[redacted]

A GUIDE TO MATHEMATICAL TABLES

Pergamon Press is also publishing Supplement No. 1
to this book, prepared from the Russian of
N.M. Burunova by D.G. Fry

A GUIDE TO MATHEMATICAL TABLES

A. V. LEBEDEV
and
R.M. FEDOROVA

English edition prepared from the Russian by
D.G.FRY B.A. (Hons.)

Pergamon Press

NEW YORK · LONDON · OXFORD · PARIS

1960

PERGAMON PRESS INC.,
122 East 55th Street, New York 22, N.Y.
1404 New York Avenue N.W., Washington 5, D.C.
P.O. Box 47715, Los Angeles, California.

PERGAMON PRESS LTD.,
4 and 5 Fitzroy Square, London W.1.
Headington Hill Hall, Oxford.

PERGAMON PRESS S.A.R.L.
24 Rue des Écoles, Paris Ve.

PERGAMON PRESS G.mb.H.
Kaiserstrasse 75, Frankfurt-am-Main.

Notice to Readers about the possibility of acquiring copies of Russian produced tables mentioned in this Guide.

Readers wishing to acquire copies of Russian produced tables mentioned in this Guide should send their order or request for a quotation to:

The Secretary,
Pergamon Institute,
Headington Hill Hall, Oxford
or
1404, New York Avenue, N.W., Washington, 5, D.C.
where their order or enquiries will receive immediate attention.

Library of Congress Card Number 60-10217

Printed in Great Britain by
PERGAMON PRINTING & ART SERVICES, LTD.,
LONDON.

TABLE OF CONTENTS

$$a_m \operatorname{ctg}^2 \frac{\pi}{n} + 2(m+1) a_{m+1} \operatorname{ctg} \frac{\pi}{n} + (m+1)(m+2) a_{m+2}; \quad I_m = \frac{\int\limits_0^\infty x^m dx}{\operatorname{sh} 2t + 2t};$$

$$J_m = \frac{\int\limits_0^\infty x^m e^{-2x} dx}{\operatorname{sh} 2t + 2t}; \qquad K_m = \frac{\int\limits_0^\infty x^m dx}{\operatorname{sh} 2t + 2t} + 2 \frac{\int\limits_0^\infty x^{m+1} dx}{\operatorname{sh} 2t + 2t} - \frac{\int\limits_0^\infty x^m e^{-2x} dx}{\operatorname{sh} 2t + 2t};$$

$$\int\limits_0^{\frac{\pi}{2}} (1 + k^{-1} \operatorname{cosec}^2 \varphi \lg \sec \varphi)^{-\frac{1}{2}} d\varphi; \quad f_r(x) = (-1)^r \left(\int\limits_x^\infty\right)^r \ln \operatorname{th} \frac{1}{2} z \, (dz)^r;$$

$$\int\limits_0^{\zeta} \left[\frac{1 - \varepsilon^2 (1 - k^2 \sin^2 t)^2}{1 - k^2 \sin^2 t}\right]^{\frac{1}{2}} dt; \qquad \int\limits_0^{\varphi} \left[\frac{\sin t}{1 + \lambda \sin t}\right]^{\frac{1}{2}} dt;$$

$$\int\limits_0^\infty \int\limits_0^\infty \frac{\sin(xu)\sin(yv)\, du\, dv}{uv\,[(u^2+v^2)^2+1]}; \; -\frac{1}{15}\left(\frac{z}{2}\right)^4 \sin\frac{2}{z} +$$

$$+ \frac{1}{15}\left(\frac{z}{2}\right)^3 \cos\frac{2}{z} - \frac{29}{30}\left(\frac{z}{2}\right)^2 \sin\frac{2}{z} - \frac{1}{6}\frac{z}{2}\cos\frac{z}{2} +$$

$$+ \frac{3}{4}\frac{z}{2}\int\limits_{2/z}^\infty \frac{\cos y\, dy}{y} + \frac{1}{6}\int\limits_{2/z}^\infty \frac{\sin y\, dy}{y} + \frac{32}{15}\left(\frac{z}{2}\right)^{\frac{3}{2}} \int\limits_0^{\frac{2}{z}} \frac{\cos y\, dy}{\sqrt{y}};$$

$$\int\limits_{-\frac{\pi}{2}}^{\frac{\pi}{2}} e^{mx} \cos^r x\, dx; \qquad \int\limits_0 e^{-(x-w^2)^2} dw; \qquad \int\limits_0^\infty f_n e^{-u\xi} \cos u\eta\, du;$$

$$\int\limits_0^\infty f_n e^{-u\xi} \sin u\eta\, du; \quad s_n = \frac{1}{\Gamma(n+1)} \int\limits_0^\infty \frac{dx}{1 + e^{x^{\frac{1}{n}}}}; \quad \sigma_n = \frac{1}{2\Gamma(n+1)} \int\limits_0^\infty \frac{dx}{\operatorname{sh} x^{\frac{1}{n}}};$$

$$J(\Omega) = -4 \int\limits_0^\infty x e^{-x^2} \cos \Omega x\, dx; \; \int\limits_0^x \frac{-\lg|1-y|\, dy}{y}; \; \int\limits_0^\infty y^s [\lg(1+y)]^n e^{-ay^2} dy;$$

$$L_n(u) = \int\limits_1^u \frac{L_{n-1}(t)}{t-1} dt; \; \frac{1}{x}\int\limits_0^x \frac{y\,dy}{e^y - 1}; \; \frac{3}{x^3}\int\limits_0^x \frac{y^3 dy}{e^y - 1}; \; \frac{12}{x^3}\int\limits_0^x \frac{y^3 dy}{e^y - 1} - \frac{3x}{e^x - 1};$$

$$\lg\left[\frac{12}{x^3}\int\limits_0^x \frac{y^3 dy}{e^y - 1} - \frac{3x}{e^x - 1}\right]; \; \int\limits_x^0 \frac{1}{y}\left[\frac{12}{y^3}\int\limits_0^x \frac{y^3 dy}{e^y - 1} - \frac{3y}{e^y - 1}\right] dy; \; \frac{3}{x^3}\int\limits_0^x \frac{y^3 dy}{e^y - 1} +$$

$$+ \int\limits_0^x \left[\frac{12}{y^4}\int\limits_0^x \frac{y^3 dy}{e^y - 1} - \frac{3}{e^y - 1}\right] dy; \quad \int\limits_{x_0}^\infty \frac{x^2 dx}{e^x - 1}; \quad \int\limits_{x_0}^\infty \frac{x^3 dx}{e^x - 1}; \quad \int\limits_{x_0}^\infty \frac{x^3}{e^x - 1} dx -$$

$$- x_0 \int\limits_{x_0}^\infty \frac{x^2}{e^x - 1} dx; \; \int\limits_x^\infty \frac{dy}{y^n (e^y - 1)}; \; \frac{15}{4\pi^4}\int\limits_0^x \frac{y^7 e^{-y}\, dy}{(1 - e^{-y})^3}; \; \frac{1}{n!}\int\limits_0^\infty x^n (e^{x-n} + 1)^{-1} dx;$$

$$\Phi_m(x) = \int\limits_1^\infty \beta^m e^{-\beta x} d\beta; \; g_n(y) = \int\limits_0^\infty \frac{e^{-x} x^n}{(x+y)^{n+1}} dx; \; \gamma_n(y) = \int\limits_0^\infty \frac{e^{-x} y^n}{(x+y)^{n+1}} dx;$$

$$S_n = 1 + \frac{1}{2^n} + \frac{1}{3^n} + \frac{1}{4^n} + \cdots; \quad S'_n = S_n - 1 = \frac{1}{2^n} + \frac{1}{3^n} + \frac{1}{4^n} + \cdots;$$

$$1 + \frac{S_n}{2^n} = 1 + \frac{1}{2^n} + \frac{1}{4^n} + \frac{1}{6^n} + \cdots; \qquad \frac{S_n}{2^n} = \frac{1}{2^n} + \frac{1}{4^n} + \frac{1}{6^n} + \cdots;$$

$$\frac{S_n - 1}{2^n} = \frac{1}{4^n} + \frac{1}{6^n} + \frac{1}{8^n} + \cdots; \quad \frac{S'_n}{n} = \frac{S_n - 1}{n} = \frac{1}{n}\left(\frac{1}{2^n} + \frac{1}{3^n} + \frac{1}{4^n} + \cdots\right);$$

$$\frac{2}{n}\,\frac{S_n - 1}{2^n} = \frac{2}{n}\left(\frac{1}{4^n} + \frac{1}{6^n} + \frac{1}{8^n} + \cdots\right); \quad \left(1 + \frac{1}{2^n} + \frac{1}{3^n} + \cdots + \frac{1}{m^n}\right)(m!)^n;$$

$$s_n = 1 - \frac{1}{2^n} + \frac{1}{3^n} - \frac{1}{4^n} + \cdots; \qquad \frac{1}{2^n}\, s_n = \frac{1}{2^n} - \frac{1}{4^n} + \frac{1}{6^n} - \frac{1}{8^n} + \cdots;$$

PART TWO

REFERENCES

INTRODUCTION

The origin of mathematical tables as an effective means of computation belongs historically to early antiquity. Attempts at the compilation of simple mathematical tables have been discovered in the literary remains of Babylon and ancient Egypt. Later, in the 16th and early 17th centuries, the computation of tables which would make it possible to make practical use of mathematical functions began with tables of trigonometric functions and logarithms.

The application of the results of observation and their painstaking improvement exerted considerable influence on the general development of the physical and mathematical sciences in addition to leading directly to the establishment of computational methods and to the computation of mathematical tables of the known functions. Many of the achievements of classical analysis were conditioned in large measure by the fact that the most important mathematicians of the time (Newton, Leibnitz, Fermat and, at a later stage, Gauss and Euler) were first class calculators.

In recent decades work on the computation of mathematical tables has been particularly extensive. The creation of powerful computing machines which considerably lighten the process of computation and the great demand of scientific establishments of all types for mathematical tables has helped to bring about a considerable increase in the number of mathematical tables published. The problems of contemporary physics, electric technology, mechanics and other branches of science have called for an extension of the tables to the field of complex arguments.

There are in existence at the present time a great many tables of the most diverse nature, and it is the purpose of this handbook to systematize all the information concerning these tables and give a description of each one.

The first part of the Handbook, consisting of 15 chapters, gives a description of the mathematica tables. Each chapter contains a number of main sub-divisions. Each sub-division contains a strictly limited group of functions. Each of these groups of functions is either given a name or related to a common formula. The description of a table almost always gives its accuracy and interval the limits of variation of the argument and a reference to the source in which the table is to be found.

The description of a table is arranged from left to right in the following sequence:

1. The accuracy of the table, i.e., the number of decimal places or significant figures;
2. The limits of variation of the argument and the interval of the table;
3. The serial number of the book or journal in the reference material to be found in the second half of the Handbook.

Let us examine this in greater detail. The description of tables of the same function is arranged in the main in accordance with the degree of accuracy, those tables which are the most accurate being placed first. This order is departe from only in exceptional cases, when, for example, tables of a given function of varying degrees of accuracy are contained in the same work.

The authors have striven to indicate the accuracy of the tables as precisely and briefly as possible. It has not always been possible, at the same time, to establish all small details. Thus, for example, no reference has been made to indicate that a given table contains two or three entries of a greater or lesser degree of accuracy than the other entries.

Special reference has usually been made when exac values are given for functions. Even, however, whe the description of a table mentions the number of decimal places the possibility is not excluded that some values will be exact. Thus, for example, in the sub-division of Chapter 9 on Legendre polynomials $P_n(x)$ almost all the tables for $P_2(x)$ contain exact values. The same can be said for certain of the tables of squares and cubes in Chapter 1. A distinction is made between the number of decimal places and the number of significant figures in describing the accuracy of the tables. Tables of logarithms, for example, are almost always compiled so as to have the same number of decimal places, while tables of antilogarithms have a constant number of significant figures.

When different parts of a table of the same function are given with varying degrees of accuracy the accuracy of each part of the table is indicated in the first sentence of the description. If the number of places in the table changes at an arbitrary value of the argument or changes several times, the minima and maxima are indicated.

Example:

5–7; 8 fig.; 8 dec. $x = 0\ (1'')\ 1^\circ 48';\ 1^\circ 48'\ (1'')\ 18^\circ 30';\ 18^\circ 30' (1'')\ 90^\circ$.

This entry shows that the table gives values for the function with 5 - 7 significant figures from 0 to $1^\circ 48'$; with 8 significant figures from $1^\circ 48'$ to $18^\circ 30'$ and with 8 decimal places from $18^\circ 30'$ to 90°.

If values such as 1°; $0^\circ,1$; 1'; 0",0001; 1^g; 1^m stand first in the sentence this indicates that the table gives values of the function corresponding to an accuracy of one degree, one tenth of a degree, one minute of arc, one ten thousandth of a second of arc, one grad, one minute of time.

A description of the argument of the table is given in the second place in the sentence. The description of the argument indicates the range of its variation and the step of the table (the interval by which the argument changes). The step of the table is given in brackets. Thus, the entry $n = 1\ (1)\ 1000$ shows that the argument n changes by unity for all the integers from 1 to 1,000. The entry $x = 0(1'')\ 6^\circ(10'')\ 45^\circ\ (30')\ 90^\circ$, however, shows that the table gives values of the function from 0 to 90°, but that the argument goes in steps of 1" from 0 to 6°, 10" from 6° to 45° and 30' from 45° to 90°.

When the step of a table varies frequently or when there is no regular step this is indicated thus: (var.). The entry ($\times 10$) shows that each successive entry in the table is 10 times larger than the preceeding one. An argument sometimes assumes only certain values within its range of variation. Examples: 1(1)50 (prime numbers) 5000 or n = 57, 73, 91, 111, 113, etc. In the first case the entry shows that at the beginning of the table the argument assumes all integers from 1 to 50, and from then on, from 50 to 5000, only prime numbers; in the second case the entry shows that the argument assumes only the values that are specified.

If the title of a sub-division contains several functions and a certain book includes a table relating to one of them, the function for which the table is given in the book is indicated in the description after the indication of accuracy (see, for example, p.132).

The third place in each entry is occupied by a reference to the publication in which the given table is to be found. The number given in square brackets is the serial number of the book or journal in the reference material which is arranged by chapters in the second part of the book.

When the reference is to a journal article the page, author and title of the article are normally indicated in addition to the name of the journal, the year of publication (in brackets), the series, and the number of the volume (in bold type) and of the issue. References to books give the name of the author and the title of the book followed, where necessary, by the volume, book or section, the publishing house and the place and year of publication.

If the book has had several editions or reprints the most recent year is chosen. Rare exceptions to this rule have been made in cases where the authors have found an earlier edition easier to use. When different editions vary considerably in content the years of these editions are noted in the references and the serial number is provided with a subscript (e.g., $/16/_2$) to indicate the year of the edition which gives the table which is being described.

If the book or journal is held in an open-access library in Moscow or Leningrad, this library is indicated after the title of the edition. (This information has been eliminated from the English edition. Translator). When a book has been issued by a Soviet publishing house in a large edition a holding source has not been indicated. Such titles are indicated in the references by a star.

The following abbreviations have been adopted in the descriptions of the tables:

exact - exact values
fig. - significant figures
dec. - decimal places
rec. - recurring decimal
(var.) - variable step
hrs - hours
m - minutes (time)
sec. - seconds (time)
g - grads
nwn - nearest whole number.

PART ONE

TABLES

Chapter I

POWERS, RATIONAL AND ALGEBRAIC FUNCTIONS[1]

This chapter contains information on tables of whole and fractional powers (including step exponential functions) of real and complex numbers, tables of reciprocal values, rational and algebraic functions and tables of the powers of certain numbers.

There are so many tables of squares and cubes and of square and cube roots in both the Russian and the foreign literature that it has not been possible to give exhaustive information on them. Such informat--ion, in any case, might not be of great interest.

The information given below is, however, an adequate account of the tables available in the Russian literature.

POSITIVE WHOLE POWERS

Squares, n^2

Squares of integers

exact	$n = 1\ (1)\ 100\,009$	[247]
exact	$n = 1\ (1)\ 100\,000$	[174] [183]
exact	$n = 1\ (1)\ 27\,000$	[162]
exact	$n = 1\ (1)\ 25\,400$	[157]
exact	$n = 1\ (1)\ 25\,200$	[115]
exact	$n = 1\ (1)\ 25\,019$	[8]
exact	$n = 1\ (1)\ 25\,000$	[109]
exact	$n = 1\ (1)\ 15\,000$	[44] [235]
exact	$n = 1\ (1)\ 12\,500$	[86] [112]
exact	$n = 1\ (1)\ 12\,000$	[123]
exact	$n = 1\ (1)\ 11\,100$	[233]
exact	$n = 1\ (1)\ 10\,000$	[71] [111] [127] [130] [201] [240] [[248]

[1] See pages 411 and 550 for the references to Chapter 1.

exact	$n = 1000\,(1)\,10\,000$	[64]
exact	$r = 1000\,(1)\,8000$	[87]
exact	$n = 1\,(1)\,3400$	[267]
exact	$n = 1\,(1)\,2000$	[13]
exact	$n = 1000\,(1)\,1899$	[38]
exact	$n = 1\,(1)\,1610$	[63]
exact	$n = 1\,(1)\,1500$	[83] [100] [139]
exact	$n = 1\,(1)\,1200$	[149]
exact	$n = 1\,(1)\,1100$	[65] [88] [269]
exact	$n = 1\,(1)\,1009$	[141]
exact	$n = 1\,(1)\,1000$	[3] [7] [9] [11] [15] [19] [20] [22] [23] [25] [30] [35] [36] [39] [46] [47] [48] [49] [50] [51] [55] [56] [57] [62] [67] [69] [70] [72] [80] [89] [90] [92] [104] [107] [110] [116] [118] [129] [136] [139] [148] [172] [176] [180] [208] [209] [210] [211] [216] [224] [226] [228] [229] [234] [236] [238] [242] [268]
exact	$n = 100\,(1)\,1000$	[96]
exact	$n = 1\,(1)\,500$	[85]
exact	$n = 1\,(1)\,120$	[61]
exact	$n = 1\,(1)\,100$	[2] [12] [16] [17] [26] [27] [29] [37] [52] [53] [63] [68] [74] [78] [84] [93] [102] [128] [131] [167] [189] [206] [207] [217] [219] [243] [244] [246]
exact	$n = 1\,(1)\,70$	[95]
exact	$n = 1\,(2)\,1518$	[94]
6 fig.	$n = 1\,(1)\,10\,000$	[5]
5 fig.	$n = 1\,(1)\,10\,000$	[33]
5 fig.	$n = 1\,(1)\,1020$	[158]
4 fig.	$n = 1\,(1)\,2000$	[215]
4 fig.	$n = 1\,(1)\,1000$	[32]

See page 406 for tables of the multiples of squares

Squares of decimal fractions

exact	$n = 1\,(0{,}01)\,10$	[212]
exact	$n = 0\,(0{,}1)\,100$	[173]

exact	$n = 0\ (0{,}1)\ 1\ (1)\ 100$	[98]
exact	$n = 0\ (0{,}01)\ 1$	[45] [142]
5—6 fig.	$n = 0\ (0{,}001)\ 10$	[40] [81]
5 fig.	$n = 0\ (0{,}001)\ 3{,}009$	[73] [77]
5 fig.	$n = 1\ (0{,}01)\ 10$	[99] [108] [132]
4 dec.	$n = 0\ (0{,}001)\ 10$	[75] [97] [142] [143] [145] [165]
4 dec.	$n = 0\ (0{,}001)\ 3{,}509$	[119] [120]
4 dec.	$n = 0{,}1\ (0{,}001)\ 1{,}009$	[61]
4 dec.	$n = 0\ (0{,}01)\ 10{,}09$	[155]
4 dec.	$n = 0\ (0{,}01)\ 10$	[4] [28] [76] [165] [182]
4 fig.	$n = 1\ (0{,}01)\ 10$	[21] [60] [82] [103] [106] [113] [131] [166] [171] [181] [218] [222]
4 fig.	$n = 10\ (0{,}1)\ 100$	[245]
4 fig.	$n = 0\ (0{,}1)\ 100$	[10]
3 dec.	$n = 0\ (0{,}001)\ 1$	[59]
3 dec.	$n = 1\ (0{,}01)\ 12$	[58] [189]
3 fig.	$n = 0\ (0{,}01)\ 10$	[66]
2—3 dec.	$n = 1\ (0{,}01)\ 10$	[206]
2 dec.	$n = 1\ (0{,}01)\ 10$	[24] [54] [167]

Squares of vulgar fractions $\left(\frac{n}{m}\right)^2$

exact	$\left\{\begin{array}{l} n = 1\ (1)\ 8;\ 11 \\ m = 2\ (1)\ 4;\ 6\ (1)\ 9\ (3)\ 15 \end{array}\right\}$	[51]
6 dec.	$\frac{n}{m} = \frac{1}{8}\left(\frac{1}{8}\right)25\frac{7}{8}$	[44]

Half squares $\frac{n^2}{2}$

5 dec.	$n = 0\ (0{,}01)\ 2$	[225]

Quarter squares $\frac{n^2}{4}$ [1]

exact	$n = 1\ (1)\ 200\,000$	[114]
exact	$n = 1\ (1)\ 100\,000$	[179]
exact	$n = 1\ (1)\ 40\,000$	[185]
exact	$n = 1\ (1)\ 30\,000$	[175]
exact	$n = 1\ (1)\ 20\,000$	[117] [203]
exact	$n = 1\ (1)\ 5100$	[205]

[1] The tables in this section are intended mainly for the multiplication of numbers by the formula:

$$ab = \frac{1}{4}(a+b)^2 - \frac{1}{4}(a-b)^2$$

Cubes, n^3

Cubes of integers

exact	$n = 1\ (1)\ 100\,000$	[174]
exact	$n = 1\ (1)\ 24\,000$	[162]
exact	$n = 1\ (1)\ 12\,500$	[86] [112]
exact	$n = 1\ (1)\ 12\,000$	[123]
exact	$n = 1\ (1)\ 10\,000$	[109] [111] [127] [130] [147] [157] [240]
exact	$n = 1\ (1)\ 3000$	[44]
exact	$n = 1001\ (1)\ 1899$	[38]
exact	$n = 1\ (1)\ 1610$	[63]
exact	$n = 1\ (1)\ 1500$	[83] [100] [139]
exact	$n = 1\ (1)\ 1300$	[233]
exact	$n = 1\ (1)\ 1200$	[149]
exact	$n = 1\ (1)\ 1100$	[65] [88] [269]
exact	$n = 1\ (1)\ 1079$	[8]
exact	$n = 1\ (1)\ 1000$	[3] [7] [11] [15] [19] [20] [22] [23] [30] [35] [46] [48] [49] [50] [51] [55] [62] [69] [70] [72] [80] [90] [92] [107] [110] [116] [118] [129] [136] [139] [141] [148] [172] [176] [208] [209] [210] [224] [226] [228] [229] [234] [236] [238] [242] [267] [268]
exact	$n = 1\ (1)\ 500$	[85]
exact	$n = 1\ (1)\ 120$	[61]
exact	$n = 1\ (1)\ 100$	[2] [12] [16] [25] [26] [27] [29] [37] [52] [53] [63] [74] [78] [84] [93] [97] [102] [128] [131] [165] [167] [189] [206] [211] [217] [219] [243] [244] [246]
fig.	$n = 1\ (1)\ 17$	[95]
6 fig.	$n = 1\ (1) 10000$	[5]
5 fig.	$n = 1\ (1) 10000$	[33]
5 fig.	$n = 1\ (1)\ 1020$	[158]
4 fig.	$n = 1\ (1)\ 2000$	[215]
4 fig.	$n = 1\ (1)\ 1000$	[32]
4 dec.	$n = 2150\ (1)\ 3400$	[267]

Cubes of decimal fractions

exact	$n = 0\ (0{,}01)\ 10$	[247]
exact	$n = 0\ (0{,}1)\ 1\ (1)\ 100$	[98]

6 fig.		$n=1\,(0{,}01)\,10$	[99] [212]
5—9 fig.		$n=0\,(0{,}001)\,10$	[81]
5 dec.	$\frac{n}{1000}$	$n=1000\,(1)\,2150$	[267]
5 dec.		$n=1\,(0{,}01)\,10\,(0{,}1)\,100$	[60]
5 fig.		$n=1\,(0{,}01)\,10$	[108] [132]
4–5 fig.		$n=1\,(0{,}01)\,12$	[189]
4 fig.		$n=1\,(0{,}001)\,2,\ 16\,(0{,}01)\,10$	[21]
4 fig.		$n=1\,(0{,}001)\,2,\ 159;\ 2{,}1\,(0{,}01)\,10$	[103]
4 fig.		$n=1\,(0{,}01)\,12$	[58]
4 fig.		$n=1\,(0{,}01)\,10$	[24] [82] [166] [171] [181] [182] [206] [218] [222]
4 fig.		$n=10\,(0{,}1)\,100$	[245]
3 fig.		$n=0\,(0{,}01)\,10$	[66]
3 dec.		$n=0\,(0{,}1)\,100$	[173]

Cubes of vulgar fractions $\left(\frac{n}{m}\right)^3$

exact	$n=1\,(1)\,8;\ 11.\ m=2\,(1)\,4;\ 6\,(1)\,9\,(3)\,15$	[51]
6 dec.	$\frac{n}{m}=\frac{1}{8}\left(\frac{1}{8}\right)25\frac{7}{8}$	[44]

Fourth powers, n^4

Fourth powers of integers

exact	$n=1\,(1)\,1099$	[122]
exact	$n=1\,(1)\,1000$	[86] [111] [112] [154] [231] [267]
exact	$n=1\,(1)\,308$	[202]
exact	$n=1\,(1)\,300$	[193]
exact	$n=1\,(1)\,100$	[141]
5 fig.	$n=1\,(1)\,100$	[118]

Fourth powers of decimal fractions

4–5 fig.	$n=1\,(0{,}01)\,12$	[189]
4 fig.	$n=1\,(0{,}01)\,12$	[58]
4 dec.	$n=0\,(0{,}1)\,8\,(0{,}2)\,10$	[159]

Multiples of fourth powers of certain numbers where $k=1\,(1)\,N$

exact $\left\{\begin{array}{llllllll} n=2\,(1)\,7; & 8; & 9; & 10; & 11; & 12; & 13;\ 14; & 15\,(1)\,19 \\ N=400; & 350; & 300; & 250; & 200; & 150; & & 100 \end{array}\right\}$ [199]

Fifth powers, n^5

Fifth powers of whole numbers

exact	$n = 1\,(1)\,1099$	[122]
exact	$n = 1\,(1)\,1000$	[111] [154] [231] [267]
exact	$n = 1\,(1)\,308$	[202]
exact	$n = 1\,(1)\,200$	[193]
exact	$n = 1\,(1)\,100$	[118] [141]

Fifth powers of decimal fractions

3—10 fig.	$n = 0{,}1;\ 0{,}102\,(0{,}002)\,0{,}15\,(0{,}005)\,0{,}4\,(0{,}01)\,1\,(0{,}02)\,2\,(0{,}05)\,5\,(0{,}1)\,10\,(0{,}2)\,30\,(0{,}5)\,50\,(1)\,99$	[236]
6—8 fig.	$n = 0{,}1\,(0{,}002)\,0{,}15\,(0{,}005)\,0{,}4\,(0{,}01)\,1\,(0{,}02)\,2\,(0{,}05)\,5\,(0{,}1)\,10\,(0{,}2)\,30\,(0{,}5)\,40$	[107]
4—5 fig.	$n = 1\,(0{,}01)\,12$	[58] [189]
4 dec.	$n = 0\,(0{,}1)\,8\,(0{,}2)\,10$	[159]

Sixth to tenth powers, n^p

exact	$p = 6.\quad n = 1\,(1)\,100$; $p = 7\,(1)\,10.\quad n = 1\,(1)\,50$	[141]

Various whole powers, n^p

Whole powers of integers

exact	$n = 1\,(1)\,120.\quad p = 1\,(1)\,50$; $n = 120\,(1)\,299.\quad p = 1\,(1)\,20$; $n = 299\,(1)\,1099.\quad p = 1\,(1)\,12$	[122]
exact	$n = 1\,(1)\,1000.\quad p = 1\,(1)\,10$	[231]
exact	$n = 57;\ 73;\ 91;\ 111;\ 133;\ 157;183;\ 211;\ 241;\ 273;\ 307;\ 343;\ 381.\quad p = 1\,(1)\,17$	[241]
exact	$n = 1\,(1)\,308.\quad p = 1\,(1)\,10$	[202]
exact	$n = 1\,(1)\,200.\quad p = 1\,(1)\,6$	[193]
exact	$n = 1\,(1)\,100.\quad p = 1\,(1)\,5$	[24]
exact	$n = 1\,(1)\,100.\quad p = 2\,(1)\,9$	[172] [238]
exact	$n = 1\,(1)\,100.\quad p = 4\,(1)\,10$	[8] [72] [86] [112]
exact	$n = 1\,(1)\,M.\quad 20 \leqslant M \leqslant 100,\ p = 3\,(1)\,6$	[45]

exact	$n = 1\ (1)\ 100.$	$p = 3\ (1)\ 6$	[108]
exact	$n = 1\ (1)\ 100.$	$p = 1\ (1)\ 10$	[60] [152] [157] [200]
exact	$n = 1\ (1)\ 100.$	$p = 11;\ 12$	[154]
exact	$n = 1\ (1)\ 11.$ $n = 11 (1)\ 31.$ $n = 31 (1)\ 50.$	$p = 1\ (1)\ 40$ $p = 1\ (1)\ 53$ $p = 1\ (1)\ 23$	[241]
exact	$n = 1\ (1)\ 25.$	$p = 1\ (1)\ 12$	[161]
exact	$n = 1\ (1)\ 10.$	$p = 11 (1)\ 20$	[186] [112]
exact	$n = 1\ (1)\ 10.$	$p = 4\ (1)\ 8$	[61]
exact	$n = 1\ (1)\ 10.$	$p = 1\ (1)\ 30$	[151]
exact	$n = 1\ (1)\ 9.$	$p = 1\ (1)\ 9$	[1] [18]
exact	$n = 2\ (1)\ 9.$	$p = 4\ (1)\ 10$	[226]
exact	$n = 2;\ 3;\ 5;\ 7.$	$p = 1\ (1)\ 25$	[214]
25 fig.	$n = 2\ (1)\ 308.$	$p = 1\ (1)\ 10$	[202]
8 fig.	$n = 1\ (1)\ 100.$	$p = 4\ (1)\ 10$	[5]
6 fig.	$n = 1\ (1)\ 50.$	$p = 4\ (1)\ 10$	[81]

Whole powers of decimal fractions

exact	$n = 1\ (0{,}01)\ 10.$	$p = 2 (1) 5$	[79]
exact	$n = 0\ (0{,}1)\ 9{,}9.$	$p = 4 (1) 9$	[247]
8 dec.	$n = 0\ (0{,}01)\ 1.$	$p = 1 (1) 11$	[177] [178] [220]
8 fig.	$n = 0\ (0{,}01)\ 1.$	$p = 1 (1) 11$	[238]
6 dec.	$n = 1{,}03;\ 1{,}04;\ 1{,}045;\ 1{,}05.$	$p = 1 (1) 25 (5) 50$	[25]
6 dec.	$n = 1{,}01;\ 1{,}02;\ 1{,}025;\ 1{,}0275;\ 1{,}03;$ $1{,}0325;\ 1{,}035;\ 1{,}0375;\ 1{,}04;\ 1{,}045;$ $1{,}05,\ 1{,}06.$	$p = 1 (1) 100$	[238]
5 fig.	$n = 1{,}1 (0{,}1) 10{,}9.$	$\pm p = 1 (1) 15$	[215]
4—5 fig.	$n = 0{,}5 (0{,}01) 1 (0{,}02) 2 (0{,}05) 5.$	$p = 2 (1) 15$	[105] [118] [163]
3—12 fig.	$n = 0 (0{,}1) 10.$	$p = 1 (1) 10$	[151]

FRACTIONAL POSITIVE POWERS

Square roots, $\sqrt{n}$

Square roots from integers

15 dec.	$n = 1 (1) 100$	[144]
12 dec.	$n = 1 (1) 10\,000$	[238]
10 dec.	$n = 1 (1) 1000$	[184]
10 dec.	$n = 1 (1) 100$	[25]

7 dec.	$n = 1\,(1)\,10000$	[111]
7 dec.	$n = 1\,(1)\,1610$	[63]
7 dec.	$n = 1\,(1)\,1200$	[149]
7 dec.	$n = 1\,(1)\,1054$	[133] [159] [160]
7 fig.	$n = 1\,(1)\,1000$	[148]
7 dec.	$n = 1\,(1)\,1000$	[146] [172] [268]
7 dec.	$n = 1\,(1)\,120$	[227]
7 dec.	$n = 1\,(1)\,100$	[69]
7; 6 dec.	$n = 1\,(1)\,1000;\ 1000\,(1)\,12\,500$	[86] [112]
7; 5 dec.	$n = 1\,(1)\,100;\ 100\,(1)\,1100$	[269]
6 fig.	$n = 1\,(1)\,10\,000$	[5]
6 fig.	$n = 1\,(1)\,1000$	[72] [176] [209] [210]
6 dec.	$n = 1\,(1)\,1000$	[35] [50] [80] [118] [267]
5 dec.	$n = 10\,010\,(1)\,25\,500$	[162]
5 fig.	$n = 100\,(1)\,1000\,(10)\,10\,000$	[180]
5 fig.	$n = 1\,(1)\,10\,000$	[33]
5 dec.	$n = 1\,(1)\,1020$	[158]
5 dec.	$n = 1\,(1)\,1000$	[7] [90]
5 fig.	$n = 10\,(1)\,1000$	[99]
5 fig.	$n = 100\,(1)\,1000$	[138]
5 dec.	$n = 1\,(1)\,100$	[18] [97] [165]
4–6 fig.	$n = 1\,(1)\,2000$	[13]
4–5 dec.	$n = 1\,(1)\,1000$	[72]
4 dec.	$n = 1\,(1)\,10\,009$	[162]
4 dec.	$n = 1\,(1)\,1000$	[236]
4 fig.	$n = 1005\,(5)\,2450\,(10)\,10\,000$	[236]
4 dec.	$n = 1\,(1)\,2000$	[215]
4 dec.	$n = 1\,(1)\,1500$	[83] [100] [139]
4 dec.	$n = 120\,(1)\,1200$	[44]
4 dec.	$n = 1\,(1)\,1100$	[65] [88]
4 dec.	$n = 1\,(1)\,1000$	[3] [19] [20] [22] [23] [35] [46] [48] [49] [51] [55] [62] [67] [70] [92] [104] [107] [116] [136] [141] [208] [224] [226] [228] [229] [234] [242]
4 fig.	$n = 1\,(1)\,1000$	[32] [36]
4 fig.	$n = 10\,(1)\,1000$	[12]
4 dec.	$n = 100\,(1)\,1000$	[25]
4 dec.	$n = 1\,(1)\,500$	[85]
4 dec.	$n = 1\,(1)\,120$	[61]

4 dec.	$n = 1\,(1)\,100$	[2] [16] [17] [29] [37] [52] [53] [63] [74] [84] [101] [102] [128] [131] [182] [189] [206] [211] [217] [219] [244] [246]
3 dec.	$n = 1\,(1)\,10\,000$	[127]
3 dec.	$n = 1\,(1)\,1000$	[15] [56] [129]
3 dec.	$n = 1\,(1)\,100$	[26] [45] [93] [166] [167] [171] [181] [218] [243] [245]
3 dec.	$n = 1\,(1)\,100\,(5)\,200\,(10)\,500\,(20)\,1000$	[78]
2—4 dec.	$n = 1\,(1)\,1000$	[110]
2 dec.	$n = 1\,(1)\,1000$	[34]
2 dec.	$n = 1\,(1)\,100$	[27]
cont. fract.	$n = 1\,(1)\,10\,000$	[197]

Square roots of decimal fractions

12 dec.	$n = 0(0{,}001)\,1(0{,}01)\,10\,(0{,}1)\,20$	[153]
7 dec.	$n = 0{,}251\,(0{,}001)\,0{,}262$	[264]
6 dec.	$n = 5{,}91\,(0{,}01)\,6{,}02$	[264]
6 dec.	$n = 23{,}9\,(0{,}1)\,25$	[264]
6 dec.	$n = 100\,(0{,}1)\,1000\,(1)\,10\,000$	[188]
5 fig.	$n = 0\,(0{,}01)\,10\,(0{,}1)\,100{,}9$	[73] [77]
5 dec.	$n = 1\,(0{,}01)\,10\,(0{,}1)\,100$	[60] [125]
5 dec.	$n = 1\,(0{,}01)\,10$	[212]
4 dec.	$n = 0\,(0{,}001)\,10{,}009$	[75]
4 dec.	$n = 1\,(0{,}01)\,10\,(0{,}1)\,100$	[43] [106] [108] [126] [132]
4 fig.	$n = 1\,(0{,}01)\,10\,(0{,}1)\,100$	[21] [42]
4 dec.	$n = 1\,(0{,}01)\,12$	[58] [189]
4 dec.	$n = 1\,(0{,}01)\,10$	[131]
4 fig.	$n = 1\,(0{,}01)\,10$	[103] [218]
4 dec.	$n = 0\,(0{,}1)\,100\,(1)\,1000$	[164]
4 dec.	$n = 0\,(0{,}1)\,100$	[173]
4 dec.	$n = 0\,(0{,}1)\,1\,(1)\,100$	[98]
3 dec.	$n = 0\,(0{,}001)\,0{,}998$	[59]
3 dec.	$n = 1\,(0{,}01)\,100$	[82]
3 dec.	$n = 1\,(0{,}01)\,10\,(0{,}1)\,100$	[4] [93] [206]
3 dec.	$n = 1\,(0{,}01)\,10$	[24] [79]
3 dec.	$n = 1\,(0{,}1)\,50\,(1)\,100$	[28] [76]
2 dec.	$n = 0\,(0{,}1)\,100$	[10]

Square roots of vulgar fractions, $\sqrt{\frac{n}{m}}$

5 dec.	$1 \leqslant n \leqslant 7.\ \ 3 \leqslant m \leqslant 12$; m & n — relatively prime.	[83] [139] [208]
4 dec.	$n = 1\,(1)\,8;\ 11.\ \ m = 2\,(1)\,4;\ 6\,(1)\,9\,(3)\,15$	[51]
3 dec.	$1 \leqslant n \leqslant 11.\ \ 2 \leqslant m \leqslant 15$; m & n — relatively prime	[62]
3 dec.	$1 \leqslant n \leqslant 7.\ \ 3 \leqslant m \leqslant 12$; m & n — relatively prime	[228]

See also p. 407.

Square roots of the type $\sqrt{10n}$

7 fig.	$n = 1\,(1)\,1000$	[148]
6 dec.	$n = 1000\,(1)\,12\,500$	[86] [112]
6 dec.	$n = 100\,(1)\,1000$	[267]
5; 4 dec.	$n = 1\,(1)\,10;\ \ 10\,(1)\,1000$	[72]
5 fig.	$n = 1000\,(1)\,10\,000$	[33]
5 dec.	$n = 1000\,(1)\,1100$	[269]
5 dec.	$n = 1\,(0{,}01)\,10$	[212]
4 dec.	$n = 1000\,(1)\,10\,000$	[5]
4 dec.	$n = 1\,(1)\,1000$	[209]
4 fig.	$n = 1\,(1)\,1000$	[32]
4 dec.	$n = 1\,(1)\,100$	[128]
3 dec.	$n = 1\,(1)\,100$	[26] [93]
3 dec.	$n = 1\,(0{,}01)\,12$	[58] [189]
3 dec.	$n = 1\,(0{,}01)\,10$	[24] [79] [131]

Square roots of powers,

$$n^{\frac{1}{2}x}$$

6 fig.	$\frac{1}{12}n = 3\frac{1}{2}\left(\frac{1}{4}\right)4\left(\frac{1}{2}\right)14\,(1)\,21$	$x = 5$	[170]
6 fig.	$n = \frac{1}{4}\left(\frac{1}{4}\right)1\,(1)\,16\,(2)\,40.$	$x = 5$	
5 fig.	$n = 1\,(1)\,1000.$	$x = 3$	

5 fig.	$12n = \frac{1}{4}\left(\frac{1}{8}\right)2\left(\frac{1}{4}\right)6\left(\frac{1}{2}\right)22\,(1)\,100$ (var.) 1200.	$x = 5$	[236]
4 dec.	$n = 0\,(0{,}001)\,1{,}49\,(0{,}01)\,12.$	$x = 3$	[156]
4 dec.	$n = 0\,(0{,}1)\,9\,(0{,}2)\,10.$	$x = 5$	[159]
4 fig.	$\frac{n}{1000} = 1\,(1)\,25.$	$x = 3$	[213]

$(10n)^{3/2}$

4 fig.	$n = 1\,(1)\,1000$	[32]

$n\sqrt{n}$

6 fig.	$n = 1\,(1)\,100$	[118]

$n^2\sqrt{n}$

5 fig.	$n = 1\,(1)\,100$	[118]

$\sqrt{\frac{2}{\pi n}}$

12 dec.	$n = 0\,(0{,}01)\,2;\quad 10\,(0{,}1)\,20$	[153]
12 dec.	$n = 2\,(1)\,10$	

Cube roots, $\sqrt[3]{n}$

Cube roots of integers

31 dec.	$n = 1$ (prime numbers) 127	[259]
10 dec.	$n = 1\,(1)\,100$	[144]
7 dec.	$n = 1\,(1)\,10000$	[111] [238]
7 dec.	$n = 1\,(1)\,1610$	[63]
7 dec.	$n = 1\,(1)\,1200$	[149]
7 dec.	$n = 1\,(1)\,1054$	[133] [159] [160]
7 dec.	$n = 1\,(1)\,1000$	[146] [172] [268]
7 dec.	$n = 1\,(1)\,120$	[227]
7 dec.	$n = 1\,(1)\,100$	[69] [90]
7; 6 dec..	$n = 1\,(1)\,1000;\ 1000\,(1)\,12\,500$	[86] [112]

7; 5 dec.	$n = 1\,(1)\,100;\ 100\,(1)\,1100$	[269]
6 fig.	$n = 1\,(1)\,10\,000$	[5]
6 dec.	$n = 1\,(1)\,1000$	[80] [118] [148] [267]
5 dec.	$n = 10\,010\,(1)\,25\,500$	[162]
5 fig.	$n = 1\,(1)\,10\,000$	[33]
5 dec.	$n = 1\,(1)\,1020$	[158]
5 dec.	$n = 1\,(1)\,1000$	[7] [35] [50] [72] [209] [210]
5 fig.	$n = 1\,(1)\,359$	[73] [77]
5 dec.	$n = 1\,(1)\,100$	[18] [165]
4 dec.	$n = 1\,(1)\,10\,009$	[162]
4 dec.	$n = 1\,(1)\,1000$	[236]
4 fig.	$n = 1005\,(5)\,2450\,(10)\,10000$	
4 dec.	$n = 1\,(1)\,2000$	[215]
4 dec.	$n = 1\,(1)\,1500$	[83] [100] [139]
4 dec.	$n = 120\,(1)\,1200$	[44]
4 dec.	$n = 1\,(1)\,1100$	[65] [88]
4 dec.	$n = 1\,(1)\,1000$	[3] [19] [20] [22] [23] [38] [46] [48] [49] [51] [55] [62] [70] [92] [107] [116] [136] [139] [141] [176] [208] [224] [226] [228] [229] [234] [242]
4 fig.	$n = 1\,(1)\,1000$	[32]
4 dec.	$n = 1\,(1)\,500$	[85]
4 dec.	$n = 1\,(1)\,120$	[61]
4 dec.	$n = 1\,(1)\,100$	[2] [12] [16] [29] [52] [63] [74] [84] [102] [128] [181] [182] [206] [211] [217] [219] [244] [246]
4 dec.	$n = 1\,(1)\,10$	[1]
3–4 dec.	$n = 1\,(1)\,10\,000$	[127]
3 dec.	$n = 1\,(1)\,1000$	[15] [110] [129]
3 dec.	$n = 1\,(1)\,100\,(5)\,200\,(10)\,500\,(20)\,1000$	[78]
3 dec.	$n = 1\,(1)\,100$	[25] [26] [45] [93] [131] [166] [167] [243] [245]
2 dec.	$n = 1\,(1)\,100$	[27]

Cube roots of decimal fractions

9 dec.	$n = 1\,(1)\,9;\ 1{,}1\,(0{,}1)\,1{,}9;\ 1{,}01\,(0{,}01)\,1{,}09;$ $1\,(0{,}001)\,1{,}009;$ $1{,}000\,000\,001\,(0{,}000\,000\,001)\,1{,}000\,000\,009$	[252]
5 dec.	$n = 1\,(0{,}01)\,10\,(0{,}1)\,100\,(1)\,1000$	[125]
5 dec.	$n = 1\,(0{,}01)\,10$	[212]
4 fig.	$n = 1\,(0{,}001)\,2{,}159$	[103]
4 dec.	$n = 1\,(0{,}01)\,10\,(0{,}1)\,100\,(1)\,1000$	[108] [126] [132] [164]
4 dec.	$n = 1\,(0{,}01)\,12$	[58] [189]
4 fig.	$n = 1\,(0{,}01)\,10$	[218]
4 dec.	$n = 0\,(0{,}1)\,100$	[173]
4 dec.	$n = 0\,(0{,}1)\,1\,(1)\,100$	[98]
3 dec.	$n = 1\,(0{,}01)\,10\,(0{,}1)\,100\,(1)\,1000$	[206]
3 dec.	$n = 1\,(0{,}01)\,10$	[79]
3 dec.	$n = 1\,(0{,}1)\,100\,(1)\,1000$	[82]

Cube roots of vulgar fractions, $\sqrt[3]{\frac{n}{m}}$

5 dec.	$1 \leqslant n \leqslant 7.\quad 3 \leqslant m \leqslant 12$; m & n — are relatively prime	[83] [139] [208]
4 dec.	$n = 1\,(1)\,8;\ 11.\quad m = 2\,(1)\,4;\ 6\,(1)\,9\,(3)\,15$	[51]
3 dec.	$1 \leqslant n \leqslant 11.\quad 2 \leqslant m \leqslant 15$; m & n — are relatively prime	[62]
3 dec.	$1 \leqslant n \leqslant 7.\quad 3 \leqslant m \leqslant 12$; m & n — are relatively prime	[228]

Cube roots of the type $\sqrt[3]{10n}$

7 fig.	$n = 1\,(1)\,1000$	[148]
6 fig.	$n = 1\,(1)\,1000$	[72]
6 dec.	$n = 100\,(1)\,1000$	[267]
5 dec.	$n = 1000\,(1)\,1100$	[269]
5 dec.	$n = 1\,(0{,}01)\,10$	[212]
5; 4 dec.	$n = 1\,(1)\,100;\ 100\,(1)\,1000$	[209]
4 fig.	$n = 1\,(1)\,1000$	[32]
3 dec.	$n = 1\,(1)\,100$	[26] [93] [131]
3 dec.	$n = 1\,(0{,}01)\,12$	[58] [189]
3 dec.	$n = 1\,(0{,}01)\,10$	[24] [79]

Cube roots of the type $\sqrt[3]{100n}$

7 fig.	$n = 1\,(1)\,1000$	[148]
6 fig.	$n = 1\,(1)\,1000$	[72]
6 dec.	$n = 100\,(1)\,1000$	[267]
5 dec.	$n = 1000\,(1)\,1100$	[269]
5 dec.	$n = 1\,(0{,}01)\,10$	[212]
4 dec.	$n = 1\,(1)\,1000$	[209]
4 fig.	$n = 1\,(1)\,1000$	[32]
3 dec.	$n = 1\,(1)\,100$	[26] [93] [131]
3 dec.	$n = 1\,(0{,}01)\,12$	[58] [189]
3 dec.	$n = 1\,(0{,}01)\,10$	[24] [79]

Cube roots of the squares of numbers,

$$n^{\frac{2}{3}}$$

31 dec.	$n = 1$ (prime numbers) 127	[259]
5 dec.	$n = 0\,(0{,}01)\,3\,(0{,}1)\,10\,(1)\,50$	[151]
3—4 dec.	$n = 1\,(0{,}1)\,31$	[255]

$$(10n)^{\frac{2}{3}}$$

4 fig.	$n = 1\,(1)\,1000$	[32]

$$(100n)^{\frac{2}{3}}$$

4 fig.	$n = 1\,(1)\,1000$	[32]

$$\frac{1}{2}(3x)^{\frac{2}{3}}$$

3 dec.	$x = m + n;\quad m = 0(1)\,5.\ \ n = \frac{1}{4};\ \frac{3}{4}$	[121]

Roots of fourth powers, $\sqrt[4]{n}$

6 dec.	$n = 1\,(1)\,100$	[267]

Roots of fifth powers $\sqrt[5]{n}$

9 dec.	$n = 1\,(1)\,9;\ 1{,}1\,(0{,}1)\,1{,}9;\ 1{,}01\ (0{,}01)\,1{,}09;$ $1{,}001\,(0{,}001)\,1{,}009;$ $1\,(0{,}000\,000\,001)\,1{,}000\,000\,009$	[252]

6 dec.	$n = 1\,(1)\,100$		[267]
4 dec.	$n = 1\,(1)\,499$		[63]

General fractional powers, n^x

Fractional powers of integers

15 dec.	$n = 2\,(1)\,9.$	$x = 0(0{,}001)0{,}01(0{,}01)0{,}99$	[232]
15 dec.	$n = 10.$	$x = 0\,(0{,}001)\,1$	[232]
6 fig.	$n = 1$ (primes) 199	$\frac{1}{x} = 4\,(1)\,10$	[162]
5 dec.	$n = 2.$	$12\,x = 1\,(1)\,11$	[207]
4—5 f.	$440 \cdot 2^x$	$12\,x = -57\,(1)\,38$	[207]
4 fig.	$n = 1\,(1)\,1000.$	$x = \frac{5}{4}$	[170]
4 fig.	$\frac{n}{1000} = 1\,(1)\,25.$	$x = 1{,}5\,(0{,}05)\,1{,}7$	[213]
4 fig.	$n = 1\,(1)\,999.$	$x = \frac{2}{3};\ \frac{3}{2}$	[32]

Fractional powers of fractions

15 dec.	$n = 0\,(0{,}01)\,10.$	$x = \pm\frac{1}{2};\ \pm\frac{1}{3};\ \pm\frac{2}{3};\ \pm\frac{1}{4};\ \pm\frac{3}{4}$	[232]
15 dec.	$n = 0\,(0{,}01)\,1.$	$x = 0\,(0{,}001)\,0{,}01\,(0{,}01)\,1$	[232]
15 dec.	$n = 10^{-3}p$, for p primes between 100 & 1000.	$x = 0\,(0{,}001)\,0{,}01\,(0{,}01)\,0{,}99$	[232]
7 dec.	$n = 0\,(0{,}001)\,N;\ 0\,(0{,}01)\,0{,}99.$ $0 < N \leqslant 0{,}299,$	$x = 0\,(0{,}01)\,0{,}99.$	[232]
7 fig.	$n = x = 0\,(0{,}01)\,0{,}05\,(0{,}05)\,1{,}5\,(0{,}1)\,3\,(0{,}2)\,4$		[118]
4—5 fig.	$n = 0{,}1;\ 0{,}5\,(0{,}1)1(0{,}2)\,3\,(0{,}5)\,5;\ 10.$	$x = 0(0{,}05)1$	[118]
4—5 fig.	$n = 0{,}5\,(0{,}1)\,1\,(0{,}2)\,3\,(0{,}5)\,5;\ 10;\ \frac{1}{10}.$	$x = 0\,(0{,}05)1$	[105] [163]
4 dec.	$n = 0\,(0{,}1)\,8\,(0{,}2)\,10.$	$x = \frac{1}{5};\ \frac{2}{5}$	[159]

ECIPROCAL VALUES (NEGATIVE WHOLE & FRACTIONAL POWERS)

Reciprocal powers of the type $\frac{1}{n}$

Reciprocals of integers

9—11 dec.	$n = 1\,(1)\,12\,500$	[86] [112]
9 dec.	$n = 1\,(1)\,1000$	[69] [118] [236]

7—8 fig.	$n = 1\,(1)\,1610$	[63]
7; 8 dec.	$n = 1\,(1)\,100;\ 100\,(1)\,1000$	[267]
7 dec.	$n = 1\,(1)\,1000$	[242] [268]
6—8 dec.	$n = 1\,(1)\,1000$	[72]
6 fig.	$n = 1\,(1)\,10\,000$	[5] [14]
6 fig.	$n = 1\,(1)\,1000$	[70]
6 dec.	$n = 1\,(1)\,1000$	[15] [80]
6 dec.	$n = 1\,(1)\,100$	[189] [246]
5—8 dec.	$n = 1\,(1)\,5000$	[180]
5—6 fig.	$n = 1\,(1)\,100\,000$	[31]
5 fig.	$n = 1\,(1)\,10\,000$	[33]
5 dec.	$n = 1\,(1)\,1100$	[269]
5 dec.	$n = 1\,(1)\,1000$	[51] [62] [92]
5 fig.	$n = 10\,(1)\,1000$	[99]
5 dec.	$n = 1\,(1)\,100$	[16] [52] [74] [84] [97] [128] [131] [165] [211]
4—5 dec.	$n = 1\,(1)\,1000$	[48]
4 fig.	$n = 100\,000\,(1)\,200\,009$	[230]
4 fig.	$n = 1\,(1)\,1000$	[32]
4 dec.	$n = 1\,(1)\,100$	[45] [61] [93] [243] [244]
4 fig.	$n = 10\,(1)\,100$	[167]
3 dec.	$n = 1\,(1)\,100$	[26]

Reciprocals of decimal fractions

6 dec.	$n = 1\,(0{,}01)\,10$	[212]
5—6 dec.	$n = 0\,(0{,}001)\,10$	[81]
5 dec.	$n = 1\,(0{,}001)\,1{,}509;\ 1{,}5\,(0{,}01)\,10$	[108]
5 dec.	$n = 1\,(0{,}01)\,10$	[60] [106]
5 fig.	$n = 1\,(0{,}01)\,10$	[132]
4—5 fig.	$n = 1\,(0{,}01)\,12$	[58] [189]
4 dec.	$n = 1\,(0{,}001)\,1{,}8\,(0{,}01)\,10$	[21]
4 dec.	$n = 1\,(0{,}01)\,10$	[24] [42] [43] [79] [82] [93] [128] [131] [206]
4 dec.	$n = 0\,(0{,}1)\,100$	[103]
3 dec.	$n = 1\,(0{,}01)\,10{,}98$	[59]

Reciprocals of the type $\frac{100}{n}$

5 dec.	$n = 1\,(1)\,100$	[2]
4 fig.	$n = 1\,(1)\,100$	[54]

Reciprocals of the type $\frac{1000}{n}$

7 fig.	$n = 1\,(1)\,1000$	[148]
6 fig.	$n = 1\,(1)\,1500$	[100] [139]
6 fig.	$n = 1\,(1)\,1000$	[19] [20] [22] [23] [38] [49] [136] [209] [210] [224]
6 fig.	$n = 0\,(0{,}1)\,1\,(1)\,100$	[98]
5—6 fig.	$n = 1\,(1)\,1100$	[65] [88]
5—6 fig.	$n = 1\,(1)\,1000$	[116] [139]
5 dec.	$n = 1\,(1)\,2000$	[215]
5 dec.	$n = 1\,(1)\,1020$	[158]
5 fig.	$n = 1\,(1)\,1000$	[55] [234]
5 fig.	$n = 1\,(1)\,100$	[222]
4—5 dec.	$n = 1\,(1)\,1500$	[83]
4—5 dec.	$n = 1\,(1)\,1000$	[228]
4 dec.	$n = 1\,(1)\,1000$	[35] [176] [226] [234]
4 dec.	$n = 1\,(1)\,500$	[85]
4 dec.	$n = 1\,(1)\,100$	[12]
4 fig.	$n = 1\,(1)\,100$	[166] [245]
4 fig.	$n = 0\,(0{,}01)\,10$	[182]
3 dec.	$n = 1\,(1)\,100$	[217] [222]

Reciprocals of the type $\frac{10\,000}{n}$

3 dec.	$n = 1\,(1)\,1000$	[67]

Reciprocals of prime numbers

The tables give the number of digits in the period of the reciprocals of prime numbers.

exact	$100\,000 < p < 112\,500$	[253]
exact	$3 \leqslant p \leqslant 19\,997$	[262]
exact	$2011 \leqslant p \leqslant 29\,989$	[263]

Reciprocals of squares, $\frac{1}{n^2}$

10 dec.	$n = 1\,(1)\,1000$	[118]
9 dec.	$n = 1\,(1)\,100$	[267]
5 dec.	$n = 1\,(1)\,100$	[97] [165]
4 fig.	$n = 1\,(1)\,1000$	[32]
4 fig.	$n = 1\,(1)\,100$	[189]

4—5 fig.	$n = 0\,(0{,}01)\,1$	[168]
3—4 dec.	$n = 0\,(0{,}1)\,7\,(0{,}5)\,9$	[186]

Reciprocals of cubes, $\frac{1}{n^3}$

9—11 dec.	$n = 1\,(1)\,100$	[118]
4 fig.	$n = 1\,(1)\,1000$	[32]

Reciprocals of square roots,

$$\frac{1}{\sqrt{n}}$$

8 dec.	$n = 1\,(1)\,100$	[146] [212]
7—8 dec.	$n = 51\,(1)\,101\,(2)\,199$	[200]
7 dec.	$n = 0\,(0{,}05)\,1\,(0{,}1)\,51{,}1$	[198]
7 dec.	$n = 1\,(1)\,1000$	[86] [112]
6; 7 dec.	$n = 1\,(1)\,100;\ 100\,(1)\,1000$	[267]
6 fig.	$n = 1\,(1)\,10\,000$	[5]
6 dec.	$n = 1\,(1)\,100$	[148]
5 fig.	$n = 1\,(1)\,1000$	[33]
5 dec.	$n = 1\,(1)\,100$	[128]
4 fig.	$n = 1\,(1)\,1000$	[32]
4 dec.	$n = 1\,(1)\,100$	[131] [189]

$$\frac{1}{\sqrt{10n}}$$

7 dec.	$n = 100\,(1)\,1000$	[267]
5 dec.	$n = 1\,(1)\,100$	[128]
4 dec.	$n = 1\,(1)\,100$	[131]

Reciprocals of cube roots, $\frac{1}{\sqrt[3]{n}}$

4 dec.	$n = 1\,(1)\,1000$	[32]

General negative powers of n^{-m}

Whole negative powers of integers

32 dec.	$n = 1\,(1)\,100.\ m = 1\,(1)\,N.\ 16 \leqslant N \leqslant 107$	[202]
10 dec.	$n = 1\,(1)\,50.\ m = 4\,(1)\,9$	[162]

7–8 fig.	$n = 50\,(1)\,101\,(2)\,199$ & 200.	$m = 1\,(1)\,3$	[200]
6–29 dec.	$n = 2\,(1)\,9$.	$m = 1\,(1)\,18$	[257]
5–8 dec.	$n = 2\,(1)\,9$.	$m = 3\,(1)\,6$	[189]

Fractional negative powers,

$$n^{-\frac{1}{2}m}$$

8 dec.	$n = 2\,(0{,}001)\,7{,}5\,(0{,}01)\,20$.	$m = 3$	[195] [196]
7 fig.	$n = 50\,(1)\,101\,(2)\,199$ & 200.	$m = 1\,(1)\,6$	[200]
4 fig.	$n = 0\,(0{,}001)\,1{,}499;\ 15\,(0{,}01)\,19{,}99$.	$m = 3$	[237]

$$n^{-n}$$

7 fig.	$n = 0\,(0{,}01)\,0{,}05\,(0{,}05)\,1{,}5\,(0{,}1)\,3\,(0{,}2)\,4$	[118]

RATIONAL FUNCTIONS

$$2 - x$$

2 dec.	$x = 0{,}001;\ 0{,}01;\ 0{,}05;\ 0{,}1\,(0{,}1)\,0{,}5$	[254]

$$3(2 - x)$$

2 dec.	$x = 0{,}001;\ 0{,}01;\ 0{,}05;\ 0{,}1\,(0{,}1)\,0{,}5$	[254]

$$x^2 + \frac{1}{2}$$

2 dec.	$x = 0\,(0{,}1)\,1\,(0{,}2)\,3$	[256]

$$x^2 + 1$$

exact	$x = 1\,(1)\,112$	[89]

$$x^2 + y^2$$

exact	$1 \leqslant x \leqslant 10.\ 1 \leqslant y \leqslant 14.$ x & y are integers.	[89]

$$x^3 - x$$

8 dec.	$x = 1\,(0{,}0001)\,1{,}1549$	[111]
7 dec.	$x = 0\,(0{,}001)\,1{,}155$	[177] [178]

$$x^3 + x$$

7; 6 dec. $x = 0\,(0{,}0001)\,1{,}1547;\ 1{,}1547\,(0{,}0001)\,3{,}28$ [249]

$$1 - x^2$$

6 dec. $x = 0\,(0{,}001)\,1$ [118] [267]

$$\left(x - \frac{1}{y}\right)^2$$

4 dec. $x = 0\,(0{,}1)\,10.\quad y = 0{,}1\,(0{,}02)\,0{,}3\,(0{,}05)\,1$ [118]

$$\frac{x}{1-x}$$

5 fig. $x = 0\,(0{,}001)\,1$ [118]

$$\frac{x}{1+x}$$

5 dec. $x = 0{,}005\,(0{,}005)\,1\,(0{,}025)\,3{,}5\,(0{,}05)\,6\,(0{,}1)\,10\,(1)\,100\,(25)\,200\,(50)\,1000$ [118]

$$\frac{1+x}{1-x}$$

4 dec. $x = 0\,(0{,}001)\,0{,}8\,(0{,}0001)\,0{,}8999$ [258]

$$\frac{x-1}{x+2}$$

6 dec. $x = 1\,(0{,}001)\,1{,}1\,(0{,}1)\,11\,(1)\,110$ [118]

$$\frac{2xy}{x+y}$$

2 dec. $x = 1\,(1)\,100.\quad y = 2\,(2)\,30\,(5)\,100$ [118]

$$\frac{x^2-1}{x^2+2}$$

6 dec. $x = 1{,}251\,(0{,}001)\,1{,}75$ [118]

ALGEBRAIC FUNCTIONS

The tables in this section include the commonest algebraic functions. It is possible to obtain values for other functions from some of them, as for example:

$$\frac{x}{\sqrt{1-x^2}} = \mathrm{tg}\,(\mathrm{arc\,sin}\,x);\quad \sqrt{1-x^2} = \sin\,(\mathrm{arc\,cos}\,x) = \mathrm{sech}\,(\mathrm{Ar\,th}\,x).$$

$\sqrt{1+x^2}$

10 dec.	$x = 0\,(0{,}001)\,1$	[91]
7 dec.	$x = 0\,(0{,}001)\,0{,}099$	[105] [163]
5 dec.	$x = 0\,(0{,}01)\,0{,}99$	
6 dec.	$x = 0\,(0{,}001)\,1$	[267]
5 dec.	$x = 0\,(0{,}01)\,1{,}02$	[236]
5 dec.	$x = 0\,(0{,}01)\,2\,(0{,}02)\,3\,(0{,}1)\,6\,(0{,}2)\,10$	[118]
4 dec.	$x = 0\,(0{,}001)\,1$	[135] [187]
3 dec.	$x = 0\,(0{,}01)\,0{,}1\,(0{,}005)\,0{,}3\,(0{,}002)\,1$	[134]

$\sqrt{1-x^2}$

10 dec.	$x = 0\,(0{,}01)\,1$	[250]
8 dec.	$x = 0\,(0{,}0001)\,1$	[169] [194]
6 dec.	$x = 0{,}998\,(0{,}00001)\,1$	[267]
6 dec.	$x = 0{,}98\,(0{,}0001)\,1$	
6 dec.	$x = 0\,(0{,}0001)\,1$	[190]
6 dec.	$x = 0\,(0{,}001)\,1$	[267]
5 dec.	$x = 0\,(0{,}01)\,1$	[118]
4 dec.	$x = 0{,}7\,(0{,}001)\,0{,}99\,(0{,}0001)\,1$	[261]
4 dec.	$x = 0\,(0{,}01)\,1$	[140]

$\sqrt{x^2-1}$

5 dec.	$x = 1\,(0{,}01)\,2\,(0{,}02)\,3\,(0{,}1)\,6\,(0{,}2)\,10$	[118]

$\sqrt{xy}$

2 dec.	$x = 1\,(1)\,100.\quad y = 2\,(2)\,30\,(5)\,100$	[118]

$\sqrt{\frac{2}{\pi}x}$

3—4 dec.	$x = 0{,}001;\ 0{,}01;\ 0{,}05;\ 0{,}1\,(0{,}1)\,0{,}5$	[254]

$\sqrt{x^2+y^2}$

2 dec.	$x = 1\,(1)\,100.\quad y = 2\,(2)\,30\,(5)\,100$	[118]
1 dec.	$x = 1\,(1)\,15.\quad y = 1\,(1)\,50$	[165]

$$\sqrt{x(1-x)}$$

8 dec. $x = 0\,(0{,}0001)\,1$ [169]
3 dec. $x = 0\,(0{,}001)\,0{,}45\,(0{,}05)\,0{,}55\,(0{,}001)\,1$ [266]

$$\frac{1}{\sqrt{1+x^2}}$$

5 dec. $x = 0\,(0{,}01)\,2\,(0{,}02)\,3\,(0{,}1)\,6\,(0{,}2)\,10$ [118]

$$\frac{1}{\sqrt{1-x^2}}$$

10 dec. $x = 0\,(0{,}005)\,0{,}1$
9 dec. $x = 0{,}1\,(0{,}001)\,0{,}199$
8 dec. $x = 0{,}2\,(0{,}001)\,0{,}9$
7 dec. $x = 0{,}9\,(0{,}0005)\,0{,}96\,(0{,}0002)\,0{,}99\,(0{,}0001)$
$0{,}995\,($ var. $)\,0{,}999\,999\,999$ [191]

$$\frac{1}{\sqrt{x(x-1)}}$$

8 dec. $x = 2\,(1)\,100$ [146]

$$\sqrt{\frac{1+x}{1-x}}$$

6 dec. $x = 0\,(0{,}01)\,1$ [118]

$$\frac{x}{\sqrt{1+x^2}}$$

5 dec. $x = 0\,(0{,}01)\,2\,(0{,}02)\,3\,(0{,}1)\,6\,(0{,}2)\,10$ [118]

$$\frac{x}{\sqrt{1-x^2}}$$

10 dec. $x = 0\,(0{,}005)\,0{,}1$
9 dec. $x = 0{,}1\,(0{,}001)\,0{,}199$
8 dec. $x = 0{,}2\,(0{,}001)\,0{,}9$
7 dec. $x = 0{,}9\,(0{,}0005)\,0{,}96\,(0{,}0002)\,0{,}99\,(0{,}0001)$
$0{,}995\,($ var. $)\,0{,}999\,999\,999$ [191]
5 dec. $x = 0\,(0{,}01)\,1$ [118]

$$\frac{\sqrt{\frac{2}{\pi}x}}{2-x}$$

3—4 dec. $x = 0{,}001;\ 0{,}01;\ 0{,}05;\ 0{,}1\,(0{,}1)\,0{,}5$ [254]

$$\frac{\sqrt{1+x^2}}{x}$$

4 dec. $x = 1\,(0{,}001)\,3\,(0{,}01)\,10\,(0{,}5)\,30\,(10)\,50$ [221]

$$\frac{x-1}{\sqrt{x}}$$

7 dec. $x = 0\,(0{,}05)\,1\,(0{,}1)\,51{,}1$ [198]

$$\frac{\frac{3}{2}\sqrt{\frac{\pi}{2}x}}{(2-x)^{3/2}}$$

3–4 dec. $x = 0{,}001;\ 0{,}01;\ 0{,}05;\ 0{,}1;\ 0{,}2\,(0{,}1)\,0{,}5$ [254]

$$\frac{x}{\sqrt{x^2+y^2}}$$

4 dec. $x = 1\,(1)\,100. \quad y = 2\,(2)\,30\,(5)\,100$ [118]

$$\frac{\sqrt{1+x^2}-1}{x}$$

6 fig. $x = 0{,}01\,(0{,}001)\,1$ [137]

$$\frac{1-\sqrt{1-x^2}}{x}$$

4 dec. $x = 0\,(0{,}001)\,0{,}8$ [124]

$$\frac{x-1}{\sqrt{x(x+1)}}$$

10 dec. $x = 2\,(1)\,25$ [200]

COMPLEX NUMBERS AND THEIR POWERS

Complex numbers

$$Re^{i\alpha} = 1 + re^{i\varphi}$$

The purpose of the table is to carry out the operations of addition and subtraction of complex numbers.

$r = 0\,(0{,}01)\,1. \quad \varphi = 0\,(1°)\,180°$ [270]

The modulus R is given to 5 decimal places and the argument α to 3 decimal places.

Powers of complex numbers

4 dec.	$n = 1\,(1)\,10.\ r = 0{,}1\,(0{,}1)\,2.\ \varphi = 10°\,(10°)\,180°$	[118]

Reciprocals of complex numbers

$$\frac{1}{1+ix} \ \& \ \frac{1}{x+i}$$

4—5 fig.	$x = 0\,(0{,}01)\,1$	[105] [150] [163]

Square roots of complex numbers

$$\sqrt{1+ix} = u + iv \ \& \ \sqrt{x+i} = U + iV$$

4—5fig.		$x = 0\,(0{,}01)\,1$	[105] [150] [163]
4 dec.	v	$x = 0\,(0{,}002)\,0{,}6\,(0{,}02)\,1{,}18$	[260]
3 dec.	v	$x = 1{,}2\,(0{,}02)\,37\,(1)\,250$	[260]
3 dec.	u	$x = 0\,(0{,}002)\,0{,}6\,(0{,}02)\,37\,(1)\,250$	[260]

$$\sqrt{1+ix} = u + iv \ \& \ \sqrt{-1+ix} = v + iu$$

4 dec.	$x = 0\,(0{,}01)\,1$	[118]

$$\sqrt{x+i} = U + iV \ \& \ \sqrt{-x+i} = V + iU$$

4 dec.	$x = 0\,(0{,}01)\,1$	[118]

VARIOUS POWERS OF CERTAIN NUMBERS

The tables here enumerated of the powers of selected constants are not exhaustive. We give belo only those tables which are closely allied in struc ture to the section of tables of the powers of arbitrary numbers. See Ch. 14 for a more detailed account of constants and their powers

Tables of exponential functions are described in Ch. 3.

$$2^n$$

exact	$n = 1\,(1)\,64$	[25] [251]
exact	$n = 1\,(1)\,45$	[172] [204] [238]
exact	$n = 1\,(1)\,40$	[5] [102] [139] [192] [228]
exact	$n = 1\,(1)\,25$	[211]
exact	$n = 1\,(1)\,24$	[23] [73] [77]
exact	$n = 1\,(1)\,20$	[24] [83] [100] [139] [228]

exact	$n = 1\,(1)\,19$	[18]
exact	$n = 7\,(1)\,17$	[45]
exact	$n = 1\,(12)\,721$	[223]
235—1205 fig.	$n = 778;\ 889;\ 971;\ 1000;\ 2000;\ 2222;\ 3000;\ 3889;\ 4001$	[265]
5 fig.	$n = 0\,(1)\,100\,(10)\,200\,(25)\,300\,(50)\,1000\,(500)\,4000\,(1000)\,10\,000\,(10\,000)\,100\,000$	[118]

3^n

exact	$n = 1\,(1)\,40$	[139]
exact	$n = 1\,(1)\,36$	[172] [238]
exact	$n = 1\,(1)\,32$	[25]
exact	$n = 1\,(1)\,30$	[5] [102] [228]
exact	$n = 1\,(1)\,24$	[23] [73] [77]
exact	$n = 1\,(1)\,20$	[24] [83] [100] [139] [228]
exact	$n = 1\,(1)\,19$	[18]
exact	$n = 7\,(1)\,17$	[45]

5^n

exact	$n = 1\,(1)\,40$	[139]
exact	$n = 1\,(1)\,32$	[25]
exact	$n = 1\,(1)\,27$	[172] [238]
exact	$n = 1\,(1)\,24$	[5] [102] [228]
exact	$n = 1\,(1)\,20$	[24]

$\left(\frac{1}{2}\right)^n$

5 dec.	$n = 1\,(1)\,20$	[83] [100] [139] [228]
5 fig.	$n = 0\,(1)\,100\,(10)\,200\,(25)\,300\,(50)\,1000\,(500)\,4000\,(1000)\,10\,000\,(10\,000)\,100\,000$	[118]

$\left(\frac{1}{3}\right)^n$

5 dec.	$n = 1\,(1)\,20$	[83] [100] [139] [228]

10^n *

4 dec.	$n = 0{,}1\,(0{,}1)\,0{,}9;\ 0{,}01;\ 0{,}001$	[45]
3 dec.	$n = 0\,(0{,}001)\,1$	[72]

10^{10^n}

6 fig.	$n = -3{,}6\,(0{,}1) - 3\,(0{,}05) - 2{,}1\,(0{,}02) - 2\,(0{,}01) - 1{,}8\,(0{,}005)\,0{,}7$	[118]

[1] See also the tables of antilogarithms in Ch. 4.

Chapter 2

TRIGONOMETRIC FUNCTIONS. VARIOUS VALUES CONNECTED WITH THE CIRCLE AND THE SPHERE[1]

This chapter contains, in the main, the characteristics of tables of natural values of trigonometric and reciprocal trigonometric functions. For the sake of convenience we give separately tables the argument of which is given:

1) in degrees, minutes and seconds, 2) in tenths of a degree, 3) in radians, 4) in grads, 5) in units of time, 6) in protractor divisions.

Tabulation of the three functions sin x, tan x (t in the text), sec x is adequate in all cases since values of cos x, cotan x (ctg in text) and cosec x may be obtained from these tables by passing to the complementary argument (for example $90^\circ - x$ in measurement by degrees).

Tables of circumferences, lengths of arc, the values of various lines in the circle, areas, volum and other geometrical tables are used widely in technical calculations. The following symbols have been used in describing these tables:

r = radius of a circle; d = diameter of a circle; l = length of arc; c = length of chord; h = height of the segment or altitude; α = the central angle.

The chapter contains the following sections:

- natural values of trigonometric functions;
- various expressions containing trigonometric functions;
- derivatives and powers of trigonometric functions
- trigonometric and reciprocal trigonometric functions of a complex variable;
- geometric quantities;
- tables for converting from one angular measure to another;
- interpolation tables.

---

[1] See pages 421 and 550 for the references to Chapter 1.

NATURAL VALUES OF TRIGONOMETRIC FUNCTIONS

$\sin x$ & $\cos x$

Argument in degrees, minutes and seconds

33 dec.	$x = 0\,(1'')\,100''\,(100'')$ 1000'' (1000'') 45°	[434]
30 dec.	$x = 1°\ (1°)\ 90°$	[388] [449]
25 dec.	$x = 1°\ (1°)\ 90°$	[364]
22 dec.	$x = 10''\ (20'')\ 34'\ 50''$	[330]
21 dec.	$x = 0\ (1'')\ 10'\ (10'')\ 90°$	[320]
17 dec.	$x = 0\ (9')\ 90°$	[140]
15 dec.	$x = 0\ (10'')90°$	[140]
15 dec.	$x = 0\ (1'')\ 1°\ (10'')$ 89° (1'') 90°	[330]
15 dec.	$x = 0\ (10'')\ 90°$	[44]
15 dec.	$x = 0\ (15')\ 90°$	[283] [284]
10 dec.	$x = 0\ (1'')\ 12'$	[400] [402]
10 dec.	$x = 0\ (10'')\ 90°$	[338]
10 dec.	$x = 0\ (1°)\ 90°$	[23] [243] [331] [346] [383]
9 dec.	$x = 0\ (10'')\ 1°$	[399]
8 dec.	$x = 0\ (1'')\ 90°$	[217] [305] [326]
8 dec.	$x = 0\ (10'')\ 90°$	[26]
8 dec.	$x = 0\ (1')\ 90°$	[246]
8 dec.	$x = 0\ (1')\ 1°$	[98]
8 dec.	$x = 0\ (1°)\ 90°$	[425]
7—8 fig.	$x = 0\ (10'')\ 5°\,50'$	[125]
7 dec.	$x = 5°\,50'\ (10'')\ 90°$	[125]
7 dec.	$x = 0\ (1'')\ 90°$	[407]
7 dec.	$x = 0\ (1'')\ 45°$	[239]
7 fig.	$x = 0\ (10'')\ 1°$	[164] [166] [399]
7 dec.	$x = 1°\,(10'')\,90°$	[164] [166] [399]
7 dec.	$x = 0\ (10'')\ 90°$	[155] [254] [262] [264] [310]
7 dec.	$x = 0\ (1')\ 90°$	[54] [99] [103] [107] [230] [312] [334] [375] [384] [389] [401] [403] [405]
7 dec.	$x = 0\ (30')\ 90°$	[98]
7 dec.	$x = 0\ (1°)\ 90°$	[104] [130]

6—7 dec.	$x = 0\ (1')\ 90°$	[8]
6 dec.	$x = 0\ (1'')\ 90°$	[120]
6 dec.	$x = 0\ (10'')\ 90°$	[80] [132] [165] [324]
6 dec.	$x = 0\,(10'')\,6°;\ 84°\,(10'')\,90°$	[124]
6 dec.	$x = 0\ (1')\ 90°$	[29] [120] [138] [456]
6; 5 dec.	$x = 0\ (1')\ 6°;\ 6°\,(1')\,84°$	[331]
6 dec.	$x = 84°\ (1')\ 90°$	
6 dec.	$x = 0\ (10')\ 90°$	[106]
6 dec.	$x = 0\ (1°)\ 360°$	[160]
5—7 dec.	$x = 0\ (10'')\ 12'\ (1')$ $89°\,48'\ (10'')\ 90°$	[222]
6—4 fig; 5 dec.	$x=0(1')4°;4°(1')86°$	[160] [190]
6—7 dec.	$x = 86°\ (1')\ 90°$	
5 dec.	$x = 0\ (10'')\ 90°$	[73] [308]
5 dec.	$x = 0\ (30'')\ 90°$	[258]
5 dec.	$x = 0\ (30'')\ 6°\ (1')\ 90°$	[299]
5—6 fig.	$x = 0\ (1')\ 6°$	[123]
5 dec.	$x = 6°\ (1')\ 90°$	
5 dec.	$x = 0\ (1')\ 90°$	[19] [35] [43] [61] [70] [74] [86] [87] [114] [122] [127] [154] [181] [183] [213] [220] [228] [236] [251] [252] [253] [286] [311] [339] [342] [344] [366] [415]
5 dec.	$x = 0\ (3')\ 90°$	[195]
5 dec.	$x = 0\ (6')\ 90°$	[146]
5 dec.	$x = 0\ (10')\ 90°$	[3] [5] [7] [16] [20] [25] [26] [40] [53] [56] [67] [69] [94] [128] [205] [209] [272] [337] [362] [370] [373] [374] [390] [459]
5 dec.	$x = 0\ (15')\ 90°$	[13]
5 dec.	$x = 0\,(1°)\,90°$	[25] [267]
4 dec.	$x = 0\ (1')\ 5°\ 20'$	[340]
5 dec.	$x = 5°\ 20'\ (1')\ 90°$	
4 dec.	$x = 0\ (1')\ 90°$	[60] [81] [127] [303]
4 dec.	$x = 0\ (6')\ 90°$	[6] [17] [32] [46] [47] [63] [97] [112] [131] [135] [188] [265] [266] [271] [336] [351]
4 dec.	$x = 0\ (10')\ 90°$	[9] [21] [24] [27] [36] [49] [64] [65] [66] [78] [81] [83] [93] [115] [118] [121] [150] [161] [214] [248] [342] [350] [352] [357] [365] [409] [417]

4 dec.	$x = 0\ (20')\ 90°$	[45]
4 dec.	$x = 0\ (1°)\ 90°$	[11] [51] [63] [91] [96] [101] [112] [116] [119] [120] [156] [200] [228] [261] [341] [414]
3 dec.	$x = 0\ (10')\ 90°$	[30] [31] [34] [52] [55] [71]
3 dec.	$x = 0\ (30')\ 90°$	[82] [85]

Argument in tenths of a degree

19 dec.	$x = 0\ (0°,625)\ 90°$	[170]
15 dec.	$x = 0\ (0°,01)\ 90°$	[170] [388]
10 dec.	$x = 0\ (0°,01)\ 90°$	[204]
9 dec.	$x = 0\ (0°,01)\ 0°,06$	[321]
8 dec.	$x = 0°,06\ (0°,01)\ 0°,58$	[321]
8 dec.	$x = 0\ (0°,01)\ 90°$	[175]
7 dec.	$x = 0\ (0°,001)\ 90°$	[321]
7 dec.	$x = 0\ (0°,01)\ 90°$	[316]
6 dec.	$x = 0\ (0°,01)\ 90°$	[111] [196] [325] [456]
6 dec. $\sin\frac{180}{x}$	$x = 1\ (1)\ 500$	[325]
5 dec.	$x = 0\ (0°,01)\ 90°$	[202] [285] [289]
5 dec.	$x = 0\ (0°,1)\ 90°$	[14] [228] [277] [339]
5 dec.	$x = 0\ (0°,1)\ 10°$	[346]
4 dec.	$x = 10°\ (0°,1)\ 90°$	[346]
4 dec.	$x = 0\ (0°,1)\ 90°$	[168] [179] [180] [192] [303]
4 dec.	$x = 0\ (0°,1)\ 4°,9$	[418]
4 fig. $\sin\frac{180}{x}$	$x = 1\ (1)\ 99$	[34]
3 dec.	$x = 0\ (0°,1)\ 90°$	[62]

Argument in radians

25 dec.	$x = m\ 10^{-n}.\ m = 1\ (1)\ 9.\ n = 4\ (1)\ 10$	[438]
23 dec.	$x = 0\ (0,001)\ 1,6\ (0,1)\ 10\ (1)\ 100$	[438]
20—25 dec.	$x = m\ 10^{-n}.\ m = 1\ (1)\ 9.\ n = 6\ (1)\ 10$	[234]
20; 18 dec.	$x = 0\ (0,00001)\ 0,001;\ 3\ (0,01)\ 9,99$	[234]
15 dec.	$x = 10\ (0,1)\ 20\ (1)\ 100$	[234]
20 dec.	$x = 1\ (1)\ 10$	[424]

15 dec.		$x = 0\ (0{,}001)\ 0{,}5$	[48]
15 dec.		$x = 0\ (0{,}1)\ 50$	[171]
15 dec.		$x = 0\ (0{,}1)\ 10$	[426]
15 dec.		$x = 20\ (0{,}2)\ 40$	[428]
15 dec.		$x = 10\ (0{,}1)\ 20\ (0{,}5)\ 50$	[427]
15 dec.		$x = m\ 10^{-n}.\ m = 1\ (1)\ 9.\ n = 1\ (1)\ 5$	[385]
12 dec.		$x = 0\ (0{,}00001)\ 0{,}00999$	
12 dec.		$x = 0{,}1\ (0{,}001)\ 2{,}999$	[234]
10 dec.		$x = 0{,}001\ (0{,}0001)\ 0{,}0999$	
11 dec.		$x = 0\ (0{,}001)\ 1{,}6$	[171] [426]
10 dec.		$x = 0\ (0{,}0001)\ 0{,}01;\ 10\ (1)\ 40$	[232]
10 dec.		$x = 0\ (0{,}01)\ 1{,}6;\ 2\ (1)\ 40$	[378]
10 dec.		$x = 0\ (0{,}1)\ 10\ (1)\ 100$	[153]
9—10 dec.		$x = \frac{n}{10};\ \frac{n}{100};\ \frac{n}{1000}.\ n = 0\ (1)\ 99$	[346]
9 dec.		$x = 0\ (0{,}0001)\ 2;\ 0\ (0{,}1)\ 10$	[379]
8 dec.		$x = 0\ (0{,}001)\ 25;\ 0\ (1)\ 100$	[385]
6 dec.		$x = 0\ (0{,}\ 001)\ 1{,}6$	[176] [456]
6 dec.		$x = 0\ (0{,}\ 01)\ 2$	[287]
6 dec.		$x = 0\ (0{,}1)\ 50\ (1)\ 100\ (100)\ 5000\ (1000)\ 10\,000$	[456]
5 dec.		$x = 0\ (0{,}0001)\ 0{,}1\ (0{,}001)\ 3\ (0{,}01)\ 6{,}3\ (0{,}1)\ 10$	[233]
5 dec.		$x = 0\ (0{,}0001)\ 0{,}1\ (0{,}001)\ 1{,}6$	[153]
5 dec.		$x = 0\ (0{,}001)\ 10$	[92]
5 dec.	$\sin x$	$x = 0{,}5\ (0{,}001)\ 1{,}6$	[185]
5 dec.		$x = 0\ (0{,}001)\ 1{,}569$	[300]
5 dec.		$x = 0\ (0{,}01)\ 11$	[346]
5 dec.		$x = 0\ (0{,}01)\ 10$	[232]
5 dec.		$x = 0\ (0{,}01)\ 1{,}75\ (0{,}05)\ 3$	[146]
5 dec.		$x = 0\ (0{,}01)\ 3{,}5\ (0{,}02)\ 5\ (0{,}1)\ 10$	[160]
5 dec.		$x = 0\ (0{,}01)\ 1{,}5\ (0{,}1)\ 10$	[61]
5 dec.		$x = 0\ (0{,}01)\ 1{,}6\ (0{,}1)\ 6$	[94] [128] [209]
5 dec.		$x = 0\ (0{,}01)\ 2$	[342]
5 dec.		$x = 0\ (0{,}01)\ 1{,}6$	[339] [381]
4 dec.		$x = 0\ (0{,}001)\ 2$	[179]
4 dec.		$x = 0\ (0{,}001)\ 1{,}57;\ 0\ (0{,}1)\ 7{,}9$	[303]
4 dec.		$x = 0\ (0{,}0029)\ 1{,}5708$	[36]
4 dec.		$x = 0\ (0{,}01)\ 6{,}28$	[81]
4 dec.		$x = 0\ (0{,}01)\ 3{,}2$	[138]
4 dec.		$x = 0\ (0{,}01)\ 1{,}6$	[21] [181] [228]
4 dec.		$x = 0\ (0{,}01)\ 1{,}59$	[301]

4 dec.	$x = 0$ (0,01) 1,58	[345]
4 dec.	$x = 0$ (0,01) 1,57	[227]
4 dec.	$x = 0{,}1$ (0,01) 1,57; 1,5 (0,1) 10,4	[192]
4 dec.	$x = 0{,}1$ (0,01) 1,57	[278]
4 dec.	$x = 0$ (0,01) 1,57 (0,1) 7,9	[63]
3 dec.	$x = 0$ (0,02) 3,2	[19]
3 dec.	$x = 0$ (var.) 1,571	[2]

Argument in grads

20 dec.	$x = 0\ (1^g)\ 100^g$	[140]
15 dec.	$x = 0\ (0^g{,}1)\ 100^g$	[178]
10 dec.	$x = 0\ (0^g{,}001)\ 0^g{,}1$	[367]
10 dec.	$x = 0\ (0^g{,}01)\ 0^g{,}5;\ 0\ (1^g)\ 100^g$	[244] [331]
10 dec.	$x = 0\ (0^g{,}1)\ 100^g$	[234] [367
8 dec.	$x = 0\ (0^g{,}01)\ 100^g$	[343]
7 dec.	$x = 0\ (0^g{,}001)\ 100^g$	[79]
7 dec.	$x = 0\ (0^g{,}001)\ 3^g\ (0^g{,}01)\ 97^g\ (0^g{,}001)\ 100^g$	[238]
6 dec.	$x = 0\ (0^g{,}001)\ 100^g$	[208] [327;
6 dec.	$x = 0\ (0^g{,}01)\ 100^g$	[124] [191] [276]
6 dec.	$x = 0\ (0^g{,}001)\ 3^g\ (0^g{,}01)\ 97^g\ (0^g{,}001)\ 100^g$	[203]
6; 5 dec.	$x = 0\ (0^g{,}01)\ 7^g;\ 7^g\ (0^g{,}01)\ 93^g$	[331]
6 dec.	$x = 93^g\ (0^g{,}01)\ 100^g$	
5 dec.	$x = 0\ (0^g{,}0001)\ 100^g$	[215]
5 dec.	$x = 0\ (0^g{,}01)\ 100^g$	[133] [149] [367] [410]
5 dec.	$x = 0\ (0^g{,}02)\ 100^g$	[159]
5 dec.	$x = 0\ (0^g{,}1)\ 100^g$	[232] [412]
5; 4 dec.	$x = 0\ (0^g{,}1)\ 10^g;\ 10^g\ (0^g{,}1)\ 100^g$	[346]
4 dec.	$x = 0\ (0^g{,}02)\ 100^g$	[372]
4 dec.	$x = 0\ (0^g{,}1)\ 100^g$	[77] [212] [224] [244] [373]
4 dec.	$x = 0\ (0^g{,}1)\ 90^g$	[209]
3 dec.	$x = 0\ (0^g{,}1)\ 100^g$	[62]
3 dec.	$x = 0\ (1^g)\ 100^g$	[351]

Argument in units of time

7 dec.	$\cos x$	$x = 0\ (0^{sec},1)\ 1^{hrs}$	[348]$_2$
7 dec.	$\sin x$	$x = 0\ (1^{sec})\ 6^{hrs}$	[309]
5—6 fig.		$x = 0\,(4^{sec})\,24^{M}$	[126] [252]
5 dec.		$x = 24^{M}\,(4^{sec})\,6^{hrs}$	
4 dec.	$\sin x$	$x = 0\ (10^{sec})\ 6^{hrs}$	[193]
4 dec.		$x = 0\ (1^{d})\ 6^{hrs}$	[331]

Argument in protractor divisions

5 dec.	$x = 0\,(0—01)\,15—00$	[88]
5 dec.	$x = 0\ (0—01)\,60—00$	[117]
4 dec.	$x = 0\ (0—01)\,15—00$	[105]
3 dec.	$x = 0\ (0—10)\ 15—00$	[38]

One large division $(1—00) = 100$ small divisions.

sin *u* & cos *u*,

where $u = f(x)$.

$$f(x) = \frac{x}{\sqrt{2}}$$

12 dec.	$x = 0\ (0,1)\ 20$	[441]

$$f(x) = \ln x$$

10 dec.	$x=0,98\ (0,001)\ 1,02\ (0,1)\ 20$	[442]

$$f(x) = \frac{\pi}{2} x^2$$

4 dec.	$x = 0\ (0,1)\ 8,5$	[160]

tg *x* & ctg *x*

Argument in degrees, minutes and seconds

30 dec.	$x = 0\,(1°)\,90°$	[449]
17 dec.	$x = 0\,(9')\,90°$	[140]
15 dec.	$x = 0\,(10'')\,90°$	[141]
11 fig.	$x = 0\,(10')\,84°10'$	[447]
10 dec.	$x = 84°\,20'(10')\,90°$	
10 dec.	$x = 0\,(10'')\,90°$	[338]
10 dec.	$x = 0\,(1°)\,90°$	[23][243][331]
8 dec.	$x = 0\,(1'')\,71°30'$	[326]
8 fig.	$x = 71°30'\,(1'')\,88°12'$	
5—7 fig.	$x = 88°12'\,(1'')\,90°$	

8 dec.	$x = 0\,(1'')\,89°25'37''$	[218]
7 dec.	$x = 89°25'38''\,(1'')\,90°$	
8 dec.	$x = 0\,(10'')\,90°$	[26]
8 fig.	$x = 0\,(1')\,90°$	[312]
8 dec.	$x = 0\,(1')\,45°$	[246]
7—8 fig.	$x = 0\,(10'')\,80°$	[125]
7 fig.	$x = 80°\,(1'')\,90°$	
7 dec.	$x = 0\,(10'')\,84°17'20''$	[155] [254]
8 fig.	$x = 84°17'30''\,(10'')\,90°$	
7 dec.	$x = 0\,(1')\,84°17'$	[255] [334] [384]
8 fig.	$x = 84°18'\,(1')\,90°$	
7—8 fig.	$x = 0\,(1')\,90°$	[230] [375] [401] [405]
7—8 fig.	$x = 0\,(1°)\,90°$	[104]
7 dec.	$x = 0\,(1'')\,90°$	[407]
7 fig.	$x = 0\,(10'')\,1°$	[164] [166] [399]
7 dec.	$x = 0\,(10'')\,84°\,(1'')\,90°$	
7 dec.	$x = 0\,(1'')\,45°$	[241]
7 dec.	$x = 45°\,(1'')\,84°18'$	[242]
6—5 dec.	$x = 84°18'\,(1'')\,90°$	
7 dec.	$x = 0\,(1')\,90°$	[103] [107] [389]
7 dec.	$x = 0\,(1°)\,90°$	[130]
6—8 fig.	$x = 0\,(1')\,90°$	[8] [384] [403]
6 dec.	$x = 0\,(10'')\,84°17'20''$	[165]
6—7 fig.	$x = 84°17'30''(10'')87°(1'')90°$	
6 fig.	$x = 87°\,(1'')\,90°$	[120]
6 dec.	$x = 0\,(10'')\,83°$	[132]
6 fig.	$x = 83°\,(10'')\,88°\,(1'')\,90°$	
6 dec.	$x = 0\,(10'')\,60°$	[80] [324]
6 fig.	$x = 60°\,(10'')\,90°$	
4—6 fig.	$x = 88°40'\,(1'')\,90°$	
6 dec. ; 6 fig.	$x = 0\,(10'')\,6°;\; 84°\,(10'')\,90°$	[124]
6 dec.	$x = 0\,(1')\,90°$	[29] [54] [138]
6 dec.	$x = 4°\,(1')\,85°$	[90]
6 dec.	$x = 0\,(10')\,90°$	[106]
6 dec.	$x = 0\,(1°)\,360°$	[160]
5—6 fig; 5 dec.	$x = 0\,(1')\,4°;\; 4°\,(1')\,86°$	[160] [190]
6—5 dec.	$x = 86°\,(1')\,90°$	
6; 5 dec.	$x = 0\,(1')\,6°;\; 6°\,(1')\,86°$	[331]
6 dec.	$x = 86°\,(1')\,90°$	
5—7 fig.	$x = 0\,(10'')\,12'\,(1')$ $89°48'\,(10'')\,90°$	[222]
5—7 fig.	$x = 0\,(1')\,86°\,(10'')$ $89°40'(1'')90°$	[60] [456]

5—6 fig.	$x = 0\,(1')\,90°$	[19] [43] [99] [127]
5 dec.	$x = 0\,(1')\,85°$	[183]
6 fig.	$x = 85°\,(1')\,90°$	
5 dec.	$x = 0\,(1')\,84°17'$	[251] [253] [366]
6 fig.	$x = 84°18'\,(1')\,90°$	
5—6 fig.	$x = 0\,(3')\,90°$	[193]
5—6 fig.	$x = 0\,(10')\,90°$	[370]
5 dec.	$x = 0\,(10'')\,63°$	[308]
5 fig.	$x = 63°\,(10'')\,82°30'\,(1'')\,90°$	
5—6 fig.	$x = 0\,(10'')\,88°\,(1'')\,90°$	[73]
5 fig.	$x = 0\,(10'')\,90°$	[208]
5 fig.; 5 dec.	$x = 0\,(1')\,6°;\ 6°\,(1')\,60°$	[123]
5 fig.	$x = 60°\,(1')\,80°\,(10'')$ $89°\,(1'')\,90°$	
5 dec.	$x = 0\,(30'')\,90°$	[258]
5 dec.	$x = 0\,(1')\,71°$	[87]
5 fig.	$x = 71°\,(1')\,80°(10'')$ $89°\,(1'')\,90°$	
5 dec.	$x = 0\,(1')71°$	[213]
3—6 fig.	$x = 71°(1')\ 79°\ (10'')$ $89°55'\,(1'')\,90°$	
5 dec.	$x = 0\,(1')45°$	[154] [181] [236] [415]
5 fig.	$x = 45°\,(1')\ 90°$	
5 dec.	$x = 0\,(1')\,90°$	[86] [252] [339]
5 fig.	$x = 0\,(1')\,90°$	[35] [61] [70] [122] [228] [286] [342] [344]
5 fig.	$x = 0\,(1')\,6°$	[220]
5 dec.	$x = 6°\,(1')\,90°$	
5 dec.	$x = 0\,(6')\,45°$	[146]
5 fig.	$x = 45°\,(6')72°\,(1')90°$	
5 fig.	$x = 0\,(10')\,90°$	[36]
5 dec.	$x = 0\,(10')\,90°$	[3] [5] [7] [16] [20] [25] [26] [40] [53] [56] [67] [69] [94] [128] [205] [209] [272] [337] [362] [373] [374] [390] [459]
5 dec.	$x = 0\,(15')\,90°$	[13]
5 dec.	$x = 0\,(1°)\,90°$	[25] [267]
4—7 fig.	$x = 0\,(1')\,90°$	[340]
4 dec.; 4 fig.	$x = 0\,(1')\,60°;\ 60°\,(1')\,90°$	[127]
4 dec.; 4—5 fig.	$x = 0\,(1')\,80°;\ 80°\,(1')\,90°$	[303]
4—5 fig.	$x = 0\,(6')\,90°$	[63]

4—5 fig.	$x = 0\,(10')\,76°\,(1')\,90°$	[27]
4—5 fig.	$x = 0\,(10')\,90°$	[9] [49] [78] [352]
4—5 fig.	$x = 0\,(20')\,90°$	[45]
4—5 fig.	$x = 0\,(1°)\,90°$	[156]
4 dec.	$x = 82°30'\,(1'')\,90°$	[208]
4 fig.	$x = 72°\,(1')\,90°$	[192]
4 fig.	$x = 0\,(6')\,81°\,(1')\,90°$	[6]
4 fig.	$x = 0\,(6')\,76°\,(1')\,90°$	[17]
4 dec.	$x = 0\,(6')\,90°$	[32] [46] [131] [135] [351]
4 fig.	$x = 0\,(6')\,90°$	[47] [97] [112] [188] [265] [266] [271] [336]
4 dec.	$x = 0\,(10')\,90°$	[64] [65] [66] [81] [83] [115] [118] [121] [161] [248] [350] [357] [365] [409] [417]
4 fig.	$x = 0\,(10')\,90°$	[21] [24] [93] [150] [214] [342] [357]
4 dec.	$x = 0\,(1°)\,90°$	[51] [91] [101] [112] [116] [119] [200] [228] [341] [414]
4 fig.	$x = 0\,(1°)\,90°$	[11] [63] [96] [261]
4 dec. 3 dec.	$x = 86°\,(10'')\,88°$ $x = 88°\,(10'')\,90°$	[456]
3—4 fig.	$x = 0\,(10')\,90°$	[30] [31] [34] [52] [55] [71]
3—4 dec.	$x = 0\,(1°)\,90°$	[1]
3 dec.	$x = 0\,(30')\,90°$	[82] [85]

Argument in tenths of a degree

10 dec.	$x = 0\,(0°{,}01)\,90°$	[170]
10 dec.	$x = 0\,(0°{,}01)\,45°$	[204]
9 dec. 8 dec.	$x = 0\,(0°{,}01)\,0°{,}06$ $x = 0°{,}06(0°{,}01)0°{,}58$	[79] [321]
8 dec. 8 fig.	$x = 0(0°{,}01)\,45°$ $x = 45°(0°{,}01)\,90°$	[175]
7 dec. 6 dec. 6—7 fig.	$x = 0\,(0°{,}001)\,45°$ $x = 45°(0°{,}001)84°$ $x = 84°(0°{,}001)\,90°$	[79] [321]
7 dec.	$x = 0\,(0°{,}01)\,90°$	[316]
6 dec. 6 fig.	$x = 0\,(0°{,}01)\,45°$ $x = 45°\,(0°{,}01)\,90°$	[111]
6 dec.	$x = 0\,(0'{,}01)\,0°10'\,(0'{,}1)\,4°$	[90]
6 dec.	$x = 0\,(0°{,}01)\,90°$	[196]
6 dec. 5 dec. 4—6 fig.	$x = 0(0°{,}01)65°$ $x = 65°\,(0°{,}01)82°$ $x = 82°\,(0°{,}01)\,90$	[325]

5—7 fig.		$x = 0\,(0^\circ,01)\,90^\circ$	[456]
5 dec.		$x = 0\,(0^\circ,01)\,45^\circ$	[285]
5 fig.		$x = 45^\circ\,(0^\circ,01)\,90^\circ$	
5 dec.		$x = 0(0^\circ,01)\,82^\circ$	[289]
4 dec.		$x = 82^\circ\,(0^\circ,01)\,90^\circ$	
5 dec.		$x = 0\,(0^\circ,01)\,45^\circ$	[202]
4 dec.		$x = 45^\circ(0^\circ,01)83^\circ$	
4—6 fig.		$x = 83^\circ\,(0^\circ,01)\,90^\circ$	
5 dec.		$x = 0\,(0^\circ,1)\,90^\circ$	[14] [277]
5 fig.		$x = 0\,(0^\circ,1)\,90^\circ$	[228] [339]
4—5 fig.		$x = 0\,(0^\circ,1)\,86^\circ\,(0^\circ,01)\,90^\circ$	[303]
4 dec.		$x = 0(0^\circ,1)80^\circ(0^\circ,01)84^\circ,99$	[180]
5 fig.		$x = 85^\circ\,(0^\circ,01)\,90^\circ$	
5; 4 dec.		$x = 0(0^\circ,1)10^\circ;\ 10^\circ(0^\circ,1)90^\circ$	[346]
4 dec.		$x = 0\,(0^\circ,1)\,90^\circ$	[168] [169]
4 dec.	$\operatorname{tg} x$	$x = 0\,(0^\circ,1)\,84^\circ$	[179]
4 fig.	$\operatorname{tg} x$	$x = 84^\circ,1\,(0^\circ,1)\,90^\circ$	
4 dec.		$x = 0\,(0^\circ,1)\,10^\circ$	[110]
4 fig.		$x = 0\,(0^\circ,1)\,79^\circ\,(0^\circ,01)\,90^\circ$	[192]
3 dec.		$x = 0\,(0^\circ,1)\,90^\circ$	[62]

Argument in radians

20 dec.		$x = 0\,(0,00001)0,001$	[234]
10 dec.	$\operatorname{tg} x$	$x = 0,001(0,0001)0,1(0,001)\ 3\,(0,01)\,10\,(0,1)\,20\,(1)\,50$	
5 dec.		$x = 1,568\,(0,0001)\,1,5707$	
8—9 fig.		$x = 0(0,0001)2$	[380]
10 dec.		$x = 2(0,1)10$	
7 dec.	$\operatorname{tg} x$	$x = 0\,(0,001)\,0,2$	[369]
5—7 fig.		$x = 0\,(0,001)\,1,6$	[456]
5 dec.	$\operatorname{tg} x$	$x = 0\,(0,0001)\,0,1\,(0,001)\ 3\,(0,01)\,6,3\,(0,1)10$	[233]
5 dec.		$x = 0\,(0,001)\,10$	[92]
5 fig.	$\operatorname{tg} x$	$x = 0\,(0,001)\,1,569$	[300]
5 dec.	$\operatorname{tg} x$	$x = 0\,(0,001)1$	[185]
5 dec.		$x = 0\,(0,01)\,3,5(0,02)5(0,1)10$	[160]
5 dec.	$\operatorname{tg} x$	$x = 1,56(0,001)1,59(0,01)10$	[232]
5 fig.		$x = 0\,(0,0029)\,1,5708$	[36]
5 dec.	$\operatorname{tg} x$	$x = 0\,(0,01)\,6$	[346]
5 fig.		$x = 0(0,01)\,1,75\,(0,05)\,3$	[146]
5 dec.	$\operatorname{tg} x$	$x = 0\,(0,01)\,1,5\,(0,1)\,10$	[61]
5 dec.	$\operatorname{tg} x$	$x = 0\,(0,01)\,1,6\,(0,1)\,6$	[94] [128] [209]

5 fig.		$x = 0\,(0,01)\,2$	[342]
5 dec.		$x = 0\,(0,01)\,1,6$	[339]
5 dec.	$\operatorname{tg} x$	$x = 0\,(0,01)\,1,6$	[381]
4 dec.	$\operatorname{tg} x$	$x = 0\,(0,01)\,1,47$	[227]
5 fig.	$\operatorname{tg} x$	$x = 1,48\,(0,01)\,1,57$	
4 dec.	$\operatorname{tg} x$	$x = 0\,(0,001)\,2$	[179]
4 fig.	$\operatorname{tg} x$	$x = 0\,(0,001)\,1,57$	[303]
4 dec.		$x = 0\,(0,01)\,6,28$	[81]
4 dec.		$x = 0\,(0,01)\,3,2$	[138]
4 dec.	$\operatorname{tg} x$	$x = 0\,(0,01)\,1,6$	[21] [181]
4 fig.		$x = 0\,(0,01)\,1,6$	[228]
4 dec. ; 4 fig.		$x = 0\,(0,01)\,1,24;\ 1,25\,(0,01)\,1,57$	[63]
4 dec.	$\operatorname{tg} x$	$x = 0,1\,(0,01)\,1,1$	[278]
3 dec.	$\operatorname{tg} x$	$x = 1,1\,(0,01)\,1,57$	
3 fig.		$x = 0\,(0,02)\,3,2$	[19]
3 dec.		$x = 0$ (var.) $1,571$	[2]

Argument in grads

20 dec.	$x = 0\,(1^g)\,100^g$	[140]
10 dec.	$x = 0(0^g,01)0^g,5;\ 0(1^g)\,100^g$	[244] [331]
8 dec.	$x = 0\,(0^g,01)\,50^g$	[343]
7 dec.	$x = 0\,(0^g,001)\,3^g(0^g,01)\ 97^g\,(0^g,001)\,100^g$	[238]
7 dec.	$x = 0\,(0^g,001)\,79^g,5$	[79]
7—5 fig.	$x = 79^g,5\ (0^g,001)100^g$	
6 dec.	$x = 0\,(0^g,001)\,74^g$	[327]
6 fig.	$x = 74^g\,(0^g,001)100^g$	
6 dec.	$x = 0\,(0^g,001)\,50^g$	[208]
6 fig.	$x = 50^g\,(0^g,001)\,100^g$	
6 dec.	$x = 0\,(0^g,001)\,3^g\,(0^g,01)\ 97^g\,(0^g,001)100^g$	[203]
6; dec.	$x = 0(0^g,01)\,7^g;\ 7^g(0^g,01)97^g$	[331]
6 dec.	$x = 97^g\,(0^g,01)\,100^g$	
6 dec.	$x = 0\,(0^g,01)\,67^g$	[124]
6 fig.	$x = 67^g\,(0^g,01)\,90^g\,(0^g,001)\,100^g$	
5 dec.	$x = 0\,(0^g,0001)\,100^g$	[215]
5 dec.	$x = 0\,(0^g,01)85^g$	[367]
5—6 fig.	$x = 85^g(0^g,01)97^g($ var. $)\,100^g$	

5 dec.		$x = 0\,(0^g,01)\,80^g$	[149]
5 fig.		$x = 80^g\,(0^g,01)\,97^g$	
2 dec.		$x = 97^g\,(0^g,01)100^g$	
5 dec.		$x = 0\,(0^g,01)\,85^g$	[398]
5 fig.		$x = 85^g\,(0^g,01)\,97^g$	
4 fig.		$x = 97^g\,(0^g,01)100^g$	
5 dec.		$x = 0\,(0^g,01)100^g$	[410]
5 dec.		$x = 0\,(0^g,02)100^g$	[159]
5 fig.		$x = 97^g(0^g,001)99^g,85$ $(0^g,0001)100^g$	[133]
5 dec.		$x = 0\,(0^g,01)85^g$;	
5 fig.		$x = 85^g\,(0^g,01)100^g$	
5 dec.		$x = 0\,(0^g,1)100^g$	[412]
4—5 dec.		$x = 0\,(0^g,1)\,100^g$	[346]
4 dec.		$x = 0\,(0^g,1)\,80^g$	[244]
4—5 fig.		$x = 80^g\,(0^g,1)\,100^g$	
4 dec.		$x = 0\,(0^g,1)\,100^g$	[77] [212] [224] [373]
4 dec.		$x = 0\,(0^g,1)\,90^g$	[209]
3—5 fig.		$x = 0\,(0^g,02)\,100^g$	[372]
3 dec.		$x = 0\,(0^g,1)100^g$	[62]
3 fig.		$x = 0\,(1^g)\,100^g$	[351]

Argument in units of time

7 dec.	tg x	$x = 0\,(0^{sec},1)\,1^{hrs}$	[348]$_1$
7 dec.	tg x	$x = 0\,(1^{sec})\,4^{hrs}$	[309]
5—7 fig.	tg x	$x = 4^{hrs}(1^{sec})\,6^{hrs}$	
5 fig.		$x = 0\,(4^{sec})\,24^M$	[126]
5 dec.		$x = 24^M\,(4^{sec})\,4^{hrs}$	
5 fig.		$x = 4^{hrs}(4^{sec})\,6^{hrs}$	
5 fig.		$x = 5^{hrs}20^M\,(1^{sec})$ $5^{hrs}52^M\,(0^{sec},1)\,6^{hrs}$	
5 dec.		$x = 0\,(4^{sec})\,6^{hrs}$	[252]
4—5 fig.	tg x	$x = 0\,(10^{sec})\,5^{hrs}(10^{sec})$ $5\ 40^M\,(1^{sec})\,6^{hrs}$	[193]
4 dec.		$x = 0\,(1^M)\,6^{hrs}$	[331]

Argument in protractor divisions

5 dec.	$x = 0\,(0-01)\,14-04$	[124]
6 fig.	$x = 14-05\,(0-01)\,15-00$	
5—6 fig.	$x = 0\,(0-01)\,60-00$	[117]
4 dec.	$x = 0\,(0-01)\,12-40$	[105]
4 fig.	$x = 12-40\,(0-01)\,15-00$	
3 dec.	$x = 0\,(0-10)\,15-00$	[38]

sec x & cosec x

Argument in degrees, minutes and seconds

17 dec.	$x = 0\,(9')\,90^\circ$	[140]
15 dec.	$x = 0\,(10'')\,90^\circ$	[142]
10 dec.	$x = 0\,(10'')\,90^\circ$	[338]
10 dec.	$x = 0\,(1')\,90^\circ$	[404]
8 dec.	$x = 0\,(1'')\,89^\circ25'37''$	[216]
10 fig.	$x = 89^\circ25'38''\,(1'')\,90^\circ$	
8 fig.	$x = 0\,(1')\,90^\circ$	[255] [312] [334] [389]
8 dec.	$x = 0\,(1')\,45^\circ$	[246]
8 fig.	$x = 0\,(1^\circ)\,90^\circ$	[104]
7 dec.	$x = 0\,(10'')\,60^\circ$	[125]
7 fig.	$x = 60^\circ\,(10'')\,80^\circ\,(1'')\,90^\circ$	
7 dec.	$x = 0\,(1')\,84^\circ15'$	[384]
8 fig.	$x = 84^\circ16'\,(1')\,90^\circ$	
7—8 fig.	$x = 0\,(1')\,90^\circ$	[8] [230] [375] [405]
7 dec.	$x = 0\,(1'')\,90^\circ$	[407]
7 dec.	$x = 0\,(1')\,90^\circ$	[103] [107] [254]
6 dec.	$x = 0\,(10'')\,83^\circ$	[132]
6—7 fig.	$x = 83^\circ(10'')\,88^\circ\,(1'')\,90^\circ$	
6 dec.	$x = 0\,(10'')\,60^\circ$	[80] [324]
6 fig.	$x = 60^\circ\,(10'')\,90^\circ$	
4—6 fig.	$x = 88^\circ40'\,(1'')\,90^\circ$	
6 fig.	$x = 0\,(1')\,90^\circ$	[120] [340]
6 dec.	$x = 0\,(1')\,90^\circ$	[138]
6 dec.	$x = 0\,(1^\circ)\,90^\circ$	[160]
6; 5 dec.	$x = 0\,(1')\,4^\circ;\; 4^\circ\,(1')\,86^\circ$	[331]
6 dec.	$x = 86^\circ\,(1')\,90^\circ$	
5—7 fig.	$x = 0\,(1')\,90^\circ$	[456]
5 dec.	$x = 0\,(1')\,84^\circ15'$	[183] [251]
6 fig.	$x = 84^\circ16'\,(1')\,90^\circ$	
5 dec.	$x = 0\,(1')\,71^\circ$	[87]
5 fig.	$x = 71^\circ\,(1')\,80^\circ\,(10'')\,89^\circ\,(1'')\,90^\circ$	

5 dec.	$x = 0(1')60^\circ$	[123]
5 fig.	$x = 60^\circ(1')80^\circ(10'')89^\circ(1'')90^\circ$	
5 fig.	$x = 0(1')90^\circ$	[35] [342] [415]
5 dec.	$x = 0(1')90^\circ$	[61] [319]
5 dec.	$x = 0(3')84^\circ15'$	[195]
5 fig.	$x = 84^\circ18'(3')90^\circ$	
5; 4 fig.	$x = 0(5')75^\circ;\ 75^\circ(5')90^\circ$	[286]
5 dec.	$x = 0(6')45^\circ$	[146]
5 fig.	$x = 45^\circ(6')72^\circ(1')90^\circ$	
4—6 fig.	$x = 0(1')90^\circ$	[70]
4 dec.	$x = 0(1')80^\circ$	[303]
4—5 fig.	$x = 80^\circ(1')90^\circ$	
4 dec.	$x = 0(1')60^\circ$	[127]
4 fig.	$x = 60^\circ(1')90^\circ$	
4—5 fig.	$x = 0(6')90^\circ$	[135] [188]
4 fig.	$x = 0(10')90^\circ$	[342]
4 dec.	$x = 0(10')90^\circ$	[64] [65] [81] [115]
4 fig.	$x = 0(1^\circ)90^\circ$	[261]
4 dec.	$x = 0(1^\circ)90^\circ$	[51] [228] [341] [414]
4 dec.	$x = 86^\circ(10'')88^\circ$	[456]
3 dec.	$x = 88^\circ(10'')90^\circ$	
3 dec.	$x = 0(30')90^\circ$	[82] [85]

Argument in tenths of a degree

10 dec.	$x = 0(0^\circ{,}01)90^\circ$	[170]
7 dec.	$x = 0(0^\circ{,}01)90^\circ$	[316]
6 dec.	$x = 0(0^\circ{,}01)90^\circ$	[196]
6 dec.	$x = 0(0^\circ{,}01)45^\circ$	[325]
5 dec.	$x = 45^\circ(0^\circ{,}01)82^\circ$	
4—6 fig.	$x = 82^\circ(0^\circ{,}01)90^\circ$	
5—7 fig.	$x = 0(0^\circ{,}01)90^\circ$	[456]
5 dec.	$x = 0(0^\circ{,}01)65^\circ$	[202]
4 dec.	$x = 65^\circ(0^\circ{,}01)83^\circ$	
4—6 fig.	$x = 83^\circ(0^\circ{,}01)90^\circ$	
5 dec.	$x = 0(0^\circ{,}01)82^\circ$	[289]
4 dec.	$x = 82^\circ(0^\circ{,}01)90^\circ$	
4—5 fig.	$x = 0(0^\circ{,}1)86^\circ$	[303]
4 fig.	$x = 86^\circ(0^\circ{,}01)90^\circ$	
4—5 fig.	$x = 60^\circ$ (var.) 90°	[41]
4 fig.	$x = 0(0^\circ{,}1)79^\circ(0^\circ{,}01)90^\circ$	[192]

Argument in radians

6—7 fig.	cosec x	$x = 0\,(0{,}001)\,1{,}6$	[456]
4 dec.	cosec x	$x = 0\,(0{,}0001)\,1$	
5 dec.		$x = 0\,(0{,}001)\,0{,}25$	[160]
5 dec.	sec x	$x = 1{,}56\,(0{,}001)\,1{,}59$	[232]
5 dec.		$x = 0\,(0{,}001)\,1{,}25$	[303]
4 dec.		$x = 1{,}25\,(0{,}001)\,1{,}57$	
4 fig.		$x = 0\,(0{,}01)\,1{,}57$	[192]
3 fig.		$x = 0{,}02\,(0{,}02)\,3{,}2$	[19]

Argument in grads

20 dec.	$x = 0\,(1^g)\,100^g$	[140]
6 dec.	$x = 0\,(0^g{,}001)\,3^g\,(0^g{,}01)\,97^g\,(0^g{,}001)\,100^g$	[203]
6 dec.	$x = 0\,(0^g{,}01)\,3^g$	[331]
5 dec.	$x = 3^g\,(0^g{,}01)\,97^g$	
6 dec.	$x = 97^g\,(0^g{,}01)\,100^g$	
4 dec.	$x = 0\,(0^g{,}1)\,80^g$	[244]
4—5 fig.	$x = 80^g\,(0^g{,}1)\,100^g$	

Argument in units of time

5 dec.	$x = 0\,(4^{\text{сек}})\,4^{\text{ч}}$	[126]
5 fig.	$x = 4^{\text{ч}}\,(4^{\text{сек}})\,6^{\text{ч}}$	
5 fig.	$x = 5^{\text{ч}}20^{\text{м}}\,(1^{\text{сек}})\,5^{\text{ч}}52^{\text{м}}\,(0^{\text{сек}}{,}1)\,6^{\text{ч}}$	
4—5 fig.	$x = 0\,(10^{\text{сек}})\,5^{\text{ч}}\,(10^{\text{сек}})\,5^{\text{ч}}40^{\text{м}}\,(1^{\text{сек}})\,6^{\text{ч}}$	[193]
4 dec.	$x = 0\,(1^{\text{м}})\,6^{\text{ч}}$	[331]

Argument in protractor divisions

4 dec. ; 4 fig. $x = 0\,(0—01)\,12—40;\ 12—40\,(0—01)\,15—00$ [105]

VARIOUS EXPRESSIONS CONTAINING TRIGONOMETRIC FUNCTIONS

$$\frac{\sin x}{x}$$

Argument in degrees, minutes and seconds

4 dec.	$x = 0\,(1°)\,180°\,(5°)\,1800°$	[295]
4 dec.	$x = 0\,(15°)\,540°$	[354]

Argument in tenths of a degree

9 dec.	$x = 0\,(0^\circ,5)\,3599^\circ,5$	[382]

Argument in radians

8 dec.	$x = 0\,(0{,}01)\,10\,(0{,}1)\,20\,(1)\,100$	[235]
7 dec.	$x = 0\,(0{,}1)\,3{,}1$	[172]
6 dec.	$x = 0\,(1)\,50$	[422]
4–5 fig.	$x = 0\,(0{,}01)\,3{,}14$	[134] [257]$_2$
4 dec.	$x = 0\,(0{,}01)\,3{,}14$	[160]
4 dec.	$x = 0\,(0{,}01)\,20\,(0{,}02)\,40\,(0{,}05)\,100$	[454]

Argument in grads

7 dec.	$x = 0\,(2^g)\,2000^g$	[263]
4 dec.	$x = 0\,(1^g)\,200^g\,(2^g)\,2000^g\,(10^g)\,8000^g$	[371]

$$\left[\frac{\sin x}{x}\right]^{(n)}$$

9 dec.	$n = 1\,(1)\,11.\;\; x = 0\,(0^\circ,5)\,3599^\circ,5$	[382]

$$\frac{\sin^2 x}{x^2}$$

Argument in degrees

4; 5 dec.	$x = 0\,(15^\circ)\,180^\circ;\;\; 180^\circ\,(15^\circ)\,540^\circ$	[354]

Argument in radians

8 dec.	$x = 0\,(0{,}01)\,10\,(0{,}1)\,20\,(1)\,100$	[235]
7 dec.	$x = 0\,(0{,}1)\,3{,}1$	[172]
6 dec.	$x = 0\,(1)\,50$	[422]
4–5 dec.	$x = 0\,(0{,}01)\,3{,}14$	[134[[257]$_2$
4 dec.	$x = 0\,(0{,}01)\,20\,(0{,}02)\,40\,(0{,}05)\,100$	[454]

Argument in grads

7 dec.	$x = 0\,(2^g)\,2000^g$	[263]
4 dec.	$x = 0\,(1^g)\,200^g\,(2^g)\,2000^g\,(10^g)\,8000^g$	[371]

$x - \sin x$

With the argument x in radians

11 dec.	$x = 0\,(0{,}00001)\,0{,}001$	[426]
7 dec.	$x = 0\,(0{,}0001)\,0{,}5$	[369]
6 dec.	$x = 0\,(0{,}000001)\,0{,}001$	[185]
5 dec.	$x = 0\,(0{,}00001)\,0{,}024$	[185]

With the argument $\sin x$

11 dec.	$\sin x = 0{,}01$ (var.) $0{,}53$	[170]
7 dec.	$\sin x = 0\,(0{,}0001)\,0{,}5$	[369]

$\sin x - x$

3 dec.	$x = 0\,(1'')\,1°$	[453]

$\frac{1}{2}(x - \sin x)$

5 dec.	$x = 0\,(1°)\,180°$	[250] [347]
5; 4 dec.	$x = 0\,(1°)\,90°;\ 91°\,(1°)\,180°$	[313[

$\frac{\cos x}{x}$

Argument in radians

8 dec.	$x = 0\,(0{,}01)\,10\,(0{,}1)\,20\,(1)\,100$	[235]
7 dec.	$x = 0\,(0{,}1)\,3{,}1$	[172]
6 dec.	$x = 0\,(1)\,50$	[422]
4—5 fig.	$x = 0\,(0{,}01)\,3{,}14$	$[257]_2$
4 dec.	$x = 0\,(0{,}01)\,20\,(0{,}02)\,40\,(0{,}05)\,100$	[454]

Argument in grads

7 dec.	$x = 0\,(2^g)\,2000^g$	[263]
4 dec.	$x = 0\,(1^g)\,200^g\,(2^g)\,2000^g\,(10^g)\,8000^g$	[371]

$1 - \cos x$

Argument in degrees

7 dec.	$x = 0\,(10'')\,125°\,59'50''$; $x = 0\,(1')\,180°$	[206]
7 dec.	$x = 0\,(1')\,90°$	[334]

6 dec. $x = 0\,(10'')\,180°$ [297]
6 dec. $x = 0\,(1')\,90°$ [144]
5; 4 dec. $x = 0\,(1')\,12°;\ 12°\,(1')\,90°$ [335]

Argument in radians

11 dec. $x = 0\,(0{,}00001)\,0{,}001$ [426]

$$m\,(1 - \cos x)$$

3 dec. $m = 10\,(10)\,100.\ x = 0\,(0°{,}01)\,50°$ [333]

$$1 - \cos\frac{x}{2}$$

5 dec. $x = 0\,(0^g{,}02)\,100^g$ [159]

$$\frac{1}{2}\,(x + \cos x)$$

4—6 dec. $x = -90°\,(1')\,90°$ [439]

$$\frac{1}{2\cos\arcsin x}$$

3 dec. $x = 0\,(0{,}01)\,0{,}4$ [432]

$$\frac{\sin x}{\sqrt{\cos x}}$$

4 dec. $x = 0\,(30')\,90°$ [160]

$$\frac{\sqrt{\cos x}}{\sin x}$$

4 dec. $x = 5°\,(5°)\,85°;\ 6°45';\ 51°50'$ [436]

$$\cos x - \sin x$$

5 dec. $x = 0\,(1')\,45°$ [223]

$$1 + \sin x + \cos x$$

5 dec. $x = 0\,(30'')\,90°$ [258]

$$\frac{1}{2}\sin x - \frac{\cos x\,(1-\cos x)}{x} \quad \& \quad \frac{(\sin 2x - \sin x)}{2x} - \frac{1}{2}\cos x$$

4 dec. $x = 0\,(0{,}1)\,7;\ \frac{\pi}{2};\ \pi;\ \frac{3\pi}{2};\ 2\pi$ [444]

$$\frac{\operatorname{tg} x}{x}$$

4—5 fig.	$\pm x = 0\,(0{,}01)\,3{,}14$	[134] [160] [257]
4 dec.	$\pm x = 0\,(0°{,}1)\,90°$	[94]

$$x - \operatorname{tg} x$$

3 dec.	$\pm x = 0\,(1'')\,3600''$	[453]
3 fig.	$\pm x = 0\,(10'')\,18000''$	

$$\operatorname{tg} x - x$$

This expression is known as the involute function; tables of these functions are used in the construction of gear wheels.

Argument in degrees, minutes and seconds

13—16 dec.	$x = 0\,(1'')\,10'$	[447]
9—5 fig.; 8 dec.	$x = 0\,(10'')\,3°;\ 3°\,(10'')\,60°$	[457]
6 dec.	$x = 0\,(1°)\,60°$	$[160]_1$
4 dec.	$x = 0\,(1')\,60°$	[34]

Argument in tenths of a degree

12; 10 dec.	$x = 0\,(0°{,}01)\,0°{,}5;\ 0°{,}5\,(0°{,}01)\,1°$	[175]
8; 7 dec.	$x = 1°\,(0°{,}01)\,37°;\ 37°\,(0°{,}01)\,60°$	
6 dec.	$x = 0\,(0°{,}01)\,90°$	[111]
6; 5 dec.	$x = 0\,(0°{,}01)\,37°;\ 37°\,(0°{,}01)\,67°{,}5$	[325]
4—6 fig.	$x = 67°{,}5\,(0°{,}01)\,90°$	
4; 3—1 dec.	$x = 0\,(0°{,}01)\,67°{,}5;\ 67°{,}5\,(0°{,}01)\,90°$	

$$\frac{1 - \operatorname{tg} x}{1 + \operatorname{tg} x}$$

5 dec.	$x = 0\,(30')\,90°$	[160]

$$\frac{3(\operatorname{tg} x - x)}{x^3}$$

4 dec.	$x = 0\,(1°)\,70°\,(0°{,}1)\,88°$	[430]
3 dec.	$x = 92°\,(1°)\,135°$	

$$\frac{3(1-2x\,\mathrm{ctg}\,2x)}{(2x)^2}$$

4 dec.	$x = 0\,(1°)\,70°\,(0°{,}1)\,88°$	[430]
3 dec.	$x = 92°\,(1°)\,135°$	

$$\frac{6(2x\,\mathrm{cosec}\,2x-1)}{(2x)^2}$$

4 dec.	$x = 0\,(1°)\,70°\,(0°{,}1)\,88°$	[430]
3 dec.	$x = 92°\,(1°)\,135°$	

$$\sec\frac{x}{2} - 1$$

5 dec.	$x = 0\,(0^g{,}02)\,100^g$	[159]

$$\mathrm{cosec}\,x - \frac{1}{x}$$

Argument in minutes and seconds

17 dec.	$x = 0\,(9')\,15°$	[140]
15 dec.	$x = 0\,(10'')\,15°$	[142]

Argument in grads

20 dec.	$x = 0\,(1^g)\,50^g$	[140]

$$\frac{1}{x} - \mathrm{ctg}\,x$$

Argument in seconds

15 dec.	$x = 0\,(10'')\,15°$	[142]

Argument in grads

20 dec.	$x = 0\,(1^g)\,50^g$	[140]

DERIVATIVES AND POWERS OF TRIGONOMETRIC FUNCTIONS

Derivatives and powers of trigonometric functions of equal argument

$$\sin^2 x$$

6 dec.	$x = 0\,(30')\,90°$	[160]
6 dec.	$x = 0\,(1°)\,90°$	[42]

5 fig.	$x = 0\,(1')\,90°$	[317]
5 dec.	$x = 0\,(10')\,90°$	[81] [123]
4 dec.	$x = 0\,(1°)\,90°$	[120]

$$\sin^2 \frac{x}{2}$$

7 dec.	$x = 0\,(10'')\,180°$	[229]
7 dec.	$x = 0\,(10'')\,120°$	[143]
5 dec.	$x = 0\,(15'')\,120°\,(30'')\ 135°\,(1')\,180°$	[310]
5 dec.	$x = 0\,(30'')\,15°\,(15'')\ 120°\,(30'')\,135°\,(1')\,180°$	[163] [198]
5 dec.	$x = 0\,(30'')\,120°\,(1')\,180°$	[177]
5—6 fig.; 5 dec.	$x = 0\,(1')\,44°;\ 44°\,(1')\,180°$	[126]
4 dec.	$x = 0\,(1')\,180°$	[65]
4 dec.	$x = 0\,(6')\,180°$	[227]
2 dec.	$x = 0\,(1^{sec})\,1200^{sec}$	[453]

$$\cos^n x$$

5 dec.	$n = 2.$	$x = 0\,(0^g{,}01)\,50^g$	[208]
6 dec.	$n = 2$	$x = 0\,(30')\,90°$	[160]
4 dec.	$n = \frac{1}{2};\ \frac{3}{2};\ 3;\ 4$	$x = 0\,(30')\,90°$	[160]
5 dec.	$n = 2.$	$x = 0\,(10')\,90°$	[81] [123]
4 dec.	$n = \frac{1}{2};\ \frac{3}{2};\ 2;\ 3;\ 4.$	$x = 0\,(30')\,90°$	[331]
4 dec.	$n = \frac{1}{2}.$	$x = 0\,(30')\,90°$	[258]

$$\mathrm{tg}^2 x$$

5 fig.	$x = 0\,(1')\,90°$	[317]

$$\frac{1}{8}\sec^3 \frac{\arcsin x}{3} \quad \& \quad -\frac{1}{8}\sec^3\left(60° \pm \frac{\arcsin x}{3}\right)$$

3 dec.	$x = 0\,(0{,}01)\,0{,}4$	[432]

$$\sin x \cos^n x$$

5 dec.	$n = 1.$	$x = 0\,(0^g{,}01)\,50^g$	[208]

5 dec.	$n = 1$; $x = 0\,(30')\,90°$	[160]
4 dec.	$n = \frac{1}{2}$; 2; 3 ; $x = 0\,(30')\,90°$	[160]

$\sin x \cos^{n-1} x$

4 dec.	$n = \frac{1}{2}$; $\frac{3}{2}$; 2; 3; 4. $x = 0\,(30')\,90°$	[331]

Derivatives of trigonometric functions of different arguments

$\sin x \ \sin y$

7 dec.	$x = 0\,(1°)\,90°$.	$y = 0\,(1°)\,90°$	[315]
4 dec.	$x = 25°\,(1°)\,90°$.	$y = 0\,(0°{,}5)\,90°$	[281]
4 dec.	$x = 0\,(1°)\,40°$.	$y = 150°\,(1°)\,180°$	[440]
2 dec.	$x = 20°\,(5°)\,90°$.	$y = 0\,(2°)\,90°$	[144]

$\sin x \cos y$

4 dec.	$x = 0\,(1°)\,40°$.	$y = 150°\,(1°)\,180°$	[440]

$\sin x \operatorname{tg} y$

4 dec.	$x = 0\,(1°)\,65°$.	$y = 0\,(10')\,50°$	[281]
2 dec.	$x = 0\,(2°)\,90°$.	$y = 0\,(10°)\,30°\,(2°)\,60°$	[288]
2 dec.	$x = 30°\,(1°)\,42°\,(2°)\,72°$; $75°\,(5°)\,90°$.	$y = 0\,(2°)\,20°\,(1°)\,70°$	[157]
2 dec.	$x = 30°\,(2°)\,88°\,(1°)\,90°$.	$y = 0\,(1°)\,8°\,(2°)\,80°\,(1°)\,90°$	[163]

$\sin x \sec y$

4 dec.	$x = 0\,(1°)\,65°$.	$y = 0\,(10')\,50°$	[281]

$\operatorname{tg} x \ \operatorname{tg} y$

4 fig.	$x = 0\,(1°)\,60°$.	$y = 0\,(4^{M})\,4^{hrs}(2^{M})\,5^{hrs}30^{M}\,(1^{M})\,6^{hrs}$	[282]
4 fig.	$x = 0\,(1°)\,65°$.	$y = 0\,(1°)\,30°\,(0°{,}5)\,75°\,(0°{,}25)\,90°$	[281]
3 fig.	$x = 0\,(1°)\,82°$.	$y = 0\,(4^{M})\,4^{hrs}(2^{M})\,6^{hrs}$.	[177]
3 dec.	$x = 0\,(1°)\,80°$.	$y = 0\,(1°)\,80°$	[335]
3 fig.	$x = 0\,(1°)\,70°$.	$y = 0\,(1°)\,90°$	[304]
3 dec.	$x = 0\,(1°)\,45°$.	$y = 0\,(5°)\,90°$	[429]
3 fig.	$x = 0\,(1°)\,60°$.	$y = 0\,(1^{M})\,6^{hrs}$	[157]
2—3 dec.	$x = 0\,(1°)\,65°$.	$y = 0\,(1°)\,90°$	[219]

$\frac{1}{2}\,\mathrm{tg}\,x\,\mathrm{tg}\,\frac{x}{2}$

2 dec.	$x = 0(1^{sec})1200^{sec}$	[453]

$\mathrm{tg}\,x\,\sec y$

4 fig.	$x = 0(0°,5)90°.\quad y = 0(1°)65°$ $x = 0(1°)65°.\quad y = 0(1°)30°(0°,5)75°(0°,25)90°$	[281]
4 fig.	$x = 0(1°)90°.\quad y = 0(1°)60°$ $x = 0(1°)29°.\quad y = 0(4^{M})4^{hrs}(2^{M})5^{hrs}30^{M}(1^{M})6^{hrs}$	[282]
3 fig.	$x = 0(1°)63°.\quad y = 0(4^{M})4^{hrs}(2^{M})6^{hrs}$ $x = 0(1°)85°.\quad y = 0(1°)82°$	[177]
3 fig.	$x = 0(1°)70°.\quad y = 0(1°)90°$	[304]
3 fig.	$x = 0(1°)60°.\quad y = 0(1^{M})6^{hrs}$ $x = 60(0°,5)90°.\ y = 0(2^{M})6^{hrs}$	[157]
2—3 dec.	$x = 0(1°)65°.\quad y = 0(1°)90°$	[219]

$\sec x\,\sec y$

4 fig.	$x = 0(1°)65°.\quad y = 0(1°)90°$	[281]
4 fig.	$x = 0(1°)60°.\quad y = 0(1°)90°$	[282]
3 dec.	$x = 0(1°)80°.\quad y = 0(1°)80°$	[335]

$m \sin x \;\; n \cos y$

5 dec.	$m = 10-n.\ n=1(1)9.\ x=90°(-1°)0.\ y=0(1°)90°$	[160]

$m \cos^2 x \;\; n \sin^2 y$

5 dec.	$m = 1(1)9.\ n=10-m.\ x=0(1°)90°.\ y=90°(-1°)0$	[160]

$m\,\mathrm{tg}\,x \;\; n\,\mathrm{ctg}\,y$

5dec.	$m = 1(1)10.\ n=10-m.\ x=0(1°)90°.\ y=90°(-1°)0$	[160]

$\sin^2 x\ \mathrm{cosec}\ 1''$

1″	$x = 0(1')45°$	[167]

$2\sin^2\frac{x}{2}\ \mathrm{cosec}\ 1''$

0″,01	$x = 0(1^{sec})16^{M}59^{sec}$	[184] [306]
0″,01; 0″,1	$x = 0(1^{sec})25^{M};\ 25^{M}(1^{sec})30^{M}$	[139]
0″,01; 0″,1	$x = 0(1^{sec})15^{M};\ 15^{M}(1^{sec})42^{M}$	[319]

0″,1	$x = 17^{M}(1^{sec})\,29^{M}59^{sec}$	[184]
0″,1	$x = 0\,(2^{sec})\,40^{M}$	[411]

$$2\sin^4 \frac{1}{2}x \operatorname{cosec} 1''$$

0″,001	$x = 0\,(10^{sec})\,30^{M}$	[137] [139]
0″,01	$x = 0\,(10^{sec})\,42^{M}$	[319]
0″,01	$x = 0\,(1^{M})\,12^{M}\,(10^{sec})\,30^{M}$	[184]
0″,01	$x = 0\,(20^{sec})\,40^{M}$	[411]

$$\cos^2 x \operatorname{cosec} 1'' \;\&\; \sin x \cos x \operatorname{cosec} 1''$$

1″	$x = 0\,(1')\,45°$	[167]

$$\operatorname{tg}^2 \frac{x}{2} \operatorname{cosec} 1''$$

0″,01	$x = 0\,(1^{sec})\,30^{M}$	[275]

Multiples of values of trigonometric functions and their derivatives

$$N \sin x \;\&\; N \cos x$$

Argument in degrees

5 dec.	$N = 1\,(1)\,10.\quad x = 0(1')\,90°$	[158]
5 dec.	$N = 1\,(1)\,9.\quad x = 0(15')\,90°$	[416]
5 dec.	$N = 1\,(1)\,9.\quad x = 0(1°)\,90°$	[279] [280] [331]
4 dec.	$N = 1\,(1)\,100.\quad x = 0(1')\,90°$	[226]
4 dec.	$N = 1\,(1)\,10.\quad x = 0(1')\,90°$	[292]
4 dec.	$N = 1\,(1)\,10.\quad x = 0(1°)\,90°$	[249]
3 dec.	$N = 1\,(1)\,100.\quad x = 0(2')\,90°$	[359]
3 dec.	$N\sin x$ $\quad N = 0\left(\frac{1}{16}\right)1\,(1)\,14.\quad x = \frac{180°}{n}$, $n = 3(1)\,26\,(2)\,32$	[313]
3 dec.	$N = 1\,(1)\,10.\quad x = 0(15')\,90°$	[355]
2 dec.	$N = 1\,(1)\,10\,(10)\,100\,(100)\,1000\,(1000)\,10\,000.\quad x = 0\,(1')\,90'$	[194]
2 dec.	$N = 1\,(1)\,500.\quad x = 0\,(10')\,90°$	[298]
1 dec.	$N = 1\,(1)\,400\,(100)\,1000$; $x = 0\,(30')\,15°\,(1°)\,75°\,(30')\,90°$	[157]
1 dec.	$N = 1\,(1)\,720.\quad x = 0\,(1°)\,90°$	[329]
1 dec.	$N = 1\,(1)\,600.\quad x = 0(1°)\,90°$	[163] [177] [335]
1 dec.	$N = 1\,(1)\,300.\quad x = 0(1°)\,90°$	[334]

Argument in seconds

6 dec. $N \cos x$ $N = \frac{2}{3}$. $\pm x = 0\,(10'')\,18\,000''$ [453]

Argument in grads

6 dec.	$N = 1.$ $x = 0\ (0^g{,}01)\,100^g$	[191]
4 dec.	$N = 2\,(1)\,9.$ $x = 0\ (0^g{,}01)\,100^g$	[191]

$N \operatorname{tg} x$

5 fig.	$N = 10\,(10)\,90.$ $x = 0\,(1')\,90°$	[57]
1 dec.	$N = 51\,(1)\,100.$ $x = 0\,(1^g)\,50^g$	[371]
in integers	$N = 102\,(2)\,500.$ $x = 0\,(1^g)\,50^g$	[371]

$\pm 15 \operatorname{tg} \frac{x}{2}$

4–5 dec. $x = 0\,(1'')\,3600''$ [453]

$x \operatorname{tg} x$

4–5 dec. $\pm x = 0\,(0{,}01)\,3{,}14$ [134] [160] [257]$_2$

$x \operatorname{tg} \frac{1}{2} \pi x$

4–5 dec. $x = 0\,(0{,}1)\,9$ [134] [257]$_2$

$N \operatorname{ctg} x$

5 fig. $N = 10\,(10)\,90.$ $x = 0\,(1')\,90°$ [57]

$\pi \operatorname{ctg} \pi x$

15 dec. $x = 0\,(0{,}01)\,0{,}5$ [197]

$m \operatorname{ctg} \frac{\pi}{n}$

7 dec. $\left\{\begin{matrix} m = \frac{1}{2} \\ m = \frac{1}{4}\,n \end{matrix}\right\} n = 3\,(1)\,12$ [334]

$x \operatorname{ctg} x$

Argument in degrees, minutes and seconds

8 fig.	$x = 0\ (10'')\ 2°$	[326]
6 fig.	$x = 0\ (0°,1)\ 3°$	[289]
6 fig.	$x = 0\ (10'')\ 1°20'$	[324]
6 fig.	$x = 0$ (var.) $1°,001$	[325]

Argument in radians

5 dec.	$x = 0\,(0{,}05)\,10$	[431]

Argument in grads

7 fig.	$x = 0(0^g{,}001)\,2^g$	[79]
6 fig.	$x = 0(0^g{,}01)\,2^g$	[327]
4 dec.	$x = 0$(var.) 2^g	[215]

Argument in units of time

8 fig.	$x = 0\,(1^{sec})\,1800^{sec}\ [7°30']$	[309]
5 fig.	$x = 0$ (var.) $1064^{sec},6\ [4°26']$	[193]

$x \operatorname{ctg} \frac{\pi}{2} x$

4—5 fig.	$x = 0\,(0{,}1)\,9$	[134] [257]$_2$

$N \sec x$

Argument in degrees

6 dec.	$N = \frac{3}{2}.\quad x = 0\,(10'')\ 5°$	[453]
4—5 dec.	$N = 1\,(1)\,9.\quad x = 0\,(1°)\,90°$	[331]
2 dec.	$N = 1\,(1)\,10.\quad x = 4°\,(2°)\,14°\,(1°)\,69°$	[335]

Argument in grads

1 dec.	$N = 51\,(1)\,100.\quad x = 0\,(1^g)\,50^g$	[371]
in integers	$N = 102\,(2)\,500.\quad x = 0\,(1^g)\,50^g$	[371]

$x \operatorname{cosec} x$

Argument in degrees

6 fig.	$x = 0\,(10'')\,1°20'$	[324]
6 fig.	$x = 0\,(0°,1)\,3°$	[289]
6 fig.	$x = 0\,($ var. $)\,1°,011$	[325]

Argument in radians

5 dec.	$x = 0\,(0{,}05)\,10$	[431]

Argument in units of time

5 fig.	$x = 0\,($ var. $)\,1055^{sec},5$	[193]

$N \sin^2 x$

5 dec.	$N = 2.\quad x = 0\,(0{,}01)\,0{,}1;\ 0\,(0{,}05)\,0{,}6$	[435]
4 dec.	$N = 50.\quad x = 0\,(0^g{,}1)\,100^g$	[412]
3 dec.	$N = 100.\ x = 0\,(1')\,11°59'$	[237]

$N \cos^2 x$

Argument in degrees

4 dec.	$N = 1\,(1)\,10.$	$x = 0\,(1')\,30°$	[291]
2 dec.	$N = 100.$	$x = 0\,(10')\,10°$	[237]
1 dec.	$N = 10\,(1)\,250.$	$x = 0\,(1')\,30°$	[260]
1 dec.	$N = 10\,(1)\,100.$	$x = 0\,(1°)\,10°\,(30')\,20°\,(20')\,30°$	[237]
	$N = 100\,(1)\,175.$	$x = 0\,(30')\,10°\,(20')\,20°$	
	$N = 175\,(1)\,250.$	$x = 0\,(30')\,5°\,(20')\,10°$	

Argument in grads

5 dec.	$N = 1\,(1)\,4$	$x = 0\,(0^g{,}02)\,31^g$	[256]
4 dec.	$N = 5\,(1)\,9$		
4 dec.	$N = 100.$	$x = 0\,(0^g{,}1)\,100^g$	[412]

$N \sin x \cos x$

Argument in degrees

5 dec.	$N = 2m.$	$m = 1\,(1)\,9.\ x = 0\,(1°)\,90°$	[160]
4 dec.	$N = 1\,(1)\,10.$	$x = 0\,(1')\,30°$	[291]
3 dec.	$N = 100.$	$x = 0\,(1')\,9°59'$	[237]
2 dec.	$N = 10\,(1)\,250.$	$x = 0\,(2')\,30°$	[260]

2 dec.	$\left\{\begin{array}{l} N = 10\,(1)\,100. \quad x = 0\,(3')\,29°57' \\ N = 100\,(1)\,175. \quad x = 0\,(2')\,19°58' \\ N = 175\,(1)\,250. \quad x = 0\,(1')\,9°59' \end{array}\right\}$	[237]

Argument in grads

5 dec.	$N = 1\,(1)\,4$	$x = 0\,(0^g,02)\,31^g$	[256]
4 dec.	$N = 5\,(1)\,9$		

RECIPROCAL TRIGONOMETRIC FUNCTIONS

arc sin x & arc cos x

Values of the functions in degrees, minutes and seconds

1″ arc sin x	$x = 0\,(0,001)\,1$	[270]
1′	$x = 0\,(0,01)\,1$	[332]
1°	$x = 0\,(0,01)\,0,99\,(0,001)\,0,999\,(0,0001)\,1$	[350]

Values of the functions in radians

20 dec.	$x = 0\,(0,00001)\,0,001$	[234]
12 dec.	$x = 0\,(0,0001)\,0,989\,(0,00001)\,1$	[376]
10 dec.	$x = 0,001\,(0,0001)\,0,0999$	[234]
7 dec.	$x = 0,1\,(0,001)\,1$	
6 dec.	$\left\{\begin{array}{l} x = 0,999\,(0,00001)\,1 \\ x = 0,99\,(0,0001)\,0,9\,(0,001)\,1 \end{array}\right\}$	[456]
6 dec.	$x = 0\,(0,001)\,1$	[199]
5 dec.	$x = 0\,(0,001)\,1$	[346]
5 dec.	$x = 0\,(0,001)\,0,89\,(0,0001)$ 0,999 (var.) 1	[349]
5 dec.	$x = 0\,(0,01)\,1$	[232] [332]
4 dec.	$x = 0\,(0,001)\,0,99\,(0,0001)\,1$	[303]
4 dec.	$x = 0\,(0,001)\,1$	[202]
4 dec.	$x = 0\,(0,01)\,0,99\,(0,001)\,1$	[63]

arc sin $(1 - x)$

13 dec.	$x = 0\,(0,00001)\,0,0005$	[376]

arc tg x

Values of the functions in degrees, minutes and seconds

0″,001	$x = 0\,(0{,}00001)\,0{,}01\,(0{,}0001)\,1$	[207]
0′,1	$x = 0\,(0{,}01)\,0{,}34$	[395]
1′	$x = 0{,}35\,(0{,}01)\,1{,}02$	[395]
1′	$x = 0\,(0{,}001)\,1{,}02$	[331]
0°,0001	$x = 0\,(0{,}001)\,0{,}099$	[257]₂
0°,001	$x = 0{,}1\,(0{,}01)\,1$	[257]₂
0°,001	$x = 1\,(0{,}001)\,3\,(0{,}01)\,10\,(0{,}5)\,30\,(10)\,50$	[356]
0°,01	$x = 0\,(0{,}001)\,1$	[302]

Values of the functions in degrees

0ᵍ,01	$x = 0\,(0{,}001)\,1{,}02$	[331]

arc tg x & arc ctg x

Values of the functions in radians

20 dec.	arc tg x	$x = m10^{-n}.\ m = 1\,(1)\,9.\ n = 1\,(1)\,6$	[437]
20 dec.		$x = 0\,(0{,}00001)\,0{,}001$	[234]
12 dec.		$x = 0\,(0{,}001)\,7\,(0{,}01)\,50\,(0{,}1)\,300\,(1)\,2000\,(10)\,10000$	[377]
11 dec.		$x = 0\,(0{,}01)\,1$	[450]
10 dec.	arc tg x	$x = 0\,(0{,}001)\,1$	[4]
10 dec.		$x = 0{,}001\,(0{,}0001)\,0{,}0999$	[234]
9 dec.		$x = 0\,(1)\,100$	[363]
8 dec.		$x = 0{,}1\,(0{,}001)\,2{,}999$	[234]
7 dec.		$x = 0\,(0{,}01)\,5\,(0{,}1)\,20\,(1)\,164$	[452]
7 dec.		$x = 3\,(0{,}01)\,10\,(0{,}1)\,20\,(1)\,50$	[234]
6 dec.		$x = 0\,(0{,}001)\,2\,(0{,}01)\,10\,(0{,}1)\,35\,(1)\,80$	[456]
5 dec.	arc tg x	$x = 0\,(0{,}001)\,1{,}1\,(0{,}01)\,10{,}09$	[346]
5 dec.		$x = 0\,(0{,}01)\,10$	[232]
5 dec.		$x = 0\,(0{,}001)\,1{,}9\,(0{,}01)\,6\,(0{,}1)\,20$ (var.) 3000	[349]
4 dec.		$x = 0\,(0{,}001)\,1$	[302]
4 dec.		$x = 0\,(0{,}001)\,1{,}5\,(0{,}01)\,5\,(0{,}1)\,15\,(1)\,60$	[303]
4 dec.		$x = 0\,(0{,}001)\,1\,(0{,}01)\,4\,(0{,}1)\,20\,(1)\,50$	[202]
4 dec.		$x = 0\,(0{,}01)\,0{,}99\,(0{,}1)\,4\,(1)\,115$	[350]
4 dec.		$x = 0\,(0{,}01)\,2\,(0{,}1)\,10\,(1)\,50$	[63]

$\frac{2}{\pi}\ \text{arc tg}\left(\frac{x}{y}\right)^{1/2}$

4 dec.	$x = 0\,(0{,}2)\,3{,}8.\quad y = 2{,}2\,(0{,}2)\,3{,}8$	[443]

arc ctg x

Values of the function in degrees

0°,0001	$x = 0\,(0{,}001)\,0{,}099)$	[257]₂
0°,001	$x = 0{,}1\,(0{,}01)\,1$	
0°,001	$x = 0{,}1\,(0{,}001)\,3\,(0{,}01)\,10\,(0{,}5)\,30\,(10)\,50$	[356]
0°,01	$x = 0\,(0{,}001)\,1$	[302]

arc ctg (k cos x)

0°,1	$k = 0\,(0{,}01)\,1\,(0{,}02)\,2\,(0{,}1)\,5\,(1)\,20$ (var.) 800; $x = 0\,(2°)\,10°\,(1°)\,65°$	[219]

arc sec x & arc cosec x

Values of the functions in radians

6 dec.	arc sec x $x = 1\,(0{,}00001)\,1{,}001\,(0{,}0001)\,1{,}01\,(0{,}001)\,2\,(0{,}01)\,10\,(0{,}1)\,35\,(1)\,80$	[456]
4 dec.	$x = 1\,(0{,}0001)\,1{,}01\,(0{,}001)\,1{,}5\,(0{,}01)\,5\,(0{,}1)\,15\,(1)\,60$	[303]

arc sin (sin x sin y)

0′,1	$x = 0\,(1°)\,90°$.	$y = 0\,(1°)\,90°$	[145]
1′	$x = 0\,(0°{,}5)\,34°\,(1°)\,90°$.	$y = 0\,(1°)\,90°$	[328]
1′	$x = 0\,(1°)\,6°\,(30')\,90°$.	$y = 0\,(1°)\,90°$	[144]
1′	$x = 0\,(1°)\,90°$.	$y = 0\,(1°)\,90°$	[268] [394] [423]
1′	$x = 0\,(1°)\,90°$.	$y = 32°\,(1°)\,39°$	[446]
0°,1	$x = 0\,(1°)\,90°$.	$y = 0\,(4^{M})\,6^{hrs}$	[221]

arc sin (sin x sec y)

0′,1	$x = 0\,(0°{,}1)\,28°{,}9$.	$y = 0$ (1° or 0°,5) $90°$	[328]
0′,1	$x = 0$ (1° or 0°,5) $90°$.	$y = 0\,(0°{,}1)\,28°{,}9$	
1′	$x = 0\,(1°)\,23°;\ 23°28'$.	$y = 0\,(1°)\,66°$	[334]
0°,1	$x = 0\,(0°{,}5)\,30°$.	$y = 0$ (var.) $77°$	[163]
0°,1	$x = 0\,(1°)\,40°$.	$y = 20°\,(1°)\,90°$	[335]
	$x = 41°\,(1°)\,48°$.	$y = 27°\,(1°)\,90°$	
0°,1	$x = 0\,(1°)\,30°$.	$y = 0\,(5°)\,10°\,(2°)\,24°\,(1°)\,74°$	[157]

arc sin (tg x tg y)

1^{M}	$x = 0\,(1°)\,40°$.	$y = 20°\,(1°)\,90°$	[335]
	$x = 41°\,(1°)\,48°$.	$y = 27°\,(1°)\,90°$	

1^M	$x = 0\,(1°)\,38°.$	$y = 0\,(5°)\,10°\,(2°)\,24°\,(1°)\,74°$	[157]
1^M	$x = 0\,(1°)\,23°;\ 23°28'.$	$y = 0\,(1°)\,60°$	[334]
1^M	$x = 0\,(2°)\,32°.$	$y = 20°\,(1°)\,88°$	[173]

arc tg (sin x tg y)

$0',1$	$x = 0\,(1°)\,90°.$	$y = 0\,(1°)\,90°$	[136] [145] [245]
$0',1$	$x = 0\,(1°)\,90°.$	$y = 25°\,(1°)\,90°$	[201] [314] [408]
$0',1$	$x = 0\,(1^M)\,6^{hrs}.$	$y = 30°\,(15')\,90°$	[329]
$0',1$	$x = 0\,(4^M)\,6^{hrs}.$	$y = 23°\,(1°)\,90°$	[360]
$1'$	$x = 0\,(1°)\,90°.$	$y = 0\,(1°)\,90°$	[268] [394] [423]
$1'$	$x = 0\,(4^M)\,6^{hrs}.$	$y = 0\,(1°)\,90°$	[221]
$1'$	$x = 0\,(1°)\,90°.$	$y = 0\,(1°)\,6°\,(30')\,90°$	[144]
$1'$	$x = 0\,(1°)\,90°.$	$y = 32°\,(1°)\,39°$	[446]
$0°,1$	$x = 0\,(1°)\,90°.$	$y = 0\,(1°)\,90°$	[136] [245]
$0°,1$	$x = 0\,(1°)\,65°.$	$y = 0\,(1°)\,90°$	[201]

TRIGONOMETRIC AND RECIPROCAL TRIGONOMETRIC FUNCTIONS OF A COMPLEX VARIABLE

sin z

The tables give the modulus and argument of the function of the variable $z = x + iy$.

4; 3 dec.

$0^g,01$

$$\left\{\begin{aligned} x &= 0\,(2^g)\,100^g \\ y &= 0\,(0{,}02)\,1 \\ \frac{1}{y} &= 1\,(-0{,}02)\,0{,}34;\ 0{,}32\,(-0{,}02)\,0{,}2 \end{aligned}\right\}$$ [231]

tg z

The tables give the modulus and argument of the function of the variable $z = x + iy$.

4 dec.

$0^g,01$

$$\left\{\begin{aligned} x &= 0\,(2^g)\,50^g \\ y &= 0\,(0{,}02)\,1 \\ \frac{1}{y} &= 1\,(-0{,}02)\,0{,}2 \end{aligned}\right\}$$ [231]

arc sin z

The tables give the real and imaginary parts u and v of the function of the variable $z = re^{i\varphi}$.

$0^g,01$ u
4 dec. v $\Big\}$ $r = 0\,(0{,}02)\,1.\quad \frac{1}{r} = 1\,(-0{,}02)\,0.\quad \varphi = 0\,(2^g)\,100^g$ [231]

0,0001	u	$r = 0\,(0{,}2)\,2{,}2.\ \varphi = 0\,(15°)\,90°$	[134] [257]$_2$
0°,01	u		
4 dec.	v		

arc tg z

The tables give the real and imaginary parts u and v of the function of the variable $z = re^{i\varphi}$.

0,0001	u	$r = 0{,}8\,(0{,}02)\,0{,}98;\ 0{,}99;\ 0{,}999;\ 1-10^{-6};\ 1-10^{-9};1$	[134] [257]$_2$
0°,01	u	$\varphi = 75°\,(5°)\,90°$	
4dec.	$\frac{1}{v}$		
0,0001	u	$r = 0\,(0{,}2)\,1.\ \varphi = 0\,(15°)\,90°$	[134] [257]$_2$
0°,01	u		
4 dec.	v		
0^g,01	u	$r = 0\,(0{,}02)\,1.\ \varphi = 0\,(2^g)\,100^g$	[231]
4 dec.	v		

GEOMETRIC QUANTITIES

Central angles

$$\frac{360°}{x}$$

0°,001; 0°,0001	$x = 1\,(1)\,90;\ 91\,(1)\,180$	[371]
1″	$x = 3\,(1)\,60$	[361]

$$\frac{180°}{x}$$

0°,0001	$x = 1\,(1)\,500$	[325]

$$\frac{\pi}{x}$$

6 dec.	$x = 1\,(1)\,500$	[325]

$$\frac{2\pi}{x}$$

4 dec.	$x = 1\,(1)\,100$	[303]

Circumference

Circumference of diameter d

7 fig.	$d = 1\,(1)\,1000$	[228]
6 fig.	$d = 1\,(0{,}1)\,100{,}9$	[293]

6 fig. $\frac{1}{2}\pi d$	$d = 0\,(0{,}1)\,100$	[420]
6 fig.	$d = 1\,(1)\,1000$	[81]
6 fig.	$d = 1\,(1)\,100$	[146]
6 dec.	$d = 1\,(1)\,100$	[25] [160]
6 dec.	$\frac{d}{2} = 1\,(1)\,100$	[331]
6 dec.	$d = 10\,(1)\,100$	[373]
6 fig.	$d = \frac{1}{64};\ \frac{1}{32};\ \frac{1}{16}\left(\frac{1}{16}\right)1\left(\frac{1}{8}\right)100$	[138]
5 dec.	$d = 0\,(0{,}001)\,1{,}009$	[413]
5 dec.	$d = 0\,(0{,}1)\,100$	[393]
5—6 fig.	$d = 1\,(1)\,1610$	[67]
5—6 fig.	$d = 1\,(1)\,1000$	[40] [138]
5 fig.	$d = 1\,(1)\,2000$	[61]
5 fig.	$d = 1\,(1)\,1500$	[94] [128] [209] [250]
5 fig.	$d = 1\,(1)\,1200$	[45]
5 fig.	$d = 1\,(1)\,1100$	[69] [99] [250]
5 dec.	$d = 1\,(1)\,1000$	[458] [459]
5 fig.	$d = 1\,(1)\,1000$	[5] [14] [16] [19] [20] [35] [53] [59] [78] [161] [205] [209] [277] [340] [362] [365] [374] [390]
5 fig.	$d = 1\,(1)\,500$	[97]
5 fig.	$d = 0\,(0{,}1)\,1\,(1)\,100$	[121] [274]
5 fig.	$d = 1\,(1)\,100$	[9] [56] [350] [357] [419]
5 fig.	$\frac{d}{2} = 1\,(1)\,100$	[120] [357]
4—6 fig.	$d = 1\,(1)\,1000$	[190]
4—5 fig.	$d = 1\,(0{,}01)\,10$	[17]
4—5 fig.	$d = 1\,(1)\,1000$	[10]
4—5 fig.	$d = 1\,(1)\,100$	[33]
4 fig.	$d = 1\,(0{,}01)\,10$	[21] [89] [93]
4 fig.	$d = 0\,(0{,}1)\,100$	[108] [131]
4 dec.	$d = 1\,(0{,}1)\,100$	[392]
4 fig.	$d = 1\,(1)\,1000$	[34] [52] [55] [66]
4 dec.	$d = 1\,(1)\,1000$	[390]
4 fig.	$d = 1\,(1)\,100$	[11] [27] [83] [96] [294]
4 fig.	$\frac{d}{2} = 1\,(1)\,100$	[271] [290] [351]
4 fig.	$d = 0\left(\frac{1}{8}\right)12$	[93]
3—4 fig.	$d = 0\,(0{,}1)\,100$	[337]
3 dec.	$\frac{d}{2} = 1\,(0{,}1)\,10\,(0{,}5)\,15$	[119]
3 dec.	$d = 1\,(1)\,1500$	[209]

3 fig.	$\frac{1}{2}\pi d$	$d = 1\,(1)\,100$	[49]
3 dec.		$d = 4\,(1)\,31$	[269]
2 dec.		$d = 1\,(0{,}1)\,17$	[31]
2 dec.		$d = 1\,(1)\,1000$	[138] [147] [160] [186] [187] [189] [269]
2 dec.		$\frac{d}{2} = 1\,(1)\,100$	[211] [261] [347]

Reciprocal of circumference of diameter d.

5—6 fig.	$d = 1\,(1)\,1610$	[67]

Length of arc $l = \frac{\pi r \alpha^\circ}{180^\circ}$

$\alpha = 1^\circ$

6 dec.	$r = 1\,(1)\,100$	[331]
5 fig.	$r = 1\,(1)\,100$	[120] [357]
4 fig.	$r = 1\,(1)\,100$	[271] [290] [351]
3 dec.	$r = 1\,(0{,}1)\,10\,(0{,}5)\,15$	[119]
2 dec.	$r = 1\,(1)\,100$	[211] [261] [347]

$r = 1$

Angle in degrees

35 dec.	$\alpha = 1''\,(1'')\,10''\,(10'')\,60''\,(1')$ $10'\,(10')\,60'$	[449]
25 dec.	$\alpha = 1^\circ\,(1^\circ)\,360^\circ;\ 1'\,(1')\,60';\ 1''\,(1'')\,60''$	[76]
15 dec.	$\alpha = 0\,(1^\circ)\,180^\circ;\ 0\,(1')\,60';\ 0\,(1'')\,60''$	[230]
11 dec.	$\alpha = 1^\circ\,(1^\circ)\,100^\circ\,(10^\circ)\,270^\circ\,(30^\circ)\,360^\circ$ $1'\,(1')\,60';\ 1''\,(1'')\,60''$	[401]
10 dec.	$\alpha = 0\,(1^\circ)\,180^\circ;\ 0\,(1')\,60';\ 0\,(1'')\,60''$	[22] [397]
10 dec.	$\alpha = 0\,(1^\circ)\,100^\circ;\ 120^\circ;\ 180^\circ;\ 200^\circ;$ $270^\circ;\ 300^\circ;\ 360^\circ$ $0\,(1')\,60';\ 0\,(1'')\,60''$	[272]
10—12 dec.	$\alpha = 1^\circ\,(1^\circ)\,100^\circ\,(10^\circ)\,150^\circ;\ 180^\circ;$ $270^\circ;\ 360^\circ;\ 0\,(1')\,60';\ 0\,(1'')\,60''$	[98]
13—7 fig; 8 dec.	$\alpha = 0\,(10'')\,5^\circ;\ 5^\circ\,(10'')\,60^\circ$	[457]

7 dec.	$\alpha = 0\,(1^\circ)\,180^\circ;\ 0\,(1')\,60';\ 0\,(1'')\,60''$	[99]
7 dec.	$\alpha = 0\,(1')\,90^\circ;\ 0\,(1'')\,60''$	[254]
6 dec.	$\alpha = 0\,(10')\,90^\circ$	[67]
6 dec.	$\alpha = 0\,(1^\circ)\,180^\circ$	[25] [85] [209]
6 dec.; 5—6 fig.	$\alpha = 0\,(1^\circ)\,180^\circ;\ 1'\,(1')\,60'$	[90] [123]
3—4 fig.	$\alpha = 1''\,(1'')\,60''$	
6 dec.	$\alpha = 0\,(1^\circ)\,10^\circ\,(10^\circ)\,180^\circ;$ $0\,(1')\,10'\,(10')\,60';$ $0\,(1'')\,10''\,(10'')\,60''$	[3]
5 dec.	$\alpha = 0\,(1^\circ)\,180^\circ$	[64] [459]
4—5 fig.	$\alpha = 0\,(1')\,60';\ 0\,(1'')\,60''$	
5 dec.	$\alpha = 0\,(0^\circ{,}1)\,180^\circ$	[346]
5 dec.	$\alpha = 0\,(10')\,180^\circ$	[94] [99] [128]
5 dec.	$\alpha = 0\,(1^\circ)\,360^\circ;\ 0\,(1')\,60';\ 0\,(1'')\,60''$	[119]
5 dec.	$\alpha = 0\,(1^\circ)\,180^\circ;\ 0\,(1')\,60';\ 0\,(1'')\,60''$	[13]
5 dec.	$\alpha = 0\,(1^\circ)\,360^\circ$	[286]
5 dec.	$\alpha = 0\,(1^\circ)\,180^\circ$	[67] [130] [252] [352]
5 fig.	$\alpha = 0\,(1^\circ)\,180^\circ$	[346]
5 dec.	$\alpha = 0\,(1^\circ)\,9^\circ;\ 1'\,(1')\,9';\ 1''\,(1'')\,9''$	[83]
4 dec.	$\alpha = 0\,(1^\circ)\,180^\circ$	[19] [20] [52] [69] [78] [94] [95] [99] [121] [128] [362] [365] [409]
4 dec.	$\alpha = 0\,(1^\circ)\,90^\circ$	[58] [156]

Angle in radians

4 fig.	$\alpha = 1\,(1)\,100$	[351] [357]
4 dec.	$\alpha = 1\,(1)\,100$	[285]
4—3 dec.	$\alpha = 1\,(1)\,100$	[271]

Length of arc in relation to the height of the segment when $d = 1$

5 dec.	$h = 0{,}001\,(0{,}001)\,0{,}5\,(0{,}05)\,1$	[67]

Length of arc in relation to the height of the segment when $c = 1$

5 dec.	$h = 0{,}005\,(0{,}005)\,0{,}1\,(0{,}001)\,0{,}5$	[67]

Length of arc in relation to the height of the segment when $h = 1$

4—5 fig.	$c = 2\,(0{,}01)\,2{,}1\,(0{,}1)\,12\,(1)\,20\,(2)\,24$	[67]

Lengths of the sides of regular polygons, of chords, radii and diameters of circles

Length of chord in relation to the size of the central angle for $r = 1$

5 fig.	$\alpha = 0\,(1°)\,180°$	[346]
5 dec.	$\alpha = 0\,(1°)\,180°$	[67] [252] [362]
4 dec.	$\alpha = 0\,(10')\,120°$	[13]
4 dec.	$\alpha = 0\,(1°)\,180°$	[19] [20] [52] [69] [78] [94] [95] [99] [121] [128] [209] [365]
4 dec.	$\alpha = 0\,(1°)\,60°$	[93]
3 dec.	$\alpha = 60°\,(1°)\,180°$	[93]
3 dec.	$\alpha = 0\,(1°)\,90°$	[112]

in relation to height of the segment when $d = 1$

5 dec.	$h = 0{,}001\,(0{,}001)\,0{,}5\,(0{,}05)\,1$	[67]

Height of segment (altitude) in relation to the size of the central angle

5 fig.	$\alpha = 0\,(1°)\,180°$	[346]
5 dec.	$\alpha = 0\,(1°)\,180°$	[67]
4 dec.	$\alpha = 0\,(1°)\,180°$	[13] [19] [20] [52] [64] [69] [78] [93] [94] [95] [99] [121] [128] [209] [362]

$\frac{1}{h}$ when $r = 1$

3—4 fig.	$\alpha = 1°\,(1°)\,180°$	[94]

Radius of a circle in relation to circumference

6 dec.	$l = 1\,(1)\,100$	[331]
3—4 fig.	$l = 1\,(1)\,100$	[347]

in relation to height of the segment for $c = 1$

5 dec.	$h = 0{,}005\,(0{,}005)\,0{,}1\,(0{,}001)\,0{,}5$	[67]

in relation to length of the chord when $h = 1$

5 dec.	$c = 2\,(0{,}01)\,2{,}1\,(0{,}1)\,12\,(1)\,20\,(2)\,24$	[67]

Diameter of circle in relation to circumference

6 dec.	$l = 1\,(1)\,100$	[25] [160]
5 fig.	$l = 1\,(1)\,1000$	[35]
4 fig.	$l = 1\,(1)\,1000$	[34]
4—3 dec.	$l = 1\,(1)\,250$	[313]

Lengths of the sides of regular polygons (inscribed & circumscribed) when $r = 1$

4 dec.	$n = 3\,(1)\,10$ (var.) 64	[64]

The relation of various lines in the circle

$$\frac{l}{h}$$

5 fig.	$\alpha = 0\,(1°)\,180°$	[346]
4 dec.	$\alpha = 0\,(1°)\,180°$	[209]
2 dec.	$\alpha = 0\,(1°)\,180°$	[99] [128] [362]

$$\frac{c}{r} = 2\sin\frac{\alpha}{2}$$

7 dec.	$\alpha = 0\,(1')\,90°$	[255] [334]
6 dec.	$\alpha = 0\,(1')\,180°$	[225]
5 dec.	$\alpha = 0\,(0°,1)\,90°$	[182]
4 dec.	$\alpha = 0\,(10')\,180°$	[211] [347]
4 dec.	$\alpha = 0\,(20')\,180°$	[278]
4 dec.	$\alpha = 0\,(0^g,1)\;200^g$	[212]

$$\frac{h}{r} = 1 - \cos\frac{\alpha}{2}$$

4 dec.	$\alpha = 0\,(10')\,180°$	[211] [347]
4 dec.	$\alpha = 0\,(0^g,1)\,200^g$	[212]

$$\frac{r}{c} = \frac{1}{2}\operatorname{cosec}\frac{\alpha}{2}$$

7 dec.	$\alpha = 0\,(1°)\,180°$	[269]
6 fig.	$\alpha = 0\,(1')\,10°\,(2')\,20°\,(10')\;30°\,(20')\,38°\,(30')\,50°$	[254]

$$\frac{l}{c} = \frac{1}{2}\,\mathrm{cosec}\,\frac{\alpha}{2}$$

7 dec.	$\alpha = 10°\,(1')\,30°\,(10')\,180°$	[361]
7 dec.	$\alpha = 0\,(1°)\,180°$	[269]

$\frac{l}{c}$ as a function of $\frac{h}{c}$

5 dec.	$h/c = 0\,(0{,}001)\,0{,}5$	[186] [187]

$$\frac{c}{h} = 2\,\mathrm{ctg}\,\frac{\alpha}{4}$$

7 dec.	$\alpha = 0\,(1°)\,180°$	[269]
5 dec.	$\alpha = 1°\,(1°)\,180°$	[67]

$$\frac{h}{c}$$

5 dec.	$\alpha = 0\,(1°)\,180°$	[67]
4 dec.	$\alpha = 1°\,(1°)\,180°$	[93]

$$\lg\frac{l}{c}$$

7 dec.	$\alpha = 0\,(1°)\,180°$	[269]
5 dec.	$\alpha = 0\,(1')\,180°$	[361]

$$\lg\frac{h}{l}$$

5 dec.	$\alpha = 0\,(1')\,180°$	[361]

$$\lg\frac{h}{c}$$

5 dec.	$\alpha = 0\,(1')\,180°$	[361]

$$\lg\frac{c}{h}$$

7 dec.	$\alpha = 0\,(1°)\,180°$	[269]

$$\lg\frac{c}{r}$$

5 dec.	$\alpha = 0\,(1')\,180°$	[361]

$\lg \frac{l}{r}$

5 dec.	$\alpha = 0\,(1')\,180°$	[361]

$\lg \frac{h}{r}$

5 dec.	$\alpha = 0\,(1')\,180°$	[361]

$\lg \frac{hr}{l^2}$

5 dec.	$\alpha = 0\,(1')\,180°$	[361]

Elements of a triangle

Altitude of equilateral triangle with side a

4–5 fig.	$a = 1\,(1)\,100$	[37]

The side of a right-angled triangle in relation to the hypotenuse and the other side

3–4 fig.	hypotenuse $= 1\,(1)\,28$; side $= 1\,(1)\,15$	[37]

Hypotenuse of a right angled triangle in relation to the sides

3–4 fig.	Length of sides $= 1\,(1)\,20$	[37]

Areas and surfaces

The area of a circle in relation to the radius

6 fig.	$r = 0{,}1\,(0{,}1)\,1\,(1)\,100$	[121]
5 fig.	$r = 0$ (var.) $10\,(0{,}25)$ $29\,(0{,}5)\,101\,(1)\,178$	[313]
6 fig.	$r = 179\,(1)\,200$	[313]
5 fig.	$r = 1{,}1\,(0{,}01)\,11{,}29$	[146]
5 fig.	$r = 1\,(1)\,100$	[120] [261] [357]
4–5 fig.	$r = 1\,(1)\,100$	[290] [347] [351]
4 dec.	$r = 0\,(0{,}1)\,100\,(1)\,600$	[393]
4 dec.	$r = 0\,(1)\,100$	[373]
4 fig.	$r = 1\,(1)\,100$	[271]

in relation to the diameter

7 fig.	$d = 1\,(1)\,1000$	[228] [458] [459]
6–7 fig.	$d = 1\,(1)\,2000$	[61]
6 fig.	$d = 1\,(1)\,1128$	[94] [128] [209] [250]
7 fig.	$d = 1129\,(1)\,1500$	
6–7 fig.	$d = 1\,(1)\,1200$	[45]
6–7 fig.	$d = 1\,(1)\,1100$	[69]
6–7 fig.	$d = \frac{1}{64};\ \frac{1}{32};\ \frac{1}{16}\left(\frac{1}{16}\right)1\left(\frac{1}{8}\right)100$	[138]
6 fig.	$d = 1\,(0{,}1)\,100{,}9$	[293]
6 dec.	$d = 0\,(0{,}1)\,100$	[393]
6 fig.	$d = 0\,(0{,}1)\,100$	[420]
6 fig.	$d = 0\,(0{,}1)\,1\,(1)\,100$	[274]
6 fig.	$d = 1\,(1)\,1000$	[5] [14] [16] [19] [20] [40] [53] [59] [78] [81] [160] [161] [190] [250] [277] [340] [362] [365] [373] [374] [390]
6 fig.	$d = 1\,(1)\,500$	[97]
6 fig.	$d = 1\,(1)\,100$	[9] [350]
5–7 fig.	$d = 1\,(1)\,2000$	[346]
5–7 fig.	$d = 1\,(1)\,1610$	[67]
5–6 fig.	$d = 0\,(0{,}1)\,100$	[337]
5–6 fig.	$d = 1\,(1)\,1100$	[99]
5–6 fig.	$d = 1\,(1)\,1000$	[5] [10] [189] [205] [269]
5–6 fig.	$d = 1\,(1)\,100$	[56]
5 dec.	$d = 0\,(0{,}001)\,1{,}009$	[413]
5 fig.	$d = 0\,(0{,}1)\,100{,}9$	[108] [182]
5 dec.	$d = 1\,(1)\,1020$	[252]
5 dec.	$d = 1\,(1)\,1000$	[362]
5 fig.	$d = 1\,(1)\,1000$	[35] [52] [55] [66]
5 fig.	$d = 1\,(1)\,100$	[11] [27] [33] [83] [96] [419]
4–6 fig.	$d = 1\,(1)\,1000$	[147]
4 fig.	$d = 1\,(0{,}001)\,1{,}13\,(0{,}01)\,10$	[17]
4 fig.	$d = 1\,(0{,}01)\,10$	[21] [58] [89] [93]
4 fig.	$d = 0\,(0{,}1)\,100$	[131]
4 dec.	$d = 1\,(0{,}1)\,100$	[391] [392]
4 dec.	$d = 1\,(1)\,1500$	[209]
4 fig.	$d = 1\,(1)\,1000$	[34]
4 fig.	$d = 1\,(1)\,100$	[294]
4 fig.	$d = 0\left(\frac{1}{8}\right)12$	[93]

2 dec.	$d=1\,(1)\,1000$	[138] [186] [187]
2 dec.	$d=1\,(1)\,100$	[357]

in relation to the circumference

2 dec.	$l=1\,(1)\,100$	[211] [347]

Area of a segment

in relation to the central angle when $r=1$

5 dec.	$\alpha=1°\,(1°)\,180°$	[19] [20] [67] [69] [78] [94] [95][99] [121] [128] [362]
5 fig.	$\alpha=0\,(1°)\,180°$	[346]
5 dec.	$\alpha=1°\,(1°)\,146°$	[93]
4 dec.	$\alpha=147°\,(1°)\,180°$	[93]
4 dec.	$\alpha=0\,(1°)\,180°$	[365] [409]

in relation to the altitude when $d=1$

5 dec.	$h=0{,}001\,(0{,}001)\,0{,}5\,(0{,}05)\,1$	[67]

in relation to the chord when $h=1$

4 dec.	$c=2\,(0{,}01)\,2{,}1\,(0{,}1)\,12\,(1)\,20\,(2)24$	[67]

in relation to the altitude when $c=1$

5 dec.	$h=0{,}005\,(0{,}005)\,0{,}1\,(0{,}001)\,0{,}5$	[67]

Area of a sector

4 dec.	$\alpha=0\,(1°)\,180°$	[209]

Areas of regular polygons (inscribed and circumscribed) when $r=1$

4 dec.	$n=3\,(1)\,10\,($ var. $)\,64$	[64]

Surface of a sphere in relation to the diameter

in integers	$\frac{d}{2}=1\,(1)\,100$	[347]
7 fig.	$d=1\,(1)\,100$	[160]
6 fig.	$d=1\,(1)\,200$	[346]

6 dec.	$d = 0\,(1)\,100$	[209]
5—6 fig.	$d = 0$ (var.) $10\,(0{,}25)\,29\,(0{,}5)\,101\,(1)\,178;\ 179\,(1)\,200$	[313]
4—5 fig.	$d = 1\,(1)\,100$	[347]
4 fig.	$d = 1\,(0{,}1)\,100$	[93]
4 dec.	$d = 0\,(0{,}1)\,100\,(1)\,600$	[393]
3 fig.	$d = 1\,(0{,}1)\,15$	[396]

Surface of spherical segment S in relation to the central angle

6 fig.	S	$\alpha = 0\,(1^\circ)\,180^\circ$	[269]
7 dec.	$\lg S$		

Surface of spheroids (prolate and oblate) ($2a$ & $2b$ — are the major & minor axes respectively)

3 fig.	$2a = m\,(0{,}1)\,15$; $2b = m - \delta\,(0{,}1)\,15 - \delta$; $\delta = 2a - 2b = 0{,}1\,(0{,}1)\,3$; $m = 1\,(0{,}5)\,7{,}5$	[396]

Volumes

Volume of a sphere in relation to the diameter

in integers	$\frac{d}{2} = 1\,(1)\,100$	[347]
7 fig.	$d = 1\,(1)\,200$	[69] [94] [99] [128] [250] [346]
7 fig.	$d = 1\,(1)\,100$	[33]
7 fig.	$\frac{d}{2} = 1\,(1)\,100$	160]
6 fig.	$d = 58\,(0{,}5)\,101\,(1)\,124$	[313]
7 fig.	$d = 125\,(1)\,200$	
6 dec.	$d = 0\,(1)\,100$	[209]
5 fig.	$d = 0$ (var.) $10\,(0{,}25)\,29\,(0{,}5)\,57{,}5$	[313]
5 fig.	$d = 1\,(1)\,1000$	[35]
4 fig.	$d = 1\,(0{,}01)\,10$	[89]
4 fig.	$d = 1\,(0{,}1)\,100$	[93]
4 dec.	$d = 0\,(0{,}1)\,100\,(1)\,600$	[393]
4 dec.	$d = 0\,(1)\,99$	[373]
3 fig.	$d = 1\,(0{,}1)\,15$	[396]
2 dec.	$d = 1\,(1)\,1000$	[409]

Volume of a spherical segment V in relation to the central angle

6 fig. V / 7 dec. $\lg V$		$\alpha = 0\,(1°)\,180°$	[269]

Volumes of spheroids (prolate and oblate)

($2a$ & $2b$ are the major & minor axes respectively)

3 fig.	$2a = m\,(0{,}1)\,15$; $2b = m - \delta\,(0{,}1)\,15 - \delta$; $\delta = 2a - 2b = 0{,}1\,(0{,}1)\,3$; $m = 1\,(0{,}5)\,7{,}5$	[396]

TABLES FOR CONVERTING FROM ONE ANGULAR MEASURE TO ANOTHER

Degrees, minutes & seconds to tenths of a degree

9 dec.	$0(0'',000001)0'',000099(0'',0001)0'',0099(0'',01)$; $0'',99\,(1'')\,59''\,(1')\,59'$	[322]
6 dec.	$0\,(1')\,60'$; $0\,(1'')\,60''$	[3] [230] [273]
5 dec.	$0\,(1')\,60'$; $0\,(1'')\,60''$; $0'',1\,(0'',1)\,1''$	[109]
5 dec.	$0\,(1')\,60'$; $0\,(1'')\,60''$	[228]
4 dec.	$0\,(1')\,59'$	[156]

Fractions of a degree to degrees, minutes & seconds

$0'',00001$; $0'',0001$; $0'',01$	$0\,(0°,00000001)\,0°,00000099$; $(0°,000001)\,0°,000099$; $(0°,0001)\,0°,0099\,(0°,01)\,0°,99$	[322]
$0'',01$; $0'',1$; $1''$	$0\,(0°,0001)0°,01(0°,001)$; $0°,1\,(0°,01)\,1°$	[88] [125] [126]
$0'',1$; $1''$	$0\,(0°,0001)\,0°,01\,(0°,001)$; $0°,1\,(0°,01)1°$	[123]
$0'$, 1	$0°,01\,(0°,01)\,1°$	[93]

Degrees, mins & secs to radians $\left(\frac{k\pi}{180};\ \frac{m\pi}{10800};\ \frac{n\pi}{648000}\right)$

35 dec.	$m = 1'\,(1')\,10'(10')\,60'$. $n = 1''\,(1'')\,10''\,(10'')\,60''$	[449]
32 dec.	$k = 1°\,(1°)\,100°$. $m = 1'\,(1')\,100'$; $n = 1''\,(1'')100''$	[323]
27 dec.	$k = 1°\,(1°)\,360°$. $m = 1'(1')\,60'$; $n = 1''\,(1'')\,60''$	[353]

25	dec.	$k = 1'\,(1')\,100.'$ $\quad m = 1'\,(1')\,100'$, $n = 1''\,(1'')\,100''$	[234]
21	dec.	$m = 1'\,(1')\,100'$	[320]
15	dec.	$k = 1^\circ\,(1^\circ)\,90^\circ$	[377]
15	dec.	$k = 1^\circ\,(1^\circ)\,90^\circ;\ 100^\circ(20^\circ)360^\circ$; $m = 1'\,(1')\,60'.\quad n = 1''\,(1'')\,60''$	[376] [380]
13	dec.	$m = 1'\,(1')\,600'$	[447]
11	dec.	$k = 1^\circ\,(1^\circ)\,360^\circ.\quad m = 1'\,(1')\,60'$, $n = 1''\,(1'')\,60''$	[402]
11	dec.	$k = 0\,(0^\circ{,}00000001)\,0^\circ{,}00000099$ $(0^\circ{,}000001)\,0^\circ{,}000099$ $(0^\circ{,}0001)\,0^\circ{,}0099\,(0^\circ{,}01)\,0^\circ{,}99\,(1^\circ)\,359^\circ$	[322]
11	dec.	$k = 1^\circ\,(1^\circ)\,100^\circ\,(10^\circ)\,270^\circ\,(30^\circ)\,360^\circ$; $m = 1'\,(1')\,60'.\quad n = 1''\,(1'')\,60''$	[153] [273] [287]
10	dec.	$k = 0\,(1^\circ)\,60^\circ.\quad m = 0\,(1')\,60'$, $n = 0\,(1'')\,60''$	[100]
10	dec.	$k = 0\,(1^\circ)\,180^\circ.\quad m = 0\,(1')\,60'$, $n = 0\,(1'')\,60''$	[22] [126]
10	dec.	$k = 1^\circ\,(1^\circ)\,180^\circ.\quad m = 0\,(1')\,60'$, $n = 0\,(1'')\,60''$	[174]
10	dec.	$k = 1^\circ\,(1^\circ)\,30^\circ\,(10^\circ)\,100^\circ$; $m = 1'\,(1')\,30'(10')\,100'$; $n = 1''\,(1'')\,30''\,(10'')\,100''$	[331]
7	dec.	$k = 0\,(1^\circ)\,180^\circ$	[125]
7	fig.	$m = 0\,(1')\,60'.\quad n = 0\,(1'')\,60''$	
7	dec.	$k = 0\,(1^\circ)\,90^\circ\,(5^\circ)\,120^\circ\,(10^\circ)\,360^\circ$; $m = 1'\,(1')\,60'.\quad n = 1''\,(1'')\,60''$	[370]
7	dec.	$k = 0\,(1^\circ)\,180^\circ.\quad m = 0\,(1')\,60'$, $n = 0\,(1'')\,60''$	[31] [36] [124] [138]
7	dec.	$m = 1'\,(1')\,5400'.\quad n = 1''\,(1'')\,60''$	[254] [255] [334]
7	dec.	$k = 1^\circ(1^\circ)\,10^\circ(10^\circ)\,90^\circ$; $m = 1'\,(1')\,10'\,(10')\,60'$; $n = 1''\,(1'')\,10''\,(10'')\,60''$	[228]
7	dec.	$k = 0\,(1^\circ)\,100^\circ\,(10^\circ)\,300^\circ$; $m = 1'\,(1')\,60'.\quad n = 1''\,(1'')\,60''$	[339]
6	dec.	$k = 1^\circ\,(1^\circ)\,180^\circ.\quad m = 0\,(1')\,60'$, $n = 0\,(1'')\,60''$	[25] [82] [85] [379] [385]
5	dec.	$k = 0\,(6')\,90^\circ$	[135] [195]
5	dec.	$k = 0\,(1^\circ)\,100^\circ\,(100^\circ)\,1000^\circ$; $m = 0\,(1')\,60'$	[95]
5	dec.	$k = 0\,(1^\circ)\,100^\circ\,(100^\circ)\,1500^\circ$	[192]

5 dec.	$k = 0\,(1°)\,360°.$ $m = 0\,(1')\,60'$ $n = 0\,(1'')\,60''$	[81]
5 dec. 5 fig.	$k = 0\,(1°)\,180°$ $m = 0\,(1')\,60'.$ $n = 0\,(1'')\,60''$	[123] [126] [214]
5 dec.	$k = 1°\,(1°)\,100°\,(100°)\,1000°;\ 90°\,(90°)\,900°$ $m = 1'\,(1')\,60'$	[63]
4 dec. 4—5 fig.	$k = 0\,(6')\,90°$ $m = 1'\,(1')\,10'.$ $n = 1''\,(1'')\,10''$	[17]
4 dec.	$k = 0\,(6')\,90°$	[46] [47] [69] [112] [131] [188] [266]
4 dec.	$k = 0\,(10')\,90°$	[209] [373]
3; 4 dec.	$k = 1°\,(1°)\,360°.$ $m = 1'\,(1')\,60'$	[19]
4 dec.	$k = 1°\,(1°)\,180°.$ $m = 1'\,(1')\,60'$	[34] [93]
4 dec.	$k = 1°\,(1°)\,90°.$ $m = 1'\,(1')\,60'$	[362]

Degrees to grads

25 dec.	1°(1°) 100°; 1′ (1′) 60′; 1″ (1″) 100″	[48]
9 dec.	0 (0°,00000001) 0°,00000099 (0°,000001) 0°,000099 (0°,0001) 0°,0099 (0°,01) 0°,99 (1°) 359°	[322]
7 dec.	0 (1°) 90°; 0 (1′) 60′; 0 (1″) 60″	[331]
6 dec.	0 (1°) 100°; 0 (1′) 60′; 0 (1″) 100″	[79]
2; 5 dec.	0 (1°) 90°; 0 (1′) 60′, 0 (1″) 60″	[77]
5 dec.	0 (1°) 100°; 0 (1′) 60′; 0 (1″) 60″	[272]
4 dec.	0 (10′) 45°	[69] [128]

Degrees to units of time

0 ,000001; 0 ,00001 0 ,001; 0 ,1; 1	0 (0°,00000001) 0°,00000099 (0°,000001) 0°,000099 (0°,0001) 0°,0099 (0°,01) 0°,99 (1°) 359°	[322]
0 ,001; 0 ,01; 0 ,1	0 (0,0001) 0,01 (0,001) 0,1 (0,01) 1	[125] [126]
1 ,1 0 ,001	0 (1°) 360°; 0 (1′) 60′ 0″ (1″) 60″	[123] [125] [126]
1 ; 1 ; 0 ,01	1° (1°) 360°; 0 (1′) 60′; 0(1″) 60″	[31]

Degress to protractor divisions

0—01: 1—00	0 (1′) 6°; 6° (6°) 90° (90°) 360°	[6]

Radians to degrees and tenths of a degree

15 dec.	0 (0,0001) 0,001 (0,001) 0,01 (0,01) 0,1 (0,1) 1 (1) 10 (10) 100	[376] [377] [380]
9 dec.	0 (0,0000000001) 0,0000000099 (0,00000001) 0,00000099 (0,000001) 0,000099 (0,0001) 0,0099 (0,01) 0,99 (1) 6	[322]
7 dec.	0 (0,00001) 0,0001 (0,0001) 0,001 (0,001) 0,01 (0,01) 0,1 (0,1) 1	[185]
6 dec.	0 (0,00001) 0,0001 (0,0001) 0,001 (0,001) 0,01 (0,01) 0,1 (0,1) 1 (1) 9	[138]
3 dec.	0 (0,01) 1 (1) 10	[95]
3 dec.	0,01 (0,01) 1 (1) 20	[63]
3 dec.	0 (0,01) 1 (1) 30	[192]

Radians to tenths of a degree

10 dec.	$m\,10^{-p}$. $m = 1\,(1)\,9$. $p = 0\,(1)\,5$	[234]
9 dec.	$m\,10^{-p}$. $m = 1\,(1)\,9$. $p = 0\,(1)\,5$	[233]
7 dec.	$m\,10^{-p}$. $m = 1\,(1)\,9$. $p = 1\,(1)\,5$	[287]
7 dec. 7 dec. 2 dec.	0 (0,1) 10 (1) 100 $m10^{-p}$. $m = 1\,(1)\,9$. $p = 1\,(1)\,5$ 0 (0,0001) 0,1 (0,001) 1,6	[153]
6 dec.	0 (0,0001) 0,001 (0,001) 0,01 (0,01) 0,1 (0,1) 1 (1) 10 (10) 100	[376] [377] [379] [380] [385]
5 dec. 2 dec.	0 (0,00001) 0,001 0,001 (0,0001) 0,1 (0,001) 3 (0,01) 10 (0,1) 20 (1) 50	[234]
2 dec.	0 (0,0001) 0,1 (0,001) 3 (0,01) 6,3 (0,1) 10	[233]
2 dec.	0,00001 (0,00001) 0,0001 (0,0001) 0,001 (0,001) 0,01 (0,01) 0,1 (0,1) 1	[19]
1 dec.	0 (0,0001) 0,001 (0,001) 0,01 (0,01) 0,1 (0,1) 1 (1) 9	[228] [339]

Radians to grads $\left(\frac{200n}{\pi}\right)$

7 dec.	$n = m\,10^{-p}$. $m = 1\,(1)\,9$. $p = 1\,(1)\,5$	[185]
6 dec.	$n = 0\,(0,01)\,1\,(1)\,100$	[331]
6 dec.	$n = 0\,(0,01)\,2\,(1)\,10$	[410]
6 dec.	$n = 0\,(0,01)\,1$	[244]

Grads into degrees, minutes and seconds

0°,000000001 до 0°,1	0 (0^g,00000001) 0^g,00000099 (0^g,000001) 0^g,000099 (0^g,0001) 0^g,0099(0^g,01) 0^g,99(1^g) 399^g	[322]
0″,001; 0″,01; 0″,1; 1″; 1′	0 (0^g,0001) 0^g,001(0^g,001) 0^g,01 (0^g,01) 0^g,1(0^g,1) 1^g(1^g) 100^g	[272]
0″,001; 0″,1; 1′	0 (0^g,0001) 0^g,01 (0^g,01) 1^g(1^g) 100^g	[124] [125] [215] [331]
0″,01; 0″,1; 1′	0 (0^g,0001) 0^g,01 (0^g,01) 1^g (1^g) 100^g	[126]
0″,1; 1′	0 (0^g,0001) 0^g,01 (0^g,01) 1^g (1^g) 100^g	[79] [123]

Grads into radians

32 dec.	1^g (1^g) 100^g	[323]
17 dec.	1^g (1^g) 100^g	[171] [380] [387]
17 dec.	1^g (1^g) 50^g	[386]
15 dec.	1^g (1^g) 100^g	[234]
10 dec.	1^g (1^g) 100^g	[80] [327]
10 dec.	1^g (1^g) 50^g (5^g) 100^g	[331]
8 dec.	1^g (1^g) 100^g	[244] [331]
7 dec.	1^g (1^g) 100^g	[124]
6 dec.	0 (0^g,0001) 0^g,01 (0^g,01) 1^g (1^g) 100^g	[410]
5—6 dec.	0 (0^g,01) 100^g	[331]
5 dec.	0 (0^g,02) 100^g	[159]

Grads into units of time

0 ,0001; 0 ,01; 1	0 (0^g,0001) 0^g,01 (0^g,01) 1^g (1^g) 100^g	[79] [125] [126]

Units of time into degrees

0°,000000001 0°,01; 1°	0 (0^{sec},000001) 0^{sec},000099 (0^{sec},0001) 0^{sec},0099 (0^{sec},01) 0^{sec},99 (1^{sec}) 59^{sec}; 0 (1^m) 59^m; 0 (1^{hrs}) 23^{hrs}	[322]

Protractor divisions into degrees

1′; 1°	0—01 (0—01) 1—00; 1—00 (1—00) 15—00 (15—00) 60—00	[6]

INTERPOLATION TABLES

We give here natural and logarithmic interpolati tables for the measurement of angles and time in divisions of sixtieths.

See Ch. 13 for interpolation in the decimal syst

$$\frac{xy}{60}$$

Rec.	$x = 1\,(1)\,60.$	$y = 0\,(1)\,160$	[162]
1 dec.	$x = 1\,(1)\,60.$	$y = 0\,(0{,}5)\,60$	[148]
0′,1	$x = 1\,(1)\,60.$	$y = 0\,(1')\,10'$	[296]

$$\frac{xy}{120}$$

Rec.	$x = 1\,(1)\,60.$	$y = 4\,(4)\,360$	[307]
1 dec.	$x = 1\,(1)\,60.$	$y = 0\,(1)\,36$	[173]

$$\frac{xy}{24^{\text{hrs}}}$$

0″,1	$x = 0\,(30^{\text{M}})\,30^{\text{hrs}}.$	$y = 0\,(30'')\,30'30''$	[335]

$$\lg \frac{60}{x}$$

3 dec.	$x = 0\,(0{,}1)\,60$	[148]

$$\lg \frac{1^\circ}{x''} = \lg \frac{3600''}{x''}$$

4 dec.	$x = 0\,(1'')\,80'$	[210]
4 dec.	$x = 0\,(1'')\,88'$	[249]

$$\lg \frac{24^{\text{hrs}}}{x}$$

5 dec.	$x = 0\,(1^{\text{M}})\,24^{\text{hrs}}$	[334]
4 dec.	$x = 0\,(1^{\text{M}})\,24^{\text{hrs}}$	[335]

$$\lg \frac{x}{3^{\text{hrs}}}$$

5 dec.	$x = 0\,(1^{\text{sec}})\,3^{\text{hrs}}$	[334]
4 dec.	$x = 0\,(1^{\text{sec}})\,3^{\text{hrs}}$	[335]

Chapter 3

EXPONENTIAL AND HYPERBOLIC FUNCTIONS

This chapter describes the following tables: exponential functions and various forms containing exponential functions; hyperbolic functions of real and complex variables and various expressions containing hyperbolic functions; expressions containing combinations of trigonometric and hyperbolic functions; inverse hperbolic functions; logarithms of exponential and hyperbolic functions; hyperbolic amplitudes (the gudermannian) and inverse hyperbolic amplitudes (the inverse gudermannian).

Chapter 7 should also be consulted for tables of $e^{\pm x^2}$ and similar tables; Chapter 1 for tables of various integrals of exponential functions. Tables of exponential power functions are described in Chapter 1.

EXPONENTIAL FUNCTIONS

Exponential functions with argument x

e^x

258 fig.	$x = 2(2)10$	[162]
42 fig.	$x = 1(1)100$	[145]
33 fig.	$x = 0(0,1)50$	
32 fig.	$x = 1(1)32$	[96]
28–29 fig.	$x = 1(1)24$	[109]
32–33 fig.	$x = 25;\ 30\ \&\ 60$	
24 dec.	$x = 0(0,001)0,099$	[68]
18 dec.	$x = 0(0,1)9,9$	
24 dec.	$x = 10(1)20$	[144]
20–22 fig.	$x = 0(0,1)32$	[140]
22–20 dec.	$x = 1(1)24$	[68]
20 fig.	$x = 1(1)100$	[170]

---

[1] See pages 438 and 550 for the references to Chapter 3.

20 dec.	$x = 1\,(1)\,10$	[129]
19 fig.	$x = 1\,(1)\,100$	[119]
18 dec.	$x = 0\,(0{,}000001)\,0{,}0001$	
18 dec.	$x = 0\,(0{,}0001)\,1$	
18 dec.	$x = 0\,(0{,}1)\,10$	[68]
17 dec.	$x = 0\,(0{,}001)\,1$	
16 dec.	$x = 0\,(0{,}1)\,3$	[161]
12 dec.	$x = 0\,(0{,}001)\,2$	
15 dec.	$x = 1\,(0{,}0001)\,2{,}5\,(0{,}001)\,5$	[119]
12 dec.	$x = 5\,(0{,}01)\,10$	
10—15 dec.	$x = 7\,(0{,}2)\,20$	[144]
20 dec.	$x = 0\,(0{,}00001)\,0{,}001$	[64]
10 dec.	$x = 0{,}001\,(0{,}0001)$ $0{,}1\,(0{,}001)\,3\,(0{,}01)\,10$	
15 dec.	$x = 10\,(0{,}1)\,20\,(1)\,41$	
33 fig.	$x = 42\,(1)\,50$	
10 fig.	$x = 0\,(0{,}001)\,10$	[170]
10 dec.	$x = 0\,(0{,}001)\,10$	[68]
9 fig.	$x = 0\,(0{,}001)\,0{,}1\,(0{,}01)$ $2\,(0{,}1)\,10\,(1)\,500$	[159]
9 fig.	$x = 0\,(0{,}001)\,0{,}01\,(0{,}01)$ $0{,}1\,(0{,}1)\,1\,(1)\,30$	[2]
9 fig.	$x = 0\,(0{,}01)\,0{,}5;\ 0\,(0{,}5)\,20$	[41]
9 fig.	$x = 11\,(1)\,100$	[63]
9 fig.	$x = 1\,(1)\,50$	[37]
8—9 fig.	$x = 0\,(0{,}02)\,16$	[6] [125]
8 fig ; 9 dec.	$x = 1\,(1)\,10;\ 11\,(1)\,100$	[63]
7—8 fig.	$x = 8\,(0{,}001)\,9{,}809$	[44]
7 fig.	$x = 0\,(0{,}01)\,10$	[75] [122] [123] [124]
6—7 fig.	$x = 0\,(0{,}001)\,6\,(0{,}01)$ $16\,(0{,}1)\,50\,(1)\,100$	[169]
6—7 fig.	$x = 0\,(0{,}01)\,7$	[40]
6—7 fig.	$x = 0\,(0{,}01)\,1{,}5\,(0{,}1)\,10$	[17]
6 dec.	$x = 0\,(0{,}001)\,1\,(0{,}002)$ $2\,(0{,}005)\,3\,(0{,}01)$ $4\,(0{,}05)\,5\,(0{,}1)\,6$	[34]
6 dec.	$x = 0\,(0{,}001)\,3\,(0{,}01)\,6$	[37]
6 fig.	$x = 0\,(0{,}01)\,1\,(0{,}1)\,9{,}9$	[76]
6 fig.	$x = 0\,(0{,}01)\,2\,(0{,}1)\,10$	[56]
6 fig.	$x = 0\,(0{,}01)\,2\,(0{,}1)\,6$	[50]
6 fig.	$x = 0\,(0{,}01)\,1{,}5\,(0{,}05)$ $2\,(0{,}1)\,6$	[36]

6 dec.	$x = 0\,(0{,}0001)\,0{,}1\,(0{,}001)$ $3\,(0{,}01)\,5{,}99$	[63]
5 dec.	$x = 6\,(0{,}01)\,6{,}3\,(0{,}1)\,10$	
6 dec.	$x = 0\,(0{,}005)\,0{,}1\,(0{,}01)$ $3\,(0{,}05)\,4\,(0{,}1)7$	[104]
5 dec.	$x = 7\,(0{,}1)\,10$	
5 dec.	$x = 0\,(0{,}01)\,10$	[62]
10 dec.	$x = 0\,(0{,}0001)0{,}01;\ 10\,(1)\,40$	
5—7 fig.	$x = 0\,(0{,}01)\,4\,(0{,}1)\,10$	]26]
5—6 fig.	$x = 0\,(0{,}001)\,10$	[20]
5—6 fig.	$x = 0\,(0{,}01)\,2\,(0{,}1)\,9{,}9$	[55]
5—6 fig.	$x = 0\,(0{,}2)\,6$	[35]
5 fig.	$x = 0\,(0{,}01)\,10{,}09$	[105]
5 dec.	$x = 0\,(0{,}01)\,10$	[117]
5 fig.	$x = 0\,(0{,}01)\,6$	[102]
5 fig.	$x = 0\,(0{,}01)\,5\,(0{,}05)\,10$	[100]
5 fig.	$x = 0\,(0{,}01)\,5{,}5\,(0{,}1)\,10$	[52]
5 fig.	$x = 0\,(0{,}01)\,4\,(0{,}1)\,10$	[5]
5 fig.	$x = 0\,(0{,}01)\,1\,(0{,}05)\,6$	[33]
5 fig.	$x = 0\,(0{,}001)0{,}009(0{,}01)$ $0{,}99\,(0{,}1)\,9{,}9$	[19]
5 fig.	$x = 0\,(0{,}01)\,3\,(0{,}05)\,4\,(0{,}1)$ $6\,(0{,}25)\,7\,(0{,}5)\,10$	[12] [47] [60]
5 fig.	$x = 0\,(0{,}01)\,2\,(0{,}1)\,8$	[114]
5 fig.	$x = 0\,(0{,}01)\,1{,}6\,(0{,}1)\,6$	[10] [15] [22] [27] [28]
5 fig.	$x = 0\,(0{,}02)\,0{,}1\,(0{,}01)$ $0{,}5\,(0{,}1)\,6$	[24]
5 fig.	$x = 0\,(0{,}1)\,5$	[48]
5 dec.	$x = 1\,(1)\,2556$	[119]
5 fig.	$x = 5\,(1)\,50\,(10)\,100\,(50)\,500$	[40]
4—15 dec.	$x = 5\,(0{,}2)\,6{,}8$	[144]
4—5 fig.	$x = 0\,(0{,}01)\,6{,}09;\ \ 60\,(1)\,200$	[53]
4—5 fig.	$x = 0\,(0{,}01)\,1\,(0{,}1)\,10{,}9$	[97]
4 dec.	$x = 0\,(0{,}001)\,1$	[86]
4—5 fig.	$x = 1\,(0{,}01)\,6\,(0{,}1)\,20$	
4 dec.	$x = 0\,(0{,}01)\,1{,}5\,(0{,}05)$ $2\,(0{,}1)\,7$	[39]
4 dec.	$x = 0\,(0{,}01)\,6$	[80]
4 dec.	$x = 0\,(0{,}01)\,0{,}99$	[85]
4 dec.	$x = 0\,(0{,}1)\,6{,}7$	[108]
4 dec.	$x = 0\,(0{,}001)\,1{,}499$	[51]
3 dec.	$x = 1{,}5\,(0{,}001)5$	

3 dec.	$x = 0\,(0{,}01)\,3$	[49]
3 fig.	$x = 0\,(0{,}1)\,1$	[107]
1—3 dec.	$x = 0\,(0{,}01)\,1\,(0{,}1)\,5{,}9$	[21]
1 dec.	$x = 0\,(0{,}1)\,15$	[144]

e^x

where

$x = m10^{-n}$.

62 dec.	$m = 1\,(1)\,9.$	$n = 1\,(1)\,10$	[145]
24—25 dec.	$m = 1\,(1)\,9.$	$n = 1\,(1)\,3$	[68]
20—23 dec.	$m = 1\,(1)\,9.$	$n = 6\,(1)\,10$	[64]
20 dec.	$m = 1\,(1)\,999.$	$n = 3;\ 6;\ 9$	[170]
20 dec.	$m = 1\,(1)\,999.$	$n = 6\,(3)\,15$	[68]
18 dec.	$m = 1\,(1)\,9.$	$n = 7\,(1)\,10$	[119]
18—12 dec.	$m = 1\,(1)\,9.$	$n = 1\,(1)\,4$	[63]

$\frac{1}{2}\,e^x$

7—9 fig.	$x = 11\,(1)\,20$	[83]
5—6 fig.	$x = 0\,(0{,}01)\,10$	[83]
3 dec.	$x = 4\,(0{,}01)\,10$	[74]
2; 1 dec.	$x = 6\,(0{,}01)\,7{,}5;\ \ 7{,}5\,(0{,}01)\,10$	[59]

e^{-x}

52 dec.	$x = 1\,(1)\,50$	[145]
62 dec.	$x = 50\,(1)\,100$	[145]
33; 38 dec.	$x = 0\,(0{,}1)\,14{,}9;\ \ 15\,(0{,}1)\,24{,}9$	[145]
43; 48 dec.	$x = 25\,(0{,}1)\,35;\ \ 35\,(0{,}1)\,50$	[145]
32 dec.	$x = 1\,(1)\,32$	[96]
30 dec.	$x = 0\,(0{,}5)\,1\,(1)\,10$	[166]
22; 16 dec.	$x = 1\,(1)\,10;\ \ 11\,(1)\,24$	[68]
20 fig.	$x = 1\,(1)\,100$	[170]
20 dec.	$x = 0\,(0{,}1)\,32$	[140]
20 dec.	$x = 1\,(1)\,10$	[129]
19 fig.	$x = 1\,(1)\,100$	[119]
18 dec.	$x = 0\,(0{,}000001)\,0{,}0001$	[119]
18 dec.	$x = 0\,(0{,}0001)\,2{,}4999$	[119]
18 dec.	$x = 0\,(0{,}1)\,37$	[160]
12 dec.	$x = 0\,(0{,}001)\,15{,}349$	[160]
14 dec.	$x = 15{,}35\,(0{,}002)\,17{,}3\,(0{,}005)\,27{,}635$	[160]

20 dec.	$x = 0\,(0{,}00001)\,0{,}001$	[64]
10 dec.	$x = 0{,}001\,(0{,}0001)\,0{,}1\,(0{,}001)$ $3\,(0{,}01)\,10$	
15 dec.	$x = 10\,(0{,}1)\,20\,(1)\,36$	
10 dec.	$x = 0\,(0{,}0001)\,0{,}01$	[62]
10 dec.	$x = 0\,(0{,}001)\,10$	[68] [170]
10 dec.	$x = 0\,(0{,}001)\,1\,(1)\,100$	[126]
10 dec.	$x = 10\,(1)\,24$	[62]
9 dec.	$x = 0\,(0{,}001)\,1\,(0{,}01)\,2$	[164]
9 fig.	$x = 0(0{,}001)\,0{,}1\,(0{,}01)\,2\,(0{,}1)\,10\,(1)\,500$	[159]
9 fig.	$x = 0\,(1)\,100$	[37] [63]
8—9 fig.	$x = 0\,(0{,}5)\,20$	[41]
8 dec.	$x = 0\,(0{,}01)\,0{,}5$	
8 fig.	$x = 0{,}001\,(0{,}001)\,0{,}01\,(0{,}01)\,0{,}1\,(0{,}1)$ $1\,(1)\,30$	[2]
7 dec.	$x = 0\,(0{,}001)\,3\,(0{,}01)\,6$	[37]
6—8 dec.	$x = 0\,(0{,}0001)\,0{,}1\,(0{,}001)3\,(0{,}01)$ $6{,}3\,(0{,}1)\,10$	[63]
6 dec.	$x = 0\,(0{,}001)\,3\,(0{,}01)\,7{,}5\,(0{,}1)\,12{,}5$	[169]
6 dec.	$x = 0(0{,}005)\,0{,}1\,(0{,}01)\,3\,(0{,}05)$ $4\,(0{,}1)\,10$	[104]
6 dec.	$x = 0\,(0{,}01)\,5{,}5\,(0{,}1)\,10$	[52]
6 dec.	$x = 0\,(0{,}001)\,1\,(0{,}002)\,2\,(0{,}005)3(0{,}01)$ $4\,(0{,}05)\,5\,(0{,}1)\,6;\ 0\,(0{,}1\pi)\,\pi$	[34]
5; 6 dec.	$x = 0\,(0{,}01)\,2{,}3;\ 2{,}31\,(0{,}01)\,4{,}6$	[100]
7 dec.	$x = 4{,}61\,(0{,}01)\,5\,(0{,}05)\,10$	
5—6 fig.	$x = 0\,(0{,}001)\,10$	[20]
5 dec.	$x = 0\,(0{,}0001)0{,}001(0{,}001)\,0{,}01\,(0{,}01)2{,}3$	[137]
6; 7 dec.	$x = 2{,}31\,(0{,}01)\,6{,}9;\ 6{,}91\,(0{,}01)\,10$	
5—6 fig.	$x = 0\,(0{,}2)\,6$	[35]
5 fig.	$x = 0\,(0{,}01)\,10{,}09$	[105]
5 dec.	$x = 0\,(0{,}01)\,10$	[62] [117]
5 dec.	$x = 0\,(0{,}01)\,6$	[102]
5 dec.	$x = 0\,(0{,}01)\,1{,}5\,(0{,}1)\,10$	[17]
5 fig.	$x = 0\,(0{,}001)\,0{,}009\,(0{,}01)\,0{,}99\,(0{,}1)\,9{,}9$	[19]
5 fig.	$x = 0\,(0{,}01)\,2\,(0{,}1)\,10$	[56]
5 dec.	$x = 0\,(0{,}01)\,1\,(0{,}1)\,9{,}9$	[76]
5 dec.	$x = 0\,(0{,}01)\,1{,}6\,(0{,}1)\,6$	[22] [27] [53]
5 dec.	$x = 0\,(0{,}01)\,2\,(0{,}1)\,6$	[50]
5 dec.	$x = 0\,(0{,}01)\,3\,(0{,}05)\,4\,(0{,}1)\,6\,(0{,}25)$ $7\,(0{,}5)\,10$	[12] [47] [60]
5 dec.	$x = 0\,(0{,}01)\,1{,}5\,(0{,}05)\,2\,(1)\,6$	[36]

5 dec.	$x = 1\,(1)\,2556$	[119]
5 fig.	$x = 5\,(1)\,50\,(10)\,100\,(50)\,500$	[40]
4—6 dec.	$x = 0\,(0{,}01)\,4\,(0{,}1)\,10$	[5] [26]
4—5 fig.	$x = 0\,(0{,}01)\,7$	[40][71]
4—5 dec.	$x = 0\,(0{,}01)\,3$	[49]
4; 5 dec.	$x = 0\,(0{,}001)\,1\,(0{,}01)\,2;\ 2\,(0{,}01)\,6$	[86]
6 dec.	$x = 6(0{,}1)\,10$	
4—5 dec.	$x = 0\,(0{,}01)\,1\,(0{,}05)\,6$	[33]
4—5 dec.	$x = 0\,(0{,}01)\,1\,(0{,}1)\,5{,}9$	[21]
4 dec.	$x = 0\,(0{,}01)\,1{,}5\,(0{,}05)\,2\,(0{,}1)\,7$	[39]
4 dec.	$x = 0\,(0{,}01)\,6$	[80]
4 dec.	$x = 0\,(0{,}001)\,0{,}1\,(0{,}01)\,3;\ 3\,(0{,}1)\,8{,}9$	[72]
4 dec.	$x = 0\,(0{,}01)\,2\,(0{,}1)\,9{,}9$	[55]
4 dec.	$x = 0\,(0{,}01)\,2\,(0{,}1)\,8$	[114]
4 dec.	$x = 0\,(0{,}02)\,0{,}1\,(0{,}01)\,0{,}5\,(0{,}1)6$	[24]
4 dec.	$x = 0\,(0{,}01)\,1{,}6\,(0{,}1)\,6$	[10]
4 dec.	$x = 0\,(0{,}01)\,1\,(0{,}1)\,6$	[15] [28]
4 dec.	$x = 0\,(0{,}01)\,0{,}99$	[85]
4 dec.	$x = 0\,(0{,}1)\,6{,}7$	[108]
4 dec.	$x = 0\,(0{,}1)\,5$	[48]
3 fig.	$x = 0\,(0{,}01)\,1\,(0{,}1)\,4{,}5\,(0{,}5)\,5\,(1)\,10$	[97]
3 fig.	$x = 0\,(0{,}1)\,1$	[107]

$$e^{-x}$$

where $x = m10^{-n}$.

63 dec.	$m = 1\,(1)\,9.$	$n = 1\,(1)\,10$	[145]
20—28 dec.	$m = 1\,(1)\,9.$	$n = 6\,(1)\,10$	[64]
20 fig.	$m = 1\,(1)\,999.$	$n = 3;\ 6;\ 9$	[170]
18 dec.	$m = 1\,(1)\,9.$	$n = 7\,(1)\,10$	[119]
18 dec.	$m = 1\,(1)\,9.$	$n = 1\,(1)\,3$	[68]
8—12 dec.	$m = 1\,(1)\,9.$	$n = 0\,(1)\,4$	[63]

$$\frac{1}{2}e^{-x}$$

5—6 fig.	$x = 0\,(0{,}01)\,8{,}09$	[83]

Exponential functions with arbitrary argument

$$e^{-\frac{1}{x}}$$

5 dec.	$x = 0\,(0{,}01)\,0{,}1\,(0{,}02)\,0{,}5\,(0{,}05)\,1\,(0{,}1)\,5\,(0{,}5)\,10\,(5)$ $50\,(25)\,100;\ 0\,(0{,}2)\,1\,(0{,}5)\,3\,(1)\,10\,(5)\,30\,(10)$ $50\,(25)\,100$	[40]

e^{x^2}

5 fig.	$x = 0(0,1)5$	[52] [82] [97]
5; 4 fig.	$x = 0(0,1)3,3;\ 3,4(0,1)5$	[115]
4 dec.	$x = 0(0,01)2$	[40]

e^{-x^2}

5 fig.	$x = 0(0,1)5$	[52]

$\frac{1}{2}e^{\beta x}$

4 dec.	$\beta x = 0(1)189$	[127]

$\frac{1}{2}e^{-\beta x}$

4 dec.	$\beta x = 0(1)189$	[127]

$e^{-\alpha(2-x)}$

5 dec. $\alpha = 2,8;\ 3.$	$x = 0,001;\ 0,01;\ 0,05;\ 0,1(0,1)0,5$	[128]

$e^{\pm(n+x)}$

3 fig.	$n = 1(1)4.\ x = 0(0,1)1$	[107]

$e^{\pm x\pi}$

25 fig.	$x = 2(1)10;\ 1,25(0,25)4;\ 1\frac{1}{6}\left(\frac{1}{6}\right)3\frac{5}{6}$	[145]
20 dec.	$x = 0(0,5)4$	[134]
10; 13 dec.	$x = 0(0,25)0,5;\ 0,75(0,25)3$	[64]
10 dec.	$x = 0(0,01)0,1(0,1)5(0,5)9;\ 10$	[104]
8—23 dec.	$x = 0(0,01)0,1(0,1)10$	[65]
5—9 fig.	$x = 0(0,01)10$	[66]
5 fig.	$x = 0(0,25)5$	[50]
5 dec.	$x = 0(0,01)0,1(0,1)10;\ 1\left(\frac{1}{6}\right)4$	[62]
4—6 fig.	$x = 0(0,1)4$	[88]
4—5 fig.	$x = 0(0,05)1$	[71]

$e^{\pm\frac{\pi x}{360}}$

23 dec.	$x = 0(1)360$	[145]
10 dec.	$x = 0(1)360$	[64]

$$e^{\frac{n}{100}}$$

9—10 dec.	$n = 0\,(1)\,99$	[105]

$$e^{\frac{1}{x}}$$

4—5 fig.	$x = 0\,(0{,}2)\,1\,(0{,}5)\,3\,(1)\,10\,(5)\,30\,(10)\,50\,(25)\,100$	[40]

$$e^{\frac{\pi x}{4}}$$

5 fig.	$x = -20\,(1)\,20$	[40] [52]

$$e^{\frac{x\sqrt{\pi}}{4}}$$

5 fig.	$x = -20\,(1)\,20$	[40] [52]

Expressions containing hyperbolic functions

$$1 - e^{-x}$$

7 dec.	$x = 0\,(0{,}1)\,16{,}9$	[91]
4 dec.	$x = 0\,(0{,}001)\,0{,}499$	[61]
4 dec.	$x = 0\,(0{,}02)\,5\,(0{,}1)\,7{,}5\,(0{,}5)\,10$	[1]

$$e^{x} - 1$$

4 dec.	$x = 0\,(0{,}001)\,0{,}499$	[61]

$$x^{n}e^{-x^2}$$

6 dec.	$n = 3;\ 4. \quad x = 0\,(0{,}1)\,3$	[165]

$$e^{-u}\sin x$$

5—6 fig.	$x = 1°\,(1°)\,90°. \quad u = 0{,}1\,(0{,}1)\,1\,(1)\,7$	[40]

$$e^{-u}\cos x$$

5—6 fig.	$x = 1°\,(1°)\,90°. \quad u = 0{,}1\,(0{,}1)\,1\,(1)\,7$	[40]

$$f(x)e^{-\frac{1}{x}}$$

5 fig.	$f(x) = x^2$	$x = 0\,(0{,}2)\,1\,(0{,}5)\,3\,(1)\,10\,(5)\,30\,(10)$ $50\,(25)\,100$	[40]
4—5 fig.	$f(x) = x^{-2}$	$x = 0\,(0{,}01)\,0{,}1\,(0{,}02)\,0{,}5\,(0{,}05)$ $1\,(0{,}1)\,5\,(0{,}5)\,10\,(5)\,50\,(25)\,100$	[40]

$$\frac{e^{-x}}{x}$$

9 dec.	$x = 0{,}1\,(0{,}001)\,1\,(0{,}01)\,2$	[84] [164]
8—10 dec.	$x = 6(0{,}01)\,15$	[32]
5—6 fig.	$x = 20\,(0{,}02)\,50$	[157]

$$\frac{e^{x}}{\sqrt{2\pi x}}$$

6 fig.	$x = 1\,(1)\,20$	[150]
6 fig.	$x = 21\,(1)\,40$	[151]

$$\frac{2}{\sqrt{\pi}}\,e^{-x}$$

27 dec.	$x = 0\,(0{,}5)\,1\,(1)\,10$	[166]

$$\frac{e^{-x}}{\sqrt{2\pi x}}$$

6 dec.	$x = 1\,(1)\,13$	[150]

$$\frac{x}{1-e^{-x}}$$

5 dec.	$x = 0\,(0{,}01)\,10$	[158]

$$\frac{x}{e^{x}-1}$$

6 dec.	$x = 0\,(0{,}005)\,3\,(0{,}01)\,8\,(0{,}05)\,15$	[139]
4—5 dec.	$x = 0\,(0{,}01)\,2\,(0{,}1)\,14$	[40]

$$\frac{x}{e^{x}-1} - \int_{x}^{0} \frac{te^{t}}{(e^{t}-1)^2}\,dt$$

4—5 dec.	$x = 0\,(0{,}01)\,7$	[40]

$$\int_x^0 \frac{te^t}{(e^t-1)^2}\,dt$$

4 dec. $x = 0\,(0{,}01)\,2\,(0{,}1)\,7$ [40]

$$\frac{1-e^{-x}}{x}$$

9 dec. $x = 0\,(0{,}001)\,0{,}1$ [164]
7 dec. $x = 0\,(0{,}1)\,16{,}9$ [91]
6 dec. $x = 0\,(0{,}05)\,5$ [163]

$$\frac{e^{-x}-1}{x}$$

9 dec. $x = 0\,(0{,}001)\,0{,}1$ [84]

$$\frac{x^2e^x}{(e^x-1)^2}$$

6 dec. $x = 0\,(0{,}005)\,3\,(0{,}01)\,8\,(0{,}05)\,15$ [139]
4—5 dec. $x = 0\,(0{,}01)\,2\,(0{,}1)\,14$ [40]

$$e^{-a}\frac{a^x}{x!}$$

6 dec. $x = 0\,(1)\,37.\quad a = 1{,}1\,(0{,}1)\,15$ [40]
6 dec. $x = 0\,(1)\,27.\quad a = 0{,}1\,(0{,}1)\,1\,(1)\,9$ [9]

$$\frac{x^{n-2}e^{-x^2}}{(n-1)!}$$

1 dec. $x = 2\,(1)\,7.\quad n = 1\,(1)\,10$ [43]

$$\frac{1}{x^5(e^{1/x}-1)}$$

4—5 fig. $x = 0{,}06\,(0{,}001)\,0{,}17\,(0{,}002)\,0{,}18\,(0{,}005)\,1\,(0{,}01)\,1{,}95$ [71]

$$\frac{1}{6\sqrt{2\pi}}\left[1-(1-x^2)\,e^{-\frac{x^2}{2}}\right]$$

4 dec. $x = 0\,(0{,}01)\,2\,(0{,}02)\,3\,(0{,}2)\,4{,}2\,(0{,}3)\,4{,}5$ [40]

$$\int_x^\infty e^{-t}dt$$

9 dec. $x = 0\,(0{,}001)\,1\,(0{,}01)\,2$ [84]

$$\int_x^\infty te^{-t}dt = (x+1)e^{-x}$$

9 dec.	$x = 0\,(0{,}001)\,1\,(0{,}01)\,2$	[84] [164]

$$\int_x^\infty t^2e^{-t}dt = (x^2+2x+2)e^{-x}$$

9 dec.	$x = 0\,(0{,}001)\,1\,(0{,}01)\,2$	[84] [164]

HYPERBOLIC FUNCTIONS

Hyperbolic sine

sh x

20 dec.	$x = 1\,(1)\,10$	[129]
15 dec.	$x = 0\,(0{,}1)\,10$	[42] [133]
15 dec.	$x = 1°\,(1°)\,90°$	[112]
20 dec.	$x = 0\,(0{,}00001)\,0{,}001$	[64]
10 dec.	$x = 0{,}001\,(0{,}0001)\,0{,}1\,(0{,}001)\,3\,(0{,}01)\,10$	[64]
15 dec.	$x = 10\,(0{,}1)\,20\,(1)\,39$	[64]
11—15 dec.	$x = 40\,(1)\,50$	[64]
9 dec.	$x = 0\,(0{,}0001)\,2\,(0{,}1)\,10$	[118]
7 dec.	$x = 0\,(0{,}0005)\,0{,}025$	[46]
7 dec.	$x = 0\,(0{,}01)\,4$	[38]
7 dec.	$x = 0\,(0{,}05)\,2{,}5$	[46]
6 dec.	$x = 0\,(0{,}0001)\,0{,}2\,(0{,}001)\,2$	[54]
6 fig.	$x = 2\,(0{,}01)\,8$	[54]
6 dec.	$x = 0\,(0{,}001)\,1\,(0{,}002)\,2\,(0{,}005)\,3\,(0{,}01)\,4\,(0{,}05)\,5\,(0{,}1)\,6$	[34]
6; 5; 4 dec.	$x = 0\,(0{,}001)\,2{,}5;\ 2{,}5\,(0{,}001)\,4;\ 4\,(0{,}001)\,6$	[169]
6 fig.	$x = 0\,(0{,}01)\,1{,}5\,(0{,}05)\,2\,(0{,}1)\,6$	[36]
6 dec.	$x = 0$ (var.) $0{,}02615$	[168]
6—5 fig.	$x = 0\,(0{,}001)\,10$	[20]
5; 6 dec.	$x = 0\,(0{,}001)\,0{,}4;\ 0{,}4\,(0{,}001)\,3\,(0{,}01)\,6$	[168]
5—6 fig.	$x = 0\,(0{,}01)\,5{,}99$	[11] [14]
5—6 fig.	$x = 0\,(0{,}01)\,4$	[40]
5—6 fig.	$x = 0\,(0{,}2)\,6$	[35]
5 dec.	$x = 0\,(0{,}01)\,2$	[50]
6 fig.	$x = 2\,(0{,}1)\,6$	[50]
5 dec.	$x = 0\,(0{,}0001)\,0{,}1\,(0{,}001)\,3\,(0{,}01)\,6{,}3\,(0{,}1)\,10$	[63]
5 dec.	$x = 0\,(0{,}005)\,0{,}1\,(0{,}01)\,3\,(0{,}05)\,4\,(0{,}1)\,10$	[104]
5 dec.	$x = 0\,(0{,}01)\,10$	[62] [117]

5 fig.	$x = 0\,(0{,}01)\,3\,(0{,}05)\,10$	[100]
5 fig.	$x = 0\,(0{,}01)\,6$	[102] [105]
5 fig.	$x = 0\,(0{,}01)\,3\,(0{,}05)\,4\,(0{,}1)\,6\,(0{,}25)\,7\,(0{,}5)\,10$	[12] [60]
5 dec.	$x = 0\,(0{,}01)\,1{,}5\,(0{,}1)\,10$	[17]
5 fig.	$x = 0\,(0{,}01)\,1{,}6\,(0{,}1)\,6$	[10]
5 fig.	$x = 0\,(0{,}01)\,1\,(0{,}1)\,6$	[15] [28]
5 fig.	$x = 0\,(0{,}02)\,0{,}1\,(0{,}01)\,0{,}5\,(0{,}1)\,6$	[24]
5 fig.	$x = 0\,(0{,}1)\,5$	[48]
6; 5; 4 dec.	$x = 0\,(0{,}01)\,2;\ 2\,(0{,}01)\,5;\ 5\,(0{,}01)\,8$	[79]
6; 5; 4 dec.	$x = 0\,(0{,}01)\,2;\ 2\,(0{,}01)\,2{,}5\,(0{,}1)\,5;\ 5\,(0{,}1)\,7{,}5$	[74]
4—6 fig.	$x = 0\,(0{,}01)\,5{,}99$	[16]
5; 4 dec.	$x = 0\,(0{,}0001)\,0{,}1\,(0{,}001)\,3;\ 3\,(0{,}01)\,6$	[37]
5; 4 dec.	$x = 0\,(0{,}001)\,1{,}509;\ 1{,}5\,(0{,}001)\,3{,}009$	[83]
5; 4 dec.	$x = 0\,(0{,}01)\,3;\ 3\,(0{,}1)\,5$	[52]
4 dec.	$x = 0\,(0{,}001)\,3$	[46]
5 fig.	$x = 3\,(0{,}01)\,8$	
4 dec.	$x = 0\,(0{,}001)\,1\,(0{,}01)\,1{,}5$	[86]
5 fig.	$x = 1{,}5\,(0{,}01)\,6\,(0{,}1)\,10$	
4 dec.	$x = 0\,(0{,}01)\,2{,}99$	[47]
5 fig.	$x = 3\,(0{,}05)\,4\,(0{,}1)\,6\,(0{,}25)\,7\,(0{,}5)\,10$	
4 dec.	$x = 0\,(0{,}01)\,3{,}39$	[97]
5 fig.	$x = 3{,}4\,(0{,}01)\,5{,}09$	
4 dec.	$x = 0\,(0{,}01)\,2{,}5\,(0{,}1)\,2{,}9$	[103]
5 fig.	$x = 3\,(0{,}1)\,7{,}5$	
4—5 fig.	$x = 0\,(0{,}01)\,5{,}99$	[13]
4—5 fig.	$x = 0\,(0{,}01)\,5{,}09$	[19]
4—5 fig.	$x = 0\,(0{,}01)\,1\,(0{,}05)\,6$	[33]
4—5 dec.	$x = 0\,(0{,}01)\,1{,}6\,(0{,}1)\,6$	[22] [27] [53]
4—5 dec.	$x = 0\,(0{,}57)\,91{,}67\,(5{,}73)\,343{,}77$	[22]
4—5 fig.	$x = 1\,(0{,}1)\,5{,}9;\ 0\,(1)\,99$	[21]
4 fig.	$x = 0\,(0{,}001)\,1\,(0{,}01)\,6\,(0{,}1)\,10$	[86]
4 dec.	$x = 0\,(0{,}01)\,5{,}99$	[3] [4]
4 dec.	$x = 0\,(0{,}01)\,5{,}09$	[23]
4 dec.	$x = 0\,(0{,}01)\,3{,}99$	[110]
4 dec.	$x = 0\,(0{,}01)\,1{,}6$	[5]
4 dec.	$x = 0\,(0{,}01)\,1$	[45]
4 dec.	$x = 0\,(0{,}02)\,3{,}98$	[81]
3; 4 dec.	$x = 0\,(0{,}001)\,2{,}009;\ 2\,(0{,}001)\,5{,}009$	[44]
5 dec.	$x = 5\,(0{,}001)\,8{,}009$	
4; 3 dec.	$x = 0\,(0{,}001)\,1{,}999;\ 2\,(0{,}001)\,3$	[51]
4; 3 dec.	$x = 0\,(0{,}01)\,3{,}39;\ 3{,}4\,(0{,}01)\,5{,}99$	[39] [111]
3—4 dec.	$x = 0\,(0{,}01)\,4;\ 4\,(0{,}01)\,6$	[59]
3—4 dec.	$x = 0\,(0{,}01)\,3$	[49]

sh πx

20—24 fig.	$x = 0\,(0{,}01)\,0{,}1\,(0{,}1)\,10$	[65]
15 dec.	$x = 0\,(0{,}0001)\,0{,}01$	[42]
15 dec.	$x = 0\,(0{,}01)\,4$	[134]
14—16 fig.	$x = 0\,(0{,}01)\,4$	[42]
10; 13 dec.	$x = 0\,(0{,}25)\,0{,}5;\ 0{,}75\,(0{,}25)\,3$	[64]
10 dec.	$x = 0\,(0{,}01)\,0{,}1\,(0{,}1)\,5\,(0{,}5)\,9;\ 10$	[104]
5 dec.	$x = 0\,(0{,}01)\ 0{,}1\ \ (0{,}1)\ 10;\ 1\left(\frac{1}{4}\right)4;\ 1\left(\frac{1}{6}\right)4$	[62]
5 fig.	$x = 0\,(0{,}1)\,4$	[88]
4—8 fig.	$x = 0\,(0{,}01)\,10$	[66]

sh $\frac{\pi}{2} x$

4—5 fig.	$x = 0\,(0{,}05)\,1$	[71]

sh $\frac{x}{\sqrt{2}}$

12 dec.	$x = 0\,(0{,}1)\,20$	[152]

sh $\frac{\pi x}{360}$

10 dec.	$x = 0\,(1)\,360$	[64]

Hyperbolic cosine

ch x

20 dec.	$x = 1\,(1)\,10$	[129]
15 dec.	$x = 0\,(0{,}1)\,10$	[42] [133]
15 dec.	$x = 1°\,(1°)\,90°$	[112]
20 dec.	$x = 0\,(0{,}00001)\,0{,}001$	[64]
10 dec.	$x = 0{,}001\,(0{,}0001)\,0{,}1\,(0{,}001)\,3\,(0{,}01)\,10$	[64]
15 dec.	$x = 10\,(0{,}1)\,20\,(1)\,39$	[64]
11—15 dec.	$x = 40\,(1)\,50$	[64]
9 dec.	$x = 0\,(0{,}0001)\,2\,(0{,}1)\,10$	[118]
7 dec.	$x = 0\,(0{,}0005)\,0{,}025$	[46]
7 dec.	$x = 0\,(0{,}01)\,4$	[38]
7 dec.	$x = 1\,(0{,}05)\,2{,}5$	[46]
6 fig.	$x = 0\,(0{,}0001)\,0{,}2\,(0{,}001)\,2\,(0{,}01)\,8$	[54]
6; 5; 4 dec.	$x = 0\,(0{,}001)\,2{,}5;\ 2{,}5\,(0{,}001)\,4;\ 4\,(0{,}001)\,6$	[169]
6 dec.	$x = 0\,(0{,}001)\,1\,(0{,}002)\,2\,(0{,}005)\,3\,(0{,}01)\,4\,(0{,}05)\,5\,(0{,}1)\,6$	[34]
6 dec.	$x = 0\,(0{,}001)\,3\,(0{,}01)\,6$	[168]

6 fig.	$x = 0\,(0{,}01)\,4$	[40]
6 fig.	$x = 0\,(0{,}01)\,1{,}5\,(0{,}05)\,2\,(0{,}1)\,6$	[36]
5 dec. 6 fig.	$x = 0\,(0{,}01)\,2$ $x = 2\,(0{,}1)\,6$	[50]
5—6 fig.	$x = 0\,(0{,}001)\,6$	[20]
5—6 fig.	$x = 0\,(0{,}01)\,5{,}99$	[4] [11] [14] [16]
5—6 fig.	$x = 0\,(0{,}2)\,6$	[35]
5 dec.	$x = 0\,(0{,}0001)\,0{,}1\,(0{,}001)\,3\,(0{,}01)$ $6{,}3\,(0{,}1)\,10$	[63]
5 dec.	$x = 0\,(0{,}005)\,0{,}1\,(0{,}01)\,3\,(0{,}05)\,4\,(0{,}1)\,10$	[104]
5 dec.	$x = 0\,(0{,}01)\,10$	[62] [117]
5 fig.	$x = 0\,(0{,}01)\,3\,(0{,}05)\,10$	[100]
5 fig.	$x = 0\,(0{,}01)\,6$	[102] [105]
5 fig.	$x = 0\,(0{,}01)\,5{,}09$	[19]
5 fig.	$x = 0\,(0{,}01)\,3\,(0{,}05)\,4\,(0{,}1)\,6\,(0{,}25)$ $7\,(0{,}5)\,10$	[12] [60]
5 dec.	$x = 0\,(0{,}01)\,1{,}5\,(0{,}1)\,10$	[17]
5 fig.	$x = 0\,(0{,}01)\,1{,}6\,(0{,}1)\,6$	[10]
5 fig.	$x = 0\,(0{,}01)\,1\,(0{,}1)\,6$	[15] [28]
5 fig.	$x = 0\,(0{,}02)\,0{,}1\,(0{,}01)\,0{,}5\,(0{,}1)\,6$	[24]
5 fig.	$x = 0\,(0{,}1)\,5$	[48]
6; 5; 4 dec.	$x = 0\,(0{,}01)\,2;\ \ 2\,(0{,}01)\,5;\ \ 5\,(0{,}01)\,8$	[79]
6; 5 dec. 4 dec.	$x = 0\,(0{,}01)\,2;\ \ 2\,(0{,}01)\,2{,}5\,(0{,}1)\,5$ $x = 5\,(0{,}1)\,7{,}5$	[74]
5; 4 dec.	$x = 0\,(0{,}0001)\,0{,}1\,(0{,}001)\,3;\ \ 3\,(0{,}01)\,6$	[37]
5; 4 dec.	$x = 0\,(0{,}001)\,1{,}509;\ \ 1{,}5\,(0{,}001)\,3{,}009$	[83]
5; 4 dec.	$x = 0\,(0{,}01)\,3;\ \ 3\,(0{,}1)\,5$	[52]
4 dec. 5 fig.	$x = 0\,(0{,}001)\,3$ $x = 3\,(0{,}01)\,8$	[46]
4 dec. 5 fig.	$x = 0\,(0{,}001)\,1\,(0{,}01)\,1{,}5$ $x = 1{,}5\,(0{,}01)\,6\,(0{,}1)\,10$	[86]
4 dec. 5 fig.	$x = 0\,(0{,}01)\,2{,}99$ $x = 3\,(0{,}05)\,4\,(0{,}1)\,6\,(0{,}25)\,7\,(0{,}5)\,10$	[47]
4 dec. 5 fig.	$x = 0\,(0{,}01)\,2{,}5\,(0{,}1)\,2{,}9$ $x = 3\,(0{,}1)\,7{,}5$	[103]
4 dec. 5 fig.	$x = 0\,(0{,}01)\,3{,}39$ $x = 3{,}4\,(0{,}01)\,5{,}09$	[97]
4—5 fig.	$x = 0\,(0{,}01)\,5{,}99$	[13]
4—5 fig.	$x = 0\,(0{,}01)\,1\,(0{,}05)\,6$	[33]
4—5 dec.	$x = 0\,(0{,}01)\,1{,}6\,(0{,}1)\,6$	[22] [27] [53]
5; 4 dec. 3 dec.	$x = 0\,(0{,}001)\,2{,}009;\ \ 2\,(0{,}001)\,5{,}009$ $x = 5\,(0{,}001)\,8{,}009$	[44]
4 dec.	$x = 0\,(0{,}01)\,5{,}99$	[3]
4 dec.	$x = 0\,(0{,}01)\,5{,}09$	[23]

4 dec.	$x = 0\,(0{,}01)\,3{,}99$	[110]
4 dec.	$x = 0\,(0{,}01)\,1{,}6$	[5]
4 dec.	$x = 0\,(0{,}01)\,1$	[45]
4 dec.	$x = 0\,(0{,}02)\,3{,}98$	[81]
4; 3 dec.	$x = 0\,(0{,}001)\,1{,}999;\ 2\,(0{,}001)\,3$	[51]
4; 3 dec.	$x = 0\,(0{,}01)\,3{,}39;\ 3{,}4\,(0{,}01)\,5{,}99$	[39] [111]
4; 3 dec.	$x = 0\,(0{,}01)\,4;\ 4\,(0{,}01)\,6$	[59]
3—4 dec.	$x = 0\,(0{,}01)\,3$	[49]
2—3 dec.	$x = 0\,(1)\,99;\ 1\,(0{,}1)\,5{,}9$	[21]

$\mathrm{ch}\,\pi x$

20—24 fig.	$x = 0\,(0{,}01)\,0{,}1\,(0{,}1)\,10$	[65]
15 dec.	$x = 0\,(0{,}0001)\,0{,}01$	[42]
15 dec.	$x = 0\,(0{,}01)\,4$	[134]
14—16 fig.	$x = 0\,(0{,}01)\,4$	[42]
10; 13 dec.	$x = 0\,(0{,}25)\,0{,}5;\ 0{,}75\,(0{,}25)\,3$	[64]
10 dec.	$x = 0\,(0{,}01)\,0{,}1\,(0{,}1)\,5\,(0{,}5)\,9;\ 10$	[104]
5 dec.	$x = 0\,(0{,}01)\,0{,}1\,(0{,}1)\,10;\ 1\left(\frac{1}{4}\right)4;\ 1\left(\frac{1}{6}\right)4$	[62]
5 fig.	$x = 0\,(0{,}1)\,4$	[88]
4—8 fig.	$x = 0\,(0{,}01)\,10$	[66]

$\mathrm{ch}\,\frac{\pi}{2}x$

4—5 fig.	$x = 0\,(0{,}05)\,1$	[71]

$\mathrm{ch}\,\frac{x}{\sqrt{2}}$

12 dec.	$x = 0\,(0{,}1)\,20$	[152]

$\mathrm{ch}\,\frac{\pi x}{360}$

10 dec.	$x = 0\,(1)\,360$	[64]

Hyperbolic tangent

$\mathrm{th}\,x$

12 dec.	$x = 0\,(0{,}01)\,1$	[89]
10 dec.	$x = 0\,(0{,}1)\,10$	[116]
8 fig.	$x = 0\,(0{,}0001)\,2$	[116]
20; 10 dec.	$x = 0\,(0{,}00001)\,0{,}001;\ 0{,}001\,(0{,}0001)\,0{,}0999$	[64]
8; 10 dec.	$x = 0{,}1\,(0{,}001)\,3;\ 3\,(0{,}01)\,10\,(0{,}1)\,12{,}3$	[64]
7 dec.	$x = 0\,(0{,}01)\,3$	[155]
6 dec.	$x = 0\,(0{,}0001)\,0{,}2\,(0{,}001)\,2\,(0{,}01)\,8$	[54]
6 dec.	$x = 0\,(0{,}001)\,3\,(0{,}01)\,6$	[169]

6 dec.	$x = 0 (0,001) 2$	[94]
6 dec.	$x = 0 (0,001) 1 (0,002) 2 (0,005) 3 (0,01) 4 (0,05) 5 (0,1) 6$	[34]
5;6 dec.	$x = 0 (0,001) 0,4;\ 0,4 (0,001) 3 (0,01) 4,99$ (var.) $7,183$	[168]
5 dec.	$x = 0 (0,0001) 0,1 (0,001) 3 (0,01) 6,3 (0,1) 10$	[63]
5 dec.	$x = 0 (0,001) 6$	[20]
5 dec.	$x = 0 (0,01) 10$	[117]
5 dec.	$x = 0 (0,01) 3 (0,05) 10$	[100]
5 dec.	$x = 0 (0,0001) 0,1 (0,001) 3 (0,01) 6$	[37]
5 dec.	$x = 0 (0,001) 1,9 (0,01) 3,09$	[83]
5 dec.	$x = 0 (0,01) 6,45$	[62]
5 dec.	$x = 0 (0,01) 6$	[102] [105]
5 fig.	$x = 0 (0,01) 3 (0,05) 4 (0,1) 6 (0,25) 7 (0,5) 10$	[12] [60]
5 dec.	$x = 0 (0,01) 4$	[40]
5 dec.	$x = 0 (0,01) 3 (0,1) 6,5$	[87]
5 dec.	$x = 0 (0,01) 3 (0,1) 5$	[52]
5 dec.	$x = 0 (0,01) 2,5 (0,1) 7,5$	[74]
5 dec.	$x = 0 (0,01) 2$	[79]
5 dec.	$x = 0 (0,01) 1,5 (0,1) 10$	[17]
5 dec.	$x = 0 (0,01) 1,5 (0,05) 2 (0,1) 6$	[36]
5 dec.	$x = 0 (0,01) 1,6 (0,1) 6$	[22] [27] [53]
4;5 dec. 6 dec.	$x = 0 (0,001) 0,999;\ 1 (0,001) 2,499$ $x = 2,5 (0,001) 3$	[51]
4;5 dec.	$x = 0 (0,01) 2;\ 2 (0,01) 6$	[59]
4;5 dec.	$x = 0 (0,02) 2,98;\ 3 (0,02) 3,98$	[81]
4;5 dec.	$x = 0 (0,01) 1,89;\ 1,9 (0,01) 5,09$	[111]
4—5 fig.	$x = 0 (1) 99;\ 1 (0,1) 59$	[21]
4 dec.	$x = 0 (0,001) 1 (0,01) 6$	[86]
4 dec.	$x = 0 (0,001) 1,8 (0,01) 3 (0,1) 4,9$	[46]
4 dec.	$x = 0 (0,01) 1,6 (0,1) 6$	[10]
4 dec.	$x = 0 (0,01) 3 (0,1) 5$	[97]
4 dec.	$x = 0 (0,01) 2,5 (0,1) 5,4$	[103]
4 dec.	$x = 0 (0,01) 3$	[49]
4 dec.	$x = 0 (0,01) 2,89$	[3] [4]]13] [14] [16][39[
4 dec.	$x = 0 (0,01) 2,39$	[19][23]
4 dec.	$x = 0 (0,01) 1,6$	[5]
4 dec.	$x = 0 (0,1) 5$	[48]

$\mathrm{th}\,\pi x$

14 dec.	$x = 0,75 (0,25) 3$	[66]
8;13 dec.	$x = 0 (0,25) 0,5;\ 0,75 (0,25) 3$	[64]
5 dec.	$x = 0 (0,01) 0,1 (0,1) 2,1$	[62]

$\text{th}\,\frac{\pi}{2}x$

4 dec.	$x = 0\,(0{,}05)\,1$	[71]

Hyperbolic cotangent

$\text{cth}\,x$

10 dec.	$x = 0\,(0{,}1)\,10$	[116]
8 dec.	$x = 0\,(0{,}0001)\,0{,}1$	[116]
8 fig.	$x = 0{,}1\,(0{,}0001)\,2$	[116]
6 fig.	$x = 0\,(0{,}001)\,0{,}05$	[169]
5 dec.	$x = 0{,}05\,(0{,}001)\,0{,}35$	[169]
6 dec.	$x = 0{,}35\,(0{,}001)\,3\,(0{,}01)\,6$	[169]
5–6 fig.	$x = 0\,(0{,}01)\,4$	[40]
5 fig.	$x = 0\,(0{,}0001)\,0{,}1\,(0{,}001)\,3\,(0{,}01)\,6$	[37]
5 fig.	$x = 0\,(0{,}01)\,3\,(0{,}1)\,5$	[52]
4–6 fig.	$x = 0\,(0{,}01)\,2{,}5\,(0{,}1)\,7{,}5$	[74]
4–5 fig.	$x = 0\,(0{,}001)\,1\,(0{,}01)\,6$	[86]
4 dec.	$x = 0\,(0{,}0001)\,1$	[169]
3 dec.	$x = 0\,(0{,}01)\,2{,}39$	[19]

$\text{cth}\,\frac{\pi}{2}x$

4–5 fig.	$x = 0\,(0{,}05)\,1$	[71]

Hyperbolic secant

$\text{sech}\,x$

5 dec.	$x = 0\,(0{,}01)\,4\,(0{,}1)\,12{,}9$	[87]
4 dec.	$x = 0\,(0{,}01)\,3$	[49]
3–5 fig.	$x = 0\,(0{,}01)\,2{,}5\,(0{,}1)\,7{,}5$	[74]

$\text{sech}\,\pi x$

14 dec.	$x = 0{,}75\,(0{,}25)\,3$	[66]

Hyperbolic cosecant

$\text{cosech}\,x$

5 dec.	$x = 0\,(0{,}01)\,4\,(0{,}1)\,12{,}9$	[87]
4 dec.	$x = 0\,(0{,}0001)\,1$	[169]
3–5 fig.	$x = 0\,(0{,}01)\,2{,}5\,(0{,}1)\,7{,}5$	[74]

Hyperbolic functions of a complex variable

Tables of the hyperbolic functions of a complex variable are constructed in various ways. One type of table for a complex variable as a rectangular system of coordinates gives the real and imaginary parts of the function, i.e. the function is also expressed in a rectangular system of coordinates. Another type of table gives values of the modulus and the argument, i.e. the function is expressed in polar coordinates. Tables also exist in which the functions of a complex variable given in polar coordinates are calculated in rectangular or polar coordinate systems.

If the function is tabulated in polar coordinates the description gives the accuracy of the modulus first followed by the accuracy of the argument.

sh z

The tables enumerated below give the real and imaginary parts of the function sh z of a complex variable given in rectangular coordinates: $z = x + iy$.

5 dec.	$x = 0\,(0{,}05)\,4.$	$y = 0\,(5^g)\,200^g$	[74]
5 dec.	$x = 0{,}2\,(0{,}2)\,1\,(1)\,3.$	$y = 0\,(5^g)50^g(10^g)\,200^g$	[104]
4—5 dec.	$x = 0\,(0{,}1)\,1{,}5.$	$y = 0\,(0{,}1)\,1{,}5$	[81]
4 dec.	$x = 0\,(0{,}1)\,3\,(0{,}2)\,4\,(0{,}5)\,7\,(1)\,10.$	$y = 0\,(15°)90°$	[154]

The following table gives the modulus and argument of the function sh z of a complex variable expressed in rectangular coordinates: $z = x + iy$.

5 dec.; 0°,001	$x = 0\,(0{,}05)\,4.$	$y = 0\,(5^g)\,200^g$	[74]

Tables of the type described below give the modulus and argument of the function sh z of a complex variable given in polar coordinates: $z = re^{i\theta}$.

6 dec.; 0°,0001	$r = 0\,(0{,}05)\,1.$	$\theta = 0\,(5°)\,45°$	[74]
5—6 fig; 0°,01	$r = 0\,(0{,}01)\,0{,}1\,(0{,}05)\,1.$	$\theta = \frac{\pi}{4}$	[148]
5 fig.; 0°,001	$r = 0\,(0{,}1)\,3.$	$\theta = 45°\,(1°)\,90°$	[74]
5 fig.; 1′	$r = 0\,(0{,}1)\,6\,(0{,}05)\,20{,}5.$	$\theta = \frac{\pi}{4}$	[74]
5 dec.; 0°,0001	$r = 0\,(0{,}01)\,0{,}5.$	$\theta = \frac{\pi}{4}$	[74]

ch z

The tables enumerated below give the real and imaginary parts of the function ch z of a complex variable given in rectangular coordinates: $z = x + iy$.

5 dec.	$x = 0\,(0{,}05)\,4.$	$y = 0\,(5^g)\,200^g$	[74]
5 dec.	$x = 0{,}2\,(0{,}2)\,1\,(1)\,3.$	$y = 0(5^g)50^g(10^g)200^g$	[104]
4—5 dec.	$x = 0\,(0{,}1)\,1{,}5.$	$y = 0\,(0{,}1)\,1{,}5$	[81]
4 dec.	$x = 0\,(0{,}1)\,3\,(0{,}2)\,4\,(0{,}5)\,7\,(1)10.$	$y = 0\,(15^\circ)\,90^\circ$	[154]

The following table gives the modulus and argument of the function ch z of a complex variable given in rectangular coordinates: $z = x + iy$.

5 dec. ; 0°,001	$x = 0\,(0{,}05)\,4.$	$y = 0\,(5^g)\,200^g$	[74]

Tables with the characteristics described below give the modulus and argument of the function ch z of a complex variable given in polar coordinates: $z = re^{i\theta}$.

6 dec. ; 0°,0001	$r = 0\,(0{,}05)\,1.$	$\theta = 0\,(5^\circ)\,45^\circ$	[74]
5—6 fig; 0°,01	$r = 0\,(0{,}01)\,0{,}1\,(0{,}05)\,1.$	$\theta = \frac{\pi}{4}$	[148]
5 dec. ; 0°,001	$r = 0\,(0{,}1)\,1.$	$\theta = 45^\circ\,(1^\circ)\,90^\circ$	[74]
5 fig.; 0°,001	$r = 1{,}1\,(0{,}1)\,3.$	$\theta = 45^\circ\,(1^\circ)\,90^\circ$	[74]
5 fig ; 1′	$r = 0\,(0{,}1)6(0{,}05)20{,}5.$	$\theta = \frac{\pi}{4}$	[74]
5 dec. ; 0°,0001	$r = 0\,(0{,}01)\,0{,}5.$	$\theta = \frac{\pi}{4}$	[74]

th z

The tables enumerated below give the real and imaginary parts of the function th z of a complex variable given in rectangular coordinates: $z = x + iy$.

5 dec.	$x = 0\,(0{,}05)\,4.$	$y = 0\,(5^g)\,200^g$	[74]
5 dec.	$x = 0{,}2\,(0{,}2)\,1\,(1)3.$	$y = 0\,(5^g)\,50^g\,(10^g)\,200^g$	[104]

The following table gives the modulus and argument of the function th z of a complex variable given in rectangular coordinates $z = x + iy$.
5 dec.

Tables with the characteristics described below tabulate the modulus and argument of the function th z of a complex variable given in polar form: $z = re^{i\theta}$.

6 dec. ; 0°,0001	$r = 0\,(0{,}05)\,1.$	$\theta = 0\,(5^\circ)\,45^\circ$	[74]
5 dec. ; 0°,001	$r = 0\,(0{,}1)\,1.$	$\theta = 45^\circ\,(1^\circ)\,90^\circ$	[74]
5 fig.; 0°,001	$r = 1{,}1\,(0{,}1)\,3.$	$\theta = 45^\circ\,(1^\circ)\,90^\circ$	[74]
5 dec. ; 0°,0001	$r = 0\,(0{,}01)\,0{,}5.$	$\theta = \frac{\pi}{4}$	[74]
4—5 fig; 1′	$r = 0\,(0{,}1)\,6\,(0{,}05)\,20{,}5.$	$\theta = \frac{\pi}{4}$	[74]

cth z

This section contains a description of tables of the modulus and argument of the function cth **z** of a complex variable given in polar form: $z = re^{i\theta}$.

5 dec.; 0°,0001	$r = 0\,(0{,}01)\,0{,}5.$	$\theta = \frac{\pi}{4}$	[74]
4—5 fig.; 1′	$r = 0\,(0{,}1)\,6\,(0{,}05)\,20{,}5.$	$\theta = \frac{\pi}{4}$	

sech z

Tables of the modulus and argument of the functio sech **z** of a complex variable given in polar form: $z = re^{i\theta}$.

5 dec. ; 0°,0001	$r = 0\,(0{,}01)\,0{,}5.$	$\theta = \frac{\pi}{4}$	[74]
4—5 fig.; 1′	$r = 0\,(0{,}1)\,6\,(0{,}05)\,20{,}5.$	$\theta = \frac{\pi}{4}$	

cosech z

The tables contain the modulus and argument of the function cosech **z** of the complex variable: $z = re^{i\theta}$.

5 dec. ; 0°,0001	$r = 0\,(0{,}01)\,0{,}5.$	$\theta = \frac{\pi}{4}$	[74]
4—5 fig; 1′	$r = 0\,(0{,}1)\,6\,(0{,}05)\,20{,}5.$	$\theta = \frac{\pi}{4}$	

$$\frac{\text{sh}\, z}{z}$$

The tables contain the modulus and argument of the function $\frac{\text{sh}\, z}{z}$ of the complex variable $z = re^{\theta}$.

6 dec ; 0°,0001	$r = 0\,(0{,}05)\,1.$	$\theta = 0\,(5°)\,45°$	[74]
5 dec ; 0°,0001	$r = 0(0{,}01)0{,}5.$	$\theta = \frac{\pi}{4}$	
5 dec ; 0°,001	$r = 0\,(0{,}1)\,1.$	$\theta = 45°\,(1°)\,90°$	
5 fig; 0°,001	$r = 1{,}1\,(0{,}1)\,3.$	$\theta = 45°\,(1°)\,90°$	

$$\frac{\text{th}\, z}{z}$$

The tables contain the modulus and argument of the function $\frac{\text{th}\, z}{z}$ of the complex variable $z = re^{i\theta}$.

6 dec. ; 0°,0001	$r = 0\,(0{,}05)\,1.$	$\theta = 0\,(5°)\,45°$	[74]
5 dec. ; 0°,0001	$r = 0(0{,}01)0{,}5.$	$\theta = \frac{\pi}{4}$	[74]
5 dec. ; 0°,001	$r = 0\,(0{,}1)\,1.$	$\theta = 45°\,(1°)\,90°$	[74]
5 fig; 0°,001	$r = 1{,}1\,(0{,}1)\,3.$	$\theta = 45°(1°)\,90°$	[74]

Expressions containing hyperbolic functions

$$\frac{\operatorname{sh} x}{x}$$

5 dec.	$x = 0\,(0{,}01)\,7$	[62]
5; 4 dec.	$x = 0\,(0{,}001)\,1;\ 1\,(0{,}001)\,3$	[46]
4 dec.	$x = 0\,(0{,}01)\,4$	[40]
3—7 dec.	$x = 0\,(0{,}01)\,0{,}25\,(0{,}01)\,0{,}75\,(0{,}01)\,2\,(0{,}1)\,4\,(0{,}1)\,6$	[73]

$$x + \frac{\operatorname{sh} 2x}{2}$$

6—7 fig.	$x = 0\,(0{,}01)\,2{,}5$	[40]

$$\frac{\operatorname{ch} x}{x}$$

5 dec.	$x = 0\,(0{,}01)\,7$	[62]
4 fig.	$x = 0\,(0{,}001)\,0{,}5$	[46]
4 dec.	$x = 0{,}5\,(0{,}001)\,3$	[46]

$$\frac{\operatorname{ch} x - 1}{x}$$

5; 4 dec.	$x = 0\,(0{,}001)\,1;\ 1\,(0{,}001)\,3$	[46]

$$\frac{\operatorname{th}\frac{x}{2}}{\frac{x}{2}}$$

4 dec.	$x = 0\,(0{,}01)\,4$	[40]

$$\frac{3\,(x - \operatorname{th} x)}{x^3}$$

4 dec.	$x = 0\,(0{,}01)\,1{,}6$	[135]

$$\operatorname{cth} x - \frac{1}{x}$$

3 dec. $x = 0\,(0{,}1)\,5$ [40]

$$\frac{1}{2}\operatorname{ch} x - \frac{1}{2x}\operatorname{sh} x + \frac{1}{x^2}(\operatorname{ch} x - 1)$$

4 fig. $x = 0\,(0{,}1)\,5\,(0{,}2)\,10$ [167]

$$\frac{16\operatorname{ch}^2 x}{55\operatorname{ch}^2 x + 25x^2 + 9}$$

6 dec. $x = 0\,(0{,}1)\,6$ [153]

$$\frac{16\operatorname{ch}^2 x}{39\operatorname{ch}^2 x + 9x^2 + 25}$$

6 dec. $x = 0\,(0{,}1)\,6$ [153]

$$\frac{3\,(2x\operatorname{ch} 2x - 1)}{(2x)^2}$$

4 dec. $x = 0\,(0{,}01)\,1{,}6$ [135]

$$\frac{6\,(1 - 2x\operatorname{cosech} 2x)}{(2x)^2}$$

4 dec. $x = 0\,(0{,}01)\,1{,}6$ [135]

$$\frac{x^2 - x\operatorname{sh} x}{2\,(\operatorname{ch} x - 1) - x\operatorname{sh} x}$$

3 dec. $x = 0\,(0{,}1)\,6$ [30]

$$\frac{x\operatorname{sh} x - x^2\operatorname{ch} x}{2\,(\operatorname{ch} x - 1) - x\operatorname{sh} x}$$

3 dec. $x = 0\,(0{,}1)\,6$ [30]

$$\frac{x^2 - x^2\operatorname{ch} x}{2\,(\operatorname{ch} x - 1) - x\operatorname{sh} x}$$

6 dec. $x = 0\,(0{,}1)\,6$ [30]

$$\frac{\frac{1}{2}\,\mathrm{ch}\,ax - \frac{1}{2ax}\,\mathrm{sh}\,ax + \frac{1}{(ax)^2}(\mathrm{ch}\,ax - 1)}{\frac{1}{2}\,\mathrm{ch}\,a - \frac{1}{2a}\,\mathrm{sh}\,a + \frac{1}{a^2}(\mathrm{ch}\,a - 1)}$$

3 dec. $x = 0\,(0{,}2)\,0{,}6\,(0{,}1)\,1.\ a = 1\,(1)\,10\,(2)\,14$ [167]

$$s = \frac{1}{2}\left[\mathrm{Ar\,sh}\,x + x\,(1 + x^2)^{\frac{1}{2}}\right]$$

5 dec. $x = 0\,(0{,}01)\,2$ [113]

EXPRESSIONS CONTAINING TRIGONOMETRIC AND HYPERBOLIC FUNCTIONS

$\mathrm{sh}\,x\ \sin x$

8 dec. $x = 0\,(0{,}001)\,3\,(0{,}01)\,10$ [64]
5 dec. $x = 0\,(0{,}01)\,10$ [62]
5 dec. $x = 0\,(0{,}01)\,3{,}5$ [40]

$\frac{1}{2}\,\mathrm{sh}\,x\ \sin x$

4—5 dec. $x = 0(0{,}001)\,0{,}04\,(0{,}01)\,6{,}3\,(0{,}1)\,10$ [31]

$\mathrm{sh}\,x\ \cos x$

8 dec. $x = 0\,(0{,}001)\,3\,(0{,}01)\,10$ [64]
5 dec. $x = 0\,(0{,}01)\,10$ [62]
5 dec. $x = 0\,(0{,}01)\,3{,}5$ [40]

$\mathrm{ch}\,x\ \sin x$

8 dec. $x = 0\,(0{,}001)\,3\,(0{,}01)\,10$ [64]
5 dec. $x = 0\,(0{,}01)\,10$ [62]
5 dec. $x = 0\,(0{,}01)\,3{,}5$ [40]

$\mathrm{ch}\,x\ \cos x$

8 dec. $x = 0\,(0{,}001)\,3\,(0{,}01)\,10$ [64]
5 dec. $x = 0(0{,}01)\,10$ [62]
5 dec. $x = 0\,(0{,}01)\,3{,}5$ [40]
4—5 dec. $x = 0\,(0{,}001)\,0{,}04\,(0{,}01)\,6{,}3\,(0{,}1)\,10$ [31]

$$\frac{1}{2}(\operatorname{ch} x \ \sin x + \operatorname{sh} x \ \cos x)$$

4—5 dec. $x = 0\,(0{,}001)\,0{,}04\,(0{,}01)\,6{,}3\,(0{,}1)\,10$ [31]

$$\frac{1}{4}(\operatorname{ch} x \ \sin x - \operatorname{sh} x \ \cos x)$$

4—5 dec. $x = 0\,(0{,}001)\,0{,}04\,(0{,}01)\,6{,}3\,(0{,}1)\,10$ [31]

cth x — ctg x

4—5 fig. $x = 0\,(1°)\,5°\,(5°)\,40°\,(2°)\,44°\,(1°)\,270°$ [149]

cosech x — cosec x

5 fig. $x = 0\,(1°)\,5°\,(5°)\,40°\,(2°)\,44°\,(1°)\,270°$ [149]

INVERSE HYPERBOLIC FUNCTIONS

Ar sh x

9 dec.	$x = 0\,(0{,}002)\,3\,(0{,}005)\,3{,}495$; $x = 3{,}5\,(0{,}005)\,6{,}5\,(0{,}01)\,15\,(0{,}02)\,35\,(0{,}05)\,65\,(0{,}1)\,150\,(0{,}2)\,350\,(0{,}5)\,650\,(1)\,1500\,(2)\,3500\,(5)\,6500\,(10)\,15000\,(20)\,22980$	[120]
7—20 dec.	$x = 0\,(0{,}00001)\,0{,}001\,(0{,}0001)\,0{,}0999\,(0{,}001)\,3\,(0{,}01)\,10\,(0{,}1)\,20\,(1)\,50$	[64]
6 dec.	$x = 0\,(0{,}001)\,2\,(0{,}01)\,10\,(0{,}1)\,35\,(1)\,80$	[169]
5 dec.	$x = 0\,(0{,}01)\,10$	[62]
5 dec.	$x = 0\,(0{,}01)\,1\,(0{,}02)\,1{,}2\,(0{,}05)\,1{,}5\,(0{,}1)\,3\,(0{,}5)\,6\,(1)\,10\,(5)\,20\,(10)\,100$	[40]
4 dec.	$x = 0\,(0{,}001)\,2\,(0{,}01)\,10$	[51]
4—3 dec.	$x = 0\,(0{,}001)\,1{,}5\,(0{,}01)\,5\,(0{,}1)\,15\,(1)\,60$	[86]
4; 3 dec.	$x = 0\,(0{,}01)\,2\,(0{,}1)\,10;\ 10\,(1)\,59$	[19]

Ar sh πx

7 dec. $x = \frac{1}{4}\left(\frac{1}{4}\right)3$ [64]

5 dec. $x = \frac{1}{4}\left(\frac{1}{4}\right)3$ [62]

$k^{-\frac{1}{2}}$ Ar sh $(kx)^{\frac{1}{2}}$

2 dec.	$x = 0\,(1)\,50\,(5)\,100\,(10)\,300$; $k = 0\,(0{,}02)\,0{,}1\,(0{,}05)\,0{,}3\,(0{,}1)\,0{,}6\,(0{,}2)\,1$	[132]

Ar ch x

9 dec.	$x = 1\,(0{,}00001)\,1{,}01\,(0{,}00005)\,1{,}025\,(0{,}0001)\,1{,}05\,(0{,}0002)\,1{,}15\,(0{,}0005)\,1{,}4\,(0{,}001)\,1{,}8\,(0{,}002)\,3{,}498$; $x = 3{,}5\,(0{,}005)\,6{,}5\,(0{,}01)\,15\,(0{,}02)\,35\,(0{,}05)\,65\,(0{,}1)\,150\,(0{,}2)\,350\,(0{,}5)\,650\,(1)\,1500\,(2)\,3500\,(5)\,6500\,(10)\,15000\,(20)\,22980$	[120]
8—7 dec.	$x = 1\,(0{,}001)\,1{,}414\,(0{,}001)\,3\,(0{,}01)\,10\,(0{,}1)\,20(1)\,50$	[64]
6 dec.	$x = 1\,(0{,}00001)\,1{,}001;\ 1\,(0{,}0001)\,1{,}01;\ 1\,(0{,}001)\,2\,(0{,}01)\,10\,(0{,}1)\,35\,(1)\,80$	[169]
5 dec.	$x = 1\,(0{,}01)\,10$	[62]
5 dec.	$x = 1\,(0{,}02)\,1{,}2\,(0{,}05)\,1{,}5\,(0{,}1)\,3\,(0{,}5)\,6\,(1)\,10\,(5)\,20\,(10)\,100$	[40]
4 dec.	$x = 1\,(0{,}001)\,2\,(0{,}01)\,10$	[51]
4 dec.; 3 dec.	$x = 1\,(0{,}0001)\,1{,}01;\ 1\,(0{,}001)\,1{,}5\,(0{,}01)\,5$; $x = 5\,(0{,}1)\,15\,(1)\,60$	[86]
4 dec.; 3—4; 3 dec.	$x = 1\,(0{,}001)\,1{,}009;\ 1\,(0{,}01)\,2$; $x = 2\,(0{,}1)\,10;\ 10\,(1)\,59$	[19]

Ar ch πx

7 dec.	$x = \frac{1}{2}\left(\frac{1}{4}\right)3$	[64]
5 dec.	$x = \frac{1}{2}\left(\frac{1}{4}\right)3$	[62]

Ar th x

9 dec.	$x = 0\,(0{,}001)\,0{,}5\,(0{,}0005)\,0{,}75\,(0{,}0002)\,0{,}9\,(0{,}0001)\,0{,}95\,(0{,}00005)\,0{,}975\,(0{,}00002)\,0{,}99\,(0{,}00001)\,1$	[120]
9—20 dec.	$x = 0\,(0{,}00001)\,0{,}001\,(0{,}0001)\,0{,}0999\,(0{,}001)\,1$	[64]
7 dec.	$x = 0\,(0{,}01)\,0{,}9\,(0{,}001)\,1$	[147]
6; 5 dec.; 5 dec.	$x = 0\,(0{,}001)\,0{,}95;\ 0{,}95(0{,}001)\,1$; $x = 0{,}99\,(0{,}0001)\,1$	[169]
5 dec.	$x = 0\,(0{,}01)\,1$	[40] [62]

4; 3 dec.	$x = 0\,(0{,}001)\,0{,}95;\; 0{,}95\,(0{,}001)\,0{,}99\,(0{,}0001)\,1$	[86]
4 dec.	$x = 0\,(0{,}001)\,1$	[51]
4 fig.	$x = 0\,(0{,}01)\,0{,}99\,(0{,}001)\,1$	[19]
4 dec.	$x = 0\,(0{,}01)\,1$	[136]
4 dec.	$x = 0\,(0{,}01)\,0{,}96$	[59]
3 dec.	$x = 0{,}97\,(0{,}01)\,1$	

Ar cth x

7 dec.	$x = 1\,(0{,}001)\,1{,}5\,(0{,}01)\,5\,(0{,}1)\,15\,(1)\,60$	[130]
5 dec.	$x = 1\,(0{,}0001)\,1{,}01$	[169]
5 dec.	$x = 1\,(0{,}001)\,1{,}05$	
6 dec.	$x = 1{,}05\,(0{,}001)\,2\,(0{,}01)\,10\,(0{,}1)\,35\,(1)\,80$	
4—5 fig.	$x = 1\,(0{,}001)\,1{,}5\,(0{,}01)\,5\,(0{,}1)\,15\,(1)\,60$	[86]
3; 3—4 dec.	$x = 1\,(0{,}001)\,1{,}009;\; 1\,(0{,}01)\,2$	[19]
4 dec.	$x = 2\,(0{,}1)\,10;\; 10\,(1)\,59$	
3 dec.	$x = 1\,(0{,}0001)\,1{,}01$	[86]

Ar sech x

4—5 fig.	$x = 0\,(0{,}001)\,1$	[86]
4 dec.	$x = 0{,}99\,(0{,}0001)\,1$	

Ar cosech x

4—5 fig.	$x = 0\,(0{,}001)\,1{,}5\,(0{,}01)\,5\,(0{,}1)\,15\,(1)\,60$	[86]

Inverse hyperbolic functions of a complex variable

The tables in this section give the real u and imaginary v parts of inverse hyperbolic functions of a complex variable $z = re^{i\theta}$. The real parts of the function are given in degrees, grads and radians.

Ar sh z

$0^g{,}01$	u	$r = 0\,(0{,}02)\,1.\quad \frac{1}{r} = 1\,(-0{,}02)\,0$	[61]
4 dec.	v	$\theta = 0\,(2^g)\,100^g$	
4 dec.	u	$r = 0\,(0{,}2)\,2{,}2.\quad \theta = 0\,(15^\circ)\,90^\circ$	[71]
$0^\circ{,}01$	u		
4 dec.	v		

Ar th z

0,01	u	$r = 0\,(0{,}02)\,1.\ \theta = 0\,(2^g)\,100^g$	[61]
4 dec.	v		
4 dec.	u	$r = 0\,(0{,}2)\,1.\ \ \theta = 0\,(15°)\,90°$	[71]
0°,01	u		
4 dec.	v		
4 dec.	u	$r = 0{,}8\,(0{,}2)\,0{,}98;\ 0{,}99;\ 0{,}999;$	[71]
0°,01	u	$1-10^{-6};\ 1-10^{-9};\ 1.$	
4 dec.	$\frac{1}{v}$	$\theta = 75°\,(5°)\,90°$	

HYPERBOLIC AMPLITUDE (GUDERMANNIAN)

The hyperbolic amplitude or gudermannian is given by the relation:

$$\varphi = \mathrm{gd}\,x = \int_0^x \mathrm{sech}\,t\,dt = 2\,\mathrm{arc\,tg}\,e^x - \frac{1}{2}\pi.$$

The hyperbolic amplitude of a real variable x is the angle φ lying between $\pm\frac{\pi}{2}$ and satisfying the equations:

$$\sin\varphi = \sin(\mathrm{gd}x) = \mathrm{th}\,x,$$
$$\cos\varphi = \cos(\mathrm{gd}x) = \mathrm{sech}\,x,$$
$$\mathrm{tg}\,\varphi = \mathrm{tg}\,(\mathrm{gd}x) = \mathrm{sh}\,x,$$
$$\mathrm{ctg}\,\varphi = \mathrm{ctg}\,(\mathrm{gd}x) = \mathrm{cosech}\,x,$$
$$\sec\varphi = \sec(\mathrm{gd}x) = \mathrm{ch}\,x,$$
$$\mathrm{cosec}\,\varphi = \mathrm{cosec}\,(\mathrm{gd}x) = \mathrm{cth}\,x.$$

It follows directly from these expressions that tables of hyperbolic amplitudes enable one to find the values of hyperbolic functions from trigonometrical tables

gd x

In degrees, minutes and seconds

0″,0001	$x = 0\,(0{,}001)\,0{,}05$	[79]
0″,001	$x = 0{,}05\,(0{,}001)\,0{,}3$	
0″,001	$x = 2\,(0{,}01)\,6$	
0″,001	$x = 6\,(0{,}01)\,9{,}99$	[64]
0″,01	$x = 0{,}3\,(0{,}001)\,2$	[79]
0″,01	$x = 0\,(0{,}001)\,3\,(0{,}01)\,6$	[37] [64]
1″	$x = 0\,(0{,}01)\,7{,}49$	[62]

0′,01	$x=0$ (var.) $0,35\,(0,01)\,2,75;\ 2,8\,(0,1)\,8,8$	[104]
1″	$x=0\,(0,01)\,3\,(0,1)\,5$	[52]

In tenths of a degree

0°,0001	$x=0(0,01)\,0,09$	[45]
0°,001	$x=0,1\,(0,01)\,1$	[45]
0°,001	$x=0\,(0,02)\,1\,(0,05)\,2\,(0,1)\,3,9$	[81]

In radians

8 dec.	$x=6\,(0,01)\,9,99$	[64]
7 dec.	$x=0\,(0,001)\,3\,(0,01)\,6$	[37] [64]
7 dec.	$x=0\,(0,01)\,3\,(0,05)\,4,8\,(0,1)\,6\,(1)\,8$	[40]
6—8 dec.	$x=0$ (var.) $0,35\,(0,01)\,2,75;\ 2,8\,(0,1)\,8,8$	[104]
6 dec.	$x=0\,(0,001)\,4,5\,(0,01)\,10$	[169]
4 dec.	$x=0\,(0,01)\,4,5\,(0,1)\,9$	[86]
4 dec.	$x=0\,(0,02)\,1\,(0,05)\,2\,(0,1)\,6\,(0,2)\,7\,(0,5)\,9$	[81]

gd πx

In degrees and minutes

0″,001	$x=2\left(\frac{1}{4}\right)3$	[64]
0″,01	$x=\frac{1}{4}\left(\frac{1}{4}\right)\frac{7}{4}$	[64]
1″	$x=\frac{1}{4}\left(\frac{1}{4}\right)\frac{9}{4}$	[62]

In radians

8 dec.; 7 dec.	$x=2\left(\frac{1}{4}\right)3;\ \frac{1}{4}\left(\frac{1}{4}\right)\frac{7}{4}$	[64]

THE INVERSE HYPERBOLIC AMPLITUDE (INVERSE GUDERMANNIAN)

The inverse hyperbolic amplitude is given by the expression:

$$x=\mathrm{gd}^{-1}x=\int_0^{\varphi}\sec\psi\,d\psi=\ln\operatorname{tg}\left(\frac{\pi}{4}+\frac{\varphi}{2}\right).$$

$$x=\int_0^{\varphi}\sec\psi\,d\psi$$

Argument in degrees and minutes

9 dec.	$\varphi=0\,(1''$ до $30'')\,1°2'30''$	[104]
9 dec.	$\varphi=0\,(1°)\,80°\,(0°,5)\,90°$	[8]
9 dec.	$\varphi=0\,(1°)\,90°$	[77] [78]
7 dec.	$\varphi=0\,(1°)\,90°$	[40] [67] [156]
6 dec.	$\varphi=1°\,(1'$ to $4')\,88°\,(2''$ до $1')\ 90°$	[104]
5 dec.	$\varphi=0\,(1')\,90°$	[97] [121]
5 dec.	$\varphi=0\,(10')\,40°\,(5')\,60°\,(3')\,72°\,(2')\,80°\,(1')\,90°$	[8]
5 dec.	$\varphi=0\,(10')\,90°$	[62]
5 dec.	$\varphi=0\,(1°)\,90°$	[57]

In tenths of a degree

12 dec.	$\varphi=0\,(0°,5)\,90°$	[77] [78]
5 dec.	$\varphi=0\,(0°,5)\,90°$	[50]

In radians

6; 5 dec.	$\varphi=0\,(0{,}001)\,1{,}2;\,1{,}2\,(0{,}001)\,1{,}4\,(0{,}0001)\,1{,}55$	[169]
4 dec.	$\varphi=1{,}55\,(0{,}0001)\,1{,}57\,(0{,}00001)\,1{,}57078\,(0{,}000001)\,1{,}570796$	[169]
4 dec.	$\varphi=0\,(0{,}01)\,1{,}57$	[59]
4 dec.	$\varphi=0\,(0{,}01)\,1$	[86]
3 dec.	$\varphi=1\,(0{,}01)\,1{,}47\,(0{,}001)\,1{,}57$	[86]

In grads

11 dec.	$\varphi=88^g\,(0^g{,}01)\,100^g$	[143]
7 dec.	$\varphi=0\ (0^g{,}01)\,100^g$	[141] [142]
5 dec.	$\varphi=0\ (0^g{,}01)\,100^g$	[97]
4 dec.	$\varphi=0\,(0^g{,}1)\,95^g$	[69]
3 dec.	$\varphi=95^g\,(0^g{,}1)\,100^g$	[69]

$$\int_0^{\varphi}\sec\psi\,d\psi+\ln(10000-100\varphi)$$

We describe here an auxiliary table of the inverse gudermannian when φ approaches a right angle.

11 dec.	$\varphi=88^g\,(0^g{,}01)\,100^g$	[143]

$$\frac{10800'}{\pi}\int_0^{\varphi}\sec\psi\,d\psi$$

Below we describe tables of the antigudermannian expressed in minutes of arc. Tables of this type are used in navigation and other related fields.

0′,001	$\varphi = 0\,(1°)\,90°$	[156]
0′,01	$\varphi = 0\,(1')\,90°$	[37]
0′,1	$\varphi = 0\,(1')\,87°\,(0',1)\,90°$	[58]
0′,1	$\varphi = 0(1')\,86°\,59'$	[98]
1′	$\varphi = 87°\,(1')\,90°$	[98]
1′	$\varphi = 0\,(1')\,78°\,59'$	[99]

$$\int_0^{\varphi}\sec\psi\,d\psi - \sin\varphi$$

6 dec.	$\varphi = 0\,(10')\,20°$	[146]
5 dec.	$\varphi = 20°\,(10')\,35°$	[146]

LOGARITHMS OF EXPONENTIAL AND HYPERBOLIC FUNCTIONS

Here we present data on tables of common logarithms of exponential and hyperbolic functions and on natural logarithm tables of certain expressions which contain exponential functions. We have not come across any natural logarithm tables of hyperbolic functions. (Translator's note. Russian lg = log /common logarithms/).

$$\lg e^x$$

10 dec.	$x = 0\,(0{,}001)\,0{,}1\,(0{,}01)\,2\,(0{,}1)\,10\,(1)\,500$	[159]
10 dec.	$x = 10\,(10)\,100\,(100)\,500$	[92]
10 dec.	$x = m\,10^{-n}.\quad m = 1\,(1)\,9.\quad n = 0\,(1)\,5$	[92]
10 dec.	$x = 1\,(1)\,100$	[37]
7 dec.	$x = 0\,(0{,}001)\,3\,(0{,}01)\,6$	[37]
7 dec.	$x = 0\,(0{,}01)\,10$	[122] [123]
5 dec.	$x = 0\,(0{,}01)\,5\,(0{,}05)\,10$	[100]
5 dec.	$x = 0\,(0{,}01)\,10$	[117]
5 dec.	$x = 0\,(0{,}01)\,6$	[102]
5 dec.	$x = 0\,(0{,}01)\,5{,}5\,(0{,}1)\,10$	[52]
5 dec.	$x = 0\,(0{,}01)\,0{,}1\,(0{,}1)\,10$	[50]
5 dec.	$x = 0\,(0{,}01)\,3\,(0{,}05)\,4\,(0{,}1)\,6\,(0{,}25)\,7\,(0{,}5)\,10$	[12] [47] [60]
5 dec.	$x = 0\,(0{,}0001)\,0{,}01\,(0{,}01)\,1\,(1)\,50$	[86]

5 dec.	$x = 0{,}0001\,(0{,}0001)\,0{,}001\,(0{,}001)\;0{,}01\,(0{,}01)\,0{,}1\,(0{,}1)\,1\,(1)\,10$	[85]
3 dec.	$x = 0\,(0{,}01)\,9{,}99$	[18]

$$\lg\left(\frac{1}{2}\,e^x\right)$$

7 dec.	$x = 4\,(0{,}01)\,10$	[74]

$$\lg\left(e^{\frac{\pi x}{4}}\right)$$

5 dec.	$x = -20\,(1)\,20$	[52]

$$\lg e^{-x}$$

10 dec.	$x = 0\,(0{,}001)\,0{,}1\,(0{,}01)\,2\,(0{,}1)\,10\,(1)\,500$	[159]
10 dec.	$x = 10\,(10)\;100\,(100)\,500$	[92]
10 dec.	$x = m10^{-n}\,.m = 1\,(1)\;9.\,n = 0\,(1)\,5$	[92]
7 dec.	$x = 0\,(0{,}0001)\,0{,}001\,(0{,}001)\,0{,}01\,(0{,}01)\,10$	[137]
5 dec.	$x = 0{,}0001\,(0{,}0001)\,0{,}001\,(0{,}001)\;0{,}01\,(0{,}01)\,0{,}1\,(0{,}1)\,1\,(1)\,10$	[85]

$$\lg e^{\pm x^2}$$

5 dec.	$x = 0\,(0{,}1)\,5$	[52]

$$\lg e^{\frac{1}{2}(x^2-1)}$$

4 dec.	$x = 1\,(0{,}01)\,3$	[25]

$$-\ln\left(1 - e^{-x}\right)$$

6 dec.	$x = 0\,(0{,}005)\,3\,(0{,}01)\,8\,(0{,}05)\,15$	[139]

$$-\lg\left[(1 + x)\,e^{-x}\right]$$

6 dec.	$x = -1\,(0{,}05)\,2{,}5$	[90] [131]

$$\lg\left\{\exp\left(\pm\frac{1}{4}\,x\sqrt{\pi}\right)\right\}$$

5 dec.	$x = 1\,(1)\,20$	[52]

$$\lg\left(\frac{e^x-1}{x}\right)$$

7 dec.	$x = 0\,(0{,}01)\,10{,}4$	[138]

$$\lg\left(\frac{1-e^{-x}}{x}\right)$$

8 dec.	$x = 0\,(0{,}1)\,1{,}5$	[91]

$\lg \operatorname{sh} x$

12 dec.	$x = 0\,(0{,}001)\,2{,}018$	[95]
10 dec.	$x = 5\,(0{,}01)\,12$	[142] [143]
10 dec.	$x = 8\,(0{,}01)\,9\,(1)\,12$	[79]
9 dec.	$x = 2\,(0{,}001)\,5$	[142] [143]
7 dec.	$x = 5\,(0{,}01)\,8$	[79]
6 dec.	$x = 0\,(0{,}0001)\,0{,}2\,(0{,}001)\,2\,(0{,}01)\,8$	[54]
6 dec.	$x = 0\,(0{,}001)\,2$	[93]
5 dec. 6 dec.	$x = 0\,(0{,}001)\,2$ $x = 2\,(0{,}01)\,5$	[79]
5—6 fig.	$x = 0\,(0{,}01)\,1{,}6\,(0{,}1)\,5{,}9$	[10]
5—6 fig.	$x = 0\,(0{,}1)\,5{,}99$	[4]
5 dec.	$x = 0\,(0{,}0001)\,0{,}1\,(0{,}001)\,3\,(0{,}01)\,6$	[37]
5 dec.	$x = 0\,(0{,}001)\,3{,}01\,(0{,}01)\,6{,}09$	[83]
5 dec.	$x = 0\,(0{,}01)\,10$	[117]
5 dec.	$x = 0(0{,}01)\,6$	[102]
5 dec.	$x = 0\,(0{,}01)5{,}09$	[106]
5 dec.	$x = 0\,(0{,}01)\,3\,(0{,}05)\,4\,(0{,}1)6\,(0{,}25)\,7\,(0{,}5)\,10$	[12] [47] [60]
5 dec.	$x = 0\,(0{,}01)\,3\,(0{,}1)\,5$	[52]
5 dec.	$x = 0\,(0{,}01)\,1{,}99\,(0{,}1)\,6$	[50]
4 dec.	$x = 0\,(0{,}001)\,1\,(0{,}01)\,6\,(0{,}1)10$	[86]
4 dec.	$x = 0(0{,}001)\,0{,}2$	[49]
4 dec.	$x = 0\,(0{,}01)\,5{,}99$	[3] [11] [14]
4 dec.	$x = 0(0{,}01)\,5{,}09$	[23] [70]
4 dec.	$x = 0(0{,}01)\,5$	[19]
4 dec.	$x = 1\,(0{,}01)\,3\,(0{,}1)\,5\,(1)\,10$	[45]
4 dec.	$x = 0\,(0{,}01)\,3$	[49]
4 dec.	$x = 0\,(0{,}1)\,5$	[48]

$\lg \operatorname{ch} x$

12 dec.	$x = 0(0{,}001)\,2{,}032$	[95]
10 dec.	$x = 5\,(0{,}01)12$	[142] [143]

10 dec.	$x = 8\,(0{,}01)\,9\,(1)\,12$	[79]
9 dec.	$x = 2\,(0{,}001)\,5$	[142] [143]
7 dec.	$x = 0\,(0{,}005)\,0{,}6$	[93]
7 dec.	$x = 5\,(0{,}01)\,8$	[79]
6 dec.	$x = 0\,(0{,}0001)\,0{,}2\,(0{,}001)\,2\,(0{,}01)\,8$	[54]
6 dec.	$x = 0\,(0{,}001)\,2$	[94]
5 dec.	$x = 0\,(0{,}001)\,2$	[79]
6 dec.	$x = 2\,(0{,}01)\,5$	
5 dec.	$x = 0\,(0{,}0001)\,0{,}1\,(0{,}001)\,3\,(0{,}01)\,6$	[37]
5 dec.	$x = 0\,(0{,}001)\,3{,}01\,(0{,}01)\,6{,}09$	[83]
5 dec.	$x = 0\,(0{,}01)\,10$	[117]
5 dec.	$x = 0\,(0{,}01)\,6$	[102]
5 dec.	$x = 0(0{,}01)\,5{,}09$	[106]
5 dec.	$x = 0\,(0{,}01)\,3\,(0{,}05)\,4\,(0{,}1)\;\;6\;\;(0{,}25)\,7\,(0{,}5)\,10$	[12] [47] [60]
5 dec.	$x = 0(0{,}01)\,3\,(0{,}1)\,5$	[52]
5 dec.	$x = 0\,(0{,}01)\,1{,}99\,(0{,}1)\,6$	[50]
4 dec.	$x = 0\,(0{,}001)\,1\,(0{,}01)\,6\,(0{,}1)\,10$	[86]
4 dec.	$x = 0\,(0{,}01)\,5{,}99$	[3] [4] [11] [14]
4 dec.	$x = 0\,(0{,}01)\,5{,}09$	[19] [23] [70]
4 dec.	$x = 1\,(0{,}01)\,3\,(0{,}1)\,5\,(1)\,10$	[45]
4 dec.	$x = 0\,(0{,}01)\,3$	[49]
4 dec.	$x = 0\,(0{,}1)\,5$	[48]
1—5 fig.	$x = 0\,(0{,}01)\,1{,}6\,(0{,}1)\,5{,}9$	[10]

lg th x

12 dec.	$x = 0(0{,}001)\,2{,}018$	[95]
10 dec.	$x = 5\,(0{,}01)\,12$	[142] [143]
10 dec.	$x = 5\,(0{,}1)\,12$	[79]
9 dec.	$x = 2\,(0{,}001)\,5$	[142] [143]
7 dec.	$x = 0\,(0{,}01)\,2$	[79]
9 dec.	$x = 2\,(0{,}1)\,4{,}9$	
7 dec.	$x = 5\,(0{,}01)\,6$	
6 dec.	$x = 0\,(0{,}0001)\,0{,}2\,(0{,}001)\,2\,(0{,}01)\,8$	[54]
5 dec.	$x = 0\,(0{,}001)\,2$	[79]
6 dec.	$x = 2\,(0{,}01)\,5$	
5 dec.	$x = 0\,(0{,}0001)\,0{,}1\,(0{,}001)\,3\,(0{,}01)\,6$	[37]
5 dec.	$x = 0\,(0{,}001)\,1{,}9\,(0{,}01)\,3{,}09$	[83]
5 dec.	$x = 0\,(0{,}01)\,1{,}99\,(0{,}1)\,6$	[50]
5 dec.	$x = 0\,(0{,}01)\,3\,(0{,}1)\,5$	[52]
4 dec.	$x = 0(0{,}001)\,1\,(0{,}01)\,6$	[86]

4 dec.	$x = 0\,(0{,}001)\,0{,}2$	[49]
4 dec.	$x = 0\,(0{,}01)\,3$	[49]
4 dec.	$x = 0\,(0{,}01)\,2{,}89$	[3] [4] [14]
4 dec.	$x = 0\,(0{,}01)\,2{,}39$	[19] [23] [70]

lg cth x

12 dec.	$x = 0\,(0{,}001)\,2{,}018$	[95]
10 dec.	$x = 5\,(0{,}01)\,12$	[142] [143]
10 dec.	$x = 5\,(0{,}1)\,12$	[79]
9 dec.	$x = 2\,(0{,}001)\,5$	[142] [143]
7 dec.	$x = 0\,(0{,}01)\,2$	[79]
9 dec.	$x = 2\,(0{,}1)\,4{,}9$	[79]
7 dec.	$x = 5\,(0{,}01)\,6$	[79]
6 dec.	$x = 0\,(0{,}0001)\,0{,}2\,(0{,}001)\,2\,(0{,}01)\,8$	[54]
6 dec.	$x = 0\,(0{,}001)\,2$	[94]
5 dec.	$x = 0\,(0{,}001)\,2$	[79]
6 dec.	$x = 2\,(0{,}01)\,5$	[79]
5 dec.	$x = 0(0{,}0001)\,0{,}1\,(0{,}001)\,3\,(0{,}01)6$	[37]
5 dec.	$x = 0\,(0{,}001)\,1{,}9\,(0{,}01)\,3{,}09$	[83]
5 dec.	$x = 0\,(0{,}01)\,3\,(0{,}1)\,5$	[52]
4 dec.	$x = 0\,(0{,}001)\,1\,(0{,}01)\,6$	[86]
4 dec.	$x = 0\,(0{,}01)\,2{,}39$	[19] [70]

$\lg \frac{\operatorname{sh} x}{x}$

12 dec.	$x = 0\,(0{,}001)\,0{,}506$	[95]
7 dec.	$x = 0\,(0{,}005)\,0{,}6$	[93]
6 dec.	$x = 0\,(0{,}001)\,0{,}499$	[94]

$\lg \frac{\operatorname{th} x}{x}$

12 dec.	$x = 0\,(0{,}001)\,0{,}506$	[95]
6 dec.	$x = 0\,(0{,}001)\,0{,}499$	[94]

Chapter 4

COMMON AND NATURAL LOGARITHMS [1]

It is common knowledge that logarithms are widely used in practical work and this explains why tables of logarithms were compiled long ago (from the beginning of the 17th century). The information given in this chapter is a comparatively complete account of the existing tables of logarithms of numbers and of trigonometric functions. Tables of logarithms for other functions (exponential, cylinder, etc.) are given in the sections where tables of these functions are described.

The chapter describes tables of:

1) logarithms of numbers (integers, fractions and primes); logarithms of complex numbers;
2) antilogarithms;
3) logarithms of certain expressions;
4) logarithms of trigonometric functions;
5) logarithms of expressions containing trigonometric functions;
6) logarithms of the sums and differences of numbers;
7) tables for computing many-decimal logarithms.

COMMON LOGARITHMS AND ANTILOGARITHMS

Logarithms of integers

102 dec.	$x = 1(1)109$	[329]
61 dec.	$x = 1(1)99$ (primes) 1097; $999\,980(1)1\,000\,021$	[52] [91] [188]
61 dec.	$x = 1(1)100$ (primes) 1097	[376]
52 dec.	$x = 1(1)1100$	[312]
50 dec.	$x = 1(1)222$	[186]
36 dec.	$x = 1(1)1000$ (primes) $10\,007$	[214]
36 dec.	$x = 1(1)1100$	[257]
23 dec.	$x = 1(1)10\,000$	[385]
21 dec.	$x = 1000(1)10\,000$	[393]

[1] See pages 445 and 550 for the references to Chapter 4.

20 dec.	$x = 10^7\,(1)\,10\,000\,200$ $10^{10}\,(1)\,10\,000\,000\,200$	[463]
20 dec.	$x = 10\,000\,(1)\,20\,000$	[469]
20 dec.	$x = 20\,000\,(1)\,30\,000$	[473]
20 dec.	$x = 30\,000\,(1)\,40\,000$	[471]
20 dec.	$x = 40\,000\,(1)\,50\,000$	[466]
20 dec.	$x = 50\,000\,(1)\,60\,000$	[467]
20 dec.	$x = 60\,000\,(1)\,70\,000$	[468]
20 dec.	$x = 70\,000\,(1)\,80\,000$	[470]
20 dec.	$x = 80\,000\,(1)\,90\,000$	[465]
20 dec.	$x = 90\,000\,(1)\,100\,000$	[463]
20; 18 dec.	$x = 1\,(1)1200;\ 101\,000\,(1)\,101\,179$	[52] [91] [188]
20 dec.	$x = 1\,(1)\,100$	[463]
15 dec.	$x = 1\,(1)\,100$	[42] [207]
14 dec.	$x = 1\,(2)\,20\,000;$ $90\,000\,(1)\,100\,000$	[178]
12 dec.	$x = 433\,300\,(1)\,434\,299$	[316]
12 dec.	$x = 1\,(1)\,1080$	[150]
12 dec.	$x = 100\,(1)\,1000$	[275]
12 dec.	$x = 10\,(1)\,100$	[104]
10 dec.	$x = 1\,(1)\,101\,000$	[415]
10 dec.	$x = 1\,(1)\,1000;\ 10\,000\,(1)\,100\,999$	[420]
10 dec.	$x = 10\,000\,(1)\,100\,000$	[215]
10 dec.	$x = 1\,(1)\,100\,000$	[323] [338] [424]
10 dec.	$x = 10\,000\,(1)\,10\,199$	[105]
10 dec.	$x = 1000\,(1)\,10\,000$	[86]
9—10 dec.	$x = 0\,(1)\,100$	[363]
9 dec.	$x = 1000\,(1)\,10\,000$	[246]
9 dec.	$x = 1\,(1)\,1000$	[5]
8 dec.	$x = 100\,000\,(1)\,110\,000$	[479]
8 dec.	$x = 1\,(1)\,200\,009$	$[10]_1$ [168]
8 dec.	$x = 12\,000\,(1)\,120\,000$	[375]
8 dec.	$x = 10\,000\,(1)\,125\,000$	[306]
8 dec.	$x = 10\,000\,(1)\,100\,000$	[325] [387]
8 dec.	$x = 1\,(1)\,20\,000$	[353]
8 dec.	$x = 100\,(1)\,1000$	[262]
10 dec. 7 dec.	$x = 1\,(1)\,1000$ $x = 20\,000\,(1)\,200\,000$	[362]
8; 7 dec. 8 dec.	$x = 1\,(1)\,1020;\ 1020\,(1)\,100\,000$ $x = 100\,000\,(1)\,108\,000$	[52] [91] [188]
7 dec. 8 dec.	$x = 1\,(1)\,100\,000$ $x = 100\,000\,(1)\,108\,000$	[24] [288] [347] [365] [418]
7 dec.	$x = 1\,(1)\,120\,000$	[327] [378] [401]

7 dec.	$x = 1\,(1)\,100\,000$	[410]
7 dec.	$x = 10\,000\,(1)\,100\,000$	[27] [184] [252] [341]
7 dec.	$x = 1\,(1)\,11\,009$	[419] [290]
7 dec.	$x = 10\,000\,(1)\,11\,000$	[182] [287] [432]
7 dec.	$x = 1\,(1)\,10\,800$	[150]
7 dec.	$x = 1\,(1)\,100;\ 1000\,(1)\,10\,499$	[25]
7 dec.	$x = 1000\,(1)\,10\,000$	[42] [207]
7 dec.	$x = 1\,(1)\,10\,000$	[7] [12] [119] [124] [130] [269] [326] [396] [421] [423]
7 dec.	$x = 2\,(1)\,134;\ 1000\,(1)\,1118$	[281]
7 dec.	$x = 100\,(1)\,1000$	[223]
7 dec.	$x = 1200\,(1)\,2400\,(2)\,6000\,(5)\,12000$	[28]
6 dec.	$x = 10\,000\,(1)\,100\,009$	[181] [280] [479]
6 dec.	$x = 1\,(1)\,100\,009$	[20]
6 dec.	$x = 1\,(1)\,21\,600$	[399]
6 dec.	$x = 1\,(1)\,20\,000$	[397]
6 dec.	$x = 1080\,(1)\,10\,800$	[150]
6 dec.	$x = 1000\,(1)\,10\,200$	[128]
6 dec.	$x = 1\,(1)\,10\,200$	[79] [127]
6 dec.	$x = 1000\,(1)\,10\,000$	[160] [199] [251] [266] [272] [279]
6 dec.	$x = 1\,(1)\,10\,000$	[313]
6 dec.	$x = 1\,(1)\,2000$	[363]
6 dec.	$x = 1\,(1)\,100$	[86] [100] [128] [272]
5; 6 dec.	$x = 1\,(1)\,10\,000;\ 10\,000\,(1)\,12\,009$	[106]
5 dec. 6 dec.	$x = 1000\,(1)\,10\,000$ $x = 10\,000\,(1)\,11\,009$	[324] [431]
5—6 dec.	$x = 1\,(1)\,10\,800$	[429]
5 dec. 6 dec.	$x = 1000\,(1)\,10\,000$ $x = 10\,000\,(1)\,11\,000$	[381]
5; 6 dec.	$x = 1\,(1)\,100;\ 1000\,(1)\,10\,999$	[384]
5—6 dec.	$x = 100\,(1)\,1200$	[248]
5 dec.	$x = 10\,000\,(1)\,100\,000$	[74] [277] [278] [371] [372]
5 dec.	$x = 4000\,(1)\,40\,009$	[192]
5 dec.	$x = 1\,(1)\,14\,500$	[298]
5 dec.	$x = 1000\,(1)\,12\,009$	[248]
5 dec.	$x = 1\,(1)\,12\,000$	[374]
5 dec.	$x = 0\,(1)\,11\,009$	[156] [233]
5 dec.	$x = 1\,(1)\,10\,800$	[34] [101] [122] [129] [262]
5 dec.	$x = 1\,(1)\,10\,109$	[147]
5 dec.	$x = 1000\,(1)\,10\,080$	[270] [271]

5 dec.	$x = 1000\,(1)\,10\,009$	[6] [169] [171] [255] [281] [432] [433]
5 dec.	$x = 1000\,(1)\,10\,000$	[2] [96] [109] [190] [357] [395]
5 dec.	$x = 100\,(1)\,10\,000$	[63] [166] [352]
5 dec.	$x = 1\,(1)\,10\,000$	[8] [9] [13] [16] [26] [29] [60] [65] [83] [[92] [99] [100] [107] [108] [126] [142] [148] [182] [203] [226] [231] [287] [290] [291] [305] [359] [364] [394] [400]
5 dec.	$x = 1000\,(1)\,2999$	[198]
5 dec.	$x = 100\,(1)\,2199$	[428]
5 dec.	$x = 0\,(1)\,2100$	[112]
5 dec.	$x = 1\,(1)\,1199$	[315]
5 dec.	$x = 100\,(1)\,1009$	[358]
5 dec.	$x = 100\,(1)\,1000$	[93] [198]
5 dec.	$x = 1\,(1)\,1000$	[18] [21] [22] [45] [58] [69] [140] [175]
5 dec.	$x = 10\,(1)\,100$	[92] [248] [283] [290] [395]
5 dec.	$x = 1003\,(3)\,3190\,(10)\,10\,000$	[84]
4; 7 dec.	$x = 100\,(1)\,999;\ 1000\,(1)\,1099$	[173]
4 dec.	$x = 1\,(1)\,11\,009$	[230]
4 dec.	$x = 1\,(1)\,10\,009$	[55] [388]
4 dec.	$x = 100\,(1)\,10\,000$	[111] [134]
4 dec.	$x = 1\,(1)\,10\,000$	[47] [73] [78] [149] [250] [277] [278] [304] [366] [423]
4 dec.	$x = 500\,(1)\,4999$	[114]
4 dec.	$x = 1\,(1)\,4100$	[356]
4 dec.	$x = 200\,(1)\,2009$	[22] [206]
4 dec.	$x = 1\,(1)\,2009$	[71] [76] [85] [135] [310]
4 dec.	$x = 100\,(1)\,2000$	[195]
4 dec.	$x = 100\,(1)\,1100$	[53] [120] [368] [373] [381] [432]
4 dec.	$x = 100\,(1)\,1000$	[4] [15] [30] [33] [34] [40] [42] [44] [49] [51] [64] [66] [75] [82] [94] [98] [101] [117] [121] [125] [138] [139 [141] [145] [147] [174] [193] [248] [281] [282] [284] [309] [357] [367] [435] [437] [438]

4 dec.	$x = 1\,(1)\,1000$	[11] [14] [17] [19] [23] [37] [38] [39] [43] [54] [57] [77] [80] [90] [102] [106] [113] [131] [133] [140] [146] [153] [230] [286] [300] [350] [363] [414]
4 dec.	$x = 1\,(1)\,100$	[29] [46] [50] [62] [70] [81] [132] [283] [358]
3 dec.	$x = 1\,(1)\,1000$	[51] [56] [72] [89] [137]
3 dec.	$x = 1\,(1)\,100$	[106] [152] [262] [395]

Logarithms of decimal fractions

11 dec.	$x = 1{,}001\,(0{,}001)\,1{,}12$	[448]
10 dec.	$x = 1{,}01(0{,}01)1{,}02(0{,}05)1{,}025\,(0{,}025)\,1{,}05\,(0{,}05)\,1{,}06$	[223]
8 dec.	$x = 0\,(0{,}001)\,0{,}999$	[479]
6 dec.	$x = 100\,(0{,}1)\,1000$	[389]
5 dec.	$x = 1\,(0{,}001)\,2{,}099\,(0{,}01)\,10$	[165]
5 dec.	$x = 0\,(0{,}01)\,10{,}09$	[248]
5 dec.	$x = 0\,(0{,}01)\,1$	[384]
5 dec.	$x = 0\,(0{,}1)\,1000$	[220]
5; 4 dec.	$x = 1\,(0{,}001)\,1{,}999;\ 2(0{,}01)\,10$	[154]
4 dec.	$x = 1\,(0{,}01)\,10$	[103]
4 dec.	$x = 10\,(0{,}1)\,100$	[37] [97] [248] [299] [352]
4 dec.	$x = 0\,(0{,}1)\,100$	[88]
3 dec.	$x = 10\,(0{,}1)\,100$	[97]

Logarithms of ordinary fractions $\lg \frac{m}{n}$

5 dec.	$m = 2\,(1)\,100.\ \frac{m}{n} \leqslant 1{,}5$	[151]

Logarithms of prime numbers

84 dec.	$1 \leqslant x \leqslant 113$	[338]
61 dec.	$113 \leqslant x \leqslant 1129$	[329]
20 dec.	$2 \leqslant x \leqslant 997$	[358]
11 dec.	$2 \leqslant x \leqslant 1811$	[223] [288]
11 dec.	$1 \leqslant x \leqslant 1063$	[394]
10 dec.	$1 \leqslant x \leqslant 15391$	[251]
10 dec.	$1 \leqslant x \leqslant 6257$	[150]
10 dec.	$1 \leqslant x < 1000$	[248] [435]
7 dec.	$2 \leqslant x \leqslant 97$	[109]

Cologarithms

(Logarithms of reciprocals)

27 dec.	$x = 11\,(1)\,110$	[408]
20 dec.	$x = 1\,(1)\,100$	[463]
6 dec.	$x = 1000\,(1)\,10\,000$	[187]
5 dec.	$x = 4000\,(1)\,40\,000$	[192]
5 dec.	$x = 1\,(1)\,10\,080$	[271]
5 dec.	$x = 1000\,(1)\,10\,080$	[270]
5 dec.	$x = 10\,(1)\,99$	[100]
5 dec.	$x = 1\,(0{,}01)\,9{,}99$	[154]
4 dec.	$x = 1000\,(1)\,10\,000$	[277] [278]
4 dec.	$x = 200\,(1)\,2000$	[345]
4 dec.	$x = 100\,(1)\,1000$	[111]

Antilogarithms

20 fig.	$0\,(0{,}0^{9}1)\,0{,}0^{7}1$; $0\,(0{,}0^{4}1)\,0{,}0^{6}1$	[463]
20 fig.	0 (0,00001) 0,00179	[91]
15 fig.	0 (0,01) 1	[42]
14 fig.	0 (0,00001) 0,0335 (0,0001) 1	[204]
13 12 fig.	0 (0,0001) 0,6999; 0,7 (0,0001) 1	[244]
11 fig.	0 (0,00001) 1	[209]
11 fig.	0 (0,0001) 1	[460]
7 fig.	0 (0,00001) 1	[224] [341] [378] [379]
7 fig.	0 (0,0001) 1	[42] [207] [208] [380]
6 fig.	0 (0,0001) 1	[192] [479]
5 fig.	0 (0,00001) 1	[74] [277] [278] [371] [372]
5 fig.	0 (0,0001) 1	[140]
5 fig.	0 (0,001) 1	[154] [165] [175] [198] [283]
4. fig.	0 (0,0001) 1	[47] [78] [134] [140] [250] [277] [278] [366]
4. fig.	0 (0,001) 0,9 (0,0001) 1	[206] [310]
4. fig.	0 (0,001) 1	[4] [11] [15] [19] [23] [30] [33] [34] [37] [43] [46] [49] [50] [51] [64] [70] [75] [76] [81] [82] [88] [98] [101] [106] [111] [117] [125] [132] [138] [141] [145] [146] [149] [153] [195] [248] [286] [303] [357]

4 fig.	0(0,01)1	[193]
3 fig.	0(0,01)1	[106] [118] [262]

Logarithms of certain expressions

$n \lg x$

5 dec.	$x = 1\,(1)\,100.\ n = 1\,(1)\,10$	[440]
4 dec.	$x = 1\,(1)\,100.\ n = 2; 3; \frac{1}{2}; \frac{1}{3}$	[279]

$\lg \lg x$

5 dec.	$x = 1$ (var.) 100 000	[175]
5 dec.	$x = 1$ (var.) 1000	[192]
4 dec.	$x = 1\,(0{,}001)\,1{,}339\,(0{,}01)\,10$	[103]

$\lg(-\lg x)$

5 dec.	$x = 0{,}001$ (var.) 1	[192]

$10 + \lg\left[\frac{x}{100 - x}\right]$

4 dec.	$x = 0\,(0{,}01)\,5\,(0{,}1)\,94{,}9$	[307]

$\lg(1 + x^2)$

7 dec.	$x = 0\,(0{,}01)\,5\,(0{,}1)\,15$	[472]

$\lg\sqrt{1 + x^2}$

3 dec.	$x = 0\,(0{,}02)\,9{,}98$	[72]

Logarithms of trigonometric functions

In this section we describe tables of the logarithms of trigonometric functions not separately for each function but simultaneously for pairs of functions: sine and cosine, tangent and cotangent, secant and cosecant.

The descriptions only relate to the tables of logarithms of the direct functions sine, tangent and secant; the only difference in the tables of the complementary functions is that the argument varies in the inverse order (this note does not apply to functions the argument of which is given in radians; in such cases the direct and complementary functions are normally given in parallel).

lg sin x & lg cos x

Argument in degrees, minutes and seconds

22 dec.	$x = 0\,(1'')\,10'\,(10')\,89°50'\,(1'')\,90°$	[338]
17 dec.	$x = 0\,(54')\,90°$	[162]
14 dec.	$x = 0\,(15')\,90°$	[294] [295]
10 dec.	$x = 0\,(1'')\,2°\,(10'')\,88°\,(1'')\,90°$	[415] [420]
10 dec.	$x = 0\,(1')\,90°$	[269] [323] [424]
10 dec.	$x = 0\,(10')\,90°$	[460]
8 dec.	$x = 0\,(1'')\,90°$	$[10]_2$ [168]
7 dec.	$x = 0\,(1'')\,90°$	[333] [377] [378]
7 dec.	$x = 0\,(1'')\,5°$	[91]
7 dec.	$x = 0\,(1'')\,1°32'$	[418]
7 dec.	$x = 0\,(1'')\,6°\,(10'')\,84°\,(1'')\,90°$	[24] [184] [410]
7 dec.	$x = 0\,(10'')\,90°$	[91] [365]
7 dec.	$x = 0\,(1'')\,5°\,(10'')\,90°$	[27] [216] [419]
7 dec.	$x = 0\,(1'')\,1°\,(10'')\,3°\,(30'')$ $87°\,(10'')\,89°\,(30'')\,90°$	[252]
7 dec.	$x = 0\,(10'')\,9°\,(1')\,81°\,(10'')\,90°$	[288]
7 dec.	$x = 0\,(10'')\,7°\,(30'')\,13°\,(1')\,44°40'$	[150]
9 dec.	$x = 44°40'\,(10'')\,45°20'$	[150]
7 dec.	$x = 45°20'\,(1')\,77°\,(30'')\,83°(10'')90°$	[150]
7 dec.	$x = 0\,(10'')\,6°\,(1')\,84°\,(10'')\,90°$	[418]
7 dec.	$x = 0\,(10'')\,13°\,(1')\,44°40'$	[122]
9 dec.	$x = 44°40'\,(10'')\,45°20'$	[122]
7 dec.	$x = 45°20'\,(1')\,77°\,(10'')\,90°$	[122]
7 dec.	$x = 0\,(1')\,90°$	[7] [25] [110] [123] [124] [130] [205] [326] [347] [353] [421] [423]
7 dec.	$x = 0\,(3')\,10°\,(2')\,80°\,(3')\,90°$	[28]
6 dec.	$x = 0\,(1'')\,5°\,(10'')\,90°$	[20] [181]
6 dec.	$x = 0\,(1'')\,1°20'$	[479]
6 dec.	$x = 0\,(1'')\,1°;\;0\,(3'')\,90°$	[276]
6 dec.	$x = 0\,(10'')\,90°$	[251]
6 dec.	$x = 0\,(1'')\,2°\,(15'')\,4°\,(30'')\,90°$	[79] [127] [128]
6 dec.	$x = 0\,(1')\,90°$	[100] [160] [272] [279] [313] [397] [389] [399]
5 dec.	$x = 0\,(1'')\,5'$	[156]
5 dec.	$x = 0\,(1'')\,1°\,(10'')\,8°\,(1')\,82°$ $(10'')\,89°\,(1'')\,90°$	[101] [231]
5 dec.	$x = 0\,(10'')\,90°$	[234] [478]
5 dec.	$x = 0\,(15'')\,90°$	[271]
5 dec.	$x = 0\,(10'')\,6°\,(1')\,84°\,(10'')\,90°$	[122] [429]

5 dec.	$x = 0\,(10'')\,4°\,(1')\,86°\,(10'')\,90°$	[156]
5 dec.	$x = 0\,(1')\,90°$	[2] [6] [8] [9] [16] [29] [34] [39] [60] [71] [99] [107] [126] [129] [171] [203] [226] [248] [255] [262] [287] [291] [298] [324] [352] [358] [359] [374] [384] [394] [395] [400]
5 dec.	$x = 0\,(1'')\,1°\,(10'')\,3°\,(1')\,90°$	[192]
5 dec.	$x = 0\,(1'')\,1°\,(10'')\,4°\,(1')\,90°$	[190]
5 dec.	$x = 0\,(6'')\,6°\,(1')\,90°$	[169]
5 dec.	$x = 0\,(1'')\,1'\,(1')\,90°$	[147]
5 dec.	$x = 0\,(3')\,90°$	[198]
5 dec.	$x = 0\,(6')\,90°$	[50] [165]
5 dec.	$x = 0\,(10')\,90°$	[69] [83] [142] [428]
5 dec.	$x = 0\,(1')\,6°\,(10')\,84°\,(1')\,90°$	[305]
4 dec.	$x = 0\,(1'')\,20'\,(10'')\,2°$	[230]
4 dec.	$x = 1°59'\,(10'')\,5°\,(1')\,90°$	[366]
4 dec.	$x = 0\,(10'')\,90°$	[53]
4 dec.	$x = 0\,(1')\,90°$	[15] [33] [55] [78] [80] [149] [153] [230] [310]
4 dec.	$x = 0\,(1')\,8°$	[356]
4 dec.	$x = 0\,(1')\,17°$	[195]
4 dec.	$x = 3°\,(1')\,87°$	[388]
4 dec.	$x = 0\,(2')\,7°$	[368]
4 dec.	$x = 0\,(2')\,5°$	[373]
4 dec.	$x = 0\,(5')\,90°$	[43]
4 dec.	$x = 0\,(1')\,14°\,(6')\,90°$	[19]
4 dec.	$x = 0\,(1')\,9°\,(6')\,90°$	[4] [85]
4 dec.	$x = 0\,(6')\,90°$	[37] [46] [49] [66] [75] [94] [132] [133] [145] [154] [193] [282] [283] [286] [299] [350] [367]
4 dec.	$x = 0\,(1')\,10°\,(10')\,80°\,(1')\,90°$	[76]
4 dec.	$x = 0\,(1')\,9°\,(10')\,81°\,(1')\,90°$	[88]
4 dec.	$x = 1°\,(1')\,8°\,(10')\,82°\,(1')\,90°$	[44]
4 dec.	$x = 0\,(1')\,6°\,(10')\,84°\,(1')\,90°$	[281]
4 dec.	$x = 0\,(1')\,5°\,(10')\,85°\,(1')\,90°$	[315]
4 dec.	$x = 0\,(10')\,90°$	[11] [30] [40] [54] [69] [97] [98] [111] [141] [358] [368] [373] [414] [437]
4 dec.	$x = 0\,(2')\,10°\,(4')\,20°\,(10')\,90°$	[82] [435]

4 dec.		$x = 0\,(5')\,3°\,(10')\,87°\,(5')\,90°$	[140]
4 dec.		$x = 14°\,(10')\,50°\,(20')\,90°$	[3]
4 dec.		$x = 0\,(1°)\,90°$	[134] [146] [248] [438]
3 dec.		$x = 0\,(10')\,90°$	[38]

Argument in tenths of a degree

14 dec.		$x = 0\,(0°{,}01)\,90°$	[179]
8 dec.		$x = 0\,(0°{,}001)\,0°{,}7$	[339]
9 dec.		$x = 0°{,}7\,(0°{,}001)\,2°{,}1$	[339]
10 dec.		$x = 2°\,(0°{,}001)\,90°$	[339]
10 dec.		$x = 0\,(0°{,}01)\,90°$	[354]
8 dec.		$x = 0\,(0°{,}001)\,3°\,(0°{,}01)\,90°$	[325]
7 dec.		$x = 0\,(0°{,}001)\,90°$	[337]
6 dec.		$x = 0\,(0°{,}001)\,90°$	[336]
6 dec.		$x = 0\,(0°{,}01)\,90°$	[328]
5 dec.		$x = 0\,(0°{,}01)\,90°$	[182] [218] [430] [432]
5 dec.		$x = 5°\,(0°{,}01)\,90°$	[381]
5 dec.		$x = 0\,(0°{,}1)\,90°$	[352]
4 dec.		$x = 0\,(0°{,}01)\,90°$	[109] [300]
4 dec.		$x = 0\,(0°{,}01)\,4°\,(0°{,}1)\,90°$	[310]
4 dec.		$x = 0\,(0°{,}01)\,11°\,(0°{,}1)\,90°$	[195]
4 dec.		$x = 0\,(0°{,}01)\,10°\,(0°{,}1)\,90°$	[189]
4 dec.		$x = 0\,(0°{,}01)\,8°\,(0°{,}1)\,82°\,(0°{,}01)\,90°$	[180]
4 dec.		$x = 0\,(0°{,}01)\,5°\,(0°{,}1)\,85°\,(0°{,}01)\,90°$	[206] [238]
4 dec.		$x = 0\,(0°{,}1)\,90°$	[17] [363]
3 dec.		$x = 0\,(0°{,}1)\,90°$	[56] [72]
3 dec.		$x = 0\,(0°{,}1)\,16°\,(1°)\,74°\,(0°{,}1)\,90°$	[123]

Argument in radians

9 dec.		$x = 0\,(0{,}001)\,0{,}5$	[52]
5 dec.		$x = 0\,(0{,}0001)\,0{,}1\,(0{,}001)\,1{,}6$	[170]
5 dec.		$x = 0\,(0{,}01)\,1{,}6$	[352]
5 dec.		$x = 0\,(0{,}001)\,1{,}569$	[308]
4 dec.		$x = 0\,(0{,}0001)\,0{,}04\,(0{,}001)\,1{,}57$	[310]
4 dec.		$x = 0\,(0{,}0029)\,1{,}5708$	[40]
4 dec.	lg sin x	$x = 0\,(0{,}001)\,0{,}2$	[195]
4 dec.		$x = 0\,(0{,}01)\,3{,}2$	[160]
4 dec.		$x = 0\,(0{,}01)\,1{,}6$	[190] [248]
4 dec.		$x = 0\,(0{,}01)\,1{,}57$	[75] [195]
4 dec.		$x = 0\,(0{,}1)\,7{,}9$	[75] [310]

Argument in grads

17 dec.	$x = 0\ (0^g{,}001)\ 3^g;\ 97^g\ (0^g{,}001)\ 100^g$	[233]
17 dec.	$x = 0\ (1^g)\ 100^g$	[162]
15 dec.	$x = 0\ (1^g)\ 100^g$	[280]
14 dec.	$x = 0\ (0^g{,}1)\ 100^g$	[188]
11 dec.	$x = 0\ (0^g{,}001)\ 0^g{,}1\ (0^g{,}1)\ 99^g{,}9\ (0^g{,}001)\ 100^g$	[176]
10 dec.	$x = 0\ (2^g)\ 100^g$	[201]
8 dec.	$x = 0\ (0^g{,}001)\ 100^g$	[375] [401]
7 dec.	$x = 0\ (0^g{,}0001)\ 3^g\ (0^g{,}001)\ 100^g$	[342]
7 dec.	$x = 0\ (0^g{,}001)\ 3^g\ (0^g{,}01)\ 97^g\ (0^g{,}001)\ 100^g$	[176] [257]
7 dec.	$x = 0\ (0^g{,}01)\ 100^g$	[188]
6 dec.	$x = 0\ (0^g{,}001)\ 20^g\ (0^g{,}01)\ 80^g\ (0^g{,}001)\ 100^g$	[280]
5 dec.	$x = 0\ (0^g{,}001)\ 3^g{,}5$	[92]
5 dec.	$x = 0\ (0^g{,}001)\ 1^g$	[290]
5 dec.	$x = 0\ (0^g{,}001)\ 1^g{,}2\ (0^g{,}01)\ 100^g$	[431]
5 dec.	$x = 0\ (0^g{,}001)\ 3^g{,}5\ (0^g{,}01)\ 100^g$	[232]
5 dec.	$x = 0\ (0^g{,}01)\ 100^g$	[92] [239] [290] [374]
5 dec.	$x = 3^g\ (0^g{,}01)\ 97^g$	[233]
5 dec.	$x = 0\ (0^g{,}01)\ 10^g\ (0^g.1)\ 100^g$	[433]
4 dec.	$x = 0\ (0^g{,}01)\ 50^g\ (0^g{,}1)\ 100^g$	[114]
4 dec.	$x = 0\ (0^g{,}01)\ 3^g\ (0^g{,}1)\ 100^g$	[261]
3 dec.	$x = 0\ (0^g{,}1)\ 100^g$	[72]

Argument in units of time

7 dec.	$x = 0\ (0^{sec}{,}1)\ 10^M$	[32]
7 dec.	$x = 0\ (1^{sec})\ 6^{hrs}$	[32] [256]
6 dec.	$x = 0\ (1^{sec})\ 6^{hrs}$	[187]
6 dec.	$x = 0\ (2^{sec})\ 6^{hrs}$	[349]
6 dec.	$x = 0\ (4^{sec})\ 60^M$	[100]
5 dec.	$x = 0\ (0^{sec}{,}1)\ 8^M\ (1^{sec})\ 6^{hrs}$	[334]
5 dec.	$x = 0\ (1^{sec})\ 6^{hrs}$	[270]
5 dec.	$x = 0\ (4^{sec})\ 6^{hrs}$	[185] [202]
5 dec.	$x = 0\ (10^{sec})\ 24^{hrs}$	[157]
5 dec.	$x = 0\ (20^M)\ 6^{hrs}$	[101]

4 dec.	$x = 0\,(4^{sec})\,6^{hrs}$	[80]
4 dec.	$x = 0\,(1^{sec})\,20^{M}\,(10^{sec})\,6^{hrs}$	[196] [206]
4 dec.	$x = 0\,(10^{sec})\,6^{hrs}$	[422]

Argument in protractor divisions

5 dec.	$x = 0$ (0—01) 15—00	[26]
4 dec.	$x = 0$ (0—01) 15—00	[101]
4 dec.	$x = 0$ (0—02) 15—00	[43]

lg sin πx

10 dec.	$x = 0\,(0{,}01)\,1$	[201]

lg tg x & lg ctg x

Argument in degrees, minutes and seconds

22 dec.	$x = 0\,(1'')\,10'\,(10')\,45°$	[338]
17 dec.	$x = 0\,(54')\,45°$	[162]
15 dec.	$x = 0\,(9')\,45°$	[162]
14 dec.	$x = 0\,(10'')\,90°$	[162]
10 dec.	$x = 0\,(1'')\,2°\,(10'')\,88°\,(1'')\,90°$	[415] [420]
10 dec.	$x = 0\,(1')\,90°$	[269] [323] [424]
8 dec.	$x = 0\,(1'')\,90°$	[10]$_2$ [168]
7 dec.	$x = 0\,(1'')\,90°$	[333] [377] [378]
7 dec.	$x = 0\,(1'')\,5°\,(10'')\,90°$	[27] [216] [419]
7 dec.	$x = 0\,(1'')\,6°\,(10'')\,84°\,(1'')\,90°$	[24] [184] [410]
7 dec.	$x = 0\,(10'')\,90°$	[91] [365]
7 dec.	$x = 0\,(1'')\,1°\,(10'')\,3°\,(30'')\,87°\,(10'')\,89°\,(30'')\,90°$	[252]
7 dec.	$x = 0\,(10'')\,6°\,(1')\,84°\,(10'')\,90°$	[418]
7 dec.	$x = 0\,(10'')\,7°\,(30'')\,13°\,(1')\,44°\,40'$	[150]
9 dec.	$x = 44°\,40'\,(10'')\,45°$	[150]
7 dec.	$x = 45°\,(1')\,77°\,(30'')\,83°\,(10'')\,90°$	[150]
7 dec.	$x = 0\,(10'')\,9°\,(1')\,81°\,(10'')\,90°$	[288]
7 dec.	$x = 0\,(1')\,90°$	[7] [25] [110] [119] [123] [124] [130] [205] [326] [347] [396] [421] [423]
7 dec.	$x = 0\,(3')\,10°\,(2')\,80°\,(3')\,90°$	[28]
7 dec.	$x = 0\,(1')\,90°$	[353]
6 dec.	$x = 0\,(1'')\,1°\,20'$	[479]
6 dec.	$x = 0\,(1'')\,5°\,(10'')\,90°$	[20] [181]
6 dec.	$x = 0\,(10'')\,90°$	[251]

6	dec.	$x = 0\,(1'')\,2°\,(15'')\,4°\,(30'')\,90°$	[79] [127] [128]
6	dec.	$x = 0\,(1')\,90°$	[160] [272] [313] [389] [397] [399]
5	dec.	$x = 0\,(1'')\,2°$	[271] [287]
5	dec.	$x = 0\,(10'')\,6°\,(1')\,84°\,(10'')\,90°$	[122] [429]
5	dec.	$x = 0\,(1'')\,5'\,(10'')\,4°\,(1')\,86°\,(10'')\,90°$	[156]
5	dec.	$x = 0\,(1'')\,1°\,(10'')\,8°\,(1')\,82°\,(10'')\,89°\,(1'')\,90°$	[101] [231]
5	dec.	$x = 0\,(15'')\,90°$	[271]
5	dec.	$x = 0\,(1'')\,1'\,(1')\,90°$	[147]
5	dec.	$x = 0\,(30'')\,6°\,(1')\,90°$	[305]
5	dec.	$x = 0\,(1')\,90°$	[2] [6] [8] [9] [16] [29] [34] [39] [60] [71] [99] [100] [107] [126] [129] [192] [203] [226] [248] [262] [287] [291] [298] [324] [352] [358] [359] [384] [395] [400]
5	dec.	$x = 0\,(3')\,90°$	[198]
5	dec.	$x = 0\,(1')\,18°\,(6')\,72°\,(1')\,90°$	[165]
5	dec.	$x = 0\,(6')\,90°$	[50]
5	dec.	$x = 0\,(1')\,6°\,(10')\,84°\,(1')\,90°$	[305]
5	dec.	$x = 0\,(10')\,90°$	[69] [83] [142] [428]
4	dec.	$x = 0\,(10'')\,90°$	[53]
4	dec.	$x = 0\,(1'')\,20'\,(10'')\,2°$	[230]
4	dec.	$x = 1°59'\,(10'')\,5°\,(1')\,85°\,(10'')\,88°1'$	[366]
4	dec.	$x = 0\,(1')\,90°$	[15] [33] [55] [78] [149] [153] [230] [310]
4	dec.	$x = 3°\,(1')\,87°$	[388]
4	dec.	$x = 0\,(1')\,18°$	[195]
4	dec.	$x = 0\,(1')\,8°$	[356]
4	dec.	$x = 0\,(2')\,7°$	[368]
4	dec.	$x = 0\,(2')\,5°$	[373]
4	dec.	$x = 0\,(5')\,90°$	[43]
4	dec.	$x = 0\,(1')\,14°\,(6')\,76°\,(1')\,90°$	[19]
4	dec.	$x = 0\,(1')\,9°\,(6')\,81°\,(1')\,90°$	[4] [85]
4	dec.	$x = 0\,(6')\,90°$	[37] [49] [66] [75] [94] [132] [133] [145] [154] [193] [282] [283] [286] [299] [350] [367]

4 dec.	$x = 0\,(1')\,10°\,(10')\,80°\,(1')\,90°$	[76]
4 dec.	$x = 1°\,(1')\,9°\,(10')\,81°\,(1')\,90°$	[88]
4 dec.	$x = 1°\,(1')\,8°\,(10')\,82°\,(1')\,89°$	[44]
4 dec.	$x = 0\,(1')\,6°\,(10')\,84°\,(1')\,90°$	[281]
4 dec.	$x = 0\,(1')\,5°\,(10')\,85°\,(1')\,90°$	[315]
4 dec.	$x = 0\,(10')\,90°$	[11] [30] [40] [54] [69] [97] [98] [111] [141] [358] [368] [373] [414] [437]
4 dec.	$x = 0\,(2')\,7°;\ 5°\,(10')\,85°;\ 83°\,(2')\,90°$	[82] [435]
4 dec.	$x = 0\,(5')\,3°\,(10')\,87°\,(5')\,90°$	[140]
4 dec.	$x = 0\,(1°)\,90°$	[134] [146] [248] [438]
3 dec.	$x = 0\,(10')\,90°$	[38]

Argument in tenths of a degree

10 dec.	$x = 0\,(0°{,}01)\,90°$	[179] [354]
10 dec.	$x = 2°\,(0°{,}001)\,88°$	[339]
9 dec.	$x = 0°{,}7\,(0°{,}001)\,2°{,}1;\ 87°{,}9\,(0°{,}001)\,89°{,}3$	[339]
8 dec.	$x = 0\,(0°{,}001)\,0°{,}7;\ 89°{,}3\,(0°{,}001)\,90°$	[339]
8 dec.	$x = 0\,(0°{,}001)\,3°;\ 0\,(0°{,}01)\,90°$	[325]
7 dec.	$x = 0\,(0°{,}001)\,90°$	[337]
6 dec.	$x = 0\,(0°{,}001)\,90°$	[336]
6 dec.	$x = 0\,(0°{,}01)\,90°$	[328]
5; 6 dec.	$x = 0\,(0°{,}01)\,5°;\ 5°\,(0°{,}01)\,90°$	[432]
5 dec.	$x = 0\,(0°{,}001)\,5°$	[355]
5 dec.	$x = 0\,(0°{,}01)\,90°$	[182] [218] [430]
5 dec.	$x = 5°\,(0°{,}01)\,85°$	[381]
5 dec.	$x = 0\,(0°{,}1)\,90°$	[352]
5 dec.	$x = 0\,(0°{,}5)\,90°$	[198]
4 dec.	$x = 0\,(0°{,}01)\,90°$	[109] [300]
4 dec.	$x = 0\,(0°{,}01)\,5°$	[66] [356]
4 dec.	$x = 0\,(0°{,}01)\,11°\,(0°{,}1)\,79°\,(0°{,}01)\,90°$	[195]
4 dec.	$x = 0\,(0°{,}01)\,10°\,(0°{,}1)\,80°\,(0°{,}01)\,90°$	[189]
4 dec.	$x = 0\,(0°{,}01)\,8°\,(0°{,}1)\,82°\,(0°{,}01)\,90°$	[180]
4 dec.	$x = 0\,(0°{,}01)\,5°\,(0°{,}1)\,85°\,(0°{,}01)\,90°$	[206] [238]
4 dec.	$x = 0\,(0°{,}01)\,4°\,(0°{,}1)\,90°$	[310]
4 dec.	$x = 0\,(0°{,}01)\,1°\,(0°{,}1)\,90°$	[432]
4 dec.	$x = 0\,(0°{,}1)\,90°$	[17] [363]
4 dec.	$x = 0\,(0°{,}1)\,5°$	[438]
3 dec.	$x = 0\,(0°{,}1)\,90°$	[56] [72]
3 dec.	$x = 0\,(0°{,}1)\,16°\,(1°)\,74°\,(0°{,}1)\,90°$	[118]

Argument in radians

5	dec.	lg tg x	$x = 0\,(0{,}001)\,1{,}569$	[308]
5	dec.		$x = 0\,(0{,}01)\,1{,}6$	[352]
4	dec.	lg tg x	$x = 0\,(0{,}0001)\,0{,}04$	[310]
4	dec.		$x = 0\,(0{,}001)\,1{,}57$	[310]
4	dec.	lg tg x	$x = 0\,(0{,}001)\,0{,}2$	[195]
4	dec.		$x = 0\,(0{,}0029)\,1{,}5708$	[40]
4	dec.	lg tg x	$x = 0\,(0{,}01)\,3{,}2$	[160]
4	dec.		$x = 0\,(0{,}01)\,1{,}6$	[190] [248]
4	dec.	lg tg x	$x = 0\,(0{,}01)\,1{,}57$	[75] [195]

Argument in grads

17	dec.	$x = 0\,(0^g{,}001)\,3^g;\ 97^g\,(0^g{,}001)\,100^g$	[233]
17	dec.	$x = 0\,(1^g)\,50^g$	[162]
11	dec.	$x = 0\,(0^g{,}001)\,0^g{,}1(0^g{,}1)99^g{,}9\,(0^g{,}001)100^g$	[176]
8	dec.	$x = 0\,(0^g{,}001)\,100^g$	[375] [401]
7	dec.	$x = 0\,(0^g{,}0001)\,3^g\,(0^g{,}001)\,100^g$	[342]
7	dec.	$x = 0\,(0^g{,}001)\,3^g\,(0^g{,}01)\,97^g\,(0^g{,}001)\,100^g$	[176] [257]
7	dec.	$x = 0\,(0^g{,}01)\,100^g$	[188]
6	dec.	$x = 0\,(0^g{,}001)\,20^g\,(0^g{,}01)\,80^g\,(0^g{,}001)\,100^g$	[280]
5	dec.	$x = 0\,(0^g{,}001)\,3^g$	[92]
5	dec.	$x = 0(0^g{,}001)1^g{,}2(0^g{,}01)98^g{,}8(0^g{,}001)100^g$	[431]
5	dec.	$x = 0\,(0^g{,}001)\,3^g\,(0^g{,}01)\,100^g$	[232]
5	dec.	$x = 0\,(0^g{,}001)\,1^g$	[290]
5	dec.	$x = 0\,(0^g{,}01)\,100^g$	[92] [239] [290] [374]
5	dec.	$x = 3^g\,(0^g{,}01)\,97^g$	[233]
5	dec.	$x = 0\,(0^g{,}01)\,10^g\,(0^g{,}1)\,100^g$	[433]
4	dec.	$x = 0\,(0^g{,}1)\,100^g$	[261]
3	dec.	$x = 0\,(0^g{,}1)\,100^g$	[72]

Argument in units of time

7	dec.	$x = 0\,(0^{sec}01)\,6^{hrs}$	[379]
7	dec.	$x = 0\,(0^{sec}1)\,10^{m}\,(1^{sec})\,6^{hrs}$	[32]
7	dec.	$x = 0\,(1^{sec})\,6^{hrs}$	[256]
6	dec.	$x = 0\,(1^{sec})\,6^{hrs}$	[187]
6	dec.	$x = 0\,(2^{sec})\,6^{hrs}$	[349]
5	dec.	$x = 0\,(0^{sec}{,}1)\,8^{m}\,(1^{sec})\,6^{hrs}$	[334]

5 dec.	$x = 0\,(1^{sec})\,6^{hrs}$	[270]
5 dec.	$x = 0\,(4^{sec})\,60^{M}$	[100]
5 dec.	$x = 0\,(4^{sec})\,6^{hrs}$	[185] [202]
5 dec.	$x = 0\,(20^{M})\,6^{hrs}$	[101]
4 dec.	$x = 0\,(4^{sec})\,6^{hrs}$	[80]
4 dec.	$x = 0\,(1^{sec})\,20^{M}\,(10^{sec})\,6^{hrs}$	[196] [206]
4 dec.	$x = 0\,(10^{sec})\,6^{hrs}$	[422]

Argument in protractor divisions

5 dec.	$x = 0\,(0{-}01)\,15{-}00$	[26]
4 dec.	$x = 0\,(0{-}01)\,15{-}00$	[101]
4 dec.	$x = 0\,(0{-}02)\,15{-}00$	[43]

lg sec x & lg cosec x

Argument in degrees, minutes and seconds

11 dec.	$x = 0\,(1'')\,10'$	[460]
10 dec.	$x = 0\,(1')\,90°$	[269] [323]
7 dec.	$x = 0\,(1')\,90°$	[119] [123] [205] [347] [353]
6 dec.	$x = 0\,(30'')\,90°$	[79] [127] [128]
6 dec.	$x = 0\,(1')\,90°$	[210]
6 dec.	$x = 0\,(1')\,79°49'$	[164]
5 dec.	$x = 0\,(30'')\,90°$	[264]
5 dec.	$x = 0\,(1')\,90°$	[34] [101] [129] [158] [192] [262] [298] [374]
5 dec.	$x = 0\,(3')\,90°$	[198]
4 dec.	$x = 0\,(6')\,90°$	[154]
4 dec.	$x = 0\,(10')\,80°\,(2')\,90°$	[435]
4 dec.	$x = 0\,(1°)\,90°$	[248]

Argument in grads

7 dec.	$x = 0\,(0^{g}{\cdot}001)\,3^{g}\,(0^{g}{\cdot}01)$ $97^{g}\,(0^{g}{\cdot}001)\,100^{g}$	[176]
5 dec.	$x = 0\,(0^{g}{\cdot}01)\,100^{g}$	[374]
4 dec.	$x = 0\,(0^{g}{\cdot}1)\,100^{g}$	[261]

Argument in units of time

5 dec.	$x = 0\,(20^{M})\,6^{hrs}$	[101]
4 dec.	$x = 0\,(4^{sec})\,6^{hrs}$	[80]

Argument in protractor divisions

4 dec.	$x = 0\,(0{-}01)\,15{-}00$	[101]

$\lg \operatorname{cosec} \frac{x}{2}$

6 dec. $x = 0\,(1')\,180°$ [164]

Logarithms of expressions containing trigonometric functions

$$S = \lg \frac{\sin x}{x} \;\&\; T = \lg \frac{tg\, x}{x}$$

Argument in degrees, minutes and seconds

22 dec.	T	$x = 0\,(1'')\,10'$	[338]
15 dec.		$x = 0\,(9')\,3°$	[162]
14 dec.		$x = 0\,(10'')\,3°$	[162]
10 dec.		$x = 0\,(10'')\,4°30'$	[411]
8 dec.		$x = 0\,(1'')\,5°$	[168]
8 dec.		$x = 0\,(5'')\,1'40''\,(10'')\,3°$	[365]
8 dec.		$x = 0\,(10'')\,8'\,(5'')\,16'40''\,(1'')\,20'\,(10'')\,2°46'40''$	[327]
8 dec.		$x = 0\,(50'')\,16'40''\,(10'')\,2°46'30''$	[24] [184]
7 dec.		$x = 0\,(1'')\,2°$	[333]
7 dec.		$x = 0\,(10'')\,5°33'20''$	[87]
7 dec.		$x = 0\,(50'')\,16'40''\,(10'')\,2°46'30''$	[419]
6 dec.		$x = 0\,(10'')\,5°33'20''$	[87]
6 dec.		$x = 0\,(50'')\,16'40''\,(10'')\,2°46'30''$	[181]
6 dec.		$x = 0\,(1')\,5°$	[279]
6 dec.		$x = 1'\,(1')\,3°4'$	[156]
6 dec.		$x = 0\,($ var. $)\,3°4'$	[479]
6 dec.		$x = 0\,(1')\,2°$	[160] [259]
6 dec.		$x = 0\,(5')\,1°40'$	[169]
5 dec.		$x = 0\,(1')\,6°$	[59]
5 dec.		$x = 0\,(1')\,4°$	[346]
5 dec.		$x = 0\,(1')\,3°$	[231]
5 dec.	S	$x = 0\,($ var. $)\,3°1'$	[255]
	T	$x = 0\,($ var. $)\,3°4'$	[255]
4 dec.		$x = 0\,(1°)\,10°$	[345]
4 dec.	S	$x = 0\,($ var. $)\,5°6',6$	[310]
	T	$x = 0\,($ var. $)\,3°26',7$	[310]

Argument in tenths of a degree

10 dec.	$x = 0\,(0°,001)\,2°,1$	[335]
8 dec.	$x = 0\,(0°,001)\,0°,3$	[339]

8 dec.		$x = 0\,(0°,01)\,3°$	[325]
7 dec.		$x = 0\,(0°,001)\,1°$	[337] [381]
6 dec.		$x = 0\,(0°,001)\,0°,4$	[336]
6 dec.		$x = 0\,(0°,01)\,5°,1$	[381] [432]
6 dec.		$x = 0\,(0°,1)\,4°$	[182]
6 dec.		$x = 0$ (**var.**) $1°,503$	[479]
5 dec.		$x = 0\,(0°,1)\,5°$	[87]
5 dec.		$x = 0°,1\,(0°,2)\,0°,9$	[206]
4 dec.		$x = 0\,(0°,1)\,10°$	[180]
4 dec.	S	$x = 0$ (**var.**) $4°,890$	[310]
	T	$x = 0$ (**var.**) $3°,602$	

Argument in radians

6 dec.		$x = 0$ (**var.**) $0,02615$	[479]
4 dec.	S	$x = 0$ (**var.**) $0,08715$	[310]
	T	$x = 0$ (**var.**) $0,06161$	

Argument in grads

17 dec.	S	$x = 0\,(1^g)\,50^g$	[162]
8 dec.		$x = 0\,(0^g,01)\,5^g$	[375]
7 dec.		$x = 0\,(0^g,001)\,3^g$	[342]
7 dec.		$x = 0\,(0^g,01)\,5^g$	[188]
6 dec.		$x = 0\,(0^g,01)\,2^g,5$	[280]
5 dec.		$x = 0\,(0^g,01)\,2^g$	[232]
5 dec.		$x = 0\,(0^g,01)\,4^g$	[346] [374]

Argument in units of time

7 dec.		$x = 0\,(10^{sec})\,20^{M}$	[256]
6 dec.		$x = 0$ (**var.**) $6^{M},01$	[479]
5 dec.		$x = 0\,(4^{sec})\,24^{M}$	[59]
5 dec.		$x = 0\,(10^{sec})\,20^{M}$	[334]

$Sh = \lg \frac{\operatorname{sh} x}{x}$ & $Th = \lg \frac{\operatorname{th} x}{x}$

6 dec.		$x = 0\,(0,001)\,0,4$	[479]
	Sh	$x = 0$ (**var.**) $0,02615$	
	Th	$x = 0$ (**var.**) $0,2621$	

$$\lg\left(\frac{\sin 2\pi x}{x}\right) \quad \& \quad \lg\left(\frac{\operatorname{tg} 2\pi x}{x}\right)$$

8 dec.	$x = 0\,(0{,}000\,001)\,0{,}012\,499$	[306]

$$\lg \frac{1}{2}(x - \sin x)$$

5 dec.	$x = 0\,(1')\,180°$	[383]

$$\lg (1 - \cos x)$$

7 dec.	$x = 0\,(15'')\,135°$	[221]
7 dec.	$x = 0\,(1')\,90°$	[211] [268]
3; 4 dec.	$x = 0\,(1')\,12°;\; 12°\,(1')\,180$	[349]

$$\lg \frac{1}{2}(1 - \cos x)$$

6 dec.	$x = 0\,(15'')\,135°\,(1')\,180°$	[267]
5; 6 dec.	$x = 0\,(1^{sec})\,45^{m};\; 45^{m}\,(1^{sec})\,2^{hr}25^{m}$	[249]
7 dec.	$x = 2^{hr}25^{m}\,(1^{sec})\,12^{hr}$	
5 dec.	$x = 0\,(15'')\,10°59'45''$	[349]
6 dec.	$x = 11°\,(15'')\,180°$	
5 dec.	$x = 0\,(30'')\,15°\,(15'')\,120°$ $(30'')\,135°\,(1')\,180°$	[202]
4 dec.	$x = 0\,(30'')\,1°$	[185]
5 dec.	$x = 1°\,(30'')\,120°\,(1')\,180°$	

$$\lg \sin \frac{x^2}{2}$$

5 dec.	$x = 0\,(15'')\,12°\,(1')\,100°\,(2')\,180°$	[298]

$$\lg \sin^2 \frac{x}{2}$$

6 dec.	$x = 0\,(15'')\,180°$	[79]
6 dec.	$x = 0\,(1^{sec})\,12^{hr}$	[153]
5 dec.	$x = 0\,(1^{sec})\,8^{hr}\,(10^{sec})\,12^{hr}$	[129] [156]
4 dec.	$x = 0\,(4^{sec})\,12^{hr}$ or $x = 0\,(1')\,90°$	[80]

$$\lg (\sec x \sec y)$$

3 dec.	$x = 0\,(1°)\,65°.\quad y = 0\,(1°)\,90°$	[212]
3 dec.	$x = 0\,(1°)\,90°.\quad y = 0\,(1°)\,90°$	[264]

$$\lg \frac{2\sin^2 \frac{x}{2}}{\sin 1''}$$

5 dec.	$x = 0\ (0^{sec},1)\ 10^{M}\ (1^{sec})\ 40^{M}$	[153]

$$\lg\left(2\sin^2 \frac{x}{2} \operatorname{cosec} 1''\right)$$

5; 6 dec.	$x = 0\ (1^{sec})\ 30^{M};\ 30^{M}\ (1^{sec})\ 1^{hr}$	[161]
5; 6 dec.	$x = 0\ (1^{sec})\ 30^{M};\ 30^{M}(1^{sec})\ 2^{hr}$	[159]
5 dec.	$x = 0\ (1^{sec})\ 30^{M}$	[191]
5 dec.	$x = 0\ (1^{sec})\ 42^{M}$	[331]

$$\lg\left(2\sin^4 \frac{x}{2} \operatorname{cosec} 1''\right)$$

5 dec.	$x = 10^{M}\ (10^{sec})\ 42^{M}$	[331]
4 dec.	$x = 20^{M}\ (10^{sec})\ 2^{hr}$	[159]
4 dec.	$x = 20^{M}\ (10^{sec})\ 1^{hr}$	[161]
4 dec.	$x = 0\ (1^{M})\ 12^{M}\ (10^{sec})\ 30^{M}$	[191]

$$\lg \frac{\sin x}{x - \sin x \cos x}$$

7 dec.	$x = 10°\ (10°)\ 540°$	[454]

$$\lg \sqrt{1 - \sin^2 x \sin^2 y}$$

5—6 dec.	$x = 25°\ (1°)\ 90°.$	$y = 0\ (1°)\ 90°$	[426]
5—6 dec.	$x = 0\ (1°)\ 90°.$	$y = 0\ (1°)\ 90°$	[155]
5 dec.	$x = 30°\ (15')\ 90°.$	$y = 0\ (1^{M})\ 6^{hr}$	[344]
5 dec.	$x = 23°\ (1°)\ 90°.$	$y = 0\ (4^{M})\ 6^{hr}$	[382]
5 dec.	$x = 25°\ (1°)\ 90°.$	$y = 0\ (1°)\ 90°$	[212]
5 dec.	$x = 0\ (1°)\ 90°.$	$y = 0\ (1°)\ 90°$	[264]

Logarithms of the sums and differences of numbers

Tables for calculating $\lg(a \pm b)$ compiled by Gauss are used to determine logarithms of the sums and differences of two numbers expressed in logarithms without reference to the numbers.

The tables can be arranged in various ways; we have adopted the arrangement proposed by Zech.

Taking logs of $a \pm b\,(a > b)$, we obtain

$$\lg(a+b) = \lg a + \lg\left(1 + \frac{1}{\frac{a}{b}}\right) = \lg a + A;$$

$$\lg(a-b) = \lg a + \lg\left(1 - \frac{1}{\frac{a}{b}}\right) = \lg a - \lg\left(\frac{1}{1 - \frac{1}{\frac{a}{b}}}\right) = \lg a - S.$$

The logarithms A and S are found from the tables by the argument

$$D = \lg a - \lg b \quad \text{or} \quad {}^*D = \lg b - \lg a.$$

The initial Gauss tables gave

$$B = \lg\left(1 + \frac{1}{\frac{a}{b}}\right) \quad \& \quad C = \lg\left(1 + \frac{a}{b}\right)$$

with the argument $A' = \lg a - \lg b$; it follows that $\lg(a + b) = \lg a + B = \lg b + C$. Direct tabulation of the differences was not considered.

In this Handbook Gauss's notation is brought into line with that of Zech:

$$A' \to D, \quad B \to A, \quad C \to (A + D).$$

Tables for calculating addition logarithms

16	dec.	A	$D = 0\,(0{,}01)\,9$	[445]
15	dec.	A	$D = 2{,}64\,(0{,}01)\,7{,}63$	[348]
14	dec.	$A,\ A+D$	$D = 0\,(0{,}00001)\,0{,}00103$	[296]
11	dec.	A	$D = 3{,}638\,(0{,}001)\,5{,}637$	[460]
9	dec.	A	${}^*D = 3\,(0{,}0005)\,3{,}3\,(0{,}001)$ $5{,}2\,(0{,}01)\,7\,(0{,}1)\,9$	[246]
7	dec.	A	$D = 0\,(0{,}0001)\,2\,(0{,}001)$ $4\,(0{,}01)\,6\,(0{,}1)\,6{,}9$	[144] [418] [439]
7	dec.	A	$D = 0\,(0{,}0001)\,2\,(0{,}001)\,4$ $(0{,}01)\,6{,}94$	[341]
7	dec.	A	${}^*D = 0\,(0{,}0001)\,2\,(0{,}001)$ $4\,(0{,}01)\,6\,(0{,}1)\,7$	[434]
7	dec.	$A+D$	$D = 0\,(0{,}0001)\,4\,(0{,}01)\,6\,(0{,}1)\,7$	[434]
7	dec.	$A,\ A+D$	$D = 0\,(0{,}0001)\,2\,(0{,}001)$ $3\,(0{,}01)\,4\,(0{,}1)\,5\,(1)\,7$	[303]
6	dec.	A	$D = 0\,(0{,}0001)\,1{,}501\,(0{,}001)$ $3{,}01\,(0{,}01)\,5\,(0{,}1)\,6$	[194]
6	dec.	A	$D = 0\,(0{,}0001)\,1{,}5\,(0{,}001)$ $2{,}5\,(0{,}01)\,5{,}09$	[35]

6 dec.	A	*D = 0 (0,0001) 1,3 (0,001) 2,8 (0,01) 5 (0,1) 6	[181]
6 dec.	A + D	D = 0 (0,0001) 0,2509	
6 dec.	A	*D = 0 (0,001) 2,8 (0,01) 5 (0,1) 6	[279]
6 dec.	A + D	D = 0 (0,0001) 0,1999	
6 dec.	A	D = 0 (0,001) 2 (0,01) 4 (0,1) 6	[245]
6 dec.	A + D	D = 0 (0,001) 2	
6 dec.	A	*D = 0(0,0001) 1 (0,001) 3	[241]
6 dec.	A + D	D = 0 (0,0001) 3	
5 dec.	A	D = 0,30103 (0,00001) 0,00001	[477]
5 dec.	A	D = 0 (0,001) 2,01 (0,01) 5	[129] [156] [169]
5 dec.	A	D = 0 (0,001) 1,65 (0,01) 3 (0,1) 5	[34] [262]
5 dec.	A	D = 0 (0,001) 1,65 (0,01) 2,35 (0,05) 3 (0,1) 5	[101]
5 dec.	A	D = 0 (0,001) 2 (0,01) 3,4 (0,05) 3,7 (0,1) 5	[99] [429]
5 dec.	A	D = 0 (0,001) 1,8 (0,01) 3; 4	[106]
5 dec.	A	*D = 0 (0,001) 2(0,01)4(0,1)5	[182]
5 dec.	A + D	D = 0 (0,001) 0,219	
5 dec.	A	*D = 0 (0,001) 2 (0,01) 4 (0,1) 5	[314]
5 dec.	A, A + D	D = 0 (0,001) 2 (0,01) 3,4 (0,1) 5	[228]
4 dec.	A	D = 0 (0,001) 0,9(0,01)2(0,1)3,9	[79] [80] [149] [206] [310]
4 dec.	A	D = 0 (0,001) 0,999 (0,01) 1,99 (0,1) 4	[366]
4 dec.	A	D = 0 (0,001) 0,9 (0,01)2(0,1)3	[53] [75] [88]
4 dec.	A	D = 0 (0,01) 3 (0,1) 4	[315]
3 dec.	A	$D = \bar{1}\,(0{,}002)\,\bar{1}{,}998;\ \bar{2}\,(0{,}02)\,\bar{2}{,}98$	[72]

Tables for calculating many-decimal logarithms and antilogarithms

16 dec.	S	D = 0 (0,01) 9	[445]
7 dec.	S	$D = 0\,(10^{-7})\,0{,}0003\,(10^{-6})\,0{,}05\,(10^{-5})\,0{,}30299$	[144] [418] [439]
7 dec.	S	D = 0,3 (0,0001) 2 (0,001) 4 (0,01) 6,94	[341]
6 dec.	S	D = 0,3 (0,0001) 1,501 (0,001) 3,01 (0,01) 5 (0,1) 6	[194]
6 dec.	S	D = 0,3 (0,0001) 0,45 (0,0001) 1,5 (0,001) 2,7 (0,01) 5	[35]
6 dec.	−S	D = 0,4 (0,0001) 1,301 (0,001) 2,81 (0,01) 5,1 (0,1) 6	[181]

6	dec.	$-S$	$D = 0{,}4\,(0{,}0001)\,0{,}5\,(0{,}001)\,2{,}8\,(0{,}01)\,5\,(0{,}1)\,5{,}9$	[279]
6	dec.	$-S$	$*D = 0{,}1001\,(0{,}0001)\,1\,(0{,}001)\,3$	[241]
5	dec.	S	$D = 0{,}30103\,(0{,}00001)\,0{,}00001$	[477]
5	dec.	S	$D = 0\,(0{,}0001)\,0{,}5\,(0{,}001)\,2\,(0{,}01)\,5$	[129] [156]
5	dec.	S	$D = 0{,}3\,(0{,}0001)\,0{,}501\,(0{,}001)\,2{,}01\,(0{,}01)\,5$	[169]
5	dec.	S	$D = 0{,}3\,(0{,}0001)\,0{,}48\,(0{,}001)\,1{,}5\,(0{,}01)\,3{,}1\,(0{,}05)\,3{,}2\,(0{,}1)\,5$	[34] [101] [262]
5	dec.	S	$D = 0{,}3\,(0{,}0001)\,0{,}509\,(0{,}001)\,1{,}5\,(0{,}01)\,3{,}4\,(0{,}05)\,3{,}7\,(0{,}1)\,5$	[99]
5	dec.	S	$D = 0\,(0{,}00001)\,0{,}01\,(0{,}0001)\,0{,}1\,(0{,}001)\,0{,}302$	[106]
5	dec.	$-S$	$D = 0{,}42\,(0{,}001)\,2{,}01\,(0{,}01)\,3{,}1\,(0{,}1)\,5$	[182]
5	dec.	S	$*D = 0\,(0{,}00001)\,0{,}051\,(0{,}0001)\,0{,}4\,(0{,}001)\,1{,}8\,(0{,}01)\,4\,(0{,}1)\,5$	[314]
4	dec.	S	$D = 0{,}2\,(0{,}001)\,1{,}21\,(0{,}01)\,3\,(0{,}1)\,3{,}9$	[206]
4	dec.	S	$D = 0{,}2\,(0{,}001)\,1\,(0{,}01)\,3\,(0{,}1)\,3{,}9$	[310]
4	dec.	S	$D = 0\,(0{,}0001)\,0{,}1\,(0{,}001)\,1\,(0{,}01)\,3{,}09$	[80]
4	dec.	S	$D = 0\,(0{,}0001)\,0{,}2999\,(0{,}001)\,1{,}099\,(0{,}01)\,1{,}99\,(0{,}1)\,3{,}9$	[366]
4	dec.	S	$D = 0\,(0{,}0001)\,0{,}1\,(0{,}001)\,0{,}5\,(0{,}01)\,2\,(0{,}1)\,3$	[88]
4	dec.	S	$D = 0\,(0{,}001)\,1\,(0{,}01)\,3\,(0{,}1)\,3{,}9$	[79] [149]
4	dec.	S	$D = 0\,(0{,}001)\,0{,}9\,(0{,}01)\,2\,(0{,}1)\,3$	[53]
4	dec.	S	$D = 0{,}3\,(0{,}001)\,1{,}1\,(0{,}01)\,3$	[75]
4	dec.	S	$D = 0\,(0{,}01)\,3\,(0{,}1)\,4$	[315]
3	dec.	S	$D = \bar{1}\,(0{,}002)\,\bar{1}{,}998;\ \bar{2}\,(0{,}02)\,\bar{2}{,}98$	[72]

Weidenbach's logarithm

Tables of Weidenbach's logarithm $W = \lg \frac{10^D + 1}{10^D - 1} = \lg \frac{1 + 10^{-D}}{1 - 10^{-D}}$ with argument D can be considered as tables of $\lg \frac{x+1}{x-1}$ with argument $\lg x$ or as tables of $\lg \frac{1+x}{1-x}$ with argument $-\lg x$ and may be used, for example, to calculate $\operatorname{tg} \frac{A-B}{2} = \frac{a-b}{a+b} \operatorname{ctg} \frac{C}{2}$.

6	dec.	$D = 0{,}00991\,(0{,}00001)\,0{,}1\,(0{,}0001)\,1{,}5\,(0{,}001)\,2{,}7\,(0{,}01)\,4\,(0{,}1)\,7$	[247]
4–5	dec.	$D = 0\,(0{,}00001)\,0{,}0001\,(0{,}000001)\,0{,}005\,(0{,}00001)\,0{,}01$	[247]

5 dec.	$D = 0{,}00001\,(0{,}00001)\,0{,}0001\,(0{,}0001)\,0{,}001\,(0{,}001)\,1\,(0{,}01)\,3\,(0{,}1)\,5$	[261]
5 dec.	$D = 0{,}382\,(0{,}001)\,2{,}002;\ 2\,(0{,}01)\,3{,}6\,(0{,}1)\,5{,}5$	[427]
4 dec.	$D = 0{,}001\,(0{,}001)\,1\,(0{,}01)\,2\,(0{,}1)\,3$	[451]
3 dec.	$*D = \bar{1}\,(0{,}002)\,\bar{1}{,}998;\ \bar{2}\,(0{,}02)\,\bar{2}{,}98$	[72]

Tables for calculating many-decimal logarithms and antilogarithms

The commonest method for calculating many-decimal logarithms is to resolve the number into a factor of the type $(1 \pm m10^{-n})$, logarithms for which are given in the tables. Tables with one, two, three and more figure arguments are distinguished in accordance with the values of m [$m = 1\,(1)\,9$, $m = 1\,(1)\,99$, $m = 1\,(1)\,999$] and so on.

In addition to logarithms of factors, many-decimal logarithms (or cologarithms) of a limited number of integers k are given.

Tables for calculating many-decimal logarithms

Four- or three-place argument

24 dec.	$1 + m10^{-n}$	$m = 1\,(1)\,999.$	$n = 3\,(3)\,15$; $k = 1\,(1)\,9$	[240]
21 dec.	$1 + m10^{-n}$	$m = 1\,(1)\,9999.$	$n = 7\,(4)\,15$; $k = 1000\,(1)\,9999$	[393]
20 dec.	$1 + m10^{-n}$	$m = 1\,(1)\,1000.$	$n = 7\,(3)\,13$	[465]
15 dec.	$1 + m10^{-n}$	$m = 1\,(1)\,999.$	$n = 5;\ 8$; $k = 1\,(1)\,999$	[392]
14 dec.	$1 + m10^{-n}$	$m = 1\,(1)\,999.$	$n = 6;\ 9$; $k = 11\,(1)\,99$	[219]
13 dec.	$1 + m10^{-n}$	$m = 1\,(1)\,1000.$	$n = 5$; $k = 100\,(1)\,999$	[163]
12 dec.	$1 + m10^{-n}$	$m = 1\,(1)\,999.$	$n = 3\,(3)\,9$; $k = 1\,(1)\,9$	[67] [242]
11 dec.	$1 + m10^{-n}$	$m = 1\,(1)\,9999.$	$n = 6$; $k = 1\,(1)\,999$	[391]
10 dec.	$1 + m10^{-n}$	$m = 1\,(1)\,10999.$	$n = 6$; $k = 1\,(1)\,999$	[343]
10 dec.	$1 + m10^{-n}$	$m = 1\,(1)\,999.$	$n = 5$; $k = 100\,(1)\,999$	[370] [372]

Two-place argument

16 dec.	$1 \pm m10^{-n}$	$m = 1\,(1)\,99.$	$n = 2\,(2)\,10$ $k = 1\,(1)\,9$	[379]
15 dec.	$1 + m10^{-n}$	$m = 1\,(1)\,99.$	$n = 3\,(2)\,9.$ $k = 10\,(1)\,99$	[261] [296]
14 dec.	$1 + m10^{-n}$	$m = 1\,(1)\,99.$	$n = 3.$ $k = 11\,(1)\,99$	[219]
12 dec.	$1 + m10^{-n}$	$m = 1\,(1)\,99.$	$n = 2\,(2)\,12$ $k = 1\,(1)\,9$	[67] [243]

Single-place argument

61 dec.	$1 \pm m10^{-n}$	$m = 1\,(1)\,9.\ n = 3\,(1)\,6.$ $k = 1\,(1)\,100($ primes $)\,1097$	[338]
52 dec.	$1 \pm 10m^{-n}$	$m = 1\,(1)\,9.\ n = 1\,(1)\,26.\ k = 1\,(1)\,9$	[475]
40 dec.		$k = 2\,(1)\,19($ primes $)\,101$ $(1)\,109\,($ primes $)\,151$	[116]
25 dec.	$1 + m10^{-n}$	$m = 1\,(1)\,9.\ n = 3\,(1)\,13$	[116]
37 dec.	$1 + m10^{-n}$	$m = 1\,(1)\,9.\ n = 1\,(1)\,22.\ k = 1(1)9$	[327]
37;36 dec.	$1 \pm m10^{-n}$	$m = 1\,(1)\,9.\ n = 2\,(1)\,19.$ $k = 1\,(1)\,1000$	[213]
33 dec.	$1 + m10^{-n}$	$m = 1\,(1)\,9.\ n = 1\,(1)17.\ k = 1(1)9$	[214]
28 dec.	$1 \pm m10^{-n}$	$m = 1\,(1)\,9.\ n = 1\,(1)14.\ k = 1(1)9$	[338]
27 dec.	$1 \pm m10^{-n}$	$m = 1\,(1)\,9.\ n = 2\,(1)\,14$ $k = 1\,(1)\,110$	[408]

Tables for the calculation of many-decimal logarithms

16 fig.	$m10^{-n}$	$m = 1\,(1)\,9.$	$n = 1\,(1)\,17$	[234]
14 fig.	$m\,10^{-n}$	$m = 1\,(1)\,999.$	$n = 7\,(3)\,13$	[204]
11 fig.	$m10^{-n}$	$m = 1\,(1)\,9.$	$n = 1\,(1)\,8$	[261]
10 fig.	$m10^{-n}$	$m = 1\,(1)\,9.$	$n = 1\,(1)\,8$	[455]

NATURAL LOGARITHMS[1]

Logarithms of integers

155 dec.	$x = 41\,(1)\,97$	[456]
137—148 dec.	$x = 2\,(1)\,11\,(1)\,31;\ 47;\ 53;\ 71;\ 97(1)113$	[409]
81 dec.	$x = 1\,(1)\,25$	[443]

---

[1] See Ch. 3 for tables of natural antilogarithms.

48 dec.	$x = 1\,(1)\,99$ (primes) 1097; 999 980 (1) 1 000 021	[52] [91] [188]
48 dec.	$x = 1\,(1)\,5000$ (primes) 10 000 (1) 10 014	[407]
48 dec.	$x = 1\,(1)\,2200$ (var.) 10 009	[369] [415]
48 dec.	$x = 1\,(1)\,146$ (primes) 10 000	[338]
48 dec.	$x = 1$ (var.) 10 009	[420]
23 dec.	$x = 1\,(1)\,10\,000$	[385]
20 dec.	$x = 1\,(1)\,1200;\ 101\,000\,(1)\,101\,179$	[52] [91] [188]
16 dec.	$x = 1\,(1)\,50\,000$	[402]
16 dec.	$x = 50\,000\,(1)\,100\,000$	[403]
11 dec.	$x = 1\,(1)\,1000$	[176]
10 dec.	$x = 1\,(1)\,1000$ (primes) 10 333	[361]
9—10 dec.	$x = 0\,(1)\,99$	[363]
8 dec.	$x = 1\,(1)\,1000$ (primes) 11 273	[251]
8 dec.	$x = 1\,(1)\,1000$ (primes) 10 000	[288] [416] [417] [418]
8 dec.	$x = 1\,(1)\,10\,000$	[167] [461]
7 dec.	$x = 1\,(1)\,20\,000$	[205]
7 dec.	$x = 1\,(1)\,1200$	[268]
7 dec.	$x = 1\,(1)\,1000$	[200]
6 dec.	$x = 1\,(1)\,21\,600$	[399]
6 dec.	$x = 1\,(1)\,1218$ (primes) 10 000	[279]
6 dec.	$x = 1\,(1)\,10$	[173]
5; 8 dec.	$x = 1\,(1)\,250;\ 250\,(1)\,1000$	[106]
5 dec.	$x = 1\,(1)\,10\,800$	[346]
5 dec.	$x = 1\,(1)\,10\,000$	[390] [425]
5 dec.	$x = 1\,(1)\,1500$	[112] [143] [156] [226] [267]
5 dec.	$x = 1\,(1)\,1100$	[120] [231] 267] [287]
5 dec.	$x = 0\,(1)\,1020$	[271]
5 dec.	$x = 1\,(1)\,1000$	[5] [8] [61] [170] [175] [190] [220] [226] [298] [353] [384]
5 dec.	$x = 1\,(1)\,409$	[398]
5 dec.	$x = 1\,(1)\,100$	[30] [203] [358]
4 dec.	$x = 150\,(1)\,500\,(10)\,1500\,(100)\,6000$	[310]
4 dec.	$x = 1\,(1)\,1199$	[83]
4 dec.	$x = 1\,(1)\,1000$	[17] [21] [22] [23] [39] [42] [68] [174] [189] [356] [388]

4 dec.	$x = 100\,(1)\,1000$	[133]
4 dec.	$x = 10\,(1)\,209$	[330]
4 dec.	$x = 1\,(1)\,100$	[282] [366] [368]
4 dec.	$x = 1\,(1)\,25$	[373]
4 dec.	$x = 1\,($ var. $)\,9909$	[18] [58]
3 dec.	$x = 1\,(1)\,100$	[55] [286] [299] [367] [438]

Logarithms of decimal fractions

16 dec.	$x = 0\,(0{,}0001)\,5$	[404]
16 dec.	$x = 5\,(0{,}0001)\,10$	[405]
20; 10 dec.	$x = 0\,(0{,}00001)\,0{,}001:\,0{,}001\,(0{,}0001)$ $0{,}1\,(0{,}001)\,3(0{,}01)10(0{,}1)20(1)50$	[254]
10 dec.	$x = 1\,(0{,}001)\,10$	[41]
7 dec.	$x = 1\,(0{,}01)\,10$	[227] [268] [292] [293] [318] [464]
7 dec.	$x = 1000\,(0{,}1)\,10\,500$	[200]
6 dec.	$x = 0{,}01\,(0{,}001)\,10$	[160]
6 dec.	$x = 1\,(0{,}001)\,10{,}009$	[480]
6 dec.	$x = 1\,(0{,}01)\,10$	[360]
5 dec.	$x = 1\,(0{,}001)\,10{,}009$	[265]
5 dec.	$x = 1\,(0{,}001)\,5{,}01\,(0{,}01)\,10{,}09$	[330]
5 dec.	$x = 0\,(0{,}01)\,10{,}09$	[398]
5 dec.	$x = 1\,(0{,}01)\,10{,}09$	[358]
5 dec.	$x = 0\,(0{,}01)\,10$	[190] [253]
5 dec.	$x = 1\,(0{,}01)\,10$	[165] [195] [198]
5 dec.	$x = 0\,(0{,}001)\,1\,(0{,}01)\,10\,(1)\,109$	[363]
5 dec.	$x = 0{,}1\,(0{,}01)\,1\,(0{,}1)\,10\,(1)\,100$	[225]
5 dec.	$x = 0\,(0{,}1)\,100$	[190]
5 dec.	$x = 1\,(0{,}1)\,10{,}9$	[198]
4 dec.	$x = 1\,(0{,}001)\,3\,(0{,}01)\,10$	[218]
4 dec.	$x = 1\,(0{,}001)\,2\,(0{,}01)\,10$	[40] [217]
4 dec.	$x = 1\,(0{,}01)\,10$	[75] [103] [111] [132] [154] [193] [309] [352]
4 dec.	$x = 0{,}001;\,0{,}01;\,0{,}05;\,0{,}1\,(0{,}1)\,0{,}5$	[441]
4 dec.	$x = 10\,(0{,}1)\,100{,}9$	[352]
4 dec.	$x = 0\,(0{,}001)\,1{,}5\,(0{,}01)\,15\,(0{,}1)$ $50\,(1)\,150\,(10)\,600$	[310]
4 dec.	$x = 0\,(0{,}01)\,10\,(1)\,1009$	[248]
4 dec.	$x = 1{,}01\,(0{,}01)\,10\,(0{,}25)\,15\,(0{,}5)$ $20\,(1)\,100\,(100)\,1000$	[71]
4 dec.	$x = 0\,(0{,}1)\,1\,(1)\,100$	[141]
3 dec.	$x = 0\,(0{,}01)\,1$	[330]

Logarithms of prime numbers

155 dec.	$11 \leqslant x \leqslant 101$	[456]
82 dec.	$1 \leqslant x \leqslant 127$	[338]
11 dec.	$1 \leqslant x \leqslant 1063$	[394]
10 dec.	$1 \leqslant x < 1000$	[435]
5 dec.	$1000 < x < 10\,000$	[170]

Cologarithms

7 dec.	$x = 0$(var.)0,001 (0,001) 1	[285]
6 dec.	$x = 1\,(1)\,1000$	[302]
5 dec.	$x = 0\,(0{,}00001)\,0{,}1$	[197]

Logarithms of complex numbers

The logarithm of a complex number $(A+Bi)$ can be calculated from the following expansions:

1. If $A \geqslant B \geqslant 0$,

$$\ln(A+Bi) = \ln A + \ln\left(1+\frac{B}{A}i\right) = \ln A + \frac{1}{2}\ln\left[1+\left(\frac{B}{A}\right)^2\right] + i\,\text{arc tg}\,\frac{B}{A}$$

2. If $A \geqslant -B \geqslant 0$,

$$\ln(A+Bi) = \ln A + \ln\left(1-\left|\frac{B}{A}\right|i\right) = \ln A + \frac{1}{2}\ln\left[1+\left(\frac{B}{A}\right)^2\right] - $$
$$- i\,\text{arc tg}\left|\frac{B}{A}\right|$$

3. If $0 \geqslant B \geqslant A$ or $0 \geqslant -B \geqslant A$,

$$\ln(A+Bi) = \ln(-1) + \ln(-A-Bi)$$

and the calculation is made as in example 1 if $B<0$ and as in example 2 if $B>0$.

4. If $|B|>|A|$,

$$\ln(A+Bi) = \ln i + \ln(B-Ai),$$

which leads to one of the cases 1 - 3.

10 dec.	$\ln x$	$x = 1\,(0{,}001)\,9{,}999$	[41]
10 dec.	$\frac{1}{2}\ln(1+x^2)$; arc tg x	$x = 0\,(0{,}001)\,1$	[41]

Logarithms of certain expressions

$x \ln 10$

5 dec.	$x = -10\,(1)\,10$	[198]
5 dec.	$x = -6\,(1)\,6$	[358]

$-2 \ln x$

4 dec.	$x = 0{,}001; 0{,}01; 0{,}05; 0{,}1\,(0{,}1)\,0{,}5$	[441]

$\ln x\, 10^{-n}$

5 dec.	$n = 1; 2.\ \ x = 0\,(1)\,100$	[358]

$\ln (1 + x\, 10^{-n})$

23 dec.	$n = 4; 8.\ \ x = 1\,(1)\,10\,000$	[385]

$\ln x^{\pm \frac{1}{2}}$

6 dec.	$x = 1\,(1)\,1000$	[302]

$\ln \ln \frac{1}{x}$

3 dec.	$x = 0\,(0{,}001)\,1$	[444]

$\ln (\pi x)$

10 dec.	$x = \frac{1}{4}\left(\frac{1}{4}\right)3$	[254]
5 dec.	$x = \frac{1}{4}\left(\frac{1}{4}\right)3$	[253]

$\ln (x + \sqrt{x^2 + y^2}$

7 dec.	$x = 1\,(1)\,16.\ \ y = 1\,(1)\,16$	[453]

$\frac{x}{\ln x - 1}$

nwn	$x = 0\,(50\,000)\,9\,000\,000$	[235]

$\frac{1}{2} \ln (1 + x^2)$

10 dec.	$x = 0\,(0{,}001)\,1$	[41]

$x^x (1 + \ln x)$ и $-x^{-x} (1 + \ln x)$

7 fig.	$x = 0\,(0{,}01)\,0{,}05\,(0{,}05)\,1{,}5\,(0{,}1)\,3\,(0{,}2)\,4$	[175]

Logarithms of trigonometric functions

$\ln \sin x$ & $\ln \cos x$

5 dec.	$x = 0\,(1')\,90°$	[386]
4 dec.	$x = 0\,(10')\,90°$	[101]

$\ln \operatorname{tg} x$ & $\ln \operatorname{ctg} x$

8 dec.	$x = 45°\,(10'')\,90°$	[412]
8 dec.	$x = 45°\,(1')\,86°\,(10'')\,90°$	[369]
7 dec.	$x = 45°\,(1')\,90°$	[317]
6 dec.	$x = 45°\,(1')\,90°$	[319]
5 dec.	$x = 45°\,(1')\,90°$	[413]
5 dec.	$x = 0\,(1')\,90°$	[386]
4 dec.	$x = 0\,(10')\,90°$	[101]

$\ln \sec x$ & $\ln \operatorname{cosec} x$

8 dec.	$x = 0\,(10'')\,90°$	[412]
8 dec.	$x = 0\,(10'')\,4°\,(1')\,86°\,(10'')\,90°$	[369]
7 dec.	$x = 0\,(1')\,90°$	[317]
7; 6 dec.	$x = 0\,(1')\,1°;\ 1°\,(1')\,90°$	[319]
5 dec.	$x = 0\,(1')\,90°$	[386] [413]
4 dec.	$x = 0\,(10')\,90°$	[101]

Tables for calculating many-decimal logarithms

These tables are similar to the tables for the calculation of many-decimal common logarithms.

137 dec.	$1 + m10^{-n}$	$m = 1\,(1)\,9.\ n = 1\,(1)\,21$	$k = 2\,(1)\,10$	[409]
137 dec.	$1 - m10^{-n}$	$m = 1\,(1)\,9.\ n = 1\,(1)\,69$		

48 dec.	$1 \pm m10^{-n}$	$m = 1(1)9.\ n = 4(1)6.\ k = 2(1)146$ (primes)9973	[338]
33 dec.	$m10^{-n} - \ln(1 + m10^{-n})$	$m = 1(1)9.\ n = 2(1)16$, $k = 2(1)99$	[260]
40 dec.		$k = 2(1)10$	[405]
25 dec.	$1 \pm m10^{-n}$	$m = 1(1)9.\ n = 1(1)13$	[405]
20 dec.	$1 \pm m10^{-n}$	$m = 1(1)9.\ n = 1(1)11.\ k = 2(1)9$	[452]
20 dec.	$1 + m10^{-n}$	$m = 1(1)9.\ n = 1(1)11.\ k = 2(1)9$	[116] [261] [296]
20 dec.	$1 + m10^{-n}$	$m = 1(1)9.\ n = 1(1)10.\ k = 2(1)9$	[393]
16 dec.	$1 + m10^{-n}$	$m = 1(1)9.\ n = 1(1)10.\ k = 2(1)9$	[365]

Chapter 5

FACTORIALS, EULER INTEGRALS AND RELATED FUNCTIONS

Two notations are employed in the literature for the gamma (factorial) function which is examined in this chapter.

$$\Gamma(1+x) = \Pi(x) = \int_0^\infty e^{-t} t^x \, dt, \quad \text{Re}(x) > 0.$$

The integral is a second order Euler integral. An integral with a variable upper limit is known as an incomplete gamma function.

If $x = n$ is a positive integer, then

$$\Gamma(n+1) = 1.2 \ldots (n-1)\, n = n!$$

$$(2n)!! = 2.4 \ldots (2n-2)\, 2n$$

$$(2n-1)!! = 1.3.5 \ldots (2n-3)(2n-1)$$

$$\Gamma\left(n + \frac{1}{2}\right) = \sqrt{\pi}\, \frac{(2n-1)!!}{2^n}.$$

The gamma function has the following properties for any argument:

$$\Gamma(x+1) = x\Gamma(x),$$

$$\Gamma(x)\,\Gamma(1+x) = \frac{\pi}{\sin \pi x}.$$

The function which represents the logarithmic derivative of $\Gamma(1+x)$ is denoted by

$$\psi(x) = \frac{\Gamma'(1+x)}{\Gamma(1+x)}.$$

Use is also made of the function

$$\varphi(x) = \psi(x) + C = \sum_{r=1}^{\infty} \frac{x}{r(r+x)} = \sum_{r=1}^{\infty}\left(\frac{1}{r} - \frac{1}{r+x}\right),$$

---

[1] See pages 403 and 550 for the references to Chapter 5.

where C is the Eular constant, $\varphi(0)=0$, $\psi(0)=-C$.
If n is a positive integer $\varphi(n)=1+\frac{1}{2}+\frac{1}{3}+\cdots+\frac{1}{n}$.

The following notation is adopted for the beta function or first order Euler integral

$$\mathrm{B}(p, q)=\int_0^1 x^{p-1}(1-x)^{q-1}\,dx, \quad \mathrm{Re}(p)>0, \quad \mathrm{Re}(q)>0.$$

The connection between beta and gamma functions is established by the formula

$$\mathrm{B}(p, q)=\frac{\Gamma(p)\Gamma(q)}{\Gamma(p+q)}.$$

The chapter consists of the following sections:

- factorials;
- logarithms of factorials;
- double factorials;
- logarithms of double factorials;
- the gamma function;
- logarithms of the gamma function;
- the generalised gamma function;
- the incomplete gamma function and related integrals;
- the psi function and its derivatives;
- the beta function;
- the incomplete beta function;
- certain constants.

FACTORIALS

$n!$

exact	$n=1\,(1)\,200$	[107]
exact	$n=1\,(1)\,120$	[130]
exact	$n=1\,(1)\,60$	[89]
exact	$n=1\,(1)\,50$	[91] [165]
exact	$n=1\,(1)\,30$	[24] [30] [168]
exact	$n=1\,(1)\,25$	[11] [62] [93]
exact	$n=1\,(1)\,20$	[5] [18] [20] [27] [35] [36] [37] [61] [63] [80] [102] [108]
exact	$n=1\,(1)\,15$	[10]
exact	$n=5\,(1)\,10\,(5)\,30$	[32]
83 fig.	$n=1\,(1)\,60$	[89]
20 fig.	$n=1\,(1)\,200$	[157]
8 fig.	$n=1\,(1)\,200$	[58]

8 fig.	$n = 1\ (1)\ 100$	[2] [3] [19] [29]
7 fig.	$n = 1\ (1)\ 1000$	[167]
6 fig.	$n = 1\ (1)\ 300$	[55]
6 fig.	$n = 1\ (1)\ 50$	[16]
6 fig.	$n = 1\ (1)\ 20$	[56] [104]
5 fig.	$n = 1\ (1)\ 100\ (10)\ 200\ (25)\ 300\ (50)\ 1000\ (500)\ 4000\ (1000)\ 10\,000\ (10\,000)\ 100\,000$	[34]
5 fig.	$n = 1\ (1)\ 100$	[40] [163]
5 fig.	$n = 1\ (1)\ 60$	[94]

$$n \cdot n! = (n+1)! - n!$$

20 fig.	$n = 2\ (1)\ 71$	[143]

$$\frac{1}{n!}$$

475 dec.	$n = 1\ (1)\ 214$	[162]
108 dec.	$n = 1\ (1)\ 74$	[135]
85 dec.	$n = 1\ (1)\ 61$	[165]
55 dec.	$n = 2\ (1)\ 43$	[89]
37 dec.	$n = 2\ (1)\ 12$	[161]
30 dec.	$n = 1\ (1)\ 25$	[11]
30 dec.	$n = 1\ (1)\ 20$	[35] [36] [37] [108]
28 dec.	$n = 13\ (1)\ 50$	[161]
23 dec.	$n = 1\ (1)\ 23$	[62]
21 dec.	$n = 1\ (1)\ 22$	[69]
21 dec.	$n = 1\ (1)\ 20$	[18] [63]
12 fig.	$n = 1\ (1)\ 50$	[87]
8 dec.	$n = 1\ (1)\ 10$	[3]
8 dec.	$n = 1\ (1)\ 9$	[82]
6 fig.	$n = 1\ (1)\ 20$	[27] [41] [56] [104]
5 fig.	$n = 1\ (1)\ 50$	[34]
5 fig.	$n = 1\ (1)\ 20$	[20] [61]

$$\left(\frac{1}{n!}\right)^2$$

105 dec.	$n = 1\ (1)\ 70$	[64]
23 dec.	$n = 1\ (1)\ 14$	[62]

$$\frac{x^n}{n!}$$

9—10 dec.	$n = 2\ (1)\ 12. \quad x = 1{,}1\ (0{,}1)\ 10{,}9$	[96]

$$\frac{\pi^n}{n!}$$

40—14 dec.	$n = 1\ (1)\ 16$	[64]

$$\frac{\left(\frac{\pi}{2}\right)^n}{n!}$$

30 dec.	$n = 1\ (1)\ 20$	[35] [37] [108]
28 dec.	$n = 1\ (1)\ 30$	[53] [54]
24 dec.	$n = 1\ (1)\ 28$	[25]
22 dec.	$n = 1\ (1)\ 26$	[66]

$$\frac{\left(\frac{\pi}{4}\right)^n}{n!}$$

24 dec.	$n = 1\ (1)\ 22$	[89]

$$\frac{1}{n!}\left(\frac{\pi}{180k}\right)^n$$

20 dec.	$k = 1;\ 60;\ 60^2.\ n = 1\ (1)\ 8$	[72]

Logarithms of factorials

$$\lg n!$$

33 dec.	$n = 1\ (1)\ 3000$	[46]
33 dec.	$n = 3050\ (50)\ 10\,000\ (10\,000)\ 100\,000$	[47]
20 dec.	$n = 1\ (1)\ 100$	[106]
19—21 dec.	$n = 1\ (1)\ 1200$	[89]
18 dec.	$n = 1\ (1)\ 1200$	[45] [89]
18 dec.	$n = 1\ (1)\ 120$	[165]
12 dec.	$n = 1\ (1)\ 1000$	[137]
10 dec.	$n = 1\ (1)\ 1200$	[58]
10 dec.	$n = 1\ (1)\ 200$	[168]
10 dec.	$n = 10\ (10)\ 1300\ (50)\ 10\,000$	[33]
8 dec.	$n = 1\ (1)\ 99$	[81] [91]
8 dec.	$n = 1\ (1)\ 30$	[99]
8 dec.	$n = 1\ (1)\ 25$	[4] [13]
8 dec.	$n = 1\ (1)\ 24$	[7] [14] [68]
7 dec.	$n = 1\ (1)\ 1000$	[60] [86]
7 dec.	$n = 1\ (1)\ 300$	[55]
7 dec.	$n = 1\ (1)\ 250$	[49]

7 dec.	$n = 1\,(1)\,20$	[102]
6 dec.	$n = 1\,(1)\,1200$	[52]
6 dec.	$n = 1\,(1)\,1000$	[167]
6 fig.	$n = 1\,(1)\,100$	[40]
6 dec.	$n = 1\,(1)\,100$	[2] [23]
6 dec.	$n = 1\,(1)\,24$	[15]
6 dec.	$n = 1\,(1)\,15$	[10]
5—6 fig.	$n = 1\,(1)\,100$	[82]
5 dec.	$n = 1\,(1)\,250$	[92]
5 dec.	$n = 1\,(1)\,199$	[61]
5 dec.	$n = 1\,(1)\,100$	[27] [41] [62]
5 dec.	$n = 1\,(1)\,60$	[94]
4 dec.	$n = 1\,(1)\,100$	[82] [163]
4 dec.	$n = 1\,(1)\,25$	[12] [93]
4 dec.	$n = 5\,(1)\,10\,(5)\,30\,(10)\,40$	[9]

$$2 \ln [(n-1)!\,(n-2)!\ldots 2!]$$

5—6 fig.	$n = 1\,(1)\,30$	[132]
4—5 fig.	$n = 1\,(1)\,30$	[131]

$$\lg\left(\frac{1}{n!}\right)$$

18 dec.	$n = 1\,(1)\,1200$	[45]

$$\lg\left(\frac{1}{n \cdot n!}\right)$$

10 dec.	$n = 1\,(1)\,71$	[143]

Double factorials

$$(2\,n)!!$$

exact	$n = 1\,(1)\,25$	[62] [91]
5 fig.	$n = 1\,(1)\,50\,(2)\,100\,(10)\,150\,(25)\,250\,(50)\,400\,(100)\,1000\,(500)\,2000\,(1000)\,5000$	[34]

$$(2n-1)!!$$

exact	$n = 1\,(1)\,25$	[62] [91]
exact	$n = 1\,(1)\,20$	[50]
5 fig.	$n = 1\,(1)\,50\,(2)\,100\,(10)\,150\,(25)\,250\,(50)\,400\,(100)\,1000\,(500)\,2000\,(1000)\,5000$	[34]

$$\frac{1}{(2n)!!}$$

5 fig.	$n = 1\ (1)\ 50\ (2)\ 100\ (10)\ 150\ (25)\ 250\ (50)\ 400\ (100)\ 1000\ (500)\ 2000\ (1000)\ 5000$	[34]

$$\frac{1}{(2n-1)!!}$$

5 fig.	$n = 1\ (1)\ 50\ (2)\ 100\ (10)\ 150\ (25)\ 250\ (50)\ 400\ (100)\ 1000\ (500)\ 2000\ (1000)\ 5000$	[34]

$$\frac{2n!!}{(2n-1)!!}$$

5 fig.	$n = 1\ (1)\ 50\ (2)\ 100\ (10)\ 150\ (25)\ 250\ (50)\ 400\ (100)\ 1000\ (500)\ 2000\ (1000)\ 5000$	[34]

$$\frac{(2n-1)!!}{(2n)!!}$$

	$n = 1\ (1)\ 15$	[75]
10 dec.	$n = 1\ (1)\ 11$	[110]
10 dec.	$n = 1\ (1)\ 10$	[28] [62] [63] [74] [109]
5—6 fig.	$n = 1\ (1)\ 50\ (2)\ 100\ (10)\ 150\ (25)\ 250\ (50)\ 400 (100)\ 1000\ (500)\ 2000\ (1000)\ 5000$	[34]
5 dec.	$n = 1\ (1)\ 20$	[15]

$$\frac{(2n-3)!!}{(2n)!!}$$

	$n = 1\ (1)\ 15$	[75]
10 dec.	$n = 1\ (1)\ 11$	[110]
10 dec.	$n = 1\ (1)\ 10$	[28] [62] [63] [74] [109]

$$\frac{(2n-1)!!}{(2n)!!\,(2n+1)}$$

10 dec.	$n = 1\ (1)\ 11$	[110]
10 dec.	$n = 1\ (1)\ 10$	[28] [62] [63] [74] [109]

$$\frac{(2n-3)!!}{(2n)!!\,(2n+1)}$$

10 dec.	$n = 1\ (1)\ 11$	[110]

10 dec.	$n = 1\ (1)\ 10$	[28] [62] [63] [74] [109]

$$2\left\{\frac{(2n-1)!!}{(2n)!!}\right\}^2$$

12 dec.	$n = 1\ (1)\ 11$	[117]

$$\frac{2}{2n+1}\left[\frac{(2n)!!}{(2n-1)!!}\right]^2$$

5 dec.	$n = 1\ (1)\ 50\ (2)\ 100\ (10)\ 150\ (25)\ 250\ (50)\ 400\ (100)\ 1000\ (500)\ 2000\ (1000)\ 5000$	[34]

$$\left[\frac{(2n-3)!!}{(2n)!!}\right]^2$$

exact	$n = 1\ (1)\ 8$	[65]

$$2\left\{\frac{(2n-1)!!}{(2n)!!}\right\}\left\{\frac{[2(n+r)-1]!!}{2(n+r)!!}\right\}$$

12 dec.	$n = 1\ (1)\ 11.\ r = 1\ (1)\ 12$	[117]

Logarithms of double factorials

$$\lg (2n)!!$$

18 dec.	$n = 1\ (1)\ 60$	[165]
8 dec.	$n = 1\ (1)\ 50$	[86] [123]

$$\lg (2n-1)!!$$

20 dec.	$n = 1\ (1)\ 50$	[106]
18 dec.	$n = 1\ (1)\ 60$	[165]
8 dec.	$n = 1\ (1)\ 50$	[86] [123]
8 dec.	$n = 1\ (1)\ 30$	[99]
8 dec.	$n = 1\ (1)\ 25$	[4] [13]
8 dec.	$n = 1\ (1)\ 24$	[7] [14] [68]
6 dec.	$n = 1\ (1)\ 24$	[15]
4 dec.	$n = 1\ (1)\ 25$	[12]

$$\lg \frac{(2n-1)!!}{2^n n!}$$

8 dec.	$n = 1\ (1)\ 15$	[99]

$$\lg \frac{(2n-1)!!}{(2n)!!}$$

8 dec.	$n = 1\ (1)\ 25$	[62] [67] [91]
7 dec.	$n = 1\ (1)\ 12$	[110]
7 dec.	$n = 1\ (1)\ 10$	[28] [74] [109]
7 dec.	$n = 1\ (1)\ 9$	[76]
5 dec.	$n = 1\ (1)\ 20$	[15]

$$\lg \frac{(2n-3)!!}{(2n)!!}$$

7 dec.	$n = 1\ (1)\ 11$	[110]
7 dec.	$n = 1\ (1)\ 10$	[28] [74] [109]
7 dec.	$n = 1\ (1)\ 9$	[76]

$$\lg \frac{(2n-3)!!}{(2n)!!\ (2n+1)}$$

7 dec.	$n = 1\ (1)\ 11$	[110]
7 dec.	$n = 1\ (1)\ 10$	[28] [74] [109]
7 dec.	$n = 1\ (1)\ 9$	[76]
5 dec.	$n = 1\ (1)\ 10$	[15]

$$\lg \frac{n(n+1)\dots(n+r-1)}{r!}$$

7 dec.	$n = \frac{1}{2}\ (1)\ \frac{9}{2}.\quad r = 1\ (1)\ 30$	[116]

$$\lg \left\{2\left[\frac{(2n-1)!!}{(2n)!!}\right]^2\right\}$$

7—10 dec.	$n = 1\ (1)\ 25$	[117]

$$\lg 2\left\{\frac{(2n-1)!!}{(2n)!!}\right\}\left\{\frac{[2(n+r)-1]!!}{[2(n+r)]!!}\right\}$$

10—7 dec.	$n = 1\ (1)\ N.\quad r = 1\ (1)\ 15$	[117]

N depends on r and is not less than 15.

THE GAMMA FUNCTION

$$\Gamma(1+x)$$

12 dec.	$x = 0\ (0{,}01)\ 1$	[38]
12 dec.	$x = 0\ (0{,}1)\ 1$	[155]
10 dec.	$x = 0\ (0{,}0001)\ 0{,}1$	[43]
10 dec.	$x = 0\ (0{,}001)\ 1$	[43]
10 fig.	$x = -11\ (0{,}01)\ 0$	[43]

8 dec.	$x = -3\,(0,01)\,0\,(0,001)\,1\,(0,01)\,3$	[63]
7 dec.	$x = 0\,(0,001)\,1,05$	[122]
7 dec.	$x = 1\,(0,001)\,2$	[34]
7 dec.	$x = -6\,(0,01)\,-3;\ 3\,(0,01)\,4$	[63]
6 dec.	$x = 1\,(0,001)\,2$	[167]
5 dec.	$x = -6\,(0,01)\,0\,(0,001)\,1\,(0,01)\,4$	[62]
5 dec.	$x = 0\,(0,001)\,1$	[49]
5 dec.	$x = 1\,(0,001)2$	[16]
5 dec.	$x = 1\,(0,002)\,2$	[10] [18] [21]
5 dec.	$x = 0\,(0,01)\,1$	[48] [98]
5 dec.	$x = 1\,(0,01)\,2$	[5] [8] [94]
5 fig.	$x = -1\,(0,1)\,10$	[122]
4—5 fig.	$x = 0\,(0,01)\,4$	[22] [26] [70]$_2$
4 dec.	$x = 0\,(0,01)\,1$	[23]
4 dec.	$x = 1\,(0,01)\,2;\ \ 0\,(0,5)\,10$	[82]
4 dec.	$x = 0\,(0,05)\,1$	[70]$_1$ [101]
4 dec.	$x = 1\,(0,05)\,2$	[133]

$$\frac{1}{\Gamma(1+x)}$$

12 dec.	$x = -2\,(0,1)\,0$	[43]
5 fig.	$x = -3\,(0,01)\,3$	[22] [34] [70]$_2$

Logarithms of the gamma function

$$\lg \Gamma(1+x)$$

20 dec.	$x=0\,(0,01)\,1$	[59]
15 dec.	$x= -11\,(0,01)\,0$	[43]
14 dec.	$x= -1\left(\frac{1}{12}\right)1$	[78] [79] [97]
12 dec.	$x=0\,(0,001)\,1$	[78] [79]
12 dec.	$x=0\,(0,01)\,1$	[83]
12 dec.	$x=1\,(0,01)\,10\,(0,1)\,100$	[43]
10 dec.	$x=0\,(0,0001)\,0,1$	[43]
10 dec.	$x=2\,(0,1)\,5\,(0,2)\,70\,(1)\,1200$	[159]
9 dec.	$x=0\,(0,01)\,1$	[76]
8; 13 dec.	$x=-1\,(0,01)\,0;\ 0\,(0,00001)\,0,001$	[63]
10; 8 dec.	$x=0,001\,(0,0001)\,0,1;\ 0,1(0,001)1\,(0,01)\,2$	[63]
8 dec.	$x=9\,(0,1)\,11$	[138]
7 dec.	$x=0\,(0,001)\,1$	[31] [39]] 60] [77] [86] [91]]124]
7 dec.	$x=1\,(0,001)\,2$	[57]
7 dec.	$x=0\,(0,01)\,49,99$	[160]
7 dec.	$x= -5\,(0,01)\,15$	[73]

6 dec.	$x=0\,(0{,}001)\,1$	[51] [111]
6 dec.	$x=1\,(0{,}001)\,2$	[42] [166]
6 dec.	$x=\frac{1}{24}\left(\frac{1}{24}\right)\frac{23}{24}$	[15]
5 dec.	$x=0\,(0{,}001)\,1$	[90]
5 dec.	$x=1\,(0{,}001)\,2$	[18]
5 dec.	$x=1\,(0{,}001)\,1{,}099;\ 1{,}1\,(0{,}01)\,2$	[21]
5 dec.	$x=0\,(0{,}01)\,1$	[48] [62] [67] [72]
5 dec.	$x=1\,(0{,}01)\,2$	[41] [94]
4 dec.	$x=1\,(0{,}01)\,2;\ 0{,}5\,(0{,}5)\,10$	[82]
4 dec.	$x=0\,(0{,}01)\,1$	[70]$_1$ [88]
4 dec.	$x=1\,(0{,}01)\,2$	[61]

$$\ln \Gamma(1+x)$$

10 dec.	$x=0\,(0{,}005)\,1$	[128]
8 dec.	$x=9\,(0{,}1)\,10{,}1$	[138]

$$\lg \frac{\Gamma(1+x)}{\sqrt{2\pi}} = \lg \Gamma(1+x) - 0{,}39908\,99342\ldots$$

10 dec.	$x=0\,(0{,}01)\,1{,}05$	[134]

Various values connected with the gamma function

$$\frac{1}{n!}\left(\frac{d}{dx}\right)^n \Gamma(1+x)$$

12 dec.	$n=2\,(1)\,20.\quad x=0;\ 0{,}1\,(0{,}1)\,1$	[155]
12 dec.	$n=0;\ 1.\quad x=0\,(0{,}1)\,1$	

$$\frac{x^x e^{-x}\sqrt{x+1}}{\Gamma(x+1)}$$

7 dec.	$x=-1\,(0{,}05)\,0\,(0{,}1)\,50{,}1$	[85]

$$\int_n^{n+1}\frac{dx}{\Gamma(x)}\,;\quad \int_1^3\frac{dx}{\Gamma(x)}\quad \&\quad \int_0^{-1}\frac{dx}{\Gamma(x)}$$

16 dec.	$n=0\,(1)\,3$	[112]

$$\int_0^x \lg \Gamma(1+t)\,dt$$

10 dec.	$x=0\,(0{,}01)\,1$	[38] [128]

$$\frac{n}{\Gamma\left(\frac{1}{n}\right)}\int_0^\infty \frac{dx}{1+e^{x^n}}$$

6 dec. $n = 1\,(1)\,20$ [150]

$$\frac{2^{1+\frac{1}{m}}-1}{2^{\frac{1}{m}}+1}\,\frac{1}{\Gamma\left(2+\frac{1}{m}\right)}\int_0^\infty \frac{dx}{e^{x^q}-1}$$

6 fig. $q = \frac{m}{m+1}$. $m = 1\,(1)\,20$ [150]

Gamma function of complex argument

$$\Gamma(x+iy)$$

5 dec. $x = 4\,(0{,}1)\,5.$ $y = 0\,(0{,}1)\,1$ [121]

$$\frac{1}{\Gamma(x+iy)} = u + iv$$

6 dec. $x = -0{,}5\,(0{,}01)\,0{,}5.$ $y = 0\,(0{,}01)\,1$ [103]

$$\ln\Gamma(x+iy) = u + iv$$

6 dec. $x = 1\,(0{,}01)\,2.$ $y = 0\,(0{,}01)\,4$ [1]

Generalised gamma function

When x is an integer the generalised gamma function $\Gamma_k(x)$ has the form $\Gamma_k(1+x) = 1^{1^k}\cdot 2^{2^k}\ldots x^{x^k}$.

The expression $\Gamma_k(x)$ for an arbitrary value of x is given in /114/. This work gives 6 place tables of values of the functions $\Gamma_1(x)$ and $\Gamma_2(x)$ for $x = 0\,(0{,}1)\,2$ and values of the function log $\Gamma_1(x$ and log $\Gamma_2(x)$ with eight decimal places for the same values of x.

The incomplete gamma function and related integrals

In this section x is replaced by the argument p in the expression $u = \frac{x}{\sqrt{p+1}}$, as is done in many works. The relation of the incomplete to the complete integral is denoted by

$$I(u,p) = \frac{\int_0^{x} e^{-t}t^p\,dt}{\int_0^\infty e^{-t}t^p\,dt}$$

7 dec. $p = -1\,(0{,}05)\,0\,(0{,}1)\,5\,(0{,}2)\,50.$ $u = 0\,(0{,}1)\ldots$ to $I(u,p) = 1$ [85]

5 dec. $p = -1\,(0{,}01)\,-0{,}75.$ $u = 0\,(0{,}1)\,6$ [85]

$$I(u, p) = \frac{1}{\Gamma(p+1)} \int_0^{u\sqrt{p+1}} e^{-t} t^p \, dt = 1 - P(\chi^2, n) = 1 - P(s, n) = 1 - \Pi(s, x)$$

5 dec.

$$\left\{\begin{array}{ll} \chi^2 = 0\,(0{,}05)\,0{,}2\,(0{,}1)\,1. & n = 0\,(0{,}05)\,0{,}2\,(0{,}1)\,6 \\ \chi^2 = 0\,(0{,}1)\,3{,}2. & n = 0\,(0{,}05)\,0{,}2\,(0{,}1)\,6 \\ \chi^2 = 3{,}2\,(0{,}2)\,7\,(0{,}5)\,10\,(1)\,35. & n = 0\,(0{,}1)\,0{,}4\,(0{,}2)\,6 \\ s = -4\,(0{,}1)\,4{,}8. & n = 6\,(0{,}5)\,11\,(1)\,32 \\ s = -4{,}5\,(0{,}1)\,4{,}8. & x = 0\,(0{,}02)\,0{,}22\,(0{,}01)\,0{,}25 \end{array}\right\}$$ [17]

$$\lg \frac{1}{x^{p+1}\Gamma(p+1)} \int_0^\infty e^{-t} t^p \, dt$$

7 dec. $p = -1\,(0{,}01) - 0{,}9.$ $x = 0\,(0{,}01)\,0{,}3$ [85]

$$x^{-n} e^{\pm x} \int_0^\infty e^{\mp t} t^n \, dt$$

12 dec. $x = 0{,}1\,(0{,}1)\,0{,}5.$ $n = 0\,(1)\,5 + 10x$ [142]

$$\int_x^\infty e^{-t} t^n \, dt$$

9 dec. $n = -2\,(1)\,2.$ $x = 0\,(0{,}001)\,1\,(0{,}01)\,2$ [163]

for $n = -1; -2$ an auxiliary table in the range $x = 0(0{,}001)\,0{,}1$ is substituted.

$$\int_1^\infty e^{-xt} t^n \, dt = \frac{1}{x^{n+1}} \int_0^\infty e^{-t} t^n \, dt$$

11 dec. $n = 0(1)\,15.$ $x = 0{,}25\,(0{,}25)\,8\,(0{,}5)\,13\,(1)\,16\,(2)\,20\,(1)\,22;\ 24$ [152]

7–8 dec. $\left\{\begin{array}{ll} n = -0{,}5\,(0{,}5)\,6. & \frac{x}{\pi} = 0{,}5\,(0{,}5)\,4;\ 0{,}75;\ 2{,}75 \\ n = -4{,}5\,(0{,}5) - 0{,}5. & \frac{x}{\pi} = 1\,(1)\,4 \end{array}\right\}$ [148]

6. dec. $\left\{\begin{array}{ll} n = 0\,(1)\,10. & x = 0{,}5\,(0{,}5)\,5\,(1)\,10\,(2)\,14 \\ n = 11\,(1)\,20. & x = 1\,(1)\,10\,(2)\,14 \end{array}\right\}$ [146]

5 dec. $n = -3\,(1) - 1.$ $x = 0{,}01\,(0{,}01)\,0{,}1\,(0{,}05)\,1\,(0{,}1)\,2\,(0{,}2)\,3\,(0{,}5)\,5;\ 6$ [154]

5 dec. $n = -8\,(1) - 2.$ $x = 0{,}01;\ 0{,}05;\ 0{,}1;\ 0{,}25;\ 0{,}5(0{,}5)3(1)6$ [141]

4–5 dec. $\left\{\begin{array}{ll} n = 0\,(1)\,6. & x = 1{,}5\,(0{,}5)\,5\,(1)\,10 \\ n = 7\,(1)\,11. & x = 3\,(1)\,10 \end{array}\right\}$ [144]

3 dec. $n = -1;\ -3.$ $x = 0{,}01;\ 0{,}05;\ 0{,}1;\ 0{,}2;\ 0{,}5\,(0{,}5)\,2;\ 5$ [153]

$$\int_{-1}^{\infty} e^{-xt} t^n \, dt = \frac{1}{x^{n+1}} \int_{-x}^{\infty} e^{-t} t^n dt$$

6 fig. $n=0(1)18$. $x=1,5\,(0,5)\,2,5\,(0,25)\,4,5;\ 5(1)8(0,5)9;\ 10\,(2)\,14$ [145]
6 fig. $n=9(1)16$. $x=3($ var. $)5\,(1)\,8\,(0,5)\,9;\ 10\,(2)\,14$

$$\int_{-1}^{+1} e^{-xt} t^n \, dt = \frac{1}{x^{n+1}} \int_{-x}^{x} e^{-t} t^n \, dt$$

11 fig. $n=0(1)8$. $x=0,25\,(0,25)\,8\,(0,5)\,13\,(1)\,16\,(2)\,20\,(1)\,22;\ 24$ [152]

$$\int_{0}^{x} e^{-t^n} \, dt$$

4 dec. $n=4$. $x=0\,(0,1)\,1$ [100]

$$n \int_{0}^{1} 10^{-t} t^{n-1} \, dt$$

11 dec. $n=1\,(1)\,29$ [134]

$$\mathrm{P}(n+1, z) = \int_{0}^{z} e^{-t} t^n \, dt$$

8 dec. $z=1;\ -1;\ i.$ $n=0\,(1)\,10$ [115]

THE PSI FUNCTION AND ITS DERIVATIVES

$$\psi(x) = \frac{d}{dx} \ln \Gamma(1+x) = \frac{\Gamma'(1+x)}{\Gamma(1+x)}$$

23 dec.	$x=-\frac{1}{2};\ \frac{1}{2};\ 10$	[59]
18 dec.	$x=0\,(0,01)\,1$	
16 dec.	$x=-0,5\,(0,5)\,99$	[43]
15 dec.	$x=-11\,(0,01)\,0$	
13 dec.	$x=0\,(0,5)\,100$	[128]
12 dec.	$x=1\,(0,02)\,19$	[43]
12 dec.	$x=0\,(0,01)\,1;\ 10\,(0,1)\,60$	[38]
10 dec.	$x=0\,(0,0001)\,0,1;\ 0\,(0,001)\,1$; $x=100\,(1)\,449$	[43]
8 dec.	$x=0\,(0,02)\,20$	[158]
8 dec.	$x=0,5\,(0,5)\,10$	[71]

7 dec.	$x=0\,(0{,}01)\,1$	[122]
7 dec.	$x=0\,(0{,}01)\,0{,}5$	[119]
4 dec.	$x=0\,(0{,}01)\,1$	[22] [34] $[70]_2$ [95]
4 dec.	$x=1\,(0{,}05)\,2$	[133]
4 dec.	$x=-1\,(0{,}1)\,10$	[122]
4 dec.	$x=0\,(0{,}05)\,1\,(1)\,10$	[101]
4 dec.	$x=0\,(1)\,10$	$[70]_1$

$$\varphi(x) = \psi(x) + C = \sum_{r=1}^{\infty} \frac{x}{r(r+x)}$$

16 dec.	$x = 0\,(0{,}1)\,60{,}3$	[129]
10 dec.	$x = 50\,(0{,}01)\,51$	[129]

$$\psi^{n-1}(x) = \left(\frac{d}{dx}\right)^{n-1} \frac{\Gamma'(1+x)}{\Gamma(1+x)}$$

12 dec.	$n = 2\,(1)\,4.$	$x = 0\,(0{,}01)\,1;\ 10\,(0{,}1)\,60$	[38]
10—19 dec.	$n = 2\,(1)\,5.$	$x = 0\,(0{,}01)\,3\,(0{,}02)\,19\,(0{,}1)\,99$	[44]
10—12 dec.	$n = 2\,(1)\,5.$	$x = -11\,(0{,}1) - 1\,(0{,}01)\,0$	[44]
8 dec.	$n = 2.$	$x = 0\,(0{,}02)\,20$	[158]
5 dec.	$n = 2.$	$x = 0\,(0{,}01)\,0{,}5$	[119]
4 dec.	$n = 2.$	$x = 0\,(0{,}01)\,1$	[22] [34] [70]
3 dec.	$n = 2.$	$x = 0\,(0{,}05)\,1$	$[70]_1$ [101]
3 dec.	$n = 2.$	$x = 1\,(0{,}05)\,2$	[133]

$$\frac{1}{\psi'(x)}$$

3 dec.	$x = 0\,(0{,}05)\,1$	$[70]_1$
3 dec.	$x = 1\,(0{,}05)\,2$	[133]

$$\frac{1}{(n-1)!}\,\psi^{(n-1)}(x) = (-1)^n \sum_{r=1}^{\infty} \frac{1}{(r+x)^n}$$

5; 3; 1 dec.	$n = 2;\ 3;\ 4.$	$x = 0\,(0{,}01)\,0{,}5$	[119]

$$\pi\,\mathrm{ctg}\,\pi x$$

This function is linked with the psi function by the relationship:

$$\psi(-x) - \psi(x-1) = \pi\,\mathrm{ctg}\,\pi x$$

15 dec.	$x = 0\,(0{,}01)\,1$	[43]

$$(-1)^n \left(\frac{d}{dx}\right)^n (\pi \operatorname{ctg} \pi x)$$

12 dec.	$n = 1$	$x = 0\,(0{,}01)\,1$	[44]
10 dec.	$n = 2;\ 3$		
9—14 dec.	$n = 4$		

$$\operatorname{Re}\left[\psi(k) - \ln k + \frac{1}{2}k^{-1}\right]$$

8 dec.	$\gamma = k^{-2} = -1\,(0{,}01)\,1$	[156]

$$\lg \psi(x)$$

10 dec.	$x = 0\,(0{,}0001)\,0{,}1$	[43]
10 dec.	$x = 0\,(0{,}001)\,1$	
10 dec.	$x = -11\,(0{,}01)\,0$	

$$\lg \psi^{(n-1)}(x)$$

10 dec.	$n = 2\,(1)\,5.\ x = -11\,(0{,}1) - 1\,(0{,}01)\,0$	[44]

THE BETA FUNCTION

$$B(p, q) = \int_0^1 x^{p-1}(1-x)^{q-1}\,dx$$

8 fig.	p и $q = 0{,}5\,(0{,}5)\,11\,(1)\,50;\ p \geqslant q$	[84]

$$\int_0^1 (1 - x^{2n})^{1/2n}\,dx$$

7 dec.	$n = 1\,(1)\,7;\ 50;\ \infty$	[140]

$$\int_0^1 (1 - x^{2n})^{1/n}\,dx$$

7 dec.	$n = 1\,(1)\,7;\ 50;\ \infty$	[140]

$$\frac{\int_0^1 (1 - x^{2n})^{3/2n}\,dx}{\int_0^1 (1 - x^{2n})^{1/2n}\,dx}$$

7 dec.	$n = 1\,(1)\,7;\ 50;\ \infty$	[140]

$$\int_0^{\frac{\pi}{2}} \sin^{n-1} x\, dx = \frac{1}{2} B\left(\frac{1}{2}, \frac{1}{2} n\right)$$

10 dec. $n = 1\,(1)\,105$ [126]

The incomplete gamma function

The integral $B_x(p, q) = \int_0^x t^{p-1}(1-t)^{q-1}\, dt$ is encountered fairly frequently in various forms. The integral has found its widest application in statistics.

$$I_x(p, q) = \frac{B_x(p, q)}{B(p, q)}$$

7 dec. p и $q = 0{,}5\,(0{,}5)\,11\,(1)\,50.$ $p \geqslant q.$ $x = 0\,(0{,}01)\,1$ [84]

$$\frac{1}{2} + \frac{1}{2} \frac{B_x(p, q)}{B(p, q)}$$

7 dec. $p = \frac{1}{2}.$ $q = 0{,}5\,(0{,}5)\,15.$ $x = 0\,(0{,}01)\,1$ [127]

CERTAIN CONSTANTS

$$\Gamma(k)$$

13 dec. $k = \frac{1}{3}$; $k = \frac{2}{3}$ [139]

$$\lg \Gamma(k)$$

14 dec. $k = \frac{1}{3}$; $k = \frac{2}{3}$ [139]

$$\Gamma(1 + x_n),$$

where x_n are the negative roots of the equation $\Gamma'(1 + x) = 0.$

9 dec. [62] [63]

$n = 1\,(1)\,4$

Chapter 6

SINE AND COSINE INTEGRALS, EXPONENTIAL AND LOGARITHMIC INTEGRALS AND RELATED FUNCTIONS

The following notation has been adopted for the principal functions examined in the present chapter

sine integral $$\mathrm{Si}\,(x)=\int_0^x \frac{\sin t}{t}\,dt;$$

cosine integral $$\mathrm{Ci}\,(x)=\int_\infty^x \frac{\cos t}{t}\,dt;$$

exponential integral

$$\mathrm{Ei}\,(x)=\int_{-\infty}^x \frac{e^t}{t}\,dt \quad \text{и} \quad -\mathrm{Ei}\,(-x)=\int_x^\infty \frac{e^{-t}}{t}\,dt;$$

logarithmic integral

$$\mathrm{li}\,(x)=\int_0^x \frac{dt}{\ln t}=\mathrm{Ei}\,(\ln x).$$

The following relationships hold for these functions:

$$\mathrm{si}\,(x)=\int_\infty^x \frac{\sin t}{t}\,dt=\mathrm{Si}\,(x)-\frac{\pi}{2},$$

$$\mathrm{Ci}\,(-x)=\mathrm{Ci}\,(x)\pm\pi i,$$

$$\mathrm{Si}\,(-x)=-\mathrm{Si}\,(x).$$

The argument x is positive in almost all cases throughout this chapter.

The integrals $\mathrm{Ei}\,(x)$ when $x>0$ and $\mathrm{li}\,(x)$ when $x>1$, are principal values, i.e.

---

[1]See pages 470 and 550 for the references to Chapter 6.

$$\mathrm{Ei}(x) = \lim_{\varepsilon\to 0}\left[\int_{-\infty}^{-\varepsilon}\frac{e^t}{t}\,dt + \int_{\varepsilon}^{x}\frac{e^t}{t}\,dt\right]$$

$$\mathrm{li}(x) = \lim_{\varepsilon\to 0}\left[\int_{0}^{1-\varepsilon}\frac{dt}{\ln t} + \int_{1+\varepsilon}^{x}\frac{dt}{\ln t}\right]$$

All the material in the chapter is grouped under the following heads:

- sine integral;
- cosine integral;
- expressions containing sine and cosine integrals;
- hyperbolic sine and cosine integrals;
- exponential integral;
- logarithmic integral;
- generalised and mixed integral functions;
- integral functions with complex arguments

The characteristics of some constants relating to integral functions are given in Chapter 15.

SINE INTEGRAL

$$\mathrm{Si}(x) = \int_0^x \frac{\sin t}{t}\,dt$$

20 dec.	$x = 1\,(1)\,10$	[36] [62]
18 dec.	$x = 0\,(0{,}01)\,1$	[51]
11 dec.	$x = 1\,(0{,}1)\,5;\ 6\,(1)\,15$	[51]
7 dec.	$x = 20\,(5)\,100\,(10)\,200\,(100)\,1000\,(1000)\,11\,000;\ 10^{5\,(1)\,8}$	[51]
11; 10 dec.	$x = 0\,(0{,}1)\,5;\ 5\,(0{,}1)\,20\,(0{,}2)\,40$	[12]
10 dec.	$x = 10\,(0{,}01)\,100$	[28]
10 dec.	$x = 0\,(0{,}01)\,1\,(0{,}1)\,7{,}5$	[62]
10 dec.	$x = 5\,(0{,}1)\,20$	[40]
10 dec.	$x = 20\,(0{,}2)\,40$	[41]
10 dec.	$x = 1\,(1)\,10$	[45]
10 dec.	$x = 0\,(0{,}001)\,10\,(0{,}1)\,20\,(0{,}2)\,40$	[32]
9 dec.	$x = 0\,(0{,}0001)\ 1{,}9999$	[31]
9 dec.	$x = 0\,(0{,}1)\,10$	[31]
9 dec.	$x = 0\,(0{,}01)\,1\,(1)\,10$	[19]
7 dec.	$x = 10^{2\,(1)\,15}$	[19]
7 dec.	$x = 0\,(0{,}0001)\,2\,(0{,}001)\,10\,(0{,}005)\,100$	[2]
7 dec.	$x = 7{,}6\,(0{,}2)\,8{,}6$	[37]
5 dec.	$x = 0\,(0{,}1)\,25;\ 50$	[43]
5 dec.	$x = 16\,(1)\,60$	[56]

5 dec. 4 dec.	$x = 0\,(0{,}01)\,1\,(0{,}1)\,5\,(1)\,15$ $x = 20\,(5)\,100\,(10)\,200\,(100)\,1000\,(1000)\,10^4;\ 10^5;\ 10^6$	[13]
4—7 dec.	$x = 0\,(0{,}01)\,1\,(0{,}1)\,6\,(1)\,15\,(5)\,100\,(10)\,200\,(100)\,1000;\ 10^{4\,(1)\,7}$	[9]
4—5 dec.	$x = 0\,(0{,}01)\,1\,(0{,}1)\,5\,(1)\,15\,(5)\,100\,(10)\,200\,(100)\,1000;\ 10^{4\,(1)\,7}$	[11]
4—5 dec.	$x = 0\,(0{,}01)\,0{,}1\,(0{,}05)\,1\,(0{,}1)\,5\,(1)\,15\,(5)\,100\,(10)\,200\,(100)\,10^3;\ 10^{4\,(1)\,7}$	[1] [5]
4—5 dec. 4 dec.	$x = 0\,(0{,}01)\,0{,}17$ $x = 0{,}18\,(0{,}01)\,1\,(0{,}1)\,5\,(1)\,15\,(5)\,100\,(10)\,200\,(100)\,10^3;\ 10^{4\,(1)\,7}$	[6] [16]

$$\mathrm{Si}\,(x) = \int_{\infty}^{x} \frac{\sin t}{t}\,dt$$

6 dec.	$x = 0\,(0{,}01)\,50$	[33]
4 dec.	$x = 0\,(0{,}01)\,1\,(0{,}1)\,6\,(1)\,15\,(5)\,100\,(10)\,200\,(100)\,10^3;\ 10^{4\,(1)\,7}$	[9]

$$\int_0^x \frac{\sin \pi t}{\pi t}\,dt = \frac{1}{\pi}\,\mathrm{Si}\,(\pi x)$$

7 dec.	$x = 0\,(0{,}01)\,10$	[18]

$$\mathrm{Si}\,(2\pi x)$$

3 dec.	$x = 0\,(0{,}001)\,0{,}5\,(0{,}01)\,10$	[4]

Extremes of the sine integral

15 dec. 15 dec.	$\mathrm{Si}\,(n\pi)$ $\mathrm{Si}\,(n\pi \pm h)$	$n = 1\,(1)\,3$ $n = 1\,(1)\,3.\ h = 0\,(0{,}0001)\,0{,}01\,(0{,}001)\,0{,}05$	[32]

The table contains values of the sine integral near to the maximum and minimum values.

7 dec.	$\mathrm{si}\,(n\pi)$	$n = 1\,(1)\,40.\ n = p\,\&\,p+1;\ p = 50\,(10)\,100\,(50)\,500\,(100)\,1000;\ 2000$	[51]
5—6 dec.	$\mathrm{Si}\,(x)$	$\frac{x}{\pi} = 1\,(1)\,15$	[9]
5 fig.	$\mathrm{Si}\,(n\pi)$	$n = 1\,(1)\,15$	[6] [16]

COSINE INTEGRAL

$$\mathrm{Ci}(x) = \int_{\infty}^{x} \frac{\cos t}{t}\, dt$$

20 dec.	$x = 1(1)10$	[36] [62]
18 dec.	$x = 0(0{,}01)1$	[51]
11 dec.	$x = 1(0{,}1)5;\ 6(1)15$	[51]
7 dec.	$x = 20(5)100(10)200(100)1000(1000)$ 11 000; $10^{5\,(1)\,8}$	[51]
10 dec.	$x = 10(0{,}01)100$	[28]
10 dec.	$x = 0(0{,}01)1(0{,}1)7{,}5$	[62]
10 dec.	$x = 5(0{,}1)20(0{,}2)40$	[12]
10 dec.	$x = 5(0{,}1)20$	[40]
10 dec.	$x = 20(0{,}2)40$	[41]
10 dec.	$x = 1(1)10$	[45]
10 dec.	$x = 0(0{,}001)9{,}999;\ 10(0{,}1)20(0{,}2)40$	[32]
9 dec.	$x = 0(0{,}0001)1{,}9999;\ 0(0{,}1)10$	[31]
9 dec.	$x = 0(0{,}01)1(1)10$	[19]
7 dec.	$x = 10^{2\,(1)\,15}$	[19]
7 dec.	$x = 0(0{,}0001)2(0{,}001)10(0{,}005)100$	[2]
7 dec.	$x = 7{,}6(0{,}2)8{,}6$	[37]
6 dec.	$x = 0(0{,}0001)0{,}05(0{,}001)1(0{,}01)50$	[33]
5 dec.	$x = 0(0{,}01)1(0{,}1)5(1)15$	[13]
4 dec.	$x = 20(5)100(10)200(100)1000(1000)$ 10 000; $10^{4\,(1)\,6}$	[13]
4—7 dec.	$x = 0(0{,}01)1(0{,}1)6(1)15(5)100(10)200$ $(100)10^3;\ 10^{4\,(1)\,7}$	[9]
4—6 fig.	$x = 0(0{,}01)0{,}1(0{,}05)1(0{,}1)5(1)15(5)$ $100(10)200(100)10^3;\ 10^{4\,(1)\,7}$	[5]
4—5 fig.	$x = 0(0{,}01)0{,}1(0{,}05)1(0{,}1)5(1)15(5)$ $100(10)200(100)10^3;\ 10^{4\,(1)\,7}$	[1]
4—5 fig.	$x = 0(0{,}01)1(0{,}1)5(1)15(5)100(10)200$ $(100)900$	[6] [11] [16]
4—3 fig.	$x = 10^{3\,(1)\,7}$	[6] [11] [16]

$\mathrm{Ci}(x) - \ln x$

If x is small this function is near to $\ln x + C$, where $C = 0{,}5772\ldots$ is the Euler constant.

11 dec.	$x = 0\,(0{,}1)\,5$	[12]
9 dec.	$x = 0\,(0{,}0001)\,0{,}01$	[31]
7 dec.	$x = 0\,(0{,}0001)\,0{,}0099$	[2]

$$\int_0^x \frac{1-\cos t}{t}\,dt = \ln x + C - \mathrm{Ci}\,(x)$$

6 dec.	$x = 0\,(0{,}001)\,10\,(0{,}01)\,50$	[8]
6 dec.	$x = 4H. \quad H = 0\,(0{,}1)\,3$	[49]
5 dec.	$x = 0\,(0{,}1)\,25;\ 50$	[43]

Ci (2πx)

3 dec.	$x = 0\,(0{,}001)\,0{,}5\,(0{,}01)\,10$	[4]

Extremes of the cosine integral

15 dec.	$\mathrm{Ci}\left(\frac{2n+1}{2}\pi\right)$	$n = 1\,(1)\,3$	[32]
15 dec.	$\mathrm{Ci}\left(\frac{2n-1}{2}\pi \pm h\right)$	$n = 1\,(1)\,3. \quad h = 0\,(0{,}0001)\,0{,}01\,(0{,}001)\,0{,}05$	[32]

These tables give values of the function Ci (x) in the vicinity of the maxima and minima.

7 dec.	$\mathrm{Ci}\left(\frac{2n+1}{2}\pi\right)$	$n = 1\,(1)\,40. \quad n = p$ и $p+1;\ p = 50\,(10)\,100\,(50)\,700$	[51]
5 fig.	$\mathrm{Ci}\left(\frac{2n+1}{2}\pi\right)$	$n = 1\,(1)\,16$	[6] [16]

EXPRESSIONS CONTAINING SINE AND COSINE INTEGRALS

The following notations for the functions are adopted in this section for the sake of brevity:

$$I\,(x) = \frac{\mathrm{Si}\,(2x) - x^{-1}\sin^2 x}{\pi} \quad \& \quad \overline{\mathrm{Ci}}\,(x) = \int_0^x \frac{(1-\cos u)}{u}\,du.$$

I (x)

5 dec.	$x = 0\,(0{,}1)\,15\,(0{,}5)\,34{,}5$	[38]

I (x + 2,4π) − I (x)

5 dec.	$\frac{x}{\pi} = -1{,}2\,(0{,}1)\,1{,}3$	[38]

$$I(x+3{,}4\pi)-I(x)$$

5 dec. $\frac{x}{\pi}=-1{,}7\,(0{,}1)\,0$ [38]

$$I_1(x+8{,}4)-I(x)$$

$$I_1(x)=\int_0^x \frac{H_1(2t)}{\pi t^2}\,dt \quad \text{и} \quad H_1(u)=\pi^{-1}\left\{2+\int_0^\pi \sin(u\sin\theta-\theta)\,d\theta\right\}$$

4 dec. $x=-4{,}2\,(0{,}1)-2{,}8$ [38]

$$\frac{\mathrm{Si}(k)-\dfrac{1-\cos k}{k}}{\pi}$$

5; 4 dec. $k=0{,}1\,(0{,}1)\,0{,}4\,(0{,}2)\,1;\ 1\,(0{,}2)\,2\,(0{,}5)\,5\,(1)\,16;\ 20\,(10)\,50;\ \pi\,(\pi)\,5\pi$ [44]

$$\tfrac{1}{2}\,[\cos x\,(\mathrm{Si}(4x)-2\mathrm{Si}(2x))+\sin x\,\overline{\mathrm{Ci}}(4x)],$$
$$\tfrac{1}{2}\,[\cos x\,(4\,\mathrm{Si}(2x)-\mathrm{Si}(4x))+\sin x\,(2\,\overline{\mathrm{Ci}}(2x)-\overline{\mathrm{Ci}}(4x)+4\ln^2 x)],$$
$$\tfrac{1}{2}\,[\cos x\,(\overline{\mathrm{Ci}}(4x)+2\,\overline{\mathrm{Ci}}(2x))-\sin x\,\mathrm{Si}(4x)],$$
$$\tfrac{1}{2}\,[\cos x\,(\overline{\mathrm{Ci}}(4x)-4\,\overline{\mathrm{Ci}}(2x))+\sin x\,(2\,\mathrm{Si}(2x)-\mathrm{Si}(4x))]$$

4 dec. $x=0\,(0{,}1)\,7;\quad \frac{\pi}{2}\left(\frac{\pi}{2}\right)2\pi$ [53]

$$\left(\frac{2}{\pi}\right)\{\mathrm{Si}(kt)-(1-k)^{-1}\,[\mathrm{Si}(t)-\mathrm{Si}(kt)+t^{-1}\,(\cos t-\cos kt)]\}$$

3 dec. $k=0\,(0{,}05)\,1.\quad t=0\,(0{,}5)\,26$ [7]

$$\left(\tfrac{1}{2}\,\pi-\mathrm{Si}(kx)\right)\cos kx+\mathrm{Ci}(kx)\,\cos kx$$

4 dec. $kx=0\,(0{,}05)\,0{,}1\,(0{,}1)\,1$ [57]

$$\int_0^z [C+\lg t-\mathrm{Ci}(t)]\,\frac{dt}{t}$$

6 dec. $z=0\,(0{,}2)\,20$ [58]

$$\int_0^z \mathrm{Si}(t)\,\frac{dt}{t}$$

6 dec. $z=0\,(0{,}2)\,20$ [58]

HYPERBOLIC SINE AND COSINE INTEGRALS

$$\mathrm{Shi}(x) = \int_0^x \frac{\mathrm{sh}\, t}{t}\, dt = \frac{\mathrm{Ei}(x) - \mathrm{Ei}(-x)}{2}$$

20 dec.	$x = 1\,(1)\,10$	[36] [62]
10 dec.	$x = 0\,(0{,}01)\,1\,(0{,}1)\,7{,}5$	[62]

$$\mathrm{Chi}(x) = \int_a^x \frac{\mathrm{ch}\, t}{t}\, dt = \frac{\mathrm{Ei}(x) + \mathrm{Ei}(-x)}{2}$$

The lower limit of a is selected so that $\mathrm{Ei}(a) + \mathrm{Ei}(-a) = 0$.

20 dec.	$x = 1\,(1)\,10$	[36] [62]
10 dec.	$x = 0\,(0{,}01)\,1\,(0{,}1)\,7{,}5$	[62]

EXPONENTIAL INTEGRAL

$$\mathrm{Ei}(x) = \int_{-\infty}^{x} \frac{e^t}{t}\, dt$$

20 dec.	$x = 1\,(1)\,10$	[36] [62]
20 dec.	$x = 10\,(1)\,20$	[46]
18 dec.	$x = 0\,(0{,}01)\,1$	[51]
11 dec.	$x = 1\,(0{,}1)\,5;\ 6\,(1)\,15$	[51]
10—11 fig.	$x = 5\,(0{,}1)\,15$	[12] [40]
10; 10—11 fig.	$x = 0\,(0{,}001)\,10;\ 10\,(0{,}1)\,15$	[32]
10 dec.	$x = 0\,(0{,}01)\,1\,(0{,}1)\,7{,}5$	[62]
10 dec.	$x = 1\,(1)\,10$	[45]
10 fig.	$x = 15\,(1)\,50$	[50]
9 dec.	$x = 0\,(0{,}0001)\,1{,}9999;\ 0\,(0{,}1)\,10$	[31]
10 dec.	$x = 5\,(0{,}2)\,14$	[46]
9 dec.	$x = 14{,}2\,(0{,}2)\,15$	[46]
8 dec.	$x = 15\,(0{,}2)\,20$	[46]
9 dec.	$x = 0\,(0{,}01)\,1\,(1)\,10$	[19]
7 fig.	$x = 0\,(0{,}0001)\,1{,}3\,(0{,}001)\,3\,(0{,}0005)\,10\,(0{,}1)\,15$	[3]

6 dec.	$x = 0\,(0{,}01)\,1\,(0{,}1)\,5\,(1)\,15$	[14]
5 dec.	$x = 0\,(0{,}01)\,1\,(0{,}1)\,3$	[13]
5–6 fig.	$x = 3{,}1\,(0{,}1)\,5\,(1)\,15$	
4–6 fig.	$x = 0\,(0{,}01)\,1\,(0{,}1)\,5\,(1)\,15$	[6] [16]
4–6 fig.	$x = 0\,(0{,}01)\,0{,}1\,(0{,}05)\,1\,(0{,}1)\,5\,(1)\,15$	[5]
4 dec.	$x = 0\,(0{,}01)\,0{,}1\,(0{,}05)\,1\,(0{,}1)\,5\,(1)\,15$	[1]
4 dec.	$x = 0\,(0{,}01)\,0{,}05;\ 0{,}1\,(0{,}1)\,3{,}9$	[15]
3; 2 dec.	$x = 4\,(0{,}1)\,6{,}9;\ 7\,(1)\,10$	

$$-\mathrm{Ei}\,(-x) = \int_x^\infty \frac{e^{-t}}{t}\,dt$$

This function is positive for positive values of x. Certain authors tabulate $\mathrm{Ei}\,(-x)$ which is negative.

Certain small tables of $\mathrm{Ei}\,(-x)$ are met with as special cases of an incomplete gamma function, but they are of little significance by comparison with the basic tables of exponential integrals and are not examined here.

20 dec.	$x = 1\,(1)\,10$	[36] [62]
18 dec.	$x = 0\,(0{,}05)\,18\,(1)\,22;\ 24;\ 25;\ 26\,(2)\,42;\ 48$	[54] [55]
18 dec.	$x = 0\,(0{,}01)\,1$	[51]
11 dec.	$x = 1\,(0{,}1)\,5;\ 6\,(1)\,15$	
13–14 dec.	$x = 5\,(0{,}1)\,15$	[12] [40]
10 dec.	$x = 0\,(0{,}01)\,1\,(0{,}1)\,7{,}5$	[62]
10 fig.	$x = 15\,(1)\,50$	[50]
9 dec.	$x = 0\,(0{,}0001)\,1{,}9999$	[31]
9 fig.	$x = 0\,(0{,}001)\,10$	[32]
14 dec.	$x = 10\,(0{,}1)\,15$	
9 fig.	$x = 0{,}1\,(0{,}001)\,1\,(0{,}01)\,2$	[60]
9 fig.	$x = 1\,(0{,}01)\,2$	[23]
9 dec.	$x = 0\,(0{,}01)\,1\,(1)\,10$	[19]
7 fig.	$x = 0\,(0{,}0001)\,1{,}3\,(0{,}001)\,3\,(0{,}0005)\,10\,(0{,}1)\,15$	[3]
6; 7 dec.	$x = 0\,(0{,}01)\,1\,(0{,}1)\,1{,}5;\ 1{,}6\,(0{,}1)\,5$	[14]
4–6 fig.	$x = 6\,(1)\,15$	
5–6 fig.	$x = 0\,(0{,}01)\,1\,(0{,}1)\,5\,(1)\,15$	[13]
5–6 fig.	$x = 20\,(0{,}02)\,50$	[59]
4 dec.	$x = 0\,(0{,}01)\,0{,}05;\ 0{,}1\,(0{,}1)\,0{,}9$	[15]
5; 6–8 dec.	$x = 1\,(0{,}1)\,6{,}9;\ 7\,(1)\,10$	
4–6 fig.	$x = 0\,(0{,}01)\,0{,}1\,(0{,}05)\,1\,(0{,}1)\,5\,(1)\,15$	[1] [5]
4–5 fig.	$x = 0\,(0{,}01)\,1\,(0{,}1)\,5\,(1)\,15$	[11]

4 dec.		$x = 0\,(0{,}01)\,1\,(0{,}1)\,1{,}5$	[6] [16]
4 fig.		$x = 1{,}6\,(0{,}1)\,5\,(1)\,15$	[6] [16]
4 dec.		$x = 0\,(0{,}1)\,0{,}2;\ 0{,}5\,(0{,}5)\,2;\ 3$	[61]

$$\mathbf{Ei}\,(x) - \ln x$$

11 dec.		$x = 0\,(0{,}1)\,5$	[12]
9 dec.		$x = 0\,(0{,}0001)\,0{,}01$	[31]
7 dec.		$x = 0\,(0{,}0001)\,0{,}0099$	[3]

$$B = \mathbf{Ei}\,(-x) - \ln x$$

11 dec.		$x = 0\,(0{,}1)\,5$	[12]
9 dec.	$-B$	$x = 0\,(0{,}0001)\,0{,}01$	[31]
9 dec.		$x = 0\,(0{,}001)\,1$	[23]
9 dec.		$x = 0\,(0{,}001)\,0{,}1$	[60]
7 dec.		$x = 0\,(0{,}0001)\,0{,}0099$	[3]

$$\int_x^\infty \frac{e^{-t}}{t^2}\,dt = \frac{e^{-x}}{x} + \mathbf{Ei}\,(-x)$$

9 dec.		$x = 0{,}1\,(0{,}001)\,1\,(0{,}01)\,2$	[60]
9 dec.		$x = 1\,(0{,}01)\,2$	[23]

$$A = \int_x^\infty \frac{e^{-t}}{t^2}\,dt - \frac{1}{x} - \ln x$$

9 dec.		$x = 0\,(0{,}001)\,0{,}1$	[60]
9 dec.	$-A$	$x = 0\,(0{,}001)\,1$	[23]

$$-\mathbf{Ei}\,(-x)\,e^x\,(x + \ln 2)$$

6 dec.		$x = 1\,(1)\,10\,(10)\,100$	[48]

$$\lg \mathbf{Ei}\,(x)$$

6 dec.		$x = 4\,(0{,}1)\,5\,(1)\,10;\ 20$	[48]

LOGARITHMIC INTEGRAL

Tables of logarithmic integrals are described separately for small and large values of the argument. The logarithmic integral for large values of the argument is used in research on the theory of numbers.

li (x)

For small values of x

8 fig.	$x = 22\,(2)\,40\,(5)\,80\,(10)\,160\,(20)\,320\,(40)$ $640\,(80)\,1280$	[24]
7 dec.	$x = 0\,(0{,}01)\,1\,(0{,}1)\,2\,(0{,}5)\,3\,(1)\,20$	[24]

For large values of x

6 dec.	$x = 10^3;\ 10^4$	[47]
5 dec.	$x = 10^5\,(10^5)\,6\times10^5;\ 10^6$	[47]
1 dec.	$x = 0\,(500\,000)\,3\,000\,000$	[42]
whole	$x = 0\,(50\,000)\,9\,000\,000;\ 10^7;\ 10^8$	[17]
whole	$x = 0\,(50\,000)\,10\,000\,000$	[22]
whole	$x = 0\,(100\,000)\,9\,000\,000$	[39]

li (10^x)

18 dec.	$x = -15\,(0{,}05)\,-10\,(0{,}02)\,-7\,(0{,}01)\,-0{,}5$	[25]
15—13 dec.	$x = 0\,(0{,}01)\,3{,}5$	[26]
14 fig.	$x = 3{,}5\,(0{,}01)\,10$	[27]

li $(x) - \ln(1-x)$

This function is finite when $x = 1$.

10 dec.	$x = 0\,(0{,}01)\,1$	[20] [21]

li $(x) - \ln|\ln x| - \ln x$

This function is finite when $x = 1$.

6 dec.	$x = 0{,}5\,(0{,}01)\,2{,}8$	[48]

li $(10^x) - \ln\ln(10^{-x})$

This function is finite when $x = 0$.

18 fig.	$x = -0{,}5\,(0{,}01)0$	[25]

$\dfrac{\mathrm{li}\,x}{x}\ln\left(\dfrac{x}{2}\right)$

6 dec.	$x = 0\,(0{,}01)\,0{,}62$	[48]

GENERALISED AND MIXED INTEGRAL FUNCTIONS

Tables of the following generalised integral functions are described in this section:

$$S(a, x) = \int_0^x \frac{\sin u}{u} dt \qquad Cs(a, x) = \int_0^x \frac{\cos u \sin x}{u} dt$$

$$C(a, x) = \int_0^x \frac{1 - \cos u}{u} dt \qquad Cc(a, x) = \int_0^x \frac{\cos u (1 - \cos x)}{u} dt$$

$$Ss(a, x) = \int_0^x \frac{\sin u \sin x}{u} dt \qquad E(a, x) = \int_0^x \frac{1 - e^{-u}}{u} dt$$

$$Sc(a, x) = \int_0^x \frac{\sin u \cos x}{u} dt \qquad Es(a, x) = \int_0^x \frac{e^{-u} \sin u}{u} dt$$

$$Ec(a, x) = \int_0^x \frac{1 - e^{-u} \cos u}{u} dt,$$

where
$$u = \sqrt{a^2 + t^2}.$$

When $a = 0$ the following hold:

$$S(0, x) = Si(x) \qquad Cs(0, x) = \frac{1}{2} Si(2x)$$

$$C(0, x) = C + \ln x - Ci(x) \qquad Ss(0, x) = \frac{1}{2}[\ln 2 + C + \ln x - Ci(2x)]$$

$$Sc(0, x) = \frac{1}{2} Si(2x) \qquad Cc(0, x) = \frac{1}{2}[\ln 2 - C - \ln x - Ci(2x)] + Ci(x),$$

where C is the Euler constant
When $a \neq 0$ the following hold:

$$Ss(a, x) = -\frac{1}{2}[Ci(z) + Ci(y)] + Ci(a),$$

$$Sc(a, x) = \frac{1}{2}[Si(z) - Si(y)],$$

$$Cs(a, x) = \frac{1}{2}[Si(z) + Si(y)] - Si(a),$$

where
$$z = u + x \;\&\; y = u - x.$$

$$S(a, x) = \int_0^\infty \frac{\sin u}{u} dt$$

6 dec. $\left\{\begin{matrix} a = 0(0{,}01)1(\text{ var. })2(0{,}05)5(0{,}1)10(0{,}2)25 \\ x = 0(0{,}01)1(\text{ var. })2(0{,}05)5(0{,}1)10(0{,}2)25 \end{matrix}\right\}$ [30]

$$C(a, x) = \int_0^\infty \frac{1 - \cos u}{u} dt$$

6 dec. $\left\{\begin{matrix} a = 0(0{,}01)1(\text{ var. })2(0{,}05)5(0{,}1)10(0{,}2)25 \\ x = 0(0{,}01)1(\text{ var. })2(0{,}05)5(0{,}1)10(0{,}2)25 \end{matrix}\right\}$ [30]

$$\mathrm{Sc}(a, x) = \int_0^x \frac{\sin u \cos t}{u}\, dt$$

6 dec. $\begin{cases} a = 0\,(0{,}01)\,1\,(\text{ var. } , 2\,(0{,}05)\,5\,(0{,}1)\,10\,(0{,}2)\,25 \\ x = 0\,(0{,}01)\,1\,(\text{ var. })\,2\,(0{,}05)\,5\,(0{,}1)\,10\,(0{,}2)\,25 \end{cases}$ [30]

$$\mathrm{Cc}(a, x) = \int_0^x \frac{\cos u\,(1 - \cos t)}{u}\, dt$$

6 dec. $\begin{cases} a = 0\,(0{,}01)\,1\,(\text{ var. })\,2\,(0{,}05)\,5\,(0{,}1)\,10\,(0{,}2)\,25 \\ x = 0\,(0{,}01)\,1\,(\text{ var. })\,2\,(0{,}05)\,5\,(0{,}1)\,10\,(0{,}2)\,25 \end{cases}$ [30]

$$\mathrm{Cs}(a, x) = \int_0^x \frac{\cos u \sin t}{u}\, dt$$

6 dec. $\begin{cases} a = 0\,(0{,}01)\,1\,(\text{ var. })\,2\,(0{,}05)\,5\,(0{,}1)\,10\,(0{,}2)\,25 \\ x = 0\,(0{,}01)\,1\,(\text{ var. })\,2\,(0{,}05)\,5\,(0{,}1)\,10\,(0{,}2)\,25 \end{cases}$ [30]

$$\mathrm{F}(x) = \int_0^x \frac{\sin t \sin (x - t)}{t}\, dt$$

4 fig. $x = 0\,(0{,}1)\,8$ [52]

$$\mathrm{G}(x) = \int_0^x \frac{\sin t \cos (x - t)}{t}\, dt$$

4 fig. $x = 0\,(0{,}1)\,8$ [52]

$$\mathrm{E}(a, x) = \int_0^x \frac{1 - e^{-u}}{u}\, dt$$

6 dec. $\begin{cases} x = 0\,(h)\,N. \quad a = 0\,(h)\,N. \quad 0{,}049 \leqslant N \leqslant 9{,}8 \\ h = 0{,}001;\ 0{,}002;\ 0{,}005;\ 0{,}01;\ 0{,}02;\ 0{,}05;\ 0{,}1;\ 0{,}2 \end{cases}$ [29]

$$\mathrm{Es}(a, x) = \int_0^x \frac{e^{-u} \sin u}{u}\, dt$$

6 dec. $\begin{cases} x = 0\,(h)\,N. \quad a = 0\,(h)\,N. \quad 0{,}049 \leqslant N \leqslant 9{,}8, \\ h = 0{,}001;\ 0{,}002;\ 0{,}005;\ 0{,}01;\ 0{,}02;\ 0{,}05;\ 0{,}1;\ 0{,}2 \end{cases}$ [29]

$$\mathrm{Ec}(a, x) = \int_0^x \frac{1 - e^u \cos u}{u} dt$$

6 dec. $\left\{\begin{array}{l} x = 0\,(h)\,N. \quad a = 0\,(h)\,N. \quad 0{,}049 \leqslant N \leqslant 9{,}8 \\ h = 0{,}001;\ 0{,}002;\ 0{,}005;\ 0{,}01;\ 0{,}02;\ 0{,}05;\ 0{,}1;\ 0{,}2 \end{array}\right\}$ [29]

$$\mathrm{Ss}(a, x) = \int_0^x \frac{\sin u \sin t}{u} dt$$

6 dec. $\left\{\begin{array}{l} a = 0\,(0{,}01)\,1\ (\text{ var.})\ 2\,(0{,}05)\,5\,(0{,}1)\,10\,(0{,}2)\,25 \\ x = 0\,(0{,}01)\,1\ (\text{ var.})\ 2\,(0{,}05)\,5\,(0{,}1)\,10\,(0{,}2)\,25 \end{array}\right\}$ [30]

$$\int_0^x \mathrm{Si}(t)\, d\, \mathrm{Ci}(t) = \int_0^x \frac{\mathrm{Si}(t) \cos t\, dt}{t}$$

5 dec. $x = k\pi. \quad k = 0{,}2\,(0{,}2)\,2$ [10]

$$\int_0^x \left(\frac{\sin t}{t}\right)^2 dt = \mathrm{Si}(2x) - \frac{\sin^2 x}{x}$$

5 dec. $\frac{x}{\pi} = 0{,}5\,(0{,}5)\,2{,}5$

4 dec. $x = 0\,(0{,}05)\,3\,(0{,}1)\,11\,(0{,}5)\,15{,}5$ [34]

$$\frac{1}{x} \int_0^x \left(\frac{\sin t}{t}\right)^2 dt$$

4 dec. $x = 0\,(0{,}05)\,3\,(0{,}1)\,11\,(0{,}5)\,15{,}5$ [34]

$$30\left[\int_0^{4H} \frac{1 - \cos x}{x} dx - \sin^2 2H\right]$$

4 dec. $H = 0{,}1\,(0{,}1)\,3$ [49]

INTEGRAL FUNCTION FOR COMPLEX ARGUMENTS

$$\mathrm{Si}(z, \alpha) = \int_0^z t^{-\alpha} \sin t\, dt$$

where $z = x + iy$.

2 dec. $x = 0\,(1)\,20. \quad y = 0\,(1)\,5. \quad \alpha = 0{,}25\,(0{,}25)\,1{,}75$ [35]

For the case $z = x$ the value $\mathrm{Si}(x, \alpha)$ is calculated in the following range:

3 dec. $x = 0\,(0{,}2)\,4\,(0{,}5)\,20. \quad \alpha = 0{,}25\,(0{,}25)\,1{,}75$ [35]

$$\mathrm{Si}(\infty, \alpha) = \int_0^\infty \frac{\sin t}{t^\alpha}\, dt$$

5 dec. $\alpha = 0{,}05\,(0{,}05)\,1{,}95;\ 1{,}99;\ 1{,}995;\ 1{,}999;\ 2$ [35]

$$\mathrm{Ci}(\infty, \alpha) = \int_0^\infty \frac{\cos t}{t^\alpha}\, dt$$

5 dec. $\alpha = 0{,}05\,(0{,}05)\,1$ [35]

Roots of $\mathrm{Si}(z, \alpha)$

The first three complex roots are calculated.

2 dec. $\alpha = 0\,(0{,}25)\,1{,}75$ [35]

Chapter 7

PROBABILITY INTEGRALS AND RELATED FUNCTIONS

We have selected the following from among the diverse notations which are used both in the Russian and the foreign literature for the most commonly used functions found in this chapter:

$$H'(x) = \frac{2}{\sqrt{\pi}} e^{-x^2},$$

$$\Phi'(x) = \frac{1}{\sqrt{2\pi}} e^{-\frac{x^2}{2}},$$

$$H(x) = \frac{2}{\sqrt{\pi}} \int_0^x e^{-t^2}\, dt,$$

$$\Phi(x) = \frac{1}{\sqrt{2\pi}} \int_{-x}^{x} e^{-\frac{t^2}{2}}\, dt.$$

In certain sections of this chapter the argument has been omitted for reasons of brevity in recording the notation of the integrals.

The connection between certain of these functions and the incomplete gamma function should also be noted. This becomes clear if one substitutes $x^2 = \frac{u}{\sqrt{2}}$ or $\frac{1}{2}x^2 = \frac{u}{\sqrt{2}}$.

The chapter contains the following sections:

- probability distribution functions and related functions;
- probability integrals and expressions containing probability integrals;
- tables in which the integral serves as the argument;
- probability integrals with complex argument;

---

[1] See pages 473 and 550 for the references to Chapter 7.

- derivatives of various orders of distribution functions;
- factors for calculating the probable error;
- logarithms of distribution functions;
- Fresnel integrals and related functions.

FUNCTIONS OF THE DISTRIBUTION OF PROBABILITY AND RELATED FUNCTIONS

See Chapter 3 for tables of exponential functions which are close to the functions in this section.

$$H'(x) = \frac{2}{\sqrt{\pi}} e^{-x^2}$$

21—26 dec.	$x = 1; 1,1; 1,25; 1, 4 (0,2) 3; 1,5; 2,5$	[186]
15 dec.	$x = 1,25 (0,01) 1,5 (0,02) 2,12; 2,17;$ $2,15 (0,05) 3 (0,1) 5,1 (0,2) 5,5; 6$	
9 dec.	$x = 0 (0,001) 1,249$	
15 dec.	$x = 0 (0,0001) 1 (0,001) 5,6$ (variable) $5,946$	[113]
8 fig.	$x = 4 (0,01) 10$	
5 dec.	$x = 0 (0,01) 3$	[1]
4 dec.	$x = 0 (0,01) 4$	[34] [41]
4 dec.	$x = 0 (0,01) 3$	[18] [65]

$$\frac{1}{2} H'(x) = \frac{1}{\sqrt{\pi}} e^{-x^2}$$

5—4 fig.	$x = 0 (0,1) 5$	[68]
4 dec.	$x = 0 (0,01) 2$	[121]
3—4 dec.	$x = 0 (0,1) 2,9$	[37]

$$\frac{1}{2} \sqrt{\pi} H'(x) = e^{-x^2}$$

24—27 dec.	$x = 1; 1,1, 1,25; 1,4 (0,2) 3$	[186]
15 dec.	$x = 3 (0,1) 5 (0,5) 6$	
9 dec.	$x = 0,05 (0,1) 4,55$	[92]
5 fig.	$x = 0 (0,1) 5$	[56] [85] [98]
5 dec.	$x = 0 (0,1) 3,4$	[111]
5 dec.	$x = 0 (0,1) 3$	[66]
4 fig.	$x = 0 (0,1) 5$	[20]

$$\frac{\rho}{\sqrt{\pi}} e^{-\rho^2 x^2}$$

The constant encountered in this and subsequent sections is such that

$$\frac{2}{\sqrt{\pi}} \int_0^{\rho} e^{-t^2} dt = \sqrt{\frac{2}{\pi}} \int_0^{\rho\sqrt{2}} e^{-\frac{t^2}{2}} dt = \frac{1}{2}.$$

5 dec.	$x = 0\,(0{,}01)\,5\,(0{,}1)\,7$	[1]

$$e^{-k^2 x^2}$$

5 dec.	$k = \frac{1}{2}\sqrt{\pi}$. $x = 0\,(0{,}1)\,3{,}9$	[111]
3 dec.	$k = 0{,}1\,(0{,}1)\,6$. $x = 0\,(0{,}5)\,4\,(1)\,8$	[109]

$$e^{k^2 x^2}$$

5 fig.	$k = \frac{1}{2}\sqrt{\pi}$. $x = 0\,(0{,}1)\,5$	[111]

$$1 - e^{-\rho^2 x^2}$$

5 dec.	$x = 0\,(0{,}1)\,5$	[1]

$$\Phi'(x) = \frac{1}{\sqrt{2\pi}} e^{-\frac{x^2}{2}}$$

15 dec.	$x = 0\,(0{,}0001)\,1\,(0{,}001)\,8{,}285$	[114]
7 dec.	$x = 0\,(0{,}01)\,4{,}5$	[93] [135]
10 dec.	$x = 4{,}5\,(0{,}01)\,6$	[93] [135]
7—10 dec.	$x = 0\,(0{,}001)\,2\,(0{,}01)\,6$	[29]
7 dec.	$x = 0\,(0{,}005)\,0{,}5$	[29]
7 dec.	$x = 0\,(0{,}01)\,4$	[71] [94] [143]
7 fig.	$x = 6\,(0{,}01)\,10$	[114]
6 dec.	$x = 0\,(0{,}01)\,5$	[192]
5 dec.	$x = 0\,(0{,}01)\,5{,}09$	[104]
5 dec.	$x = 0\,(0{,}01)\,4{,}99$	[60] [101]
5 dec.	$x = 0\,(0{,}01)\,4{,}07$	[9]
5 dec.	$x = 0\,(0{,}01)\,3{,}99$	[2] [45] [74]
5 dec.	$x = 0\,(0{,}01)\,3\,(0{,}1)\,5$	[44]
5 fig.	$x = 0\,(0{,}01)\,2\,(0{,}1)\,4$	[16]
5 dec.	$x = 0\,(0{,}1)\,4{,}8$	[122]
5 dec.	$x = 0\,(0{,}1)\,4$	[57]
5 dec.	$x = 5\,(1)\,50\,(10)\,100\,(50)\,500$	[29]
4 dec.	$x = 0\,(0{,}01)\,4{,}4\,(0{,}05)\,4{,}7\,(0{,}1)\,5$	[96]

4 dec.	$x = 0(0{,}01)\,3{,}99\,(0{,}05)\,4{,}95$	[61]
4 dec.	$x = 0\,(0{,}01)\,3{,}99$	[5] [100] [102]
4 dec.	$x = 0\,(0{,}01)\,3\,(0{,}1)\,4{,}9$	[38] [124]
4 dec.	$x = 0\,(0{,}01)\,3\,(0{,}1)\,3{,}9$	[55]
4 fig.	$x = 0\,(0{,}01)\,3$	[8]
3 dec.	$x = 0\,(0{,}01)\,2;3$	[103]

$$\frac{1}{\sqrt{2\pi}}\left(1 - e^{-\frac{x^2}{2}}\right)$$

7 dec.	$x = 0\,(0{,}1)\,5$	[93] [137]

$$\sqrt{2\pi}\,\Phi'(x) = e^{-\frac{x^2}{2}}$$

10 dec.	$x = -7\,(0{,}1)\,6{,}6$	[32]
5 dec.	$x = 0\,(0{,}01)\,4{,}09$	[9]
5 dec.	$x = 0\,(0{,}01)\,3\,(0{,}1)\,5$	[44] [58]
4 dec.	$x = 0\,(0{,}1)\,4{,}5$	[115]

$$\frac{1}{x}\,\frac{1}{\sqrt{2\pi}}\,e^{-\frac{x^2}{2}}$$

4 fig.	$x = 0\,(0{,}01)\,2\,(0{,}02)\,3\,(0{,}2)\,4{,}2\,(0{,}3)\,4{,}5$	[29]

$$\frac{1}{2}\,\frac{\rho}{\sqrt{\pi}}\left[e^{-\rho^2(x-a)^2} + e^{-\rho^2(x+a)^2}\right]$$

3 dec.	$x = 0\,(0{,}2)\,5.$ $\quad a = 0\,(0{,}2)\,4$	[1]

PROBABILITY INTEGRALS AND EXPRESSIONS CONTAINING PROBABILITY INTEGRALS

Integrals containing $e^{\pm t^2}$

$$H(x) = \frac{2}{\sqrt{\pi}}\int_0^x e^{-t^2}\,dt = \frac{1}{\sqrt{\pi}}\int_{-x}^{x} e^{-t^2}\,dt$$

19—22 dec.	$x = 1;\ 1{,}25;\ 1{,}4;\ 1{,}5;\ 1{,}6;\ 1{,}8;\ 2;\ 2{,}5;\ 3$	[186]
15 dec.	$x = 0\,(0{,}0001)\,1\,(0{,}001)\,5{,}6$ (var.) $5{,}946$	[113]
15 dec.	$x = 1\,(0{,}001)\,1{,}5\,(0{,}002)\,3\,(0{,}1)\,5\,(0{,}5)\,6$	[186]
9 dec.	$x = 0\,(0{,}001)\,1{,}249$	[186]

7; 11 dec.	$x = 0\ (0{,}01)\ 3{,}46;\ 3{,}46\ (0{,}01)\ 4$ $(0{,}1)\ 4{,}8$	[27] [40] [41] [117]
7—11 dec.	$x = 0{,}01\ (0{,}01)\ 3{,}84\ (0{,}02)\ 4\ (0{,}1)\ 4{,}6$ $(0{,}2)\ 5$	[29]
7 dec.	$x = 0\ (0{,}001)\ 2{,}499\ (0{,}01)\ 3{,}79$	[9]
7 dec.	$x = 0\ (0{,}01)\ 3$	[51] [89]
7 dec.	$x = 0\ (0{,}01)\ 2{,}97;\ 3$	[50]
7 dec.	$x = 0\ (0{,}01)\ 2$	[133] [154] [164]
6—7 dec.	$x = 0\ (0{,}001)\ 2{,}5\ (0{,}01)\ 3{,}79$	[82]
6; 7 dec.	$x = 0\ (0{,}01)\ 2{,}5;\ 2{,}5\ (0{,}01)\ 3$	
8 dec.	$x = 3\ (0{,}01)\ 3{,}35$	[192]
9; 10 dec.	$x = 3{,}35\ (0{,}01)\ 3{,}74;\ 3{,}74\ (0{,}01)\ 4$	
6—7 dec.	$x = 0\ (0{,}1)\ 3{,}9$	[68]
6 dec.	$x = 0\ (0{,}1)\ 3{,}5$	[79]
5 dec.	$x = 0\ (0{,}001)\ 1{,}8$	
6 dec.	$x = 1{,}8\ (0{,}001)\ 2{,}5$	[10]
7 dec.	$x = 2{,}5\ (0{,}01)\ 3$	
5—6 dec.	$x = 0\ (0{,}001)\ 1{,}6\ (0{,}01)\ 3{,}19$	[46]
5—6 dec.	$x = 0\ (0{,}01)\ 3$	[1] [42] [62]
5—6 dec.	$x = 0\ (0{,}01)\ 2{,}6\ (0{,}05)\ 3{,}6\ (0{,}1)\ 5$ $(0{,}5)\ 6$	[35]
5 dec.	$x = 0\ (0{,}001)\ 2\ (0{,}01)\ 3$	[97]
5 dec.	$x = 0\ (0{,}001)\ 1{,}5\ (0{,}002)\ 2{,}5$ $(0{,}01)\ 3{,}09$	[95]
5 dec.	$x = 0\ (0{,}001)\ 1{,}5\ (0{,}01)\ 3{,}09$	[47]
5 dec.	$x = 0\ (0{,}01)\ 3\ (0{,}1)\ 3{,}3$	[56]
5 dec.	$x = 0\ (0{,}01)\ 3{,}09$	[9] [106] [110 [120]
5 dec.	$x = 0(0{,}01)2{,}99$	[22] [37]
5 dec.	$x = 0\ (0{,}01)\ 2{,}4\ (0{,}1)\ 3{,}2$	[25]
5 dec.	$x = 0\ (0{,}01)\ 2$	[14] [39] [92]
5 dec.	$x = 0\ (0{,}02)\ 3$	[77]
5 dec.	$x = 0\ (0{,}05)\ 2$	[23]
5 dec.	$x = 0\ (0{,}01)\ 0{,}1\ (0{,}1)\ 3$	[153]
4—6 dec.	$x = 0\ (0{,}01)\ 2{,}99$	[18] [65]
4—6 dec.	$x = 0\ (0{,}01)\ 2{,}5\ (0{,}1)\ 3{,}4$	[59]
4—5 dec.	$x = 0\ (0{,}001)\ 1{,}5\ (0{,}01)\ 3{,}09$	[108]
4; 5 dec.	$x = 0\ (0{,}01)2;\ 2\ (0{,}01)\ 3$	[87]
4 dec.	$x = 0\ (0{,}001)\ 1{,}509;\ 1{,}5\ (0{,}01)\ 2{,}89$	[17]
4 dec.	$x = 0\ (0{,}001)\ 1{,}5\ (0{,}01)\ 2{,}89$	[12] [34] [41] [172]
4 dec.	$x = 0\ (0{,}025)\ 2{,}875$	[149]
4 dec.	$x = 0\ (0{,}01)\ 2{,}69$	[85]
4 dec.	$x = 0\ (0{,}01)\ 2{,}55$	[160]

4 dec.	$x = 0\,(0{,}01)\ 2{,}5\ (0{,}1)\ 2{,}9$	[86]
4 dec.	$x = 0\,(0{,}01)\ 2\ (0{,}1)\ 2{,}9$	[64]
4 dec.	$x = 0\,(0{,}01)\ 2$	[121]
3 dec.	$x = 0\,(0{,}01)\ 2\ (0{,}1)\ 2{,}9$	[28]

$$H(u) = \frac{2}{\sqrt{\pi}} \int_0^u e^{-t^2}\, dt$$

17 dec.	$u = \sqrt{n\pi}.\quad n = 1$	[189]
15 dec.	$u = \sqrt{n\pi}.\ n = 2\,(1)\ 11$	[189]
5 dec.	$u = \rho x.\ x = 0\,(0{,}01)\ 3{,}5\ (0{,}1)\ 5$	[14]
5 dec.	$u = \rho \frac{\Delta}{r}.\ \frac{\Delta}{r} = 0\,(0{,}01)\ 4\ (0{,}1)\ 5$	[133]

$$1 - H(x) = \frac{2}{\sqrt{\pi}} \int_x^\infty e^{-t^2}\, dt$$

8 fig.	$x = 4\,(0{,}01)\ 10$	[113]
5 dec.	$x = 0\,(0{,}1)\ 1{,}5;\ 2;\ 2{,}5$	[190]

$$2x^2 \cdot \frac{2}{\sqrt{\pi}} \int_x^\infty e^{-t^2}\, dt$$

5 dec.	$x = 0\,(0{,}1)\ 1{,}5\ (0{,}5)\ 2{,}5$	[190]

$$\frac{1}{2} H(x) = \frac{1}{\sqrt{\pi}} \int_0^x e^{-t^2}\, dt$$

6 dec.	$x = 0\,(0{,}1)\ 3$	[20] [49]

$$\frac{1}{2} H(\rho x) = \frac{1}{\sqrt{\pi}} \int_0^{\rho x} e^{-t^2}\, dt$$

8—13 dec.	$x = 0\,(0{,}01)\ 9\ (0{,}1)\ 10$	[162]
4—13 dec.	$x = 0\,(0{,}01)\ 5\ (0{,}1)\ 10$	[156]
4—5 dec.	$x = 0\ (0{,}05)\ 5{,}95$	[58]

$$\frac{1}{2} \sqrt{\pi}\, H(x) = \int_0^x e^{-t^2}\, dt$$

10 dec.	$x = 0\,(0{,}01)\ 4{,}52$	[92]

$$\frac{1}{2}\sqrt{\pi}\,[1-H(x)]=\int_x^{\infty} e^{-t^2}\,dt$$

15—21 dec.	$x=3\,(0,1)\,5\,(0,5)\,6$	[186]
11 dec.	$x=0\,(0,001)\,3\,(0,01)\,4,8$	[81]
11—14 dec.	$x=3\,(0,01)\,4,5$	[170]
10 dec.	$x=0\,(0,01)\,0,5$	[76]
10 dec.	$e^{-x^2}=0\,(0,01)\,0,8$	[76]
8—11 dec.	$x=0\,(0,01)\,3$	[75] [154]
4 dec.	$x=0\,(0,05)\,2,6$	[90]

$$H(\rho x)=\frac{2}{\sqrt{\pi}}\int_0^{\rho x} e^{-t^2}\,dt$$

7 dec.	$x=0\,(0,01)\,7$	[182]
5 dec.	$x=0\,(0,01)\,3,5\,(0,1)\,5$	[39]
5 dec.	$x=0\,(0,01)\,3,4\,(0,1)\,5$	[133]
5 dec.	$x=0\,(0,01)\,3\,(0,1)\,5,9$	[56]
4 dec.	$x=0\,(0,01)\,3,5\,(0,1)\,6,1$	[67]
4 dec.	$x=0\,(0,01)\,3,5\,(0,1)\,5,9$	[85]]86]

$$\frac{2\rho}{\sqrt{\pi}}\int_0^{x} e^{-\rho^2 t^2}\,dt$$

5—6 dec.	$x=0\,(0,01)\,4\,(0,1)\,7$	[1]
5 dec.	$x=0\,(0,01)\,3,1\,(0,1)\,5,4$	[9]

$$\varphi(x)=\frac{\rho}{2e\sqrt{\pi}}\int_{x-l}^{x+l} e^{-\rho^2 t^2}\,dt$$

4 dec.	$x=0\,(0,2)\,5.\;\; l=0\,(0,2)\,5$	[1]

$$\int_0^{x} e^{t^2}\,dt$$

6 dec.	$x=0\,(0,01)\,2$	[158] [159]
7—9 fig.	$x=2\,(0,01)\,4$	[158] [159]
6 dec.	$x=0\,(0,01)\,2$	[177]
4; 3; 2 dec.	$x=0\,(0,01)\,1;\;1\,(0,01)\,1,8;\;1,8\,(0,01)\,1,99$	[18] [29] [65]

$$\frac{2}{\sqrt{\pi}}\int_0^{x} e^{t^2}\,dt$$

4—5 fig.	$x=0\,(0,1)\,2,6\,(0,2)\,7,4$	[107]

$$\frac{1}{\sqrt{\pi}}e^{-x^2}+x\frac{2}{\sqrt{\pi}}\int_0^x e^{-t^2}dt$$

4 fig. $x=0\ (0{,}1)\ 1\ (0{,}2)\ 3$ [132]

$$\frac{2\rho}{\sqrt{\pi}}\int_0^x e^{-\rho^2t^2}dt-2x\frac{\rho}{\sqrt{\pi}}e^{-\rho^2x^2}$$

5 dec. $x=0\ (0{,}1)\ 5$ [1]

$$\left(2-\frac{2}{\sqrt{\pi}}\int_x^\infty e^{-t^2}dt\right)(1+2x^2)+\frac{2x}{\sqrt{\pi}}e^{-x^2}$$

5 dec. $x=0\ (0{,}1)\ 1{,}5\ (0{,}5)\ 2{,}5$ [190]

$$\frac{1}{\sqrt{\pi}}\left[(1+x^2)e^{-x^2}+2x\left(\frac{3}{2}+x^2\right)\int_0^x e^{-t^2}dt\right]$$

3 fig. $x=0\ (0{,}1)\ 1\ (0{,}2)\ 3$ [132]

$$\frac{1-H(x)}{H'(x)}=e^{x^2}\int_x^\infty e^{-t^2}dt$$

10 dec. $x=-0{,}2\ (0{,}05)\ 3{,}8\ (0{,}1)\ 6{,}3$ [105]

$$\frac{2x[1-H(x)]}{H'(x)}=2xe^{x^2}\int_x^\infty e^{-t^2}dt$$

15 dec. $x=3\ (0{,}1)\ 5\ (0{,}5)\ 6$ [186]

$$e^{x^2}\int_0^x e^{-t^2}dt$$

5 fig. $x=0\ (0{,}001)\ 3\ (0{,}0002)\ 5$ [7]

$$xe^{x^2}\int_0^x e^{-t^2}dt$$

5 dec. $x=0{,}7\ (0{,}01)\ 1{,}2$ [191]

$$e^{-x^2}\int_0^x e^{t^2}dt$$

12 dec.	$x = 3\,(1)\,16$	[130]
10 dec.	$x = -0{,}2\,(0{,}05)\,4\,(0{,}1)\,6{,}5\,(0{,}5)\,12{,}5$	[105]
6; 8 dec.	$x = 0\,(0{,}01)\,1{,}99;\ 2\,(0{,}01)\,4\,(0{,}05)\,4{,}95$	[161]
9 dec.	$x = 5\,(0{,}05)\,7{,}5\,(0{,}1)\,10\,(0{,}2)\,12$	[161]
6 dec.	$x = 0\,(0{,}1)\,1;\ \frac{1}{2}x^2 = 0{,}6\,(0{,}2)\,6;\ 16;\ 25;\ 100$	[184]
5 dec.	$x = 0\,(0{,}001)\,2\,(0{,}01)\,10$	[7]
4—5 dec.	$x = 0\,(0{,}2)\,12$	[88]
4 dec.	$x = 0\,(0{,}01)\,3{,}6;\ 4;\ 4{,}5;\ 5;\ 6;\ 8;\ 10$	[163]
4 dec.	$x = 0\,(0{,}1)\,10$	[188]

$$xe^{-x^2}\int_0^x e^{t^2}dt$$

5 dec.	$x = 0{,}1;\ 0{,}5;\ 0{,}8;\ 1;\ 1{,}2\,(0{,}05)\,2{,}2$	[191]

$$\frac{1}{x}\,e^{-x^2}\int_0^x e^{t^2}dt$$

4 dec.	$x = 0\,(0{,}01)\,3{,}6;\ 4;\ 4{,}5;\ 5;\ 6;\ 8;\ 10$	[163]

$$\left(2xe^{x^2}\int_x^\infty e^{-t^2}dt\right)^{-1}$$

8 dec.	$x = 0{,}05;\ 0{,}1\,(0{,}1)\,2\,(0{,}5)\,4{,}5$	[190]

$$1 + \frac{1}{2}x^{-2} - \left[2xe^{x^2}\int_0^\infty e^{-t^2}dt\right]$$

8 dec.	$x = 0{,}05;\ 0{,}1\,(0{,}1)\,2\,(0{,}5)\,4{,}5$	[190]

$$\int_0^\infty \frac{e^{-u^2}du}{u+x}$$

4 dec.	$x = 0\,(0{,}02)\,2\,(0{,}05)\,3\,(0{,}1)\,10$	[183]

Integrals containing $e^{\pm\frac{t^2}{2}}$

$$\Phi(x) = \sqrt{\frac{2}{\pi}}\int_0^x e^{-\frac{t^2}{2}}dt = \frac{1}{\sqrt{2\pi}}\int_{-x}^{x} e^{-\frac{t^2}{2}}dt$$

15 dec.	$x = 0\,(0{,}0001)\,1\,(0{,}001)\,7{,}8\,($ var. $)\,8{,}285$	[114]
6; 7; 8 dec.	$x = 0\,(0{,}01)\,2{,}5;\ 2{,}5\,(0{,}01)\,4;\ 4\,(0{,}01)\,4{,}5$	[192]
9; 10; 11 dec.	$x = 4{,}5\,(0{,}01)\,5;\ 5\,(0{,}01)\,5{,}5;\ 5{,}5\,(0{,}01)\,6$	
5 dec.	$x = 0\,(0{,}01)\,4{,}09$	[6]
5 dec.	$x = 0\,(0{,}01)\,3{,}99\,(0{,}1)\,4{,}5$	[9]
4—6 dec.	$x = 0\,(0{,}01)\,3\,(0{,}1)\,6$	[124]
4 dec.	$x = 0\,(0{,}01)\,2\,(0{,}05)\,4$	[3]
4 dec.	$x = 0\,(0{,}01)\,2\,(0{,}1)\,4$	[16]

$$\frac{1}{2}\Phi(x) = \frac{1}{\sqrt{2\pi}}\int_0^x e^{-\frac{t^2}{2}}\,dt$$

7—10 dec.	$x = 0\,(0{,}001)\,2\,(0{,}01)\,6$	[29]
6 dec.	$x = 0\,(0{,}004)\,6{,}468$	[112]
6 dec.	$x = 0\,(0{,}01)\,4{,}99$	[101]
5 dec.	$x = 0\,(0{,}001)\,2{,}5\,(0{,}01)\,4\,(0{,}1)\,5$	[44]
5 dec.	$x = 0\,(0{,}01)\,5{,}09$	[104]
5 dec.	$x = 0\,(0{,}01)\,4{,}99$	[60]
5 dec.	$x = 0\,(0{,}01)\,3{,}99$	[45] [74]
5 dec.	$x = 0\,(0{,}1)\,2;\ 2{,}2;\ 2{,}5;\ 3{,}4$	[36]
4—7 dec.	$x = 0\,(0{,}01)\,4{,}4\,(0{,}05)\,4{,}7\,(0{,}1)\,5$	[96]
4—7 dec.	$x = 0\,(0{,}01)\,3{,}2\,(0{,}1)\,4\,(0{,}5)\,5$	[58]
4—6 dec.	$x = 0\,(0{,}01)\,2\,(0{,}02)\,3\,(0{,}2)\,4\,(0{,}5)\,5$	[116]
4; 5 dec.	$x = 0\,(0{,}01)\,2\,(0{,}02)\,2{,}98;\ 3\,(0{,}2)\,3{,}4$	[31]
6 dec.	$x = 3{,}6\,(0{,}2)\,4;\ 4{,}5$	
4 dec.	$x = 0\,(0{,}01)\,3{,}99$	[102]
4 dec.	$x = 0\,(0{,}01)\,2\,(0{,}02)\,3\,(0{,}2)\,4\,(0{,}5)\,5$	[5] [8]
4 dec.	$x = 0\,(0{,}01)\,2;\ 3$	[103]
4 dec.	$x = 0\,(0{,}02)\,1{,}5\,(0{,}05)\,2\,(0{,}1)\,3\,(0{,}2)\,4$	[21]

$$\sqrt{2}\,\Phi(x) = \frac{2}{\sqrt{\pi}}\int_0^x e^{-\frac{t^2}{2}}dt$$

5 dec.	$x = 0\,(0{,}01)\,3{,}99$	[2]

$$\sqrt{\frac{\pi}{2}}\,\Phi(x) = \int_0^x e^{-\frac{t^2}{2}}dt$$

5 dec.	$x = 0\,(0{,}1)\,4{,}5$	[115]

$$\frac{1}{2}[1+\Phi(x)] = \frac{1}{\sqrt{2\pi}}\int_{-\infty}^{x} e^{-\frac{t^2}{2}}\,dt$$

7 dec.	$x = 0\,(0{,}01)\,4{,}5$	[93] [135]
10 dec.	$x = 4{,}5\,(0{,}01)\,6$	
7 dec.	$x = 0\,(0{,}01)\,4$	[71]
7 dec.	$x = 0\,(0{,}1)\,5{,}4$	[168]
5 dec.	$x = 0\,(0{,}1)\,4{,}4$	[122]
5 dec.	$x = 0\,(0{,}1)\,4$	[57]
5 dec.	$x = 0\,(0{,}1)\,0{,}4\,(0{,}05)\,0{,}7\,(0{,}01)\,0{,}8\,(0{,}05)\,1{,}1;\ 1{,}5\,(0{,}5)\,3{,}5$	[69]
4 dec.	$x = 0\,(0{,}05)\,3{,}95$	[53]
4 dec.	$x = -2{,}9\,(0{,}1)\,-2\,(0{,}01)\,2\,(0{,}1)\,2{,}9$	[38]

$$1-\Phi(x) = \sqrt{\frac{2}{\pi}}\int_{x}^{\infty} e^{-\frac{t^2}{2}}\,dt$$

7 fig.	$x = 6\,(0{,}01)\,10$	[114]
7 dec.	$x^2 = 1\,(1)\,30$	[93] [134]
5 dec.	$x^2 = 0\,(0{,}01)\,1\,(0{,}1)\,10$	[122]
5 dec.	$x = 0\,(0{,}01)\,3{,}99\,(0{,}1)\,4{,}9$	[9]
5 dec.	$x = 0\,(0{,}1)\,4{,}5$	[122]
4 dec.	$x = 0\,(0{,}01)\,4$	[57]
4 dec.	$x = 0\,(0{,}1)\,4{,}1$	[72]

$$\frac{1}{2}[1-\Phi(x)] = \frac{1}{\sqrt{2\pi}}\int_{x}^{\infty} e^{-\frac{t^2}{2}}\,dt$$

7 dec.	$x = 0\,(0{,}01)\,4$	[94] [143]
7–10 dec.	$x = 0\,(0{,}5)\,5$	[100]
5 fig.	$x = 6\,(1)\,10$	
4 dec.	$x = 0\,(0{,}01)\,3{,}99$	

$$\sqrt{\frac{\pi}{2}}[1-\Phi(x)] = \int_{x}^{\infty} e^{-\frac{t^2}{2}}\,dt$$

10 dec.	$x = -7\,(0{,}1)\,6{,}6$	[32]

$$\frac{1}{\sqrt{\pi}}\int_{-\infty}^{x} e^{-\frac{t^2}{2}}\,dt$$

5 dec.	$x = 0\,(0{,}01)\,3{,}99\,(0{,}1)\,4{,}4$	[9]

$$\frac{\frac{1}{2}[1-\Phi(x)]}{\Phi'(x)} = e^{\frac{x^2}{2}} \int_x^\infty e^{-\frac{t^2}{2}} dt$$

24 dec.	$x = 0\,(0{,}1)\,10$	[33]
12 dec.	$x = 0\,(0{,}01)\,10$	[33]
5 dec.	$x = 0\,(0{,}01)\,4\,(0{,}05)\,5\,(0{,}1)\,10$	[94] [145]
3 dec.	$x = 0\,(0{,}01)\,0{,}1(0{,}1)\,3$	[93]
3 dec.	$-x = 0\,(0{,}01)\,0{,}1(0{,}1)\,3$	[94] [140]

$$\frac{\frac{1}{2}x[1-\Phi(x)]}{\Phi'(x)} - 1 = -1 + xe^{\frac{x^2}{2}} \int_x^\infty e^{-\frac{t^2}{2}} dt$$

23 dec.	$x = 0\,(0{,}1)\,10$	[33]
10 dec.	$x = 0\,(0{,}01)\,10$	[33]

$$\frac{\Phi'(x)}{\frac{1}{2}[1-\Phi(x)]} = \left(e^{\frac{x^2}{2}} \int_x^\infty e^{-\frac{t^2}{2}} dt\right)^{-1}$$

15 dec.	$x = 0\,(0{,}1)\,10$	[33]

$$x + \frac{\Phi'(x)}{\frac{1}{2}[1-\Phi(x)]} = x + \left(e^{\frac{x^2}{2}} \int_\infty^x e^{-\frac{t^2}{2}} dt\right)^{-1}$$

3 dec.	$x = -4\,(0{,}2)\,3$	[109]

Various integrals

$$P(t,n) = \Pi(t,x) = \mathrm{P}(\chi^2,n) = \frac{1}{2^{\frac{n-2}{2}}\Gamma\left(\frac{n}{2}\right)} \int_\chi^\infty x^{n-1} e^{-\frac{x^2}{2}} dx$$

5 dec.	$t = -4\,(0{,}1)\,4{,}8.$	$n = 6\,(0{,}5)\,11\,(1)\,32$	[11]
5 dec.	$t = -4{,}5\,(0{,}1)\,4{,}8.$	$x = 0\,(0{,}02)\,0{,}22\,(0{,}01)\,0{,}25$	[11]
5 dec.	$\chi^2 = 0\,(0{,}1)\,3{,}2.$	$n = 0\,(0{,}05)\,0{,}2\,(0{,}1)\,6$	[11]
5 dec.	$\chi^2 = 3{,}2\,(0{,}2)7(0{,}5)10(1)35.$	$n = 0\,(0{,}1)\,0{,}4\,(0{,}2)\,6$	[11]
4 dec.	$\chi = 1\,(1)\,30.$	$n = 1\,(1)\,29$	[5]

$$S(x) = \frac{\Gamma\left(\frac{n}{2}\right)}{\sqrt{\pi(n-1)}\,\Gamma\left(\frac{n-1}{2}\right)} \int_{-\infty}^x \left(1 + \frac{t^2}{n-1}\right)^{-\frac{n}{2}} dt$$

5 dec.	$n-1 = 1\,(1)\,20.$	$x = 0\,(0{,}1)\,4\,(0{,}2)\,8$	[150]
5 dec.	$n-1 = 20(1)24;\ 30;\ 40;\ 60;\ 120;\ \infty.$	$x = 0(0{,}05)2(0{,}1)4{,}5$	[150]
3 dec.	$n = 2\,(1)\,20;\ \infty.$	$x = 0\,(0{,}1)\,6$	[5]

$$\frac{2}{\sqrt{\pi}}\left\{[1+2x^2(1-\theta^2)]\int_x^\infty e^{-t^2}dt - xe^{-x^2}\right\}$$

5 dec. $\theta = 0\,(0{,}1)\,0{,}4;\ 1.\quad x = 0\,(0{,}1)\,1{,}5;\ 2;\ 2{,}5$ [190]

$$\frac{4}{\sqrt{\pi}}\int_x^\infty (t-x)\,e^{-t^2}\,dt$$

4 dec. $x = 0\,(0{,}01)\,1\,(0{,}02)\,2{,}18$ [167]

$$\frac{4}{\sqrt{\pi}}\int_x^\infty (t-x)^2 e^{-t^2}\,dt$$

4 dec. $x = 0\,(0{,}01)\,1\,(0{,}02)\,2$ [167]

$$e^{-x^2}f(x\cos\theta) = \frac{2}{\sqrt{\pi}}e^{x^2(1+\cos^2\theta)}\int_{-x\cos\theta}^\infty (t+x\cos\theta)\,e^{-t^2}dt$$

2 dec. $x = 0{,}1\,(0{,}1)\,1{,}8.\quad \theta = 0\,(5°)\,360°$ [109]

$$F(x) = \sqrt{\pi}x + 2\int_x^\infty (t-x)\,e^{-t^2}dt$$

3 dec. $x = 0\,(0{,}05)\,2$ [109]

$$\psi(x) = \sqrt{\pi}\left(\frac{1}{2}+x^2\right) - 2\int_x^\infty (t-x)^2\,e^{-t^2}\,dt$$

7 fig. $x = 0\,(0{,}05)\,1{,}5\,(0{,}1)\,10\,(0{,}2)\,20{,}4$ [175]
6 dec. $x = 0\,(0{,}1)\,3$ [185]
3 dec. $x = 0\,(0{,}1)\,3$ [66]

$$\int_0^w e^{-y^2}dy\int_0^y e^{x^2}dx$$

10 dec. $w = -0{,}2\,(0{,}05)\,3{,}5\,(0{,}1)\,6$ [105]

$$\int_0^w e^{y^2} dy \int_y^\infty e^{-x^2} dx$$

10 dec. $w = -0{,}2\,(0{,}05)\,3{,}5\,(0{,}1)\,6$ [105]

$$\frac{\frac{1}{\sqrt{\pi}}\left[e^{-x^2} + 2x\int_0^x e^{-t^2} dt\right]}{\left\{2\left[\left(x^2+\frac{1}{2}\right)^2 - \frac{1}{\pi}\left(e^{-x^2} + 2x\int_0^x e^{-t^2} dt\right)^2\right]\right\}^{1/2}}$$

4 — 5 fig. $x = 0\,(0{,}1)\,1\,(0{,}2)\,3$ [132]

$$\frac{\int_{-x}^\infty (t+x)^2 e^{-t^2} dt}{\int_{-x}^\infty (t+x) e^{-t^2} dt}$$

3 dec. $x = -1\,(0{,}1)\,1{,}6$ [48]

3 dec. $x = -1\,(0{,}1)\,1{,}7$ [109]

$$\frac{1}{n!}\int_x^\infty (t-x)^n e^{-\frac{t^2}{2}} dt$$

10 dec. $n = 0\,(1)\,11.$ $x = -7\,(0{,}1)\,X$; $n = 12\,(1)\,17.$ $x = -5\,(0{,}1)\,X$; $n = 18\,(1)\,21.$ $x = -2{,}5\,(0{,}1)\,X$ [32]

10 dec. $n = 0\,(1)\,11.$ $x = -10(1)-7(0{,}1)X$; $n = 12\,(1)\,15.$ $x = -5\,(0{,}1)\,X$; $n = 16\,(1)\,21.$ $x = -2{,}5\,(0{,}1)X$ [152]

$X \leqslant 6{,}6$

$$L' = e^{-\frac{x^2}{2}}\left(\int_x^\infty e^{-\frac{t^2}{2}} dt\right)^{-1}$$

15 dec. $x = 0\,(0{,}1)\,10$ [33]

$$\frac{1}{2}L'' = \frac{1}{2}e^{-x^2}\left(\int_x^\infty e^{-\frac{t^2}{2}} dt\right)^{-2}$$

14 dec. $x = 0\,(0{,}1)\,10$ [33]

$$\frac{1}{2}e^{x^2}\int_x^\infty e^{-\frac{t^2}{2}}dt\int_x^\infty (t-x)^2 e^{-\frac{t^2}{2}}dt$$

9 dec.	$x = -7\,(0{,}1)\,0$	[32]
8 dec.	$x = 0\,(0{,}1)\,0{,}9$	
7 dec.	$x = 1\,(0{,}1)\,1{,}9$	
6 dec.	$x = 2\,(0{,}1)\,3$	
3—5 dec.	$x = 3\,(0{,}1)\,5$	

$$\psi_1 = e^{x^2}\int_x^\infty e^{-\frac{t^2}{2}}dt\int_x^\infty (t-x)^2 e^{-\frac{t^2}{2}}dt - 1$$

3 dec.	$x = 0\,(0{,}01)\,0{,}1\,(0{,}1)\,3$	[93]
3 dec.	$-x = 0\,(0{,}01)\,0{,}1\,(0{,}1)\,3$	[94] [140]

$$\psi_3 = \sqrt{2\pi}\left(\int_x^\infty e^{-\frac{t^2}{2}}dt\right)^{-1}$$

3 dec.	$x = 0\,(0{,}01)\,0{,}1\,(0{,}1)\,3$	[93]
3 dec.	$-x = 0\,(0{,}01)\,0{,}1\,(0{,}1)\,3$	[94] [140]

$$\frac{1}{2\pi\sqrt{1-r^2}}\int_h^\infty\int_k^\infty \exp\left\{-\frac{1}{2(1-r^2)}(x^2+y^2-2rxy)\right\}dx\,dy$$

In all the following tables the range of variation of h and k is 0(0,1)2,6.

8 dec.	$-r = 0{,}65\,(0{,}05)\,0{,}75$	[147]
7 dec.	$-r = 0{,}8\,(0{,}05)\,1$	[141]
7 dec.	$-r = 0\,(0{,}05)\,0{,}6$	[147]
6 dec.	$r = 0\,(0{,}05)\,1$	[146]
4 dec.	$r = 0{,}8\,(0{,}05)\,1$	[93] [139]

$$\int_{-\infty}^{\infty}\left[1-\left(\frac{1}{\sqrt{2\pi}}\int_{-\infty}^{x} e^{-\frac{t^2}{2}}dt\right)^n-\left(\frac{1}{\sqrt{2\pi}}\int_x^\infty e^{-\frac{t^2}{2}}dt\right)^n\right]dx$$

5 dec.	$n = 2\,(1)\,1000$	[94] [144]

$$\int_0^x y^{-1/2} e^{-\pi^2/y}\, dy = 2\sqrt{x}\, e^{-\pi^2/x} - 4\pi \int_{\frac{\pi}{\sqrt{x}}}^{\infty} e^{-y^2}\, dy$$

11 dec. $x = 0{,}4\,(0{,}1)\,2{,}5$ [161]

Moments

These integrals can also be expressed by the incomplete gamma function.

$$m_n(x) = \frac{1}{\sqrt{2\pi}(n-1)!!} \int_0^x t^n e^{-\frac{t^2}{2}}\, dt$$

7 dec. $n = 11$ & 12. $x = 0\,(0{,}1)\,5$ [94]
7 dec. $n = 1\,(1)\,10$. $x = 0\,(0{,}1)\,5$ [93] [137]

$$1 - \sqrt{2\pi}\frac{\mu_{n-1}(x\sqrt{n})}{2^{1/2\,n-1}\left(\frac{1}{2}n-1\right)!}$$

$$\mu_n(x) = \frac{1}{\sqrt{2\pi}} \int_0^x t^n e^{-\frac{t^2}{2}}\, dt.$$

4 dec.ч. $n = 1\,(1)\,10; 12; 15; 19; 24; 30$. $x = 0\,(0{,}1)\,4$ [72]

TABLES IN WHICH THE INTEGRAL SERVES AS THE ARGUMENT

Values of the abscissa and deviations of x

Argument $H = \frac{2}{\sqrt{\pi}} \int_0^x e^{-t^2}\, dt.$

9 dec. $H = 0\,(0{,}1)\,1$ [186]

Argument $1 - H = \frac{2}{\sqrt{\pi}} \int_x^{\infty} e^{-t^2}\, dt$

5 fig. $1 - H = 0\,(0{,}01)\,1$ [56]

Argument $\frac{1}{2}(1+H)=\frac{1}{2}+\frac{1}{\sqrt{\pi}}\int_0^x e^{-t^2}\,dt$

4 dec.	$\left\{\begin{array}{l}\frac{1}{2}(1+H)=0{,}5\,(0{,}01)\,0{,}83\,(0{,}0025)\,0{,}97\,(0{,}001)\,1\\ \frac{1}{2}(1+H)=\frac{1}{2}\left(\frac{1}{64}\right)1\end{array}\right\}$	[52]

Argument $\frac{1}{2}H(\rho x)=\frac{1}{\sqrt{\pi}}\int_0^{\rho x} e^{-t^2}\,dt$

7—6 dec.	$\frac{1}{2}H(\rho x)=0\,(0{,}001)\,0{,}45\,(0{,}0001)\,0{,}4999$	[162]

Argument $H(\rho x)=\frac{2}{\sqrt{\pi}}\int_0^{\rho x} e^{-t^2}\,dt$

9 dec.	$H(\rho x)=0\,(0{,}1)\,1$	[186]

Argument $50\,H(\rho x)=\frac{100}{\sqrt{\pi}}\int_0^{\rho x} e^{-t^2}\,dt$

2 dec.	$50\,H(\rho x)=0\,(0{,}1)\,45\,(0{,}01)\,49{,}99$	[156]

Argument $\Phi=\sqrt{\frac{2}{\pi}}\int_0^x e^{-\frac{t^2}{2}}\,dt$

dec.	$\Phi=0\,(0{,}01)\,0{,}8$	[93] [135]

Argument $\frac{1}{2}\Phi=\frac{1}{\sqrt{2\pi}}\int_0^x e^{-\frac{t^2}{2}}\,dt$

6 dec.	$\frac{1}{2}\Phi=0\,(0{,}001)\,0{,}499$	[73]
3 dec.	$\frac{1}{2}\Phi=0\,(0{,}01)\,0{,}5$	[58] [63]

Argument $1-\Phi=\sqrt{\frac{2}{\pi}}\int_x^\infty e^{-\frac{t^2}{2}}\,dt$

6 dec.	$1-\Phi=0\,(0{,}01)\,1$	[54] [55]

Argument $\frac{1}{2}(1+\Phi)=\frac{1}{\sqrt{2\pi}}\int_{-\infty}^{x} e^{-\frac{t^2}{2}}\,dt$

10 dec.	$\frac{1}{2}(1+\Phi)=0{,}5\,(0{,}001)\,1$	[93] [94] [136] [148]
8 dec.	$\frac{1}{2}(1+\Phi)=0{,}5\,(0{,}0001)\,1$	[72]
6; 5 dec.	$\frac{1}{2}(1+\Phi)=0{,}5\,(0{,}001)\,0{,}95;$ $0{,}95\,(0{,}001)\,0{,}99\,(0{,}0001)\,0{,}9989$	[192]
4 dec.	$\frac{1}{2}(1+\Phi)=0{,}9989\,(0{,}0001)\,1$	[192]
4 dec.	$\frac{1}{2}(1+\Phi)=0\,(0{,}01)\,1$	[38]
3 dec.	$\frac{1}{2}(1+\Phi)=0{,}9^k0\,(10^{-(k+1)})\,0{,}9^k8$, $k=4\,(1)\,7$, where e.g. $0{,}9^40=0{,}99990$. $\frac{1}{2}(1+\Phi)=0{,}9^k5\,(10^{-(k+1)})\,0{,}9^{k+1}$. $k=8;\ 9$	[192]

Argument $\frac{1}{2}(1-\Phi)=\frac{1}{\sqrt{2\pi}}\int_{x}^{\infty} e^{-\frac{t^2}{2}}\,dt$

5 dec.	$\frac{1}{2}(1-\Phi)=0\,(0{,}001)\,0{,}5$	[93] [138]
4 dec.	$\frac{1}{2}(1-\Phi)=0\,(0{,}001)\,0{,}5$	[96]
4 dec.	$\frac{1}{2}(1-\Phi)=0\,(0{,}005)\,0{,}5$	[100]

Values of the ordinate

Argument $\Phi=\sqrt{\frac{2}{\pi}}\int_{0}^{x} e^{-\frac{t^2}{2}}\,dt$

7 dec.	$\Phi=0\,(0{,}01)\,0{,}8$	[93] [135]

Argument $\frac{1}{2}\Phi=\frac{1}{\sqrt{2\pi}}\int_{0}^{x} e^{-\frac{t^2}{2}}\,dt$

6 dec.	$\frac{1}{2}\Phi=0\,(0{,}001)\,0{,}499$	[73]
3 dec.	$\frac{1}{2}\Phi=0\,(0{,}01)\,0{,}5$	[58] [63]

Argument $\frac{1}{2}(1+\Phi)=\frac{1}{\sqrt{2\pi}}\int_{-\infty}^{x} e^{-\frac{t^2}{2}}\,dt$

10 dec.	$\frac{1}{2}(1+\Phi)=0{,}5\,(0{,}001)\,1$	[94] [148]
8 dec.	$\frac{1}{2}(1+\Phi)=0{,}5\,(0{,}0001)\,1$	[72]
6 dec.	$\frac{1}{2}(1+\Phi)=0{,}5\,(0{,}001)\,0{,}99\,(0{,}0001)\,1$	[192]
5 dec.	$\frac{1}{2}(1+\Phi)=0\,(0{,}001)\,1$	[94] [142]

Argument $\frac{1}{2}(1-\Phi)=\frac{1}{\sqrt{2\pi}}\int_{x}^{\infty} e^{-\frac{t^2}{2}}\,dt$

9—14 dec.	$\frac{1}{2}(1-\Phi)=10^{-4};\ 10^{-p};\ 5\cdot 10^{-p}$ $p=5\,(1)\,9$	[72]
5 dec.	$\frac{1}{2}(1-\Phi)=0\,(0{,}001)\,0{,}5$	[93] [138]
4 dec.	$\frac{1}{2}(1-\Phi)=0\,(0{,}001)\,0{,}5$	[96]

PROBABILITY INTEGRALS WITH COMPLEX ARGUMENT

$$w(z)=e^{-z^2}\int_0^z e^{x^2}\,dx$$

The values of the function are calculated in rectangular coordinates.

$$w(z)=u(\rho,\theta)+iv(\rho,\theta),$$

where ρ is the modulus and θ the argument of z.

5 dec.	$\theta=0.\ \rho=0\,(0{,}001)\,2\,(0{,}01)\,10$; $\theta=2°{,}5\,(\text{ var. })\,45°.\ \rho=k\,(\text{ var.})\,5$, $0\leqslant k\leqslant 3{,}5$	[7]
5 fig.	$\theta=90°.\ \rho=0\,(0{,}001)\,3\,(0{,}0002)\,5$	[7]

$$w(z)=e^{-z^2}\left(1+\frac{2i}{\sqrt{\pi}}\int_0^z e^{t^2}\,dt\right)$$

The tables give values of the real u and imaginary v parts of the function of complex argument $z=x+iy$

6 dec.	$x=0\,(0{,}02)\,3.\ y=0\,(0{,}02)\,3$; $x=3\,(0{,}1)\,5.\ y=0\,(0{,}1)\,2{,}9$; $x=0\,(0{,}1)\,5.\ y=3\,(0{,}1)\,5$	[15]

$$\frac{2}{\eta} e^{z^2} \int_z^\infty e^{-t^2} dt$$

The real and imaginary parts of the integral are given in the table. Here

$$z = \frac{1 - iu}{\eta}.$$

3 fig.	$u = 0\,(0{,}2)\,4.\ \eta = 2;\ 1;\ \frac{1}{2}\ \&\ \frac{1}{10}$	[30]

$$\frac{2}{\sqrt{\pi}} \mathrm{Re}\left\{e^{z^2} \int_z^\infty e^{-t^2} dt\right\}$$

$$z = x + iy.$$

4 dec.	$x = 0\,(0{,}5)\,1{,}5.$	$y = 0\,(0{,}2)\,2$	[173]
4 dec.	$x = 0.$	$y = 0\,(0{,}2)\,3$	[88]
3—4 dec.	$x = 0{,}5;\ 1{,}5.$	$y = 0\,(0{,}2)\,2\,(2)\,10$	[88]
3—4 dec.	$x = 1.$	$y = 0\,(0{,}2)\,10$	[88]
3 fig.	$x = 2;\ 10.$	$\frac{y}{x} = 0\,(0{,}2)\,4$	[88]
3 fig.	$x = 0\,(0{,}01)\,0{,}2.$	$y = 0\ \left(\frac{1}{4}\right)\,5$	[131]
2 fig.	$x = 0\,(0{,}1)\,0{,}5.$	$y = 0\ \left(\frac{1}{4}\right)\,2\,\left(\frac{1}{2}\right)\,5$	[131]

DERIVATIVES OF VARIOUS ORDERS OF DISTRIBUTION FUNCTION

$$\Phi^{(2)}(x) = -x\Phi'(x)$$

4 dec.	$x = 0\,(0{,}01)\,2\,(0{,}02)\,3\,(0{,}2)\,4{,}2\,(0{,}3)\,4{,}5$	[29]

$$\Phi^{(3)}(x) = (x^2 - 1)\,\Phi'(x)$$

4 dec.	$x = 0\,(0{,}01)\,3{,}99;\ 4\,(0{,}05)\,4{,}95$	[61]
4 dec.	$x = 0\,(0{,}01)\,2\,(0{,}02)\,3\,(0{,}2)\,4{,}2\,(0{,}3)\,4{,}5$	[29]

$$\Phi^{(4)}(x) = (3x - x^3)\,\Phi'(x)$$

4 dec.	$x = 0\,(0{,}01)\,3{,}99;\ 4\,(0{,}05)\,4{,}95$	[61]
4 dec.	$x = 0\,(0{,}01)\,2\,(0{,}02)\,3\,(0{,}2)\,4{,}2\,(0{,}3)\,4{,}5$	[29]

$$\Phi^{(5)}(x) = (x^4 - 6x^2 + 3)\,\Phi'(x)$$

4 dec.	$x = 0\,(0{,}01)\,3{,}99;\ 4\,(0{,}05)\,4{,}95$	[61]
4 dec.	$x = 0\,(0{,}01)\,2\,(0{,}02)\,3\,(0{,}2)\,4{,}2\,(0{,}3)\,4{,}5$	[29]

$$\frac{1}{3!}\left[\Phi^{(3)}(x)+\frac{1}{\sqrt{2\pi}}\right]=\frac{1}{3!\sqrt{2\pi}}\left[\frac{d^2}{dx^2}(e^{-\frac{x^2}{2}})+1\right]$$

4 dec.	$x = 0\,(0{,}01)\,2\,(0{,}02)\,3\,(0{,}2)\,4{,}2;\ \infty$	[31]
4 dec.	$x = 0\,(0{,}1)\,3\,(0{,}2)\,4{,}2;\ \infty$	[43]

$$2^{-(n-1)}\left(\frac{d}{dx}\right)^n H(x)=2^{-(n-1)}\frac{d^n}{dx^n}\left(\frac{2}{\sqrt{\pi}}\int\limits_0^x e^{-t^2}dt\right)$$

4 dec.	$n = 1\,(1)\,6.$	$x = 0\,(0{,}01)\,4$	[34] [41]
4 dec.	$n = 1\,(1)\,6.$	$x = 0\,(0{,}01)\,3$	[18] [65]
4 dec.	$n = 1\,(1)\,6.$	$x = 0\,(0{,}01)\,2{,}99$	[12] [17]

$$\Phi^{(n+1)}(x)=\frac{d^n}{dx^n}\left[\frac{1}{\sqrt{2\pi}}e^{-\frac{x^2}{2}}\right]$$

7 dec.	$n = 1\,(1)\,6.$	$x = 0\,(0{,}01)\,4$	[71]
6 dec.	$n = 0\,(1)\,4.$	$x = 0\,(0{,}004)\,6{,}468$	[112]
	$n = 5\,(1)\,10.$	$x = 0\,(0{,}004)\,8{,}236$	
	$n = 11\,(1)\,15.$	$x = 0\,(0{,}002)\,9{,}610$	
	$n = 16\,(1)\,20.$	$x = 0\,(0{,}002)\,10{,}902$	
6 fig.	$n = 1\,(1)\,10.$	$x = 0\,(0{,}01)\,12$	[80]
5 dec.	$n = 1\,(1)\,6.$	$x = 0\,(0{,}1)\,4$	[57]
5 dec.	$n = 2\,(1)\,4.$	$x = 0\,(0{,}01)\,4{,}99$	[101]
5 dec.	$n = 2\,(1)\,8.$	$x = 0\,(0{,}01)\,4{,}99$	[60]
4—5 dec.	$n = 3.$	$x = 0\,(0{,}01)\,3\,(0{,}1)\,6$	[124]
4 dec.	$n = 4.$	$x = 0\,(0{,}01)\,3\,(0{,}1)\,6$	
4 dec.	$n = 3;\ 4.$	$x = 0\,(0{,}01)\,3\,(0{,}1)\,4{,}9$	[38]
4 dec.	$n = 3\,(1)\,6.$	$x = 0\,(0{,}05)\,3{,}95$	[53]

$$(-1)^{n-1}\sqrt{2\pi}\,\Phi^{(n)}(x)=(-1)^{n-1}\frac{d^{n-1}}{dx^{n-1}}(e^{-\frac{x^2}{2}})$$

10 dec.	$n = 1\,(1)\,3.$	$x = -7\,(0{,}1)\,6{,}6$	[32]
9; 8; 7; 6 dec.	$n = 4;\ 5;\ 6;\ 7.$	$x = -7\,(0{,}1)\,6{,}6$	
4; 3; 2; 1 dec.; integers	$n = 1;\ 2;\ 3;\ 4;\ 5.$	$x = 0\,(0{,}1)\,4{,}5$	[115]

$$(-1)^{n-1}\frac{1}{\sqrt{n!}}\Phi^{(n)}(x)=(-1)^{n-1}\frac{1}{\sqrt{n!}}\frac{d^{n-1}}{dx^{n-1}}\left(\frac{1}{\sqrt{2\pi}}e^{-\frac{x^2}{2}}\right)$$

7 dec.	$n = 0\,(1)\,19.$	$x = 0\,(0{,}1)\,4$	[94] [143]

FACTORS FOR CALCULATING THE PROBABLE ERROR

$$\frac{0{,}67449}{\sqrt{n}}$$

4 dec. $n = 2\,(1)\,100$ [87]

$$\frac{0{,}67449}{\sqrt{n-1}}$$

5 dec. $n = 2\,(1)\,100$ [104]
4 dec. $n = 2\,(1)\,100$ [29] [61]

$$\frac{0{,}67449}{\sqrt{n\,(n-1)}}$$

4 dec. $n = 2\,(1)\,100$ [29] [61] [87]

$$\frac{0{,}84535}{\sqrt{n-1}}$$

4 dec. $n = 2\,(1)\,100$ [29]

$$\frac{0{,}84535}{\sqrt{n\,(n-1)}}$$

5 dec. $n = 2\,(1)\,100$ [104]
4 dec. $n = 2\,(1)\,100$ [61]

$$\frac{0{,}84535}{n\sqrt{n-1}}$$

dec. $n = 2\,(1)\,100$ [29] [61]

LOGARITHMS OF DISTRIBUTION FUNCTIONS

$$\lg\left[\frac{2}{\sqrt{\pi}}e^{-x^2}\right]$$

16 dec. $x = 1\,(0{,}001)\,1{,}5\,(0{,}002)\,3$ [186]

$$\lg\left[\frac{\rho}{\sqrt{\pi}}e^{-\rho^2x^2}\right]$$

5 dec. $x = 0\,(0{,}01)\,5\,(0{,}1)\,7$ [1]

$$\lg e^{\pm x^2}$$

5 dec. $x = 0\,(0{,}1)\,5$ [56]

$$\lg\left[e^{-\frac{x^2}{2}}\right]$$

8 dec.	$x^2 = 1\,(1)\,100$	[93] [134]

$$\lg\left[2x\frac{1}{\sqrt{2\pi}}e^{-\frac{x^2}{2}}\right]$$

8 dec.	$x^2 = 1\,(1)\,100$	[93] [134]

$$\lg\frac{1}{2}\sqrt{\pi}\,[1 - H(x)]$$

7 dec.	$x = 0\,(0{,}01)\,3$	[75] [154]

$$\pm\lg\frac{1}{2}[1 - \Phi(x)]$$

12 dec.	$x = 0\,(0{,}1)\,10$	[33]
8 dec.	$x = 0\,(0{,}01)\,10$	[33]
5 dec.	$x = 5\,(1)\,50\,(10)\,100\,(50)\,500$	[93]

$$-\ln\frac{1}{2}[1 - \Phi(x)]$$

24 dec.	$x = 0\,(1)\,10$	[33]
16 dec.	$x = 0\,(0{,}1)\,10$	[33]

$$\lg\frac{1 - H(x)}{H'(x)} = \lg\left[e^{x^2}\int_x^\infty e^{-t^2}dt\right]$$

10 dec.	$x^2 = 3\,(1)\,16$	[130]
7 dec.	$x = 10^t.\ t = 0\,(0{,}001)\,1$	[130]
7 dec.	$x = -0{,}12\,(0{,}001)\,1{,}01$	[130]
7 dec.	$x = 0\,(0{,}01)\,3$	[75]
7 dec.	$x = 10^t.\ t = 0\,(0{,}01)\,1$	[26]
7 dec.	$x = 0\,(0{,}01)\,1$	[26]

$$\lg\left[2xe^{x^2}\int_x^\infty e^{-t^2}dt\right]$$

8 dec.	$x = 0{,}05;\ 0{,}1\,(0{,}1)\,2\,(0{,}5)\,4{,}5$	[190]

$$\lg\frac{1 - \Phi(x)}{\sqrt{2\pi e}\,\Phi'(x)} = \lg\left[\sqrt{\frac{2}{\pi}}e^{1/_2x^2 - 1}\int_x^\infty e^{-\frac{t^2}{2}}dt\right]$$

4 dec.	$x = 1\,(0{,}01)\,3$	[39]

$$\lg\left[\frac{2}{\sqrt{\pi}} e^{x^2} \int_{-x}^{\infty} (t+x)\, e^{-t^2} dt\right]$$

4 dec.	$x = -1{,}2\,(0{,}1)\,2$	[48] [109]

FRESNEL INTEGRALS AND RELATED FUNCTIONS

Fresnel integrals

$$C(u) = \int_0^u \cos\left(\frac{1}{2}\pi t^2\right) dt = \frac{1}{2}\int_0^u J_{-1/2}(t)\, dt$$

7 dec.	$u = 0\,(0{,}001)\,25$	[13]
7 fig.	$u = 0\,(0{,}001)\,0{,}101$	
7 dec.	$u = 0{,}02\,(0{,}02)\,1$	[4]
6 dec.	$u = 0{,}5\,(0{,}5)\,50$	
5 dec.	$u = 0\,(0{,}01)\,20$	[118]
5 dec.	$u = 0\,(0{,}02)\,1$	[17]
5 dec.	$u = 0\,(0{,}1)\,1\,(0{,}5)\,50$	[12] [17]
5 dec.	$u = 0\,(0{,}01)\,0{,}1$	[166] [176]
4 dec.	$u = 0{,}15\,(0{,}05)\,1\,(0{,}1)\,5{,}5$	
4 dec.	$u = 8\,(0{,}02)\,16$	[24]
4 dec.	$u = 8\,(0{,}02)\,15{,}98$	[91]
4 dec.	$u = 0\,(0{,}1)\,20$	[24] [91]
4 dec.	$u = 0\,(0{,}1)\,8{,}5$	[17] [18] [29] [65] [126]
4 dec.	$u = 0\,(0{,}1)\,5{,}5$	[165]
4 dec.	$u = 0\,(0{,}1)\,5$	[49] [83] [99] [155]
3 dec.	$u = 0\,(0{,}1)\,5{,}5$	[19]

$$S(u) = \int_0^u \sin\left(\frac{1}{2}\pi t^2\right) dt = \frac{1}{2}\int_0^u J_{1/2}(t)\, dt$$

7 dec.	$u = 0\,(0{,}001)\,25$	[13]
7 fig.	$u = 0\,(0{,}001)\,0{,}58$	
7 dec.	$u = 0{,}02\,(0{,}02)\,1$	[4]
6 dec.	$u = 0{,}5\,(0{,}5)\,50$	
5 dec.	$u = 0\,(0{,}01)\,20$	[118]
5 dec.	$u = 0\,(0{,}02)\,1$	[17]
5 dec.	$u = 0\,(0{,}1)\,1\,(0{,}5)\,50$	[12] [17]

5 dec.	$u = 0\,(0{,}01)\,0{,}1$	[166] [176]
4 dec.	$u = 0{,}15\,(0{,}05)\,1\,(0{,}1)\,5{,}5$	
4 dec.	$u = 8\,(0{,}02)\,16$	[24]
4 dec.	$u = 8\,(0{,}02)\,15{,}98$	[91]
4 dec.	$u = 0\,(0{,}1)\,20$	[24] [91]
4 dec.	$u = 0\,(0{,}1)\,8{,}5$	[17] [18] [29] [65] [126]
4 dec.	$u = 0\,(0{,}1)\,5{,}5$	[165]
4 dec.	$u = 0\,(0{,}1)\,5$	[49] [83] [99] [155]
3 dec.	$u = 0\,(0{,}1)\,5{,}5$	[19]

$$\frac{1}{\sqrt{2}}\,C(u)$$

3 dec.	$u = 0\,(0{,}1)\,8{,}5$	[187]

$$\frac{1}{\sqrt{2}}\,S(u)$$

3 dec.	$u = 0\,(0{,}1)\,8{,}5$	[187]

The extremes $C(u)$

The table gives the maxima and minima of this integral.

6 dec.	$u = \left(n - \frac{1}{2}\right)\pi.\ n = 1\,(1)\,16$	[119]

The extremes $S(u)$

The maxima and minima of $S(u)$. are given.

6 dec.	$u = n\pi.\ n = 1\,(1)\,16$	[119]

$$C\left(\sqrt{\frac{2x}{\pi}}\right)$$

7 dec.	$x = 0\,(0{,}02)\,1$	[4] [119]
6 dec.	$x = 0\,(0{,}1)\,20$	[151]
6 dec.	$x = 0\,(0{,}1)\,1\,(0{,}5)\,50$	[123]
6 dec.	$x = 0\,(0{,}5)\,50$	[4] [119]
5 dec.	$x = 0\,(0{,}02)\,1\,(0{,}5)\,50$	[62]
4 dec.	$x = 0\,(0{,}1)\,1\,(0{,}5)\,50$	[18] [65]

$$S\left(\sqrt{\frac{2x}{\pi}}\right)$$

7 dec.	$x = 0\,(0{,}02)\,1$	[4] [119]
6 dec.	$x = 0\,(0{,}1)\,20$	[151]
6 dec.	$x = 0\,(0{,}1)\,1\,(0{,}5)\,50$	[123]
6 dec.	$x = 0\,(0{,}5)\,50$	[4] [119]
5 dec.	$x = 0\,(0{,}02)\,1\,(0{,}5)\,50$	[62]
4 dec.	$x = 0\,(0{,}1)\,1\,(0{,}5)\,50$	[18] [65]

$$C^2(u)$$

4 fig.	$u = 0\,(0{,}1)\,20$	[91]
4 dec.	$u = 8\,(0{,}02)\,15{,}98$	[91]

$$S^2(u)$$

4 fig.	$u = 0\,(0{,}1)\,20$	[91]
4 dec.	$u = 8\,(0{,}02)\,15{,}98$	[91]

$$C^2(u) + S^2(u)$$

4dec.	$u = 0\,(0{,}1)\,8{,}5$	[29]
4dec.	$u = 0\,(0{,}1)\,3$	[171]

$$\frac{1}{2}\left[C^2(u) + S^2(u)\right]$$

4dec.	$u = 0\,(0{,}1)\,8{,}5$	[126]

$$c_1^2(u) = \left[C(u) - \frac{1}{2}\right]^2$$

4fig.	$u = 0\,(0{,}1)\,20$	[91]

$$s_1^2(u) = \left[S(u) - \frac{1}{2}\right]^2$$

4fig.	$u = 0\,(0{,}1)\,20$	[91]

$$\sqrt{C^2(u) + S^2(u)}$$

4fig.	$u = 0\,(0{,}1)\,20$	[91]

$$\sqrt{c_1^2(u) + s_1^2(u)}$$

4fig.	$u = 0\,(0{,}1)\,20$	[91]

$$\frac{1}{2}\left[\frac{1}{2}+\boldsymbol{C}(\boldsymbol{u})\right]^2+\frac{1}{2}\left[\frac{1}{2}+\boldsymbol{S}(\boldsymbol{u})\right]^2$$

4 dec. $u = 0\,(0{,}01)\,0{,}1\,(0{,}05)\,1\,(0{,}1)\,5{,}5$ [176]

$$\frac{1}{2}\left[\frac{1}{2}-\boldsymbol{C}(\boldsymbol{u})\right]^2+\frac{1}{2}\left[\frac{1}{2}-\boldsymbol{S}(\boldsymbol{u})\right]^2$$

4 dec. $u = 0\,(0{,}05)\,1$ [176]

$$\frac{1}{2}\cos\frac{\pi}{2}\boldsymbol{u}^2+\frac{1}{2}\sin\frac{\pi}{2}\boldsymbol{u}^2-\boldsymbol{C}(\boldsymbol{u})\cos\frac{\pi}{2}\boldsymbol{u}^2-\boldsymbol{S}(\boldsymbol{u})\sin\frac{\pi}{2}\boldsymbol{u}^2$$

12 dec. $u = -0{,}06\,(0{,}02)\,3{,}5;\; 3\,(0{,}05)\,5{,}15$ [105]

$$\frac{1}{2}\cos\frac{\pi}{2}\boldsymbol{u}^2-\frac{1}{2}\sin\frac{\pi}{2}\boldsymbol{u}^2+\boldsymbol{C}(\boldsymbol{u})\sin\frac{\pi}{2}\boldsymbol{u}^2-\boldsymbol{S}(\boldsymbol{u})\cos\frac{\pi}{2}\boldsymbol{u}^2$$

12 dec. $u = -0{,}06\,(0{,}02)\,3{,}5;\; 3\,(0{,}05)\,5{,}15$ [105]

Auxiliary functions

Fresnel integrals can be expressed in terms of Lommel functions.

$$C\left(\sqrt{\frac{2x}{\pi}}\right)=\frac{1}{\sqrt{2}}U_{1/2}(2x;0)\cos x+\frac{1}{\sqrt{2}}U_{3/2}(2x;0)\sin x=$$

$$=\frac{1}{2}+\frac{1}{\sqrt{2}}V_{1/2}(2x;0)\sin x+\frac{1}{\sqrt{2}}V_{3/2}(2x;0)\cos x.$$

$$S\left(\sqrt{\frac{2x}{\pi}}\right)=\frac{1}{2}U_{1/2}(2x;0)\sin x-\frac{1}{\sqrt{2}}U_{3/2}(2x;0)\cos x=$$

$$=\frac{1}{2}-\frac{1}{\sqrt{2}}V_{1/2}(2x;0)\cos x+\frac{1}{\sqrt{2}}V_{3/2}(2x;0)\sin x.$$

$$\boldsymbol{M}(\boldsymbol{u}^2)=\frac{1}{\sqrt{2}}\boldsymbol{V}_{1/2}(\boldsymbol{u}^2\boldsymbol{\pi};0)\ \&\ \boldsymbol{N}(\boldsymbol{u}^2)=-\frac{1}{\sqrt{2}}\boldsymbol{V}_{3/2}(\boldsymbol{u}^2\boldsymbol{\pi};0)$$

5 dec. $u^2 = 0\,(0{,}01)\,0{,}3\,(0{,}02)\,2\,(0{,}1)\,9\,(1)\,30$ [166]
4 dec. $u = 0\,(0{,}05)\,1{,}5\,(0{,}1)\,8{,}5\,(0{,}5)\,13$ [161]

$$\boldsymbol{M}(\boldsymbol{v})=-\frac{1}{2}\sqrt{\boldsymbol{\pi}}\,\boldsymbol{V}_{3/2}(2\boldsymbol{v}^2;0)\ \&\ \boldsymbol{N}(\boldsymbol{v})=\frac{1}{2}\sqrt{\boldsymbol{\pi}}\,\boldsymbol{V}_{1/2}(2\boldsymbol{v}^2;0)$$

5 dec. $v = 0\,(0{,}1)\,6$ [128]

$$M(\beta) = -\frac{1}{2}\sqrt{\pi}\, V_{3/2}[(2\beta+1)\pi; 0] \ \& \ N(\beta) = \frac{1}{2}\sqrt{\pi}\, V_{1/2}[(2\beta+1)\pi; 0]$$

6 dec. $\beta = 0(0,1)8,9$ [129]

$$\sqrt{\frac{\pi}{2y}}\, V_{1/2}(y; 0); \quad \sqrt{\frac{\pi}{2y}}\, V_{3/2}(y; 0) \ \& \ M_1^2 = \frac{\pi}{2y}[V_{1/2}^2(y; 0) + V_{3/2}^2(y; 0)]$$

6 dec. $y = 0(1)60$ [123]

$$M_2^2 = \frac{1}{4}[V_{1/2}^2(y; 0) + V_{3/2}^2(y; 0)] = \frac{1}{2}\left[\frac{1}{2} - C\left(\sqrt{\frac{y}{\pi}}\right)\right]^2 + \frac{1}{2}\left[\frac{1}{2} - S\left(\sqrt{\frac{y}{\pi}}\right)\right]^2$$

6 dec. $y = 2x = 0(1)60(2)100$ [123]

4 dec. $u = \sqrt{\frac{2x}{\pi}} = 0(0,1)8,5$ [126]

4 dec. $u = \sqrt{\frac{2x}{\pi}} = 0(0,1)5,5$ [165]

$$\frac{1}{2} V_{1/2}(y; 0) \ \& \ \frac{1}{2} V_{3/2}(y; 0)$$

6 dec. $y = 2x = 0(1)60(2)100$ [123]

4 dec. $u = \sqrt{\frac{2x}{\pi}} = 0(0,1)8,5$ [126]

4 dec. $u = \sqrt{\frac{2x}{\pi}} = 0(0,1)5,5$ [165]

$$m^2 = \frac{\pi}{2y}[U_{1/2}^2(y; 0) + U_{3/2}^2(y; 0)] = \frac{\pi}{y}\left[C^2\left(\sqrt{\frac{y}{\pi}}\right) + S^2\left(\sqrt{\frac{y}{\pi}}\right)\right],$$

6 dec. $y = 0(1)60$ [123]

$$\sqrt{\frac{\pi}{2y}}\, U_{1/2}(y; 0) \ \& \ \sqrt{\frac{\pi}{2y}}\, U_{3/2}(y; 0)$$

6 dec. $y = 0(1)60$ [123]

$$N(y) = \sqrt{\frac{\pi}{2y}}\, U_{3/2}(y; 0); \quad \frac{N(y)}{\sqrt{y}}$$

5 dec. $y = 0(1)60$ [180]

$$\frac{N(y)}{y}$$

6 dec. $y = 0(1)60$ [180]

$$m_1^2 = \left\{\frac{1}{2}\cos\left(\frac{1}{2}y + \frac{1}{4}\pi\right) - \frac{1}{2}U_{1/2}(y;0)\right\}^2 + \left\{\frac{1}{2}\sin\left(\frac{1}{2}y + \frac{1}{4}\pi\right) + \frac{1}{2}U_{1/2}(y;0)\right\}^2 =$$
$$= \frac{1}{2}\left[\frac{1}{2} + C\left(\sqrt{\frac{y}{\pi}}\right)\right]^2 + \frac{1}{2}\left[\frac{1}{2} + S\left(\sqrt{\frac{y}{\pi}}\right)\right]^2$$

6 dec. $y = 0\,(1)\,60\,(2)\,100$ [123]

$$\frac{1}{2}\{1 - C(\sqrt{m-q}) - C(\sqrt{m+q})\}^2 + \frac{1}{2}\{1 - S(\sqrt{m-q}) - S(\sqrt{m+q})\}^2$$

4 dec. $m = q = 0\,(0{,}1)\,4{,}2$
4 dec. $m = 3.\quad q = 0\,(0{,}1)\,5{,}5$
4 dec. $m = 1{,}5.\quad q = 0\,(0{,}1)\,7$
4 dec. $m = 0{,}5.\quad q = 0\,(0{,}3)\,1{,}8;\ 2\,(0{,}1)\,8$ [126]

$$F_1(u) = \cos\frac{1}{2}\pi u^2 + \sin\frac{1}{2}\pi u^2 - \pi u\,\{C(u) + S(u)\}$$

4 dec. $u = 0\,(0{,}1)\,3{,}9$
3 dec. $u = 4\,(0{,}1)\,5$ [127]

$$F_2(u) = -3\pi^2\,\{c_4(u) + s_4(u)\}$$

$$c_1(u) = \int_u^\infty \cos\frac{1}{2}\pi t^2 dt, \quad c_n(u) = \int_u^\infty c_{n-1}(t)\,dt;$$
$$s_1(u) = \int_u^\infty \sin\frac{1}{2}\pi t^2 dt, \quad s_n(u) = \int_u^\infty s_{n-1}(t)\,dt$$

4 dec. $u = 0\,(0{,}1)\,1{,}4$
3 dec. $u = 1{,}5\,(0{,}1)\,2$ [127]

Chapter 8

ELLIPTIC INTEGRALS AND ELLIPTIC FUNCTIONS

The following notations are employed in the description of tables of elliptic integrals in this chapter:

incomplete elliptic integral of the first kind

$$F(k, \varphi) = \int_0^{\varphi} \frac{d\psi}{\sqrt{1-k^2\sin^2\psi}},$$

incomplete elliptic integral of the second kind

$$E(k, \varphi) = \int_0^{\varphi} \sqrt{1-k^2\sin^2\psi}\, d\psi,$$

incomplete elliptic integral of the third kind

$$\Pi(k, \varphi, n) = \int_0^{\varphi} \frac{d\psi}{(1+n\sin^2\psi)\sqrt{1-k^2\sin^2\psi}}.$$

Elliptic integrals within the limits 0 to $\frac{\pi}{2}$ are known as complete elliptic integrals.

The modulus of the elliptic integral is denoted by $k = \sin\theta$, the complementary modulus by $k' = \sqrt{1-k^2}$ the angle of the modulus by θ and the parameter of the integral of the third kind by n.

In addition, the notation $F(k^2, \varphi)$ is used for incomplete elliptic integrals of the first kind and $E(k^2, \varphi)$ for incomplete elliptic integrals of the second kind when k^2 is the argument of the tabulation. When these integrals are tabulated in terms of θ, the corresponding notations $F(\theta, \varphi)$ and $E(\theta, \varphi)$ are employed. The notations K, $K(k)$, $K(k^2)$, $K(\theta)$, $K' = K(k')$ and E, $E(k)$, $E(k^2)$, $E(\theta)$, $E' = E(k')$ are employed for complete elliptic integrals of the first and second kinds.

---

[1]See pages 481 and 550 for the references to Chapter 8.

The following notations are used in describing tables of theta functions

$$\vartheta_1(x) = 2q^{\frac{1}{4}}(\sin x - q^2 \sin 3x + q^6 \sin 5x - q^{12} \sin 7x + \ldots).$$

$$\vartheta_2(x) = 2q^{\frac{1}{4}}(\cos x + q^2 \cos 3x + q^6 \cos 5x + q^{12} \cos 7x + \ldots).$$

$$\vartheta_3(x) = 1 + 2q \cos x + 2q^4 \cos 4x + 2q^9 \cos 6x + \ldots$$

$$\vartheta_4(x) = 1 - 2q \cos 2x + 2q^4 \cos 4x - 2q^9 \cos 6x + \ldots$$

Here, as usual, $q = e^{-\pi \frac{K'}{K}}$ is the Jacobian paramete[r]. Jacobian elliptic functions can be expressed through a theta function. If

$$\operatorname{sn} u = \frac{1}{\sqrt{k}} \frac{\vartheta_1(x)}{\vartheta_4(x)}; \quad \operatorname{cn} u = \sqrt{\frac{k'}{k}} \frac{\vartheta_2(x)}{\vartheta_4(x)}; \quad \operatorname{dn} u = \sqrt{k'} \frac{\vartheta_3(x)}{\vartheta_4(x)}.$$

Jacobi used the following notations for the theta function

$$\Theta(u) = \vartheta_4(x); \ \Theta_1(u) = \vartheta_3(x); \ \mathrm{H}(u) = \vartheta_1(x); \ \mathrm{H}_1(u) = \vartheta_2(x).$$

In describing theta functions and Jacobian elliptic functions the arguments x and u of these functions will be used in the sense indicated above

The material of the chapter is arranged in the following sections:

- moduli of the integrals;
- complete elliptic integrals;
- expressions containing complete elliptic integrals of the first and second kinds;
- incomplete elliptic integrals;
- expressions containing incomplete elliptic integrals of the first and second kinds;
- elliptic integrals of the third kind;
- elliptic functions;
- values of Jacobi's parameter $q = e^{-\pi \frac{K'}{K}}$;
- theta functions.

MODULI OF THE INTEGRALS

Values of the modulus and related magnitudes

Here we give only the characteristics of certain tables of the angle of the modulus θ, the moduli k and k' and the square of the modulus k^2 as functio[n] of each other.

See the section "Squares" in Chapter 1 for table[s] of k^2 as a function of k, the section "Square root[s]" in Chapter 1 for tables of k as a function of k^2, the section "Sines of angles" in Chapter 2 for tab[les] of k as a function of θ and the section "Inverse trigonometrical functions" in Chapter 2 for tables of θ as a function of k.

$k^2(\theta)$

5 dec. $\theta = 0\,(1°)70°(0°,5)80°(0°,2)89°(0°,1)90°$ [2] [24]

$k\,(k^2)$

12 dec. $k^2 = 0\,(0,001)\,1$ [25]

$k'\,(k^2)$

4 dec. $k^2 = 0\,(0,01)0,99;\ 0,9(0,001)0,999$ [55]

$$(k_1')^2 = \frac{4k'}{(1+k')^2}$$

7 dec. $k^2 = 0(0,001)\,1$ [87]

$\theta\,(k^2)$

varying accuracy (in fractions of a second) $\left\{\begin{array}{l} k^2 = 0{,}000\,000\,000\,001\,(\times 10)0{,}01(0{,}01)1 \\ k^2 = 0{,}999(9\cdot 10^{-n})0{,}999\,999\,999\,999 \\ n = 4\,(1)\,12 \end{array}\right\}$ [2]

$\ln\left(\frac{4}{k'}\right)$

4 dec. $k^2 = 0{,}7(0{,}01)1$ [13] [29]

$$\lg k_n = \lg\frac{1-k'_{n-1}}{1+k'_{n-1}},$$

where $k_0 = k,\ k_0' = \sqrt{1-k_0^2},\ k_n = \dfrac{1-k'_{n-1}}{1+k'_{n-1}},\ k_n' = \sqrt{1-k_n^2}$.

14 dec. $n = 0\,(1)\,m.\quad 2 \leqslant m \leqslant 4.\quad \theta = 0(0°,1)15°(0°,5)45°$ [31] [33]

$$\lg k_n' = \lg\sqrt{(1-k_n^2)}$$

14 dec. $n = 0\,(1)\,m.\quad 1 \leqslant m \leqslant 3.\quad \theta = 0\,(0°,1)15°(0°,5)45°$ [31] [33]

COMPLETE ELLIPTIC INTEGRALS

Complete elliptic integrals of the first kind

Tables with argument θ

$K(\theta)$

12 dec.	$\theta = 0\,(1°)\,90°$	[32] [33]
9–10 dec.	$\theta = 5°\,(5°)\,80°\,(1°)\,90°$	[14]
7 dec.	$\theta = 0(1°)70°(0°,5)80°(0°,2)$ $89°(0°,1)90°$	[2] [17]
7 dec.	$\theta = 0\,(1°)\,90°$	[30]
6 dec.	$\theta = 0\,(1°)\,90°$	[4] [46] [89] [90]
5 dec.	$\theta = 86°\,(1')\,90°$	[62]
5 fig.	$\theta = 0\,(20')\,90°$	[8]
5 dec.	$\theta = 0\,(1°)70°(30')80°(12')$ $89°\,(6')\,90°$	[9] [22] [24] [34] [40]
5 dec.	$\theta = 0\,(1°)90°$	[18]
4 dec.	$\theta = 0(1°)70°(30')80°(12')$ $89°(6')89°30'(3')90°$	[44]
4 dec.	$\theta = 0(1°)45°(0°,5)80°\,(0°,2)$ $89°(0°,1)90°$	[49]
4 dec.	$\theta = 0(1°)65°(0°,5)80°(0°,2)$ $89°(0°,1)90°$	[15] [23]
4 dec.	$\theta = 0\,(1°)70°(30')80°\,(12')$ $89°(6')90°$	[13] [29]
4 dec.	$\theta = 0\,(1°)\,90°$	[1] [5] [10] [11] [20] [28] [41]
3 dec.	$\theta = 0(1°)50°(30')70°(12')$ $82°(6')89°(2')89°40'(1')90°$	[3]

$$\frac{1}{M(\theta)} = \frac{2}{\pi} K(\theta)$$

6 dec.	$\theta = 0\,(0°,1)\,90°$	[68]

$$\frac{1}{\ln(4\sec\theta)} K(\theta)$$

7–11 dec.	$\theta = 86°(1')90°$	[62]

$\lg K(\theta)$

12 dec.	$\theta = 0\,(0°,1)\,90°$	[32] [33]
8 dec.	$\theta = 45°\,(0°,1)\,90°$	[89] [90]

8 dec.	$\theta = 80°(0°,1)90°$	[46]
7 dec.	$\theta = 0(0°,5)90°$	[16] [21] [43]
7 dec.	$\theta = 0(1°)70°(0°,5)80°(0°,2)89°(0°,1)90°$	[2]
7 dec.	$\theta = 0(1°)90°$	[31]
6 dec.	$\theta = 0(1°)90°$	[20]
5 dec.	$\theta = 0(0°,1)90°$	[53] [58]
4 dec.	$\theta = 0(1^g)100^g$	[27]

$$\lg\left[\frac{1}{M(\theta)}\right] = \lg\left[\frac{2}{\pi}K(\theta)\right]$$

14 dec.	$\theta = 0(0°,1)15°(0°,5)45°$	[32] [33]
4 dec.	$\theta = 0(1^g)100^g$	[27]

$$\lg K(\theta) - \lg\ln\left(\frac{4}{k'}\right)$$

4 dec.	$\theta = 50^g(1^g)100^g$	[27]

Tables with argument $\operatorname{tg}\theta$

$$K(\operatorname{tg}\theta)$$

7 dec.	$\operatorname{tg}\theta = 0,1(0,1)1(0,5)3(1)5(2,5)12,5$	[89] [90]

$$\lg K(\operatorname{tg}\theta)$$

7 dec.	$\operatorname{tg}\theta = 0,1(0,1)1(0,5)3(1)5(2,5)12,5$	[89] [90]

Tables with argument k^2

$$K(k^2)$$

12 dec.	$k^2 = 0,835(0,001)1$	[25]
10 dec.	$k^2 = 0(10^{-7})10^{-5}(10^{-5})0,00249$	[25]
10 dec.	$k^2 = 0(0,001)0,834$	[25]
10 dec.	$\lambda = -\lg k'^2 = 1(0,005)2(0,01)6$	[93]
9 dec.	$k^2 = 0(0,01)1$	[81]
8 dec.	$k'^2 = 0(10^{-7})10^{-5}(10^{-5})0,003$; $k^2 = 0,0025(10^{-5})0,003$	[25]
7 dec.	$k^2 = 0(0,001)1$	[7]
7 dec.	$k^2 = 0(0,01)1$	[38] [39]

6 dec.	$k^2 = 0{,}000\,000\,000\,001(\times 10)0{,}01(0{,}01)1$ $k^2 = 0{,}999(0{,}0001)0{,}9999(0{,}00001)$ $0{,}99999(0{,}000\,001)0{,}999999$ $(0{,}0000001)0{,}9999999\,(9\cdot 10^{-n})$ $0{,}999\,999\,999\,999.\ n = 4(1)12$	[2]
6 dec.	$k^2 = 0\,(0{,}001)\,1$	[87]
5 dec.	$k^2 = 0{,}000\,001\,(0{,}000\,001)0{,}00001$ $k^2 = 0{,}0001\,(0{,}0001)\,0{,}003$ $k^2 = 0\,(0{,}01)\,0{,}5$	[40]
5 dec.	$k^2 = 0\,(0{,}001)\,1$	[24]
4 dec.	$k^2 = 0\,(0{,}001)\,0{,}997\,(10^{-5})$ $0{,}99999\,(10^{-7})\,1$	[19]
4 dec.	$k^2 = 0\,(0{,}01)\,1$	[13] [29]

The following tables of values of the complete elliptic integral of the first kind K are calculated for values of a complex modulus.

modulus 10 dec. argument 7 dec.	$x = 0\,(0{,}1)\,2.$	$y = 0\,(0{,}1)\,2$	[69]
modulus 10 dec. argument 7 dec.	$x = 1{,}9\,(0{,}01)\,2.$	$y = 0\,(0{,}01)\,0{,}1$	[69]
modulus 10 dec. argument 1—3 dec.	$x = 0\,(0{,}1)2.$	$y = 0\,(0{,}1)\,2$	[69]

$$M\,(k^2) = \frac{\pi}{2}\,K\,(k^2)$$

12 dec.	$k^2 = 0{,}835\,(0{,}001)\,1$	[25]
9 dec.	$k^2 = 0\,(0{,}01)\,1$	[81]
8 dec.	$k^2 = 0{,}997\,(10^{-5})\,0{,}99999\,(10^{-7})\,1$	[25]

$$K = K_1 \ln\left(\frac{4}{k'}\right) - K_2$$

where K_1 & K_2 are certain power series in k'^2.

13 dec.	$k'^2 = 0\,(0{,}00001)\,0{,}0001$	[75]
12 dec.	$k'^2 = 0\,(0{,}0001)\,0{,}001$	[75]
10 dec.	$k'^2 = 0\,(0{,}001)\,0{,}17$	[75]

$$K = \beta \lg\left(\frac{4}{k'}\right)$$

$$\beta \to \ln 10 \text{ for } k^2 \to 1$$

7 dec.	$k^2 = 0{,}95\,(0{,}001)\,1$	[7]

$$h = K - \ln\left(\frac{4}{k'}\right)$$

4 dec. $k^2 = 0{,}7\,(0{,}01)\,1$ [13] [29]

$$\frac{K}{K'}$$

6 dec. $k^2 = 0{,}000\,000\,000\,001\,(\times 10)\,0{,}01\,(0{,}01)\,1$; $k^2 = 0{,}999\,(0{,}0001)\,0{,}9999(0{,}00001)\,0{,}99999\,(0{,}000\,001)\,0{,}999\,999(0{,}0000001)\,0{,}9999999\,(9\cdot 10^{-n})\,0{,}999\,999\,999\,999.\ n = 4\,(1)\,12$ [2]

5 dec. $k^2 = 0{,}000001\,(0{,}000001)\,0{,}00001$; $k^2 = 0{,}0001\,(0{,}0001)\,0{,}003$; $k^2 = 0\,(0{,}01)\,0{,}5$ [40]

5 dec. $k^2 = 0\,(0{,}1)\,1$ [35]

$$\frac{K'}{K}$$

10 dec. $k^2 = 0\,(0{,}001)\,0{,}5$ [25]

8 dec. $k^2 = 0\,(10^{-7})\,10^{-5}\,(10^{-5})\,0{,}003$ [25]

5 dec. $k^2 = 0{,}000001\,(0{,}000001)\,0{,}00001$; $k^2 = 0{,}0001\,(0{,}0001)\,0{,}003$; $k^2 = 0\,(0{,}01)\,0{,}5$ [40]

Tables with argument k

$$K(k)$$

10 dec. $k = 0\,(0{,}01)\,1$ [76]

$$b_0 = \frac{4}{\pi}K(k)$$

11 dec. $k = 0\,(0{,}01)\,0{,}75\,(0{,}005)\,1$ [74]

$$M(k) = \frac{\pi}{2}\frac{1}{K(k)}$$

10 dec. $k = 0\,(0{,}01)\,1$ [76]

$$\lg b_0 = \lg\left[\frac{4}{\pi}K(k)\right]$$

7 dec. $k = 0\,(0{,}005)\,0{,}75$ [45]

$$\lg \frac{K}{2\pi} + 2\lg(1+\sqrt{k'})$$

8 dec.	$k = 0{,}4\,(0{,}01)\,0{,}6\,(0{,}001)\,0{,}849$	[73]

$$\lg\left[\frac{4K}{\pi}\cos\frac{\theta}{2}\sqrt[4]{\cos\theta}\right]$$

7 dec.	$k = 0{,}45\,(0{,}01)\,0{,}75$	[57]

$$k'K$$

Tables for the complex modulus $k^2 = -e^{ix}$.

modulus 10 dec. argument 7 dec.	$x = 0\,(0{,}1)\,1{,}9\,(0{,}01)\,2$	[69]

Complete elliptic integrals of the second kind

Tables with argument θ

$$E(\theta)$$

12 dec.	$\theta = 0\,(1°)\,90°$	[32] [33]
9—10 dec.	$\theta = 5°\,(5°)\,80°\,(1°)\,90°$	[14]
7 dec.	$\theta = 0(1°)70°(0°{,}5)80°(0°{,}2)89°(0°{,}1)90°$	[2] [17]
6 dec.	$\theta = 0\,(1°)\,90°$	[8] [46] [89] [90]
5 dec.	$\theta = 0\,(20')\,90°$	[8]
5 dec.	$\theta = 0\,(1°)\,70°\,(30')\,80°\,(12')\,89°\,(6')\,90°$	[9] [22] [24] [34] [40]
5 dec.	$\theta = 0\,(1°)\,90°$	[18]
4 dec.	$\theta = 0\,(1°)\,70°\,(30')\,80°\,(12')\,89°\,(6')$ $89°30'\,(3')\,90°$	[44]
4 dec.	$\theta = 0(1°)45°(0°{,}5)80°(0°{,}2)89°(0°{,}1)90°$	[49]
4 dec.	$\theta = 0\,(1°)\,70°\,(30')\,80°\,(12')\,89°\,(6')\,90°$	[13] [29]
4 dec.	$\theta = 0(1°)65°(0°{,}5)80°(0°{,}2)89°(0°{,}1)90°$	[15] [23]
4 dec.	$\theta = 0\,(1°)\,90°$	[1] [5] [10] [11] [20] [28] [41]
3 dec.	$\theta = 0\,(1°)\,90°$	[3]

$$\frac{2}{\pi}E(\theta)$$

6 dec.	$\theta = 0\,(0°{,}1)\,90°$	[68]

$$\frac{2}{\pi k'^2} E(\theta)$$

8 dec.	$\theta = 0\,(0°{,}1)\,50°$	[56]

$$\lg E(\theta)$$

12 dec.	$\theta = 0\,(0°{,}1)\,90°$	[32] [33]
8 dec.	$\theta = 45°\,(0°{,}1)\,90°$	[89] [90]
8 dec.	$\theta = 80°\,(0°{,}1)\,90°$	[46]
7 dec.	$\theta = 0\,(0°{,}5)\,90°$	[16] [21] [43]
7 dec.	$\theta = 0\,(1°)70°(0°{,}5)80°(0°{,}2)89°(0°{,}1)\,90°$	[2]
7 dec.	$\theta = 0\,(1°)\,90°$	[31]
6 dec.	$\theta = 0\,(1°)\,90°$	[20]
5 dec.	$\theta = 0\,(0°{,}1)\,90°$	[53] [58]
4 dec.	$\theta = 0\,(1^g)\,100^g$	[27]

Tables with argument $\operatorname{tg}\theta$

$$E(\operatorname{tg}\theta)$$

7 dec.	$\operatorname{tg}\theta = 0{,}1\,(0{,}1)\,1\,(0{,}5)\,3\,(1)\,5\,(2{,}5)\,12{,}5$	[89] [90]

$$\lg E(\operatorname{tg}\theta)$$

7 dec.	$\operatorname{tg}\theta = 0{,}1\,(0{,}1)\,1\,(0{,}5)\,3\,(1)\,5\,(2{,}5)\,12{,}5$	[89] [90]

Tables with argument k^2

$$E(k^2)$$

10 dec.	$\lambda = -\lg k'^2 = 1\,(0{,}005)\,2\,(0{,}01)\,6$	[93]
10 dec.	$k^2 = 0\,(0{,}001)\,1$	[26]
9 dec.	$k^2 = 0\,(0{,}01)\,1$	[81]
7 dec.	$k^2 = 0\,(0{,}001)\,1$	[7]
7 dec.	$k^2 = 0\,(0{,}01)\,1$	[6] [38] [39]
6 dec.	$k^2 = 0\,(0{,}001)\,1$	[87]
5–6 dec.	$k^2 = 0{,}000\,000\,000001(\times 10)\;0{,}01(0{,}01)1$	[2]
5 dec.	$k^2 = 0\,(0{,}001)\,1$	[19] [24]
4 dec.	$k^2 = 0\,(0{,}01)\,1$	[13] [29]

$$E = E_1 \ln\left(\frac{4}{k'}\right) + E_2$$

The table is intended for the calculation of $E(k^2)$ when k^2 approximates to 1. The values of E_1 and E_2 which are power series in k'^2 are tabulated.

12 dec.	$k'^2 = 0\,(0{,}00001)\,0{,}0001$	[75]
11 dec.	$k'^2 = 0\,(0{,}00001)\,0{,}0001$	
10 dec.	$k'^2 = 0\,(0{,}001)\,0{,}1$	

Tables with argument k

$$E(k)$$

10 dec. $k = 0\,(0{,}01)\,1$ [76]

Expressions containing complete elliptic integrals of the first and second kind

$$\frac{E}{K}$$

7 dec. $k^2 = 0\,(0{,}1)\,1$ [83]

$$e^{\frac{iy}{2}}\left[1 - \frac{E'}{K'}\right]$$

The tables give the real and imaginary part of this expression for a complex modulus.

10 dec. $x = 0\,(0{,}1)\,2.\ y = 2\,(-0{,}1)\,0$ [69]

$$D = \frac{K - E}{k^2}$$

4 dec. $k^2 = 0\,(0{,}01)\,1$ [13] [29]

$$b_1 = \frac{4}{\pi k}(K - E)$$

11 dec. $k = 0\,(0{,}01)\,0{,}9\,(0{,}005)\,1$ [74]

$$M(n) = (n+1)\int_0^1 k^n K\,dk$$

10 dec. $n = -0{,}9\,(0{,}1)\,2$ [70]

$$\int_{\alpha}^{1}(K-E)\,dk$$

10 dec. $\alpha=0\,(0{,}01)\,1$ [84]

$$\int_{\alpha}^{1}(K-E)\,k^{-3}dk$$

10 dec. $\alpha=0\,(0{,}01)\,1$ [84]

$$(-1)^n k'^2\int_0^{\frac{\pi}{2}}\frac{\cos 2n\varphi}{(1-k^2\sin^2\varphi)}\,d\varphi$$

4 dec. $\left\{\begin{matrix} n=0\,(1)\,3. \ k^2=0\,(0{,}01)\,0{,}99; \\ 0{,}9\,(0{,}001)\,0{,}999\end{matrix}\right\}$ [55]

$$B=K-D=K-\frac{1}{k^2}(K-E)$$

4 dec. $k^2=0\,(0{,}01)\,1$ [13] [29]

$$C=\frac{1}{k^2}(D-B)=\frac{1}{k^4}\,[2\,(K-E)-k^2K]$$

4 dec. $k^2=0\,(0{,}01)\,1$ [13] [29]

$$C+2-\ln\left(\frac{4}{k'}\right)=\frac{1}{k^4}\,[2\,(K-E)-k^2K]+2-\ln\left(\frac{4}{k'}\right)$$

4 dec. $k^2=0{,}7\,(0{,}01)\,1$ [13] [29]

$$D+1-\ln\left(\frac{4}{k'}\right)=\frac{1}{k^2}(K-E)+1-\ln\left(\frac{4}{k'}\right)$$

4 dec. $k^2=0{,}7\,(0{,}01)\,1$ [13] [29]

$$4\pi\left[\left(\frac{2}{k}-k\right)K-\frac{2}{k}\,E\right]$$

5—6 fig. $k^2=0\,(0{,}001)\,1$ [88]
5—6 dec. $\theta=0\,(6')\,90^\circ$ }
4 dec. $\theta=89^\circ\,(1')\,90^\circ$ } [42]

$$2\pi k\left[\left(\frac{2}{k}-k\right)K-\frac{2}{k}\,E\right]$$

4 fig. $k^2=0\,(0{,}01)\,1$ [85]

$$\frac{4\pi}{1000}\left[\left(\frac{2}{k}-k\right)K-\frac{2}{k}E\right]$$

5—6 dec. $k^2 = 0\,(0{,}005)\,1$ [86]

$$\pi k\left(\frac{2-k^2}{1-k^2}E-2K\right)$$

5—6 fig. $k^2 = 0\,(0{,}001)\,1$ [88]

$$\frac{2}{3}\left[\frac{K+(\mathrm{tg}^2\theta-1)E}{\sin\theta}-\mathrm{tg}^2\theta\right]$$

4 dec. $\mathrm{tg}\,\theta = 0\,(0{,}01)\,1.$ $\mathrm{ctg}\,\theta = 0\,(0{,}01)\,1$ [63]

$$\frac{4\pi}{3}\left[\frac{K+(\mathrm{tg}^2\theta-1)E}{\sin\theta}-\mathrm{tg}^2\theta\right]$$

This formula is used for calculating the inductance of thin-walled cylindrical coils

4—5 fig. $\mathrm{tg}\,\theta = 0\,(0{,}01)\,1.$ $\mathrm{ctg}\,\theta = 0\,(0{,}01)\,1$ [13] [29]

$$\frac{8\pi}{3}\left[\frac{K+(\mathrm{tg}^2\theta-1)E}{\sin\theta}-\mathrm{tg}^2\theta\right]$$

4 dec. $\mathrm{tg}\,\theta = 0{,}2\,(0{,}1)\,1\,(0{,}2)\,4$ [46]

$$\frac{4\pi\,\mathrm{tg}\,\theta}{3}\left[\frac{K+(\mathrm{tg}^2\theta-1)E}{\sin\theta}-\mathrm{tg}^2\theta\right]$$

4—5 fig. $\mathrm{tg}\,\theta = 0\,(0{,}01)\,1.$ $\mathrm{ctg}\,\theta = 0\,(0{,}01)\,1$ [13] [29]

$$\frac{4\,\mathrm{ctg}\,\theta}{3\pi}\left[\frac{K+(\mathrm{tg}^2\theta-1)E}{\sin\theta}-\mathrm{tg}^2\theta\right]$$

6 dec. $\left\{\begin{array}{l}\mathrm{tg}\,\theta = 0\,(0{,}01)\,1\,(0{,}05)\,2\,(0{,}1)\,5\,(0{,}5)\,10\\ \theta = 0\,(1^\circ)\,90^\circ\end{array}\right\}$ [66]

6 dec. $k^2 = 0\,(0{,}001)\,1.$ $\mathrm{tg}\,\theta,\ \mathrm{ctg}\,\theta = 0\,(0{,}001)\,1$ [88]

$$\frac{16\pi}{3(1-\alpha)^2}\left[\int_\alpha^1 (K-E)\,dk-\alpha^3\int_\alpha^1\frac{K-E}{k^3}\,dk\right.$$

7fig. $\alpha = 0\,(0{,}05)\,0{,}9$ [54]

4fig. $\alpha = 0;\ 0{,}05\,(0{,}01)\,0{,}95;\ 0{,}99$ [54]

6dec. $\alpha = 0\,(0{,}01)\,1$ [84]

$$\lg\left(\frac{b_1}{k}\right) = \lg\left[\frac{4}{\pi}\,\frac{K-E}{k^2}\right]$$

7 dec. $k = 0\,(0{,}005)\,0{,}75$ [45]

$$\lg\left[\left(\frac{2}{k} - k\right)K - \frac{2}{k}\,E\right]$$

7 dec. $\theta = 60°\,(0°{,}1)\,90°$ [36] [89] [90]

$$\lg\left\{4\pi\left[\left(\frac{2}{k} - k\right)K - \frac{2}{k}\,E\right]\right\}$$

6 dec. $k^2 = 0\,(0{,}001)\,1$ [88]

$$\lg\left\{\frac{4\pi}{1000}\left[\left(\frac{2}{k} - k\right)K - \frac{2}{k}\,E\right]\right\}$$

6 dec. $k^2 = 0\,(0{,}005)\,1$ [86]

$$\lg\left[k\left(\frac{2-k^2}{1-k^2}E - 2K\right)\right] = \lg\{\sin\theta\,[(1+\sec^2\theta)\,E - 2K]\}$$

8 dec. $\theta = 55°\,(0°{,}1)\,70°$ [91]
7 dec. $\theta = 55°(0°{,}1)\,70°$ [77]

$$\lg\left[\pi k\left(\frac{2-k^2}{1-k^2}E - 2K\right)\right]$$

6 dec. $k^2 = 0\,(0{,}001)\,1$ [88]

$$\lg\left\{\frac{4\pi\,\mathrm{ctg}\,\theta}{3}\left[\frac{K + (\mathrm{tg}^2\theta - 1)\,E}{\sin\theta} - \mathrm{tg}^2\theta\right]\right\}$$

6 dec. $k^2 = 0\,(0{,}001)\,1.\ \mathrm{tg}\,\theta,\ \mathrm{ctg}\,\theta = 0\,(0{,}001)\,1$ [88]

$$\lg\left\{\frac{4\,\mathrm{ctg}\,\theta}{3\pi}\left[\frac{K + (\mathrm{tg}^2\theta - 1)\,E}{\sin\theta} - \mathrm{tg}^2\theta\right]\right\}$$

6 dec. $\theta = 0\,(1°)\,90°$ [66]

INCOMPLETE ELLIPTIC INTEGRALS

Incomplete elliptic integrals of the first kind

$$F(\theta,\ \varphi)$$

12 dec.	$\theta = 0\,(1°)\,90°.$	$\varphi = 45°$	[32] [33]
10 dec.	$\theta = 0\,(1°)\,45°.$	$\varphi = 0\,(1°)\,90°$	
10 dec.	$\theta = 89°.$	$\varphi = 0\,(0°{,}5)\,90°$	
9 dec.	$\theta = 45°\,(1°)\,90°.$	$\varphi = 0\,(1°)\,90°$	

6 dec.	$\theta = 0\,(15°)\,45°\,(5°)\,75°\,(3°)\,78°\,(2°)\,84°\,(1°)\,90°$.	$\varphi = 0\,(5°)\,75°\,(3°)\,78°\,(2°)\,84°\,(1°)\,90°$	[2]
6 dec.	$\theta = 0\,(5°)\,90°$.	$\varphi = 0\,(1°)\,90°$	[4]
5 dec.	$\theta = 1°\,(1°)\,60°$. $\theta = 61°\,(1°)\,90°$.	$\varphi = 0\,(5°)\,90°$; $\varphi = 0\,(5°)\,50°\,(2°)\,88°\,(1°)\,90°$	[12]
5 dec.	$\theta = 0\,(1°)\,90°$.	$\varphi = 0\,(1°)\,90°$	[9]
5 dec.	$\theta = 0\,(3°)\,90°$.	$\varphi = 0\,(1°)\,90°$	[17]
5 dec.	$\theta = 0\,(5°)\,90°$.	$\varphi = 0\,(1°)\,90°$	[18] [22] [24] [34] [58]
5 dec.	$\theta = 5°\,(5°)\,90°$.	$\varphi = 0\,(1°)\,90°$	[8]
4—5 dec.	$\theta = 5°\,(5°)\,90°$.	$\varphi = 0\,(1°)\,90°$	[40]
4 dec.	$\theta = 0\,(5°)\,90°$.	$\varphi = 0\,(1°)\,90°$	[13] [29] [49]
4 dec.	$\theta = 5°\,(5°)\,90°$.	$\varphi = 0\,(1°)\,90°$	[44]
4 dec.	$\theta = 0\,(10°)\,90°$.	$\varphi = 0\,(10°)\,90°$	[1] [5] [10] [11]
4 dec.	$\theta = 0\,(10^g)\,100^g$.	$\varphi = 0\,(10^g)\,100^g$	[27]

$F(k^2, \varphi)$

5 dec.	$k^2 = 0\,(0{,}01)\,1$.	$\varphi = 0\,(1°)\,90°$	[7]

$\frac{1}{K} F(\theta, \varphi)$

4 dec.	$\theta = 0\,(10^g)\,100^g$.	$\varphi = 0\,(10^g)\,100^g$	[27]

$\lg F(\theta, \varphi)$

4 dec.	$\theta = 90^g\,(1^g)\,100^g$. $\theta = 0\,(10^g)\,100^g$.	$\varphi = 90^g\,(1^g)\,100^g$; $\varphi = 0\,(10^g)\,100^g$	[27]

Incomplete elliptic integrals of the second kind

$E(\theta, \varphi)$

12 dec.	$\theta = 0\,(1°)\,90°$.	$\varphi = 45°$	[32] [33]
10 dec.	$\theta = 89°$.	$\varphi = 0\,(0°{,}5)\,90°$	[32] [33]
10 dec.	$\theta = 0\,(1°)\,45°$.	$\varphi = 0\,(1°)\,90°$	[32] [33]
9 dec.	$\theta = 45°\,(1°)\,90°$.	$\varphi = 0\,(1°)\,90°$	[32] [33]
6 dec.	$\theta = 0\,(5°)\,90°$.	$\varphi = 0\,(1°)\,90°$	[4]
6 dec.	$\theta = 0\,(15°)\,45°\,(5°)\,75°\,(3°)\,78°\,(2°)\,84°\,(1°)\,90°$.	$\varphi = 0\,(5°)\,75°\,(3°)\,78°\,(2°)\,84°\,(1°)\,90°$	[2]

5 dec.	$\theta = 0\,(1°)\,90°$.	$\varphi = 0\,(5°)\,90°$	[12]
5 dec.	$\theta = 0\,(3°)\,90°$.	$\varphi = 0\,(3°)\,90°$	[17]
5 dec.	$\theta = 0\,(5°)\,90°$.	$\varphi = 0\,(1°)\,90°$	[9] [18] [22] [24] [34] [58]
5 dec.	$\theta = 5°\,(5°)\,90°$.	$\varphi = 0\,(1°)\,90°$	[8]
4 dec.	$\theta = 0\,(5°)\,90°$.	$\varphi = 0\,(1°)\,90°$	[13] [29] [49]
4 dec.	$\theta = 5°\,(5°)\,90°$.	$\varphi = 0\,(1°)\,90°$	[40] [44]
4 dec.	$\theta = 0\,(10°)\,90°$.	$\varphi = 0\,(10°)\,90°$	[1] [5] [10] [11]
4 dec.	$\theta = 0\,(10^g)\,100^g$.	$\varphi = 0\,(10^g)\,100^g$	[27]

$$E(k^2,\ \varphi)$$

5 dec.	$k^2 = 0\,(0{,}01)\,1$.	$\varphi = 0\,(1°)\,90°$	[7]

$$E(k',\ \sin\varphi)$$

11 dec.	$k' = 0\,(0{,}01)\,1$.	$\sin\varphi = 0\,(0{,}01)\,1$	[48]

$$\lg E(\theta,\ \varphi)$$

4 dec.	$\theta = 0\,(10^g)\,100^g$.	$\varphi = 0\,(10^g)\,100^g$	[27]

Expressions containing incomplete elliptic integrals of the first and second kind

$$\frac{\cos\varphi\cos\theta}{\sin^3\theta\sin^2\alpha}\,[F(k,\ \theta) - E(k,\ \theta)],$$

$$\frac{\cos\varphi\cos\theta}{\sin^3\theta\sin^2\alpha\cos^2\alpha}\left[E(k,\ \theta) - \cos^2\alpha\, F(k,\ \theta) - \frac{\sin^2\alpha\sin\theta\cos\theta}{\cos\varphi}\right],$$

$$\frac{\cos\varphi\cos\theta}{\sin^3\varphi\cos^2\alpha}\left[\frac{\sin\theta\cdot\cos\varphi}{\cos\theta} - E(k,\ \theta)\right]$$

$a \geqslant b \geqslant c \geqslant 0$ are semiaxes of the ellipsoid

$$\cos\varphi = \frac{b}{a};\quad \cos\theta = \frac{c}{a};\quad \sin\alpha = \frac{\sin\varphi}{\sin\theta} = k;\quad 0 \leqslant \varphi \leqslant \frac{\pi}{2}$$

$$0 \leqslant \theta \leqslant \frac{\pi}{2};\quad 0 \leqslant \alpha \leqslant \frac{\pi}{2}$$

5 dec.	$\frac{c}{a} = 0{,}017452$ (var.) $0{,}98481$.	$0{,}039018 \leqslant \frac{b}{a} \leqslant 1$	[79]
4 dec.	$\frac{b}{a} = 0{,}1\,(0{,}1)\,1$.	$0{,}0175 \leqslant \frac{c}{a} \leqslant 1$	[79]

Lemniscate case $\left(\theta = 45^\circ \text{ or } k^2 = \frac{1}{2}\right)$

In addition to the tables in this section, any table of $F(\theta, \varphi)$ and $E(\theta, \varphi)$ when $\theta = 45^\circ$ can be used to find the values required.

$F(\theta, \varphi)$

12 dec. $\varphi = 0\,(0^\circ{,}5)\,90^\circ$ [32] [33]

$E(\theta, \varphi)$

12 dec. $\varphi = 0\,(0^\circ{,}5)\,90^\circ$ [32] [33]

ELLIPTIC INTEGRALS OF THE THIRD KIND

An incomplete elliptic integral of the third kind is a function of three variables. When $0 < n < k^2$ it can be expressed in the terms of theta functions and Jacobian zeta functions of real argument. When n lies outside the range $(0, k^2)$ the integral can not be presented as dependent only on the functions of two of the real variables. The complete integral may however be expressed through complete and incomplete integrals of the first and second kind.

Tables of elliptic integrals of the third kind have not been calculated in direct form. The function

$$\Lambda(\alpha, \beta, \varphi) = \int_0^{\varphi} \frac{\cos^2\alpha \sin\beta \cos\beta \sqrt{1 - \cos^2\alpha \sin^2\beta}}{\cos^2\alpha \cos^2\beta + \sin^2\alpha \cos^2\psi} \frac{d\psi}{\sqrt{1 - \sin^2\alpha \cdot \sin^2\psi}},$$

dependent on three variables (the angle of the modulus α, the parametric angle β and the amplitude φ) has been studied. When $\varphi = \frac{\pi}{2}$ the corresponding full integral can be denoted by $\Lambda(\alpha, \beta)$.

The integral $\Lambda(\alpha, \beta, \varphi)$ is connected with the elliptic integral of the third kind $\Pi(\alpha, n, \varphi)$ by the relationship: $\Lambda(\alpha, \beta, \varphi) = Q \cdot \Pi(\alpha, n, \varphi)$ where

$$Q = \frac{\cos^2\alpha \sin\beta \cos\beta}{\sqrt{1 - \cos^2\alpha \sin^2\beta}}.$$

$$\Lambda_0(\alpha, \beta) = \frac{2}{\pi} \int_0^{\frac{\pi}{2}} \frac{\cos^2\alpha \sin\beta \cos\beta \sqrt{1 - \cos^2\alpha \sin^2\beta}}{\cos^2\alpha \cos^2\beta + \sin^2\alpha \cos^2\psi} \frac{d\psi}{\sqrt{1 - \sin^2\alpha \sin^2\psi}}$$

The integral which we have examined is connected with $\Lambda(\alpha, \beta)$ by the equation

$$\Lambda_0(\alpha, \beta) = \frac{2}{\pi} \Lambda(\alpha, \beta).$$

The function $\Lambda_0(\alpha, \beta)$ has been calculated for the case $k^2 < n < 1$ (the case $n < 0$ may be reduced to this case).

6 dec.	$\alpha = 0\,(1°)\,90°.\quad \beta = 0\,(1°)\,90°$	[68]

An auxiliary table with the following characteristics is given in the same work.

6 dec.	$\alpha = 0\,(0°{,}1)\,5°{,}9.\quad \beta = 80°\,(1°)\,89°$	[68]

ELLIPTIC FUNCTIONS

Jacobian elliptic functions

$\varphi = \mathrm{am}\, u$

0″,000 0001	$\theta = 0\,(0°{,}1)\,45°.$	$u = \frac{1}{10} K$	[32] [33]
1′	$\theta = 0\,(5°)\,80°\,(1°)\,89°.$	$x = \frac{\pi u}{2K} = 0\,(1°)\,90°$	[14]
0°,01	$\theta = 45°;\ 75°.$	$x = 0\,(1°)\,90°$	[59]
	$\theta = 15°.$	$x = 0\,(3°)\,90°$	
0°,001	$\theta = 45°.$	$x = 0\,(1°)\,90°$	[61]
1′	$\frac{K}{K'} = 2;\ 4.$	$x = 0\,(1°)\,90°$	
0°,01	$\frac{K}{K'} = \frac{1}{\sqrt{2}};\ \sqrt{2};\ \frac{3}{\sqrt{2}};\ 2\sqrt{2};\ 3;\ 2\sqrt{3};\ 4;\ 3\sqrt{2};\ 5;\ 3\sqrt{3}.$	$x = 0\,(1°)\,90°$	[60]
0°,01	$\frac{K}{K'} = 2.$	$x = 0\,(1°)\,90°$	[59]

$\mathrm{sn}\, u$

12 dec.	$\frac{u}{K} = 0\left(\frac{1}{90}\right)1.$	$\theta = 1°\,(1°)\,89°$	[50]
5 dec.	$u = 0\,(0{,}01)\,2.$	$k^2 = 0\,(0{,}1)\,1$	[6] [12] [39]
	$u = 2\,(0{,}01)\,2{,}5.$	$k^2 = 0{,}6\,(0{,}1)\,1$	
	$u = 2{,}5\,(0{,}01)\,3.$	$k^2 = 0{,}9;\ 1$	
	$u = 3\,(0{,}1)\,6{,}5.$	$k^2 = 1$	
5 dec.	$u = 0\,(0{,}01)\,2.$	$k^2 = 0\,(0{,}1)\,0{,}5$	[38]
	$u = 0\,(0{,}01)\,2{,}5.$	$k^2 = 0{,}6\,(0{,}1)\,0{,}8$	
	$u = 0\,(0{,}01)\,3.$	$k^2 = 0{,}9$	
5 dec.	$\frac{u}{K} = 0{,}1\,(0{,}1)\,1.$	$k^2 = 0\,(0{,}1)\,1$	[35]

5 fig.	$u = 0\,(0{,}1)\,2{,}5.$	$k^2 = 0\,(0{,}1)\,1$	[2]
4 dec.	$u = 0\,(0{,}05)\,1.$	$k = 0{,}9585$	[51]

cn u

12 dec.	$\frac{u}{K} = 0\left(\frac{1}{90}\right)1.$	$\theta = 1°\,(1°)\,89°$	[50]
5 dec.	$u = 0\,(0{,}01)\,2.$	$k^2 = 0\,(0{,}1)\,1$	[6] [12] [39]
	$u = 2\,(0{,}01)\,2{,}5.$	$k^2 = 0{,}6\,(0{,}1)\,1$	
	$u = 2{,}5\,(0{,}01)\,3.$	$k^2 = 0{,}9;\ 1$	
	$u = 3\,(0{,}01)\,3{,}5.$	$k^2 = 1$	
5 dec.	$u = 0\,(0{,}01)\,2.$	$k^2 = 0\,(0{,}1)\,0{,}5$	[38]
	$u = 0\,(0{,}01)\,2{,}5.$	$k^2 = 0{,}6\,(0{,}1)\,0{,}8$	
	$u = 0\,(0{,}01)\,3.$	$k^2 = 0{,}9$	
5 dec.	$u = 0\,(0{,}1)\,2{,}5.$	$k^2 = 0\,(0{,}1)\,1$	[2]
4 dec.	$u = 0\,(0{,}05)\,1.$	$k = 0{,}9585$	[51]

dn u

12 dec.	$\frac{u}{K} = 0\left(\frac{1}{90}\right)1.$	$\theta = 1°\,(1°)\,89°$	[50]
5 dec.	$u = 0\,(0{,}01)\,2.$	$k^2 = 0\,(0{,}1)\,1$	[6] [12] [39]
	$u = 2\,(0{,}01)\,2{,}5.$	$k^2 = 0{,}6\,(0{,}1)\,1$	
	$u = 2{,}5\,(0{,}01)\,3.$	$k^2 = 0{,}9;\ 1$	
	$u = 3{,}5\,(0{,}01)\,4.$	$k^2 = 1$	
5 dec.	$u = 0\,(0{,}01)\,2.$	$k^2 = 0\,(0{,}1)\,0{,}5$	[38]
	$u = 0\,(0{,}01)\,2{,}5.$	$k^2 = 0{,}6\,(0{,}1)\,0{,}8$	
	$u = 0\,(0{,}01)\,3.$	$k^2 = 0{,}9$	
5 fig.	$u = 0\,(0{,}1)\,3.$	$k^2 = 0{,}1\,(0{,}1)\,1$	[2]
4 dec.	$u = 0\,(0{,}05)\,1.$	$k = 0{,}9585$	[51]

sn $(u + iv)$

2—3 dec. $\theta = 15°;\ 20°;\ 75°.\quad u = 0\left(\frac{K}{10}\right)K.\quad v = 0\left(\frac{K'}{10}\right)K'$ [92]

Jacobian function $Z(u)$

$$Z(u) = \frac{M\vartheta_4'(x)}{\vartheta_4(x)},\ \text{where}\ x = Mu = \frac{\pi u}{2K}.$$

Note that for $k^2 = 1 \quad Z(u) = \text{th}\, u$.

10 dec.	$\theta = 45°$.	$x = 0\,(1°)\,90°$	[61]
10 dec.	$\theta = 0\,(5°)\,80°\,(1°)\,89°$.	$x = 0\,(1°)\,90°$	[14]
7 dec.	$u = 0\,(0{,}01)\,2$. $u = 2\,(0{,}01)\,2{,}5$. $u = 2{,}5\,(0{,}01)\,3$.	$k^2 = 0\,(0{,}1)\,1$ $k^2 = 0{,}7\,(0{,}1)\,1$ $k^2 = 0{,}9;\ 1$	[39]
7 dec.	$u = 0\,(0{,}01)\,2$. $u = 0\,(0{,}01)\,2{,}5$. $u = 0\,(0{,}01)\,3$.	$k^2 = 0\,(0{,}1)\,0{,}6$ $k^2 = 0{,}7;\ 0{,}8$ $k^2 = 0{,}9;\ 1$	[83]
7 dec. 6 fig.	$\theta = 15°$. $\theta = 45°;\ 75°$.	$x = 0\,(3°)\,90°$ $x = 0\,(1°)\,90°$	[59]

Weierstrassian elliptic functions

Weierstrassian function $\wp(u, g_2, g_3)$

The doubly periodic Weierstrassian function with periods 2ω and $2\omega'$ is defined as

$$\wp(u) = \wp(u|\omega,\omega') = -\frac{d^2 \ln \sigma(u)}{du^2}.$$

The function $z = \wp(u)$ is connected with the elliptic integral in the normal Weierstrassian form

$$u = \int_z^\infty \frac{dt}{\sqrt{4t^3 - g_2 t - g_3}},$$

where g_2 and g_3 are invariants of the function $\wp(u)$.

For the equiharmonic case ($g_2 = 0;\ g_3 = 1$) in which the function $\wp$ satisfies the equation

$$\wp'^2(u;\ 0;\ 1) = 4\wp^3(u;\ 0;\ 1) - 1,$$

there are the following tables of the function $\wp(u)$:

5 dec.	$r = \frac{180u}{\omega} = 0\,(1)\,240$	[82]
4 dec.	$r = \frac{180u}{\omega} = 0\,(1)\,240$	[12] [13] [29]

Tables exist for the case, slightly different from the first, in which $\wp'^2(u) = 4\wp(u) - 4$:

5—7 fig.	$r = 0\,(1)\,180$	[80]

A table for imaginary argument has been computed for the values of the argument

4 dec.	$u = 0{,}04i\,(0{,}04i)\,1{,}36i$	[51]

Derivatives of the Weierstrassian function $\wp'(u; 0; 1)$

5 dec.	$r = 0(1)240$	[82]
4 dec.	$r = 0(1)240$	[12] [13] [29]

Weierstrassian elliptic function $\zeta(u)$

5 dec.	$r = 0(1)240$	[82]
4 dec.	$r = 0(1)240$	[12] [13] [29]

Weierstrassian elliptic function $\sigma(u)$

5 dec.	$r = 0(1)240$	[82]
4 dec.	$r = 0(1)240$	[12] [13] [29]

VALUES OF THE JACOBIAN PARAMETER $q = e^{-\pi \frac{K'}{K}}$

Powers of q

$q(\theta)$

14—15 dec.	$\theta = 0(5°)80°(1°)90°$	[14]
8 dec.	$\theta = 0(1°)90°$	[71]
7 dec.	$\theta = 0(1°)27°$	[30]
6 dec.	$\theta = 0(1°)80°(0°,5)90°$	[2]
6 dec.	$\theta = 28°(1°)79°$	[30]
5 fig.	$\theta = 0(20')90°$	[8]
5 dec.	$\theta = 80°(1°)90°$	[30]

$q(k^2)$

10 dec.	$k^2 = 0(0,01)1$	[67]
8 dec.	$k^2 = 0(0,001)1$	[7]
8 dec.	$k^2 = 0(0,001)0,5$	[25]
8 dec.	$k^2 = 0(0,01)1$	[6] [38] [39]
5 dec.	$k^2 = 0(0,001)0,5$	[24]

Inverse values $\frac{1}{q}$

$\frac{1}{q(\theta)}$

6—7 fig.	$\theta = 15°; 45°; 75°$	[59]

Powers of q

$q^4(k^2)$

8 dec.	$k^2 = 0\,(0{,}001)\,0{,}5$	[25]
5 fig.	$k^2 = 0\,(0{,}001)\,0{,}01\,(0{,}01)\,0{,}5$	[2]

$10q^4(k^2)$

8 dec.	$k^2 = 0\,(0{,}001)\,0{,}01\,(0{,}01)\,0{,}5$	[2]

$q^{\frac{1}{4}}(k^2)$

8 dec.	$k^2 = 0\,(0{,}001)\,0{,}5$	[25]
5 dec.	$k^2 = 0\,(0{,}001)\,0{,}01\,(0{,}01)\,0{,}5$	[2]

$q^{\frac{9}{4}}(k^2)$

10 dec.	$k^2 = 0\,(0{,}001)\,0{,}5$	[25]

$q^{\frac{25}{4}}(k^2)$

11 dec.	$k^2 = 0{,}3\,(0{,}001)\,0{,}5$	[25]

Logarithms of q

$\lg q\,(\theta)$

10 dec.	$\theta = 0\,(1^\circ)\,45^\circ$	[72]
8 dec.	$\theta = 0\,(1^\circ)\,90^\circ$	[37] [71]
5 dec.	$\theta = 0\,(0^\circ{,}1)\,90^\circ$	[65]
5 dec.	$\theta = 0\,(5')\,90^\circ$	[16] [34] [43]
5 dec.	$\theta = 0\,(1^\circ)\,90^\circ$	[47] [49]
4 dec.	$\theta = 0\,(5')\,90^\circ$	[12] [13] [29]
4 dec.	$\theta = 0\,(0^\circ{,}1)\,90^\circ$	[58]
4 dec.	$\theta = 0\,(1^g)\,100^g$	[27]

$\lg q\,(k^2)$

10 dec.	$k^2 = 0\,(0{,}001)\,1$	[25]
8 dec.	$k^2 = 0\,(10^{-7})\,10^{-5}\,(10^{-5})\,0{,}003$	[25]
7 dec.	$k^2 = 0\,(0{,}001)\,1$	[87]

6 dec.	$k^2 = 0{,}000000000001\ (\times 10)\ 0{,}01\ (0{,}01)\ 1$; $k^2 = 0{,}999\ (0{,}0001)\ 0{,}9999\ (0{,}00001)\ 0{,}99999\ (0{,}000001)\ 0{,}999999\ (0{,}0000001)\ 0{,}9999999\ (9\cdot 10^{-n})\ 0{,}999999999999$; $n = 8\ (1)\ 12$	[2]

$$\lg \left[q(\theta) \operatorname{ctg}^2 \frac{\theta}{2} \right]$$

10 dec.	$\theta = 0\ (1°)\ 45°$	[72]

$$\lg \left[\frac{q(k^2)}{k^2} \right]$$

7 dec.	$k^2 = 0\ (0{,}001)\ 0{,}05$	[87]

$$\lg \left[\frac{16 q(k^2)}{k^2} \right]$$

4 dec.	$k^2 = 0\ (1^g)\ 50^g$	[27]

$$\lg \frac{1}{q(\theta)}$$

10 dec.	$\theta = 0\ (0°,1)\ 45°\ (1°)\ 90°$	[64]
5 dec.	$\theta = 0\ (1°)\ 80°\ (0°,5)\ 90°$	[2]
5 dec.	$\theta = 0\ (1°)\ 90°$	[30]

$$\ln \left(\frac{k^2}{q(k^2)} \right)$$

7 dec.	$k^2 = 0\ (0{,}001)\ 0{,}05$	[87]

$$\lg \lg \frac{1}{q(\theta)}$$

14 dec.	$\theta = 0\ (0°,1)\ 15°$	[52]
12 dec.	$\theta = 15°\ (0°,1)\ 45°$	[52]
10 dec.	$\theta = 0\ (1°)\ 45°$	[72]
7 dec.	$\theta = 0\ (0°,5)\ 90°$	[16] [43]
7 dec.	$\theta = 0\ (1°)\ 80°\ (0°,5)\ 90°$	[2]

$$\lg \ln \left(\frac{1}{q(k^2)} \right)$$

7 dec.	$k^2 = 0\ (0{,}001)\ 1$	[87]

THETA FUNCTIONS

The argument x met with in certain tables has the value $x = \frac{\pi u}{2K}$ and $2x + 2x_1 = 1$. The notations of the functions conform with those used in this handbook.

Theta functions and their combinations

$\vartheta_1(x)$

4 fig.	$\theta = 0(9°)81°$.	$2x = 0(0,1)1$	[2] [12] [13] [29]

$\vartheta_2(x)$

4 fig.	$\theta = 0(9°)81°$.	$2x_1 = 0(0,1)1$	[2] [12] [13] [29] [83]

$\vartheta_3(x)$

4 fig.	$\theta = 0(9°)81°$.	$2x_1 = 0(0,1)1$	[2] [12] [13] [29]

$\vartheta_4(x)$

4 fig.	$\theta = 0(9°)81°$.	$2x = 0(0,1)1$	[2] [12] [13] [29]

$\vartheta_n(ix)$

4 dec.	$n = 1(1)4$. $\theta = 9°$.	$\frac{x}{\pi} = 0(0,05)1$	[78]

$$A(x) = \frac{\vartheta_1(x)}{\vartheta_2(0)}$$

10dec.	$\theta = 0(5°)80°(1°)89°$.	$x = 0(1°)90°$	[14]
10dec.	$\theta = 45°$.	$x = 0(1°)90°$	[61]
6 fig.	$\theta = 15°$.	$x = 0(3°)90°$	[59]
6 fig.	$\theta = 45°;\ 75°$.	$x = 0(1°)90°$	[59]

$$B(x) = \frac{\vartheta_2(x)}{\vartheta_2(0)}$$

10dec.	$\theta = 0(5°)80°(1°)89°$	$x = 0(1°)90°$	[14]
10dec.	$\theta = 45°$.	$x = 0(1°)90°$	[61]
6 fig.	$\theta = 15°$.	$x = 0(3°)90°$	[59]
6 fig.	$\theta = 45°;\ 75°$.	$x = 0(1°)90°$	[59]

$$C(x)=\frac{\vartheta_3(x)}{\vartheta_4(0)}$$

10 dec.	$\theta=0(5°)80°(1°)89°$.	$x=0(1°)90°$	[14]
10 dec.	$\theta=45°$.	$x=0(1°)90°$	[61]
6 fig.	$\theta=15°$.	$x=0(3°)90°$	[59]
6 fig.	$\theta=45°;\ 75°$.	$x=0(1°)90°$	[59]

$$D(x)=\frac{\vartheta_4(x)}{\vartheta_4(0)}$$

10 dec.	$\theta=0(5°)80°(1°)90°$.	$x=0(1°)90°$	[14]
10 dec.	$\theta=45°$.	$x=0(1°)90°$	[61]
6 fig.	$\theta=15°$.	$x=0(3°)90°$	[59]
6 fig.	$\theta=45°;\ 75°$.	$x=0(1°)90°$	[59]

Logarithms of theta functions

$$\lg\vartheta_1(x)$$

4 dec.	$\theta=0(10^g)90^g$.	$x=0(10^g)100^g$	[27]
4 dec.	$\theta=0(9°)90°$.	$x=0(9°)90°$	[2]

$$\lg\vartheta_2(x)$$

4 dec.	$\theta=0(10^g)90^g$.	$x=0(10^g)100^g$	[27]
4 dec.	$\theta=0(9°)90°$.	$x_1=0(9°)90°$	[2]

$$\lg\vartheta_3(x)$$

4 dec.	$\theta=0(10^g)90^g$.	$x=0(10^g)100^g$	[27]
4 dec.	$\theta=0(9°)90°$.	$x_1=0(9°)90°$	[2]

$$\lg\vartheta_4(x)$$

4 dec.	$\theta=0(10^g)90^g$.	$x=0(10^g)100^g$	[27]
4 dec.	$\theta=0(9°)90°$.	$x=0(9°)90°$	[2]

Derivatives of theta functions

$$\pi\vartheta_n'(ix)$$

4 dec.	$n=1(1)4.\ \theta=9°$.	$\frac{x}{\pi}=0(0{,}05)1$	[78]

Logarithmic derivatives of theta functions

$$\frac{d}{dx}\ln\vartheta_n(x)=\frac{\vartheta_n'(x)}{\vartheta_n(x)}$$

4 dec.	$n=1(1)4$.	$\theta=0(9°)81°$.	$x=0(0{,}1)1$	[2] [13] [29]

Logarithms of logarithmic derivatives of theta functions

4 dec.	$n=1(1)4$.	$\theta=0(10^g)90^g$.	$x=0(10^g)100^g$	[27]
4 dec.	$n=1(1)4$.	$\theta=0(9°)90°$.	$x=0(9°)90°$	[2]

Theta functions of zero

In this section theta functions and their derivatives with respect to x are written without notation of the argument x. This implies those values of the functions which correspond to $x=0$.

Values of theta functions of zero are given by the following rapidly convergent series:

$$\vartheta_1'=2\pi q^{1/4}(1-3q^2+5q^6-7q^{12}+\ldots),$$

$$\vartheta_2=2q^{1/4}(1+q^2+q^6+q^{12}+\ldots),$$

$$\vartheta_3=1+2q+2q^4+2q^9+\ldots,$$

$$\vartheta_4=1-2q+2q^4-2q^9+\ldots.$$

Natural values of theta functions of zero

ϑ_2

10—12 dec.	$\theta=0(5°)80°(1°)89°$	[14]
8 dec.	$k^2=0(0{,}001)0{,}5$	[25]
6 dec.	$k^2=0(0{,}001)0{,}01(0{,}01)0{,}5$	[2]
5 dec.	$k^2=0(0{,}001)0{,}5$	[24]

ϑ_3

8 dec.	$k^2=0(0{,}001)0{,}5$	[25]
6 dec.	$k^2=0(0{,}001)0{,}01(0{,}01)0{,}5$	[2]
5 dec.	$k^2=0(0{,}001)0{,}5$	[24]

ϑ_4

10—12 dec.	$\theta = 0\,(5°)\,80°\,(1°)\,89°$	[14]
8 dec.	$k^2 = 0\,(0{,}001)\,0{,}5$	[25]
6 dec.	$k^2 = 0\,(0{,}001)\,0{,}01\,(0{,}01)\,0{,}5$	[2]
5 dec.	$k^2 = 0\,(0{,}001)\,0{,}5$	[24]

Derivatives of theta functions of zero

$\pi\vartheta_1'$

8 dec.	$k^2 = 0\,(0{,}001)\,0{,}5$	[25]
6 dec.	$k^2 = 0\,(0{,}001)\,0{,}01\,(0{,}01)\,0{,}5$	[2]
5 dec.	$k^2 = 0\,(0{,}001)\,0{,}5$	[24]

$$\frac{\pi^2\vartheta_1'''}{\vartheta_1'};\quad \frac{\pi^2\vartheta_2''}{\vartheta_2};\quad \frac{\pi^2\vartheta_3''}{\vartheta_3};\quad \frac{\pi^2\vartheta_4''}{\vartheta_4}$$

8 dec.	$k^2 = 0\,(0{,}001)\,0{,}5$	[25]

Logarithms of theta functions of zero

$\lg\vartheta_2$ и $\lg\vartheta_4$

7 dec.	$k^2 = 0\,(0{,}001)\,0{,}5$	[87]

$\lg\vartheta_3$

9 dec.	$\theta = 0\,(0°{,}5)\,90°$	[16] [43]
7 dec.	$k^2 = 0\,(0{,}001)\,0{,}5$	[87]

Logarithms of derivatives of theta functions of zero

$\lg(\pi\vartheta_1')$

7 dec.	$k^2 = 0\,(0{,}001)\,0{,}5$	[25]

$$\lg\left(\frac{\pi\vartheta_1'}{\sqrt{k}}\right);\quad \lg\left(\frac{-\pi^2\vartheta_3''}{k^2\vartheta_3}\right);\quad \lg\left(\frac{\pi^2\vartheta_4''}{k^2\vartheta_4}\right)$$

7 dec.	$k^2 = 0\,(0{,}001)\,0{,}05$	[87]

$$\lg\left(\frac{\pi^2\vartheta_1'''}{\vartheta_1'}\right);\quad \lg\left(\frac{\pi^2\vartheta_2''}{\vartheta_2}\right);\quad \lg\left(\frac{\pi^2\vartheta_3''}{\vartheta_3}\right);\quad \lg\left(\frac{\pi^2\vartheta_4''}{\vartheta_4}\right)$$

7 dec.	$k^2 = 0\,(0{,}001)\,0{,}5$	[25]

Chapter 9

LEGENDRE FUNCTIONS AND POLYNOMIALS[1]

The solutions of the Legendre differential equation

$$(1-x^2)\frac{d^2y}{dx^2}-2x\frac{dy}{dx}+n(n+1)y=0.$$

are known as spherical Legendre functions of order n.

The general solution of this equation has the form:

$$y=AP_n(x)+BQ_n(x),$$

where A and B are arbitrary constants; $P_n(x)$ are Legendre functions of the first kind, $Q_n(x)$ are Legendre functions of the second kind.

When n is a positive integer or zero, Legendre functions of the first kind are polynomials of order n:

$$P_n(x)=\frac{1}{2^n}\sum_k(-1)^k\frac{(2n-2k)!}{(n-2k)!\,(n-k)!\,k!}x^{n-2k},$$

where the sum is taken over k integer from 0 to $\frac{n-1}{2}$ for odd values of n and to $\frac{n}{2}$ for even values of n. These polynomials are known as Legendre polynomials.

The associated Legendre functions $P_n^m(x)$ and $Q_n^m(x)$ are solutions of the differential equation:

$$(1-x^2)\frac{d^2y}{dx^2}-2x\frac{dy}{dx}+\left[n(n+1)-\frac{m^2}{1-x^2}\right]y=0.$$

The functions $P_n^m(x)$, $P_{nm}(x)$ and $P^{nm}(x)$ are distinguished from each other by a normalizing factor. $R_n^m(x)$ and $F_n^m(x)$ are also normalized associated Legendre functions (see references /15/, /17/ and /33/ of this chapter).

---

[1] See pages 485 and 550 for the references to this chapter.

Other tables of polynomials are described in Chapter 11.

This chapter contains information on tables grouped under the following heads:

- Legendre polynomials and Legendre functions of the first kind;
- Legendre functions of the second kind;
- Associated Legendre functions of the first kind
- Associated Legendre functions of the second kinc
- Logarithms of Legendre functions;
- Roots of Legendre functions;
- Various expressions containing Legendre functior

LEGENDRE POLYNOMIALS AND LEGENDRE FUNCTIONS OF THE FIRST KIND

$P_n(x)$

The first seven Legendre polynomials have the for

$$P_0(x) = 1,$$

$$P_1(x) = x,$$

$$P_2(x) = \frac{1}{2}(3x^2 - 1),$$

$$P_3(x) = \frac{1}{2}(5x^3 - 3x),$$

$$P_4(x) = \frac{1}{8}(35x^4 - 30x^2 + 3),$$

$$P_5(x) = \frac{1}{8}(63x^5 - 70x^3 + 15x),$$

$$P_6(x) = \frac{1}{16}(231x^6 - 315x^4 + 105x^2 - 5).$$

Tables of Legendre polynomials with argument x are described below.

exact	$x = 0(0,01)1.$	$n = 1(1)8$	[21] [44]
exact	$x = 0(0,01)1.$	$n = 1(1)7$	[51]
10 dec.	$x = 1,1.$	$n = 1(1)20$	[63]
10 dec.	$x = 1,2;\ 2;\ 3.$	$n = 1(1)10$	[63]
7—11 dec.	$x = 0(0,01)1.$	$n = 1(1)7$	[35]
7—8 dec.	$x = 0(0,01)6.$	$n = 1(1)12$	[11]
7 dec.	$x = 0(0,01)1.$	$n = 1(1)5$	[79]
6 dec.	$x = 0(0,001)1.$	$n = 1(1)16$	[47]
6 dec.	$x = 0(0,1)1.$	$n = 1(1)4$	[33]
5 dec.	$x = 0(0,1)1.$	$n = 5(1)7$	[33]
5 dec.	$x = 0(0,01)1.$	$n = 1(1)10$	[20] [29]
5 dec.	$x = 0(0,01)1.$	$n = 1(1)7$	[14]
5; 4 dec.	$x = 0(0,01)1.$	$n = 1(1)7;\ 8(1)10$	[16]

4 dec.	$x=0\,(0{,}01)\,1.$	$n=1\,(1)\,7$	[4] [5] [6] [7] [12] [24] [30]
4 dec.	$x=0\,(0{,}01)\,1.$	$n=1\,(1)\,8$	[37]
4 dec.	$x=0\,(0{,}05)\,1.$	$n=1\,(1)\,7$	[1] [3]

$P_n(\cos\theta)$

In trigonometric form the first five Legendre polynomials appear as:

$$P_0(\cos\theta)=1,$$

$$P_1(\cos\theta)=\cos\theta,$$

$$P_2(\cos\theta)=\frac{1}{4}(3\cos 2\theta+1),$$

$$P_3(\cos\theta)=\frac{1}{8}(5\cos 3\theta+3\cos\theta),$$

$$P_4(\cos\theta)=\frac{1}{64}(35\cos 4\theta+20\cos 2\theta+9).$$

Tables of Legendre polynomials with argument $\cos\vartheta$. are given below.

10 dec.	$\theta=0\,(1°)\,90°.$	$n=1\,(1)\,8$	[21] [45]
7 dec.	$\theta=0\,(5°)\,90°.$	$n=1\,(1)\,20$	[66]
6 dec.	$\theta=0\,(10')\,90°.$	$n=1\,(1)\,32$	[48]
6 dec.	$\theta=10°\,(10°)\,170°;175°.$	$n=0{,}1\,(0{,}1)\,2$	[80]
6 dec.	$\theta=0\,(5°)\,90°.$	$n=1\,(1)\,4$	[33]
5 dec.	$\theta=0\,(5°)\,90°.$	$n=5\,(1)\,7$	[33]
5 dec.	$\theta=0\,(1°)\,90°.$	$n=1\,(1)\,32$	[55] [56]
5 dec.	$\theta=0\,(1°)\,90°.$	$n=1\,(1)\,10$	[4] [20]
5 dec.	$\theta=0\,(1°)\,90°.$	$n=1\,(1)\,7$	[5]
5; 4 dec.	$\theta=0\,(1°)\,90°.$	$n=1\,(1)7;8\,(1)\,10$	[16]
4 dec.	$\theta=0\,(1°)\,90°.$	$n=1\,(1)\,8$	[37]
4 dec.	$\theta=0\,(1°)\,90°.$	$n=1\,(1)\,7$	[6] [10] [12] [18] [24] [30] [34] [64]

$P_n(\operatorname{ch} x)$

4 dec.	$x=0\,(0{,}1)\,3$	$n=-\frac{1}{2};\ \frac{1}{2};\ \frac{3}{2}.$	[76]

$P_n'(x)$

10 dec.	$x=1{,}1.$	$n=0\,(1)\,20$	[63]
10 dec.	$x=1{,}2;\ 2;\ 3.$	$n=0\,(1)\,10$	[63]

$$P_n^{(m)}(x) = \frac{d^m P_n(x)}{dx^m}$$

exact	$x = 0\,(0{,}01)\,1.$	$m = 1\,(1)\,n.$	$n = 2\,(1)\,8$	[44]
4 dec.	$x = 0\,(0{,}01)\,1.$	$m = 1\,(1)\,n.$	$n = 2\,(1)\,8$	[37]

$$P_n^{(m)}(\cos\theta) = \left(\frac{d}{d\cos\theta}\right)^m P_n(\cos\theta)$$

8—11 fig.	$\theta = 0\,(1^\circ)\,90^\circ.$	$m = 1\,(1)\,n-1.$	$n = 2\,(1)\,8$	[46]
4 dec.	$\theta = 0\,(1^\circ)\,90^\circ.$	$m = 1.$	$n = 2\,(1)\,7$	[5]
4; 3 dec.	$\theta = 0\,(1^\circ)\,90^\circ.$	$m = 1.$	$n = 1\,(1)\,6;\ 7$	[72]
3; 4 dec.	$x = 0\,(1^\circ)\,90^\circ.$	$m = 1.$	$n = 2\,(1)\,7$	[4]

$$P_n'(\operatorname{ch} x)$$

4 dec.	$x = 0\,(0{,}1)\,3$	$n = -\frac{1}{2};\ \frac{1}{2};\ \frac{3}{2}.$	[76]

LEGENDRE FUNCTIONS $Q_n(x)$ OF THE SECOND KIND

When $|x| \leqslant 1$, the second solution $Q_n(x)$ of a Legendre equation of order n (n is a positive integer) is defined so that

$$Q_n(x) = P_n(x)\,\operatorname{Ar\,th} x - Z_n(x),$$

where $Z_n(x)$ is a definite polynomial of degree If $|x| > 1$ it is normally assumed that

$$Q_n'(x) = P_n(x)\,\operatorname{Ar\,cth} x - Z_n(x).$$

$Q_0(x)$ is equal to Ar th$x$ in the first case and to Ar cth$x$ in the second.

Coefficients of polynomials $P_n(x)$ & $Z_n(x)$

5 dec.	$n = 0\,(1)\,10$	[68]

$$2Z_n(x)$$

4 dec.	$x = 0\,(0{,}01)\,1.$	$n = 0\,(1)\,8$	[37]

$$Q_n(x)$$

10 dec.	$x = 1{,}1.$	$n = 0\,(1)\,20$	[63]
10 dec.	$x = 1{,}2;\ 2;\ 3.$	$n = 0\,(1)\,10$	[63]
5 dec.	$x = 0\,(0{,}01)\,1.$	$n = 0\,(1)\,7$	[77]
4 dec.	$x = 1{,}1\,(0{,}001)\,1{,}3\,(0{,}01)\,3{,}2\,(0{,}1)\,6\,(0{,}5)\,10.$	$n = 1$	[70]

$2Q_n(x)$

4 dec.	$x = 0(0{,}01)1.$	$n = 0(1)8$	[37]

$Q_n(\operatorname{ch} x)$

4 dec.	$x = 0(0{,}1)3.$	$n = -\frac{1}{2};\ \frac{1}{2};\ \frac{3}{2}$	[76]

$Q'_n(x)$

10 dec.	$x = 1{,}1.$	$n = 0(1)20$	[63]
10 dec.	$x = 1{,}2;\ 2;\ 3.$	$n = 0(1)10$	[63]
4 dec.	$x = 1{,}1(0{,}001)1{,}62(0{,}01)3{,}2(0{,}1)6(0{,}5)10$	$n = 1$	[70]

$Q'_n(\operatorname{ch} x)$

4 dec.	$x = 0(0{,}1)3.$	$n = -\frac{1}{2};\ \frac{1}{2};\ \frac{3}{2}$	[76]

ASSOCIATED LEGENDRE FUNCTIONS OF THE FIRST KIND

Associated Legendre functions of argument x

$P_{nm}(x)$

3—17 dec.	$x = 0(0{,}01)1.$	$m = 1(1)8.$	$n = 1(1)8$	[45]
2—16 dec.	$x = 0(0{,}01)1.$	$m = 1(1)n.$	$n = 1(1)8$	[46]
8—13 fig.	$x = 0(0{,}001)1.$	$m = 1(1)n.$	$n = 1(1)10$	[28]
5—6 fig.	$x = 0(0{,}01)1.$	$m = 1(1)n.$	$n = 1(1)8$	[29]
5—6 fig.	$x = 0(0{,}01)1.$	$m = 1.$	$n = 9$	[29]
4 dec.	$x = 0(0{,}01)1.$	$m = 1(1)n.$	$n = 1(1)8$	[37]

$P_n^m(x)$

exact	$x = 1(0{,}1)2.$	$m = 0;\ 2;\ 4.$	$n = 4(1)10$	[36]
6 fig.	$x = 1(0{,}1)10.$	$m = 0(1)4.$	$n = 1(1)10$	[36]
6; 5 dec.	$x = 0(0{,}1)1.$	$m = 1(1)n.$	$n = 1(1)3;\ 4(1)6$	[33]

$(x^2-1)^{-\frac{1}{2}m} P_n^m(x)$

exact	$x = 1(0{,}1)2.$	$m = 1;\ 3.$	$n = 2(1)10$	[36]

$$\sqrt{x^2-1}\,P^m_{n+\frac{1}{2}}(x)$$

4—7 dec.	$x=1\,(0{,}1)\,2.$	$m=1;\ 3.$	$n=-1\,(1)\,4$	[36]

$$P^m_{n+\frac{1}{2}}(x)=\frac{\Gamma(n+m+3/2)}{\pi\Gamma(n+3/2)}\int_0^\pi\left[x+\sqrt{x^2-1}\cos t\right]^{n+\frac{1}{2}}\cos mt\,dt$$

6 fig.	$x=1{,}6\,(0{,}1)\,10.$	$m=0\,(1)\,4.$	$n=-1\,(1)\,4$	[36]
5—6 fig.	$x=1\,(0{,}1)\,1{,}6.$	$m=0\,(1)\,4.$	$n=-1\,(1)\,4$	

$$\frac{d}{dx}P^m_n(x)$$

exact	$x=1\,(0{,}1)\,2.$	$m=0;\ 2;\ 4.$	$n=4\,(1)\,10$	[36]
6 fig.	$x=1\,(0{,}1)\,10.$	$m=0\,(1)\,4.$	$n=1\,(1)\,10$	

$$\frac{d}{dx}P^m_{n+\frac{1}{2}}(x)$$

7 dec.	$x=1\,(0{,}1)\,10.$	$m=0\,(1)\,3.$	$n=-1$	[36]
6 dec.	$x=1\,(0{,}1)\,10.$	$m=4.$	$n=-1$	
6 dec.	$x=1\,(0{,}1)\,10.$	$m=0\,(1)\,4.$	$n=0\,(1)\,4$	

$$\sqrt{x^2-1}\,\frac{d}{dx}P^m_{n+\frac{1}{2}}(x)$$

4—7 dec.	$x=1\,(0{,}1)\,2.$	$m=1;\ 3.$	$n=-1\,(1)\,4$	[36]

$$(x^2-1)^{-\left(\frac{1}{2}m-1\right)}\frac{d}{dx}P^m_n(x)$$

exact	$x=1\,(0{,}1)\,2.$	$m=1;\ 3.$	$n=2\,(1)\,10$	[36]

$$i^{-n}P^m_n(ix)$$

6 fig.	$x=0\,(0{,}1)\,10.$	$m=0\,(1)\,4.$	$n=1(1)10;\ m\leqslant n$	[36]

$$i^{-n}\frac{d}{dx}P^m_n(ix)$$

6 fig.	$x=0\,(0{,}1)\,10.$	$m=1\,(1)\,4.$	$n=0(1)10;\ m\leqslant n$	[36]

Associated Legendre functions with argument

$$P_{nm}(\cos\theta)$$

up to 11 fig.	$\theta=0\,(1°)\,90°.$	$m=1\,(1)\,n.$	$n=1\,(1)\,8$	[46]
4 dec.	$\theta=0\,(1°)\,90°.$	$m=1\,(1)\,n.$	$n=1\,(1)\,8$	[37]

$P_n^m(\cos\theta)$

4—11 fig.	$\theta = 0(1°)90°$.	$m = 0(1)5$.	$n = -0{,}5(0{,}1)10$	[81]
6 fig.	$\theta = 0(1°)90°$.	$m = 0(1)4$.	$n = 1(1)10$	[36]
6 dec.	$\theta = 0(5°)90°$.	$m = 1(1)n$.	$n = 1(1)3$	[33]
5 dec.	$\theta = 0(5°)90°$.	$m = 1(1)n$.	$n = 4(1)6$	[33]
4 dec.	$\theta = 0(5°)90°$.	$m = 1(1)n$.	$n = 1(1)4$	[41]

$P^{nm}(\cos\theta)$

10 dec. $\theta = 0(5°)90°$. $m = 1(1)n$. $n = 1(1)7$ [41]

$F_n^m(\cos\theta)$

The table gives coefficients of the Fourier series for these functions

7 dec. $m = 0(1)n$. $n = 0(1)12$ [15]

$R_{nm}(\cos\theta)$

The table gives coefficients of the Fourier series for these functions.

14 dec. $m = 0(1)n$. $n = 0(1)20$ [17]

$$Y_n^m = \frac{m}{n}\frac{P_n^m(\cos\theta)}{\sin\theta}$$

5—6 dec.	$\begin{cases}\theta = 0(5°)90° \\ x = \cos\theta = 0(0{,}1)1\end{cases}$	$m = 0(1)n$.	$n = 1(1)6$	[33]
4 dec.	$\theta = 0(5°)90°$.	$m = 0(1)n$.	$n = 1(1)4$	[41]

$$Y^{nm} = \frac{m}{n}\frac{P^{nm}(\cos\theta)}{\sin\theta}$$

8 dec. $\theta = 0(5°)90°$. $m = 0(1)n$. $n = 1(1)7$ [41]

$\frac{d}{d\theta}P_n^m(\cos\theta)$

6 fig. $\theta = 0(1°)90°$. $m = 1(1)4$. $n = 1(1)10$ [36]

$$X_n^m = \frac{1}{n}\frac{dP_n^m(\cos\theta)}{d\theta}$$

5—6 dec.	$\begin{cases}\theta = 0(5°)90° \\ x = \cos\theta = 0(0{,}1)1\end{cases}$	$m = 0(1)n$.	$n = 1(1)6$	[33]
4 dec.	$\theta = 0(5°)90°$.	$m = 0(1)n$.	$n = 1(1)4$	[41]

$$X^{nm} = \frac{1}{n} \frac{dP^{nm}(\cos\theta)}{d\theta}$$

8 dec. $\theta = 0\,(5°)\,90°$. $m = 0\,(1)\,n$. $n = 1\,(1)\,7$ [41]

ASSOCIATED LEGENDRE FUNCTIONS OF THE SECOND KIND

$$(-1)^m Q_n^m(x)$$

6 fig. $x = 1\,(0{,}1)\,10$. $m = 0\,(1)\,4$. $n = 0\,(1)\,10$ [36]

$$(-1)^m Q_{n+\frac{1}{2}}^m(x)$$

6 fig. $x = 1\,(0{,}1)\,10$. $m = 0\,(1)\,4$. $n = -1\,(1)\,4$ [36]

$$(-1)^{m+1} \frac{d}{dx} Q_n^m(x)$$

6 fig. $x = 1\,(0{,}1)\,10$. $m = 0\,(1)\,4$. $n = 0\,(1)\,10$ [36]

$$(-1)^{m+1} \frac{d}{dx} Q_{n+\frac{1}{2}}^m(x)$$

6 fig. $x = 1\,(0{,}1)\,10$. $m = 0\,(1)\,4$. $n = -1\,(1)\,4$ [36]

$$i^{n+2m+1} Q_n^m(ix)$$

6 fig. $x = 0\,(0{,}1)\,10$. $m = 0\,(1)\,4$. $n = 0\,(1)\,10$ [36]

$$i^{n+2m\pm1} \frac{d}{dx} Q_n^m(ix)$$

6 fig. $x = 0\,(0{,}1)\,10$. $m = 0\,(1)\,4$. $n = 0\,(1)\,10$ [36]

LOGARITHMS OF LEGENDRE FUNCTIONS

Logarithms of Legendre polynomials

$$\lg P_n(x)$$

6 dec. $x = 0\,(0{,}1)\,1$. $n = 1\,(1)\,7$ [33]

$$\lg P_n(\cos\theta)$$

6 dec. $\theta = 0\,(5°)\,90°$. $n = 1\,(1)\,7$ [33]
4 dec. $\theta = 0\,(5°)\,90°$. $n = 1\,(1)\,4$ [41]

Logarithms of associated Legendre functions

$$\lg P_n^m(x)$$

6 dec.	$x = 0(0{,}1)1.$	$m = 1(1)n.$	$n = 1(1)6$	[33]

$$\lg P_n^m(\cos\theta)$$

6 dec.	$\theta = 0(5°)90°.$	$m = 1(1)n.$	$n = 1(1)6$	[33]
4 dec.	$\theta = 0(5°)90°.$	$m = 1(1)n.$	$n = 0(1)4$	[41]

$$\lg R_n^m(\cos\theta)$$

5 dec.	$\theta = 0(5°)90°.$	$m = 0(1)n.$	$n = 1(1)5$	[74]

$$\lg X_n^m(\cos\theta) = \lg \frac{1}{n} \frac{dP_n^m(\cos\theta)}{d\theta}$$

5—6 dec.	$\left\{\begin{array}{l}\theta = 0(5°)90° \\ x = \cos\theta = 0(0{,}1)1\end{array}\right\}$	$m = 0(1)n.$	$n = 1(1)6$	[33]
4 dec.	$\theta = 0(5°)90°.$	$m = 0(1)n.$	$n = 1(1)4$	[41]

$$\lg \frac{m}{n} \frac{P_n^m(\cos\theta)}{\sin\theta}$$

5—6 dec.	$\left\{\begin{array}{l}\theta = 0(5°)90° \\ x = \cos\theta = 0(0{,}1)1\end{array}\right\}$	$m = 0(1)n.$	$n = 1(1)6$	[33]

$$\lg \frac{m}{n} \cdot \frac{P_n^m(\cos\theta)}{d\theta}$$

4 dec.	$\theta = 0(5°)90°.$	$m = 0(1)n.$	$n = 1(1)4$	[41]

ROOTS OF LEGENDRE FUNCTIONS

Roots of polynomials $P_n(x)$

16 dec.	$n = 2(1)7$	[2]
16 dec.	$n = 7(1)9$	[73]
16 dec.	$n = 2(1)7$	[22]

15 dec.	$n = 2(1)16$	[52]
16—15 dec.	$n = 2(1)7$	[51]
13 dec.	$n = 12$	[50]
12 dec.	$n = 7$	[75]
10 dec.	$n = 17$	[40]
7 dec.	$n = 2(1)8$	[37]
7 dec.	$n = 2(1)7$	[34]
7 dec.	$n = 2(1)6$	[33]
6 dec.	$n = 2(1)40$	[58]
4 dec.	$n = 2(1)10$	[29]

Roots of polynomials $P_n(\cos\theta)$

These roots may easily be obtained from the roots of polynomials $P_n(x)$ in the preceeding section.

0″, 000 001	$n = 17$	[40]
0″, 000 1	$n = 2(1)7$	[78]
1″	$n = 2(1)6$	[33]
1′	$n = 2(1)10$	[29]
1′	$n = 3(1)7$	[66]

Roots of $P'_n(\cos\theta)$

5 fig.	50 roots	[59]

Roots of $\frac{d}{d\theta} P'_n(\cos\theta)$

5 fig.	50 roots	[59]

Roots of the function $P_{nm}(x)$

10 dec.	$m = 1.$	$n = 8(1)10$	[60]
8 dec.	$m = 1.$	$n = 3(1)7$	[60]
8 dec.	$m = 1.$	$n = 8;\ 9$	[57]
7 dec.	$m = 1(1)n.$	$n = 2(1)8$	[37]
4 dec.	$m = 1(1)n-2.$	$n = 3(1)8$	[29]
4 dec.	$m = 1.$	$n = 9$	[29]

Roots of $P'_{nm}(x)$

7 dec.	$m = 1(1)n-1.$	$n = 2(1)6$	[33]

Roots of $P_{nm}(\cos\theta)$

0″, 0001	$m = 1(1)n-2.$	$n = 3(1)7$	[78]
1″	$m = 1(1)n-2.$	$n = 3(1)6$	[33]
1′	$m = 1(1)n-2.$ $m = 1.$	$n = 3(1)8$ $n = 9$	[29]

Roots of $P'_{nm}(\cos\theta)$

1″	$m = 1(1)n-1.$	$n = 2(1)6$	[33]

VARIOUS EXPRESSIONS CONTAINING LEGENDRE FUNCTIONS

Certain tables of integrals containing Legendre functions are described in this section. The value of the magnitude ϵ_m which is encountered in the section is equal to 1 if $m = 0$ and to 2 if m is a positive integer.

$$\int_0^x P_{nm}(t)\,dt$$

5 dec.	$x = 0(0{,}01)1.$ $m = 0(1)n.$ $n = 0(1)8$ pairs of values $(n, m) = (9; 0), (10; 0), (9; 1)$	[29]

$$\int_{-1}^{+1} x^r P_n(x)\,dx$$

6 fig.	$n = 0(1)10.$ $r = n(2)N$ $N = 9$ n odd , $N = 10$ n even	[68]

$$M = \frac{1}{4}\int_{-1}^{+1} R_n^m(x)\cos p\theta\,dx \;\&\; \lg M$$

6 dec.	$x = \cos\theta.$ n, m & p to 12	[65]

$$N = \frac{1}{2\varepsilon_m}\int_{-1}^{+1} R_n^m(x)\sin p\theta\,dx \;\&\; \lg N$$

6 dec.	$x = \cos\theta.$ n, m & p to 12	[65]

$$\int_1^\infty e^{-\alpha t} t^m Q_n(t)\,dt$$

9 fig.	$n = 1$	$m = 0(1)7.$	$\alpha = 0{,}5;\ 1(0{,}25)5{,}5;\ 6(0{,}25)6{,}5(0{,}5)11$ (var.) 24	[71]
		$m = 8(1)15.$	$\alpha = 1{,}2(0{,}5)11$ (var.) 24	
9 fig.	$n = 0$	$m = 0(1)8.$	$\alpha = 0{,}5;\ 1(0{,}25)5{,}5;\ 6(0{,}25)6{,}5(0{,}5)11$ (var.) 24	
		$m = 9(1)15.$	$\alpha = 1{,}2(0{,}5)11$ (var.) 24	
6 fig.	$n = 0$	$m = 0(1)10.$	$\alpha = 0{,}5(0{,}5)5(1)10(2)14$	[68]
		$m = 11(1)20$	$\alpha = 1(1)10(2)14$	
3—5 fig.	$n = 1$	$m = 0(1)5.$	$\alpha = 1{,}5(0{,}5)5(1)10$	[67]
		$m = 6(1)11$	$\alpha = 3(1)10$	

$$\int_{x_0}^0 [P'_n(t)]^2\,dt$$

5 fig.	$\theta = 90°(5°)130°$	$x_0 = \cos\theta$	[59]
4 fig.	$\theta = 90°(5°)145°$	$x_0 = \cos\theta$	

$$\int_0^1 P_m(u)\,P_n(u)\,du$$

exact	$m = 0(1)5.$	$n = 0(1)5$	[49]
5 dec.	$m = n = 0(0{,}05)1$		[59]

Chapter 10

CYLINDER FUNCTIONS

The solutions of the linear differential equation

$$y'' + \frac{1}{x} y' + \left(1 - \frac{\nu^2}{x^2}\right) y = 0. \qquad (1)$$

are called cylinder functions.

The general solution of equation (1) has the form:

$$y = C_1 J_\nu(x) + C_2 Y_\nu(x),$$

where

$$J_\nu(x) = \sum_{k=0}^{\infty} \frac{(-1)^k \left(\frac{x}{2}\right)^{\nu+2k}}{\Gamma(k+1)\,\Gamma(k+\nu+1)}, \qquad Y_\nu(x) = \frac{J_\nu(x)\cos\nu\pi - J_{-\nu}(x)}{\sin\nu\pi}$$

are cylinder functions of the first and second kind respectively (Bessel and Neumann functions).

For $\nu = n$, an integer

$$J_n(x) = \sum_{k=0}^{\infty} \frac{(-1)^k \left(\frac{x}{2}\right)^{n+2k}}{k!\,(n+k)!},$$

$$Y_n(x) = \frac{2}{\pi} J_n(x)\left(\ln\frac{x}{2} + C\right) - \frac{1}{\pi}\sum_{k=0}^{n-1} \frac{(n-k-1)!}{k!}\left(\frac{x}{2}\right)^{2k-n} -$$

$$- \frac{1}{\pi}\sum_{k=0}^{\infty} \frac{(-1)^k}{k!\,(n+k)!}\left(\frac{x}{2}\right)^{2k+n}\left\{1 + \frac{1}{2} + \cdots + \frac{1}{n+k} + 1 + \frac{1}{2} + \cdots + \frac{1}{k}\right\},$$

with the additional relationships

$$J_{-n}(x) = (-1)^n J_n(x); \quad Y_{-n}(x) = (-1)^n Y_n(x).$$

Cylinder functions of order half an odd integer are widely used. Note among them spherical Bessel or Stokes functions

$$\sqrt{\frac{\pi}{2x}}\, J_{\pm\left(n+\frac{1}{2}\right)}(x)$$

---

[1] See pages 489 and 550 for the references to Chapter 10.

and Riccati-Bessel functions

$$S_n(x)=\sqrt{\frac{\pi x}{2}}\,J_{n+\frac{1}{2}}(x) \;\&\; C_n(x)=(-1)^n\sqrt{\frac{\pi x}{2}}\,J_{-n-\frac{1}{2}}(x).$$

Lommel's functions of two variables are expressed through cylinder functions of the first kind

$$U_\nu(y,z)=\sum_{m=0}^{\infty}(-1)^m\left(\frac{y}{z}\right)^{\nu+2m}J_{\nu+2m}(z),$$

$$V_\nu(y,z)=\cos\frac{1}{2}\left(y+\frac{z^2}{y}+\pi\nu\right)+U_{-\nu+2}(y,z).$$

Cylinder functions of the third kind (Hankel functions) $H_\nu^{(1)}(x)$ and $H_\nu^{(2)}(x)$ are defined by the expressions:

$$H_\nu^{(1)}(x)=J_\nu(x)+iY_\nu(x) \;\&\; H_\nu^{(2)}(x)=J_\nu(x)-iY_\nu(x).$$

The general solution of the modified Bessel equation

$$y''+\frac{1}{x}y'-\left(1+\frac{\nu^2}{x^2}\right)y=0,$$

obtained from (1) by the substitution of ix for is

$$y=C_1I_\nu(x)+C_2K_\nu(x),$$

where

$$I_\nu(x)=\sum_{k=0}^{\infty}\frac{\left(\frac{x}{2}\right)^{2k+\nu}}{\Gamma(k+\nu)\Gamma(k+\nu+1)},\quad K_\nu(x)=\frac{\pi}{2}\,\frac{I_{-\nu}(x)-I_\nu(x)}{\sin\nu\pi}$$

are respectively cylinder functions of the first and second kind of the imaginary argument; when $\nu=n$ (n is an integer) $K_n(x)=\lim\limits_{\nu\to n}K_\nu(x)$.

Cylinder functions of argument $xi^{3/2}$ are of great value

$$J_\nu(xi^{\pm 3/2})=J_\nu(xe^{\pm 3/4\pi i})=\mathrm{ber}_\nu x\pm i\,\mathrm{bei}_\nu x$$

$$i^{\mp\nu}K_\nu(xi^{\pm 1/2})=i^{\mp\nu}K_\nu(xe^{\pm 1/4\pi i})=\mathrm{ker}_\nu x\pm i\,\mathrm{kei}_\nu x$$

$$H_\nu^{(1)}(xi^{3/2})=\mathrm{her}_\nu x+i\,\mathrm{hei}_\nu x,$$

where

$\mathrm{ber}_\nu x$ & $\mathrm{bei}_\nu x$, $\mathrm{ker}_\nu x$ & $\mathrm{kei}_\nu x$, $\mathrm{her}_\nu x$ & $\mathrm{hei}_\nu x$ are Thomson functions of the first, second and third kinds.

A number of in mathe.atical physics are reduced to the solutions of which are functions closely related to cylinder functions, such as Airy functions (solutions of $y''=xy$)

$$\mathrm{Ai}(x)=\frac{1}{3}x^{1/2}\left\{I_{-1/3}\left(\frac{2}{3}x^{3/2}\right)-I_{1/3}\left(\frac{2}{3}x^{3/2}\right)\right\},$$

$$\mathrm{Bi}(x)=\left(\frac{1}{3}x\right)^{1/2}\left\{I_{-1/3}\left(\frac{2}{3}x^{3/2}\right)+I_{1/3}\left(\frac{2}{3}x^{3/2}\right)\right\},$$

Struve functions

$$H_n(x) = \frac{2\left(\frac{x}{2}\right)^n}{\sqrt{\pi}\,\Gamma\left(n+\frac{1}{2}\right)} \int_0^{\frac{\pi}{2}} \sin(x\cos\theta)\sin^{2n}\theta d\theta,$$

Lommel-Weber functions

$$\Omega_n(x) = \frac{1}{\pi}\int_0^{\pi} \sin(x\sin\theta - n\theta)\,d\theta.$$

The chapter contains the following sections:
cylinder functions of the first and second kind of real argument;
various expression containing cylinder functions of the first and second kind of real argument;
Riccati-Bessel functions;
spherical Bessel or Stoke's functions;
Lommel functions of two variables;
cylinder functions of the third kind;
cylinder functions of the first and second kind of imaginary argument;
various expressions containing cylinder functions of the first and second kind of imaginary argument;
cylinder functions of complex argument;
cylinder functions of imaginary order & argument;
Thomson functions; their moduli & arguments;
various expressions containing Thomson functions;
Airy, Struve and Lommel-Weber functions;
zeros of cylinder functions;
values of functions of the type $J_n(j_{m,s})$, $Y_n(y_{m,s})$, $Y_n(z_{m,s})$;
integrals of cylinder functions.

CYLINDER FUNCTIONS OF THE FIRST AND SECOND KIND OF REAL ARGUMENT

Cylinder functions of the first kind of real argument (Bessel functions)

$J_0(x)$ & $J_1(x)$

35—47 dec.		$x = 0(0{,}01)0{,}05$; $0{,}1(0{,}1)0{,}5$	[38]
23—104 dec.		$x = 1$; 2; $10(10)50$ и 100	[38]
21 dec.		$x = 0(0{,}1)6$	[253]
18 dec.		$x = 0(0{,}0001)0{,}0055$	[39]
17—20 dec.	$J_1(x)$	$x = 1(1)5$	[39]
12 dec.		$x = 25(0{,}01)25{,}51$	[39]

18 dec.		$x = 0\,(0{,}001)\,25\,(0{,}01)\,99{,}99$	[66]
18 dec.		$x = 0\,(0{,}001)\,0{,}11$	[38]
16; 12 dec.		$x = 0{,}12\,(0{,}01)\,0{,}5;\ 0{,}5\,(0{,}01)\,25{,}1$	[38]
15 dec.	$J_1(x)$	$x = 0{,}1\,(0{,}1)\,0{,}5;\ 15{,}5\,(0{,}5)\,25$	[38]
	$J_0(x)$	$x = 15{,}5\,(1)\,24{,}5$ & 25	[38]
14 dec.		$x = 9;\ 10$	[213]
12 dec.		$x = 0\,(0{,}01)\,15{,}5$	[4] [34] [89]
10 dec.		$x = 0\,(0{,}001)\,16\,(0{,}01)\,25$	[23]
10 dec.		$x = 0\,(0{,}01)\,10$	[62]
10 dec.		$x = 0\,(0{,}01)\,3{,}2$	[88]
10 dec.		$x = 6{,}5\,(0{,}5)\,16$	[128]
7 dec.		$x = 0\,(0{,}001)\,16\,(0{,}01)\,25$	[6]
7 dec.		$x = 0\,(0{,}02)\,16$	[2] [86]
6 dec.		$x = 0\,(0{,}1)\,20$	[36] [42] [295]
6 dec.		$x = 0\,(0{,}1)\,15$	[58]
6 dec.	$J_1(x)$	$x = 20\,(0{,}1)\,31\,(0{,}2)\,41$	[181]
6 dec.		$x = 0\,(1)\,10;\ 16;\ 20$	[110]
6 dec.		$x = 0\,(0{,}2)\,6$	[128]
6 dec.		$x = 0\,(1)\,12$	[190]
6; 4 dec.		$x = 0\,(0{,}01)\,4;\ 4\,(0{,}1)\,15$	[32]
5 dec.		$x = 0\,(0{,}01)\,16$	[9] [13] [37]
5 dec.		$x = 0\,(0{,}01)\,15{,}09$	[87]
5 dec.		$x = 0\,(0{,}01)\,15{,}5$	[60]
5 dec.		$x = 0\,(0{,}01)\,10$	[8]
5 dec.		$x = 0\,(0{,}1)\,15$	[28]
5 dec.	$J_0(x)$	$x = 0\,(0{,}1)\,1{,}5;\ 2\,(0{,}1)\,2{,}5;$ $3{,}5\,(0{,}1)\,4{,}3;\ 5{,}2\,(0{,}1)\,5{,}6;$ $6{,}7\,(0{,}1)\,7{,}3$	[108]
4 dec.		$x = 0\,(0{,}01)\,15{,}5$	[14] $[41]_{1,2,3}$
4 dec.		$x = 0\,(0{,}02)\,16$	[3]
4 dec.		$x = 0\,(0{,}1)\,16$	[7] [20] [44]
4 dec.		$x = 0\,(0{,}1)\,15$	[21] [25] [26] [55]
4 dec.		$x = 0\,(0{,}1)\,13{,}4$	[52]
4 dec.		$x = 0\,(0{,}1)\,10$	[1]
4 dec.		$x = 0\,(0{,}1)\,0{,}2\,(0{,}2)\,8$	[48]
4 dec.	$J_0(x)$	$x = 0\,(0{,}2)\,10$	[194]
4 dec.	$J_0(x)$	$x = 0\,(0{,}2)\,8$	[37]
4 dec.		$x = 0{,}1;\ 0{,}4\,(0{,}1)\,0{,}6;\ 0{,}8\,(0{,}1)\,1{,}2;$ $1{,}5\,(0{,}5)\,3\,(1)\,6$	[240]
4 fig.		$x = 1\,(1)\,29$	$[41]_{2,3}$
4 fig.		$x = 1\,(1)\,24$	$[41]_1$
4 dec.		$x = 25{,}3$ (var.) $39{,}9$	[161]

$J_n(x)$

97—48 dec.	$n = 0(1)40.$	$x = 2$	
98—46 dec.	$n = 0(1)32.$	$x = 1$	
47—20 dec.	$n = 0(1)N-1;$ $8 \leqslant N \leqslant 15.$	$x = 0(0{,}01)0{,}05;$ $0{,}1(0{,}1)0{,}5$	
65—30 dec.	$n = 0(1)50.$	$x = 10$	[38]
45—20 dec.	$n = 0(1)56.$	$x = 20$	
36—18 dec.	$n = 0(1)69;\ 82.$	$x = 30;\ 40$	
31—15 dec.	$n = 0(1)88.$	$x = 50$	
24—10 dec.	$n = 0(1)135.$	$x = 100$	
18—20 dec.	$n = 0(1)2;\ 3(1)6.$	$x = 0(0{,}0001)0{,}0055$	[39]
18 dec.	$n = 2;\ 3.$	$x = 0(0{,}001)25(0{,}01)99{,}99$	[67]
18 dec.	$n = 0(1)N-1;$ $17 \leqslant N \leqslant 61.$	$x = 1(1)24$	[4] [34]
15 dec.	$n = 0(1)11.$	$x = 0(0{,}001)0{,}5$	[27]
11 dec.	$n = 0(1)30.$	$x = 0(0{,}01)10{,}5$	
15 dec.	$n = x = 0(1)100$		[77]
10 dec.	$n = 4(1)6.$	$x = 0(0{,}001)25(0{,}01)99{,}99$	[68]
10 dec.	$n = 7(1)9.$	$x = 0(0{,}001)25(0{,}01)99{,}99$	[69]
10 dec.	$n = 10(1)12.$	$x = 0(0{,}001)25(0{,}01)99{,}99$	[70]
10 dec.	$n = 13(1)15.$	$x = 0(0{,}001)25(0{,}01)99{,}99$	[71]
10 dec.	$n = 16(1)27.$	$x = 0(0{,}01)99{,}99$	[72]
10 dec.	$n = 28(1)39.$	$x = 0(0{,}01)99{,}99$	[73]
10 dec.	$n = 40(1)51.$	$x = 0(0{,}01)99{,}99$	[74]
10 dec.	$n = 52(1)63.$	$x = 0(0{,}01)99{,}99$	[75]
10 dec.	$n = 64(1)78.$	$x = 0(0{,}01)99{,}99$	[76]
10 dec.	$n = 79(1)96.$	$x = 0(0{,}01)99{,}99$	[77]
	$n = 97(1)114.$	$x = 0(0{,}05)99{,}95$	
	$n = 115(1)135.$	$x = 0(0{,}1)99{,}9$	
	$n = 0(1)135.$	$x = 100$	
10 fig.	$n = 0(1)20.$	$x = 0(0{,}1)25$	[96]
10 dec.	$n = 0(1)13.$	$x = 6{,}5(0{,}5)16$	[128]
8 fig.	$n = 0(1)13.$	$x = 0(0{,}01)14{,}99$	[11]
8 dec.	$n = 2(1)20.$	$x = 0(0{,}1$ or $0{,}01)$ $10(0{,}1)20$	[96]
8 dec.	$n = 935(1)1035.$	$x = 1000$	[166]
8 dec.	$n = 967;\ 968;$ $981(1)1000.$	$x = 1000$	[110]
7 dec.	$n = 2(1)5.$	$x = 0{,}1(0{,}1)5$	[2] [86]
7 dec.	$n = N;\ N+1.$	$x = X(0{,}2)X+2$	[36] [42]
	$N = 5\ 8\ 11\ 13\ 15\ 18\ 20\ 23\ 25\ 27$		
	$X = 0\ 2\ 4\ 6\ 8\ 10\ 12\ 14\ 16\ 18$		

7 dec.	$n = x = 1\,(1)\,50$		[2] [86]
7 dec.	$n = N;\ N+1.$ $N = 5\ \ 8\ \ 11\ \ 13\ \ 15\ \ 18\ \ 20\ \ 23\ \ 25\ \ 27$ $X = 0\ \ 1\ \ 2\ \ 3\ \ 4\ \ 5\ \ 6\ \ 7\ \ 8\ \ 9$	$x = X(0,1)X + 1$	[295]
6 dec.	$n = 0\,(1)\,120.$	$x = 0\,(0,1)\,124,9$	[11]
6 dec.	$n = 0\,(1)\,13.$	$x = 0\,(0,2)\,6$	[128]
6 dec.	$n - x = -1\,(0,1)\,1.$	$x = 1\,(1)\,20$	[222]
6 dec.	$n = 0\,(1)\,N-1$ $8 \leqslant N \leqslant 36.$	$x = 0\,(1)\,10;\ 16$ и 20	[110]
6 dec.	$n = 0\,(1)\,20.$	$x = 1\,(1)\,12$	[2] [86]
6 dec.	$n = 100\,(1)\,109.$	$x = 109$	[208]
6 dec.	$n = x;\ x \pm 1.$	$x = 1\,(1)\,20$	[221]
6 dec.	$n = 0\,(1)\,20.$	$x = 0\,(1)\,12$	[190]
6 dec.	$n = x;\ x - 1.$	$x = 0\,(1)\,10$	[134]
6 dec.	$n = x;\ x - 1.$	$x = 1\,(1)\,50\,(5)\,100\,(10)$ $200\,(20)\,400\,(50)\,1000$ (var.) 10^6	[130]
5 dec.	$n = 2\,(1)\,5.$ $n = 0\,(1)\,20.$	$x = 0\,(0,1)\,5$ $x = 1\,(1)\,12$	[9] [13]
5 dec.	$n = x = 1\,(1)\,50$		[13]
5 dec.	$n = 0\,(1)\,12.$	$x = 0\,(1)\,20$	[28]
4—21 dec.	$n = 0\,(1)\,16$ до $60.$	$x = 1\,(1)\,24$	[9]
4 fig.	$n = 0\,(1)\,7.$	$x = 0\,(0,1)\,9,9$	[115]
4 dec.	$n = 2\,(1)\,4.$	$x = 0\,(0,1)\,4,9$	[44]
4 dec.	$n = 0\,(1)\,3.$ $n = 0\,(1)\,8.$	$x = 0,1;\ 0,4\,(0,1)\,0,6;$ $0,8\,(0,1)\,1$ $x = 1,1;\ 1,2;\ 1,5\,(0,5)$ $3\,(1)\,6$	[240]
4 dec.	$n = 2.$	$x = 0\,(0,1)\,0,2\,(0,2)\,8$	[48]
4 fig.	$n = 0\,(1)\,N-1;$ $17 \leqslant N \leqslant 61.$	$x = 1\,(1)\,24$	[13] [14] $[41]_{1,2,3}$
4 fig.	$n = 0\,(1)\,N-1;$ $42 \leqslant N \leqslant 45.$	$x = 25\,(1)\,29$	[13] $[41]_3$
4 dec.	$n = 2\,(1)\,9.$ $n = 10\,(1)\,17.$	$x = 0\,(1)\,24$ $x = 4\,(1)\,29$	[20]
4 dec.	$n = 2\,(1)\,10.$	$x = 1\,(1)\,24$	[55]

$J_n(ms)$

5 dec.	$m = 1;\ \pi.$	$n = 0\,(1)\,10.$	$s = 1\,(1)\,40$	[165]
	$m = \begin{cases} 2\,(1)\,10 \\ 2\pi\,(\pi)\,5\pi \end{cases}$	$n = 0\,(1)\,10.$	$s = 1\,(1)\,20$	

$J_{2n}(n)$

8; 10; 12 dec.	$n=10(1)14;\ 15(1)19;\ 20\quad 21$	[112]

$J_0(\pi x)$

6 dec.	$x=1(1)50$	[158]
5 dec.	$x=1(1)50$	[37]
3 dec.	$180x=0(10)600$	[187]
3 dec.	$180x=600(10)1040$	[188]

$J_n(2\sqrt{x})$

8 dec.	$n=0\left\{\begin{array}{l} x=0(0{,}02)1{,}5(0{,}05)3(0{,}1)13(0{,}2)\\ 45(0{,}5)115(1)410\end{array}\right.$	[305]
8; 9 dec.	$n=1\left\{\begin{array}{l} x=0(0{,}02)1{,}5(0{,}05)3(0{,}1)13(0{,}2)\\ 45(0{,}5)115(1)125;\ 125(1)410\end{array}\right.$	[305]

$J_\nu(x)\left(\nu=m+\frac{1}{2},\ m\right.$ is an integer)

12 dec.	$\left\{\begin{array}{l}\nu=\mu+\frac{1}{2}(1)M+\frac{1}{2}\\ -1\leqslant\mu\leqslant-23\\ 11\leqslant M\leqslant 44\end{array}\right\}$	$x=1(1)20$	[136]
12; 9 dec.	$\nu=\pm\frac{1}{2};\ \pm\frac{3}{2}$	$\left\{\begin{array}{l}x=0(0{,}01)1{,}99;\\ 2(0{,}01)9{,}89\end{array}\right.$	[38]
12 dec.	$\nu=\pm\frac{1}{2};\ \pm\frac{3}{2}$	$\left\{\begin{array}{l}x=9{,}9(0{,}01)\\ 10(0{,}1)20(1)100\end{array}\right.$	[38]
6 dec.	$\nu=\pm\frac{1}{2}.$	$x=0(0{,}02)20$	[136]
6 dec.	$\nu=-6\frac{1}{2}(1)6\frac{1}{2}.$	$x=1(1)50$	[2] [86]
	$\nu=7\frac{1}{2}(1)18\frac{1}{2}.$	$x=1(1)20$	
6 dec.	$\nu=-6\frac{1}{2}(1)6\frac{1}{2}.$	$x=1(1)50$	[191]
	$\nu=7\frac{1}{2}(1)34\frac{1}{2}.$	$x=1(1)20$	
6 dec.	$\nu=\pm x;\ \pm(x-1).$	$x=\frac{1}{2}(1)9\frac{1}{2}$	[134]
6 dec.	$\nu=x\pm\frac{1}{2}.$	$x=1(1)20$	[222]
5 dec.	$\nu=\pm\frac{1}{2};\ \pm\frac{3}{2}.$	$\left.\begin{array}{l}x=0(0{,}01)1(0{,}1)\\ 10(1)100\end{array}\right\}$	[37]
5 dec.	$\pm\nu=\frac{1}{2}(1)\frac{13}{2}.$	$x=1(1)50$	[9] [13]
	$\pm\nu=7\frac{1}{2}(1)18\frac{1}{2}.$	$x=2(1)20$	

4 fig.	$\nu = -6\frac{1}{2}(1)6\frac{1}{2}$.	$x = 1(1)50$	[14] [41]$_{1,2,3}$
	$\nu = \frac{1}{2}(1)6\frac{1}{2}$ до $18\frac{1}{2}$.	$x = 1(1)24$	
4 dec.	$\nu = \pm\frac{1}{2}$.	$x = 0(0{,}1)15$	[15]
4 dec.	$\nu = -2\frac{1}{2}(1)2\frac{1}{2}$.	$x = 0(0{,}2)8$	[100]
4 dec.	$\nu = -6\frac{1}{2}(1)6\frac{1}{2}$.	$x = 0(1)24$	[20]

$J_\nu(x)\left(\nu = m \pm \frac{1}{3},\ m \text{ is an integer}\right)$

10 dec.	$\nu = \frac{1}{3}$.	$x = 0(0{,}001)0{,}6(0{,}01)25$	[65]$_1$
	$\nu = -\frac{1}{3}$.	$x = 0(0{,}001)0{,}8(0{,}01)25$	
10 dec.	$\nu = \frac{2}{3}$.	$x = 0(0{,}001)0{,}5(0{,}01)25$	[65]$_1$
	$\nu = -\frac{2}{3}$.	$x = 0(0{,}001)0{,}9(0{,}01)25$	
7 dec.	$\nu = \frac{1}{3}$.	$x = 0(0{,}02)16$	[2] [86]
7 dec.	$\nu = -\frac{1}{3}$.	$x = 2(2)16$	[86]
5 dec.	$\nu = \frac{1}{3}$.	$x = 0(0{,}02)16$	[13]
5 dec.	$\nu = \pm\frac{1}{3}$.	$x = 0(0{,}1)10$	[37]
5 dec.	$\nu = \pm\frac{2}{3}$.	$x = 0(0{,}1)10$	[290]
4 dec.	$\nu = \frac{1}{3}$.	$x = 0(0{,}02)14(0{,}04)16$	[3]
4 dec.	$\nu = \pm\frac{2}{3}$.	$x = 0(0{,}02)1$	[225]
4 dec.	$\nu = \pm\frac{1}{3}$.	$x = 0(0{,}05)2(0{,}2)8$	[256]
4 dec.	$\nu = \frac{1}{3}$.	$x = 0(0{,}1)16$	[7]
4 dec.	$\nu = \pm\frac{2}{3}$.	$x = 0(0{,}1)15$	[2]
4 dec.	$\nu = \pm\frac{1}{3};\ \pm\frac{2}{3}$.	$x = 0(0{,}1)15$	[15]

$J_\nu(x)\left(\nu = \pm\frac{1}{4};\ \pm\frac{3}{4}\right)$

10 dec.	$\nu = \frac{1}{4}$.	$x = 0(0{,}001)0{,}6(0{,}01)25$	[65]$_1$
	$\nu = -\frac{1}{4}$.	$x = 0(0{,}001)0{,}8(0{,}01)25$	
	$\nu = \frac{3}{4}$.	$x = 0(0{,}001)0{,}5(0{,}01)25$	
	$\nu = -\frac{3}{4}$.	$x = 0(0{,}001)0{,}9(0{,}01)25$	

4 dec.	$\nu = \pm\frac{1}{4};\ \pm\frac{3}{4}.$	$x = 0(0{,}1)15$	[2] [15] [102]
4 dec.	$\nu = \pm\frac{1}{4};\ \pm\frac{3}{4}.$	$x = 0(0{,}1)8$	[290]
4 dec.	$\nu = \pm\frac{1}{4};\ \pm\frac{3}{4}.$	$x = 0(0{,}2)8$	[14] $[41]_{2,3}$ [102]

$J_\nu(\nu)$ & $J_{\nu-1}(\nu)$

6 dec.	$\nu = 0\left(\frac{1}{4}\right)10$	[134]

$J_\nu(x)\left(\nu = \pm\frac{1}{6};\ \pm\frac{5}{6}\right)$

4 dec.	$\nu = \pm\frac{1}{6};\ \pm\frac{5}{6}.$	$x = 0(0{,}1)15$	[2] [15]

Cylinder functions of the second kind of real argument (Neumann functions)

$Y_0(x)$ & $Y_1(x)$

14—17 dec.	$x = 16(1)25$	[39]
10 dec.	$x = 16(0{,}01)25{,}51$	[39]
8 dec.	$x = 0(0{,}01)25$	[23]
8 fig.	$x = 0(0{,}0001)0{,}05(0{,}001)1$	[64]
7 dec.	$x = 0(0{,}02)16$	[2] [86]
5 dec.	$x = 0(0{,}01)16$	[37]
5 dec.	$x = 0(0{,}01)10$	[8]
5 dec.	$x = 0(0{,}02)16$	[13]
4 dec.	$x = 0(0{,}01)15{,}99$	[14] $[41]_{2,3}$
4 dec.	$x = 0(0{,}02)16$	[3]
4 dec.	$x = 0(0{,}1)16$	[20] [44]
4 dec.	$x = 0(0{,}1)10{,}2$	$[41]_1$
4 dec.	$x = 0(0{,}1)0{,}2(0{,}2)8$	[48]

$Y_n(x)$

16—19 fig.	$n = 0(1)10.$	$x = 0{,}1(0{,}1)5$	[9]
13—17 dec.	$n = 0(1)15$	$x = 16(1)25$	[39]
15—2 dec.	$n = 16(1)30$	$x = 16(1)25$	[39]
10 fig.	$n = 0(1)20.$	$x = 0(0{,}1)25$	[96]
8 fig.	$n = 2(1)20.$	$x = 0(0{,}1$ or $0{,}01)$ $10(0{,}1)20$	[96]

7–8 fig.	$n = 0(1)10.$	$x = 0{,}1(0{,}1)5$	[2] [86]
7 dec.	$n = 0(1)13.$	$x = 6(1)12$	
7 dec.	$n = x = 1(1)50$		
6 dec.	$n = x;\ x - 1.$	$x = 0(1)10$	[134]
5–6 fig.	$n = 0(1)10.$	$x = 0{,}1(0{,}1)5$	[13]
5 dec.	$n = 0(1)13.$	$x = 6(1)12$	
5 dec.	$n = x = 1(1)50$		
4 dec.	$n = 2.$	$x = 0(0{,}1)0{,}2(0{,}2)8$	[48]

$Y_\nu(x)$ & $Y_{-\nu}(x)$

6 dec.	$\nu = x;\ x - 1.$	$x = \frac{1}{4}\left(\frac{1}{2}\right)9\frac{3}{4}$	[134]
4 dec.	$\nu = \pm\frac{1}{4};\ \pm\frac{3}{4}.$	$x = 0(0{,}1)8$	[290]

$Y_\nu(x)\left(\nu = \pm\frac{1}{3};\ \pm\frac{2}{3}\right)$

7 dec.	$\nu = \frac{1}{3}.$	$x = 0(0{,}02)16$	[2] [86]
	$\nu = -\frac{1}{3}.$	$x = 2(2)16$	
5 dec.	$\nu = \frac{1}{3}.$	$x = (0{,}02)16$	[13]
5 dec.	$\nu = \pm\frac{1}{3}.$	$x = 0(0{,}1)10$	[37]
5 dec.	$\nu = \pm\frac{2}{3}.$	$x = 0(0{,}1)10$	[290]
4 dec.	$\nu = \frac{1}{3}.$	$x = 0(0{,}02)14(0{,}04)16$	[3]

$G_0(x) = -\frac{\pi}{2}Y_0(x)$ & $G_1(x) = -\frac{\pi}{2}Y_1(x)$

21 dec.		$x = 0(0{,}1)6$	[253]
14 dec.		$x = 9$ и 10	[213]
10 dec.		$x = 6{,}5(0{,}5)15{,}5$	[128]
7 dec.		$x = 0(0{,}01)16$	[126]
7 dec.		$x = 0(0{,}1)16$	[124] [205]
5 dec.		$x = 0(0{,}1)6(0{,}5)16$	[127]
4 dec.		$x = 0(0{,}01)16$	[103]
4 dec.	$G_0(x)$	$x = 0(0{,}01)1(0{,}1)9{,}9$	[60]
4 dec.		$x = 0(0{,}01)1(0{,}1)10{,}3$ или $10{,}2$	[13] [41], [199]
4 dec.		$x = 0(0{,}01)1(0{,}1)5$	[9]

$2G_0(x)$ & $2G_1(x)$

4 fig.	$x = 0{,}1;\ 0{,}4(0{,}1)0{,}6;\ 0{,}8(0{,}1)1{,}2;\ 1{,}5(0{,}5)3(1)6$	[241]

$G_n(x)$

6 dec.	$n = x;\ x-1.$	$x = 1(1)50(5)100(10)200(20)$ 400(50)1000 (var.)10^6	[130]
6 dec.	$n = x;\ x-1.$ $n = 100(1)107.$	$x = 6(1)13$ $x = 104$	[208]
5 dec.	$n = 0(1)13.$	$x = 0(0,1)6(0,5)16$	[127]
3—4 fig.	$n = 0(1)4.$ $n = 0(1)6.$ $n = 7(1)10.$	$x = 0,1;\ 0,4(0,1)0,6;\ 0,8(0,1)1$ $x = 1,1;\ 1,2;\ 1,5(0,5)3(1)6$ $x = n-4(1)6$	[240]

$$N_n(x) = \frac{\pi}{2}Y_n(x) + (\ln 2 - C)J_n(x)$$

$N_0(x)$ & $N_1(x)$

14 dec.	$x = 9$ и 10	[213]
10 dec.	$x = 0(0,1)6(0,5)15,5$	[128]
7 dec.	$x = 0(0,1)16$	[124] [205]
6 dec.	$x = 0(0,02)15,5$	[127]
6 dec.	$x = 0(0,2)6(0,5)15,5$	[128]
4 dec.	$x = 0(0,01)1(0,1)10,2$	[9] [13] [41]$_1$ [184]

$N_n(x)$

6 dec.	$n = x;\ x-1$	$x = 1(1)50(5)100(10)200(20)$ 400(50)1000 (var.)10^6	[130]
6 dec.	$n = 0(1)13.$	$x = 0(0,2)6(0,5)15,5$	[128]

Derivatives of cylinder functions of the first and second kind of real argument

$2J'_n(x)$

20; 17 dec.	$n = 1(1)27.$	$x = 1;\ 2;\ 3(1)5$	[39]
4 dec.	$n = 1(1)6.$	$x = 1,2;\ 1,5(0,5)3(1)5$	[240]
3 dec.	$n = 1(1)3.$	$x = 0,5;\ 0,6;\ 0,8;\ 0,9;\ 1$	[240]

$J'_n(n)$

7 dec.	$n = 1(1)50$	[2] [86]
5 dec.	$n = 1(1)50$	[13]

$J_0^{(p)}(x)$

$(16-p)$ dec.	$p = 1(1)10.$ $p = 1(1)9.$	$x = 0,1(0,1)0,5$ $x = 16(0,5)25$	[38]

$$J_\nu^{(p)}(x)$$

15—8 dec. $\pm\nu=\frac{1}{3};\frac{2}{3};\frac{1}{4};\frac{3}{4}$. $p=1\,(1)\,15$. $x=10\,(0{,}1)\,25$ [177]

$$\frac{\partial}{\partial\nu}[J_\nu(x)]$$

6 dec.	$\nu=\pm\frac{1}{2};\pm\frac{3}{2}$.	$x=0\,(0{,}1)\,20$	[138]
5 dec.	$\frac{\partial}{\partial\nu}[J_\nu(x)]$ } $\nu-x=-1\,(0{,}1)\,1$.	$x=1\,(1)\,20$	[221]
4 dec.	$\frac{\partial^2}{\partial\nu^2}[J_\nu(x)]$ } $\nu-x=-1\,(0{,}1)\,1$.	$x=1\,(1)\,20$	[221]

$$\frac{1}{2n!}\frac{d^{n-1}}{dx^{n-1}}[J_{\pm\frac{1}{2}}(x)]$$

6 dec. $n=1\,(1)\,6$. $x=1\,(1)\,50$ [191]

$$Y_n^{(p)}(x)$$

7 dec.	$p=1$.	$n=x=1\,(1)\,50$		[2] [86]
5 dec.	$p=1$.	$n=x=1\,(1)\,50$		[13]
$(16-p)$ dec.	$p=1\,(1)\,14$.	$n=0$.	$x=15\,(1)\,25$	[39]

$$-2G_n'(x)$$

4 fig.	$n=1\,(1)\,3$.	$x=0{,}5;\ 0{,}6;\ 0{,}8\,(0{,}1)\,1$	[240]
	$n=1\,(1)\,6$.	$x=1{,}2;\ 1{,}5\,(0{,}5)\,3\,(1)\,5$	
	$n=7;\ 8$.	$x=4$	
	$n=7\,(1)\,10$.	$x=5$	

Various expressions containing cylinder functions of the 1st & 2nd kind of real argument

$$\Lambda_n(x)=\frac{n!\,J_n(x)}{\left(\frac{1}{2}x\right)^n}$$

6 dec.	$n=1$.	$x=0\,(0{,}1)\,20$	[190] [296]
4—5 dec.	$n=0\,(1)\,8$.	$x=0\,(0{,}02)\,9{,}98$	[14] [301]
4 dec.	$n=1$.	$x=0\,(0{,}2)\,12$	[18] [50] [273] [281]
4; 5 dec.	$n=1$	$180\frac{x}{\pi}=0\,(15)\,195;\ 210\,(15)\,1125$	[50] [59] [273]
4 dec.	$10\,\Lambda_1(x)$	$180\frac{x}{\pi}=0\,(40)\,600$	[187]
4 dec.	$1-\Lambda_1(x)$	$x=0\,(0{,}1)\,0{,}2\,(0{,}2)\,4\,(0{,}5)\,10\,(1)\,15$	[48]

$$-xJ_1(x)$$

5 dec.	$x = 0(0,1)1,5;\ 3,5(0,1)4,3;\ 6,7(0,1)7,3$	[108]

$$e^{-x}J_n(x)$$

7 dec.	$n = 2(1)5.$	$x = 0,1(0,1)5$	[2]

$$n!\,J_n(x)$$

4 dec.	$n = \pm\frac{1}{3};\ \pm\frac{2}{3}.$	$x = 0(0,2)8$	[99]

$$\{J_0(x)\}^2$$

4 dec.	$x = 0(0,2)10$	[194]

$$\left[\frac{2}{x}J_1(x)\right]^2$$

6 dec.	$x = 1(0,02)15,2$	[109]
6 dec.	$x = 0(0,1)20$	[190] [296]
4; 5 dec.	$180\frac{x}{\pi} = 0(15)150;\ 165(1)1125$	[50] [59] [273]
4 dec.	$x = 0(0,2)12$	[50] [273]
4 dec.	$180\frac{x}{\pi} = 0(40)600$	[187]

$$n^{\frac{1}{3}}J_n(n)\ \&\ n^{\frac{2}{3}}J_n'(n)$$

7 dec.	$n = 1(1)50$	[2] [86]
5 dec.	$n = 1(1)50$	[13]

$$\frac{xJ_1(x)}{J_0(x)}$$

2—5 dec.	$x = 0(0,01)0,1(0,05)1,5$ (var.) $14,8$	[248]

$$\frac{1}{6}[J_0(x) + 2J_0(2x) + 3J_0(3x)]$$

6 dec.	$x = 0(0,2)7,4$	[296]

$$\frac{4}{3}\left\{2\frac{J_1(x)}{x} - \frac{1}{2}\frac{J_1\left(\frac{1}{2}x\right)}{\frac{1}{2}x}\right\}$$

6 dec.	$x = 0(0,2)20$	[296]

$$\left(-\frac{1}{2}\right)^{\frac{1}{2}}\left(\frac{1}{4}\right)!\,x^{-\frac{1}{4}}\,J_{\frac{1}{4}}\left(\frac{1}{2}x\right)$$

3 dec. $x = 0(0{,}2)8(1)12$ [41]$_1$ [51]

$$2^{-\frac{1}{2}}\left(-\frac{1}{4}\right)!\,x^{\frac{1}{4}}\,J_{-\frac{1}{4}}\left(\frac{1}{2}x\right)$$

3 dec. $x = 0(0{,}2)8(1)12$ [41]$_1$ [51]

$$\lg J_0(\pi x)$$

6 dec. $x = 1(1)50$ [158]

$$-n^{\frac{1}{3}}\,Y_n(n)\ \&\ n^{\frac{2}{3}}\,Y_n'(n)$$

7 dec. $n = 1(1)50$ [2] [86]
5 dec. $n = 1(1)50$ [13]

$$(\sqrt{x})^{-n}\,Y_n(2\sqrt{x})$$

8 dec. $n = 0$ $\begin{cases} x = 0(0{,}02)1{,}5(0{,}05)3(0{,}1)13(0{,}2)45(0{,}5) \\ 115(1)410 \end{cases}$ [305]
8; 9 dec. $n = 1$ $\begin{cases} x = 0(0{,}02)1{,}5(0{,}05)3(0{,}1)13(0{,}2)45(0{,}5) \\ 115(1)125;\ 125(1)410 \end{cases}$ [305]

Auxiliary functions

For large values of the argument

$$A_n(x)\ \&\ B_n(x)$$

$$A_0(x) = J_0(x)\sin x - Y_0(x)\cos x \qquad B_0(x) = J_0(x)\cos x + Y_0(x)\sin x$$
$$A_1(x) = -J_1(x)\cos x - Y_1(x)\sin x \qquad B_1(x) = J_1(x)\sin x + Y_1(x)\cos x$$

8 dec. $n = 0;\ 1.$ $x = 25(0{,}1)50(1)150(10)1150;\ 1000(100)6000$ [23]

$$U_\nu(x),\ V_\nu(x)\ \&\ \frac{V_\nu(x)}{U_\nu(x)}$$

$$J_{\pm\nu}(x) = U_\nu(x)\cos\left(x \mp \frac{1}{2}\nu\pi - \frac{1}{4}\pi\right) - V_\nu(x)\sin\left(x \mp \frac{1}{2}\nu\pi - \frac{1}{4}\pi\right)$$
$$Y_{\pm\nu}(x) = U_\nu(x)\sin\left(x \mp \frac{1}{2}\nu\pi - \frac{1}{4}\pi\right) + V_\nu(x)\cos\left(x \mp \frac{1}{2}\nu\pi - \frac{1}{4}\pi\right)$$

10 dec. $\nu=\pm\frac{1}{3};\ \pm\frac{2}{3};\ \pm\frac{1}{4};\ \pm\frac{3}{4}$. $x=25($ var. $)\,30\,000$ [65]$_1$

4 dec. $\nu=\frac{1}{3}$. $x=0\,(0{,}05)\,2\,(0{,}2)\,8$ [256]

$$Q_n(x)=\sqrt{\tfrac{1}{2}\pi x}\,V_n(x)=\sqrt{\tfrac{1}{2}\pi x\,\{J_n^2(x)+Y_n^2(x)\}}\,\sin\alpha_n=$$
$$=\sqrt{\frac{\pi x}{2}}\left\{Y_n(x)\cos\left(x-\frac{1}{2}n\pi-\frac{1}{4}\pi\right)-J_n(x)\sin\left(x-\frac{1}{2}n\pi-\frac{1}{4}\pi\right)\right\}$$

The argument α_n is defined in the section "Cylinder functions of the third kind" (see p. 263).

9—10 dec. $Q_n(x)$ $n=0\,(1)\,6$. $x=10\,(10)\,100\,(100)\,1000$ [123]

9—10 dec. $Q_0(x)$
8 dec. $\lg|Q_0(x)|$ } $x=10\,(10)\,100\,(100)\,1000$ [122]

For small values of the argument

$C_n(x)$ & $D_n(x)$

$$Y_0(x)=C_0(x)+D_0(x)\lg x \qquad Y_1(x)=\frac{1}{x}C_1(x)+D_1(x)\lg x.$$

8 dec. $n=0;\,1.$ $x=0\,(0{,}0001)\,0{,}005$ [64]

8 dec. $n=0;\,1.$ $x=0\,(0{,}01)\,0{,}5$ [23]

$\sqrt{2\pi x}$, $\sqrt{\dfrac{2}{\pi x}}$, $\ln\frac{1}{2}x+C$

12 dec. $\sqrt{\dfrac{2}{\pi x}}$ $x=0\,(0{,}01)\,2;\ 10\,(0{,}1)\,20$ [38]

8 dec. $\left\{\ln\frac{1}{2}x+C\right\}$ $x=0{,}2\,(0{,}2)\,6$ [41]$_1$

6 dec. $\sqrt{2\pi x}$ $x=1\,(1)\,40$ [215]

RICCATI-BESSEL FUNCTIONS

$S_n(x)=\sqrt{\dfrac{\pi x}{2}}\,J_{n+\frac{1}{2}}(x)$ & $S_n'(x)$

7 dec. $n=0\,(1)\,N$ $\begin{cases}x=1\,(1)\,10 & [127]\\ x=0{,}1\,(0{,}1)\,0{,}9 & [133]\\ x=1{,}1\,(0{,}1)\,1{,}9 & [131]\end{cases}$

In each case $S_{N+1}(x)<\frac{1}{2}10^{-7}$.

Typical values

x	0,1	0,2	0,5	1	1,5	2	3	4	6	8	10
N	3	4	6	7	9	10	12	14	17	20	23

6 dec. $n=0(1)6.$ $x=0(0{,}05)10$ [310]

6 dec. $\begin{cases} n=1(1)N;\ 5\leqslant N\leqslant 420. & x=1(1)6(2)10(5)100\ (10)200(50)400 \\ n=1(1)N;\ 6\leqslant N\leqslant 436. & x=1{,}2(\ \text{var.}\ )640 \end{cases}$ [35]

4 dec. $n=0(1)4.$ $x=0(0{,}1)5{,}6$ [272]

$$C_n(x)=(-1)^n\sqrt{\frac{\pi x}{2}}\,J_{-n-\frac{1}{2}}(x)\ \&\ C'_n(x)$$

7 fig. $n=0(1)N$ $\begin{cases} x=1(1)10 & [127] \\ x=0{,}1(0{,}1)0{,}9 & [133] \\ x=1{,}1(0{,}1)1{,}9 & [131] \end{cases}$

Typical values :	x	0,1	0,2	0,5	1	1,5	2	3	4	6	8	10
	N	3	4	6	7	9	10	12	14	17	20	23

7 dec. $n=1(1)N;\ 5\leqslant N\leqslant 420$ $\{x=1(1)6(2)10(5)\ 100(10)200(50)400\}$ [35]

$$\frac{xS'_n(x)}{S_n(x)}$$

4 dec. $n=0(1)4.$ $x=0(0{,}1)5{,}6$ [272]

$$\lg|S_n(x)|,\ \lg|C_n(x)|,\ \lg|S'_n(x)|,\ \lg|C_n(x)|$$

7 dec. $n=0(1)N'$ $\begin{cases} x=1(1)10 & [127] \\ x=0{,}1(0{,}1)0{,}9 & [133] \\ x=1{,}1(0{,}1)1{,}9 & [131] \end{cases}$

B [131] & [133] $N'=N$ (see $S_n(x)$ & $C_n(x)$), in [127] $N'<N$.

Typical values :	x	1	2	4	6	8	10
	N'	4	6	9	11	14	17

$$E_n^2(x)=S_n^{2}(x)+C_n^2(x),\ E_n'^2(x)=S_n'^2(x)+C_n'^2(x),\ \lg E_n^2(x),\ \lg E_n'^2(x)$$

7 fig. $n=0(1)N''.$ $x=0{,}1(0{,}1)2(1)10$ [127] [131] [133]

N'' is such that for $n>N''$ $E_n^2(x)=C_n^2(x)$ & $E_n'^2(x)=C_n'^2(x)$ with tabulation accuracy.

Typical values :	x	0,1	0,5	1	2	4	6	8	10
	N''	1	2	3	5	8	10	12	15

Re(A_n), Re(P_n), Im(A_n), Im(P_n)

$$A_n = \left[\frac{(-1)^{n+\frac{1}{2}}(2n+1)}{n(n+1)}\right]\left[\frac{S_n'(m\alpha)S_n(\alpha) - mS_n'(\alpha)S_n(m\alpha)}{S_n'(m\alpha)\Phi_n(\alpha) - m\Phi_n'(\alpha)S_n(m\alpha)}\right],$$

$$P_n = \left[\frac{(-1)^{n+\frac{3}{2}}(2n+1)}{n(n+1)}\right]\left[\frac{mS_n(\alpha)S_n'(m\alpha) - S_n(m\alpha)S_n'(\alpha)}{m\Phi_n(\alpha)S_n'(m\alpha) - S_n(m\alpha)\Phi_n'(\alpha)}\right],$$

$\Phi_n(\alpha) = S_n(\alpha) + iC_n(\alpha).$

6 dec. $\left\{\begin{array}{l} m = 1,2;\ 1,33;\ 1,4;\ 1,44;\ 1,5;\ 1,6.\ 1 \leqslant n \leqslant 421 \\ \alpha = 1(\text{ var. })10(5)100(10)200(50)400 \end{array}\right\}$ [301]

$$f_n(ix) = i^{-(n+1)} e^{ix} \sqrt{\frac{\pi x}{2}}\,[J_{n+\frac{1}{2}}(x) + i(-1)^n J_{-n-\frac{1}{2}}(x)],$$

$$F_n(ix) = ix(2n+1)^{-1}[nf_{n-1}(ix) + (n+1)f_{n+1}(ix)],$$

$$\varphi_n(ix) = \frac{f_n(ix)}{F_n(ix)}$$

The real and imaginary parts of these functions are given in the table.

4—5 fig. $\left\{\begin{array}{l} n = 1(1)5.\ x = 0,1(0,05)0,2(0,1)1(0,25)4(0,5) \\ 6(1)10(5)20;\ 30 \end{array}\right\}$ [303]

SPHERICAL BESSEL OR STOKE'S FUNCTIONS

$\sqrt{\frac{\pi}{2x}}\,J_\nu(x)\left(\nu = m + \frac{1}{2}, m\right.$ is an integer)

8—10; 7 fig.	$\pm\nu = \frac{1}{2}(1)\frac{25}{2}$.	$x = 0(0,01)10;\ 10(0,1)25$	[79]
7—8 fig.	$\pm\nu = \frac{27}{2}$.	$x = 0(0,01)10;\ 10(0,05)$ 10,5(0,1)25	
8—10; 7 fig.	$\pm\nu = \frac{29}{2}(1)\frac{43}{2}$.	$x = 0(0,01)10;\ 10(0,1)25$	[80]
7 fig.	$\pm\nu = \frac{45}{2}(1)\frac{61}{2}$.	$x = 10(0,1)25$	
5 fig.	$\pm\nu = \frac{1}{2}(1)\frac{5}{2}$.	$x = 0(0,1)0,2(0,2)8$	[48]

$$x^{-\nu+\frac{1}{2}}\sqrt{\frac{\pi}{2x}}\,J_\nu(x)$$

8—10 fig. $\nu = -\frac{1}{2}$; $\pm\nu = \frac{3}{2}(1)\frac{27}{2}$. $x = 0(0,01)x_n$. $0,5 \leqslant x_n \leqslant 2,5$ [79]

$$\Lambda_\nu(x) = \frac{\Gamma(\nu+1)}{\left(\frac{x}{2}\right)^\nu} J_\nu(x)$$

$$\sqrt{\frac{\pi}{2x}}\, J_\nu(x) = C_\nu x^{\nu-\frac{1}{2}} \Lambda_\nu(x), \quad \text{где } C_\nu = \frac{\sqrt{\pi}}{2^{\nu+\frac{1}{2}}\,\Gamma(\nu+1)}.$$

9 dec. $\nu = \frac{1}{2}\left(\frac{1}{2}\right)\frac{41}{2}(1)\frac{61}{2}$. $x = 0\,(0{,}1)\,10$

7 fig. $\nu = \frac{1}{2}(1)\frac{61}{2}$. $x = 10\,(0{,}1)\,25$

$-\nu = \frac{29}{2}(1)\frac{33}{2}$. $x = 0\,(0{,}1)\,9{,}5\,(0{,}05)\,10\,(0{,}1)\,25$

$-\nu = \frac{35}{2}(1)\frac{61}{2}$. $x = 0\,(0{,}1)\,25$ [80]

$$\Psi_n(x) = (-1)^n (2n+1)!!\sqrt{\frac{\pi}{2x^{2n+1}}}\, J_{-n-\frac{1}{2}}(x),$$

$$\psi_n(x) = (2n+1)!!\sqrt{\frac{\pi}{2x^{2n+1}}}\, J_{n+\frac{1}{2}}(x)$$

5–6 fig. $n = 0\,(1)\,3$. $8x = 8;\ 12\,(2)\,18;\ 21\,(3)\,27$ [254]

5 fig. $n = 0\,(1)\,4$ or 6. $x = 1,\ 25$ [207]

$$D_n = \left[\left\{\frac{d}{dx}\left(\sqrt{\frac{\pi}{2x}}\, J_{n+\frac{1}{2}}(x)\right)\right\}^2 + \left\{\frac{d}{dx}\left(\sqrt{\frac{\pi}{2x}}\, J_{-n-\frac{1}{2}}(x)\right)\right\}^2\right]^{1/2}$$

note esp. $D_0 = \frac{\sqrt{1+x^2}}{x^2}$ & $D_1 = \frac{\sqrt{4+x^4}}{x^3}$.

4 fig. $n = 0\,(1)\,4$. $x = 0\,(0{,}1)\,0{,}2\,(0{,}2)\,5$ [48]

$$\delta_n = \operatorname{arc\,tg}\left\{\frac{\frac{d}{dx}\left(\sqrt{\frac{\pi}{2x}}\, J_{n+\frac{1}{2}}(x)\right)}{\frac{d}{dx}\left(\sqrt{\frac{\pi}{2x}}\, J_{-n-\frac{1}{2}}(x)\right)}\right\}$$

0°, 1 $n = 0\,(1)\,4$. $x = 0\,(0{,}1)\,0{,}2\,(0{,}2)\,5$ [48]

LOMMEL'S FUNCTIONS OF TWO VARIABLES[1]

$$U_\nu(y, z) = \sum_{m=0}^{\infty} (-1)^m \left(\frac{y}{z}\right)^{\nu+2m} J_{\nu+2m}(z),$$

$$V_\nu(y, z) = \cos\frac{1}{2}\left(y + \frac{z^2}{y} + \pi\nu\right) + U_{-\nu+2}(y, z).$$

[1] Tables of the Lommel function for z = 0 are given in Ch. 7.

For integer order $\nu = n$

$$U_\nu(y, z) = U_n(y, z),\ V_n(y, z) = \sum_{m=0}^{\infty} (-1)^m \left(\frac{y}{z}\right)^{-n-2m} J_{-n-2m}(z),$$

where

$$U_0(z, z) = V_0(z, z) = \frac{1}{2}\,[J_0(z) + \cos z],\quad U_1(z, z) = -V_1(z, z) = \frac{1}{2}\sin z,$$

$$U_2(z, z) = V_2(z, z) = \frac{1}{2}\,[J_0(z) - \cos z].$$

$$\frac{2}{y}\,U_1(y, z),\ \frac{2}{y}\,U_2(y, z),\ M^2 = \frac{4}{y^2}\,[U_1^2(y, z) + U_2^2(y, z)]$$

6 dec. $y = \pi(\pi)10\pi.\ \ z = 0(1)12$ [190]

$$pM^2$$

4 dec. $p = 1(1)4.\ \ y = p\pi.\ \ z = 0(1)12$ [196]

$$\frac{2}{z}\,U_1(z, z),\ \frac{2}{z}\,U_2(z, z),\ M^2 = \frac{4}{z^2}\,[U_1^2(z, z) + U_2^2(z, z)]$$

6 dec. $z = 0(0{,}5)12$ [190]

$$\frac{2}{z}\,V_1(z, z),\ \frac{2}{z}\,V_0(z, z),\ M_1^2 = \frac{4}{z^2}\,[V_1^2(z, z) + V_2^2(z, z)]$$

6 dec. $z = 0(0{,}5)12$ [190]

$$\frac{2}{y}\,V_0(y, z),\ -\frac{2}{y}\,V_1(y, z),\ M_1^2 = \frac{4}{y^2}\,[V_1^2(y, z) + V_0^2(y, z)]$$

6 dec. $y = \pi(\pi)8\pi.\ \ z = 0(1)12$ [190]

$$\sqrt{\frac{\pi}{2y}}\,U_{1/2}(y, z),\ \sqrt{\frac{\pi}{2y}}\,U_{3/2}(y, z),\ M^2 = \frac{\pi}{2y}\,[U_{1/2}^2(y, z) + U_{3/2}^2(y, z)]$$

6 dec. $y = 3(3)30.\ \ z = 0(1)12$ [191]

$$\sqrt{\frac{\pi}{2y}}\,V_{1/2}(y, z),\ \sqrt{\frac{\pi}{2y}}\,V_{3/2}(y, z),\ M_1^2 = \frac{\pi}{2y}\,[V_{3/2}^2(y, z) + V_{1/2}^2(y, z)]$$

6 dec. $y = 3(3)12.\ \ z = 0(1)12$ [191]

CYLINDER FUNCTIONS OF THE THIRD KIND (HANKEL FUNCTIONS)

$$|H_n^{(1)}(x)| = \sqrt{J_n^2(x) + Y_n^2(x)}$$

7 dec.	$n = 0;\ 1;\ \frac{1}{3}.$	$x = 0\,(0{,}02)\,16$	[2] [86]
5 dec.	$n = 0;\ 1;\ \frac{1}{3}.$	$x = 0\,(0{,}02)\,16$	[13]
4 fig.	$\lvert 2H_1^{(1)}(x)\rvert$	$x = 0\,(0{,}1)\,0{,}2\,(0{,}2)\,5$	[48]

$$\pi^2 J_n^2(x) + Y_n^2(x)$$

4 fig. $\begin{cases} n = 0\,(1)\,4. & x = 0{,}1;\ 0{,}4\,(0{,}1)\,0{,}6;\ 0{,}8\,(0{,}1)\,1 \\ n = 0\,(1)\,6. & x = 1{,}1;\ 1{,}2;\ 1{,}5\,(0{,}5)\,3\,(1)\,6 \\ n = 7;\,8;\,9\ \&\ 10. & x = 4\,(1)\,6;\ 5\ \&\ 6;\ 6 \end{cases}$ [240]

$$\lg \sqrt{\frac{1}{2}\pi x\,\{J_\nu^2(x) + Y_\nu^2(x)\}}$$

7 fig. $\nu = 0\left(\frac{1}{2}\right)6\,\frac{1}{2}.$ $x = 10\,(10)\,100\,(100)\,1000$ [122]

$$\lg \sqrt{x\,\{J_\nu^2(x) + Y_\nu^2(x)\}}$$

7 dec. $\nu = 0\left(\frac{1}{2}\right)6\,\frac{1}{2}.$ $x = 10\,(10)\,100\,(100)\,1000$ [122]

$$\sqrt{J_n'^2(x) + Y_n'^2(x)}$$

4 fig. $n = 1\,(1)\,4.$ $x = 0\,(0{,}1)\,0{,}2\,(0{,}2)\,5$ [48]

$$4\{\pi^2 J_n'^2(x) + Y_n'^2(x)\}$$

4 fig. $\begin{cases} n = 0\,(1)\,3. & x = 0{,}5;\ 0{,}6;\ 0{,}8\,(0{,}1)\,1 \\ n = 0\,(1)\,6. & x = 1{,}1;\ 1{,}2;\ 1{,}5\,(0{,}5)\,3\,(1)\,5 \\ n = 7\ \&\ 8;\ 7\,(1)\,10. & x = 4;\ 5 \end{cases}$ [240]

$$\arg H_n^{(1)}(x) = \operatorname{arc\,tg}\left\{\frac{Y_n(x)}{J_n(x)}\right\}$$

0″,01	$n = 0;\ 1;\ \frac{1}{3}.$	$x = 0\,(0{,}02)\,16$	[2] [86]
1 sec.	$n = 0;\ 1;\ \frac{1}{3}.$	$x = 0\,(0{,}02)\,16$	[13]

$$\operatorname{arc\,tg}\left\{\frac{J'_n(x)}{Y'_n(x)}\right\}$$

0°,1	$n = 0\,(1)\,4.\quad x = 0\,(0{,}1)\,0{,}2\,(0{,}2)\,5$	[48]

Functions of the argument

$$\alpha_n = \frac{1}{2}\left(n + \frac{1}{2}\right)\pi - x + \operatorname{arc\,tg}\left\{\frac{Y_n(x)}{J_n(x)}\right\}$$

9 dec.	$-\sin\alpha_0$	$x = 10\,(10)\,100\,(100)\,1000$	[122]
8 dec.	$\lg(-\sin\alpha_0)$		
8 dec.	$\lg(-x\sin\alpha_0)$		
7 dec.	$\lg\left(\frac{8x\alpha_\nu}{4\nu^2-1}\right)$	$\nu = 0\left(\frac{1}{2}\right)6\,\frac{1}{2}.\quad x = 10\,(10)\,100\,(100)\,1000$	[123]

CYLINDER FUNCTIONS OF THE FIRST AND SECOND KIND OF IMAGINARY ARGUMENT

Cylinder functions of the first kind of imaginary argument

$I_0(x)$ & $I_1(x)$

21 dec.		$x = 0\,(0{,}1)\,6$	[252]
18 dec.		$x = 6\,(1)\,11$	
12 fig.		$x = 0\,(0{,}2)\,6$	[4] [34] [119]
12; 10 fig.		$x = 0\,(1)\,6;\ 7\,(1)\,10\,(10)\,100$	[271]
10 dec.		$x = 0\,(0{,}01)\,10$	[62]
9 dec.	$I_0(x)$	$x = 0\,(0{,}001)\,5{,}1$	[121]
9 dec.	$I_1(x)$	$x = 0\,(0{,}001)\,5{,}1$	[120]
9 dec.		$x = 0\,(0{,}01)\,5{,}1$	[4] [34]
8 dec.		$x = 0\,(0{,}001)\,10$	[10]
8; 7 dec.		$x = 0\,(0{,}001)\,4;\ 4\,(0{,}001)\,5$	[23]
6 fig.		$x = 5\,(1)\,15;\ 16\,(1)\,37$	[215] [217]
5—6 fig.		$x = 0\,(0{,}01)\,10$	[8]
5 fig.		$x = 0\,(0{,}1)\,6$	[279]
5 fig.		$x = 0\,(0{,}1)\,0{,}2\,(0{,}2)\,8$	[48]
5 fig.		$x = 0\,(0{,}2)\,5$	[182]
5 fig.	$I_0(x)$	$x = 12\,(2)\,20\,(10)\,100\,(100)\,1000$	[90]
4—6 fig.		$x = 0\,(1)\,15$	[98]
4—5 fig.		$x = 0\,(0{,}01)\,9{,}99$	[14] $[41]_{2,3}$
4—5 fig.		$x = 0\,(0{,}1)\,10$	[1] [21]
4 dec.		$x = 0\,(0{,}1)\,16$	[7]
4 dec.	$I_0(x)$	$x = 0\,(0{,}1)\,5{,}9$	[26]

4 dec.		$x = 0\,(0{,}2)\,8$	[37]
4 dec.		$x = 0\,(0{,}2)\,6$	[9] [52]
4 dec.	$I_0(x)$	$x = 0\,(0{,}1)\,6\,(1)\,11$	[41]$_1$
4 dec.	$I_1(x)$	$x = 0\,(0{,}01)\,5{,}1\,(0{,}1)\,6\,(1)\,11$	[41]$_1$

$I_n(x)$

15 dec.	$I_n\left(\frac{3n}{4}\right)$	$n = 4\,(4)\,24$	[177]
12 fig.		$n = 0\,(1)\,10$; $x = 0\,(0{,}2)\,6$	[4] [34] [119]
10—12 fig.		$n = 11$; $x = 0\,(0{,}2)\,6$	[4] [34] [119]
12 fig.		$n = 0\,(1)\,11.$ $x = 0\,(1)\,6$	[271]
10 fig.		$n = 0\,(1)\,23$ (var) $54.$ $x = 7\,(1)\,10\,(10)\,100$	[271]
10 fig.		$n = 0\,(1)\,20.$ $x = 0\,(0{,}1)\,25$	[96]
8 fig.		$n = 2\,(1)\,20.$ $x = 0\,(0{,}1$ or $0{,}01)\,10\,(0{,}1)\,20$	[96]
6 fig.		$n = 0\,(1)\,6.$ $x = 5\,(1)\,15;\ 16\,(1)\,37$	[215] [217]
5 dec.		$n = 1\,(1)\,21.$ $x = 0\,(0{,}01)\,1\,(0{,}02)\,5$; $n = 22.$ $x = 1\,(0{,}02)\,5$	[84]
5 fig.		$n = 0\,(1)\,11.$ $x = 0\,(0{,}1)\,6$	[279]
5 fig.		$n = 0\,(1)\,11.$ $x = 0\,(0{,}2)\,5$	[28]
5 fig.		$n = 0\,(1)\,10.$ $x = 0\,(0{,}2)\,5$	[182]
5 fig. 4		$n = 0\,(1)\,2.$ $x = 0\,(0{,}1)\,0{,}2\,(0{,}2)\,8$	[48]
4—6 fig.		$n = 0\,(1)\,11.$ $x = 0\,(0{,}2)\,6$	[9] [13] [14] [41]$_{1,2,3}$
4—6 fig.		$n = 0\,(1)\,9.$ $x = 0\,(1)\,15$	[98]

$I_\nu(x)\left(\nu = m + \frac{1}{2},\ m\right.$ is an integer$\left.\right)$

7 fig.	$\nu = -\frac{1}{2}\,(1)\,4\frac{1}{2}.$	$x = 0{,}5\,(0{,}5)\,10$	[231]
4—6 fig.	$\nu = \pm\frac{1}{2}.$	$x = 0\,(0{,}1)\,15$	[15]

$I_\nu(x)\left(\nu = \pm\frac{1}{3},\ \pm\frac{2}{3}\right)$

10 dec.	$\nu = \pm\frac{1}{3}.$	$x = 0\,(0{,}001)\,0{,}6$ or $0{,}8\,(0{,}01)\,25$ or 13	[65]$_2$
10 dec.	$\nu = \pm\frac{2}{3}.$	$x = 0\,(0{,}001)\,0{,}5$ or $1\,(0{,}01)\,25$ or 13	[65]$_2$
4—6 fig.		$x = 0\,(0{,}1)\,12$	[2] [15]
4 dec.		$x = 0\,(0{,}1)\,16$	[7]
4 dec.		$x = 0\,(0{,}2)\,4$	[14] [41]$_{2,3}$
4—5 fig.		$x = 4\,(0{,}2)\,8$	[104]

$I_\nu(x)\left(\nu=\pm\frac{1}{4};\pm\frac{3}{4}\right)$

10 dec.	$\nu=\pm\frac{1}{4}$. $x=0\,(0{,}001)\,0{,}6$ or $0{,}8\,(0{,}01)\,25$ or 13	[65]₂
10 dec.	$\nu=\pm\frac{3}{4}$. $x=0\,(0{,}001)\,0{,}5$ or $1\,(0{,}01)\,25$ or 13	
4—6 fig.	$x=0\,(0{,}1)\,15$	[2] [15]

$I_\nu(x)\left(\nu=\pm\frac{1}{6};\pm\frac{5}{6}\right)$

4—6 fig.	$x=0\,(0{,}1)\,15$	[15]

$I_0(2\sqrt{x})$

8; 7 dec.	$x=0\,(0{,}02)\,1;\ 1\,(0{,}02)\,1{,}5\,(0{,}05)\,6{,}2$	[305]

Cylinder functions of the second kind of imaginary argument

$K_0(x)$ & $K_1(x)$

21 dec.	$x=0\,(0{,}1)\,6$	[4] [34] [252]
18 dec.	$x=7\,(1)\,11$	
9—16 dec.	$x=6{,}1\,(0{,}1)\,12$	
8—9 fig.	$x=0\,(0{,}01)\,5$	[23]
8 fig.	$x=0\,(0{,}001)\,10$	[10]
7 fig.	$x=0\,(0{,}0001)\,0{,}033\,(0{,}001)\,1$	[64]
7 dec.	$x=0\,(0{,}1)\,5$	[86]
5—6 fig.	$x=0\,(0{,}01)\,10$	[8]
5 fig.	$x=0\,(0{,}2)\,5$	[182]
4 fig.	$x=0\,(0{,}1)\,12$	[41]₁
4 dec.	$x=0\,(0{,}1)\,0{,}9$	[198]

$K_n(x)$

10 fig.	$n=0\,(1)\,20$.	$x=0\,(0{,}1)\,25$	[96]
8 fig.	$n=2\,(1)\,20$.	$x=0\,(0{,}1$ or $0{,}01)\,10\,(0{,}1)\,20$	
7—8 fig.	$n=0\,(1)\,10$.	$x=0{,}1\,(0{,}1)\,5$	[2] [86]
5—6 fig.	$n=0\,(1)\,10$.	$x=0{,}1\,(0{,}1)\,5$	[13]
5 dec.	$n=0\,(1)\,23$.	$x=0\,(0{,}01)\,1\,(0{,}02)\,5$	[84]
5 fig.	$n=2\,(1)\,10$.	$x=0{,}2\,(0{,}2)\,5$	[4] [34]
5 fig.	$n=0\,(1)\,10$.	$x=0\,(0{,}2)\,5$	[182]

$K_\nu(x)\left(\nu=m+\frac{1}{2},\ m\right.$ is an integer)

7 fig.	$\nu=-\frac{1}{2}\,(1)\,2$.	$x=0{,}5\,(0{,}5)\,10\,(1)\,25$	[231]
5 fig.	$\nu=\frac{1}{2}\,(1)\,2\frac{1}{2}$.	$x=0\,(0{,}1)\,5;\ 6\,(2)\,10$	[227]

$$K_\nu(x)\left(\nu=\pm\frac{1}{4},\ \pm\frac{3}{4}\right)$$

5 fig.	$x=0\,(0{,}1)\,5;\ 6\,(2)\,10$	[227]

$$\frac{2}{\pi}K_0(x)=iH_0^{(1)}(ix)\ \&\ \frac{2}{\pi}K_1(x)=-H_1^{(1)}(ix)$$

4–6 fig.	$x=0{,}1\,(0{,}1)\,12$	[9] [13]
4–5 fig.	$x=0\,(0{,}01)\,15{,}99$	[14] [41]$_{2,3}$
4 fig.	$x=0\,(0{,}2)\,6$	[41]$_1$

$$K_0(2\sqrt{x})$$

8; 9dec.	$x=0\,(0{,}02)\,1{,}5\,(0{,}05)\,2{,}5;$ $2{,}5\,(0{,}5)\,6{,}2$	[305]

Various expressions containing cylinder functions of the first & second kind of imaginary argument

$$x^m I_n(x)$$

8 fig.	$m=-\frac{1}{2}\,(1)-\frac{21}{2}.$	$n=\frac{1}{2}\,(1)\,\frac{21}{2}.$	$x=0\,(0{,}1)\,5$	[97]
5 dec.	$m=\frac{1}{2}.$	$n=0\,(1)\,22.$	$x=5\,(0{,}04)\,25$	[84]

$$\left(\frac{1}{2}x\right)^{-1/3} I_{1/3}(x)$$

6 dec.	$x=0\,(0{,}02)\,1{,}2$	[2] [17]

$$e^{-x}I_\nu(x)$$

10 dec.	$\nu=\pm\frac{1}{4};\ \pm\frac{1}{3};\ \pm\frac{2}{3};\ \pm\frac{3}{4}.$	$x=25\,(0{,}1)\,50\,(1)\,500\,(10)\,5000\,(100)\,10000\,(200)\,30000$	[65]$_2$
9 dec.	$\nu=0;\ 1.$	$x=20\,(1)\,50$	[269]
8 fig.	$\nu=\frac{1}{2}\,(1)\frac{21}{2}.$	$x=5\,(0{,}1)\,10$	[97]
8 dec.	$\nu=0;\ 1.$	$x=5\,(0{,}01)\,10\,(0{,}1)\,20$	[23]
7 dec.	$\nu=0;\ 1.$	$x=0\,(0{,}02)\,16$	[2] [86]
7 dec.	$\nu=2(1)5.$	$x=0\,(0{,}1)\,5$	[86]
6 dec.	$\nu=\frac{1}{3}.$	$x=1{,}2\,(0{,}05)\,4\,(0{,}1)16$	[2] [17]
6 dec.	$\nu=0;\ 1.$	$x=16{,}08\,($ var. $)\,50$	[270]
5 dec.	$\nu=0;\ 1.$	$x=0\,(0{,}02)\,16$	[13]
5 dec.	$\nu=2(1)5.$	$x=0\,(0{,}1)\,5$	[13]
4 dec.	$\nu=0;\ 1.$	$x=0\,(0{,}1)\,16$	[7]
4 dec.	$\nu=\pm\frac{2}{3}.$	$x=12{,}1\,(0{,}1)\,15$	[2]

4 dec.	$\nu = \frac{1}{3}; \frac{2}{3}$. $x = 12{,}1\,(0{,}1)\,15$	[15]

$$e^{-x}\,[I_0(x) + I_1(x)]$$

4 dec.	$\frac{1}{2}x = 0\,(0{,}1)\,6\,(0{,}2)\,10\,(0{,}5)\,18\,(1)\,25$	[106]

$$2xe^{-x^2-y^2}I_0(2xy)$$

4 dec.	$x = 0\,(0{,}2)\,6{,}6$. $y^2 = 0;\ \frac{1}{4};\ 1;\ 4;\ 6;\ 7;\ 9$	[146]

$$\lg e^{-x}I_n(x)$$

5 dec.	$n = 1\,(1)\,6\,(2)\,10$.	$x = 0\,(0{,}2)\,6;\ 10\,(10)\,40$	[146]
5; 4 dec.	$n = 0$	$x = 0\,(0{,}1)\,2\,(0{,}2)\,4\,(0{,}5)\,6\,(1)\,20;\ 25\,(5)\,65$	

$$\lg I_0(x) \ \&\ \lg \frac{I_1(x)}{x}$$

6 dec.	$x = 0\,(0{,}01)\,10$	[19]

$$\psi(x) = e^{-x}\{(1 + 2x)\,I_0(x) + 2xI_1(x)\}$$

3 dec.	$\psi(x)$	$\sqrt{2x} = 0\,(0{,}1)\,2$	[61]
2 dec.	$\frac{1}{2}\sqrt{\pi}\psi(x)$	$2x = 0\,(1)\,10\,(5)\,20\,(10)\,100;\ 150$	

$$I_1(2\sqrt{x})\,/\,\sqrt{x}$$

7 dec.	$x = 0\,(0{,}02)\,1{,}5\,(0{,}05)\,6{,}2$	[305]

$$\sqrt{\frac{1}{2}\pi x}\,e^{-x}\,[I_{\nu-1}(x) + I_\nu(x)]$$

4 dec.	$\nu = 1\left(\frac{1}{2}\right)2\,\frac{1}{2}$.	$\sqrt{2x} = 0\,(0{,}1)\,2$	[61]
4 dec.	$\nu = 1\left(\frac{1}{2}\right)2$.	$\sqrt{2x} = 0\,(0{,}1)\,2$	[186]

$$\sqrt{x(1 - x^2)\frac{I_1(x)}{I_0(x)}}$$

4 dec.	$x = 0\,(0{,}1)\,1$	[53] [198]
4 dec.	$x^2 = 0{,}05;\ 0{,}1\,(0{,}1)\,0{,}9$	[238]

$$\frac{J_n(x)}{I_n(x)}$$

6 dec.	$n = 2; 3.\quad x = 1\,(1)\,15$	[217]

$$e^{-2\sqrt{x}} I_0(2\sqrt{x}) \;\&\; e^{-2\sqrt{x}} I_1(2\sqrt{x})/\sqrt{x}$$

8; 9 dec.	$e^{-2\sqrt{x}} I_0(2\sqrt{x})$ $x = 6{,}2\,(0{,}1)\,13\,(0{,}2)\,36\,(0{,}5)\,115;\ 115\,(1)\,160\,(5)\,410$; $e^{-2\sqrt{x}} I_1(2\sqrt{x})/\sqrt{x}$ $x = 6{,}2\,(0{,}1)\,13\,(0{,}2)\,36\,(0{,}5)\,65;\ 65\,(0{,}5)\,115\,(1)\,160\,(5)\,410$	[305]

$$x^m K_n(x)$$

5 dec.	$m = 1.\quad n = 0; 1.\quad x = 0\,(0{,}1)\,4{,}5$	[189]
5 dec.	$m = \frac{1}{2}.\quad n = 0\,(1)\,23.\quad x = 5\,(0{,}04)\,25$	[84]

$$\frac{2}{\pi} x^m K_n(x)$$

8 fig.	$m = n = \frac{1}{2}\,(1)\,\frac{21}{2}.\quad x = 0\,(0{,}1)\,5$	[97]
4 dec.	$m = n = 1.\quad x = 0\,(0{,}5)\,4{,}5$	[189]

$$e^x K_\nu(x)$$

9 dec.	$\nu = 0; 1.\quad x = 20\,(1)\,50$	[269]
8 fig.	$\frac{2}{\pi} e^x K_\nu(x)\quad \nu = \frac{1}{2}\,(1)\,\frac{21}{2}.\quad x = 5\,(0{,}1)\,10$	[97]
8 dec.	$\nu = 0; 1.\quad x = 5\,(0{,}01)\,10\,(0{,}1)\,20$	[23]
7 dec.	$\nu = 0; \frac{1}{3}; 1.\quad x = 0\,(0{,}02)\,16$	[2] [86]
5 dec.	$\nu = 0; \frac{1}{3}; 1.\quad x = 0\,(0{,}02)\,16$	[13]
4 dec.	$\nu = 0; 1.\quad x = 0\,(0{,}1)\,16$	[7]
4 dec.	$\nu = 0; 1\quad x = 0{,}01\,(0{,}01)\,0{,}1\,(0{,}1)\,3\,(0{,}2)\,6\,(0{,}5)\,10\,(1)\,20;\ \sqrt{x} = 0{,}01\,(0{,}01)\,0{,}1;\ 5\,(1)\,10$	[228]

$$K'_\nu(x)$$

5 fig.	$\nu = \frac{1}{4};\ \frac{3}{4}.\quad x = 0\,(0{,}1)\,5;\ 6\,(2)\,10$	[227]

$$T_\nu(x) = \frac{\left(\frac{1}{2}x\right)^\nu K_\nu(x)}{\sqrt{\pi}\,\Gamma\left(\nu - \frac{1}{2}\right)}; \quad \lg T_\nu(x) \qquad \frac{K_{\nu-1}(x)}{K_\nu(x)}$$

6 fig. $\left\{\begin{array}{l} \nu = 0\left(\frac{1}{2}\right)5\frac{1}{2}.\ x = 0\,(0,1)\,4\,(0,5)\,16\,(2)\,40\,(5)\,120 \\ \nu = 6\left(\frac{1}{2}\right)11\frac{1}{2}.\ x = 0\,(0,5)\,16\,(2)\,40\,(5)\,120 \end{array}\right\}$ [49] [116]

$$2\sqrt{n^2 + x^2}\, I_n(x)\, K_n(x)$$

5 dec. $\left\{\begin{array}{l} n = 1.\ x = 0{,}02;\ 0{,}1;\ 0{,}3\,(0{,}3)\,0{,}9;\ 1\,(1)\,10\,(3)\,16 \\ n = 3.\ x = 0{,}1;\ 0{,}2\,(0{,}2)\,1\,(0{,}5)\,5 \end{array}\right\}$ [223]

$$-2\frac{d}{dx}\left\{\frac{1}{x}\frac{d}{dx}\left[I_1\left(\frac{1}{2}bx\right)K_1\left(\frac{1}{2}bx\right)\right]\right\},$$

where $b = i^{1/2}$

4 dec. $x = 0{,}2\,(0{,}2)\,1\,(0{,}5)\,2$ [267]

$$K_1(2\sqrt{x})/\sqrt{x}$$

6; 7 dec. $x = 0(0{,}02)0{,}04;\ 0{,}06(0{,}02)0{,}28$
8; 9 dec. $x = 0{,}3\,(0{,}02)\,1{,}5\,(0{,}05)\,2{,}5;\ 2{,}5\,(0{,}05)6{,}2$ } [305]

$$e^{2\sqrt{x}} K_0(2\sqrt{x}) \qquad e^{2\sqrt{x}} K_1(2\sqrt{x})/\sqrt{x}$$

8 dec. $e^{2\sqrt{x}} K_0(2\sqrt{x})$ $\left\{\begin{array}{l} x = 6{,}2\,(0{,}1)\,13\,(0{,}2)\,36\,(0{,}5) \\ \quad 115\,(1)\,160\,(5)\,410 \end{array}\right.$

8 dec.
9 dec. } $e^{2\sqrt{x}} K_1(2\sqrt{x})/\sqrt{x}$ $\left\{\begin{array}{l} x = 6{,}2\,(0{,}1)\,13\,(0{,}2)\,36\,(0{,}5)\,40 \\ x = 40\,(0{,}5)\,115\,(1)\,160\,(5)\,410 \end{array}\right.$ } [305]

$$\lg K_0(x)$$

7; 6 dec. $x = 2\,(0{,}001)\,4;\ 4\,(0{,}001)\,12$ [230]

Auxiliary functions

For large values of the argument

$$\sqrt{2\pi x}\, e^{-x} I_n(x)$$

7 dec. $n = 0;\ 1.$ $\frac{1}{x} = 0\,(0{,}001)\,0{,}22$
12 dec. $n = 0;\ 1.$ $x = 5;\ 6;\ 8;\ 10$ } [241]

7 dec. $n = 0;\ 1.$ $\frac{1}{x} = 0\,(0{,}01)\,0{,}1$ [151]

6 dec. $n = 0; 1.$ $x = 10\,(0{,}1)\,50\,(1)\,200$ (var.) 10^6 [19]

$$\lg\,[\sqrt{2\pi x}\, e^{-x} I_1(x)]$$

5 dec. $x = 11\,(0{,}1)\,21\,(1)\,109$ [300]

For small values of the argument

$$E_n(x)\ \&\ F_n(x)$$

$$K_0(x) = E_0(x) + F_0(x)\lg x, \quad K_1(x) = \frac{E_1(x)}{x} + F_1(x)\lg x$$

8 dec. $n = 0; 1.$ $x = 0\,(0{,}01)\,0{,}5$ [23]
7 dec. $n = 0; 1.$ $x = 0\,(0{,}001)\,0{,}03$ [64]

CYLINDER FUNCTIONS OF COMPLEX ARGUMENT

Cylinder functions of the first kind of complex argument

$$J_n(re^{i\varphi}) = U_n(r,\varphi) + iV_n(r,\varphi)$$

10 dec. $\begin{Bmatrix} U_0, V_0 \\ U_1, V_1 \end{Bmatrix}$ $r = 0\,(0{,}01)\,10.$ $\varphi = 0\left(\frac{\pi}{36}\right)\frac{\pi}{2}$ [62]

7 dec. $\begin{cases} U_0, V_0 & r = 0\,(0{,}05)\,10. \quad \varphi = 0\left(\frac{\pi}{36}\right)\frac{\pi}{2} \\ U_1, V_1 & r = 0\,(0{,}02)\,10. \quad \varphi = \frac{\pi}{4} \end{cases}$ [2]

4 dec. $\begin{Bmatrix} U_0, V_0 \\ U_1, V_1 \end{Bmatrix}$ $r = 0\,(0{,}2)\,8.$ $\varphi = \frac{\pi}{16}\left(\frac{\pi}{16}\right)\frac{\pi}{2}$ [13] [37] [302]

4 dec. U_1, V_0 $r = 0\,(0{,}2)\,8.$ $\varphi = \frac{\pi}{2} - 0{,}001$

$$\frac{z^{-1}J_1(z)}{J_0(z)}$$

$$z = x + iy$$

4 dec. $x = 0{,}5\,(0{,}1)\,2.$ $y = 0\,(0{,}1)\,1$ [250]

Cylinder functions of the second kind of complex argument

$$Y_n(re^{i\varphi}) = U_n(r,\varphi) + iV_n(r,\varphi)$$

10 dec. $\begin{Bmatrix} U_0, V_0 \\ U_1, V_1 \end{Bmatrix}$ $r = 0\,(0{,}01)\,10.$ $\varphi = 0\left(\frac{\pi}{36}\right)\frac{\pi}{2}$ [63]

$$Y_0(re^{i\varphi}) - \frac{2}{\pi}J_0(re^{i\varphi})\lg r\ \&\ Y_1(re^{i\varphi}) - \frac{2}{\pi}J_1(re^{i\varphi})\lg r + \frac{2}{\pi}e^{-\varphi i}$$

10 dec. $r = 0\,(0{,}01)\,0{,}5.$ $\varphi = 0\left(\frac{\pi}{36}\right)\frac{\pi}{2}$ [63]

Cylinder functions of the third kind of complex argument

Tables of the real and imaginary parts of the solutions $h_1(z)$ & $h_2(z)$ of Stoke's equation

$$\frac{d^2u}{dz^2} + zu = 0,$$

are expressed in terms of the Hankel function of order $\frac{1}{3}$:

$$h_1(z) = \left(\frac{2}{3} z^{3/2}\right)^{1/3} H^{(1)}_{1/3}\left(\frac{2}{3} z^{3/2}\right),$$

$$h_2(z) = \left(\frac{2}{3} z^{3/2}\right)^{1/3} H^{(2)}_{1/3}\left(\frac{2}{3} z^{3/2}\right),$$

tables of other functions are included.

$$\mathrm{Re}(h_n),\ \mathrm{Re}(h'_n),\ \mathrm{Jm}(h_n),\ \mathrm{Jm}(h'_n)$$

$$z = x + iy$$

8 dec. $\left\{\begin{matrix} n = 1;2. \quad x = M(0,1)N. \quad y = 0(0,1)6 \\ -6 \leqslant M \leqslant -1. \quad 1 \leqslant N \leqslant 6 \end{matrix}\right\}$ [78]

$$\frac{H_1^{(2)}(z)}{H_1^{(2)}(z) + iH_0^{(2)}(z)}$$

$$z = re^{i\theta}$$

7 dec. $\left\{\begin{matrix} r = 0(0{,}01)0{,}3(0{,}02)0{,}5(0{,}05)1(0{,}5)4(1)10. \ \theta = -5°(5°)30° \\ r = 0(0{,}002)0{,}1(0{,}01)0{,}3\ (\ \text{var.}\ )\ 2(0{,}5)10(10)50; \\ 100; \infty. \ \theta = 0 \end{matrix}\right.$ [153]

CYLINDER FUNCTIONS OF IMAGINARY ORDER AND IMAGINARY ARGUMENT

The functions $F_\nu(v)$ and $G_\nu(v)$ are solutions of the differential equation

$$v^2y'' + vy' + (v^2 - \nu^2)y = 0,$$

where v & ν are real; this is a special case of the equation $z^2u'' + zu' + (z^2 - p^2)u = 0$, in which $z = iv$, $p = i\nu$.

$$F_\nu(v) = \frac{\pi}{2}\,\frac{I_{i\nu}(v) + I_{-i\nu}(v)}{\mathrm{sh}\,\nu\pi} = \frac{\pi}{\mathrm{sh}\,\nu\pi}\,\mathrm{Re}\,I_{i\nu}(v),$$

$$G_\nu(v) = \frac{i\pi}{2}\,\frac{I_{i\nu}(v) - I_{-i\nu}(v)}{\mathrm{sh}\,\nu\pi} = -\frac{\pi}{\mathrm{sh}\,\nu\pi}\,\mathrm{Im}\,I_{i\nu}(v) =$$

$$= K_{i\nu}(v) = \frac{1}{2}\pi i e^{-\frac{1}{2}\nu\pi} H^{(1)}_{i\nu}(iv).$$

$F_\nu(e^x)$

4—5 fig.	$\nu = 0{,}2\,(0{,}2)\,10.$	$x = -0{,}49\,(0{,}01)\,2{,}5$	[47]

$G_\nu(e^x)$

4—5 fig.	$\nu = 0{,}2\,(0{,}2)\,10.$	$x = -0{,}49\,(0{,}01)\,0{,}5$	[47]
	$\nu = 1{,}2\,(0{,}2)\,2.$	$x = -0{,}49\,(0{,}01)\,1$	
	$\nu = 2{,}2\,(0{,}2)\,4.$	$x = -0{,}49\,(0{,}01)\,1{,}5$	
	$\nu = 4{,}2\,(0{,}2)\,7.$	$x = -0{,}49\,(0{,}01)\,2$	
	$\nu = 7{,}2\,(0{,}2)\,10.$	$x = -0{,}49\,(0{,}1)\,2{,}5$	

THOMSON FUNCTIONS

Thomson functions of the first kind

ber x & bei x

21 dec.	$x = 0\,(0{,}1)\,6$	[253]
10 dec.	$x = 0\,(0{,}01)\,10$	[62]
9 dec.	$x = 0\,(0{,}1)\,10$	[5] [30] [125] [155] [277]
9 dec.	$x = 0\,(0{,}2)\,6$	[4] [34] [120]
6 dec.	$x = 0\,(0{,}02)\,4{,}6$	[174]
5—6 dec.	$x = 1\,(1)\,10$	[276]
5 dec.	$x = 0\,(0{,}01)\,1\,(0{,}02)\,1{,}98$	[9] [13]
5 fig.	$x = 0\,(0{,}1)\,20$	[29] [278]
5 dec.	$x = 0\,(0{,}2)\,6$	[150]
5—3 fig.	$x = 0\,(0{,}01)\,6\,(0{,}1)\,10$	[14] $[41]_{2,3}$
5; 4 dec.	$x = 0\,(0{,}01)\,1\,(0{,}02)\,2;\ 2{,}1\,(0{,}1)\,6$	[37]
4 fig.	$x = 0\,(0{,}01)\,21$	[83]
4 fig.	$x = 0\,(0{,}1)\,10$	[20] [44]
4 dec.	$x = 0\,(0{,}1)\,6$	[9] [13] $[41]_1$
4 fig.	$x = 0\,(0{,}1)\,10\,(5)\,20;\ 30$	[21]
4 dec.	$x = 0\,(0{,}2)\,8$	[37]
4 dec.	$x = 0\,(0{,}2)\,6$	[216]
4 fig.	$x = 1\,(1)\,30$	[56] [202] [243]
4 dec.	$x = 0\,(0{,}5)\,6\,(2)\,10\,(5)\,20$	$[41]_1$ [154] [180] [298]
4 dec.	$x = 0\,(0{,}5)\,6\,(2)\,10\,(5)\,20$ и 30	[43]

For $\nu=0$ the order is omitted; for example $\mathrm{ber}_0\,x = \mathrm{ber}\,x$

$\mathrm{ber}'x$ & $\mathrm{bei}'x$

9 dec.	$x = 0\,(0{,}1)\,10$	[5] [30] [125] [155] [277]
5 dec.	$x = 0\,(0{,}01)\,1\,(0{,}02)\,1{,}98$	[9] [13]
5 fig.	$x = 0\,(0{,}1)\,20$	[29] [278]
5 dec.	$x = 0\,(0{,}2)\,6$	[150]
5—3 fig.	$x = 0\,(0{,}01)\,6\,(0{,}1)\,10$	[14] $[41]_{2,3}$
5; 4 dec.	$x = 0\,(0{,}01)\,1\,(0{,}02)\,2;\ 2{,}1\,(0{,}1)\,6$	[37]
5—3 fig.	$x = 0\,(0{,}1)\,10\,(5)\,20;\ 30$	[21]
4 fig.	$x = 0\,(0{,}01)\,21$	[83]
4 fig.	$x = 0\,(0{,}1)\,10$	[20] [44]
4 dec.	$x = 0\,(0{,}1)\,6$	[9] [13]
4 dec.	$x = 0\,(0{,}5)\,6\,(2)\,10\,(5)\,20$	$[41]_1$ [154] [180] [298]
4 fig.	$x = 0\,(0{,}5)\,6\,(2)\,10\,(5)\,20;\ 30$	[56]
4 dec.	$x = 0\,(0{,}5)\,6\,(2)\,10\,(5)\,20;\ 30$	[43]

$\mathrm{ber}_1 x$ & $\mathrm{bei}_1 x$

10 dec.	$x = 0\,(0{,}01)\,10$	[62]
5—6 fig.	$x = 1\,(1)\,10$	[30] [276] [277]
4 dec.	$x = 0\,(0{,}2)\,8$	[37]
4 dec.	$x = 0\,(0{,}2)\,6$	[216]
4 fig.	$x = 1\,(1)\,10$	[44]

$-\sqrt{2}\,\mathrm{ber}_1 x$ & $\sqrt{2}\,\mathrm{bei}_1 x$

21 dec.	$x = 0\,(0{,}1)\,6$	[253]
6 dec.	$x = 0\,(0{,}02)\,4{,}6$	[174]
4 dec.	$x = 0\,(0{,}1)\,6$	$[41]_1$

$\mathrm{ber}_n x$ & $\mathrm{bei}_n x$

5—6 fig.	$n = 1\,(1)\,5.\quad x = 1\,(1)\,10$	[5] [30] [276] [277]
4 fig.	$n = 2.\quad x = 0\,(0{,}01)\,21$	[83]
4 fig.	$n = 1\,(1)\,5.\quad x = 1\,(1)\,10$	[44]

$\mathrm{ber}'_n x$ & $\mathrm{bei}'_n x$

5—6 fig.	$n = 1\,(1)\,5.\quad x = 1\,(1)\,10$	[5] [30] [276] [277]
4 fig.	$n = 1\,(1)\,5.\quad x = 1\,(1)\,10$	[44]

Thomson functions of the second and third kinds

ker x & kei x

7; 8; 9 dec.	$x = 0\,(0{,}1)\,3{,}8;\ 3{,}9\,(0{,}1)\,6{,}4;\ 6{,}5\,(0{,}1)\,10$	[2] [5] [30] [128] [155] [277]
5 dec.	$x = 0\,(0{,}01)\,1\,(0{,}02)\,1{,}98$	[13]
4 fig.	$x = 0\,(0{,}01)\,21$	[83]
4 dec.	$x = 0\,(0{,}01)\,1{,}22\,(0{,}02)\,2{,}4$	[2]
4 fig.	$x = 1\,(1)\,30$	[56] [202] [243]
4 fig.	$x = 0\,(0{,}1)\,10$	[20] [44]

ker′x & kei′x

7; 8; 9 dec.	$x = 0\,(0{,}1)\,3{,}8;\ 3{,}9\,(0{,}1)\,6{,}4;\ 6{,}5\,(0{,}1)\,10$	[2] [5] [30] [128] [155] [277]
5 dec.	$x = 0\,(0{,}01)\,1\,(0{,}02)\,1{,}98$	[13]
4 fig.	$x = 0\,(0{,}01)\,21$	[83]
4 fig.	$x = 0\,(0{,}1)\,10$	[20] [44]

$\mathrm{ker}_n x$ & $\mathrm{kei}_n x$

5—6 fig.	$n = 1\,(1)\,5.$	$x = 1\,(1)\,10$	[5] [30] [277]
4 fig.	$n = 2$	$x = 0\,(0{,}01)\,21$	[83]
4 fig.	$n = 1\,(1)\,5.$	$x = 1\,(1)\,10$	[44]

$\mathrm{ker}'_n x$ & $\mathrm{kei}'_n x$

6 — 5 fig.	$n = 1\,(1)\,5.$	$x = 1\,(1)\,10$	[30] [277]
4 fig.	$n = 1\,(1)\,5.$	$x = 1\,(1)\,10$	[44]

$\mathrm{her}\,x = \frac{2}{\pi}\,\mathrm{kei}\,x$ & $\mathrm{hei}\,x = -\frac{2}{\pi}\,\mathrm{ker}\,x$

5 dec.	$x = 0\,(0{,}01)\,1\,(0{,}02)\,2\,(0{,}2)\,6$	[37]
5 dec.	$x = 0\,(0{,}01)\,1\,(0{,}02)\,1{,}98$	[13]
4 fig.	$x = 0\,(0{,}2)\,6$	[9] [13] $[41]_1$ [216]
3—4 fig.	$x = 0\,(0{,}01)\,6\,(0{,}1)\,10$	[14] $[41]_{2,3}$

$\mathrm{her}'x = \frac{2}{\pi}\,\mathrm{kei}'x$ & $\mathrm{hei}'x = -\frac{2}{\pi}\,\mathrm{ker}'x$

5 dec.	$x = 0\,(0{,}01)\,1\,(0{,}02)\,2\,(0{,}2)\,6$	[37]
5 dec.	$x = 0\,(0{,}01)\,1\,(0{,}02)\,1{,}98$	[13]
3—5 fig.	$x = 0\,(0{,}01)\,6\,(0{,}1)10$	[14] $[41]_{2,3}$

$$\mathrm{her}_1 x = \frac{2}{\pi}\,\mathrm{kei}_1 x \quad \& \quad \mp \mathrm{hei}_1 x = \pm \frac{2}{\pi}\,\mathrm{ker}_1 x$$

5—6 fig.	$x = 0\,(0{,}2)\,6$	[41]$_1$
4 fig.	$x = 0\,(0{,}2)\,6$	[216]

Moduli and phases of Thomson functions

The cylinder functions $J_n(xi^{3/2})$ & $K_n(xi^{1/2})$ can be expressed in the form:

$$J_n(xi^{3/2}) = M_n(x)\,e^{i\theta_n(x)},$$

$$K_n(xi^{1/2}) = N_n(x)\,e^{i\{\varphi_n(x)+\frac{1}{2}n\pi\}},$$

where $M_n(x) = \sqrt{\mathrm{ber}_n^2 x + \mathrm{bei}_n^2 x}$ & $N_n(x) = \sqrt{\mathrm{ker}_n^2 x + \mathrm{kei}_n^2 x}$ are the moduli of Thomson functions while $\theta_n(x) = \mathrm{arc\,tg}\,\dfrac{\mathrm{bei}_n x}{\mathrm{ber}_n x}$ & $\varphi_n(x) = \mathrm{arc\,tg}\,\dfrac{\mathrm{kei}_n x}{\mathrm{ker}_n x}$ are their phases.

$$\mathrm{ber}_n x = M_n(x)\cos\theta_n(x), \qquad \mathrm{bei}_n x = M_n(x)\sin\theta_n(x),$$
$$\mathrm{ker}_n x = N_n(x)\cos\varphi_n(x), \qquad \mathrm{kei}_n x = N_n(x)\sin\varphi_n(x).$$

Note that

$$\mathrm{ber}'^2 x + \mathrm{bei}'^2 x = \mathrm{ber}_1^2 x + \mathrm{bei}_1^2 x \qquad \theta_1(x) - 45^\circ = \mathrm{arctg}\,\frac{\mathrm{bei}'\,x}{\mathrm{ber}'\,x}$$

$$\mathrm{ker}'^2 x + \mathrm{kei}'^2 x = \mathrm{ker}_1^2 x + \mathrm{kei}_1^2 x \qquad \varphi_1(x) - 45^\circ = \mathrm{arctg}\,\frac{\mathrm{kei}'\,x}{\mathrm{ker}'\,x}$$

Moduli of Thomson functions

$M_n(x)$

6—5 dec.	$n = 0;\,1.\ x = 0\,(0{,}1)\,10$	[134]
5 fig.	$n = 0;\,1.\ x = 0\,(0{,}1)\,20$	[29] [278]
5 fig.	$n = 0;\,1.\ x = 0\,(0{,}1)\,10$	[235] [275]
5 fig.	$n = 0;\,1.\ x = 0\,(0{,}2)\,6$	[216]
4—5 fig.	$n = 0;\,1.\ x = 0\,(0{,}1)\,10$	[14] [41]$_{2,3}$
4 fig.	$n = 0.\ x = 0\,(0{,}05)\,2\,(0{,}1)\,4\,(0{,}5)\,6\,(1)\,12\,(2)\,20\,(5)\,45$; $n = 1.\ x = 0\,(0{,}05)\,2{,}7\,(0{,}1)\,3{,}8$; $4\,(0{,}5)\,8\,(1)\,12\,(2)\,20\,(5)\,50; 4{,}25$	[44]
4 fig.	$n = 0;1.\ x = 0\,(0{,}05)\,1{,}7\,(0{,}1)\,3\,(0{,}2)\,5\,(0{,}5)\,6\,(1)\,12\,(2)\,20\,(5)\,45$	[20]

$\lg\{\sqrt{x}\,M_0(x)\}$

4 dec.	$x = 0\,(0{,}05)\,2\,(0{,}1)\,4\,(0{,}5)\,6\,(1)\,12\,(2)\,20\,(5)\,45$	[44]

$$\lg\{\sqrt{x}\,M_1(x)\}$$

4 dec.	$x = 0\,(0{,}05)\,2{,}7\,(0{,}1)\,3{,}8;\ 4\,(0{,}5)\,8\,(1)\,12\,(2)\,20\,(5)\,50;\ 4{,}25$	[44]

$$\ln\{\sqrt{2\pi x}\,M_0(x)\}$$

4 dec.	$x = 1\,(1)\,30$	[202] [243]

$$N_n(x)$$

5 fig.	$n = 0;\ 1.$	$x = 0\,(0{,}1)\,10$	[220]

$$\frac{2}{\pi}N_n(x)$$

4—5 fig.	$n = 0;\ 1.$	$x = 0\,(0{,}1)\,10$	[14] $[41]_{2,3}$
4 fig.	$n = 0;\ 1.$	$x = 0\,(0{,}2)\,6$	[216]

$$\ln\left\{\sqrt{\frac{2x}{\pi}}\,N_0(x)\right\}$$

4 dec.	$x = 1\,(1)\,30$	[202] [243]

Phases of Thomson functions

$$\theta_0(x)$$

0°,00001		$x = 0\,(0{,}1)\,10$	[134]
0,0001		$x = 1\,(1)\,30$	[202] [243]
0°,001		$x = 0\,(0{,}1)\,20$	[29] [278]
0°,001		$x = 0\,(0{,}1)\,10$	[235] [275]
0°,01		$x = 0\,(0{,}05)\,2\,(0{,}1)\,4\,(0{,}5)\,6\,(1)\,12\,(2)\,20\,(5)\,45$	[44]
1″		$x = 0\,(0{,}2)\,6$	[216]
0°,01		$x = 0\,(0{,}05)\,1{,}7\,(0{,}1)\,3\,(0{,}2)\,4\,(0{,}5)\,6\,(1)\,12\,(2)\,20\,(5)\,45$	[20]
0°,01	$-\theta_0(x)$	$x = 0\,(0{,}1)\,10$	[14] $[41]_{2,3}$

$$\theta_1(x)$$

0°,01	$x = 0\,(0{,}05)\,2{,}7\,(0{,}1)\,3{,}8;\ 4\,(0{,}5)\,8\,(1)\,12\,(2)\,20\,(5)\,50;\ 4{,}25$	[44]

$0^\circ{,}01$	$x = 0\,(0{,}05)\,1{,}7\,(0{,}1)\,3\,(0{,}2)\,4\,(0{,}5)$ $6\,(1)\,12\,(2)\,20\,(5)\,45$	[20]

$\theta_1(x) - 45^\circ$

$0^\circ{,}001$	$x = 0\,(0{,}1)\,20$	[29] [278]

$\theta_1(x) - 180^\circ$

$0^\circ{,}00001$	$x = 0\,(0{,}1)\,10$	[134]
$0^\circ{,}001$	$x = 0\,(0{,}1)\,10$	[235] [275]
$1''$	$x = 0\,(0{,}2)\,6$	[216]

$200^g - \theta_1(x)$

$0^g{,}01$	$x = 0\,(0{,}1)\,10$	[14] $[41]_{2,3}$

$\theta_0(x) - \theta_1(x) + 135^\circ$

$0^\circ{,}00001$	$x = 0\,(0{,}1)\,10$	[134]

$\theta_1(x) - \theta_0(x) - 200^g$

$0^g{,}01$	$x = 0\,(0{,}1)\,10$	[14] $[41]_{2,3}$

$\varphi_n(x)$

0,0001	$n = 0.$	$x = 1\,(1)\,30$	[202] [243]
$0^\circ{,}001$	$n = 0; 1.$	$x = 0\,(0{,}1)\,10$	[220]

$\varphi_1(x) - \varphi_0(x);\; -\varphi_0(x) - 100^g;\; -\varphi_1(x) - 100^g$

$0^g{,}01$	$x = 0\,(0{,}1)\,10$	[14] $[41]_{2,3}$

$\varphi_0(x) + 90^\circ;\; \varphi_1(x) + 90^\circ$

$1''$	$x = 0\,(0{,}2)\,6$	[216]

Various expressions containing Thomson functions

$\mathrm{ber}^2\, x + \mathrm{bei}^2\, x$

6 fig.	$x = 0\,(0{,}2)\,10$	[132]
4 fig.	$x = 0\,(1)\,30$	[56] [202] [243]
2—6 dec.	$x = 0\,(0{,}2)\,6$	[150]

$\mathbf{ber}'^2 x + \mathbf{bei}'^2 x$

6 fig.	$x = 0\,(0{,}2)\,10$	[132]
4 fig.	$x = 0\,(1)\,30$	[56] [202] [243]
2—6 dec.	$x = 0\,(0{,}2)\,6$	[150]

$\mathbf{ber}\,x\, \mathbf{ber}'\, x + \mathbf{bei}\, x\, \mathbf{bei}'\, x$

6 fig.	$x = 0\,(0{,}2)\,10$	[132]
4 fig.	$x = 0\,(1)\,30$	[56] [202] [243]
2—6 dec.	$x = 0\,(0{,}2)\,6$	[150]

$\mathbf{ber}\, x\, \mathbf{bei}'\, x - \mathbf{bei}\, x\, \mathbf{ber}'\, x$

6 fig.	$x = 0\,(0{,}2)\,10$	[132]
4 fig.	$x = 0\,(1)\,30$	[56] [202] [243]
2—6 dec.	$x = 0\,(0{,}2)\,6$	[150]

$-\dfrac{\mathbf{ber}_3 x \pm \mathbf{bei}_3 x}{\sqrt{2}}$

4 fig.	$x = 0\,(0{,}01)\,21$	[83]

$\mathbf{ker}^2 x + \mathbf{kei}^2 x$

6 fig.	$x = 0\,(0{,}2)\,10$	[132]
4 fig.	$x = 0\,(1)\,30$	[33] [56] [202] [243]

$\mathbf{ker}'^2 x + \mathbf{kei}'^2 x$

6 fig.	$x = 0\,(0{,}2)\,10$	[132]
4 fig.	$x = 0\,(1)\,30$	[33] [56] [202] [243]

$\mathbf{ker}\, x\, \mathbf{ker}'\, x + \mathbf{kei}\, x\, \mathbf{kei}'\, x$

6 fig.	$x = 0\,(0{,}2)\,10$	[132]
4 fig.	$x = 0\,(1)\,5$	[202] [243]

$\mathbf{ker}\, x\, \mathbf{kei}'\, x - \mathbf{kei}\, x\, \mathbf{ker}'\, x$

6 fig.	$x = 0\,(0{,}2)\,10$	[132]
4 fig.	$x = 0\,(1)\,5$	[202] [243]

$\mathbf{ber}'\, x\, \mathbf{ker}'\, x + \mathbf{bei}'\, x\, \mathbf{kei}'\, x$

6 fig.	$x = 0\,(0{,}2)\,10$	[132]
4 fig.	$x = 0\,(1)\,30$	[33] [56] [202] [243]

$$\text{bei}' x \, \text{ker}' x - \text{ber}' x \, \text{kei}' x$$

6 fig.	$x = 0\,(0,2)\,10$	[132]
4 fig.	$x = 0\,(1)\,30$	[33] [56] [202] [243]

$$-\frac{\text{ker}_3 x \pm \text{kei}_3 x}{\sqrt{2}}$$

4 fig.	$x = 0\,(0,01)\,21$	[84]

$$\frac{\text{ber}'^2 x + \text{bei}'^2 x}{\text{ber}^2 x + \text{bei}^2 x}$$

5 dec.	$x = 0\,(0,2)\,10;\ \infty$	[132]
4 dec.	$x = 0\,(1)\,30;\ \infty$	[202] [243]

$$\frac{\text{ber}\, x\, \text{ber}' x + \text{bei}\, x\, \text{bei}' x}{\text{ber}^2 x + \text{bei}^2 x}$$

6 dec.	$x = 3\,(1)\,6$	[247]
5 dec.	$x = 0\,(0,2)\,10;\ \infty$	[132]
4 dec.	$x = 3\,(0,2)\,7$	[203] [244]
4 dec.	$x = 0\,(1)\,30;\ \infty$	[202] [243]

$$\frac{\text{ber}\, x\, \text{bei}' x - \text{bei}\, x\, \text{ber}' x}{\text{ber}^2 x + \text{bei}^2 x}$$

6 dec.	$x = 3\,(1)\,6$	[247]
5 dec.	$x = 0\,(0,2)\,10;\ \infty$	[132]
4 dec.	$x = 0\,(1)\,30;\ \infty$	[202] [243]

$$\frac{x}{4}\,\frac{\text{ber}\, x\, \text{bei}' x - \text{bei}\, x\, \text{ber}' x}{\text{ber}^2 x + \text{bei}^2 x}$$

5 dec.	$x = 0\,(0,1)\,10$	[226]

$$\frac{2}{x}\,\frac{\text{ber}\, x\, \text{ber}' x + \text{bei}\, x\, \text{bei}' x}{\text{ber}^2 x + \text{bei}^2 x}$$

4—5 dec	$x = 0\,(0,5)\,6$ (var.) 20	[31]
4 fig.	$x = 0\,(0,5)\,5.$	[229]

$$\frac{2}{x}\,\frac{\text{ber}\, x\, \text{bei}' x - \text{bei}\, x\, \text{ber}' x}{\text{ber}^2 x + \text{bei}^2 x}$$

4—5 dec.	$x = 0\,(0,5)\,6$ (var.) 20	[31]

$$\frac{x}{4}\,\frac{\operatorname{ber} x \operatorname{ber}' x + \operatorname{bei} x \operatorname{bei}' x}{\operatorname{ber}^2 x + \operatorname{bei}^2 x}$$

3 fig.	$x = 0\,(0{,}5)\,5$	[229]

$$\frac{\operatorname{ber} x \operatorname{ber}' x + \operatorname{bei} x \operatorname{bei}' x}{\operatorname{ber}'^2 x + \operatorname{bei}'^2 x}$$

6 dec.	$x = 5\,(1)\,8$	[247]
5 dec.	$x = 0\,(0{,}2)\,10;\ \infty$	[132]
5 dec.	$x = 0\,(0{,}1)\,5\,(0{,}2)\,10\,(0{,}5)\,15$ (var.) $100;\ \infty$	[287]
4 dec.	$x = 0\,(1)\,30;\ \infty$	[202] [243]
3 dec.	$x = 0\,(0{,}5)\,6\,(2)\,10\,(5)\,20$	[298]

$$\frac{\operatorname{ber} x \operatorname{bei}' x - \operatorname{bei} x \operatorname{ber}' x}{\operatorname{ber}'^2 x + \operatorname{bei}'^2 x}$$

6 dec.	$x = 5\,(1)\,8$	[247]
5 dec.	$x = 0\,(0{,}2)\,10;\ \infty$	[132]
5 dec.	$x = 0\,(0{,}1)\,5\,(0{,}2)\,10\,(0{,}5)\,15$ ( ) $100;\ \infty$	[287]
4 dec.	$x = 0\,(1)\,30;\ \infty$	[202] [243]
4 dec.	$x = 0\,(0{,}5)\,6\,(2)\,10\,(5)\,20$	[33] [154] [180]
4 dec.	$x = 0{,}5\,(0{,}5)\,2;\ 5{,}5;\ 6\,(2)\,10\,(5)\,20\,(10)\,50$	[200] [242]
3 dec.	$x = 0\,(0{,}5)\,6\,(2)\,10\,(5)\,20$	[298]

$$\frac{x}{2}\,\frac{\operatorname{ber} x \operatorname{ber}' x + \operatorname{bei} x \operatorname{bei}' x}{\operatorname{ber}'^2 x + \operatorname{bei}'^2 x}$$

6 dec.	$x = 0\,(0{,}1)\,10$	[134]
5 dec.	$x = 0\,(0{,}1)\,10$	[226]
2—3; 3—4 dec.	$\frac{x}{2\sqrt{2}} = 0\,(0{,}1)\,0{,}9$; $x = 0{,}5\,(0{,}5)\,6\,(2)\,10\,(5)\,20$	$[41]_1$

$$\frac{x}{2}\,\frac{\operatorname{ber} x \operatorname{bei}' x - \operatorname{bei} x \operatorname{ber}' x}{\operatorname{ber}'^2 x + \operatorname{bei}'^2 x}$$

6 dec.	$x = 0\,(0{,}1)\,10$	[134]
5 dec.	$x = 0\,(0{,}1)\,5\,(0{,}2)\,10$ (var.) $100;\ \infty$	[33] [56]
5 dec.	$x = 0\,(0{,}1)\,5\,(0{,}2)\,10\,(0{,}5)\,15$ (var.) $100;\ \infty$	[287]
5 dec.	$x = 0\,(0{,}1)\,10\,(5)\,20;\ 30;\ 50;\ 100$	[21]
4 dec.	$x = 0\,(0{,}5)\,6\,(2)\,10\,(5)\,20;\ 30$	[43]

4 dec. $x = 0\,(0{,}5)\,6\,(2)\,10\,(5)\,20$ [33] [154] [180]

2—3; 3—4 dec. $\left\{\begin{array}{l} \frac{x}{2\sqrt{2}} = 0\,(0{,}1)\,0{,}9 \\ x = 0{,}5\,(0{,}5)\,6\,(2)\,10\,(5)\,20 \end{array}\right\}$ [41]₁

$$\frac{4}{x}\,\frac{\operatorname{ber} x \operatorname{ber}' x + \operatorname{bei} x \operatorname{bei}' x}{\operatorname{ber}'^2 x + \operatorname{bei}'^2 x}$$

5 dec. $x = 0\,(0{,}1)\,5\,(0{,}2)\,10$ (var.) 100; ∞ [33] [56]

5 dec. $x = 0\,(0{,}1)\,5\,(0{,}2)\,10\,(0{,}5)\,15$ (var.) 100; ∞ [287]

$$\frac{4}{x}\,\frac{\operatorname{ber} x \operatorname{bei}' x - \operatorname{bei} x \operatorname{ber}' x}{\operatorname{ber}'^2 x + \operatorname{bei}'^2 x} - \frac{8}{x^2}$$

4 fig. $x = 0\,(0{,}5)\,5$ [229]

$$\frac{x}{2}\,\frac{\operatorname{ber} x \operatorname{bei}' x - \operatorname{bei} x \operatorname{ber}' x}{\operatorname{ber}'^2 x + \operatorname{bei}'^2 x}$$

3 fig. $x = 0\,(0{,}5)\,5$ [229]

$$\frac{\operatorname{ber}' x}{\operatorname{ber}'^2 x + \operatorname{bei}'^2 x} \quad \& \quad \frac{\operatorname{bei}' x}{\operatorname{ber}'^2 x + \operatorname{bei}'^2 x}$$

4 fig. $x = 0\,(0{,}5)\,6\,(2)\,10\,(5)\,20$ [298]

$$x\,\frac{\operatorname{ber} x \operatorname{ber}' x + \operatorname{bei} x \operatorname{bei}' x}{\operatorname{ber}'^2 x + \operatorname{bei}'^2 x}$$

2—3 dec. $x^2 = \frac{1}{2}\left(\frac{1}{2}\right)3$ [81]

$$x\,\frac{\operatorname{ber} x \operatorname{bei}' x - \operatorname{bei} x \operatorname{ber}' x}{\operatorname{ber}'^2 x + \operatorname{bei}'^2 x}$$

2—3 dec. $x^2 = \frac{1}{2}\left(\frac{1}{2}\right)3$ [81]

$$\frac{x}{2}\sqrt{\frac{\operatorname{ber}^2 x + \operatorname{bei}^2 x}{\operatorname{ber}'^2 x + \operatorname{bei}'^2 x}}$$

6 dec. $x = 0\,(0{,}1)\,10$ [134]

$$\sqrt{\frac{\operatorname{ber}^2 x + \operatorname{bei}^2 x}{\operatorname{ber}'^2 x + \operatorname{bei}'^2 x}} \quad \& \quad \sqrt{\frac{\operatorname{kei}^2 x + \operatorname{kei}^2 x}{\operatorname{ker}'^2 x + \operatorname{kei}'^2 x}}$$

4—5 fig. $x = 0\,(0{,}1)\,10$ [14] [41]₂,₃

$$\left(\mathbf{bei}\,x-\frac{2}{\pi}\,\mathbf{ker}\,x\right)+i\left(\mathbf{ber}\,x+\frac{2}{\pi}\,\mathbf{kei}\,x\right)=Y_0\,(xi^{1/2})$$

5; 4 dec.	$x = 0\,(0{,}01)\,1\,(0{,}02)\,2;\ 2{,}2\,(0{,}2)\,6$	[37]
4 dec.	$x = 0\,(0{,}2)\,6$	[9] [13] [41]$_1$

$$\left(\mathbf{bei}'x-\frac{2}{\pi}\,\mathbf{ker}'\,x\right)+i\left(\mathbf{ber}'\,x+\frac{2}{\pi}\,\mathbf{kei}'\,x\right)=\frac{d}{dx}\,Y_0\,(xi^{1/2})$$

5; 4 dec.	$x = 0\,(0{,}01)\,1\,(0{,}02)\,2;\ 2{,}2\,(0{,}2)\,6$	[37]

AIRY FUNCTIONS

$$\mathrm{Ai}\,(x)=\frac{1}{\pi}\int_0^\infty \cos\left(\frac{1}{3}\,t^3+xt\right)dt=\frac{1}{3}\,x^{1/2}\left\{I_{-1/3}\left(\frac{2}{3}\,x^{3/2}\right)-I_{1/3}\left(\frac{2}{3}\,x^{3/2}\right)\right\}$$

$$\mathrm{Bi}\,(x)=\frac{1}{\pi}\int_0^\infty\left\{\exp\left(-\frac{1}{3}\,t^3+xt\right)+\sin\left(\frac{1}{3}\,t^3+xt\right)\right\}dt=\left(\frac{1}{3}\,x\right)^{1/2}\left\{I_{-1/3}\left(\frac{2}{3}\,x^{3/2}\right)+\right.$$
$$\left.+I_{1/3}\left(\frac{2}{3}\,x^{3/2}\right)\right\},$$
$$\mathrm{Ai}\,(-x)=\frac{1}{3}\,x^{1/2}\left\{J_{-1/3}\left(\frac{2}{3}\,x^{3/2}\right)+J_{1/3}\left(\frac{2}{3}\,x^{3/2}\right)\right\},$$
$$\mathrm{Bi}\,(-x)=\left(\frac{1}{3}\,x\right)^{1/2}\left\{J_{-1/3}\left(\frac{2}{3}\,x^{3/2}\right)-J_{1/3}\left(\frac{2}{3}\,x^{3/2}\right)\right\}.$$

$\mathbf{Ai}\,(x)$ & $\mathbf{Ai}'\,(x)$

8 dec.	$x = -20\,(0{,}01)\,2$	[46]
4—5 fig.	$x = -9\,(0{,}02)\,9$	[12]

$\mathbf{Bi}\,(x)$ & $\mathbf{Bi}'\,(x)$

8 dec.	$x = -10\,(0{,}1)\,2{,}5$	[46]
4—5 fig.	$x = -9\,(0{,}02)\,9$	[12]

$\left(\frac{h^n}{n!}\right)\mathbf{Bi}^{(n)}\,(x)$

10—9 dec.	$h = 0{,}1.\quad n = 1\,(1)\,8.\quad x = -10\,(0{,}1)\,2{,}5$	[46]

$\frac{\mathbf{Ai}'\,(x)}{\mathbf{Ai}\,(x)}$

7 dec.	$x = 0\,(0{,}1)\,25\,(1)\,75$	[46]

$\frac{\mathbf{Bi}'\,(x)}{\mathbf{Bi}\,(x)}$

7 dec.	$x = 0\,(0{,}1)\,10$	[46]

lg Ai (x)

8 dec.	$x = 0\,(0{,}1)\,25\,(1)\,75$	[46]

lg Bi (x)

8 dec.	$x = 0\,(0{,}1)\,10$	[46]

$$W(n) = \int_0^\infty \cos \frac{1}{2}\pi (w^3 - nw)\, dw$$

5 dec.	$n = -4\,(0{,}2)\,4$	[282]
5 dec.	$n = -5{,}6\,(0{,}2)\,5{,}6$	[107] [283]
5 dec.	$n = -5{,}6; -5\,(1)\,-1; -0{,}2\,(0{,}2)\,5{,}6$	[273]
4 dec.	$n = -4\,(0{,}4)\,4$	[195] [273]

$$\left(\frac{3}{\pi^2}\right)^{1/3} W(n)$$

5 dec.	$n = -4\,(0{,}2)\,4$	[201]

$W^2(n)$

5 dec.	$n = -4\,(0{,}2)\,4$	[282]
4 dec.	$n = -5{,}6; -5\,(1)-1; -0{,}2\,(0{,}2)\,5{,}6$	[273]
3—6 dec.	$n = -4$ (var.) $8{,}48$	[197]
3 dec.	$n = -2\,(0{,}2)\,13{,}6$	[40] [50] [159]

$$\int_0^x \cos \frac{1}{2}\pi (w^3 - nw)\, dw$$

5 dec.	$n = -4\,(0{,}2)\,4$	$x = 1; 1{,}26; 1{,}44; 1{,}58; 1{,}70; 1{,}92; 2$	[282]
4 dec.	$n = -4\,(0{,}4)\,4.$	$x = 1; 1{,}26; 1{,}44$	[195] [273]

Zeros Ai (x) & Ai′ (x)

8 dec.	a_s & a'_s	$s = 1\,(1)\,50$	[46]

Zeros Bi (x) & Bi′ (x)

8 dec.	b_s & b'_s	$s = 1\,(1)\,20$	[46]

Zeros $W(n)$

4 dec.	First 50 zeros	[284]
4 dec.	s zeros; $s = 1(1)\,10\,(2)14\,(3)20(5)50$	[273]
4 dec.	First 10 zeros	[40] [50] [159]

Zeros $W'(n)$

4 dec. First 10 zeros [40] [50] [129] [159] [273]]284]

$\max W^2(n)$

3 dec. First 10 max [40] [159]

$\mathrm{Ai}(a_s') \,\&\, \mathrm{Ai}'(a_s)$

where $\mathrm{Ai}(a_s) = \mathrm{Ai}'(a_s') = 0$.

8 dec. $s = 1(1)50$ [46]

$\mathrm{Bi}(b_s') \,\&\, \mathrm{Bi}'(b_s)$

where $\mathrm{Bi}(b_s) = \mathrm{Bi}'(b_s') = 0$.

8 dec. $s = 1(1)20$ [46]

Auxiliary functions

Airy functions can be expressed in terms of trigonometric functions in the form:

$$\mathrm{Ai}(x) = F(x)\sin\varphi(x), \quad \mathrm{Ai}'(x) = G(x)\sin\psi(x),$$
$$\mathrm{Bi}(x) = F(x)\cos\varphi(x), \quad \mathrm{Bi}'(x) = G(x)\cos\psi(x),$$

where $F(x), G(x), \varphi(x), \psi(x)$ — are auxiliary functions.

$F(x) \,\&\, G(x)$

8dec. $-x = 0(0{,}1)30(1)80$ }
7dec. $x = 0(0{,}1)2{,}5$ } [46]

$\varphi(x) \,\&\, \psi(x)$

$0^{\circ}{,}000001$ $-x = -2{,}5(0{,}1)30(1)80$ [46]

STRUVE AND LOMMEL-WEBER FUNCTIONS

$$\mathrm{H}_n(x) = \frac{2\left(\frac{1}{2}x\right)^n}{\sqrt{\pi}\,\Gamma\left(n+\frac{1}{2}\right)} \int_0^{\frac{\pi}{2}} \sin(x\cos\theta)\sin^{2n}\theta\, d\theta,$$

$$\Omega_n(x) = \frac{1}{\pi}\int_0^{\pi} \sin(x\sin\theta - n\theta)\, d\theta,$$

$$L_n(x) = \sum_{m=0}^{\infty} (-1)^m \left(\frac{x}{2}\right)^{2m+1+n} / \,\Gamma\left(m+\frac{3}{2}\right)\Gamma\left(m+n+\frac{3}{2}\right),$$

where

$$L_0(x) = -iH_0(ix) = -i\Omega_0(ix).$$

Struve functions $\mathbf{H}_n(x)$

$\mathbf{H}_0(x) = \Omega_0(x)$

7 dec.	$x = 0(0{,}02)16$	[2] [86]
6 dec.	$x = 0(0{,}02)16$	[135]
4 dec.	$x = 0(0{,}01)14{,}99$	[14] $[41]_{2,3}$
4 dec.	$x = 0(0{,}02)16$	[3]
4 dec.	$x = 0(0{,}1)15{,}9$	[44]

$\mathbf{H}_1(x) = \Omega_1(x) + \frac{2}{\pi}$

7 dec.	$x = 0(0{,}02)16$	[2] [86]
4 dec.	$x = 0(0{,}01)14{,}99$	[14] $[41]_{2,3}$
4 dec.	$x = 0(0{,}02)16$	[3]
4 fig.	$x = 0(0{,}1)15{,}9$	[44]

$\mathbf{H}_n(x)$

7 dec.	$n = 0; -1; -2.$	$x = 0(0{,}1)10$	[167]
4 dec.	$n = 2(1)4.$	$x = 0(0{,}1)10$	[171]

$\frac{2\mathbf{H}_1(x)}{x}$ & $\frac{3\pi\mathbf{H}_1(x)}{2x^2}$

4 dec.	$x = 0(0{,}1)0{,}2(0{,}2)4(0{,}5)10(1)15$	[48]
4 dec.	$x = 0(0{,}4)18$	[249]

$x\,\mathbf{H}_{-2}(x)$

7 dec.	$x = 0(0{,}1)2$	[167]

$F_{\pm n}(n)$ & $F_{\pm(n-1)}(n)$,

where

$$F_n(x) = \int_0^{\infty} \exp(-x\,\mathrm{sh}\,\theta - n\theta)\,d\theta.$$

6 dec.	$n = 0\left(\frac{1}{4}\right)10$	[134]

$$\mathbf{L}_n(x)$$

8—10 fig.	$n = 0; -1; -2.$ $x = 0{,}1\,(0{,}1)\,10$	[167]
6—7 fig.	$n = 0.$ $x = 0{,}02;\ 0{,}1;\ 0{,}5;\ 1(1)12$	[218]

$$x\mathbf{L}_{-2}(x)$$

8 dec.	$x = 0\,(0{,}1)\,10$	[167]

$$\mathbf{L}_0'(x) = \frac{2}{\pi} - \mathbf{H}_1(ix) = -\Omega_1(ix)$$

6—7 fig.	$x = 0{,}02;\ 0{,}1;\ 0{,}5;\ 1\,(1)\,12$	[218]

$$\frac{1}{2}\pi\,[I_0(x) - \mathbf{L}_0(x)]$$

3—5 dec.	$x = 0\,(0{,}1)\,1\,(0{,}2)\,13{,}6\,(0{,}4)\,16$	[292]

$$-\frac{1}{2}\pi\,[I_0'(x) - \mathbf{L}_0'(x)]$$

3—5 dec.	$x = 0\,(0{,}1)\,1\,(0{,}2)\,13{,}6\,(0{,}4)\,16$	[292]

Lommel-Weber functions $\Omega_n(x)$

$$\Omega_1(x) = \mathbf{H}_1(x) - \frac{2}{\pi}$$

6 dec.	$x = 0\,(0{,}02)\,16$	[135]
4 dec.	$x = 0\,(0{,}01)\,14{,}99$	[14] $[41]_{2,3}$

$$\Omega_{1/2}(x)$$

6 dec.	$x = 0\,(0{,}02)\,20$	[136]

$$\Omega_{-1/2}(x) = -\Omega_{1/2}(-x)$$

6 dec.	$x = 0\,(0{,}02)\,20$	[136]

$$\Omega_n(n)\ \&\ \Omega_{n-1}(n)$$

6 dec.	$n = 0\left(\frac{1}{4}\right)10$	[134]

$$\Omega_{2n-1}(2n) - \Omega_{2n+1}(2n)$$

6 dec.	$n = 1\,(1)\,15$	[257]

ZEROS OF CYLINDER FUNCTIONS

Zeros of cylinder functions of the first and second kind of real argument

$$J_n(j_{n,s}) = Y_n(y_{n,s}) = N_n(y_s^n) = J_n'(j_{n,s}')=0$$

$j_{0,s}$

10 dec.	$s=1$ (1) 150	[23] [140]
10 dec.	$s=1$ (1) 40	[4] [34] [139]
10 dec.	$s=1$ (1) 10	[89]
8 dec.	$s=1$ (1) 40	[6] [58]
7 dec.	$s=1$ (1) 40	[86] [213]
7 dec.	$s=1$ (1) 6	[148]
6 dec.	$s=1$ (1) 12	[233]
6 dec.	$s=1$ (1) 10	[234]
5 dec.	$s=1$ (1) 60	[37]
5 dec.	$s=1$ (1) 40	[3] [11]
5 dec.	$s=1$ (1) 16	[13]
5 dec.	$s=1$ (1) 10	[87]
5; 4 dec.	$s=1$ (1) 10; 11 (1) 25	[60]
4 dec.	$s=1$ (1) 40	[13] [14] [21] $[41]_{1,2,3}$
4 dec.	$s=1$ (1) 12	[28]
4 dec.	$s=1$ (1) 10	[26] [32] [133]
3 dec.	$s=1$ (1) 9	[95]

$\frac{j_{0,s}}{\pi}$

4 dec.	$s=1$ (1) 12	[284]

$j_{1,s}$

16 dec.	$s=1$ (1) 50	[4] [34] [45]
10 dec.	$s=1$ (1) 150	[23] [140]
8 dec.	$s=1$ (1) 50	[58]
8 dec.	$s=0$ (1) 40	[6]
7 dec.	$s=1$ (1) 40	[86] [213]
6 dec.	$s=1$ (1) 15	[32]
6 dec.	$s=1$ (1) 6	[190] [274] [296]
5 dec.	$s=1$ (1) 60	[37]
5 dec.	$s=1$ (1) 39	[11]
5 dec.	$s=1$ (1) 32	[60]
5 dec.	$s=1$ (1) 10	[87]
4 dec.	$s=1$ (1) 50	[13] [14] $[41]_{1,2,3}$

4 dec.	$s=1\ (1)\ 40$	[3] [21]
4 dec.	$s=1\ (1)\ 12$	[28]
4 dec.	$s=1\ (1)\ 10$	[26] [133]
3 dec.	$s=1\ (1)\ 9$	[95]

$$\frac{j_{1,s}}{\pi}$$

6 dec.	$s=1\ (1)\ 6$	[190] [274] [296]
4 dec.	$s=1\ (1)\ 12$	[284]

$$\frac{j_{1,s}}{2\pi}$$

3 dec.	$s=1\ (1)\ 9$	[50]
3 dec.	$s=1\ (1)\ 6$	[273]

$$\left(\frac{1}{2}j_{1,s}\right)^2$$

6 dec.	$s=1\ (1)\ 5$	[149]
5 dec.	$s=1\ (1)\ 4$	[206]

$$\frac{180\,j_{1,s}}{\pi}$$

1 dec.	$s=1\ (1)\ 6$	[59]

$$\left(\frac{1}{2}j_{n,s}\right)^2$$

5 dec.	$\begin{Bmatrix} n=2.\ s=1\ (1)\ 4 \\ n=3.\ s=1\ (1)\ 3 \end{Bmatrix}$	[206]

$$\frac{j_{2,s}}{\pi}$$

5 dec.	$s=1\ (1)\ 5$	[87] [190] [274] [296]
4 dec.	$s=1\ (1)\ 3$	[48]
3 dec.	$s=1\ (1)\ 9$	[258]

$$\frac{j_{2,s}}{2\pi}$$

3 dec.	$s=1\ (1)\ 9$	[50] [273]

$$\frac{180\, j_{2,s}}{\pi}$$

1 dec.	$s = 1\ (1)\ 5$		[59]
in integers	$s = 6\ (1)\ 10$		[59]

$$j_{n,s}$$

7 dec.	$n = 1\ (1)\ 5.$	$s = 1\ (1)\ 40$	[2] [86]
6 fig.	$n = 10;\ 100\ \&\ 1000.$	$s = 1\ (1)\ 5$	[209]
5 fig.	$n = 0\ (1)\ 5.$	$s = 1\ (1)\ 9$	[28] [55]
5 dec.	$n = 0\ (1)\ 115.$	$1 \leqslant s \leqslant 40$	[11]
5 dec.	$n = 2\ (1)\ 5.$	$s = 1\ (1)\ 40$	[13]
5 dec.	$n = 0\ (1)\ 19.$	$s = 1\ (1)\ 8$	[160]
5 dec.	$n = 1000.$	$s = 1$	[37] [289]
5 dec.	$n = 2.$	$s = 1\ (1)\ 5$	[87] [190] [296]
4—5 dec.	$n = 0\ (1)\ 19.$	$s = 1\ (1)\ 8$	[20]
4 dec.	$n = 0 (1) 10\ (5)\ 20\ (10)\ 50.$	$s = 1\ (1)\ 10$	[133]
4 dec.	$n = 2.$	$s = 1\ (1)\ 6$	[14] $[41]_{1,2,3}$
3 dec.	$n = 0\ (1)\ 10.$	$s = 1\ (1)\ 10$	[82]
3 dec.	$n = 0\ (1)\ 5.$	$s = 1\ (1)\ 10$	[224]
3 dec.	$n = 0\ (1)\ 5.$	$s = 1\ (1)\ 9$	[4] [28] [34] [95]
3 dec.	$n = 75;\ 100\ (100)\ 500.$	$s = 1\ (1)\ 10$	[133]
3; 2 dec.	$n = 750;\ 1000.$	$s = 1\ (1)\ 10$	[133]
2—3 dec.	$0 \leqslant n \leqslant 10.$	$s = 1\ (1)\ 5$	[54]

$$j_{\nu,s}\left(\nu = \pm\frac{1}{3};\ \pm\frac{2}{3}\right)$$

12 dec.	$\nu = \pm\frac{1}{3};\ \pm\frac{2}{3}.$	$s = 1\ (1)\ 8$	[178]
12 dec.	$\nu = \pm\frac{1}{3};\ \pm\frac{2}{3}.$	$s = 1\ (1)\ 30$	[65]
9 dec.	$\nu = -\frac{1}{3}.$	$s = 1$	[37] [289]
7 dec.	$\nu = \frac{1}{3}.$	$s = 1\ (1)\ 40$	[2] [86]
7 dec.	$\nu = \frac{2}{3}.$	$s = 1$	[105]
5 dec.	$\nu = \frac{1}{3}.$	$s = 1\ (1)\ 40$	[13]
4 dec.	$\nu = \frac{1}{3}.$	$s = 1\ (1)\ 40$	[3]
4 dec.	$\nu = \pm\frac{1}{3}.$	$s = 1$	[4] [34] [214]
3 dec.	$\nu = \pm\frac{1}{3};\ \pm\frac{2}{3}.$	$s = 1\ (1)\ 5$	[15]
2 dec.	$\nu = \frac{1}{3};\ \frac{2}{3}.$	$s = 1;\ 2$	[99] [214]
	$\nu = -\frac{1}{3};\ -\frac{2}{3}.$	$s = 1\ (1)\ 3$	[99] [214]

$$j_{\nu,s}\left(\nu=\pm\frac{1}{4};\ \pm\frac{3}{4}\right)$$

12 dec.	$\nu=\pm\frac{1}{4};\ \pm\frac{3}{4}.$	$s=1\ (1)\ 8$	[177]
10—12 fig.	$\nu=\pm\frac{1}{4};\ \pm\frac{3}{4}.$	$s=1\ (1)\ 30$	[65]
3dec.	$\nu=\pm\frac{1}{4};\ \pm\frac{3}{4}.$	$s=1\ (1)\ 5$	[15]
2dec.	$\nu=\frac{1}{4};\ \frac{3}{4}.$	$s=1;\ 2$	[102] [214]
	$\nu=-\frac{1}{4};\ -\frac{3}{4}.$	$s=1\ (1)\ 3$	

$$j_{\nu,s}\left(\nu=\pm\frac{1}{6};\ \pm\frac{5}{6}\right)$$

4 dec.	$\nu=\pm\frac{1}{6}.$	$s=1$	[4] [34] [214]
4 dec.	$\nu=\pm\frac{1}{6};\ \pm\frac{5}{6}.$	$s=1\ (1)\ 5$	[15]

$$j_{\nu,s}$$

8—12 fig.	$\nu=-\frac{43}{2}(1)\frac{39}{2}.$	$1\leqslant s\leqslant r.\ r=1\ (1)\ 7$	[80]
5 dec.	$\nu=-1\ (0{,}1)\ 1.$	$s=1$	[137]
4 dec.	$\nu=-\frac{1}{2}\left(\frac{1}{40}\right)1.$	$s=1;\ 2$	[4] [34] [214]
4 dec.	$\nu=-0{,}4\ (0{,}01)\ 1{,}49.$	$s=1$	[14] $[41]_{2,3}$
	$\nu=-0{,}4\ (0{,}01)\ 0{,}99.$	$s=2$	
3 dec.	$\nu=0\ (0{,}1)\ 1.$	$s=1$	[105]

$$j'_{\nu,s}$$

8—9 fig.	$\nu=-\frac{39}{2}\ (1)\ \frac{45}{2}.$	$1\leqslant s\leqslant r.\ r=1\ (1)\ 8$	[80]
8dec.	$\nu=0;\ 1.$	$s=1\ (1)\ 40$	[6]
5dec.	$\nu=0\ (1)\ 22.$	$s=1\ (1)\ 9$	[160]
4dec.	$\nu=0\ (1)\ 22.$	$s=1\ (1)\ 9$	[20]
4dec.	$\frac{j'_{\nu,1}}{\sqrt{2\nu}};\ j'_{\nu,2}\ \nu=-\frac{1}{2}\left(\frac{1}{40}\right)\frac{1}{2}$		[214]
3dec.	$\nu=1.$	$s=1\ (1)\ 6$	[53] [81]
	$\nu=2;\ 3.$	$s=1\ (1)\ 3$	
3dec.	$\nu=1.$	$s=1\ (1)\ 4$	[285]

$$\frac{j^2_{\nu s1}}{4(\nu+1)}$$

4 dec.	$\nu = -1\left(\frac{1}{40}\right)0$	[214]

$\lg j_{0,s}$

10 dec.	$s = 1\ (1)\ 150$	[140]
10 dec.	$s = 1\ (1)\ 40$	[139]

$\lg j_{1,s}$

10 dec.	$s = 1\ (1)\ 150$	[140]

$y_{0,s}$

8 dec.	$s = 1\ (1)\ 50$	[23]
7 dec.	$s = 1\ (1)\ 40$	[86]
5 dec.	$s = 1\ (1)\ 40$	[37]
5 dec.	$s = 1\ (1)\ 16$	[13]
5 dec.	$s = 1\ (1)\ 10$	[245]
4 dec.	$s = 1\ (1)\ 40$	[3]

$y_{1,s}$

8 dec.	$s = 1\ (1)\ 50$	[23]
7 dec.	$s = 1\ (1)\ 40$	[86]
5 dec.	$s = 1\ (1)\ 40$	[37]
5 dec.	$s = 1\ (1)\ 10$	[245]
4 dec.	$s = 1\ (1)\ 40$	[3]

$y_{n,s}$

7 dec.	$n = 0\ (1)\ 5.$	$s = 1\ (1)\ 40$	[2] [86]
5 dec.	$n = 0\ (1)\ 5.$	$s = 1\ (1)\ 40$	[13]
3 dec.	$n = 100.$	$s = 1\ (1)\ 5$	[209]

$y_{1/3,s}$

7 dec.	$s = 1\ (1)\ 40$	[2] [86]
5 dec.	$s = 1\ (1)\ 40$	[13]
4 dec.	$s = 1(1)40$	[3]

y^n_s

5 dec.	$n = 0;\ 1;\ 2.$	$s = 1\ (1)\ 40$	[245]
3 dec.	$n = 0;\ 1.$ $n = 1.$	$s = 1\ (1)\ 4$ $s = 1\ (1)\ 3$	[14][41]$_1$[297]
3 dec.	$n = 100.$	$s = 1\ (1)\ 5$	[209]

Zeros of various expressions containing cylinder functions of the first and second kinds of real argument

$xJ_0(x) - kJ_1(x) = (2-k)J_1(x) - xJ_2(x)$ or $(2-k)J_0(x) - kJ_2(x)$

4 dec. $k = \frac{1}{2}; \frac{2}{3}$. $s = 1\ (1)\ 10$ [101]

4 dec. $k = \frac{2}{3}$. $s = 1\ (1)\ 8$ [141]

3 dec. $k = \frac{1}{2}; \frac{2}{3}$. $s = 1\ (1)\ 4$ [285]

$$J_{-\left(n+\frac{1}{2}\right)}(kx)\,J_{n+\frac{1}{2}}(x) - J_{n+\frac{1}{2}}(kx)\,J_{-\left(n+\frac{1}{2}\right)}(x)$$

6 fig. $k = 0{,}2;\ 0{,}3; 0{,}4\ (0{,}2)\ 0{,}8.$ $n = 1(1)5;\ 1(1)6;\ 1(1)15$; $s = 1$ [306]

$$J_0(x)\,Y_1(kx) - J_1(kx)\,Y_0(x)$$

6 fig. $k = 0$ (var.) $3\ (1)\ 10;\ 20.$ $s = 1$
4—5 dec. $k = 1\ (0{,}01)\ 2;\ 2{,}1;\ 3\ (1)\ 10;\ 20.$ $s = 1$ [172]

$$Y_1(x)\,J_0(kx) - J_1(x)\,Y_0(kx)$$

4 fig. $\frac{1}{k} = 0{,}05$ (var.) $1.$ $s = 1$ [172]

4 dec. $k = (1{,}1)^{1/2};\ 1{,}06\ (0{,}02)\ 1{,}1\ (0{,}05)\ 1{,}3\ (0{,}1)\ 1{,}5;\ 2(1)5.$ $s = 1(1)6$ [304]

$$Y_2(x)\,J_0(kx) - J_2(x)\,Y_0(kx)$$

3 dec. $k^2 = 1{,}1;\ 1{,}2;\ 1{,}5;\ 2;\ 2{,}25;\ 3{,}4;\ 9;\ 16;\ 25.$ $s = 1\ (1)\ 4$ [204]

$$Y_n(x)\,J_n(kx) - J_n(x)\,Y_n(kx)$$

where $x_{n,s}$ are the zeros of this expression.

6 dec. $x_{n,s};\ (k-1)\,x_{n,s}$ $k = 1\frac{1}{2}\left(\frac{1}{2}\right)4;\ 1\left(\frac{1}{2}\right)4.$ $n = 0.$ $s = 1\ (1)\ 5$ [164]

5 dec. $x_{n,1}$ $k = 0{,}2;\ 0{,}3\ (0{,}1) 0{,}5;\ 0{,}6;\ 0{,}8.$ $n = 1\ (1)\ 5;\ 1(1)6;\ 1\ (1)\ 8;\ 1(1)12$ [307]

5—6 dec. $x_{n,s}$ $k = 10.$ $n = 1.$ $s = 1\ (1)\ 8$
4—5 dec. $x_{n,s}$ $k = 5.$ $n = 1.$ $s = 1\ (1)\ 7$
4 dec. $x_{n,s}$ $k = 1{,}2;\ 1{,}5;\ 2;\ 3.$ $n = 1.$ $s = 1\ (1)\ 6$ [156]

3—4 dec. $(k-1)x_{n,s}$ $1 \leqslant k \leqslant 12{,}287$; ∞. $n=0;1$. $s=1(1)4$ [14] $[41]_{1,2,3}$ [297]

3 dec. $x_{n,s}$ $k=1{,}5;\ 2{,}5;\ 10;\ 20;\ \infty$. $n=0(1)6$. $s=1(1)6$ [54]

3 dec. $\begin{cases} \left(\frac{3}{2}x_{n,s}\right)^2;\ (3x_{n,s})^2 & k=2\sqrt{2},\ 3\sqrt{3};\ 3\sqrt{3}.\quad n=\frac{1}{3}.\ s=1;\ 2 \\ (8\ x_{n,s})^2;\ (2x_{n,s})^2 & k=9;\ 4.\quad n=\frac{1}{4}.\ s=1;\ 2 \end{cases}$ [290]

4—2 dec. $x_{n,s};\ (k-1)x_{n,s}$ $k=1{,}2;\ 1{,}5;\ 2$. $n=0\left(\frac{1}{2}\right)2\frac{1}{2}$. $s=1(1)6$ [14] $[41]_{1,2,3}$ [297]

2 dec. $x_{n,s}$ $k=8;\ 4$. $n=\frac{1}{3};\ \frac{1}{4}$. $s=1(1)3;\ 1(1)2$ [143]

1—4 dec. $(k-1)x_{n,1}$ $k=1;\ 1{,}2;\ 1{,}5;\ 2(1)11;\ 19;\ 39;\ \infty$. $n=0\left(\frac{1}{2}\right)2\frac{1}{2}$ [14] $[41]_{1,2,3}$ [297]

$$x^2 J_0(x) + kJ_2(x)$$

4 dec. $k=0;0{,}5;1(1)6(2)10$. $s=1(1)4$ [48]

$$J_{-1/3}(x) \pm J_{1/3}(x)$$

7 dec. $s=1\ (1)\ 40$ [2] [86]

$$\frac{1}{2}(3x_s)^{2/3}$$

where x_s are zeros $J_{2/3}(x) - J_{-2/3}(x)$ & $J_{1/3}(x) + J_{-1/3}(x)$.

3 dec. $s=0(1)5$ [22]

$$F(x) - F(kx)$$

$$F(x) = \frac{3xY_0(x) - 2Y_1(x)}{3xJ_0(x) - 2J_1(x)}.$$

4 dec. $k=\frac{1}{4};\ \frac{1}{2};\ \frac{3}{4}$. $s=1(1)10$ [101]

$$10x\frac{J_1(x)}{J_0(x)} = \frac{J_1(3y/2)Y_1(y) - J_1(y)Y_1(3y/2)}{J_1(3y/2)Y_0(y) - J_0(y)Y_1(3y/2)}y$$

$y = [10x^2 + 0{,}09\pi^2 n^2]^{1/2}$.

2 dec. $n=1;\ 3;\ 5$. $s=1(1)4$ [309]

Series containing cylinder functions of the first and second kind of real argument

$$\sum_{s=1}^{\infty} e^{-j_{0,s}^2 x}$$

4 dec. $x = 0(0{,}02)\,0{,}1\,(0{,}1)\,1$ [81] [239]

$$\sum_{s=1}^{\infty} \frac{e^{-j_{0,s}^2 x}}{j_{0,s}^2}$$

4 dec. $x = 0(0{,}02)\,0{,}1\,(0{,}1)\,1$ [81] [239]

$$C(x) = \sum_{s=1}^{\infty} \frac{2e^{-j_{0,s}^2 x}}{j_{0,s} J_1(j_{0,s})} \quad \& \quad \lg C(x)$$

5 dec. $x = 0(0{,}001)\,0{,}4$ [147]

$$\sum_{s=1}^{\infty} \frac{J_0(j_{0,s} r)}{j_{0,s}^p J_1(j_{0,s})} e^{-j_{0,s} x}$$

4 dec. $\left.\begin{cases} p = 1. & x = 0\,(0{,}05)\,1{,}75 \\ p = 2. & x = 0\,(0{,}05)\,1{,}5 \\ p = 3;\ 4. & x = 0\,(0{,}05)\,1 \end{cases}\right\} r = 0(0{,}1)0{,}9$ [157]

$$\sum_{s=1}^{\infty} \frac{e^{-j_{1,s}^2 x}}{j_{1,s}^2} \quad \& \quad \sum_{s=1}^{\infty} \frac{e^{-j_{1,s}^2 x}}{j_{1,s}^2 J_0(j_{1,s})}$$

5 dec. $x = 0\,(0{,}005)\,0{,}01\,(0{,}01)\,0{,}1\,(0{,}02)\,0{,}2\,(0{,}05)\,0{,}5;\ 1$ [280]

$$\sum_{s=3}^{\infty} (j_{n,s})^{-2p}$$

5 fig. $n = 0;1.\quad p = 2\,(1)\,10$ [247]

Zeros of cylinder functions of the imaginary argument and expressions containing them

Zeros $x_{\nu,s}$ of the function $I_{-\nu}(x)$

3 dec. $\nu = 1\frac{1}{4};\ 1\frac{1}{2};\ 1\frac{3}{4}.\ s = 1$ $[41]_{2,3}$

Zeros $x_{\nu,s}$ of $\frac{dI_{-\nu}(x)}{dx}$

3 dec. $\nu = -1/3;\ -2/3.$ $s = 1$ [104]
2 dec. $\nu = -1/3;\ -2/3;\ -1/4;\ -3/4.$ $s = 1$ [15]

Zeros $x_{n,s}$ of $J_{n+1}(x)\,I_n(x) + J_n(x)\,I_{n+1}(x)$

4; 3 dec. $n = 0; 1\,(1)\,3.\ s = 1\,(1)\,10$ [246]
4 dec. $\frac{x_{n,s}}{\pi}$ / 3 dec. $x_{n,s}$ } $n = 2.\ s = 1\,(1)\,10$ [258]

3—4 dec. $n = 0;\ 1\ \&\ 2;\ 3.\ s = 1\,(1)\,5;\ 1\,(1)\,4;\ 1\,(1)\,3$ [219]
3 dec. $n = 0;\ 1\ \&\ 2;\ 3.\ s = 1\,(1)\,5;\ 1\,(1)\,4;\ 1\,(1)\,3$ [93]
2—3 dec. $n = 0;\ 1;\ 2.$ $s = 1\,(1)\,5;\ 1\,(1)\,3;\ 1\,(1)\,2$ [14]
[41]$_{1,2,3}$

Zeros $x_{n,s}$ of $J_{n+p}(2\sqrt{x})\,I_n(2\sqrt{x}) - J_n(2\sqrt{x})\,I_{n+p}(2\sqrt{x})$

3 dec. $p = 1.\ n = 0;\ 1;\ 2;\ 3.\ s = 1\,(1)\,4;\ 1\,(1)\,3;\ 1\,(1)\,4;\ 1\,(1)\,3$ } [206]
3 dec. $p = 2.\ n = 0;\ 1.$ $s = 1\,(1)\,3$ }

Zeros $x_{n,s}$ of $\frac{(m-1)\left[x\frac{dJ_n(x)}{dx} - n^2J_n(x)\right] - x^2J_n(x)}{(m-1)\left[x\frac{dI_n(x)}{dx} - n^2I_n(x)\right] + x^2I_n(x)}$

4 dec. $m = 0{,}33\,(0{,}01)\,0{,}35.$ $n = 0.$ $s = 1\,(1)\,7$ [241]
3 dec. $m = \frac{1}{4}.$ $n = 0\ \&\ 1;\ 2\ \&\ 3.$ $s = 1\,(1)\,9;\ 1\,(1)\,10$ } [246]
3 dec. $m = \frac{1}{3}.$ $n = 0\ \&\ 1;\ 2\ \&\ 3.$ $s = 1\,(1)\,4;\ 1\,(1)\,5$ }
3 dec. $m = 0{,}33$ $\begin{cases} n = 0\ \&\ 1;\ 2\,(1)\,10. & s = 1\,(1)\,10;\ 1\,(1)\,11 \\ n = 11;\ 12. & s = 1\,(1)\,2;\ 1 \end{cases}$ [241]
3—2 dec. $m = 0{,}33.$ $n = 0\ \&\ 1;\ 2\,(1)\,6.$ $s = 1\,(1)\,10;\ 1\,(1)\,11$ [224]

Zeros of cylinder functions of complex argument and expressions containing them

$z_{n,s}$

where $Y_n(z_{n,s}) = 0.$

6 dec. $n = 0;\ 1.$ $s = 1;\ 2;\ 3$ } [63] [179]
5 dec. $n = 0;\ 1.$ $s = 1\,(1)\,15$ }

$z'_{1,s}$

where
$Y'_1(z'_{1,s}) = 0.$

9 dec.	$s = 1(1)3$	[63] [179]
5 dec.	$s = 1(1)15$	

$$\frac{d}{dz} H_1^{(2)}(z) = 0$$

$$z = x + iy = re^{i\theta}$$

$x = 0{,}5012.$	$y = 0{,}6435$	[14] $[41]_{1,2,3}$
$r = 0{,}8156.$	$\theta = 52°{,}085 = 0{,}9091 = 57^g{,}87$	[301]

Zeros $t_{n,s}$ of $J_n(it)J'_n(t) - J_n(t)J'_n(it)$

1—3 dec.	$n = 0;\ 1\ \&\ 2;\ 3;\ 4;\ 5(1)7$ $s = 0(1)4;\ 0(1)3;\ 0(1)2;\ 0\ \&\ 1;\ 0$	[152]

Zeros of Thomson functions

ber x, **bei** x, **ber′** x, **bei′** x, **ker** x, **kei** x, **ker′** x, **kei′** x

5 dec.	$s = 1(1)10$	[137]

Values of
functions of type $J_n(j_{m,s})$, $Y_n(y_{m,s})$, $Y_n(z_{m,s})$

$J_0(j_{1,s})$

16 dec.	$s = 1(1)50$	[4] [34] [45]
10 dec.	$s = 1(1)150$	[23] [140]
8 dec.	$s = 1(1)50$	[58]
7 dec.	$s = 1(1)40$	[213]
6 dec.	$s = 1(1)15$	[32]
5 dec.	$s = 1(1)60$	[37]
5 dec.	$s = 1(1)32$	[60]
5 dec.	$s = 1(1)9$	[87]
4 dec.	$s = 1(1)50$	[13] [14] $[41]_{1,2,3}$ [301]
4 fig.	$s = 1(1)40$	[21]

$J_1(j_{0,s})$

10 dec.	$s = 1(1)150$	[23] [140]
8 dec.	$s = 1(1)40$	[4] [34] [58] [139]
7 dec.	$s = 1(1)40$	[213]

6 dec.	$s=1(1)12$	[233]
5 dec.	$s=1(1)60$	[37]
4 fig.	$s=1(1)40$	[13] [14] [21] [41]$_{1,2,3}$
4 dec.	$s=1(1)10$	[32]

$$J_1(j_{2,s})$$

4 dec.	$s=1(1)4$	[48]

$$J_\nu(j'_{\nu,s})$$

6—9 fig.	$\nu=-\frac{39}{2}(1)\frac{45}{2}.\ 1\leqslant s\leqslant r.\ r=1(1)7$	[80]

$$J'_\nu(j_{\nu,s})$$

7—9 fig.	$\nu=-\frac{43}{2}(1)\frac{39}{2}.\ 1\leqslant s\leqslant r.\ r=1(1)8$	[80]

$$Y_0(y_{1,s}) \quad \& \quad Y_1(y_{0,s})$$

8 dec.	$s=1(1)50$	[23]

$$2\,[J_1^2(j_{0,s})]^{-1},\ 2\,[J_0^2(j'_{0,s})]^{-1},\ 2\,[(1-j'^{-2}_{1,s})J_1^2(j'_{1,s})]^{-1}$$

7 fig.	$s=1(1)40$	[6]

$$J_1(j'_{0,s}x)$$

7 dec.	$s=1(1)40.\ x=0(0{,}01)1$	[6]

$$J_0(j'_{0,s}x) \ \& \ J_1(j'_{1,s}x)$$

7 dec.	$s=1(1)40.\quad x=0(0{,}01)1$	[6]

$$J_0(j_{0,s}x)$$

7 dec.	$s=1(1)40.\quad x=0(0{,}01)1$	[6]
5 fig.	$s=1(1)10.\quad x=0(0{,}01)1$	[264]

$$\frac{2}{j_{2,s}}J_1(j_{2,s})$$

6 dec.	$s=1(1)5$	[190] [274] [296]
5 dec.	$s=1(1)10$	[59]

$$\frac{4}{j_{2,s}^2} J_1^2(j_{2,s})$$

6 dec.	$s = 1\,(1)\,5$	[190] [274] [296]
5 dec.	$s = 1\,(1)\,10$	[59]
5 dec.	$s = 1\,(1)\,9$	[50] [273]
5 dec.	$s = 1\,(1)\,5$	[87]

$$\lg|J_0(j_{1,s})|$$

9 dec.	$s = 1\,(1)\,150$	[140]

$$\lg|J_1(j_{0,s})|$$

9 dec.	$s = 1\,(1)\,150$	[140]
7 dec.	$s = 1\,(1)\,40$	[139]

$$\frac{2J_1(x_s)\,J_0(kx_s)}{x_s\{[J_0(kx_s)]^2 - [J_1(x_s)]^2\}}$$

where x_s are the first six roots of the equation $J_0(kx)\,Y_1(x) - Y_0(kx)\,J_1(x) = 0.$

4 dec.	$k = (1{,}1)^{\frac{1}{2}}$; 1,15 (0,05) 1,3 (0,1) 1,5; 2 (1) 5	[304]

$$J_n(x_{n,1}),\quad Y_n(x_{n,1}),\quad J_n(kx_{n,1}),\quad Y_n(kx_{n,1}),$$

$$\frac{2}{\pi^2 x_{n,1}^2}\left\{\frac{J_n^{2'}(kx_{n,1})}{J_n^2(x_{n,1})} - 1\right\}$$

where $x_{n,1}$ is the first root of the equation $Y_n(kx)\,J_n(x) - J_n(kx)\,Y_n(x) = 0.$

7—9 dec.	$\begin{cases} k = 0{,}2;\ 0{,}3\,(0{,}1)\,0{,}5;\quad 0{,}6;\ 0{,}8 \\ n = 1\,(1)\,5;\ 1\,(1)\,6;\ 1\,(1)\,8;\ 1\,(1)\,12 \end{cases}$	[307]

$$J_{\pm\left(n+\frac{1}{2}\right)}(x_{n,1}),\quad J_{\pm\left(n+\frac{1}{2}\right)}(kx_{n,1}),$$

$$\frac{2}{(\pi x_{n,1})^2}\left\{\frac{J^2_{n+\frac{1}{2}}(kx_{n,1})}{J^2_{n+\frac{1}{2}}(x_{n,1})} - 1\right\}$$

where $x_{n,1}$ is the first root of the equation

$$J_{-\left(n+\frac{1}{2}\right)}(kx)\,J'_{n+\frac{1}{2}}(x) - J_{n+\frac{1}{2}}(kx)\,J_{-\left(n+\frac{1}{2}\right)}(x) = 0.$$

7 dec.	$\begin{cases} k = 0{,}2;\ 0{,}3;\ 0{,}4(0{,}2)0{,}8 \\ n = 1(1)5;\ 1(1)\,6;\ 1\,(1)\,15 \end{cases}$	[306]

$$Y_0(z_{1,s}),\quad Y_1'(z_{0,s}),\quad Y_1(z_{0,s}),\quad Y_1'(z_{1,s})$$

9 dec. $s = 1(1)3$ [63] [179]

$$Y_0(z_{1,s}'),\quad Y_1(z_{1,s}')$$

9 dec. $s = 1(1)4$ [63] [179]

$$A_{m,n}(z_{m,n}),\ B_{m,n}(z_{m,n}),\ C_{m,n}(z_{m,n})$$

where $z_{m,n}$ are roots of $J_m(iz)J_m'(z) - J_m(z)J_m'(iz) = 0$.

$$A_{m,n}(z_{m,n}) = l_m J_m^2(z_{m,n})J_m^2(iz_{m,n})\, i^{2m}$$

4 fig. $l_m = 1$. $m = 1; 2; 3(1)5$. $n = 0(1)3$; 0 & 1; 0 [152]

$$A_{0,n}(z_{m,n})$$

4 fig. $l_0 = 2$. $n = 0(1)3$ [152]

$$B_{m,n}(z_{m,n}) = \frac{2i^m J_m(iz_{m,n})J_{m-1}(z_{m,n})}{z_{m,n}}$$

4 dec. $m = 0$ & 1; 2; $3(1)5$. $n = 0(1)3$; 0 & 1; 0 [152]

$$C_{m,n}(z_{m,n}) = \frac{J_m(z_{m,n})}{z_{m,n}J_{m-1}(z_{m,n})}$$

4 dec. $m = 0; 1$. $n = 0(1)3$ [152]

$$2mC_{m,n}(z_{m,n})$$

4 dec. $m = 0; 1; 3(1)5$. $n = 0(1)3$; 0 & 1; 0 [152]

INTEGRALS OF CYLINDER FUNCTIONS

$$\int_0^x J_n(t)\,dt$$

10 dec. $n = 0$. $x = 0(0{,}01)10$ [162]

7 dec. $\frac{1}{2}\int_0^x J_0(t)\,dt$ $x = 0{,}02(0{,}02)1$ [2][86]

5 dec. $\frac{1}{2}\int_0^x J_0(t)\,dt$ $x = 0{,}02(0{,}02)1$ [9][13]

5 dec.	$n = 0(1)8.$	$x = 0(0{,}01)10$	[311]
4 dec.	$n = 0.$	$x = 0(0{,}02)1$	[3]
4 dec.	$n = 0.$	$x = 0(0{,}2)15(1)24;\ \infty$	[294]

In /2/ and /86/ values of the first 16 extrema of the function $\frac{1}{2}\int_0^x J_0(t)\,dt$ are given to 7 decimals. In /13/ these values are given to 5 decimals.

$$\mathrm{Ji}_0(x) = \int_\infty^x \frac{J_0(t)}{t}\,dt \quad \& \quad F(x) = \mathrm{Ji}_0(x) + \ln\frac{1}{2}x$$

10 dec.	$\mathrm{Ji}_0(x)$	$x = 0(0{,}1)10(1)22$	[163]
12 dec.	$F(x)$	$x = 0(0{,}1)3$	[163]

$$F^{(n)}(x)\,/\,n!$$

12 dec. $n = 0(1)13.$ $x = 10(1)21$ [163]

$$2^{-n}\,\mathrm{Ji}_{0,n}(x) = \underbrace{\int_0^x dt \cdots \int_0^x}_{n} J_0(t)\,dt$$

8 dec. $n = 1(1)7.$ $x = 0(1)24$ [169]

$$\int_0^\infty \frac{J_1(t)}{t}\,dt$$

4 dec. $x = 0(0{,}1)14;\ 15$ [92]

$$\int_0^\infty \frac{J_0(nr)}{n+k}\,dn = kr - \frac{(kr)^3}{1^2\cdot 3^2} + \frac{(kr)^5}{1^2\cdot 3^2\cdot 5^2} - \cdots + \frac{\pi}{2}K_0(kr)$$

$$\frac{1}{k}\int_0^\infty \frac{nJ_1(nr)}{n+k}\,dn = \frac{\pi}{2}K_1(kr) - \left\{1 - \frac{(kr)^2}{1^2\cdot 3} + \frac{(kr)^4}{1^2\cdot 3^2\cdot 5} - \cdots\right\}$$

4 dec. $kr = 0(0{,}05)0{,}1(0{,}1)10$ [255]

$$n\int_0^1 J_n(nx)\,dx$$

7 dec. $n = 1(2)23$ [211]

$$\int_0^\infty t^n J_n(t)\,dt$$

4 dec. $n = 1(1)4.$ $x = 0(0{,}1)10$ [171]

$$\int_0^1 x^p J_0^m(j_{0,s}x) J_1^n(j_{0,s}x)\,dx$$

$J_0(j_{0,s}) = 0.$

5 dec. $p = 0(1)10.$ $s = 1(1)10 \begin{cases} m = 1;\ 0;\ 2;\ 0;\ 1 \\ n = 0;\ 1;\ 0;\ 2;\ 1 \end{cases}$ [312]

$$\int_0^1 x^p J_m(j_{0,r}x) J_n(j_{0,s}x)\,dx$$

5 dec. $p = 0(1)5.$ $r = s = 1(1)5 \begin{cases} m = 0;\ 0;\ 1 \\ n = 0;\ 1;\ 1 \end{cases}$ [312]

$$\Phi_n(t) = \int_0^\infty J_0[2\sqrt{ut}]\,J_n(u)\,du \ \&\ \Phi_n'(t)$$

5 dec. $n = 0(1)7.$ $t = 0(1)24$ [169]

$$\psi_n(t) = \Phi_{2n}(t) + \Phi_{2n+2}(t)$$

5 dec. $n = 0(1)2.$ $t = 0(1)24$ [169]

$$\int_0^x Y_0(t)\,dt$$

10 dec. $x = 0(0{,}01)10$ [162]

7 dec. $\frac{1}{2}\int_0^x Y_0(t)\,dt$ $x = 0{,}02\,(0{,}02)\,1$ [2] [86]

5 dec. $\frac{1}{2}\int_0^x Y_0(t)\,dt$ $x = 0{,}02\,(0{,}02)\,1$ [9][13]

4 dec. $x = 0(0{,}02)1$ [3]

In /2/ and /86/ values of the first 16 extrema of the function $\frac{1}{2}\int_0^x Y_0(t)\,dt$ are given to 7 decimals. In /13/ these values are given to 5 decimals.

$$\int_0^\infty \frac{e^{-tx^2}}{J_0^2(x) + Y_0^2(x)}\,\frac{dx}{x}$$

3 dec. $t = 0(0{,}01)1(0{,}1)10(1)100(10)1000$ [251]

$$\int_0^x I_0(t)\,dt$$

8 dec.	$x = 0(0{,}1)20(1)25$	[265]
7 dec.	$x = 0(0{,}1)6$	[2]
7 dec.	$x = 0(0{,}1)16$	[16]
6 fig.	$x = 0(0{,}1)5{,}2(0{,}2)16$	[292]

$$e^{-x}\int_0^x I_0(t)\,dt$$

9 fig.	$x = 15(1)25$	[265]
7 dec.	$x = 0(0{,}1)16$	[2][16]
6 fig.	$x = 0(0{,}1)5{,}2(0{,}2)16$	[292]

$$\frac{1}{\pi}\int_0^\infty \frac{t^m dt}{I_n^2(t)}$$

7 dec.	$m = 2;\ 4.\qquad n = 0;\ 1$	[261]

$$T(m, n, r) = 2r^{n-m+1}e^{-r^2}\int_0^\infty t^{m-n}e^{-t^2}I_n(2rt)\,dt$$

5 dec.	$m = -\frac{1}{2}\left(\frac{1}{2}\right)1.n = -2\left(\frac{1}{2}\right)2.r = 0(0{,}2)4;5;6;10;25;50$	[286]
3 dec.	$\left\{\begin{matrix} 2T(2, 0, x) \\ \frac{4x}{\sqrt{\pi}}\,T(2, 0, x)\end{matrix}\right\} x = 0\,(0{,}1)\,1\,(0{,}2)\,3$	[146]

$$\int_0^x K_0(t)\,dt$$

7 fig.	$x = 0{,}02;\ 0{,}1;\ 0{,}5;\ 1\,(1)\,12$	[218]
6 fig.	$x = 0(0{,}02)0{,}04;\ 0{,}1(0{,}1)2{,}6(0{,}2)13{,}6(0{,}4)16$	[292]
6 dec.	$x = 0(0{,}1)4(0{,}5)18$	[117]

$$\int_x^\infty K_0(t)\,dt$$

7—8 fig.	$x = 0(0{,}1)12$	[2][16]

$$e^x\int_x^\infty K_0(t)\,dt$$

7 dec.	$x = 0(0{,}1)16$	[2][16]

$$\int_0^x tK_1(t)\,dt = \int_0^x K_0(t)\,dt - xK_0(x)$$

6 dec. $x = 0(0,1)4(0,5)18$ [117]
5 dec. $x = 0(0,1)4,5;\ \infty$ [189]

$$\mathrm{Ki}_n(x) = \int_x^\infty \mathrm{Ki}_{n-1}(t)dt = \int_x^\infty \frac{(t-x)^{n-1}}{(n-1)!} K_0(t)\,dt$$

9 dec. $n = 1(1)16.$ $x = 0(0,05)0,2(0,1)2;\ 3$ [223]
7—8 fig. $n = 1.$ $x = 0(0,1)12$ [16]
4 dec. $n = 2.$ $x = 0,25(0,01)1(0,05)7$ [192]

$$\int_0^x \frac{\left(\frac{1}{2}t\right)^n K_n(t)}{\sqrt{\pi}\left(n-\frac{1}{2}\right)!}\,dt$$

6 dec. $\left\{\begin{array}{ll} n = 0\left(\frac{1}{2}\right)5\frac{1}{2}. & x = 0(0,1)4(0,5)18 \\ n = 6\frac{1}{2}\left(\frac{1}{2}\right)11\frac{1}{2}. & x = 0(0,5)18 \end{array}\right\}$ [117]

$$\int_0^x e^t K_0(t)dt$$

6 dec. $\sqrt{x} = 0,01(0,01)0,1$
4 dec. $x = 0,02(0,01)0,1(0,1)3(0,2)6(0,5)10(1)20$ [228]
3 dec. $\sqrt{x} = 5(1)10$

$$\int_{-\infty}^x 2e^{-y} K_0(2e^{-\frac{1}{2}y})\,dy$$

5—8 dec. $x = -3,2(0,1)10,6$ [118]

$$I_{2n} = \int_0^\infty \frac{K_0(m)m^{2n}}{I_0(m)}\,dm$$

5 fig. $n = 0(1)6$ [263]

$$\left|\int_0^x e^{\frac{1}{2}\pi t^2} J_0(\alpha'\sqrt{t})\,dt\right|^2$$

4 dec. $\alpha' \frac{10\sqrt{10}}{2\pi} = 1;\ 10;\ 20;\ 30.$ $x = 0(0,1)3$ [210]

$$\mathrm{Jc}(\lambda, x) = \int_0^\infty J_0(\lambda t)\cos t\,dt$$

6; 8 dec. $\lambda = 0{,}1\,(0{,}1)\,1$. $x = 0\,(0{,}02)\,2;\ 0\,(0{,}1)\,5$ [175]

$$\mathrm{Js}(\lambda, x) = \int_0^\infty J_0(\lambda t)\sin t\,dt$$

6; 8 dec. $\lambda = 0{,}1\,(0{,}1)\,1$. $x = 0\,(0{,}02)\,2;\ 0\,(0{,}1)\,5$ [175]

$$\mathrm{Yc}(\lambda, x) = \int_0^\infty Y_0(\lambda t)\cos t\,dt$$

6 dec. $\lambda = 0{,}1\,(0{,}1)\,1$. $x = 0\,(0{,}02)\,2\,(0{,}1)\,5$ [175]

$$\mathrm{Ys}(\lambda, x) = \int_0^\infty Y_0(\lambda t)\sin t\,dt$$

6 dec. $\lambda = 0{,}1\,(0{,}1)\,1$. $x = 0\,(0{,}02)\,2\,(0{,}1)\,5$ [175]

$$\mathrm{Cc}(\lambda, x) = \mathrm{Yc}(\lambda, x) - \frac{2}{\pi}\ln\lambda x\,\mathrm{Jc}(\lambda, x)$$

6 dec. $\lambda = 0\,(0{,}1)\,1$. $x = 0\,(0{,}02)\,X$. $X/_{\lambda=0} = 2$ to $X/_{\lambda=1} = 0{,}2$ [175]

$$\mathrm{Cs}(\lambda, x) = \mathrm{Ys}(\lambda, x) - \frac{2}{\pi}\ln\lambda x\,\mathrm{Js}(\lambda, x)$$

6 dec. $\lambda = 0\,(0{,}1)\,1$. $x = 0\,(0{,}02)\,X$. $X/_{\lambda=0} = 2$ to $X/_{\lambda=1} = 0{,}2$ [175]

$$F_1(H) = \int_0^H J_1(x)\frac{\cos x}{x}\,dx \quad \& \quad F_2(H) = \int_0^H J_2(x)\frac{\sin x}{x}\,dx$$

6 dec. $H = 0\,(0{,}1)\,3$ [228]

$$H_{e,a} = 2F_2(H)\cos H \quad \& \quad H_{e,s} = 2F_1(H)\sin H$$

4 dec. $H = 0{,}1\,(0{,}1)\,3$ [228]

$$R_a \approx 64\,|H_{e,a}|^2, \quad R_s \approx 80\,|H_{e,s}|^2 \quad \& \; R_a + R_s$$

4 dec. $H = 0{,}1\,(0{,}1)\,3$ [228]

$$I_p = \frac{1}{\pi}\int_0^{2x}[1 - I_0^2(t) - I_1^2(t)]\frac{dt}{\sqrt{4x^2 - t^2}}$$

4 dec. $x = 20\,(5)\,50\,(10)\,100\,(50)\,300\,(100)\,1000\,(500)\,3000$ [113]

Integrals of Struve functions

$$\overline{\mathbf{H}}_n = \int_0^x \mathbf{H}_n(t)\,dt$$

5—6 dec. $n = 0;\ 1.\quad x = 0(0{,}1)10$ [170]

$$\mathbf{D}_n = \int_0 t^n\,\mathbf{H}_n(t)\,dt$$

4 dec. $n = 0(1)4.\ x = 0(0{,}1)10$ [171]

$$\overline{\mathbf{L}}_n = \int_0^x \mathbf{L}_n(t)\,dt$$

6 dec. : 6 fig. $n = 0;\ 1.\quad x = 0(0{,}1)10$ [170]

$$\frac{1}{\pi}\int_x^\infty \frac{\mathbf{H}_1(2t)}{t^2}\,dt$$

4 dec. $x = 0(0{,}5)2(1)5(2)11$ & 15 [34]
4 dec. $x = 0(0{,}1)4(0{,}2)7(0{,}4)15$ [94]
4 dec. $x = 0(0{,}5)3(1)7(2)11$ & 15 [145]

$$\frac{1}{\pi}\int_0^x \frac{\mathbf{H}_1(2t)\,dt}{t^2}$$

4 dec. $x = 0(0{,}1)15$ [114]

Integrals of Airy functions

$$\int_0^x \mathbf{Ai}(t)\,dt$$

7 dec. $x = 0(0{,}1)7{,}5$ [308]

$$\int_0^x \mathbf{Bi}(t)\,dt$$

7 dec. $x = 0(0{,}1)2$ [308]

$$\int_0^x \mathbf{Ai}(-t)\,dt \quad \& \quad \int_0^x \mathbf{Bi}(-t)\,dt$$

7 dec. $x = 0(0{,}1)10$ [308]

Chapter 11

CERTAIN SPECIAL FUNCTIONS AND INTEGRALS

The material for the chapter is arranged in the following sections:

polynomials;
Riemann zeta function;
Mathieu and Lamé functions;
hypergeometric and confluent hypergeometric functions;
certain integrals;
various functions.

Information on Legendre polynomials and function is given in Chapter 9.

POLYNOMIALS

Chebyshev polynomials

Chebyshev polynomials, which are used in many mathematical studies, can be defined by

$$C_n(x) = 2\cos n\theta,$$

$$S_n(x) = \frac{\sin(n+1)\theta}{\sin\theta},$$

$$x = 2\cos\theta.$$

These polynomials satisfy the recurrence relationship

$$C_{n+1} = xC_n - C_{n-1}.$$

The explicit expression of the first four polynomials is

$$\begin{array}{ll} C_0 = 2 & S_0 = 1 \\ C_1 = x & S_1 = x \\ C_2 = x^2 - 2 & S_2 = x^2 - 1 \\ C_3 = x^3 - 3x & S_3 = x^3 - 2x \\ C_4 = x^4 - 4x^2 + 2 & S_4 = x^4 - 3x^2 + 1 \end{array}$$

---

[1] See pages 503 and 550 for the references to Chapter 11.

Chebyshev polynomials used in a form somewhat different from that accepted here are written:

$$T_n(x) = \frac{1}{2} C_n(2x); \quad T_n^*(x) = \frac{1}{2} C_n(4x-2)$$
$$U_n(x) = S_n(2x); \quad U_n^*(x) = S_n(4x-2)$$

$$C_n(x) \,\&\, S_n(x)$$

12 dec. $x = 0(0{,}001)2.$ $n = 2(1)12$ [134]

The coefficients of the polynomials $C_n(x)$, orthogonal for the interval $(-1; +1)$ to n = 12 are given in /25/ and orthogonal for the interval (0;1) to n = 10 in /64/. The coefficients of the polynomials $S_n(x)$ orthogonal for the interval (-1; +1) to n = 12 are given in /25/.

$$y_n(x) = 2^n T_n\left(\frac{x}{2}\right) + 2^{n-2} T_{n-2}\left(\frac{x}{2}\right) + 2^{n-4} T_{n-4}\left(\frac{x}{2}\right) + \ldots$$

2—10 dec. $x = -2(0{,}1)0.$ $n = 1(1)10.$ [74]

Chebyshev-Hermite polynomials

Polynomials $H_n(x) = (-1)^n e^{x^2} \left(\frac{d}{dx}\right)^n e^{-x^2}$

8 dec. $x = 0(0{,}01)4.$ $n = 4$
6 dec. $x = 0(0{,}01)4.$ $n = 3$ [5]
4 dec. $x = 0(0{,}01)4.$ $n = 2$

$$e^{-\frac{1}{2}x^2} H_n(x) = (-1)^n e^{\frac{1}{2}x^2} \frac{d^n}{dx^n} e^{-x^2}$$

5 fig. $x = 0(0{,}04)1(0{,}1)4(0{,}2)7(0{,}5)8.$ $n = 0(1)11$ [62]

Polynomials $h_n(x) = (-1)^n e^{\frac{1}{2}x^2} \frac{d^n}{dx^n} e^{-\frac{1}{2}x^2}$

exact $x = 0(0{,}01)4.$ $n = 1(1)6$ [19]

Polynomials $\frac{h_n(x)}{\sqrt{(n+1)!}} = \frac{1}{\sqrt{(n+1)!}} (-1)^n e^{\frac{1}{2}x^2} \frac{d^n}{dx^n} e^{-\frac{1}{2}x^2}$

4 dec. $n+1 = 2(1)25$ [24]

x is selected so that

$$\frac{1}{\sqrt{2\pi}} \int_x^\infty e^{-\frac{1}{2}t^2} dt = 0{,}05(0{,}01)0{,}4.$$

Expressions containing Chebyshev-Hermite polynomials

$$r'(r, k) = r - \frac{r^3}{2\cdot 3} h_2(k) + \frac{r^5}{2\cdot 4\cdot 5} h_4(k) - \frac{r^7}{2\cdot 4\cdot 6\cdot 7} h_6(k) + \ldots$$

4; 3 dec. $r = 0{,}01\,(0{,}01)\,0{,}9;\ 0{,}91\,(0{,}01)\,1$ [24]

k is selected so that

$$\frac{1}{\sqrt{2\pi}} \int_k^\infty e^{-\frac{1}{2}t^2}\, dt = 0{,}09\,(0{,}01)\,0{,}25.$$

Coefficients of polynomials $h_n(x)$

$n = 1\,(1)\,8\,(2)\,16$ [24]

$n = 0\,(1)\,10$ [27]

Roots of polynomials $H_n(x)$

9–12 dec. $n = 1\,(1)\,10$ [50]

7 dec. $n = 2\,(1)\,9$ [42]

Roots of polynomials $h_n(x)$

6 dec. $n = 1\,(1)\,27$ [38]

Jacobian polynomials

Jacobian polynomials may be presented in the following form:

$$G_n(p, q, x) = \sum_{j=0}^{n} (-1)^j C_n^j \frac{(q+n-1)^{(j)}}{(p+2n-1)^{(j)}} x^{n-j}$$

7 dec. $x = 0\,(0{,}01)\,1.\ q = 0{,}1\,(0{,}1)\,1.\ p = 1{,}1\,(0{,}1)\,3.\ n = 1\,(1)\,5.$ [4]

Here all the roots are given with seven decimals and the coefficients of the polynomials with seven significant figures.

Mittag-Leffler polynomials

Mittag-Leffler polynomials $g_n(x)$ are defined by the following expression

$$(1+t)^x (1-t)^{-x} = \sum_0^\infty g_n(x)\, t^n, \quad |t| < 1$$

exact $x = 0\,(1)\,10 \quad n = 0\,(1)\,10.$ [97]

Laguerre polynomials

Polynomials $L_n(x) = e^x \frac{d^n}{dx^n}(x^n e^{-x})$

3—5 fig. $x = 0(0{,}01)0{,}1(0{,}1)18(0{,}2)20(0{,}5)21(1)26(2)30.$ $n = 0(1)5$ [34]

Polynomials $L_n^{(\alpha)}(x) = \sum_{m=0}^{n} \binom{n+\alpha}{n-m} \frac{(-x)^m}{m!}$

Four formulas given in /135/ fully describe the asymptotic behaviour of the polynomial $L_n^{(\alpha)}(x)$ for $n \to \infty$. . The formulas are established for the cases:

1. x varies around the origin.
2. $0 < x < \nu$.
3. x is close to ν.
4. $x > \nu$.

$\nu = 4n + 2\alpha + 2.$

A table of the polynomial $l_{10}(x) = e^{-\frac{1}{2}x} L_{10}^{(0)}(x)$. has been compiled for the verification of formulas.

4 dec. $\left\{\begin{array}{l} 1)\ x = 0(0{,}1)1 \\ 2)\ x = 0{,}5(0{,}5)3(1)6(2)36 \\ 3)\ x = 34(2)50 \\ 4)\ x = 46(2)52 \end{array}\right\}$ [135]

Roots of Laguerre polynomials $L_n(x)$

12 dec. $x_i^{(n)}$ $n = 1(1)15$ [51]

Weight factors (Christoffel numbers) of Laguerre polynomials

$$\alpha_i^{(n)} = \frac{1}{L_n'(x_i^{(n)})} \int_0^\infty \frac{e^{-x} L_n(x)}{x - x_i^{(n)}} dx = \frac{1}{x_i^{(n)}} \left[\frac{n!}{L_n'(x_i^{(n)})}\right]^2$$

12 fig. $n = 1(1)15$ [51]

Products of exponential and weight factors

$$\alpha_i^{(n)} e^{x_i^{(n)}}$$

12 fig. $n = 1(1)15$ [51]

Neumann polynomials

Neumann polynomials $O_n(t)$ are defined as the coefficients of $\varepsilon_n J_n(z)$ in the expansion of $\frac{1}{t-z}$ into series of Bessel functions of integer order, i.e.

$$\frac{1}{t-z} = \sum_{n=0}^{\infty} \varepsilon_n J_n(z) O_n(t).$$

The function $O_n(t)$ has the form:

$$O_n(t) = \frac{1}{4} \sum_{m=0}^{\leqslant \frac{1}{2}n} \frac{n(n-m-1)!}{m!\left(\frac{1}{2}t\right)^{n-2m+1}}; \quad n \geqslant 1$$

and is a polynomial of $\frac{1}{t}$ of degree $n+1$.

$$O_0(t) = \frac{1}{t}$$

Coefficients of Neumann polynomials

exact	$n = 0(1)15$	[79]
	$n = 0(1)5$	[2] [33]

Schläfli's polynomials

Schläfli's polynomials are defined as

$$S_n(t) = \sum_{m=0}^{< \frac{1}{2}n} \frac{(n-m-1)!}{m!} \left(\frac{1}{2}t\right)^{-n+2m}; \quad n \geqslant 1.$$

$$S_0(t) = 0.$$

Schläfli's polynomials are connected with Neuman polynomials by

$$\frac{1}{2} n S_n(t) = t O_n(t) - \cos^2 \frac{1}{2} n\pi.$$

Coefficients of Schläfli's polynomials

exact	$n = 1(1)12$	[79]
	$n = 1(1)6$	[2] [33]

Bernoulli polynomials

Polynomials of the type

$V_n(x) = x^n - \frac{1}{2} n x^{n-1} + C_n^2 B_1 x^{n-2} - C_n^4 B_2 x^{n-4} + \ldots + V_n(0)$, where $V_{2n}(0) = (-1)^{n-1} B_n$, $V_{2n+1}(0) = 0$ & $V_1(0) = -\frac{1}{2}$, are calle Bernoulli polynomials.

Polynomials related to Bernoulli polynomials are sometimes used. They include

$$A_n(x) = \frac{1}{n} V_n(x),$$

$$S_{n-1}(x-1) = \frac{1}{n}[V_n(x) - V_n(0)] = A_n(x) - A_n(0).$$

$V_n(x) - V_n(0)$

10 dec.	$n = 2(1)5.\ x = 0(0{,}01)1$ $n = 6(1)8.\ x = 0(0{,}01)1$	[13]
6 dec.	$n = 1(1)6.\ x = -4(1)-1\left(\frac{1}{4}\right)2(1)5$	[26]

$V_n(x)$

Values of the polynomial coefficients are given.

exact	$n = 0(1)15$	[13]
exact	$n = 0(1)10$	[29]
exact	$n = 0(1)8$	[115]
exact	$n = 0(1)6$	[20]

$S_{n-1}(x)$

Values of the polynomial coefficients are given.

exact	$n-1 = 1(1)50$	[12]
exact	$n-1 = 1(1)25$	[13] [117]

$\frac{1}{n} V_n(x) = A_n(x)$

Values of the polynomial coefficients are given.

exact	$n = 1(1)8$	[26] [78] [114]

Values of the coefficients for the powers $\left(x - \frac{1}{2}\right)$ are given in the expressions of these polynomials.

exact	$n = 1(1)4$	[78]

$2^{n+1} S_n(x)$

Values of the coefficients for the powers $(2x+1)$ are given in the expressions of these polynomials.

exact	$n = 1(1)25$	[13] [117]
exact	$n = 1(1)11$	[75]

$$V_{2n}(x) - V_{2n}(0) \ \& \ \frac{V_{2n+1}(x) - V_{2n+1}(0)}{2x-1}$$

Values of the coefficients for the powers $(x - x^2)$ are given in the expressions of these polynomials.

exact	$n = 3(1)18$	[29]

Euler polynomials

Polynomials of type

$$U_n(x) = V_n(x) - 2^n V_n\left(\frac{1}{2}x\right) = \frac{1}{2}nx^{n-1} - C_n^2(2^2-1)B_1 x^{n-2} + \\ + C_n^4(2^4-1)B_2 x^{n-4} - \ldots + (-1)^n U_n(0),$$

where $U_{2n}(0) = (-1)^n (2^{2n}-1) B_n, U_{2n+1}(0) = 0 \ \& \ U_1(0) = \frac{1}{2}$, are called Euler polynomials.

Many polynomials are connected with Euler polynomials, for example:

$$U_n(x) = nA_n'(x) = \frac{1}{2}nE_{n-1}(x).$$

$$E_n(x)$$

exact	$x = 0(0{,}01)1$	$n = 2(1)5.$	[13]
10 dec.	$x = 0(0{,}01)1$	$n = 6(1)8.$	

$$U_n(x)$$

Values of the coefficients of the polynomials are given

exact	$n = 1(1)8$	[115]

$$\frac{1}{2}E_{n-1}(x)$$

Values of the coefficients of the polynomials are given

exact	$n = 1(1)8$	[78] [115]

Coefficients of the powers $(x - \frac{1}{2})$ are given in expressions of these polynomials.

exact	$n = 1(1)8$	[78]

Various orthogonal polynomials

Polynomials $\xi'_r(x)$

$\xi'_r(x)$ are polynomials of x of power r, which satisfy the equation

$$\sum_{x=1}^{n} \xi'_i(x)\xi'_j(x) = 0; \quad i \neq j.$$

$n = 3(1)75. \qquad r = 1(1)5$ [15]

$n = 3(1)26. \qquad r = 0(1)n-1$ [14]

Integrals of orthogonal polynomials

$$I_r = \int_0^{n-1} \xi'_r(x, n)\,dx \ \& \ J_r = \int_{-\frac{1}{2}}^{n-\frac{1}{2}} \xi'_r(x,n)\,dx$$

$n \leqslant 26. \qquad r \leqslant 14$ [14]

Riemman zeta function

The Riemann zeta function $\zeta(s)$, where $s = \sigma + it$ has been the subject of a great many studies.
When $\sigma > 1$ the Riemann zeta function may be defined by

$$\zeta(s) = \sum_1^{\infty} n^{-s}, \text{ or } \quad \zeta(s) = \frac{1}{\Gamma(s)} \int_0^{\infty} \frac{x^{s-1}\,dx}{e^x - 1}.$$

Note that when $\sigma > 0$

$$\zeta(s) = \lim_{n\to\infty} \left\{ \left(\sum_1^{n} n^{-s}\right) - \frac{n^{1-s}}{1-s} \right\}.$$

When $\sigma < 0$, $\zeta(s)$ may be found from Riemann's functional equation

$$\cos\frac{1}{2}\pi s\Gamma(s)\zeta(s) = 2^{s-1}\pi^s\zeta(1-s).$$

The function is zero when s is an even negative integer, and all its non-trivial zeros lie in the critical strip $0 < \sigma < 1$.
Riemann's hypothesis that all the zeros of $\zeta(s)$ in the strip $0 < \sigma < 1$ have a real part equal to $1/2$, i.e. lie on the line $\sigma = \frac{1}{2}$ has never been proved or

disproved. See Chapter 13. for details of certain numerical series connected with $\zeta(s)$ investigated by Euler.

The majority of the tables given below are calculated for real values of s.

$\zeta(s)$

31 dec.	$s = \frac{1}{2}$	[84]
10 fig.; 10 dec.	$s = -24\,(0{,}1) - 16{,}5; -16{,}4\,(0{,}1)\,24$	[73]
6 dec.	$s = \frac{1}{m}$. $m = 1\,(1)\,20$	[96]
6 dec.	$s = 1 + \frac{1}{m}$. $m = 1\,(1)\,20$	[96]
4 fig.	$s = -23{,}9\,(0{,}1)\,23{,}9$	[18]
3—9 dec.	$s = 1\,(0{,}1)\,20{,}9$	[6]
3—4 dec.	$s = -10{,}9\,(0{,}1)\,10{,}9$	[7]

$\zeta(s) - 1$

4 fig.	$s = -23{,}9\,(0{,}1)\,23{,}9$	[18]

$(s-1)\zeta(s)$

10; 11 dec.	$s = -2\,(0{,}1)\,0;\ 0\,(0{,}1)\,4$	[73]

$\frac{\zeta'(s)}{\zeta(s)}$

7 dec.	$s = -6\,(0{,}1)\,7$	[36]

Generalised Riemann function $\zeta(s, a) = \sum_{n=0}^{\infty} (n+a)^{-s}$

10 dec.	$s = \frac{1}{2}$. $a = 1\,(0{,}01)\,2\,(0{,}02)\,5\,(0{,}05)\,10$	[123]
5—6 fig.	$s = 2\,(2)\,8$. $a = 3\left(\frac{1}{12}\right)4;\ 3\,(0{,}05)\,4$	[136]

Coefficients in the expansion of $(s-1)\zeta(s)$

This section describes tables of coefficients C_n of the expression $(s-1)\zeta(s) = 1 + \sum_{n=1}^{\infty} C_n (s-1)^n$

16 dec.	$n = 1\,(1)\,16$	[84]
9 dec.	$n = 1\,(1)\,9$	[56]

Coefficients of expressions connected with

$$\xi(t)=\frac{1}{2}s(s-1)\pi^{-\frac{1}{2}s}\Gamma\left(\frac{1}{2}s\right)\zeta(s)$$

Here $s=\frac{1}{2}+it$. The function $\xi(t)$ is real when t is real and is an even function of t.

$$\ln\xi(t)$$

This section describes coefficients for t^{2n} in the expansion of the function $\ln\xi(t)$.

22—30 dec.	$n=0(1)9$	[35] [83]
16 dec.	$n=0(1)6$	[84]

$$\frac{\xi(t)}{\xi(0)}$$

Coefficients for t^{2n} in the expansion of the function $\frac{\xi(t)}{\xi(0)}$ are given here.

16 dec.	$n=0(1)7$	[84]

Roots α_n of the function $\xi(t)$

$0<t\leqslant 200\cdot$	$n=1(1)79$	[82]
$0<t\leqslant 300\cdot$	$n=1(1)138$	[129]
$0<t\leqslant 390\cdot$	$n=1(1)195$	[105]
$0<t\leqslant 1468\cdot$	$n=1(1)1041$	[106]
$0<t<50\cdot$	$n=1(1)10$	[18] [35] [83]
$50<t<100\cdot$	$n=11(1)29$	[129]

MATHIEU AND LAME FUNCTIONS

Mathieu functions

Mathieu functions are periodic solutions of the differential equation

$$\frac{d^2y}{dx^2}+(a\pm 16q\cos 2x)\,y=0 \qquad (*)$$

or of the equation

$$\frac{d^2y}{dx^2} + (a \pm b\cos x)\,y = 0$$

with period π or 2 .

The notations $\theta = 8q$ and $\alpha = \frac{1}{4}a$ are sometimes used for convenience. Values of the parameter a for which such solutions exist are called characteristi numbers. There are infinitely many characteristic numbers $a_0(q), a_1(q), a_2(q), \ldots; b_0(q)\, b_1(q), b_2(q), .$for each value of q They give rise to periodic solutions $ce_0(x, q), ce_1(x, q), ce_2(x, q), \ldots;$ $se_1(x, q), se_2(x, q), \ldots,$ or $ce_0(x, \theta), \ldots,$ $se_1(x, \theta), \ldots,$ which when $q = 0$ become equal to

$$\cos x,\ \cos 2x, \ldots;\ \sin x,\ \sin 2x, \ldots$$

The solution of ce_0 is usually taken to be equal to 1 or $\frac{1}{\sqrt{2}}$.

The authors of [30] use the equation

$y'' + (b - c^2\cos^2 x)\,y = 0,$ which can

be reduced to equation(*), if one assumes that

$$c^2 = 32q = 4\theta \;\&\; b = a + 16q = a + 2\theta.$$

A Mathieu equation of type

$$\frac{d^2y}{dx^2} + (b - s\cos^2 x)\,y = 0. \qquad (**)$$

is investigated in /32/.

Characteristic numbers

The numbers $\alpha = \frac{1}{4}a$, $\beta = \frac{1}{4}b$. are sometimes tabulate instead of the characteristic numbers a and b.

The equation $\frac{d^2F}{d\theta^2} + (\alpha + \beta\cos\theta)F = 0$ is investigated in /111/. Assuming $F = y$, $\theta = 2x$ & $\beta = 4q$, this equation can be transformed to $y'' + (4a + 16q\cos 2x)\,y = 0$.

The characteristic numbers given below relate to the last two forms for recording the Mathieu equati

10 dec.	$a_{0(1)3}$.	$b_{1(1)4}$.	$q = 2; 4, 5; 8; 12{,}5; 18,$ т. е. $\sqrt{32q} = 8\,(4)\,24$	[100]
7 dec.	$a_{0(1)5}$.	$b_{1(1)6}$.	$\theta = 8q = 0\,(1)\,10\,(2)\,20\,(4)\,40$	[101]
7 dec.	$a_{1(2)7}$.	$b_{1(2)7}$.	$\theta = 8q = 0\,(0{,}1)\,2{,}3$	[76]
6 dec.	$a_{0(1)5}$.	$b_{1(1)6}$.	$q = 0\,(0{,}5)\,5$	[99]
5 dec.	$\alpha_0; \alpha_1$.	$\beta_1; \beta_2$.	$4q = 0\,(0{,}2)\,4\,(0{,}4)\,10\,(1)\,16\,(2)\,20$	[111]
	$\alpha_{2(1)5}$.	$\beta_{3(1)6}$.	$4q = 0\,(0{,}2)\,4\,(0{,}4)\,6\,(2)\,20$	

5 dec. $\alpha_{0(1)2}\cdot\ \beta_1;\ \beta_2.\quad q=0(0,1)1(0,2)4;\ 5(5)30$ (var.) 200 [130]

5 dec. $a_{0(1)2}\cdot\ b_1;\ b_2.\quad q=0(0,1)1$ [98]

5 fig. $a_{0(1)4}+2\theta$ & $b_{1(1)4}+2\theta$ for $c=2\sqrt{\theta}=0(0,2)4,\ 4;\ 0,1$ и $0,5(1)4,5$ [30]

The characteristic numbers $be_r(s)$ correspond to even periodic solutions of equation (**).

8 dec. [32]

$r=0.$	$s=0(0,2)20(0,5)37(1)100$
$r=1.$	$s=0(0,2)15(0,5)50(1)100$
$r=2.$	$s=0(0,2)40(0,5)77(1)100$
$r=3.$	$s=0(0,2)13(0,5)56(1)100$
$r=4(1)6.$	$s=0(0,5)100$
$r=7;\ 8.$	$s=0(1)100$
$r=9(1)15.$	$s=0(2)100$

The characteristic numbers $bo_r(s)$ correspond to odd periodic solutions of equation (**).

8 dec. [32]

$r=1(1)4.$	$s=0(0,5)100$
$r=5;\ 6.$	$s=0(0,1)100$
$r=7.$	$s=0(1)100$
$r=8(1)15.$	$s=0(2)100$

The functions $ce_{0(1)5}(x,q)$ & $se_{1(1)6}(x,q)$

5 dec. $\theta=8q=0(1)10.\quad x=0(1^\circ)90^\circ$ [101]

Coefficients for the expansion of Mathieu functions into Fourier series

7 dec. $ce_{0(1)5}.\quad se_{1(1)6}\quad \theta=8q=0(1)10(2)20(4)40$ [101]

5 dec. $ce_{0(1)2}.\quad se_1;\ se_2\quad q=0(0,1)1(0,2)4;\ 5(5)30$ (var.) 200 [130]

5 dec. $ce_{0(1)4}.\quad se_{1(1)4}\quad c=2\sqrt{\theta}=0(0,2)4,4;\ 0,1$ и $0,5(1)4,5$ [30]

For $ce_n(x,q)=\sum_k A_{k,n}\cos kx$ & $se_n(x,q)=\sum_k B_{k,n}\sin kx$, $A_{k,n}$ & $B_{k,n}$ are given with the following characteristics.

7 dec. $n=1(2)7.\quad k=1(2)17.\quad \theta=0(0,1)2,3$ [76]

Coefficients $kA_{k,n}$ & $kB_{k,n}$ of Fourier series for calculating the derivative with respect to x of Mathieu functions $ce_n(x,q)$ & $se_n(x,q)$.

7 dec. $n=1(2)7.\quad k=1(2)17.\quad \theta=0(0,1)2,3$ [76]

Coefficients $De_k(s)$ of the expansion of Mathieu functions into Fourier series:

$$Se_{2r}(s, x) = \frac{1}{A} ce_{2r}(x, q) = \sum_{k=0}^{\infty} De_{2k}(s) \cos 2kx$$

$$\& \; Se_{2r+1}(s, x) = \frac{1}{A} ce_{2r+1}(x, q) = \sum_{k=0}^{\infty} De_{2k+1}(s) \cos (2k+1)x$$

10fig. [32]

- $r=0.\ k=0(1)13.\ s=0(0{,}2)8(0{,}5)20(1)100;\ 0(0{,}5)20(1)40(2)100$
- $r=1.\ k=0(1)13.\ s=0(0{,}2)8(0{,}5)15(1)40(2)100; 0(0{,}2)20(1)40(2)100$
- $r=2.\ k=0(1)13.\ s=0(1)60(2)100;\ 0(1)100$
- $r=3.\ k=0(1)13.\ s=0(2)100$
- $r=4.\ k=0(1)13.\ s=0(4)68(2)100;\ 0(4)100$
- $r=5.\ k=0(1)14.\ s=0(5)100;\ 0(10)100$
- $r=6.\ k=0(1)14.\ s=0(10)100$
- $r=7.\ k=0(1)15.\ s=0(10)100$

Coefficients $Do_k(s)$ for the expansion of Mathieu functions into Fourier series:

$$So_{2r}(x, s) = \frac{1}{B} se_{2r}(x, q) = \sum_{k=1}^{\infty} Do_{2k}(s) \sin 2kx \quad \&$$

$$So_{2r+1}(x, s) = \frac{1}{B} se_{2r+1}(x, q) = \sum_{k=0}^{\infty} Do_{2k+1}(s) \sin (2k+1)x.$$

10fig. [32]

- $r=0.\ k=0(1)12.\quad s=0(0{,}5)10(1)40(2)100$
- $r=1.\ k=0(1)13.\quad s=0(1)40(2)100; 0(1)20(2)68(4)100$
- $r=2.\ k=0(1)12.\quad s=0(1)14(2)40(4)100;\ 0(2)60(4)100$
- $r=3.\ k=0(1)12.\quad s=0(2)100;\ 0(4)100$
- $r=4.\ k=0(1)12.\quad s=0(4)100$
- $r=5.\ k=0(1)14.\quad s=0(5)100;\ 0(10)100$
- $r=6.\ k=0(1)14.\quad s=0(10)100$
- $r=7.\ k=0(1)15.\quad s=0(10)100$

Roots of Mathieu functions $ce_n(x,q)$ & $se_n(x,q)$

$0^\circ{,}0001 \quad \left\{\begin{array}{ll} ce_n(x). & n=2(1)5 \\ se_n(x) & n=3(1)6 \end{array}\right\} \; 0 = 8q = 0(1)10(2)20(4)40$ [101]

Associated factors for calculating modified Mathieu functions[1]

Associated factors $s^{\frac{1}{2}R} g_{e,R}(s)$

$R = 2r$ or $2r+1$

[1] Notations taken from /32/.

7—8 fig.	$R=0.$	$s=0\,(0{,}2)\,4\,(0{,}5)\,10\,(1)\,34\,(2)\,100$	[32]
	$R=1.$	$s=0\,(0{,}5)\,20\,(1)\,40\,(2)\,100$	
	$R=2;\ 3.$	$s=0\,(0{,}5)\,15\,(1)\,50\,(2)\,100$	
	$R=4;\ 5.$	$s=0\,(1)\,100$	
	$R=6\,(1)\,9.$	$s=0\,(2)\,100$	
	$R=10.$	$s=0\,(5)\,100$	
	$R=11\,(1)\,15.$	$s=0\,(10)\,100$	

Associated factors $s^{\frac{1}{2}R}g_{o,R}(s)$
$R=2r$ or $2r+1$

7—8 fig.	$R=1.$	$s=0\,(0{,}5)\,10\,(1)\,62\,(2)\,100$	[32]
	$R=2;\ 3;\ 4.$	$s=0\,(1)\,100$	
	$R=5\,(1)\,9.$	$s=0\,(2)\,100$	
	$R=10;\ 11.$	$s=0\,(5)\,100$	
	$R=12\,(1)\,15.$	$s=0\,(10)\,100$	

Associated factors $s^r f_{e,r}(s)$

7—9 dec.	$r=0.$	$s=0\,(0{,}2)\,59\,(0{,}5)\,100$	[32]
	$r=1.$	$s=0\,(0{,}2)\,59\,(0{,}5)\,100$	
	$r=2.$	$s=0\,(0{,}2)\,25\,(0{,}5)\,100$	
	$r=3.$	$s=0\,(0{,}2)\,20\,(0{,}5)\,35\,(1)\,100$	
	$r=4\,(1)\,6.$	$s=0\,(1)\,100$	
	$r=7;\ 8;\ 9.$	$s=0\,(2)\,100$	
	$r=10\,(1)\,15.$	$s=0\,(5)\,100$	

Associated factors $s^r f_{o,r}(s)$

7—9 dec.	$r=1.$	$s=0\,(0{,}2)\,55\,(0{,}5)\,100$	[32]
	$r=2.$	$s=0\,(0{,}2)\,25\,(0{,}5)\,100$	
	$r=3.$	$s=0(0{,}2)\,10(0{,}5)\,67\,(1)\,100$	
	$r=5.$	$s=0\,(1)\,100$	
	$r=6\,(1)\,9.$	$s=0\,(1)\,100$	
	$r=10\,(1)\,15.$	$s=0\,(5)\,100$	

Auxiliary functions for interpolation of associated factors

$$f_{e,o}+\frac{1}{\pi}\ln s$$

7—8 dec. $s=0\,(0{,}2)\,10$ [32]

$$s\left(f_{e,1}+\frac{1}{\pi}\ln s\right)$$

7 dec. $s=0\,(0{,}2)\,3$ [32]

$$s^2\left(f_{e,2} + \frac{1}{\pi}\ln s\right)$$

8 dec. $s = 0(0,2)4$ [32]

$$s\left(f_{o,1} - \frac{1}{\pi}\ln s\right)$$

8 dec. $s = 0(0,2)5$ [32]

$$s^2\left(f_{o,2} - \frac{1}{\pi}\ln s\right)$$

7 dec. $s = 0(0,2)4$ [32]

Lamé functions

Information is here given concerning tables of characteristic numbers for periodic solutions of the Lamé equation

$$\frac{d^2y}{du^2} = [n(n+1)k^2\mathrm{sn}^2u - h]\,y.$$

4 dec. $n = -\frac{1}{2};\ 0(1)25.$ $k^2 = 0{,}1;\ 0{,}5;\ 0{,}9$ [102]

4 dec. $n = \frac{1}{2}(1)\frac{9}{2}.$ $k^2 = 0{,}1;\ 0{,}5;\ 0{,}9$ [103]

HYPERGEOMETRIC & CONFLUENT HYPERGEOMETRIC FUNCTIONS

Hypergeometric functions

The hypergeometric function $y = F(\alpha, \beta, \gamma, x)$ defined by the series

$$F(\alpha, \beta, \gamma, x) = 1 + \frac{\alpha\beta}{1\cdot\gamma}x + \frac{\alpha(\alpha+1)\beta(\beta+1)}{1\cdot 2\gamma(\gamma+1)}x^2 + \frac{\alpha(\alpha+1)(\alpha+2)\beta(\beta+1)(\beta+2)}{1\cdot 2\cdot 3\gamma(\gamma+1)(\gamma+2)}x^3 + \cdots,$$

satisfies the Gauss differential equation

$$x(1-x)\frac{d^2y}{dx^2} + [\gamma - (\alpha+\beta+1)x]\frac{dy}{dx} - \alpha\beta y = 0.$$

Many of the functions previously encountered can be expressed by a hypergeometric function with specific values of α, β & γ.

The calculation of the hypergeometric function $F(\alpha, \beta, \gamma, x)$ presents great difficulties since it requires the compilation of a table with four arguments.

$$F\left(1; 1; \frac{1}{2}n + \frac{1}{2}; x^2\right)$$

$n = 3(1)25;\ 50;\ 100;\ 200;\ 400$ [47]

$$F\left(2; 2; \frac{1}{2}n + \frac{3}{2}; x^2\right)$$

$n = 3(1)25;\ 50;\ 100;\ 200;\ 400$ [47]

$$2F(-n; 1; n+1; x) - 1$$

6 dec. $n = 0(1)10.\quad x = -1(0{,}1)1$ [46]

$$\frac{2\pi}{u}\left\{(1+\sigma)^3 F\left(\frac{3}{2}; \frac{3}{2}; 1; \sigma^2\right) - 1\right\}$$

It is assumed that $\sigma = (u - \sqrt{u^2 - 1})^2$.

4 dec. $u = 1{,}4(0{,}01)3{,}1(0{,}1)5$ [93]

$$\ln F\left(\frac{1}{6}; \frac{5}{6}; 2; \sin^2 \frac{1}{2}x\right)$$

7 dec. $x = 0(1°)90°$ [80]
7 dec. $x = 90°(1°)180°$ [81]

$$\ln F\left(-\frac{1}{6}; \frac{7}{6}; 2; \sin^2 \frac{1}{2}x\right)$$

7 dec. $x = 0(1°)90°$ [80]
7 dec. $x = 90°(1°)180°$ [81]

Confluent hypergeometric functions

The confluent hypergeometric function
$_1F_1(\alpha, \gamma, x)$ or $M(\alpha, \gamma, x)$
is defined by the series

$$_1F_1(\alpha, \gamma, x) = M(\alpha, \gamma, x) = 1 + \frac{\alpha}{\gamma}\frac{x}{1!} + \frac{\alpha(\alpha+1)}{\gamma(\gamma+1)}\frac{x^2}{2!} + \frac{\alpha(\alpha+1)(\alpha+2)}{\gamma(\gamma+1)(\gamma+2)}\frac{x^3}{3!} + \cdots$$

and satisfies the differential equation

$$x\frac{d^2y}{dx^2} + (\gamma - x)\frac{dy}{dx} - \alpha y = 0.$$

This equation has two linearly independent solutions

$$M(\alpha, \gamma, x) \ \& \ x^{1-\gamma} M(\alpha-\gamma+1; \quad 2-\gamma; x).$$

Various functions (exponential integral, probability integral, Hermite polynomials, etc.) can be expressed by means of confluent hypergeometric functions.

$M(\alpha, \gamma, x)$

exact	$\alpha = -3(0,5)4.\ \gamma = 1(1)7.\ x = 1(1)6(2)10$	[85]
11 dec.	$\alpha = -1(0,2)1.\ \gamma = 0,2(0,2)1.\ x = 0,1;\ 0,2(0,2)1$	[113]
8 fig.	$\alpha = -1(0,1)1.\ \gamma = 0,1(0,1)1.\ x = 1(1)10$	[95]
7 fig.	$\alpha = \frac{1}{2}n.\ n = 3(2)201.\ \gamma = \frac{1}{2}.\ x = 0(0,01)0,1$	[31]
7 fig.	$\alpha = 1(1)40.\ \gamma = 3.\ x = 0,02(0,02)0,1(0,1)1(1)10(10)50;\ 100;\ 200$	[128]
7 fig.	$\alpha = 1(1)50.\ \gamma = 4.\ x = 0,02(0,02)0,1(0,1)1(1)10(10)50;\ 100;\ 200$	[128]
6 fig.	$\alpha = 0,001;\ 0,01;\ 0,05;\ 0,1;\ 0,2;\ 0,25;\ 0,3(0,1)1.\ \gamma = \frac{1}{2}\left(\frac{1}{2}\right)2.\ x = 0,5(0,5)8$	[67]
5; 6 dec.	$\alpha = -4\left(\frac{1}{2}\right)4.\ \gamma = \pm\frac{1}{2};\ \pm\frac{3}{2}.\ x = 0(0,02)0,08;\ 0,15(0,1)0,95$	[49]
5 dec.; 6—7 fig.	$\alpha = -4\left(\frac{1}{2}\right)4.\ \gamma = \pm\frac{1}{2};\ \pm\frac{3}{2}.\ x = 0(0,1)1(0,2)3(0,5)8$	[48]
5 dec.; 6—7 fig.	$\alpha = -4\left(\frac{1}{2}\right)4.\ \gamma = 1(1)4.\ x = 0(0,02)0,1(0,05)1(0,1)2(0,2)3(0,5)8$	[49]
4 fig.	$\alpha = -3\left(\frac{1}{2}\right)4.\ \gamma = 1(1)7.\ x = 1(1)6(2)10$	[85]
4 dec.	$\alpha = 1,3;\ 0,65;\ 0,325.\ \gamma = 3;\ 2;\ 1,5.\ x = 0,02(0,02)0,1(0,05)1(0,1)2,5$	[58]
4 dec.	$\alpha = 0,65;\ 0,5.\ \gamma = 2$; $\alpha = 1,65;\ 1,5.\ \gamma = 3$; $\alpha = -0,5;\ -0,35.\ \gamma = 1$; $x = kp^2.\ k = -2\left(\frac{1}{2}\right)2.\ p = 0(0,1)1$	[57]
	$\alpha = 0,325.\ \gamma = 1,5$; $\alpha = 1,325.\ \gamma = 2,5$; $\alpha = -0,675.\ \gamma = 0,5$; $\alpha = 0,825.\ \gamma = 1,5$; $x = kp^4.\ k = -2\left(\frac{1}{2}\right)2.\ p = 0(0,1)1$	[57]

$$e^{-\frac{\alpha x}{\gamma}} M(\alpha, \gamma, x)$$

4—5 dec. $\alpha = 0{,}001;\ 0{,}01;\ 0{,}05;\ 0{,}1;\ 0{,}2\,(0{,}5)\,0{,}3(0{,}1)1$; $\gamma = 0{,}5(0{,}5)2.\ x = 0{,}1\,(0{,}1)\,1\,(0{,}5)\,8$ [54]

$$\frac{1}{\operatorname{ch}\sqrt{(2n-1)\,x}}\, M\left(\frac{1}{2}n;\ \frac{1}{2};\ x\right)$$

6 dec. $x = 0\,(0{,}01)\,0{,}1.\quad n = 43\,(2)\,201$ [31]

$$\frac{1}{\sqrt{2nx}} \ln M\left(\frac{1}{2}n;\ \frac{1}{2};\ x\right)$$

6 dec. $x{=}0(0{,}01)0{,}6(0{,}05)2(0{,}2)7(1)45(5)100.\ n{=}3(2)21$ [31]
6 dec. $x{=}0(0{,}01)0{,}6(0{,}1)2(0{,}2)7(1)45(5)100.\ n{=}23(2)201$

$$H(m,\ \alpha,\ x) = e^{-ix} M(m + 1 - i\alpha;\ 2m + 2;\ 2ix)$$

7 fig. $x = 0\,(1)\,10.\ \alpha = 0\,(1)\,10.\ m = 0\,(1)\,3$ [65]

A table of the derivative $\frac{dH(m,\ \alpha,\ x)}{dx}$ with the same characteristics is given.

$$T(m,\ n,\ r) = \frac{r^{2n-m+1} e^{-r^2}\Gamma\left(\frac{1}{2}m + \frac{1}{2}\right) M\left(\frac{1}{2}m + \frac{1}{2};\ n + 1;\ r^2\right)}{\Gamma(n+1)}$$

$m = -\frac{1}{2}\left(\frac{1}{2}\right)1.\qquad n = -2\left(\frac{1}{2}\right)2.\qquad 2 \leqslant r \leqslant 5$ [131]

WHITTAKER FUNCTIONS

Whittaker functions, which are a solution of the equation

$$\frac{d^2W}{dz^2} + \left(-\frac{1}{4} + \frac{k}{z} - \frac{l(l+1)}{z^2}\right) W = 0,$$

can be expressed by means of the confluent hypergeometric function. Solutions which are regular for z = 0 are denoted by $M_{k,\,l+\frac{1}{2}}(z)$ and coincide with $W_{k,l+\frac{1}{2}}(z)$ for $z \to \infty$

$$W_{k,l+\frac{1}{2}}(z) = e^{-\frac{z}{2}} z^k u_{k,l}(z),$$

$$W'_{k,l+\frac{1}{2}}(z) = e^{-\frac{z}{2}} z^k v_{k,l}(z).$$

The values $M_{k,\frac{1}{2}}(x)$ & $M'_{k,\frac{1}{2}}(x)$ are given with the followi characteristics:

6 dec. $x = 0(0{,}1)5.$ $k = 0(0{,}05)1$ [119]

The values $u_k(x)$ & $v_k(x)$ are given with the following characteristics:

6 dec. $x = 5(0{,}2)25.$ $k = 0{,}05(0{,}05)0{,}95$ [119]

See Chapter 12 also for solutions of Whittaker equations.

CERTAIN INTEGRALS

$$\int_0^X x^x dx$$

6 fig. $X = 0(0{,}01)0{,}05(0{,}05)1{,}5(0{,}1)3(0{,}2)4$ [11]

$$\int_0^X x^{-x} dx$$

6 fig. $X = 0(0{,}01)0{,}05(0{,}05)1{,}5(0{,}1)3(0{,}2)4$ [11]

$$\int_{-1}^{1} \frac{t^{2n+1}}{t-y} dt$$

4 dec. $n = 0(1)2.$ $y = 0(0{,}1)0{,}9;\ 0{,}95$ [87]

$$\int_{-1}^{1} \frac{t^{2n+2}}{1-yt} dt$$

4 dec. $n = 0(1)2.$ $y = 0(0{,}1)0{,}9;\ 0{,}95$ [87]

$$\sqrt{2\pi}\int_0^y \frac{dx}{\ln(x\sqrt{2\pi})}$$

7 fig. $y = 0(0{,}005)0{,}5$ [11]

$$\int_0^\varphi \sin^{\frac{1}{3}} x\, dx$$

8 dec. $\varphi = 0(1°)90°$ [72]

$$\frac{4}{3}\sin^{-\frac{4}{3}}\varphi\int_0^{\varphi}\sin^{\frac{1}{3}}x\,dx$$

8 dec. $\varphi = 0(30')90°$
7 dec. $\varphi = 90°(30')180°$ [72]

$$\int_0^{\varphi}\sin^{2p}x\,dx$$

6 dec. $\varphi = 5°44'21''$(var.)$90°$. $p = 1(1)5$ [3]

$$\int_0^{\infty}\sin(x^n)\,dx$$

7 dec. $\pm n = 2(1)10$ [77]

$$\int_0^{1}x^k\sin n\pi x\,dx$$

10 dec. $k = 0(1)10$. $n = 0(1)100$ [66]

$$\int_0^{1}x^k\cos n\pi x\,dx$$

10 dec. $k = 0(1)10$. $n = 0(1)100$ [66]

$$\frac{1}{\pi}\int_0^{\pi}(1-b\cos x)^{-\frac{3}{2}}\,dx$$

6 dec. $b = 0(0{,}001)0{,}809$; $0{,}8(0{,}001)1$; $0{,}995(0{,}0001)1$ [60]

$$\frac{1}{\pi}(1-b)\int_0^{\pi}(1-b\cos x)^{-\frac{3}{2}}\,dx$$

6 dec. $b = 0{,}995(0{,}0001)1$; $0{,}8(0{,}001)1$ [60]

$$\frac{1}{\pi}\int_0^{\pi}(1-b\cos x)^{-\frac{3}{2}}\cos x\,dx$$

6 dec. $b = 0{,}995(0{,}0001)1$; $0(0{,}001)0{,}809$; $0{,}8(0{,}001)1$ [60]

$$\frac{1}{\pi}(1-b)\int_0^{\pi}(1-b\cos x)^{-\frac{3}{2}}\cos x\,dx$$

3 dec. $b = 0{,}995\,(0{,}0001)\,1;\ 0{,}8\,(0{,}001)\,1$ [60]

$$\frac{1}{\pi}\left[\int_0^{\pi}(1-b\cos x)^{-\frac{3}{2}}dx-\int_0^{\pi}(1-b\cos x)^{-\frac{3}{2}}\cos x\,dx\right]$$

6 dec. $b = 0{,}995\,(0{,}0001)\,1;\ 0{,}8\,(0{,}001)\,1$ [60]

$$\frac{1}{\pi}\int_0^{\infty}\sin\left(uz+\frac{1}{3}u^3\right)du$$

7 dec. $z = 0\,(0{,}1)\,10$ [121]

$$s(z)=\frac{1}{z}\int_0^1 \sin zx\,(1-x^2)^{\frac{1}{2}}\,dx$$

4—5 dec. $z = 0\,(0{,}5)\,5\,(1)\,10;\ 12$ [104]

$$\mathrm{cl}\,(x) = -\int_0^x \ln\left(2\sin\frac{1}{2}t\right)dt$$

16 dec. $x = 0\,(1°)\,180°$ [22] [71]

$$a_m=\frac{n}{\pi m!\,\sin^m\frac{\pi}{n}}\int_0^{\frac{\pi}{n}}\left(\cos x-\cos\frac{\pi}{n}\right)^m dx$$

5 dec. $m = 0\,(1)\,6.\ n = 3;\ 4;\ 6$ [59]

The same characteristics apply to tables of the functions:

$$a_m \operatorname{ctg}\frac{\pi}{n}+(m+1)\,a_{m+1}\quad \&$$

$$a_m \operatorname{ctg}^2\frac{\pi}{n}+2\,(m+1)\,a_{m+1}\operatorname{ctg}\frac{\pi}{n}+(m+1)\,(m+2)\,a_{m+2}.$$

$$I_m=\frac{\int_0^{\infty}x^m dx}{\operatorname{sh}2t+2t}$$

5fig. $m = 1\,(2)\,15$ [37]

$$J_m = \frac{\int_0^\infty x^m e^{-2x} dx}{\operatorname{sh} 2t + 2t}$$

1—5 fig. $m = 1\,(2)\,15$ [37]

$$K_m = \frac{\int_0^\infty x^m dx}{\operatorname{sh} 2t + 2t} + 2\,\frac{\int_0^\infty x^{m+1} dx}{\operatorname{sh} 2t + 2t} - \frac{\int_0^\infty x^m e^{-2x} dx}{\operatorname{sh} 2t + 2t}$$

5—6 fig. $m = 1\,(2)\,15$ [37]

$$\int_0^{\frac{\pi}{2}} (1 + k^{-1} \operatorname{cosec}^2 \varphi \lg \sec \varphi)^{-\frac{1}{2}} d\varphi$$

$k = 0{,}05\,(0{,}05)\,0{,}5\,(0{,}1)\,1{,}5$ (var.) 5 [118]

$$f_r(x) = (-1)^r \left(\int_x^\infty\right)^r \ln \operatorname{th} \frac{1}{2} z\,(dz)^r$$

4 dec. $\left\{ \begin{array}{c} r = 0;2;4.\ x = 0\,(0{,}01)\,0{,}1\,(0{,}02)\,0{,}3\,(0{,}05)\,0{,}5\,(0{,}1) \\ 1\,(0{,}2)\,3\,(0{,}5)\,5\,(1)\,10 \end{array} \right\}$ [107]

$$\int_0^\zeta \left[\frac{1 - \varepsilon^2 (1 - k^2 \sin^2 t)^2}{1 - k^2 \sin^2 t}\right]^{\frac{1}{2}} dt$$

3—4 dec. $\left\{ \begin{array}{l} k = \sin \frac{\varphi_0}{2};\ \varphi_0 = 10°;\ 45°;\ 90°; 135°; 170° \\ \zeta = 10°\,(10°)\,90° \\ \varepsilon^2 = 0{,}1;\ 0{,}25;\ 0{,}5;\ 0{,}75;\ 0{,}9;\ 1 \end{array} \right\}$ [70]

$$\int_0^\varphi \left[\frac{\sin t}{1 + \lambda \sin t}\right]^{\frac{1}{2}} dt$$

4 dec. $\lambda = 0\,(0{,}1)\,0{,}5.\quad \varphi = 0\,(2°)\,90°$ [68]

$$\int_0^\infty \int_0^\infty \frac{\sin(xu) \sin(yv)\, du\, dv}{uv\,[(u^2 + v^2)^2 + 1]}$$

3 dec. $x = y = 0\,(0{,}5)\,8\,(1)\,12;\ \infty$ [126]

$$-\frac{1}{15}\left(\frac{z}{2}\right)^4 \sin\frac{2}{z} + \frac{1}{15}\left(\frac{z}{2}\right)^3 \cos\frac{2}{z} - \frac{29}{30}\left(\frac{z}{2}\right)^2 \sin\frac{2}{z} - \frac{1}{6}\,\frac{z}{2} \cos\frac{z}{2} +$$

$$+ \frac{3}{4}\,\frac{z}{2} \int_{2/z}^\infty \frac{\cos y\, dy}{y} + \frac{1}{6} \int_{2/z}^\infty \frac{\sin y\, dy}{y} + \frac{32}{15}\left(\frac{z}{2}\right)^{3/2} \int_0^{\frac{2}{z}} \frac{\cos y\, dy}{\sqrt{y}}$$

3 dec. $z = 0\,(0{,}1)\,1\,(0{,}2)\,3\,(0{,}5)\,5$ }
2 dec. $z = 6\,(1)\,10$ } [44]

$$\int_{-\frac{\pi}{2}}^{\frac{\pi}{2}} e^{mx}\cos^r x\,dx$$

7 fig. $m = r\operatorname{tg}\alpha$. $\alpha = 0\,(5°)\,45°$. $r = 1\,(1)\,10\,(2)\,20\,(5)\,50$ [11]

$$\int_0^\infty e^{-(x-w^2)^2}\,dw$$

4 dec. $x = 0\,(-0{,}1)-3{,}2;\ 0\,(0{,}1)\,3\,(0{,}2)\,8$ [16]

$$\int_0^\infty f_n e^{-u\xi}\cos u\eta\,du \quad \& \quad \int_0^\infty f_n e^{-u\xi}\sin u\eta\,du$$

The following functions are denoted by f_n in these integrals:

$$f_1 = 1 - u\sqrt{2}\,[(1+u^4)^{\frac{1}{2}} - u^2]^{\frac{1}{2}}$$

$$\text{и} \quad f_2 = 2u^2 - u\sqrt{2}\,[(1+u^4)^{\frac{1}{2}} + u^2]^{\frac{1}{2}}.$$

3—4 dec. $n = 1;\,2.\ \xi,\ \eta = 0;\ 0{,}5;\ 1\,(1)\,5;\ 10$ [120]

$$s_n = \frac{1}{\Gamma(n+1)}\int_0^\infty \frac{dx}{1+e^{x^{1/n}}} \quad \& \quad \sigma_n = \frac{1}{2\Gamma(n+1)}\int_0^\infty \frac{dx}{\operatorname{sh} x^{1/n}}$$

16 dec. $n = 1\,(1)\,35$ [96]

$$J(\Omega) = -4\int_0^\infty x e^{-x^2}\cos\Omega x\,dx$$

Ω	0	1	1,57	1,85	2	2,5	3	4	5	6	10	$\to\infty$
$J(\Omega)$	—2	—1,05	—3,2	0	0,16	0,48	0,56	0,4	0,2	0,12	0,05	$\to\frac{4}{\Omega^2}$

[8]

$$\int_0^x \frac{-\lg|1-y|\,dy}{y}$$

9dec. $x = -1\,(0{,}01)\,1$ [90]

$$\int_0^\infty y^s \{\lg(1+y)\}^n e^{-ay^2}\, dy$$

6 dec. $a = 0{,}1\,(0{,}1)\,1{,}5$ $\left\{\begin{array}{ll} s = 0. & n = 1;\ 2 \\ s = 1. & n = 1\,(1)\,3 \end{array}\right\}$ [94]

$$L_n(u) = \int_1^u \frac{L_{n-1}(t)}{t-1}\, dt$$

The following relationships are used to define these functions:

$L_0(u) = \frac{u-1}{u}$; $L_1(u) = \ln u$; $L_2(u) = \int_1^u \frac{\ln t}{t-1}\, dt$; $L_3(u) = \int_1^u \frac{L_2(t)}{t-1}\, dt$ etc.

12 dec. $n = 2.$ $u = 1 + x.\ x = 0\,(0{,}01)\,0{,}5$ [22]
12 dec. $-L_2(1-x)$ $x = 0\,(0{,}01)\,0{,}5$ [22]
9 dec. $n = 3.$ $u = 1\,(1)\,100$ [28]
9 dec. $n = 2.$ $u = 2\,(1)\,100$ [28]
7 dec. $n = 2.$ $u = 0\,(0{,}01)\,2\,(0{,}02)\,6$ [88]

$$\frac{1}{x}\int_0^x \frac{y\, dy}{e^y - 1}$$

3 fig. $x = 0\,(0{,}2)\,2\,(0{,}5)\,3\,(1)\,10\,(2)\,16;\ 20$ [39] [127]

$$\frac{3}{x^3}\int_0^x \frac{y^3 dy}{e^y - 1}$$

6 fig. $x = 0\,(0{,}01)\,24$ [61]
5 dec. $x = 0\,(0{,}01)\,2\,(0{,}1)\,16$ [11]

$$\frac{12}{x^3}\int_0^x \frac{y^3 dy}{e^y - 1} - \frac{3x}{e^x - 1}$$

6 fig. $x = 0\,(0{,}01)\,24$ [61]
5 dec. $x = 0\,(0{,}01)\,2\,(0{,}1)\,16\,(1)\,30$ [11]

$$\lg\left[\frac{12}{x^3}\int_0^x \frac{y^3 dy}{e^y - 1} - \frac{3x}{e^x - 1}\right]$$

5 dec. $x = 0\,(0{,}01)\,2{,}09$ [53]

$$\int_x^0 \frac{1}{y}\left[\frac{12}{y^3}\int_0^x \frac{y^3dy}{e^y-1} - \frac{3y}{e^y-1}\right]dy$$

4—5 dec. $x = 0(0{,}01)2(0{,}1)15$ [11]

$$\frac{3}{x^3}\int_0^\infty \frac{y^3dy}{e^y-1} + \int_0^x\left[\frac{12}{y^4}\int_0^x \frac{y^3dy}{e^y-1} - \frac{3}{e^y-1}\right]dy$$

4—5 dec. $x = 0(0{,}01)2(0{,}1)15$ [11]

$$\int_{x_0}^\infty \frac{x^2\,dx}{e^x-1}$$

4—5 dec. $x_0 = 0;\ 0{,}2;\ 0{,}3;\ 0{,}6;\ 1(0{,}5)5(1)11$ [110] [132]

$$\int_{x_0}^\infty \frac{x^3\,dx}{e^x-1}$$

3—4 dec. $x_0 = 0;\ 0{,}2;\ 0{,}3;\ 0{,}6;\ 1(0{,}5)5(1)11$ [110] [132]

$$\int_{x_0}^\infty \frac{x^3}{e^x-1}\,dx - x_0\int_{x_0}^\infty \frac{x^2}{e^x-1}\,dx$$

3—4 dec. $x_0 = 0;\ 0{,}2;\ 0{,}3;\ 0{,}6;\ 1(0{,}5)5(1)11$ [132]

$$\int_x^\infty \frac{dy}{y^n(e^y-1)}$$

6 dec. $x = 0(0{,}1)10.$ $n = 0;\ \frac{1}{4};\ \frac{1}{2};\ \frac{3}{4};\ 1;\ 2$ [86]

$$\frac{15}{4\pi^4}\int_0^x \frac{y^7e^{-y}\,dy}{(1-e^{-y})^3}$$

5—4 dec. $x = 0(0{,}5)3(0{,}2)5(0{,}1)9$ (var.) ∞ [43]
2; 1 dec. $x = 0(0{,}1)5;\ 5{,}1(0{,}1)12(1)16;\ \infty$ [133]

$$\frac{1}{n!}\int_0^\infty x^n(e^{x-\eta}+1)^{-1}\,dx$$

7 dec. $n = 1(1)4.$ $\eta = -4(0{,}1)0$ [109]

$$\Phi_m(x) = \int_1^\infty \beta^m e^{-\beta x} d\beta$$

6—7 fig. $x = 1\,(0{,}1)\,6.\; m = -\frac{9}{2};\; -\frac{7}{2};\; -\frac{5}{2} \ldots;\; \frac{17}{2};\; -4\,(1)\,9$ [9]

$$g_n(y) = \int_0^\infty \frac{e^{-x} x^n}{(x+y)^{n+1}} dx$$

13—15 dec. $n = 0\,(1)\,14.$ $y = 1$ [52]
4—8 dec. $n = 0\,(1)\,6.$ $y = 0\,(0{,}01)\,0{,}1\,(0{,}05)\,0{,}45$

A table of values of $g_n(y) = 0.$ is introduced here.

$$\Upsilon_n(y) = \int_0^\infty \frac{e^{-x} y^n}{(x+y)^{n+1}} dx$$

13—15 dec. $n = 0\,(1)\,14.$ $y = 1$ [52]

$$\int_0^\infty \frac{e^{-x} x^{2n}}{(x^2+z^2)^{n+1}} dx$$

8 dec. $n = 0\,(1)\,10.$ $z = 1;\; 2$ [52]

$$\int_0^\infty \frac{e^{-x} z^{2n}}{(x^2+z^2)^{n+1}} dx$$

8 dec. $n = 0\,(1)\,10.$ $z = 1;\; 2$ [52]

$$\int_0^\infty \frac{e^{-x} x z^{2n}}{(x^2+z^2)^{n+1}} dx$$

8 dec. $n = 0\,(1)\,10.$ $z = 1;\; 2$ [52]

$$\int_0^\infty \frac{e^{-x} x^{2n+1}}{(x^2+z^2)^{n+1}} dx$$

8 dec. $n = 0\,(1)\,10.$ $z = 1;\; 2$ [52]

After-effect function $\Psi = \int_{-\infty}^{+\infty} e^{-y^2t^2-t-xe^{-t}}\,dt$

3 fig. $\left\{ y = 0\,(0{,}05)\,1.\ x = 0{,}001;\ 0{,}003;\ 0{,}008;\ 0{,}02;\ 0{,}05;\ 0{,}01;\ 0{,}2;\ 0{,}4;\ 0{,}7;\ 1;\ 1{,}5;\ 3;\ 5 \right\}$ [11]

$$\psi(x) = \int_{-\infty}^{\infty} \exp(t - e^t - e^{-t-x})\,dt$$

5 dec. $x = -3\,(0{,}5)\,10{,}5$ [41]

The reciprocal function of $\psi(x)$

2 dec. $\left\{ \psi(x) = 0{,}0002\,(0{,}0001)\,0{,}001\,(0{,}001)\,0{,}01\,(0{,}01)\,0{,}1\,(0{,}1)\,0{,}9\,(0{,}01)\,0{,}99\,(0{,}001)\,0{,}999\,(0{,}0001)\,0{,}9997 \right\}$ [41]

$$\varphi(x) = e^{-x} \int_{-\infty}^{\infty} \exp(-e^t - e^{-t-x})\,dt$$

5 dec. $x = -3\,(0{,}5)\,10{,}5$ [41]

$$\mathrm{Hi}(z) = \frac{1}{\pi} \int_0^{\infty} \exp\left(uz - \frac{1}{3}u^3\right) du$$

7 dec. $z = 0\,(0{,}1)\,10$ [121]

$$\int_x^{\infty} t^{1-n} \exp\left(-\frac{1}{4}t^2\right) dt$$

4 fig. $n = 2;\ 3.$ $x = 0\,(0{,}1)\,0{,}4\,(0{,}2)\,4\,(1)\,6$ [108]

$$\frac{1}{2} x^n \exp\left(\frac{1}{4}x^2\right) \int_x^{\infty} t^{1-n} \exp\left(-\frac{1}{4}t^2\right) dt$$

3–5 fig. $n = 2;\ 3.$ $x = 0\,(0{,}1)\,0{,}4\,(0{,}2)\,4\,(1)\,6$ [108]

$$\int_0^{\infty} y^3 \exp\left(-y^2 + \frac{ix}{y}\right) dy$$

5 dec. $x = 0\,(0{,}1)\,0{,}4\,(0{,}2)\,8\,(0{,}5)\,20$ [69]

$$g(x_1, x_1) = \frac{x_1}{\sqrt{\pi}} \int_{x_1}^{\infty} x^{-1} (x^2 - x_1^2)^{-\frac{1}{2}} e^{-(x-x_1)^2} dx$$

4 dec. $x_1 = 2\,(0{,}5)\,5\,(1)\,10$ [45]

$$4\pi^{-3/2} \int_0^{\infty} e^{-(\tau+1)r^2} \left\{ \frac{4}{\pi} \left[1 - re^{-r^2} \int_0^{r} e^{z^2} dz \right]^2 + r^2 e^{-2r^2} \right\}^{-1} dr$$

4 dec. $\tau = 0;\ 0{,}1;\ 0{,}2\ (0{,}2)\ 2$ [124]

$$\frac{1}{2} \int_0^{\infty} e^{-(\tau+1)r} \left\{ \left[1 + \frac{1}{2} re^{-r} \int_{-2}^{\infty} z^{-1} e^{-z} dz \right]^2 + \frac{1}{4} \pi^2 r^2 e^{-2r} \right\}^{-1} dr$$

4 dec. $\tau = 0;\ 0{,}1;\ 0{,}2\,(0{,}2)\,2$ [124]

$$\frac{2}{3} F_{3/2}(\eta) = \frac{2}{3} \int_0^{\infty} \frac{x^{3/2}\, dx}{e^{x-\eta} + 1}$$

6 dec. $\eta = -4\,(0{,}1)\,4$
5 dec. $\eta = 4\,(0{,}1)\,20$ [91]

$$F;\ \frac{1}{10}\frac{dF}{d\eta};\ \frac{1}{100}\frac{d^2F}{d\eta^2};\ \frac{1}{1000}\frac{d^3F}{d\eta^3},$$

$$F = F_{1/2}(\eta) = \frac{d}{d\eta} \left\{ \frac{2}{3} F_{3/2}(\eta) \right\}$$

6 dec. $\eta = -4\,(0{,}1)\ 4$
5 dec. $\eta = 4\,(0{,}1)\,20$ [91]

$$g_n^{(2)} = \frac{1}{n!} \int_0^1 \int_0^t t(t-1)\ldots(t-n+1)\,(dt)^2$$

exact $n = 1\,(1)\,20$ [89]

$$g_n^{(k)} = \frac{1}{n!} \underbrace{\int_0^1 \int_0^t \cdots \int_0^t}_{k} t(t-1)\ldots(t-n+1)\,(dt)^k$$

9—12 dec. $n = 1\,(1)\,20.\qquad k = 2\,(1)\,6$ [89]

$$H_n^{(k)} = \frac{1}{n!}\underbrace{\int_0^1\int_0^t\ldots\int_0^t}_{k} t(t+1)\ldots(t+n-1)(dt)^k$$

7—11 dec. $n = 1(1)20.$ $k = 2(1)6$ [89]

$$\int_0^x e^{-t^2}\,dt = \begin{cases} A + ax + bx^3 + cx^5 & 0 \leqslant x \leqslant 1 \\ A + \frac{a}{x} + \frac{b}{x^3} + \frac{c}{x^5} & x > 1 \end{cases}$$

6 dec. A

7 dec. $\left\{\begin{matrix} \lg a \\ \lg b \\ \lg c \end{matrix}\right\}$ $0 \leqslant x \leqslant 3$ [40]

$$\int_0^1 \sin n\pi x\,dx$$

10 dec. $n = 1(1)100$ [66]

$$\int_0^1 x^m \sin n\pi x\,dx$$

10 dec. $n = 1(1)100.$ $m = 1(1)10$ [66]

$$\int_0^1 x^m \cos n\pi x\,dx$$

10 dec. $n = 1(1)100.$ $m = 0(1)10$ [66]

VARIOUS FUNCTIONS

Stirling formulas

$$\sigma = \sqrt{2\pi n}\left(\frac{n}{e}\right)^n$$

7 dec. $\left\{\begin{matrix} n = 1(1)100(10)200(25)300(50)1000 \\ (500)4000(1000)10000(10000)100000 \end{matrix}\right\}$ [11]

$$\frac{\sigma}{n!} = \frac{1}{n!}\sqrt{2\pi n}\left(\frac{n}{e}\right)^n$$

5 fig. $\left\{\begin{matrix} n = 1(1)100(10)200(25)300(50)1000 \\ (500)4000(1000)10000(10000)100000 \end{matrix}\right\}$ [11]

Chaplygin functions

Values connected with the solution of Chaplygin's equation in the form presented below are given in /17/:

$$\tau(1-\tau)\frac{d^2Y_k}{d\tau^2} + [(k+1)-(k+1-\beta)\tau]\frac{dY_k}{d\tau} + \frac{1}{2}\beta k(k+1)Y_k = 0.$$

Four-decimal tables contain the solution of the Chaplygin equation Y_k and its derivatives $\frac{dY_k}{d\tau}$ for

$$M^2 = 5\tau(1-\tau)^{-1};\quad \beta = 2{,}5;\quad k = -15(0{,}5)15;\quad \tau = 0(0{,}01)0{,}5.$$

Values connected with the solution of Chaplygin's equation in the form presented below are given in /21/:

$$\tau(1-\tau)\frac{d^2Y_k}{d\tau^2} + [(k+1+\beta)\tau-(k+1)]\frac{dY_k}{d\tau} - \frac{1}{2}\beta k(k-1)Y_k = 0.$$

Y_k denotes the partial solution and $\overline{Y}_{-k}$ the second independent solution of Chaplygin's equation. Six-figure tables contain the values

$$Y_k,\quad \frac{dY_k}{dw},\quad \overline{Y}_{-k}, \frac{d\overline{Y}_{-k}}{dw},\quad L_k = w^kY_k,\quad L'_k = \frac{d}{dw}(w^kY_k),\quad \overline{L}'_{-k} = \frac{d}{dw}(w^k\overline{Y}_{-k}),$$

$$\Psi_k = \rho(kL_k - k^{-1}wL'_k),\quad \overline{\Psi}_{-k} = \rho(k\overline{L}_{-k} - k^{-1}w\overline{L}'_{-k})$$

for $\rho = (1-w^2)^{\frac{5}{2}};\quad k = 2(1)10;\quad w = 0{,}01(0{,}01)0{,}5;\quad 6^{-\frac{1}{2}};$

$$\beta = \frac{5}{2};\quad \tau = w^2.$$

Ballistic functions

This section only contains information on certain recently published tables of certain magnitudes connected with ballistic problems.

$$\xi(\theta) = \int_0^\theta \sec^3 t\, dt$$

7 fig. $\theta = 0(1')87°$ [10]

$$\boldsymbol{s}(\xi) = \sin\theta(\xi)$$

A function reciprocal to $\theta(\xi)$ is denoted by

$$\xi(\theta) = \int_0^\theta \sec^3 t\,dt.$$

9 dec. $\xi = 0\,(0{,}01)\,50$ [10]

$$\boldsymbol{c}(\xi) = \cos\theta(\xi)$$

8 dec. $\xi = 0\,(0{,}01)\,50$ [10]

$$\boldsymbol{X}(\beta, b) = \int_1^\beta \boldsymbol{c}\,[b(1-t)]\,t^{-1}\,dt$$

8 dec. $b = 0{,}1\,(0{,}1)\,2.$ $\beta = 0\,(0{,}02)\,3.$ $\beta(b-1) \leqslant 2$ [10]

$$\boldsymbol{Y}(\beta, b) = \int_1^\beta \boldsymbol{s}\,[b(1-t)]\,t^{-1}dt$$

8 dec. $b = 0{,}1\,(0{,}1)\,2.$ $\beta = 0\,(0{,}02)3.$ $\beta(b-1) \leqslant 2$ [10]

$$\boldsymbol{T}(\beta, b) = \int_1^\beta \boldsymbol{c}\,[b(1-t)]\,t^{-\frac{1}{2}}dt$$

8 dec. $b = 0{,}1\,(0{,}1)\,2.$ $\beta = 0\,(0{,}02)\,3.$ $\beta(b-1) \leqslant 2$ [10]

The function $D(u)$

4—5 fig. $u = 50\,(2)\,400\,(10)\,1000$ [1]

The function $T(u)$

3fig. $u = 50\,(2)\,400\,(10)\,1000$ [1]

Chapter 12

SOLUTIONS OF CERTAIN EQUATIONS[1]

The present chapter contains descriptions of tables which are solutions of certain algebraic, transcendental and differential equations. The greater part of the differential equations are obtained in connection with the solution of technical problems of various types. In this chapter we have only described solutions for equations which are close to the standard differential equations of mathematical physics.

The roots of special functions are given additionally in the relevant chapters.

The material contained in the chapter is classified in the following sections:

algebraic equations;
equations containing power and exponential functions;
equations containing trigonometric functions;
solutions of differential equations.

ALGEBRAIC EQUATIONS

Tables of squares and square roots listed in Chapter 1 are normally used in the solution of quadratic equations. Certain authors have also compiled small tables of the discriminants of quadratic equations.

The tables in Chapter 1 may also be used to find the real roots of cubic equations since, by substitution of the variable, any cubic equation may be reduced to one of the standard forms:

$$x^3 \pm x - c = 0, \qquad x^3 - c = 0.$$

Solutions of the equation $x^3 \pm x - c = 0$ may be found by inverse interpolation in tables of $x^3 \pm x$. Everett interpolation coefficients $E_2 = \frac{1}{6}(n^3 - n)$

---

[1] See pages 510 and 550 for the references to Chapter 12.

(Chapter 13) may be used in the same way, as may the associated Legendre functions $P_{3,2}(x) = 15(x - x^3)$ (Chapter 9). Other substitutions of the variable are also used to reduce a given equation to a polynomial for which tables exist.

Solutions of cubic equations are also directly tabulated. Such tables are enumerated below.

The discriminant $\sqrt{\frac{p^2}{4} - q}$ of a quadratic equation

6 dec. $p = -10(1)10.\ q = -10(1)10$ [6]

The discriminant $\sqrt{\frac{q^2}{4} + \frac{p^3}{27}}$ of a cubic equation

6 dec. $p = -10(1)10.\ q = -10(1)10$ [6]

Solutions of a cubic equation $x^3 + px + q = 0$

152 dec.	$p = -2.\quad q = -5$	[29] [71]
101 dec.	$p = -2.\quad q = -5$	[38]
4—5 dec.	$\begin{cases} p = -9{,}9(0{,}1)1(0{,}05)2(0{,}02)2{,}8(0{,}01)3 \\ q = 2 \end{cases}$	[4] [13]
	$p = -q \begin{cases} p = 8\cdot10^{-9}\ (\text{var.})\ 300 \\ p = -6{,}75\ (\text{var.}) - 8\cdot10^{-9} \\ p = -6{,}75\ (\text{var.}) - 300 \end{cases}$	[3]
3 dec.	$p = -30(1)30.\quad x = 0(0{,}1)5$	[2]
	$p = \pm 1.\ q = 0(0{,}02)0{,}3(0{,}01)0{,}37\ (\text{var.})1100$	[14]

Solutions of $x^{4n} + 4nx^{2n} = 1$

6 dec. $n = 1(1)4$ [39]

Solutions of $B\sum_{k=0}^{n}\frac{y^k}{k!} = \frac{y^n}{n!}$

12 dec. $B=1.\quad n=3(1)24$ [80]

2 dec. $\begin{cases} B = 0{,}001(0{,}001)0{,}005;\ 0{,}01;\ 0{,}02;\ 0{,}05 \\ n = 1(1)260 \end{cases}$ [8]

EQUATIONS CONTAINING POWER & EXPONENTIAL FUNCTIONS

Solutions of $x = e^{\pi \operatorname{ctg} x}$

3 fig. [44]

Solutions of $e^{\frac{3x}{4}} = 1 + x$ or $\ln(1+x) = \frac{3}{4}x$

The **value** of the root of this equation is given to 5 decimals in /5/.

Solutions of

The **value** of the root of this equation is given to 4 decimals in /5/.

Solutions of $ue^{-u} = e^{-t-1}$

Solutions are given to this equation for $u < 1$ & $u > 1$. **in this work.**

7—8 dec. $t = 0\,(0{,}02)\,1\,(0{,}1)\,11$ [58]

Solutions of $t = x^t$

Tables of the solutions λ and μ of this equation when $1 < \lambda < e < \mu < \infty$. are given in the work.

λ is given for

$x = 0\,(0{,}01)\,0{,}06;\ e^{-e} = 0{,}06599;\ 0{,}1\,(0{,}1)\,1;\ 1{,}2;\ e^{\frac{1}{e}} = 1{,}447$ [66]

μ is given for $x = 1;\ 1{,}2;\ e^{\frac{1}{e}};\ e.$

Solutions of $t = x^{x^t}$

A table of the solutions $t = \sigma$ and τ of this equation when $0 < \sigma < e^{-1} < \tau < 1$ is given in the work

4 dec. $x = 0\,(0{,}01)\,0{,}06;\ e^{-e};\ 0{,}1\,(0{,}1)\,1;\ 1{,}2;\ e^{\frac{1}{e}} = 1{,}447$ [66]

Solutions of $x^{-\alpha} + (x-1)^{-\alpha} = 1$

Solutions of this equation with three decimals for values of $\alpha = 0{,}5\,(0{,}1)\,3$ are given in /55/.

The same work gives (to three decimals) powers of $x_i^{-\alpha}$ of the solutions of this equation for the values

$$\alpha = 0{,}5\,(0{,}1)\,3.$$

EQUATIONS CONTAINING TRIGONOMETRIC FUNCTIONS

The majority of the equations considered here are intimately connected with Bessel functions of order $\frac{n}{2}$, where n is an odd integer.

$$\sin z = z$$

Values of the first ten roots of the equation sin z = z having positive real and imaginary parts are given to six decimals in /47/.

$$\sin z = -z$$

A six-place table of values of the first ten roots in the first quadrant is given in /48/.

$$\operatorname{sh} z + z = 0$$

A six-place table of the first eleven roots is given in /23/.

$$\sin (z - q) = m \sin^4 z.$$

3 dec. $q = 1°\,(1°)\,36°$. $1° \leqslant z \leqslant 212°33'$. $m > 0$ [28]

Solutions of the equation sin(z - q) = $m$sin^{4}z are given in /1/ through the solutions of the equation $y = t(1+y)^4$, where

$$y = \frac{\operatorname{tg} x}{\operatorname{tg} q}; \qquad t = \frac{m \sin^4 q \cos^3 x}{\operatorname{tg} q}.$$

A four-decimal table contains values of log t as a function of log y:

$$\lg y = 7{,}9\,(0{,}01)\,9{,}25$$

$$\cos z = z$$

One real root and six complex roots having positive real parts are given to eight and four decimals respectively in /54/.

One real root and three complex roots having negative real parts are given to five decimals and five significant figures respectively in /46/.
The root $x = 42°20'47'',3$ (or 0,739085133 radians) of this equation is given in /5/.

$\operatorname{ch} x \cdot \cos x = 1$

The subsequent roots of the equation are denoted below by

7 dec.	x_n	$n = 1(1)5$	[5] [17] [19]
4 dec.	x_n	$n = 1(1)5$	[12]
4 dec.	$\frac{1}{2} x_n$	$n = 1(1)4$	$[13]_2$
4 dec.	x_n	$n = 1(1)3$	$[13]_1$

$\operatorname{ch} x \cos x = -1$

6 dec.	x_n	$n = 1(1)6$	[5] [17] [19]
4 dec.	x_n	$n = 1(1)6$	[12]
4 dec.	$\frac{1}{2} x_n$	$n = 1(1)5$	$[13]_2$
4 dec.	x_n	$n = 1(1)4$	$[13]_1$

$\operatorname{tg} x = x$ or $J_{3/2}(x) = 0$

10 dec.	x_n	$n = 1;\ 2$	[63]
6 dec.	x_n	$n = 1(1)16$	[22]
6 dec.	x_n	$n = 1(1)8$	[21]
5—7 dec.	x_n	$n = 1(1)3$	[42]
5 fig.	$\frac{2\pi}{x_n}$	$n = 1(1)6$	[18] [56]
4; 6 dec.	x_n	$n = 1(1)16;\ 17(1)35$	[19]
4 dec.	x_n	$n = 1(1)36$	[12]
4 dec.	x_n	$n = 1(1)17$	[4]
4 dec.	x_n	$n = 1(1)16$	[5] $[13]_2$
4 dec.	x_n	$n = 1(1)6$	$[13]_1$
4 dec.	$\frac{x_n}{\pi}$	$n = 1(1)7$	[16] [20]
4 dec.	$\frac{x_n}{\pi}$	$n = 1(1)6$	[17] [30] [56] [57]

$\operatorname{tg} x = \lambda x$

4—6 dec.	x_n	$\lambda = 4;\ \frac{8}{5}.\quad n = 1(1)4;\ 1(1)3$	[42]
4 dec.	x_n	$\lambda = \operatorname{tg}\theta.\ \theta = -90°(5°)90°.\ n = 1(1)4$	[59]

$$2\alpha x + \operatorname{tg} x = 0$$

/61/ gives a table with 5-7 figures of the first 16 positive roots for $\alpha = 2; \frac{1}{2}; \frac{1}{3}; \frac{1}{6}; \frac{1}{30}$.

$$\operatorname{tg} x = \frac{x}{1-x^2} \text{ or } \frac{d}{dx}\left[x^{\frac{1}{2}} J_{3/2}(x)\right] = 0$$

4 dec.	$\frac{x_n}{\pi}$	$n = 1\,(1)\,6$	[57]
4 fig.	x_n	$n = 1\,(1)\,6$	[5] [24]
4 fig.	$\frac{x_n}{2\pi}$	$n = 1\,(1)\,4$	[12] [19]

$$\operatorname{tg} x = \frac{2x}{2-x^2} \text{ or } \frac{d}{dx}\left[x^{-\frac{1}{2}} J_{3/2}(x)\right] = 0$$

5 fig.	$\frac{2\pi}{x_n}$	$n = 1\,(1)\,6$	[18] [56]
4 dec.	$\frac{x_n}{\pi}$	$n = 1\,(1)\,6$	[57]

$$\operatorname{tg} x = \frac{3x}{3-x^2} \text{ or } J_{5/2}(x) = 0$$

4 dec.	x_n	$n = 1\,(1)\,6$	[13]
4 dec.	$\frac{x_n}{\pi}$	$n = 1\,(1)\,6$	[5] [57]

$$\operatorname{tg} x = \frac{4x}{4-3x^2}$$

4 dec.	$\frac{x_n}{\pi}$	$n = 1\,(1)\,6$	[5] [57]

$$\operatorname{tg} x = \frac{8x}{8-7x^2}$$

4 dec.	$\frac{x_n}{\pi}$	$n = 1\,(1)\,6$	[5] [57]

$$\operatorname{tg} x = \frac{6x - x^3}{6-3x^2} \text{ or } \frac{d}{dx}\left[x^{\frac{1}{2}} J_{5/2}(x)\right] = 0$$

4 dec.	$\frac{x_n}{\pi}$	$n = 1\,(1)\,6$	[57]

$$\operatorname{tg} x = \frac{9x - x^3}{9 - 4x^2} \text{ or } \frac{d}{dx}\left[x^{-\frac{1}{2}} J_{5/2}(x)\right] = 0$$

5 fig.	$\frac{2\pi}{x_n}$	$n = 1\,(1)\,4$	[18] [56]
4 dec.	x_1		[18] [56]

$$\operatorname{tg} x = \frac{12x - x^3}{12 - 5x^2} \text{ or } \frac{d}{dx}\left[x^{-3/2} J_{5/2}(x)\right] = 0$$

4 dec.	$\frac{x_n}{\pi}$	$n = 1\,(1)\,6$	[57]

$$\operatorname{tg} x = \frac{480x - 92x^3 + 5x^5}{480 - 252x^2 + 25x^4 - x^6}$$

4 dec.	$\frac{x_n}{\pi}$	$n = 1\,(1)\,6$	[57]

$$\operatorname{ctg} x = \frac{1}{\pi} \ln R$$

Values of x are given with an accuracy of up to $1'$. $R = 1\ (0{,}01)\ 2$ [44]

$$\operatorname{ctg} x = \lambda x$$

4 dec.	x_n	$\lambda = \operatorname{tg} \theta.$	$\theta = 0\,(5°)\,90°.$	$n = 1\,(1)\,4.$	[15]
4 dec.	x_n	$\lambda = \operatorname{tg} \theta.$	$\theta = -90°\,(5°)\,90°.$	$n = 1\,(1)\,4.$	[59]
4 dec.	x_n	$\lambda = \frac{1}{u}.$	$u = 0;\ 0{,}1;\ 0{,}5;\ 1\,(1)\,10;\ \infty$	$n = 1\,(1)\,7$	[68]
$0'',1$	x_n	$\lambda = 1.$	$n = 1$		[5]

$$\operatorname{cth} x = x$$

22 dec.	x_n	$n = 1$	[65]
6 dec.	x_n	$n = 1$	$[13]_2$

$$\operatorname{ctg} x = -\frac{x}{1 - x^2} \text{ or } \frac{d}{dx}\left[x^{\frac{1}{2}} J_{-3/2}(x)\right] = 0$$

4 fig.	x_n	$n = 1\,(1)\,4$	[5] [24]

$$\sec x = 1 + x^2$$

6 dec.	x_n	$n = 1\,(1)\,6$	[5]

SOLUTIONS OF DIFFERENTIAL EQUATIONS

Solution of a system of nonlinear differential equations:

$$\left.\begin{aligned} (\eta-\varphi)\varphi' &= \frac{1}{\gamma}\frac{f'}{\psi}-\frac{3}{2}\varphi \\ \frac{\psi'}{\psi} &= \frac{\varphi'+\frac{2\varphi}{\eta}}{\eta-\varphi} \\ 3f+\eta f'+\frac{\gamma\psi'}{\psi}(-\eta+\varphi)f-\varphi f' &= 0 \end{aligned}\right\}$$

where γ = const.

A solution of this system of equations, which describes the formation of the shock wave of a very powerful explosion, is given in /76/. For the initial conditions $\gamma=1{,}4$; $f(1)=1{,}167$; $\varphi(1)=0{,}833$; $\psi(1)=6$ and when $\eta=1(-0{,}02)0{,}5$ values of the functions f, φ and ψ are given to three decimals. For the initial conditions $\gamma=1{,}666$; $f(1)=1{,}250$; $\varphi(1)=0{,}750$; $\psi(1)=4$ and when $\eta=1$; 0,95; 0,9; 0,8; 0,7; 0,5; 0 values of the functions f and φ are given to three significant figures and tables of the function ψ to two decimals. For the initial conditions $\gamma=1{,}30$; $f(1)=1{,}130$; $\varphi(1)=0{,}869$; $\psi(1)=7{,}667$ and when η = 1(-0,02)0,9(-0,05)0,4 values the functions f, φ, ψ are given to three decimals.

Solutions of $F''(z)+\left\{-\frac{1}{4}+\frac{\nu}{z}+\frac{1-p^2}{4z^2}\right\}F(z)=0$

Solutions $m_{i\tau}(i\zeta)$ of this equation are given in the table with parameter ζ.

4—5 dec. $\tau=-3(0{,}5)3$. $\zeta=1(1)6$ [75]

Solutions of $3\theta\frac{d^2\theta}{d\gamma^2}+\left(\frac{d\theta}{d\gamma}\right)^2-e^{\gamma}=0$

/49/ gives the solution for $\theta(0)=0$; $\left.\frac{d\theta}{d\gamma}\right|_0=1$.

Solutions of $\frac{d^2\varphi}{d\xi^2}+\left(a+b\frac{e^{-\xi}}{\xi}\right)\varphi=0$

Solutions for the condition $\varphi(0)$ = 0 and certain values of φ are given for five pairs of values a and b in /53/.

Solutions of $\varphi''=x^{-\frac{1}{2}}\varphi^{3/2}$

Solutions for $\varphi(0)=1$ & $\varphi(\infty)=0$ are given in /50/ and /51/.

Solutions of $y'' - xy' + ny = 0$

The table gives values of the positive roots of

$$h_n(x) = (-1)^n e^{\frac{x^2}{2}} \frac{d^n}{dx^n} \int_0^x e^{-\frac{1}{2}(x^2-u^2)} du$$, the Hermite function of the second kind.

x_n $n = 1(1)6$ [25]

Solutions of $y'' + \frac{2}{x}y' + y^n = 0$

The following tables of solutions exist for the initial conditions

$n = 0;\ 1;\ 2;\ 2{,}5;\ 3;\ 4;\ 5$	[31]
$n = 0(0{,}5)3;\ 4;\ 4{,}5;\ 4{,}9;\ 5;\ 6$	[11]
$n = 2;\ 2{,}5;\ 3$	[10]
$n = 3{,}25$	[27]
$n = 5(0{,}5)8$	[36]

For the initial conditions $y(0) = 1$, $y'(0) = 0$ values of y, y' to seven decimals for $n = 1(0{,}5)4{,}5$ are given in /7/. Values of y and y' for $n = 5$ are given for the same initial conditions in /7/ to 5 decimals.

Solutions for the equation $n = 3$ are given in /40/ and /41/ for boundary conditions.

Solutions of $y'' + \frac{2}{x}y' + e^y = 0$

For the initial conditions $y(0) = 0;\ y'(0) = 0$ solutions of the equation are given in /10/, /11/ and /31/.

The equation is transformed by the substitution of $x = \frac{1}{z}$ to $\frac{d^2y}{dz^2} = -\frac{e^y}{z^4}$. Values of $\frac{dy}{dz}$ and e^y are given in /45/ for the initial conditions

$y(\infty) = 0;\ \left.\frac{dy}{dz}\right|_\infty = 0$

Solutions of $\frac{1}{x^2}\frac{d}{dx}\left(x^2\frac{dy}{dx}\right) = -(y^2 - c)^{3/2}$

Solutions are given for the initial conditions

$y(0) = 1;\ \left.\frac{dy}{dx}\right|_0 = 0.$

$c = 0{,}01;\ 0{,}02;\ 0{,}05;\ 0{,}1(0{,}1)0{,}6;\ 0{,}8$ [9]

Solutions of $\frac{d^2T(x)}{dx^2} - ixT(x) = 1$

A table is given of the real and imaginary parts of the solution of the equation $T(x)$ and its derivative $\frac{dT(x)}{dx}$ in the terms of Lommel functions

$T(x) = -\frac{2}{3}(ix)^{\frac{1}{2}} S_{0\frac{1}{3}}\left[\frac{2}{3}(ix)^{3/2}\right]$

3 dec. $x = 0(0,5)5$ [32]

Solutions of $\frac{d^2U_a(\xi)}{d\xi^2} + (\xi^2 + a)U_a(\xi) = 0$

4 dec. $\xi = 0(0,1)3.$ $a = 1(1)3;\ -3(1)-1$ [33]

Solutions of $\frac{d}{d\xi}\left[(1-\xi^2)\frac{dX}{d\xi}\right] + (k^2\xi^2 + \Lambda)X = 0$

Solutions of the equation for spheroidal wave functions of zero order are given in /35/. The solutions of the equation in Legendre polynomials $P_m(\xi)$ is denoted through $X_n(\xi)$. Tables of this characteristic function are given in /35/ to six decimals for the following values:

$n = 0.$	$m = 0(2)10.$	$k^2 = -10(1)10$
$n = 1.$	$m = 1(2)11.$	$k^2 = -10(1)10$
$n = 2.$	$m = 0(2)10.$	$k^2 = 0(1)10$
$n = 3.$	$m = 1(2)11.$	$k^2 = 0(1)10$
$n = 4.$	$m = 0(2)12.$	$k^2 = 0(1)10$
$n = 5.$	$m = 1(2)11.$	$k^2 = 0(1)10$
$n = 6.$	$m = 0(2)14.$	$k^2 = 0(1)10$

Tables of the characteristic value Λ_n are also given to six decimals:

6 dec. $n = 0;\ 1;\ 2.$ $k^2 = -10(1)10$
6 dec. $n = 3(1)6.$ $k^2 = 0(1)10$

Solutions of $\frac{d^2x}{dt^2} + \frac{1}{2}\frac{dx}{dt}\left|\frac{dx}{dt}\right| + 2b\frac{dx}{dt} + x = 0$

Solutions are given in /43/, /73/, /74/.

Solutions of $\frac{d}{dx}[x^{-\frac{1}{2}}J_\nu(x)] = 0$

4 fig. $\frac{2\pi}{x_n}$ $n = 1(1)3.$ $\nu = \frac{7}{2}(1)\frac{13}{2}$ [18] [56]

Solutions of $\frac{d^2}{dx^2}\left(x^3 \frac{d^2u}{dx^2}\right) + xu = 0$[1]

Four solutions of this equation are given in the tables.

$$u_1 = \frac{1}{\sqrt{x}} \operatorname{bei}'(2\sqrt{x}),$$

$$u_2 = -\frac{2}{\sqrt{x}} \operatorname{ber}'(2\sqrt{x}),$$

$$u_3 = \frac{1}{\sqrt{x}} [2(1-\gamma) \operatorname{bei}'(2\sqrt{x}) - \frac{1}{2}\pi \operatorname{ber}'(2\sqrt{x}) - 2 \operatorname{kei}'(2\sqrt{x})],$$

$$u_4 = \frac{1}{\sqrt{x}} [-2\gamma \operatorname{ber}'(2\sqrt{x}) + \frac{1}{2}\pi \operatorname{bei}(2\sqrt{x}) - 2 \operatorname{ker}'(2\sqrt{x})].$$

9 dec.	$x = 0(0{,}5)10(1)30$	[70]
8 dec.	$x = 0(0{,}5)2(1)30$	[52]
8 dec.	$x = 0(0{,}5)10(1)30$	[69]

Solutions of $y''' + yy'' = \beta(y'^2 - 1)$

The tables give the values $y; y'; y''$ for $\beta = 1$

4 dec.	$x = 0(0{,}1)4{,}3$	[67]

Solutions of $\frac{\partial t}{\partial \theta} = \alpha \left[\frac{\partial^2 t}{\partial r^2} + \frac{1}{r}\frac{\partial t}{\partial r}\right]$

The equation is resolved for the boundary conditions

$$\theta = 0; \quad t = t_0; \quad r_i \leqslant r \leqslant r_0$$

$$\theta > 0 \quad \begin{cases} t = t_i \\ \frac{\partial t}{\partial r} + \frac{f}{k}(t - t_0) = 0; \; r = r_0 \end{cases}$$

Values of the temperature relationship $\frac{t-t_0}{t_i-t_0}$. are given in /72/ with four decimals. θ varies from 0 to 2.

Solutions of $\frac{\partial t}{\partial \theta} = \alpha \frac{\partial^2 t}{\partial x^2}$

The equation is resolved for the conditions

$$\theta = 0; \quad t = t_0; \quad x \geqslant 0$$

$$\theta > 0; \quad t = t_i; \quad x = 0$$

Values are given for the temperature relationship $\frac{t-t_0}{t_i-t_0}$.

4 dec.	$\frac{x}{2\sqrt{\alpha\theta}} = 0(0{,}25)2$	[72]

[1] See Chapter 10 for the functions ber, bei, ker, kei.

Solutions of $2f'''(\eta)+f(\eta)f''(\eta)=0$

Solutions of the equation for a laminar boundary layer for steady flow of a viscous incompressible liquid in parallel streams of varying density and viscosity are given in /64/. The solutions depend on the velocity ratio of the two streams $\lambda=\frac{U_2}{U_1}$ and the derivative $\rho\mu$ of the corresponding density and viscosity ratios.

The boundary layer equations for the upper and lower streams

$$u\frac{\partial u}{\partial x}+v\frac{\partial u}{\partial y}=\nu_1\frac{\partial^2 u}{\partial y^2}$$

$$u\frac{\partial u}{\partial x}+v\frac{\partial u}{\partial y}=\nu_2\frac{\partial^2 u}{\partial y^2}$$

are converted to the form

$$2\frac{d^3 f_1(\eta_1)}{d\eta_1^3}+f_1(\eta_1)\frac{d^2 f_1(\eta_1)}{d\eta_1^2}=0$$

$$2\frac{d^3 f_2(\eta_2)}{d\eta_2^3}+f_2(\eta_2)\frac{d^2 f_2(\eta_2)}{d\eta_2^2}=0$$

Values of $f_1(\eta_1)$, $f_1'(\eta_1)$ & $f_1''(\eta_1)$ to four decimals are given for the following cases

$\rho\mu=1.$	$\lambda=0.$	$\eta_1=0($ var. $)6{,}497;\ \infty$
$\rho\mu=1.$	$\lambda=0{,}501.$	$\eta_1=0($ var. $)6{,}001;\ \infty$
$\rho\mu=5{,}965\cdot10^4.$	$\lambda=0.$	$\eta_1=0(0{,}4)6{,}8;\ \infty$
$\rho\mu=10.$	$\lambda=0.$	$\eta_1=0($ var. $)6{,}764;\ \infty$
$\rho\mu=100.$	$\lambda=0.$	$\eta_1=0($ var. $)7{,}101$

Values of $f_2(\eta_2)$, $f_2'(\eta_2)$ & $f_2''(\eta_2)$ to four decimals are given for the following cases

$\rho\mu=1.$	$\lambda=0.$	$\eta_2=-14{,}980($ var. $)-0{,}286;\ -\infty$
$\rho\mu=1.$	$\lambda=0{,}501.$	$\eta_2=-7{,}602($ var. $)-0{,}4;\ -\infty$
$\rho\mu=5{,}965\cdot10^4.$	$\lambda=0.$	$\eta_2=-60(10)-30(5)-20(2)0;\ -\infty$
$\rho\mu=10.$	$\lambda=0.$	$\eta_2=-19{,}608($ var. $)0;\ -\infty$
$\rho\mu=100.$	$\lambda=0.$	$\eta_2=-24{,}647($ var. $)0;\ -\infty$

Chapter 13

SUMS AND VALUES CONNECTED WITH FINITE DIFFERENCES

This chapter is devoted, in the main, to tables connected with the interpolation, differentiation and integration of functions using numerical methods.

Only tables of interpolation coefficients are indicated from among the many interpolation tables. In this chapter, however, as in all other chapters, there are no data on differences, which are computed in various tables.

Information is given here on the sums of certain numerical series and derivatives. In setting out the material on interpolation, numerical differentiation and integration, diagonal finite differences of functions are denoted by Δ^k, central differences by δ^k, the arithmetic mean of adjacent differences by μ, the step of the argument by x and the number of mesh points in the interval by n.

The tables are classified in the following manner:

binomial coefficients;
interpolation coefficients;
numerical differentiation and integration;
sums of series.

BINOMIAL COEFFICIENTS

$$C_n^m = \frac{n!}{m!\,(n-m)!}$$

exact	$m = 1\,(1)\,12.$	$n = 6\,(1)\,132$	[86]
exact	$m = 1\,(1)\,10.$	$n = 4\,(1)\,110$	[113]
exact	$m \leqslant n.$	$n = 1\,(1)\,60$	[80]
exact	$m \leqslant n.$	$n = 1\,(1)\,50$	[81] [149]
exact	$m = 1\,(1)\,10.$	$n = 1\,(1)\,50$	[26]
exact	$m \leqslant n.$	$n = 1\,(1)\,40$	[117]

[1] See pages 514 and 550 for the references to Chapter 13.

exact	$m \leqslant n.$	$n = 1\,(1)\,25$	[36]
exact	$m = 1\,(1)\,10.$	$n = 1\,(1)\,20$	[9] [45]
exact	$m = 1\,(1)\,15.$	$n = 1\,(1)\,15$	[4] [90]
exact	$m \leqslant n.$	$n = 1\,(1)\,15$	[2] [12] [48] [50]
exact	$m \leqslant n.$	$n = 1\,(1)\,12$	[10] [18] [69]
8 fig.	$m \leqslant n.$	$n = 1\,(1)\,100$	[41]
8 fig.	$m \leqslant 25.$	$n = 1\,(1)\,50$	[1]
7 dec.	$m = 1\,(1)\,6.$	$n = 0\,(0{,}01)\,100$	[106]
7 dec.	$m = 1\,(1)\,6.$	$n = 0\,(0{,}01)\,1$	[46]
5 dec.	$m = 1\,(1)\,6.$	$n = 0\,(0{,}01)\,1$	[22] [81]

$$\frac{1}{n} C_n^m$$

dec.	$m < n.$	$n = 1\,(1)\,53$	[107]

$$C_{n+m-1}^m = \frac{n\,(n+1)\ldots(n+m-1)}{m!}$$

dec.	$m = 1\,(1)\,7.$	$n = 5\,(1)\,20$	[44]
dec.	$m = 1\,(1)\,12.$	$n = 1\,(1)\,30$	[58]
	$m = 1\,(1)\,10.$	$n = 51 - m$	[26]

$$C_{n+1}^2 = \frac{1}{2}\, n\,(n+1)$$

dec.		$n = 1\,(1)\,20000$	[53]
dec.		$n = 1\,(1)\,5000$	[16]

$$\lg C_n^m$$

7 dec.	$m \leqslant n.$	$n = 1\,(1)\,64$	[43]
7 dec.	$m \leqslant n.$	$n = 1\,(1)\,40$	[117]

Partitions $A_n^m = \dfrac{n!}{(n-m)!}$

exact	$m \leqslant n.$	$n = 1\,(1)\,25$	[81]
exact	$m = 1\,(1)\,25.$	$n = 26\,(1)\,50$	[81]
exact	$m \leqslant n.$	$n = 1\,(1)\,15$	[48]

$$\lg A_n^{50} = \lg \frac{n!}{(n-50)!}$$

10 dec.		$n = 50\,(50)\,10\,000$	[21]

INTERPOLATION COEFFICIENTS

Interpolation coefficients for the Lagrange formula

The Lagrange interpolation formula enables interpolation to be carried out without the use of differences.

The four-point formula for example has the form:

$$f(a+px) = A_{-1}(p) f(a-x) + A_0(p) f(a) + A_1(p) f(a+x) + A_2(p) f(a+2x),$$

or more briefly:

$$y_n = A_{-1}(p)\, y_{-1} + A_0(p)\, y_0 + A_1(p)\, y_1 + A_2(p)\, y_2,$$

where the $A_i(p)$ are Lagrange interpolation coefficients

Lagrange interpolation coefficients for an even number of interpolation points

Coefficients for 4-point interpolation

10 dec.	$p = -1\,(0{,}001)\,0\,(0{,}0001)\,1\,(0{,}001)\,2$	[102]
10 dec.	$p = 0\,(0{,}001)\,1$	[55]
9 dec.	$p = -1\,(0{,}002)\,0$	[65]
7 dec.	$p = -1\,(0{,}01)\,2$	[152]
7 dec.	$p = 0\,(0{,}01)\,1$	[24]
6 dec.	$p = -1\,(0{,}001)\,2$	[7]
5 dec.	$p = 0\,(0{,}01)\,1$	[73][74][75][76][171]
4 dec.	$p = -1\,(0{,}1)\,2$	[102]
common fractions	$p = -1\left(\frac{1}{12}\right)2$	[102]
common fractions	$p = \frac{1}{2};\ 1\frac{1}{2}$ etc.	[168]

Coefficients for 6-point interpolation

10 dec.	$p = -2\,(0{,}01)\,0\,(0{,}001)\,1\,(0{,}01)\,3$	[102]
10 dec.	$p = 0\,(0{,}01)\,1$	[55]
8 dec.	$p = -2\,(0{,}01)\,3$	[152]
8 dec.	$p = -2\,(0{,}1)\,3$	[102]
6 dec.	$p = 0\,(0{,}2)\,1$	[150]
6 dec.	$p = -2\,(0{,}2)\,3$	[168]
common fractions	$p = -2\left(\frac{1}{12}\right)3$	[102]
common fractions	$p = \frac{1}{2};\ 1\frac{1}{2}$ etc.	[168]

Coefficients for 8-point interpolation

exact	$p = 0\,(0{,}1)\,1$	[168]
11 dec.	$p = 0\,(0{,}1)\,1$	[55]
10 dec.	$p = -3\,(0{,}1)\,0\,(0.001)\,1\,(0{,}1)\,4$	[102]
8 dec.	$p = -3\,(0{,}2)\,4$	[168]
common fractions	$p = -3\left(\frac{1}{12}\right)4$	[102]
	$p = \frac{1}{2};\ 1\frac{1}{2}$ etc.	[168]

Coefficients for 10-point interpolation

10 dec.	$p = -4\,(0{,}1)\,5$	[102]
10 dec.	$p = -4\,(0{,}2)\,5$	[168]
common fractions	$p = \frac{1}{2};\ 1\frac{1}{2}$	[168]

Lagrange interpolation coefficients for an odd number of interpolation points

Coefficients for 3-point interpolation

9 dec.	$\pm p = -1\,(0{,}0001)\,1$	[102]
5 dec.	$\pm p = 0\,(0{,}01)\,1$	[171]
3 dec.	$\pm p = 0\,(0{,}1)\,1;\ 2;\ 3$	[131]
3 dec.	$\pm p = 0\,(0{,}1)\,1$	[36] [133]
3 dec.	$\pm p = 0\,(0{,}1)\,1$	[102]
common fractions	$\pm p = 0\left(\frac{1}{12}\right)1$	[102]

Coefficients for 5-point interpolation

exact	$\pm p = 0\,(0{,}1)\,0{,}5;\ 1{,}5$	[168]
11 dec.	$\pm p = 0{,}25\,(0{,}5)\,1{,}75$	[134]
10 dec.	$\pm p = 0\,(0{,}001)\,2$	[102]
10 dec.	$\pm p = 0\,(0{,}001)\,1$	[94] [100]
7 dec.	$\pm p = 0\,(0{,}1)\,1;\ 3;\ 4$	[131]
7 dec.	$\pm p = 0\,(0{,}001)\,2$	[99]
7 dec.	$\pm p = 0\,(0{,}1)\,2$	[102] [133]
7 dec.	$\pm p = 0\,(0{,}1)\,1$	[36]
6 dec.	$\pm p = -2\,(0{,}001)\,2$	[7]
common fractions	$\pm p = 0\left(\frac{1}{12}\right)2$	[102]

Coefficients for 7-point interpolation

exact	$\pm p = 0(0,1)0,5(1)2,5$	[168]
16 dec.	$\pm p = 0,25(0,5)2,75$	[134]
10 dec.	$\pm p = 0(0,001)1(0,01)3$	[102]
10 dec.	$\pm p = 0(0,1)1;\ 4;\ 5$	[131]
10 dec.	$\pm p = 0(0,1)3$	[102] [133]
7 dec.	$\pm p = 0(0,1)1$	[36]
common fractions	$\pm p = 0\left(\frac{1}{12}\right)3$	[102]

Coefficients for 9-point interpolation

exact	$\pm p = 0(0,1)0,5(1)3,5$	[168]
10 dec.	$\pm p = 0(0,1)4$	[102]
5 dec.	$\pm p = 0(0,001)1$	[100]

Coefficients for 11-point interpolation

exact	$\pm p = 0(0,1)0,5(1)4,5$	[168]
10 dec.	$\pm p = 0(0,1)5$	[102]

Newton interpolation coefficients

Tables of Newton interpolation coefficients are described below:

$$f(a+tx) = f(a) + t\Delta f(a) + \frac{t(t-1)}{2!}\Delta^2 f(a) + \frac{t(t-1)(t-2)}{3!}\Delta^3 f(a) + \cdots$$

11 dec.	$t = 0(0,01)1.$	$n = 3(1)7$	[32]
7 dec.	$t = 0(0,001)1.$	$n = 3$	[30] [96]
7 dec.	$t = 0(0,01)1.$	$n = 3(1)7$	[17][36][105] [106]
6 dec.	$t = 0(0,01)1.$	$n = 3(1)6$	[56]
6; 5; 4; 3 dec.	$t = 0(0,01)1.$	$n = 3;\ 4;\ 5;\ 6$	[87]
5 dec.	$t = 0(0,01)1.$	$n = 3(1)7$	[81]
5 dec.	$t = 0(0,01)1.$	$n = 3(1)6$	[43] [54] [66] [83]
5; 4 dec.	$t = -0,5(0,01)0,5.$	$n = 3;4(1)6$	[64]
5; 4 dec.	$t = 0(0,01)1.$	$n = 3;4(1)6$	[110]
5; 4 dec.	$t = 0(0,01)1.$	$n = 3;\ 4;\ 5$	[19] [171]
4—7 fig.	$t = 0(0,001)1.$	$n = 3$	[92]

4 dec.	$t = 0(0{,}1)1.$	$n = 3(1)8$	[83]
4 dec.	$t = -1(0{,}1)1.$	$n = 3(1)7$	[27]
4; 3 dec.	$t = 0(0{,}01)1.$	$n = 3; 4; 5$	[35]
3 dec.	$t = 0(0{,}01)1.$	$n = 3;\ 4$	[31]
3 dec.	$t = 0(0{,}01)1.$	$n = 3;\ 4$	[82]

Gauss interpolation coefficients

Tables of Gauss interpolation coefficients are given in this section:

$$f(a + tx) = f(a) + t\Delta f(a) + \frac{t(t-1)}{2!}\Delta^2 f(a - x) + \\ + \frac{(t+1)t(t-1)}{3!}\Delta^3 f(a - x) + \dots$$

exact	$t = 0(0{,}01)1.$	$n = 3(1)7$	[30]
7—10 dec.	$t = 0(0{,}001)1.$	$n = 3(1)7$	[30]
5; 4 dec.	$t = -0{,}25(0{,}01)0{,}25.$	$n = 3;\ 4(1)7$	[64]
4 dec.	$t = -1(0{,}1)1.$	$n = 3(1)7$	[27]

Stirling interpolation coefficients

Information is here given on tables of Stirling interpolation coefficients:

$$f(a + tx) = y_0 + t\mu\delta y_0 + \frac{t^2}{2!}\delta^2 y_0 + \frac{t(t^2-1)}{3!}\mu\delta^3 y_0 + \frac{t^2(t^2-1)}{4!}\delta^4 y_0 + \dots,$$

in which the central even and mean odd differences are included. Coefficients for odd differences are the same as in the Gauss formula. They may also be given as coefficients for double mean odd differences:

$$\frac{t}{2};\ \frac{1}{2}\frac{t(t^2-1)}{3!};\ \frac{1}{2}\frac{t(t^2-1)(t^2-4)}{5!}\ \text{etc.}$$

Stirling coefficients for even and mean odd differences

exact	$\pm t = 0(0{,}01)0{,}55.$	$n = 3(1)7$	[52]
exact	$\pm t = 0(0{,}1)3.$	$n = 3(1)6$	[52]
12 dec.	$\pm t = 0(0{,}01)1.$	$n = 3(1)7$	[32]
6 dec.	$\pm t = 0(0{,}1)3.$	$n = 7(1)13$	[52]
5 dec.	$\pm t = 0(0{,}01)1.$	$n = 3(1)6$	[43]
5 dec.	$\pm t = 0(0{,}01)1.$	$n = 3(1)6$	[83]
5; 4 dec.	$\pm t = 0(0{,}01)\ 0{,}5.$	$n = 3;\ 4;\ 5$	[19]

5; 4 dec.	$\pm t = 0(0{,}01)0{,}25.$	$n = 3;\ 4(1)7$	[64]
4 fig.	$\pm t = 0(0{,}01)0{,}5.$	$n = 3(1)5$	[171]
4 dec.	$\pm t = 0(0{,}1)1.$	$n = 3(1)7$	[27]

Stirling coefficients for even and double mean odd differences

5; 4 dec.	$t = 0(0{,}01)1.$	$n = 4;\ 5;\ 6$	[47]
5; 4 dec.	$t = 0(0{,}01)1.$	$n = 3;\ 4(1)6$	[79]

Bessel interpolation coefficients

This section lists tables of the Bessel interpolation coefficients

$$f(a+tx) = \mu y_{\frac{1}{2}} + \left(t - \frac{1}{2}\right)\delta y_{\frac{1}{2}} + \frac{t(t-1)}{2!}\mu\delta^2 y_{\frac{1}{2}} + \ldots,$$

for odd central and mean even differences. Coefficients for even differences can also be tabulated as coefficients for double mean even differences:

$$\frac{1}{2}\frac{t(t-1)}{2!};\quad \frac{1}{2}\frac{(t+1)t(t-1)(t-2)}{4!}\quad \text{etc.}$$

Bessel coefficients for odd and mean even differences

exact	$t = 0(0{,}01)1.$	$n = 3(2)7$	[30]
10—11 dec.	$t = 0(0{,}01)1.$	$n = 3;\ 7$	[32]
10 dec.	$t = 0(0{,}001)1.$	$n = 3;\ 5(2)9$	[118]
8 dec.	$t = 0(0{,}001)1.$	$n = 3(2)7$	[30]
6; 5; 4 dec.	$t = 0(0{,}01)1.$	$n = 3(1)5$	[15]
5 dec.	$t = 0(0{,}01)1.$	$n = 3;\ 6$	[43] [54] [66] [83]
5; 4 dec.	$t = 0(0{,}01)1.$	$n = 3(1)5$	[19] [51] [87]
5; 4 dec.	$t = 0(0{,}01)1.$	$n = 3;\ 4(1)6$	[110]
5; 4 dec.	$t = 0{,}25(0{,}01)0{,}75.$	$n = 3(1)7$	[64]
4 dec.	$t = -1(0{,}1)1.$	$n = 3(1)7$	[27]

Bessel coefficients for odd and double mean even differences

exact	$t = 0(0{,}1)1.$	$n = 3(1)6$	[150]
8 dec.	$t = 0(0{,}01)1.$	$n = 7$	[74]

7; 6 dec.	$t=0(0,1)1.$	$n=7(1)10$	[150]
6 dec.	$t=0(0,001)1.$	$n=3$	[74]
5; 4 dec.	$t=0(0,001)1.$	$n=4;\ 5$	[74]
5; 4 dec.	$t=0(0,01)1.$	$n=3(1)5$	[79]
4 dec.	$t=0(0,001)1.$	$n=3$	[11]

Everett interpolation coefficients

Here we give information on tables of the coefficients

$$\frac{p(p^2-1)}{3!},\quad \frac{p(p^2-1)(p^2-4)}{5!},\ \text{etc.}$$

which exist for the second, fourth and higher order differences in Everett's formula:

$$y_p = qy_0 + \frac{q(q^2-1)}{3!}\delta^2 y_0 + \frac{q(q^2-1)(q^2-4)}{5!}\delta^4 y_0 + \dots +$$
$$+ py_1 + \frac{p(p^2-1)}{3!}\delta^2 y_1 + \frac{p(p^2-1)(p^2-4)}{5!}\delta^4 y_1 + \dots,$$

where $q = 1 - p$.

Central differences of even order of two adjacent rows are used in Everett's formula:

exact	$p=0(0,01)1.$	$n=3(1)9$	[169]
exact	$p=0(0,1)1.$	$n=3(1)13$	[169]
11 dec.	$p=0(0,1)1.$	$n=7$	[55]
11 dec.	$p=0(0,1)1.$	$n=7$	[169]
10—13 dec.	$p=-5(0,1)6.$	$n=3(1)13$	[169]
10 dec.	$p=0(0,01)1.$	$n=3;\ 5;\ 7$	[32]
10; 8 dec.	$p=0(0,001)1.$	$n=3;\ 5$	[30]
10 dec.	$p=0(0,001)1.$	$n=3(1)9$	[169]
9—13 dec.	$p=0(0,2)1.$	$n=3(1)17$	[169]
9 dec.	$p=0\left(\frac{1}{12}\right)1.$	$n=3;\ 5;\ 7$	[126]
7 dec.	$p=0(0,001)1.$	$n=3$	[71]]104]
7; 13 dec.	$p=0(0,01)1.$	$n=3;\ 5$	[30]
7; 10; 8 dec.	$p=0(0,01)1.$	$n=3;\ 5;\ 7$	[74]
7 dec.	$p=0(0,01)1.$	$n=3$	[25]
7; 6 dec.	$p=0(0,001)1.$	$n=3;\ 5$	[101]
6 dec.	$p=0(0,001)1.$	$n=3$	[92]
6 dec.	$p=0(0,001)1.$	$n=3$	[93] [96] [97] [103]
5 dec.	$p=0(0,01)1.$	$n=3;\ 5;\ 7$	[43]
5 dec.	$p=0(0,01)1.$	$n=3$	[73] [74]
4 dec.	$p=0(0,001)1.$	$n=3$	[11]
4 dec.	$p=0(0,01)1.$	$n=3$	[72]
4 fig.	$p=0(0,01)1.$	$n=3;\ 5$	[172]

Taylor interpolation coefficients $\frac{t^k}{k!}$

Taylor's well-known formula

$$f(a+tx) = f(a) + txf'(a) + \frac{t^2}{2!}x^2 f''(a) + \cdots$$

is sometimes used for interpolation.

8 dec.	$t = 0(0{,}01)1.$	$k = 2(1)6$	[24]

NUMERICAL DIFFERENTIATION AND INTEGRATION

Numerical differentiation

Numerical differentiation formulae are used to find the derivatives of functions if the relationship between $f(x)$ and x is given by the table.

These formulae are obtained by the differentiation of interpolation formulae.

Coefficients for calculating first derivatives by Newton's formula

exact	$t = 0(0{,}01)1.$	$n = 3(1)8$	[32]
5 dec.	$t = 0(0{,}01)1.$	$n = 3(1)6$	[43] [83]

Coefficients for calculating first derivatives by Stirling's formula

5 dec.	$t = 0(0{,}01)1.$	$n = 3(1)6$	[32] [43] [83]
5; 4 dec.	$t = 0(0{,}01)0{,}25.$	$n = 4;\ 5(1)7$	[64]

Coefficients for calculating second derivatives by Stirling's formula

5; 4 dec.	$\pm t = 0(0{,}01)0{,}25.$	$n = 5;\ 6;\ 7$	[64]

Coefficients for calculating first derivatives by Bessel's formula

5 dec.	$t = 0(0{,}01)1.$	$n = 3(1)5$	[19]
5 dec.	$t = 0(0{,}01)1.$	$n = 3(1)6$	[32] [43] [83]
5 dec.	$t = 0{,}25(0{,}01)0{,}75.$	$n = 3(1)7$	[64]

Coefficients for calculating second derivatives by Bessel's formula

5; 4 dec.	$t = 0{,}25(0{,}01)0{,}75.$	$n = 5(1)7$	[64]

Coefficients for calculating first derivatives by **Everett's** formula

10 dec.	$p = 0(0{,}01)1.$	$n = 3(2)7$	[32]
5 dec.	$p = 0(0{,}01)1.$	$n = 3(2)7$	[43]

Coefficients for calculating first derivatives without differences by the formula

$$f'(a+px) \sim \frac{1}{xc(n)} \sum_{i=-\left[\frac{n-1}{2}\right]}^{i=\left[\frac{n}{2}\right]} C_i^{(n)}(p) f(a+ix)$$

4 dec.	$p = 0(0{,}01)2.$	$n = 4(1)7$	[85]

Numerical integration

Interpolation formulae which permit approximate representation of a given function may also be used to find an approximate value for the integral of the function.

Problems of this type arise when the function is determined experimentally or when its indefinite integral is unknown.

Tables of the coefficients of certain well-known numerical integration formulas are given in this section.

Coefficients of Cote's numerical integration formula

Equidistant abscissae are selected in this case. The coefficients are all rational.

exact	$n = 1(1)20$	[167]
exact	$n = 1(1)12$	[70]
exact	$n = 1(1)10$	[8] [20] [29] [59] [67] [91] [119] [120]
exact	$n = 2(1)8;\ 10$	[19]
exact	$n = 1(1)6;\ 8;\ 10$	[139]
exact	$n = 7;\ 9$	[135]
exact	$n = 1(1)8$	[121]
exact	$n = 1(1)6;\ 8$	[108]
exact	$n = 1(1)7$	[68]

Coefficients of the formula for forward integration

This formula allows one to express the value of the integral $\int^{(n+1)x} y\,dt$ through the values of the ordinates $y_0, y_1, y_2, \ldots, y_n$. The integral involves the initial but not the final ordinate of the range of integration, i.e. the formula is of "closed" type "on the left" and of "open" type "on the right". The formula is of great importance in the numerical solution of differential equations.

$n = 1\,(1)\,5$ [139]

Coefficients of Steffensen's formula for forward integration

Steffensen's formulas are of "open" type with equidistant abscissae.

exact	$n = 3\,(1)\,12$	[9]
exact	$n = 2\,(1)\,10$	[121]

Coefficients of Maclaurin's formula for numerical integration

Maclaurin's formula allows the integral $\int_0^{nx} y\,dt$ to be expressed in terms of the ordinates $y_{\frac{1}{2}}; y_{\frac{3}{2}}, \ldots y_{n-\frac{1}{2}}$ or, which is the same thing, $\int_{\frac{1}{2}x}^{\left(n+\frac{1}{2}\right)x} y\,dt$ in terms of the ordinates $y_1, y_2, \ldots, y_n$.

exact	$n = 1\,(1)\,10$	[70]
exact	$n = 1\,(1)\,5$	[68]
exact	$n = 5$	[78]

Abscissae of Chebyshev's formula for numerical integration

P.L.Chebyshev proposed an integration formula of the type

$$\int_a^b f(x)\,dx = \frac{b-a}{n}\{f(x_1) + f(x_2) + \ldots + f(x_n)\}.$$

The abscissae $x_1, x_2, \ldots, x_n$ are determined to give maximum accuracy. The range is usually taken to be $(-1, +1)$, $\left(-\frac{1}{2}, +\frac{1}{2}\right)$.

8 dec.	$n=2(1)5.$	$(-1,+1)$	[68]
8 dec.	$n=2(1)7;\ 9.$	$\left(-\frac{1}{2},+\frac{1}{2}\right)$	[70]
6 dec.	$n=2(1)7.$	$(-1,+1)$	[128]
6 dec.	$n=2(1)5.$	$(-1,+1)$	[84] [127]
6 dec.	$n=2(1)5.$	$\left(-\frac{1}{2},+\frac{1}{2}\right)$	[109]
5 dec.	$n=2(1)7;\ 9.$	$(-1,+1)$	[170]
5 dec.	$n=2(1)5.$	$(-1,+1)$	[50]

Coefficients of Lubbock's summation formula

This section contains information on coefficients of Lubbock's summation formula:

$$f(a)+f\left(a+\frac{x}{n}\right)+\ldots+f\left[a+\left(m-\frac{1}{n}\right)x\right]+f(a+mx)=n\,[f(a)+$$
$$+f(a+2x)+\ldots+f(a+mx)]-\frac{n-1}{2}[f(a+mx)+f(a)]-$$
$$-\frac{n^2-1}{12n}\{\Delta f_{m-1}[a+(m-1)x]-\Delta f_0[a+(m-1)x]\}+$$
$$+\frac{n^2-1}{24n}\{\Delta^2 f_{m-2}[a+(m-2)x]+\Delta^2 f_0[a+(m-2)x]\}-\frac{(n^2-1)(19n^3-1)}{720n^3}\times$$
$$\times\{\Delta^3 f_{m-3}[a+(m-3)x]-\Delta^3 f_0[a+(m-3)x]\}-\frac{(n^2-1)(9n^2-1)}{480n^3}\times$$
$$\times\{\Delta^4 f_{m-4}[a+(m-4)x]+\Delta^4 f_0[a+(m-4)x]\}-\ldots,$$

linking two sums obtained from division of the integration interval into m equal parts and mn equal parts.

12 dec.	$n=2(1)100.$	1st–6 th differences	[33]
10 dec.	$n=2(1)11.$	1st–6 th differences	[91]
8 dec.	$n=2(1)12.$	1st–3 rd differences	[43]
6 dec.	$n=2(1)11.$	1st–6 th differences	[136]

SUMS OF SERIES

Sums of negative powers of integers

$$S_n=1+\frac{1}{2^n}+\frac{1}{3^n}+\frac{1}{4^n}+\ldots$$

50 dec.	$n=1(1)167$	[63]
32 dec.	$n=2(1)107$	[33] [166]
32 dec.	$n=2(1)100$	[80]
32 dec.	$n=2(1)70$	[111]
24 dec.	$n=24(2)38$	[159]
22 24 dec.	$n=2(2)12$	[155]
22 dec.	$n=3$	[129]

20 dec.	$n = 2(2)12$	[54]
16 dec.	$n = 2(1)35$	[23] [34] [37] [61] [62] [77] [138] [140] [158]
16 dec.	$n = 2(1)20$	[32]
16; 17; 21 dec.	$n = 2(1)17;\ 18(1)23;\ 24(1)40$	[115]
16 dec.	$n = 2(1)16$	[39] [151]
16 dec.	$n = 2(1)10$	[57]
10 dec.	$n = 2(1)20$	[31]
10 dec.	$n = 2(1)11$	[14] [48]
6 dec.	$n = -\frac{1}{m};\ n = -1 - \frac{1}{m}.\ m = 1(1)20$	[154]

$$S'_n = S_n - 1 = \frac{1}{2^n} + \frac{1}{3^n} + \frac{1}{4^n} + \dots$$

32 dec.	$n = 2(1)107$	[33] [166]
32 dec.	$n = 2(1)100$	[80]
32 dec.	$n = 2(1)70$	[111]
24 dec.	$n = 24(2)38$	[159]
22—24 dec.	$n = 2(2)12$	[155]
22 dec.	$n = 3$	[129]
20 dec.	$n = 2(2)12$	[54]
16; 17; 21 dec.	$n = 2(1)17;\ 18(1)23;\ 24(1)40$	[115]
16 dec.	$n = 2(1)35$	[23] [34] [37] [61] [62] [77] [138] [140] [158]
16 dec.	$n = 2(1)20$	[32]
16 dec.	$n = 2(1)16$	[39] [151]
16 dec.	$n = 2(1)10$	[57]
10 dec.	$n = 2(1)20$	[31]
10 dec.	$n = 2(1)11$	[34] [48]

$$1 + \frac{S_n}{2^n} = 1 + \frac{1}{2^n} + \frac{1}{4^n} + \frac{1}{6^n} + \dots$$

32 dec.	$n = 2(1)100$	[80]
23 dec.	$n = 2(2)48$	[38]
16; 17; 21 dec.	$n = 2(1)17;\ 18(1)23;\ 24(1)40$	[115]

$$\frac{S_n}{2^n} = \frac{1}{2^n} + \frac{1}{4^n} + \frac{1}{6^n} + \dots$$

50 dec.	$n = 1(1)167$	[63]
32 dec.	$n = 2(1)100$	[80]
23 dec.	$n = 2(2)48$	[38]
16; 17; 21 dec.	$n = 2(1)17;\ 18(1)23;\ 24(1)40$	[115]

$$\frac{S_n-1}{2^n}=\frac{1}{4^n}+\frac{1}{6^n}+\frac{1}{8^n}+\dots$$

32 dec.	$n=2(1)50$	[80]

$$\frac{S'_n}{n}=\frac{S_n-1}{n}=\frac{1}{n}\left(\frac{1}{2^n}+\frac{1}{3^n}+\frac{1}{4^n}+\dots\right)$$

10 dec.	$n=1(2)11$	[59]

$$\frac{2}{n}\,\frac{S_n-1}{2^n}=\frac{2}{n}\left(\frac{1}{4^n}+\frac{1}{6^n}+\frac{1}{8^n}+\dots\right)$$

20 dec.	$n=2(2)30$	[6] [28] [38]

$$\left(1+\frac{1}{2^n}+\frac{1}{3^n}+\dots+\frac{1}{m^n}\right)(m!)^n$$

	$m=4.$ $n=1(1)20$	[165]
	$m=6.$ $n=1(1)16$	

$$s_n=1-\frac{1}{2^n}+\frac{1}{3^n}-\frac{1}{4^n}+\dots$$

50 dec.	$n=1(1)167$	[63]
32 dec.	$n=1(1)107$	[166]
32 dec.	$n=1(1)100$	[33]
22—24 dec.	$n=2(2)12$	[155]
16; 17; 21 dec.	$n=2(1)17;\ 18(1)23;\ 24(1)40$	[115]
16 dec.	$n=1(1)35$	[154]
12 dec.	$n=2(2)10$	[89]
10 dec.	$n=1(1)20$	[31]
6 dec.	$n=\frac{1}{m}.$ $m=1(1)20$	[154]
5 dec.	$n=1+\frac{1}{m}.$ $m=1(1)20$	

$$\frac{1}{2^n}s_n=\frac{1}{2^n}-\frac{1}{4^n}+\frac{1}{6^n}-\frac{1}{8^n}+\dots$$

50 dec.	$n=1(1)167$	[63]

$$\frac{2}{n}s_n=\frac{2}{n}\left[1-\frac{1}{2^n}+\frac{1}{3^n}-\frac{1}{4^n}+\dots\right]$$

20 dec.	$n=2(2)18$	[130]
5 dec.	$n=2(2)16$	[157]

$$s'_n = 1 - s_n = \frac{1}{2^n} - \frac{1}{3^n} + \frac{1}{4^n} - \ldots$$

50 dec.	$n = 1\,(1)\,167$	[63]
32 dec.	$n = 1\,(1)\,107$	[166]
32 dec.	$n = 1\,(1)\,100$	[80]
20 dec.	$n = 2\,(2)\,18$	[77]
16 dec.	$n = 2\,(1)\,34$	[77]

$$s_n - 1 + \frac{1}{2^n} = \frac{1}{3^n} - \frac{1}{4^n} + \frac{1}{5^n} - \ldots$$

32 dec.	$n = 1\,(1)\,53$	[80]

$$\frac{s_n}{2^{2n-2}} = \frac{1}{2^{2n-2}}\left(1 - \frac{1}{2^n} + \frac{1}{3^n} - \frac{1}{4^n} + \ldots\right)$$

18 dec.	$n = 1\,(2)\,19$	[148]

$$\frac{s'_n}{2^n\,(n+1)} = \frac{1}{2^n\,(n+1)}\left(\frac{1}{2^n} - \frac{1}{3^n} + \frac{1}{4^n} - \ldots\right)$$

22 dec.	$n = 2\,(2)\,34$	[156]

$$U_n = 1 + \frac{1}{3^n} + \frac{1}{5^n} + \frac{1}{7^n} + \ldots$$

32 dec.	$n = 2\,(1)\,67$	[33] [166]
32 dec.	$n = 2\,(1)\,50$	[80]
23 dec.	$n = 2\,(2)\,44$	[38]
22—24 dec.	$n = 2\,(2)\,12$	[155]
26; 17; 21 dec.	$n = 2\,(1)\,17;\ 18\,(1)\,23;\ 24\,(1)\,40$	[115]
16 dec.	$n = 2\,(1)\,35$	[154]
16 dec.	$n = 2\,(1)\,23$	[77]
12 dec.	$n = 2\,(2)\,24$	[164]
10 dec.	$n = 2\,(1)\,20$	[31]
5; 4 dec.	$n = 1 + \frac{1}{m}\,.\ m = 1\,(1)\,18;\ 19;\ 20$	[154]

$$U_n - 1 = \frac{1}{3^n} + \frac{1}{5^n} + \frac{1}{7^n} + \ldots$$

32 dec.	$n = 2\,(1)\,67$	[33] [166]
32 dec.	$n = 2\,(1)\,50$	[80]
23 dec.	$n = 2\,(2)\,44$	[38]

22—24 dec.	$n = 2(2)12$	[155]
16; 17; 21 dec.	$n = 2(1)17;\ 18(1)23;\ 24(1)40$	[115]
16 dec.	$n = 2(1)35$	[154]
16 dec.	$n = 2(1)23$	[77]
12 dec.	$n = 2(2)24$	[164]

$$\frac{2U_n}{n} = \frac{2}{n}\left(1 + \frac{1}{3^n} + \frac{1}{5^n} + \frac{1}{7^n} + \ldots\right)$$

20 dec.	$n = 2(2)24$	[6] [28]

$$\frac{2}{n}(U_n - 1) = \frac{2}{n}\left(\frac{1}{3^n} + \frac{1}{5^n} + \frac{1}{7^n} + \ldots\right)$$

20 dec.	$n = 2(2)38$	[6] [28]

$$u_n = 1 - \frac{1}{3^n} + \frac{1}{5^n} - \frac{1}{7^n} + \ldots$$

50 dec.	$n = 1(1)105$	[63]
32; 31; 29 dec.	$n = 2;\ 4;\ 6$	[147]
32; 30; 28 dec.	$n = 2;\ 4;\ 6$	[33]
28—32 dec.	$n = 2;\ 4;\ 6$	[147]
25 dec.	$n = 2$	[124]
23; 21 dec.	$n = 3;\ 5(2)11$	[146]
20 dec.	$n = 2$	[142] [156] [162]
18 dec.	$n = 1(1)38$	[33] [146]
16; 15 dec.	$n = 4;\ 6$	[145]
14 dec.	$n = 2$	[123]
13 dec.	$n = 2$	[88]
12 dec.	$n = 3(2)25$	[164]
9 dec.	$n = 2$	[141]
7; 8 dec.	$n = 2;\ 4;\ 6$	[14]
7 dec.	$n = 3(2)13$	[40]

$$\frac{2}{n}u_n = \frac{2}{n}\left(1 - \frac{1}{3^n} + \frac{1}{5^n} - \frac{1}{7^n} + \ldots\right)$$

28—16 dec.	$n = 1(2)19$	[130]

$$u_n' = 1 - u_n = \frac{1}{3^n} - \frac{1}{5^n} + \frac{1}{7^n} - \ldots$$

50 dec.	$n = 1(1)105$	[63]
33; 31; 29 dec.	$n = 2;\ 4;\ 6$	[147]
32 dec.	$n = 1(1)53$	[80]

28—32 dec.	$n = 2; 4; 6$	[147]
23; 21 dec.	$n = 1; 3; 5\,(2)\,11$	[146]
18 dec.	$n = 1\,(1)\,38$	[146]
13 dec.	$n = 2\,(1)\,20$	[77]

$$\frac{2}{n}u_n' = \frac{2}{n}\left(\frac{1}{3^n} - \frac{1}{5^n} + \frac{1}{7^n} - \dots\right)$$

25—18 dec.	$n = 1\,(2)\,17$	[77] [130]
18 dec.	$n = 2\,(2)\,34$	[146]

$$\frac{u_n'}{2^{n-2}} = \frac{1}{2^{n-2}}\left(\frac{1}{3^n} - \frac{1}{5^n} + \frac{1}{7^n} - \dots\right)$$

21 dec.	$n = 2\,(2)\,28$	[148]

$$(n-1)\frac{u_n'}{2^{n-2}} = \frac{n-1}{2^{n-2}}\left(\frac{1}{3^n} - \frac{1}{5^n} + \frac{1}{7^n} - \dots\right)$$

20 dec.	$n = 2\,(2)\,28$	[148]

$$u_n'' = 1 - \frac{1}{3^n} + \frac{1}{5^n} - u_n = \frac{1}{7^n} - \frac{1}{9^n} + \frac{1}{11^n} - \dots$$

33; 31; 29 dec.	$n = 2; 4; 6$	[147]

$$u_n''' = \frac{1}{11^n} - \frac{1}{13^n} + \frac{1}{15^n} - \dots$$

33; 31; 29 dec.	$n = 2; 4; 6$	[147]

$$\frac{u_n'}{2^{n-2}} = \frac{1}{2^{n-2}}\left(\frac{1}{5^n} - \frac{1}{7^n} + \frac{1}{9^n} - \dots\right)$$

21 dec.	$n = 2\,(2)\,20$	[148]

$$(n-1)\frac{u_n'}{2^{n-2}} = \frac{n-1}{2^{n-2}}\left(\frac{1}{5^n} - \frac{1}{7^n} + \frac{1}{9^n} - \dots\right)$$

20 dec.	$n = 2\,(2)\,22$	[148]

$$P_n = 1 + \frac{1}{3^n} - \frac{1}{5^n} - \frac{1}{7^n} + + - - \dots$$

20 dec.	$n = 2$	[148]
19 dec.	$n = 2$	[148]
19 dec.	$n = 2$	[33]

$$P_2 - 1 - \frac{1}{3^2} + \frac{1}{5^2} + \frac{1}{7^2} = \frac{1}{9^2} + \frac{1}{11^2} - - + + \dots$$

20 dec.		[148]

$$1 + \frac{1}{3^2} - P_2 = \frac{1}{5^2} + \frac{1}{7^2} - - + + \dots$$

20 dec.		[148]

$$\sum \frac{1}{n^2}$$

9 dec.	$n = 10\,(10)\,1000$	[22]

$$1 + \frac{1}{5^2} + \frac{1}{9^2} + \dots$$

20 dec.		[142] [156]
9 dec.		[33] [141]

$$\frac{1}{3^2} + \frac{1}{7^2} + \frac{1}{11^2} + \dots$$

20 dec.		[142] [156]
9 dec.		[33] [141]

Sums and products containing prime numbers only

$$\sum_n = \frac{1}{2^n} + \frac{1}{3^n} + \frac{1}{5^n} + \frac{1}{7^n} + \frac{1}{11^n} + \dots$$

50 dec.	$n = 1\,(1)\,167$	[63]
24 dec.	$n = 2\,(1)\,80$	[33] [159]
24 dec.	$n = 2$	[143]
15 dec.	$n = 2\,(2)\,36$	[38] [125]
15 dec.	$n = 2\,(1)\,35$	[138] [158]
15 dec.	$n = 3$	[160]

$$\frac{2}{n} \sum_n = \frac{2}{n} \left(\frac{1}{2^n} + \frac{1}{3^n} + \frac{1}{5^n} + \frac{1}{7^n} + \frac{1}{11^n} + \dots \right)$$

24 dec.	$n = 2\,(2)\,74$	[159]

$$\gamma_n = \frac{1}{3^n} - \frac{1}{5^n} + \frac{1}{7^n} + \frac{1}{11^n} - \frac{1}{13^n} - \frac{1}{17^n} + \dots$$

14 dec.	$n = 2$	[162]
12 dec.	$n = 1\,(2)\,25$	[161]
7 dec.	$n = 1\,(2)\,13$	[40]

$$\Gamma_n = \frac{1}{2^n} + \frac{1}{5^n} - \frac{1}{7^n} + \frac{1}{11^n} - \frac{1}{13^n} + \dots$$

12 dec. $n = 1\,(2)\,39$ [163]

$$\Gamma'_n = \Gamma_n - \frac{1}{2^n} = \frac{1}{5^n} - \frac{1}{7^n} + \frac{1}{11^n} - \frac{1}{13^n} + \dots$$

12 dec. $n = 1\,(2)\,17$ [163]

$$\frac{1}{2}\left(\Gamma_1 - \gamma_1 - \frac{1}{6}\right) = \frac{1}{5} - \frac{1}{7} + \frac{1}{17} - \frac{1}{19} + \frac{1}{29} - \frac{1}{31} + \dots$$

12 dec. [163]

$$\frac{1}{2}\left(\Gamma_2 + \gamma_1 - \frac{5}{6}\right) = \frac{1}{11} - \frac{1}{13} + \frac{1}{23} - \frac{1}{37} + \frac{1}{47} + \frac{1}{59} - \frac{1}{61} - \dots$$

12 dec. [163]

$$P(p) = \left(1 - \frac{1}{2}\right)\left(1 - \frac{1}{3}\right)\left(1 - \frac{1}{5}\right)\dots\left(1 - \frac{1}{p}\right)$$

7 dec. $p = 2$ (primes) 9973 [144]
6 dec. $p \leqslant 1229$ [60]

$$\lg P(p)$$

7 dec. $p = 2$ (primes) 9973 [144]

Various sums and products

$$\sum \frac{1}{n}$$

8 dec. $n = 1\,(1)\,200;\ 10\,(10)\,1000$ [22]

$$p_1 = \frac{\pi}{2\sqrt{2}} = 1 + \frac{1}{3} - \frac{1}{5} - \frac{1}{7} + + - - \dots$$

31 dec. [33] [148]

$$q_1 = \frac{1}{\sqrt{2}} \ln(\sqrt{2} + 1) = 1 - \frac{1}{3} - \frac{1}{5} + \frac{1}{7} + - - + \dots$$

21 dec. [148]
20 dec. [33]

$$q_1 - 1 + \frac{1}{3} + \frac{1}{5} = \frac{1}{7} + \frac{1}{9} - - + + \ldots$$

18 dec. [148]

$$q_1 - 1 + \frac{1}{3} + \frac{1}{5} - \frac{1}{7} = \frac{1}{9} - \frac{1}{11} - \frac{1}{13} + + - - \ldots$$

21 dec. [148]

$$1 - \frac{1}{3} - q_1 = \frac{1}{5} - \frac{1}{7} - \frac{1}{9} + + - - \ldots$$

21 dec. [148]

$$\frac{1}{n}\left(1 + \frac{1}{3} + \frac{1}{5} + \ldots + \frac{1}{2n-1}\right)$$

exact $n = 1\,(1)\,14$ [49]

$$\Sigma x^n$$

exact $x = 1\,(1)\,50.$ $n = 1\,(1)\,5$ [1]

$$a x^n$$

3–7 fig. $a = 1\,(1)\,9.$ $x = 0\,(0{,}001)\,1.$ $n = 1\,(1)\,7$ [5]

$$P_0(x) = 1 - \frac{1^2 \cdot 3^2}{2!\,(8x)^2} + \frac{1^2 \cdot 3^2 \cdot 5^2 \cdot 7^2}{4!\,(8x)^4} - \ldots$$

4 dec. $x = 6\,(0{,}1)\,7$ [116]

$$P_1(x) = 1 + \frac{3 \cdot 5}{2!\,(8x)^2} - \frac{3 \cdot 5 \cdot 21 \cdot 45}{4!\,(8x)^4} + \ldots$$

4 dec. $x = 6\,(0{,}1)\,7$ [116]

$$Q_0(x) = -\frac{1}{8x} + \frac{1^2 \cdot 3^2 \cdot 5^2}{3!\,(8x)^3} - \frac{1^2 \cdot 3^2 \cdot 5^2 \cdot 7^2 \cdot 9^2}{5!\,(8x)^5} + \ldots$$

4 dec. $x = 6\,(0{,}1)\,7$ [116]

$$Q_1(x) = \frac{3}{8x} - \frac{3 \cdot 5 \cdot 21}{3!\,(8x)^3} + \frac{3 \cdot 5 \cdot 21 \cdot 45 \cdot 77}{5!\,(8x)^5} - \ldots$$

4 dec. $x = 6\,(0{,}1)\,7$ [116]

$$M_n = \frac{0!}{n!} + \frac{1!}{(n+1)!} + \frac{2!}{(n+2)!} + \ldots = \frac{1}{(n-1)(n-1)!}$$

24--41 dec. $n = 2\,(1)\,21$ [115]

$$K_n = \frac{0!}{n!} - \frac{1!}{(n+1)!} + \frac{2!}{(n+2)!} - \frac{3!}{(n+3)!} + \ldots$$

18—20 dec. $n = 2\,(1)\,10$ [115]

$$N_n = \frac{0!}{n!} + \frac{2!}{(n+2)!} + \frac{4!}{(n+4)!} + \ldots$$

18 dec. $n = 2\,(1)\,11$ [115]

$$Q_n = \frac{1!}{(n+1)!} + \frac{3!}{(n+3)!} + \frac{5!}{(n+5)!} + \ldots$$

18 dec. $n = 2\,(1)\,11$ [115]

$$f(x) = 2x \sum_{n=0}^{\infty} (-1)^n \left[(2n+1)^2 - x^2\right]^{-1}$$

4 fig. $x = 0\,(0{,}05)\,1$ [122]

$$F(x) = (1-x)\left[\frac{1}{2x} f(x) + \frac{1}{2} \ln 2 - \frac{1}{1+x}\right]$$

4 fig. $x = -1\,(0{,}05)\,1$ [122]

$$C_n^r p^r q^{n-r}$$

7 dec. $p = 0\,(0{,}01)\,0{,}5.$ $q = 1 - p.$ $n = 2\,(1)\,49.$ $r = 1\,(1)\,n-1$ [95]

$$\sum_{s=r}^{n} C_n^s p^s q^{n-s}$$

7 dec. $p = 0\,(0{,}01)\,0{,}5.$ $q = 1 - p.$ $n = 2\,(1)\,49.$ $r = 1\,(1)\,n$ [95]

$$f_k(x) = \sum_{n=1}^{\infty} (2n-1)^{-k} \cos (2n-1)\,x$$

4 dec. $k = 1\,(2)\,5.$ $x = m\frac{\pi}{180}.$ $m = 0\,(1°)\,180°$ [13]

$$f_k(x) = \sum_{n=1}^{\infty} (2n-1)^{-k} \sin(2n-1)x$$

4 dec. $k = 2\,(2)\,6$. $x = m\frac{\pi}{180}$. $m = 0\,(1°)\,180°$ [13]

$$F_k(x) = \sum_{n=1}^{\infty} n^{-k} \cos nx$$

4 dec. $k = 1\,(2)\,5$. $x = m\frac{\pi}{180}$. $m = 0\,(1°)\,180°$ [13]

$$F_k(x) = \sum_{n=1}^{\infty} n^{-k} \sin nx$$

4 dec. $k = 2\,(2)\,6$. $x = m\frac{\pi}{180}$. $m = 0\,(1°)\,180°$ [13]

$$\sum_{n=1}^{\infty} \frac{\sin^2 n\pi x}{x^2 n^3}$$

3—7 dec. $x = 0(0{,}01)\,1$ [153]

$$G(x, z) = \sum_{n=1}^{\infty} n^{-1} \sin(nx + n^{-1}z)$$

5 dec. $\left\{\begin{array}{l} z = 0;\ 3{,}15012;\ 4{,}14543;\ 6{,}30024;\ 8{,}29086; \\ \quad 8{,}45505;\ 9;\ 12{,}60048;\ 16{,}91010;\ 18;\ 18{,}90072; \\ \quad 25{,}20096;\ 27;\ 36.\ x = -\pi\left(\frac{\pi}{36}\right)\pi \end{array}\right\}$ [137]

$$\sum_{m=0}^{k} \frac{a^m e^{-a}}{m!}$$

6 dec. $a = 0{,}1\,(0{,}1)\,1\,(1)\,3$. $k = 0\,(1)\,15$ [3]

$$K(x) = \sum_{k=-\infty}^{\infty} (-1)^k e^{-2k^2x^2}$$

6 dec. $x = 0{,}28\,(0{,}01)\,2{,}5\,(0{,}05)\,3$ [3]

$$1 - 2\sum_{\nu=1}^{\infty} (-1)^{\nu-1} e^{-2\nu^2 z^2}$$

6 dec. $z = 0{,}28\,(0{,}01)\,2{,}5\,(0{,}05)\,3$ [114]

$$P_m = \sum_{t=1}^{\infty} \frac{(\ln t)^m}{t!}$$

10 fig. $m = 1\,(1)\,15$ [132]

$$\frac{1}{m!} \sum_{t=1}^{\infty} \frac{(\ln t)^m}{t!}$$

10 fig. $m = 1\,(1)\,15$ [132]

$$\sum_{k=1}^{m} \frac{P_k}{k!}$$

10 dec. $m = 1\,(1)\,15$ [132]

$$2 \sum_{r=1}^{x} (2r-1)^n$$

exact $n = 2\,(2)\,10.\quad x = 1\,(1)\,10$ [52]
exact $n = 2\,(2)\,6.\quad x = 1\,(1)\,25$ [112]

$$4\left\{x \sum_{r=1}^{x} (2r-1)^4 - \left[\sum_{r=1}^{x} (2r-1)^2\right]^2\right\}$$

exact $x = 1\,(1)\,25$ [112]

$$4\left\{\sum_{r=1}^{x} (2r-1)^6 \sum_{r=1}^{x} (2r-1)^2 - \left[\sum_{r=1}^{x} (2r-1)^2\right]^4\right\}$$

exact $x = 1\,(1)\,25$ [112]

$$\frac{\sum_{r=1}^{x} (2r-1)^4}{\sum_{r=1}^{x} (2r-1)^2}$$

exact $x = 1\,(1)\,25$ [112]

Logarithms of sums

$$\lg S_n = \lg\left(1 + \frac{1}{2^n} + \frac{1}{3^n} + \dots\right)$$

15 dec. $n = 2\,(1)\,35$ [138]

$$\ln S_n = \ln\left(1+\frac{1}{2^n}+\frac{1}{3^n}+\dots\right)$$

50 dec.	$n = 1\,(1)\,167$	[63]
24 dec.	$n = 2\,(1)\,80$	[33] [159]
15 dec.	$n = 2\,(1)\,34$	[138] [158]

$$\ln U_n = \ln\left(1+\frac{1}{3^n}+\frac{1}{5^n}+\frac{1}{7^n}+\dots\right)$$

14 dec.	$n = 4;\ 6;\ 12;\ 20$	[162]
12 dec.	$n = 2\,(4)\,22$	[161]

$$\lg u_n = \lg\left(1-\frac{1}{3^n}+\frac{1}{5^n}-\frac{1}{7^n}+\dots\right)$$

18 dec.	$n = 1\,(1)\,38$	[33]

$$-\ln u_n = -\ln\left(1-\frac{1}{3^n}+\frac{1}{5^n}-\frac{1}{7^n}+\dots\right)$$

20 dec.	$n = 2$	[162]
14 dec.	$n = 10;\ 14;\ 22;\ 26$	[162]
12 dec.	$n = 6$	[162]
12 dec.	$n = 1\,(2)\,25$	[161]
12 dec.	$n = 6$	[162]

$$\ln G_n = \ln\left(1+\frac{1}{2^n}+\frac{1}{4^n}+\frac{1}{5^n}+\frac{1}{7^n}+\frac{1}{8^n}+\dots\right)$$

12 dec.	$n = 2\,(4)\,14$	[163]

$$-\ln g_n = -\ln\left(1-\frac{1}{2^n}+\frac{1}{4^n}-\frac{1}{5^n}+\frac{1}{7^n}-\frac{1}{8^n}+\dots\right)$$

12 dec.	$n = 1(2)\,11$	[163]

$$\ln W_n = \ln\left(1+\frac{1}{5^n}+\frac{1}{7^n}+\frac{1}{11^n}+\frac{1}{13^n}+\frac{1}{17^n}+\dots\right)$$

12 dec.	$n = 2\,(4)\,14$	[163]

$$-\ln h_n = -\ln\left(1-\frac{1}{5^n}+\frac{1}{7^n}-\frac{1}{11^n}+\frac{1}{13^n}-\frac{1}{17^n}+\dots\right)$$

12 dec.	$n = 1\,(2)\,17$	[163]

Chapter 14

MATHEMATICAL CONSTANTS

This chapter gives a most complete description of constants involving π , the base e of natural logarithms and the modulus M for conversion from natural to common logarithms.

The wealth of possible combinations in which these constants are encountered has made it necessary to unite them into groups and to use fractional indices in the entries.

The constants described in the chapter are grouped in the following sections:

constants involving π;
constants involving e;
constants involving modulus M;
various constants involving π, e and M;
Bernoulli numbers;
Euler numbers;
Euler's constants C and γ;
algebraic constants
certain values of trigonometric functions;
certain values of hyperbolic functions
certain values of integrals;
logarithms of various constants;
special numbers and miscellaneous constants.

See Chapter 1 for the powers of various constants. Particular values of certain special functions are to be found in the corresponding chapters.

Constants involving π

π

2035 dec.	[187]
707 dec.	[97] [222] [248]
411 dec.	[145]
140 dec.	[138]

[1] See pages 521 and 550 for the references to Chapter 14.

70 dec.	[140]
52 dec.	[88]
26 dec.	[100]
24 dec.	[6] [69]
15 dec.	[19]
10 dec.	[1]
8 dec.	[13] [20]
7 dec.	[22]
6 dec.	[20] [21]

$$\frac{1}{\pi}$$

2037 dec.	[188]
253 dec.	[209]
140 dec.	[153]
70 dec.	[140]
52 dec.	[97]
36 dec.	[36]
24 dec.	[69] [100]
15 dec.	[19]
10 dec.	[1]
8 dec.	[16] [20]
7 dec.	[21] [22]

$$n\pi$$

441 dec.	$n = \frac{1}{4}$	[145]
51 dec.	$n = \frac{1}{4}; \frac{1}{6}$	[153]
48 dec.	$n = \frac{1}{4}$	[97]
35 dec.	$n = 1\,(1)\,9$	[140]
32 dec.	$n = 1\,(1)\,9$	[97]
31 dec.	$n = \sqrt{2}; \frac{1}{\sqrt{2}}$	[134]
31 dec.	$n = \frac{1}{2\sqrt{2}}$	[43] [195]
30 dec.	$n = 1\,(1)\,100$	[79]
30 dec.	$n = 2;\ 4;\ 36;\ \frac{2}{3};\ \frac{4}{3}$	[134]
24 dec.	$n = 1\,(1)\,9$	[61]
13 dec.	$\left\{\begin{array}{l} n = 2;\ 3;\ 4;\ 6;\ 12;\ 18;\ 24;\ 36;\ 108; \\ \quad 144;\ \frac{2}{3};\ \frac{3}{4};\ \frac{4}{3};\ \sqrt{2};\ \frac{1}{\sqrt{2}} \end{array}\right\}$	[84]
10 dec.	$n = 2$	[1]

9 dec.	$n = \sqrt{2};\ \frac{1}{2}$	[23]
9 dec.	$n = \frac{1}{360}$	[21]
8 dec.	$n = 2;\ 3;\ 4;\ \frac{1}{2};\ \frac{1}{3};\ \frac{1}{4};\ \frac{1}{6}$	[16]
8 dec.	$n = \frac{1}{2};\ \frac{1}{3};\ \frac{1}{4};\ \frac{1}{6}$	[20]
7 dec.	$n = \sqrt{2};\ \frac{1}{2};\ \frac{1}{3};\ \frac{1}{4}$	[22]
7 dec.	$n = \frac{1}{4};\ \frac{1}{6}$	[21]
6 dec.	$n = 1\,(1)\,100$	[57]
6 dec.	$n = 10\,(1)\,100$	[1] [53]
6 dec.	$n = 2$	[20]
6 dec.	$n = 2;\ 3;\ \frac{1}{2};\ \frac{1}{3}$	[21]
6 dec.	$n = \frac{1}{\sqrt{2}}$	[22]
5—7 dec.	$n = 1\,(1)\,2000$	[105]
5 dec.	$n = 1\,(1)\,1020$	[73]
5 fig.	$n = 1\,(1)\,1000$	[120]
5 dec.	$n = 4;\ \frac{1}{360}$	[20]
5 dec.	$n = 4$	[21]
2 dec.	$n = 1\,(1)\,1000$	[41]

$$\frac{n\pi}{m}$$

36 dec.	$n = 1.$	$m = 180\ k.$	$k = 1;\ 60;\ 60^2$	[36]
35 dec.	$n = 1\,(1)\,9.$	$m = 180$		[140]
35 dec.	$n = 1\,(1)\,60\,(60)\,600\,(600)\,3600.$	$m = 648000$		[246]
35 dec.	$n = 1.$	$m = 1\,(1)\,9$		[140]
30 dec.	$n = 1.$	$m = 2;\ 3;\ 4;\ 6;\ 8;\ 12;\ 24;\ 180;\ 360;\ 10800;\ 648000$		[134]
24 dec.	$n = 1\,(1)\,9.$	$m = 4$		[61]
24 dec.	$n = 1\,(1)\,9.$	$m = 6$		[61]
24 dec.	$n = 1\,(1)\,9.$	$m = 12$		[61]
23 dec.	$n = 1.$	$m = 180\ k.$	$k = 1;\ 60;\ 60^2$	[64]
16 dec.	$n = 1\,(1)\,9.$	$m = 2$		[70]
15 dec.	$n = 1\,(1)\,100.$	$m = 2$		[122] [123] [124] [130] [131]
15 dec.	$n = 1.$	$m = 180\ k.$	$k = 1;\ 60;\ 60^2$	[29] [142]

13 dec.	$n=1$.	$m=2; 3; 4; 6; 8; 9; 12; 18; 36; 108; 144; 216; 648$	[84]
10 dec.	$n=1$.	$m=1(1)9$	[142]
10 dec.	$n=\frac{1}{2}; 1; 2$.	$m=1(1)200$	[140]
8 dec.	$n=4$.	$m=3$	[16]
6 dec.	$n=1$.	$m=1(1)500$	[98]
6 dec.	$n=1$.	$m=1(1)100$	[28]
6 dec.	$n=2$.	$m=1(1)100$	[100]
6 dec.	$n=1(1)10$.	$m=1(1)10$	[100]
6 dec.	$n=1(1)9$.	$m=1(1)7; 16; 24; 32; 180$	[85]
6 dec.	$n=3; 4$.	$m=2; 3; 5$	[21]
4 dec.	$n=2$.	$m=1(1)100$	[87]

$$\frac{n}{\pi}$$

134 dec.	$n=180; 200$	[204]
131 dec.	$n=648000$	[204]
36 dec.	$n=2; 4$	[36]
35 dec.	$n=1(1)9$, $n=180\,k$. $k=1(1)9$	[140]
35 dec.	$n=4$	[50]
34—31 dec.	$n=180\,k$. $k=1; 60; 60^2$	[36]
32 dec.	$n=1(1)9$	[97]
30 dec.	$n=2; 4; 6; 180; 360; 10800; 648000; \sqrt{2}$	[134]
30 dec.	$n=180$	[94]
13 dec.	$n=\sqrt{2}$	[84]
12 dec.	$n=1(1)100$	[79]
8 dec.	$n=2; 3; 4; 6$	[16]
8 dec.	$n=3$	[20]
7 dec.	$n=2; 3; 4$	[21]
6 dec.	$n=10(1)99$	[1] [53] [119]
6 dec.	$n=4; 6$	[21]
5 dec.	$n=1(1)999$	[8]
5 dec.	$n=4$	[20]
4 dec.	$n=360$	[20] [21]
3 dec.	$n=180$	[20]

$$\frac{n}{m\pi}$$

51 dec.	$n=1$.	$m=4$	[153]
35 dec.	$n=1$.	$m=1(1)9$	[140]

35 dec.	$n=1$.	$m=2$	[50]
30 dec.	$n=1$.	$m=2;\ 4;\ 6;\ \sqrt{2}$	[134]
30 dec.	$n=1$.	$m=2$	[94]
24 dec.	$n=1\,(1)\,9$.	$m=4$	[61]
20 dec.	$n=\left(\frac{1}{6}+\ln 2\right);\ \ln 2$.	$m=4$	[241]
13 dec.	$n=1$.	$m=2;\ 3;\ 4;\ 6;\ 18;\ 24;\ 36;\ 48;\ 72;\ 96;\ 144;\ \sqrt{2}$	[84]
10 dec.	$n=1$.	$m=1\,(1)\,9$	[142]
10 dec.	$n=1$.	$m=2$	[1]
8 dec.	$n=1$.	$m=2;\ 3;\ 4$	[16]
8 dec.	$n=1$.	$m=4$	[21]
8 dec.	$n=3$.	$m=4$	[20]
7 dec.	$n=1$.	$m=2;\ 3$	[21]
7 dec.	$n=3$.	$m=4$	
7 dec.	$n=108$.	$m=0{,}15\ (0{,}05)\,1\,(0{,}1)\,2\ (0{,}25)\,4\,(0{,}5)\,10\ (1)\ 70$	[74]
6 dec.	$n=1\,(1)\,100$.	$m=2$	[100]
6 dec.	$n=1\,(1)\,10$.	$m=1\,(1)\ 10$	

$m\pi^n$

261 dec.	$m=1$.	$n=2$	[209]
217 dec.	$m=1$.	$n=2$	[259]
140 dec.	$m=1$.	$n=\frac{1}{2}$	[153]
66 dec.	$m=1;\ 2;\ \frac{1}{6}$.	$n=-\frac{1}{2}$	
60 dec.	$m=1$.	$n=\frac{1}{2}$	[97]
51 dec.	$m=2$.	$n=\frac{1}{2}$	[153]
51 dec.	$m=1$.	$n=\frac{1}{3};\ \frac{2}{3};\ -2$	
51 dec.	$m=\sqrt[3]{36}$.	$n=\frac{1}{3}$	
51 dec.	$m=1$.	$n=\frac{1}{2}$	[97]
51 dec.	$m=\sqrt[3]{6}$.	$n=\frac{2}{3}$	[153]
51 dec.	$m=\frac{1}{6}$.	$n=-2$	
50 dec.	$m=\sqrt[3]{6}$.	$n=-\frac{1}{3}$	
38—41 dec.	$m=1$.	$n=2\,(1)\,16$	[63]
35 dec.	$m=\frac{1}{\sqrt{2}}$.	$n=-\frac{1}{2}$	[50]

32 dec.	$m=1.$	$-n=1\,(1)\,32$	[97]
31 dec.	$m=1.$	$n=-\frac{1}{2}$	
30 dec.	$m=\frac{1}{2}n.$	$n=1\,(1)\,20$	[37] [136]
30 dec.	$m=1.$	$n=\pm 2$	[134]
30 dec.	$m=2;\ 1;\ \frac{1}{2};\ \frac{1}{8}.$	$n=\pm\frac{1}{2}$	
30 dec.	$m=\frac{1}{6}.$	$n=-\frac{1}{2}$	
30 dec.	$m=\frac{1}{2}.$	$n=\frac{1}{2}$	[255]
30 dec.	$m=\sqrt{2};\frac{1}{\sqrt{2}}.$	$n=\frac{1}{2};\ -\frac{1}{2}$	[134]
28 dec.	$m=2^{-k-\frac{1}{2}}.\ k=2;\ 3;\ 6;\ 7;\ 9;\ 14;\ 15;\ 18;\ 22;\ 24;\ 25;\ 33;\ 34;\ 37$	$n=-\frac{1}{2}$	[50]
25—30 dec.	$m=\left(\frac{1}{2}\right)^n.$	$n=1\,(1)\,20$	[36]
24 dec.	$m=2\,(2)\,18.$	$n=-\frac{1}{2}$	[61]
24—27 dec.	$m=1.$	$-n=1\,(1)\,12$	[205]
22—24 dec.	$m=1.$	$n=1\,(1)\,12$	
21 dec.	$m=\frac{1}{\sqrt{6}}.$	$n=\frac{1}{3}$	[61]
21 dec.	$m=\sqrt[3]{6}.$	$n=-\frac{1}{3}$	
21—22 dec.	$m=1.$	$n=1\,(1)\,12$	[76]
20 dec.	$m=1.$	$-n=1\,(1)\,12$	
20 dec.	$m=(180k)^{-n}.\ k=1;\ 60;\ 60^2$	$n=1\,(1)\,11$	
15 dec.	$m=1.\ n=0(0{,}001)1;$	$\pm n=\frac{1}{4};\ \frac{1}{3};\ \frac{1}{2};\ \frac{2}{3};\ \frac{3}{4}\ \&\ 1(1)12$	[125]
15 dec.	$m=1.$	$n=\pm 2;\ \pm\frac{1}{2};\ \pm\frac{1}{3}$	[19]
15 dec.	$m=\sqrt{k}.\ k=1;\ 2;\ 3;\ 5;\ 6.$	$n=\frac{1}{2}$	[132]
15 dec.	$m=\sqrt{2};\ \frac{1}{\sqrt{2}}.$	$n=\pm\frac{1}{2}$	[19]

13 dec.	$m = 1.$	$n = 1;\ 2;\ 4$	[84]
13 dec.	$m = \sqrt[3]{36};\ \sqrt[3]{\frac{3}{4}}.$	$n = \frac{1}{3}$	
13 dec.	$m = \sqrt[3]{6}.$	$n = \frac{2}{3}$	
13 dec.	$m = 6\sqrt{10};\ \sqrt{10}.$	$n = -\frac{1}{2}$	
11 dec.	$m = \sqrt[3]{36}.$	$n = \frac{1}{3}$	[134]
10 dec.	$m = 1.$	$n = \pm 2;\ \pm\frac{1}{2}$	[1]
10 dec.	$m = \sqrt{2}.$	$n = \frac{1}{2}$	
10 dec.	$m = \frac{1}{\sqrt{2}}.$	$n = \pm\frac{1}{2}$	
10 dec.	$m = \sqrt[3]{\frac{4}{3}}.$	$n = \frac{1}{3}$	[87]
10 dec.	$m = \sqrt[3]{\frac{3}{4}}.$	$n = -\frac{1}{3}$	
9 dec.	$m = 1.$	$n = \frac{1}{2};\ 2;\ 3$	[23]
9 dec.	$m = \frac{1}{\sqrt{2}}.$	$n = \frac{1}{2}$	
9 dec.	$m = 1.$	$n = \frac{1}{2}$	[20]
8 dec.	$m = 1.$	$n = \frac{1}{2};\ 2.$	[13]
8 dec.	$m = 1.$	$n = \pm\frac{1}{2};\ \pm\frac{1}{3};\ \pm 2;\ \pm 3$	[16]
8 dec.	$m = 1.$	$n = \frac{1}{2};\ \frac{2}{3}$	[20]
8 dec.	$m = \sqrt{3}.$	$n = -\frac{1}{2}$	
8 dec.	$m = \sqrt[3]{6}.$	$n = \frac{2}{3}$	[134]
8 dec.	$m = \frac{1}{\sqrt{2}};\ \frac{1}{2};\ \sqrt{2};\ 2.$	$n = \pm\frac{1}{2}$	[16]
8 dec.	$m = \sqrt[3]{\frac{3}{4}}.$	$n = -\frac{1}{2}$	
7 dec.	$m = 1.$	$n = \frac{1}{2};\ \frac{1}{3};\ \pm 2;\ -3;\ \frac{3}{2};\ \frac{4}{3}$	[22]
7 dec.	$m = \frac{1}{4}.$	$n = 2$	

7 dec.	$m = 4.$	$n = 2$	[20] [21]
7 dec.	$m = 1.$	$n = -\frac{1}{2}; -\frac{1}{3}; -2; -3$	
7 dec.	$m = \frac{1}{2}; \sqrt{2}; \frac{1}{\sqrt{2}}; \sqrt{3}.$	$n = -\frac{1}{2}$	[21]
7 dec.	$m = \frac{1}{\sqrt[3]{6}}.$	$n = \frac{1}{3}$	
7 dec.	$m = \sqrt[3]{\frac{3}{4}}.$	$n = -\frac{1}{3}$	
6 dec.	$m = 1\,(1)\,100.$	$n = 2$	[34]
6 dec.	$m = \frac{1}{4};\ 1;\ 4.$	$n = 2$	[20] [21] [22]
6 dec.	$m = 1.$	$n = 3;\ \pm 4;\ -5$	[22]
6 dec.	$m = 1\,(1)\,100.$	$n = -2$	[34]
6 dec.	$m = k\sqrt{2}.\quad k = 1\,(1)\,100.$	$n = \frac{1}{2}$	
6 dec.	$m = \sqrt{k}.\ k = 1\,(1)\,80.$	$n = \frac{1}{2}$	[199]
6 dec.	$m = 1;\ 2;\ \sqrt{2};\ 2\sqrt{2};\ \frac{1}{\sqrt{2}}.$	$n = \frac{1}{2}$	[21]
6 dec.	$m = \sqrt{2};\ \frac{1}{\sqrt{2}}.$	$n = \frac{1}{2}$	
6 dec.	$m = \sqrt[3]{2};\ \frac{1}{\sqrt[3]{2}};\ \frac{1}{\sqrt[3]{4}}.$	$n = \frac{1}{3}$	[22]
6 dec.	$m = \sqrt{2};\ \sqrt{3}.$	$n = -\frac{1}{2}$	
6 dec.	$m = \sqrt[3]{2};\ \sqrt[3]{3}.$	$n = -\frac{1}{3}$	
6 dec.	$m = 1.$	$n = \frac{1}{3};\ \frac{2}{3}$	[20] [21]
6 dec.	$m = 2.$	$n = -\frac{1}{2}$	[21]
6 dec.	$m = \frac{1}{\sqrt{2}}k.\ k = 1\,(1)100.$	$n = -\frac{1}{2}$	[34]
5 dec.	$m = 1.$	$n = 1\,(1)\,16$	[64]

5 dec.	$m=1.$	$n=-\frac{1}{2}$; 3; 4	[20]
5 dec.	$m=\sqrt{2}.$	$n=\frac{1}{2}$	[20]
5 dec.	$m=\frac{1}{4}.$	$n=2$	[20]
5 dec.	$m=1.$	$n=3;\ 4;\ 5$	[21]
5 dec.	$m=2;\ 3.$	$n=2$	[21]
5 dec.	$m=1.$	$n=5$	[22]
4 dec.	$m=\sqrt{4k}.\ k=1(1)999.$	$n=-\frac{1}{2}$	[7]

$$\frac{\pi^{12}-1}{\pi-1} \text{ и } \frac{1-\pi^{-12}}{1-\pi^{-1}}$$

24 dec.	[205]

Constants involving *e*

e

2010 dec.	[187]
808 dec.	[169]
707 dec.	[150] [254]
606 dec.	[169]
404 dec.	[169]
225 dec.	[117] [171]
223 dec.	[185] [189]
205 dec.	[146] [223]
187 dec.	[215]
137 dec.	[111] [151] [159] [253]
113 dec.	[170]
110 dec.	[224]
105 dec.	[97] [154]
72 dec.	[97]
52 dec.	[97]
25 dec.	[11]
24 dec.	[5] [69]
15 dec.	[19]
9 dec.	[23]
8 dec.	[13] [16]
6 dec.	[21] [22]
5 dec.	[20]

$\frac{1}{e}$

477 dec.		[254]
105 dec.		[97] [154] [189]
72 dec. 52 dec.		[97]
15 dec.		[19]
8 dec.		[16]
7 dec.		[21]
6 dec.		[20] [22]

e^n

290 fig.	$n = 10$	[210]
120 dec.	$n = 0{,}1$	[189]
116 dec.	$n = 100$	[254]
72 dec. 32 fig.	$n = \frac{1}{2}$ $n = 1\,(1)\,32$	[97]
15 dec.	$n = \frac{1}{2};\ 2$	[19]
8 dec.	$n = \frac{1}{2};\ 2$	[16]
6 dec.	$n = \frac{1}{2};\ 2$	[21] [22]
6 dec.	$n = \frac{1}{3}$	[22]
5 dec.	$n = \frac{1}{2};\ 2$	[20]
5 dec. 4 dec.	$n = 3;\ 4$ $n = 5$	[21]

e^{-n}

294 dec.	$n = 10$	[210]
258 dec.	$n = 10$	[254]
120 dec.	$n = 0{,}1$	[189]
83 dec.	$n = 100$	[254]
72 dec. 32 dec.	$n = \frac{1}{2}$ $n = 1\,(1)\,32$	[97]
15 dec.	$n = \frac{1}{2};\ 2$	[19]
8 dec.	$n = \frac{1}{2};\ 2$	[16]

7 dec.	$n = \frac{1}{2};\ 2$	[21]
6 dec.	$n = 2$	[22]

$$\frac{n}{k} e^{\pm 1}$$

32 dec.	$n = 1\,(1)\,9.\quad k = 1$	[97]
9 dec.	$n = 1\,(1)\,10.\ k = 1\,(1)\,10$	[100]

$$\frac{1}{2\sqrt{2}} e$$

23 dec.	[255]

$$(e - 1)^{-1}$$

12 dec.	[228]

$$e_n(p)$$

$e_2(p) = e^{e^p};\ e_3(p) = e^{e_2(p)}$ etc.

4 fig.	$n = 2;\ 3;\ 4.\qquad p = -4\,(1)\,5$	[203]

Constants involving modulus M

$$M$$

330 dec.	[210]
325 dec.	[135]
282 dec.	[97]
272 dec.	[167] [226]
263 dec.	[97] [225]
205 dec.	[146] [223]
105 dec.	[93]
102 dec.	[219]
81 dec.	[157]
61 dec.	[116]
59 dec.	[111] [215]
42 dec.	[47]
40 dec.	[133]
25 dec.	[11]
24 dec.	[5]
20 dec.	[60]

15 dec.	[19]
8 dec.	[13] [16]
7 dec.	[21]
6 dec.	[20] [21] [22]
5 dec.	[22]

nM

100 dec.	$n = 2;\ 4;\ 8;\ \frac{1}{2};\ \frac{1}{3};\ \frac{1}{4}$	[93]
70 dec.	$n = 1\,(1)\,9$	[10]
61 dec.	$n = 1\,(1)\,100$	[97]
40 dec.	$n = 1\,(1)\,100$	[46]
40 dec.	$n = 1\,(1)\,10$	[47]
34 dec.	$n = 10\,(1)\,20\,(2)\,52\,(4)\,96$	[93]
30 dec.	$n = \frac{1}{2}k^2.\ k = 10\,(1)\,20\,(2)\,52\,(4)\,96$	
24 dec.	$n = 1\,(1)\,100$	[10] [256]
24 dec.	$n = 1\,(1)\,9$	[69]
21 dec.	$n = 1\,(1)\,100$	[96]
18 dec.	$n = 1\,(1)\,9$	[67]
16 dec.	$n = 1\,(1)\,9$	[70]
12 dec.	$n = 11\,(1)\,110$	[133]
12 dec.	$n = 10\,(1)\,100$	[14]
11 dec.	$n = 1\,(1)\,100$	[175]
10 dec.	$n = -500\,(1)\,500$	[253]
10 dec.	$n = 1\,(1)\,100$	[3] [30] [31] [38] [58] [110]]121]
8 dec.	$n = 1\,(1)\,10\,(10)\,100$	[5] [12] [13]
8 dec.	$n = 1\,(1)\,100$	[34] [57] [60] [69] [103]
7 dec.	$n = 1\,(1)\,100$	[11]
7 dec.	$n = -1;\ \pm 2;\ 3;\ 4;\ 5;\ \pm\frac{1}{2}$	[21]
7 dec.	$n = \frac{1}{2}$	[20]
6 dec.	$n = 0\,(0{,}001)\,0{,}999$	[258]
6 dec.	$n = 1\,(1)\,24$	
6 dec.	$n = 1\,(1)\,10$	[33]
6 dec.	$n = -1;\ 2$	[20]
5 dec.	$n = 0\,(0{,}0001)\,0{,}01$	
5 dec.	$n = 0\,(0{,}01)\,1$	[87]
5 dec.	$n = 1\,(1)\,50$	
5 dec.	$n = 10\,(1)\,99$	[53]
5 dec.	$n = -1;\ \pm 2;\ \frac{1}{2};\ \frac{1}{3}$	[22]

$$\frac{n}{M}$$

137 dec.	$n = 20(10)90;\ 110$	[135]
70 dec.	$n = 1(1)9$	[10]
61 dec.	$n = 1(1)100$	[97]
40 dec.	$n = 1(1)10$	[47]
33 dec.	$n = 1(1)30$	[68]
24 dec.	$n = 1(1)100$	[256]
24 dec.	$n = 1(1)9$	[69]
23 dec.	$n = 1(1)100$	[10]
21 dec.	$n = 1(1)100$	[96]
20 dec.	$n = 1(1)20$	[83] [100]
20 dec.	$n = 1(1)10$	[126] [127] [128] [129]
20 dec.	$n = 1(1)9$	[70] [115]
18 dec.	$n = 1(1)9$	[67]
12 dec.	$n = 11(1)110$	[133]
12 dec.	$n = 10(1)99$	[14]
11 dec.	$n = 1(1)100$	[175]
10 dec.	$n = 1(1)100$	[3] [30] [38] [58] [110] [121]
8 dec.	$n = 1(1)10(10)100$	[5] [12] [13]
8 dec.	$n = 1(1)100$	[34] [57] [60] [69] [103]
8 dec.	$n = 1(1)9$	[114]
7 dec.	$n = 1(1)100$	[11]
6 dec.	$n = 0(0{,}001)1(1)24$	[258]
6 dec.	$n = 1(1)10$	[33]
5 dec.	$n = 0(0{,}0001)0{,}01;\ 0(0{,}01)1;1(1)10$	[87]
5 dec.	$n = 10(1)99$	[53]
5 dec.	$n = 1(1)50$	[141]

$$M^n$$

32 dec.	$n = 1(1)32$	[97]

$$M^{-n}$$

272 dec.	$n = 1$	[97]
49 dec.	$n = 1$	[256]
42 dec.	$n = 1$	[47]
32 fig.	$n = 1(1)32$	[97]
25 dec.	$n = 1$	[11]
20 dec.	$n = 1$	[60]
15 dec.	$n = 1$	[19]
6 dec.	$n = 1$	[22]

$$\lg \frac{e}{2\sqrt{2}} = M - \frac{3}{2} \lg 2$$

23 dec.	[255]

$$\lg \frac{e}{\sqrt{2\pi}} = M - \frac{1}{2} \lg 2\pi$$

23 dec.	[255]

Various constants involving π, e & M

$$e^{\pi}$$

52 dec.	[148]
40 dec.	[97]
28 dec.	[149] [183]
8 dec.	[16]
5 dec.	[21]

$$e^{-\pi}$$

55 dec.	[148]
51 dec.	[56]
50 dec.	[97] [149]
8 dec.	[16] [21]

$$e^{\pi\sqrt{n}}$$

39 dec.	$n = 163$	[186]
24 fig.	$n = 67$	[59]
19 fig.	$n = 58$	[238]
18 fig.	$n = 43$	[59]
15 fig.	$n = 37$	[238]

$$ke^{-n\pi}$$

35 dec.	$k = 2.$	$n = 4$	[56] [149]
27 dec.	$k = 2.$	$n = 9;\ 16$	[56] [149]
15 dec.	$k = 1.$	$n = 4;\ 9$	[229]

$$e^{\frac{\pi}{n}}$$

54 dec.	$n = \pm 2$	[148]
50 dec.	$n = -4$	[56] [149]

48 dec.	$n = -4$	[97]
46 dec.	$n = 4$	
40 dec.	$n = \pm 100$	[63]
34 dec.	$n = 2$	[56]
28 dec.	$n = 2$	[149] [183]
8 dec.	$n = \pm 2$	[16]
7 dec.	$n = -2;\ -4$	[21]
6 dec.	$n = 2;\ 4$	

$ke^{\frac{n\pi}{4}}$

57 dec.	$k = 1.$	$n = -9$	[56] [148]
32 dec.	$k = 2.$	$n = -25$	
28 dec.	$k = 2.$	$n = -49$	[56] [149]
17 dec.	$k = 1.$	$n = -9$	[229]
15 dec.	$k = 1.$	$n = -25$	
4 fig.	$k = 1.$	$n = 9;\ 81$	[118]

$\sqrt{\frac{1}{2}\pi e}$

9 dec.	[228]

$\frac{e}{\sqrt{2\pi}}$

23 dec.	[255]

π^e

28 dec.	[149] [183]

kM

20 dec.	$k = \frac{\pi}{200}$	[77]
10 dec.	$k = \pi;\ \pi^2$	[99]
7 dec.	$k = \pm\pi;\ -\frac{\pi}{2};\ \pm\frac{\pi}{4}$	[21]
6 dec.	$k = \frac{\pi}{2}$	
5 dec.	$k = n\pi.\ n = 0\,(0{,}25)\,5$	[54]

$\frac{1}{kM}$

20 dec.	$k = 4\pi$	[241]

Bernoulli numbers

B_n

exact	$n = 91\,(1)\,110$	[165]
exact	$n = 98$	[152]
exact	$n = 1\,(1)\,90$	[17] [44] [97]
exact	$n = 1\,(1)\,62$	[44] [160] [162] [179]
exact	$n = 1\,(1)\,50$	[100]
exact	$n = 1\,(1)\,32$	[104]
exact	$n = 1\,(1)\,31$	[176]
exact	$n = 1\,(1)\,25$	[147] [164] [198]
exact	$n = 1\,(1)\,22$	[64]
exact	$n = 1\,(1)\,21$	[62]
exact	$n = 1\,(1)\,20$	[86] [180]
exact	$n = 1\,(1)\,18$	[65]
exact	$n = 17$	[51] [196]
exact	$n = 1\,(1)\,15$	[46] [80] [91] [92] [197]
exact	$n = 1\,(1)\,13$	[76] [239]
exact	$n = 1\,(1)\,11$	[48] [49]
exact	$n = 1\,(1)\,10$	[23] [24] [35] [40] [42] [75] [101] [106] [168]
exact	$n = 1\,(1)\,5$	[32]
13 dec.	$n = 1\,(1)\,10$	[168]
13 dec.	$n = 1\,(1)\,8$	[80]
9 fig.	$n = 1\,(1)\,250$	[44] [252]
9 dec.	$n = 1\,(1)\,20$	[34]
9 fig.	$n = 1\,(1)\,15$	[76]
9 dec.	$n = 1\,(1)\,14$	[53]
7 fig.	$n = 1\,(1)\,35$	[49]
5—6 fig.	$n = 1\,(1)\,20$	[42]
5 fig.	$n = 1\,(1)\,60$	[103]
4 fig.	$n = 1\,(1)\,250$	[97]

$\lg B_n$

20 dec.	$n = 1\,(1)\,40$	[163]
10 dec.	$n = 1\,(1)\,250$	[43] [97] [252]
10 dec.	$n = 1\,(1)\,18$	[78] [164] [198]
10 dec.	$n = 1\,(1)\,15$	[76]
8 dec.	$n = 1\,(1)\,31$	[70]
8 dec.	$n = 1\,(1)\,30$	[100]
8 dec.	$n = 2m - 1 = 1\,(2)\,19$	[109]
7 dec.	$n = 1\,(1)\,200$	[58]
7 dec.	$n = 1\,(1)\,35$	[49]
5 dec.	$n = 1\,(1)\,60$	[103]

Euler numbers

E_n

exact	$n = 1\,(1)\,50$	[43] [242]
exact	$n = 1\,(1)\,32$	[240]
exact	$n = 1\,(1)\,30$	[97]
exact	$n = 1\,(1)\,27$	[236]
exact	$n = 1\,(1)\,14$	[91] [104] [108]
exact	$n = 10$	[178]
exact	$n = 1\,(1)\,10$	[92] [101]
exact	$n = 9$	[176]
exact	$n = 1\,(1)\,9$	[40] [42] [173]
exact	$n = 1\,(1)\,8$	[26] [51] [76] [87] [197] [234]
exact	$n = 1\,(1)\,5$	[100]
1—71 fig.	$n = 1\,(1)\,30$	[97]
10 fig.	$n = 1\,(1)\,250$	[43]
7 fig.	$n = 1\,(1)\,35$	[49]

$\lg E_n$

12 dec.	$n = 1\,(1)\,250$	[43]
10 dec.	$n = 1\,(1)\,50$	[97]
7 dec.	$n = 1\,(1)\,35$	[49]

Euler's constants C & γ

Euler's constant C is defined as the difference of two integrals

$$C = \int_0^1 \frac{1-e^{-t}}{t}\,dt - \int_1^\infty \frac{e^{-t}}{t}\,dt = 0{,}57721\,56649\,01532\,86061\,\ldots$$

The second constant is connected with the first equation $\gamma = e^C$.

C

263 dec.	[97] [161] [225]
101 dec.	[220]
99 dec.	[218]
80 dec.	[217]
59 dec.	[216]
50 dec.	[90]
41 dec.	[177]
40 dec.	[55] [257]

21 dec.	[112]
20 dec.	[126] [127] [128] [129]
16 dec.	[40]
15 dec.	[196]
9 dec.	[23]

$$\ln 2 - C$$

25 dec.	[227]

$$\frac{C + \ln\left(\frac{4}{\pi}\right)}{4\pi}$$

20 dec.	[241]

$$\gamma = e^c \ \& \ \frac{1}{\gamma} = e^{-c}$$

20 dec.	[52]

$$\lg C$$

20 dec.	[250]

$$\ln C$$

20 dec.	[126] [127] [128] [129]

Algebraic constants

$$n^m$$

exact		$\left\{\begin{array}{l} n = 2.\ m = 120 \\ n = 3.\ m = 70 \\ n = 5.\ m = 60 \end{array}\right\}$	[97]
1001 fig.	$450! \cdot n^m$	$n = 10.\ m = -111$	[245]

$$m\sqrt{n}$$

544 dec.	$m = 1.\ n = 2$	[213]
1316 dec.	$m = 1.\ n = 3;\ \frac{1}{3}$	[213]
110 dec.	$m = 1.\ n = 2$	[45]

66 dec. 64 dec.	$m = 1.\ n = 3$ $m = 1.\ n = 2$	[97]
30 dec.	$m = 1.\ n = 2;\ \frac{1}{2}$	[94]
24 dec.	$m = 1.\ n = 2;\ 3;\ 5;\ 15;\ \frac{1}{2};\ \frac{3}{2};\ \frac{5}{2};\ \frac{15}{2};\ 5 \pm \sqrt{5}$	[59] [192]
24 dec.	$m = \sqrt{2}.\ n = 5 \pm \sqrt{5}$ $m = \sqrt{3}.\ n = 5 \pm \sqrt{5}$ $m = \sqrt{6}.\ n = 5 \pm \sqrt{5}$	[59]
24 dec.	$m = 1.\ n = 10 \pm 2\sqrt{5}$ $m = 1.\ n = 15 \pm 3\sqrt{5}$ $m = 1.\ n = 30 \pm 6\sqrt{5}$	[192]
15 dec.	$m = 1.\ n = 2;\ 3;\ 10$	[19] [29]
12 dec.	$m = 1.\ n = 2;\ 3;\ 5;\ 15;\ \frac{1}{2};\ \frac{3}{2};\ \frac{5}{2};\ \frac{15}{2};\ 5 \pm \sqrt{5}$	[137]
12 dec.	$m = \sqrt{2}.\ n = 5 \pm \sqrt{5}$ $m = \sqrt{3}.\ n = 5 \pm \sqrt{5}$ $m = \sqrt{6}.\ n = 5 \pm \sqrt{5}$	[137]
10 dec.	$m = 1.\ n = 2;\ 3$	[1] [20]
10 dec.	$m = 1(1)9.\ n = 2;\ 3$	[4] [53] [119]
10 dec.	$m = 1.\ n = 5;\ \frac{1}{2};\ \frac{1}{3};\ \frac{1}{5}$	[1]
10 dec.	$m = 1.\ n = \frac{1}{2}$	[20]
10 dec.	$m = 1(1)9.\ n = 5$	[4] [119]
7 dec. 6 dec.	$m = 1.\ n = \frac{1}{2};\ \frac{1}{3};\ \frac{1}{5};\ \frac{1}{6};\ \frac{1}{7};\ \frac{1}{8};\ \frac{1}{10}$ $m = 1.\ n = 2;\ 3;\ 5;\ 6;\ 7;\ 8;\ 10$	[21]
6 dec.	$m = 1(1)100.\ n = 2;\ 3;\ 5;\ 10$	[34]
5 dec.	$m = 1.\ n = 2$	[19]

$$m\sqrt[3]{n}$$

710 dec. 709 dec. 478 dec.	$m = 1.\ n = 2$ $m = 1.\ n = 3$ $m = 1.\ n = 4;\ 9$	[260]

56 dec.	$m=1.\ n=2;\ 4$	[190]
56 dec.	$m=1.\ n=6;\ 36$	[153]
50 dec.	$m=1.\ n=2$	[191]
15 dec.	$m=1.\ n=2;\ 10;\ 100$	[19] [29]
10 dec.	$m=1.\ n=10;\ 100;\ 0{,}1;\ 0{,}01$	[1]
7 dec.	$m=1.\ n=4;\frac{1}{2};\ \frac{1}{3};\ \frac{1}{4};\ \frac{1}{5};\ \frac{1}{6};\ \frac{1}{7};\ \frac{1}{9};\ \frac{1}{10}$	[21]
6 dec.	$m=1(1)100.\ n=2;\ 3;\ 5;\ 10;\ 100$	[34]
6 dec.	$m=1.\ n=2;\ 3;\ 4;\ 5;6;7;9;10;100$	[21]
5 dec.	$m=1.\ n=10000$	[21]

$$\frac{m}{\sqrt{n}}$$

6 dec.	$m=1(1)100.\ n=2;\ 3;\ 5$	[34]
6 dec.	$m=2(1)19.\ \ n=2;\ 3;\ 5$	[1]

$$\frac{m}{\sqrt[3]{n}}$$

6 dec.	$m=1(1)100.\ n=2;\ 3;\ 5;\ 10;\ 100$	[34]

Certain values of trigonometric functions

$$\sin\frac{0{,}1}{\sqrt{2}}\ \ \&\ \ \cos\frac{0{,}1}{\sqrt{2}}$$

18 dec. [200]

$$\sin 0{,}1\ \ \&\cos 0{,}1$$

120 dec. [189]

$$\sin\frac{1}{\sqrt{2}}\ \ \&\ \ \cos\frac{1}{\sqrt{2}}$$

20 dec. [200]

$$\sin 1\ \&\ \ \cos 1$$

477 dec.	[254]
105 dec.	[97] [154] [189]
52 dec.	[97]

sin 10 & cos 10

284 dec.		[210]
212 dec.		[254]

cos 20

284 dec.		[210]
212 dec.		[254]

sin 100 & cos 100

72 dec.		[254]

sin 200 & cos 200

72 dec.		[254]

$\sin\left(\frac{\pi}{3}\right)$

316 dec.		[213]

$\sin(\ln \pi)$ & $\cos(\ln \pi)$

13 dec.		[201]

$\sin(\ln k)$ & $\cos(\ln k)$

15 dec.	$k = 2$	[201]
14 dec.	$k = 5;\ 10;\ 100$	
13 dec.	$k = 1000$	

tg 1

11 dec.		[228]

$\operatorname{arc\,tg} \frac{1}{k}$

709 dec.	$k = 239$	[222]
709 dec.	$k = 5;\ 239$	[221]
609 dec.	$k = 5;\ 239$	[111]
530 dec.	$k = 5;\ 239$	[214]
335 dec.	$k = 577$	[210]
330 dec.	$k = 2449;\ 4999;\ 8749$	
261 dec.	$k = 2;\ 3;\ 5;\ 7$	[155]
250 dec.	$k = 3;\ 5;\ 7;\ 239$	[158]
216 dec.	$k = 10081$	[210]
215 dec.	$k = 451$	

Certain values of hyperbolic functions

sh 1 & ch 1

105 dec. [97] [154]
52 dec. [97]
24 dec. [172]

$$\text{sh}\,\frac{1}{\sqrt{2}}\ \&\ \text{ch}\,\frac{1}{\sqrt{2}}$$

20 dec. [200]

$$\text{sh}\,\frac{0{,}1}{\sqrt{2}}\ \&\ \text{ch}\,\frac{0{,}1}{\sqrt{2}}$$

18 dec. [200]

$$\text{sh}\,\frac{\pi}{6}\ \&\ \text{ch}\,\frac{\pi}{6}$$

54 dec. [148]

th 1

11 dec. [228]

Certain values of integrals

Si 1 & Ci 1

37 dec. [257]
35 dec. [154]

Si 2 & Ci 2

43 dec. [202]

shi 1 & chi 1

37 dec. [257]
35 dec. [154]

$$\text{li}\, e = \text{Ei}\, 1\ \&\ \text{li}\left(\frac{1}{e}\right) = \text{Ei}\,(-1)$$

37 dec. [257]
35 dec. [154]

$$\text{Ei}\,(\pm 2)$$

43 dec. [202]

Logarithms of various constants

$n \lg m$

274 dec.	$n = 1.$	$m = \frac{10}{9};\ \frac{25}{24};\ \frac{81}{80};\ \frac{50}{49};\ \frac{126}{125}$	[225]
230 dec.	$n = 1.$	$m = 2;\ 3;\ 5;\ 7;\ 17$	[209]
110 dec.	$n = 1.$	$m = 71;\ 113$	[209]
60 dec.	$n = \pm 1.$	$m = \frac{1}{7};\ \frac{1}{5};\ \frac{1}{3};\ \frac{1}{2};\ 2;\ 3;\ 5;\ 7$	[116]
42 dec.	$n = 1.$	$m = 2;\ 5$	[47]
12 dec.	$n = 1\,(1)\,9.$	$m = 2$	[100]
8 dec.	$n = 1\,(1)\,24.$	$m = 2$	[5] [12] [13] [15] [69]
8 dec.	$n = 1\,(1)\,25.$	$m = 2;\ 3$	[2] [11]
8 dec.	$n = 1\,(1)\,30.$	$m = 2$	[109]
7 dec.	$n = \frac{1}{2}.$	$m = 2;\ 3;\ 6;\ 7;\ 8;\ 10$	[21]
7 dec.	$n = -\frac{1}{2}.$	$m = 2;\ 3;\ 5;\ 6;\ 7;\ 8;\ 10$	[21]
7 dec.	$n = \frac{1}{3}.$	$m = 2; 3; 5; 6; 7; 9; 10; 100; 10000$	[21]
7 dec.	$n = -\frac{1}{3}.$	$m = 2;\ 3;\ 4;\ 5;\ 6;\ 7;\ 9;\ 10$	[21]
6 dec.	$n = \frac{1}{2}.$	$m = 5$	[21]
4 dec.	$n = 1\,(1)\,25.$	$m = 2$	[102]

$\lg m\pi^n$

214 dec.	$m = 1.$	$n = 1$	[209]
214 dec.	$m = 2.$	$n = \frac{1}{2}$	[209]
61 dec.	$m = 1.$	$n = \frac{1}{2};\ 1$	[97]
40 dec.	$m = 2.$	$n = \frac{1}{2}$	[46]
36 dec.	$m = 2.$	$n = \frac{1}{2}$	[47]
32 dec.	$m = \frac{1}{2};\ 1;\ 2.$	$n = \frac{1}{2}$	[133]

32 dec.	$m = 1.$	$n = 1(1)9$	[97]
30 dec.	$m = \sqrt{2};\ \frac{1}{\sqrt{2}}.$	$n = 1$	[134]
24 dec.	$m = 1.$	$n = 1$	[5] [69]
15 dec.	$m = 1.$	$n = 1$	[100]
10 dec.	$m = \frac{1}{3}.$	$n = 1$	[20]
9 dec.	$m = \frac{1}{6};\ \frac{1}{2}.$	$n = 1$	[20]
9 dec.	$m = 1.$	$n = \frac{1}{2}$	[20]
8 dec.	$m = 1.$	$n = 1$	[13] [16] [20]
8 dec.	$m = 1.$	$n = 2$	[13] [20]
8 dec.	$m = 1.$	$n = 3;\ 4$	[20]
8 dec.	$m = \frac{1}{4};\ 2;\ 4.$	$n = 1$	[20]
8 dec.	$m = 1.$	$n = \frac{1}{2}$	[13]
8 dec.	$m = 1.$	$n = \frac{1}{3};\ \frac{2}{3};\ -1$	[20]
8 dec.	$m = \frac{1}{2};\ \frac{3}{4};\ 3;\ 4.$	$n = -1$	[20]
7 dec.	$m = 1;\ 2.$	$n = 1$	[1] [21]
7 dec.	$m = 2.$	$n = 1$	[1]
7 dec.	$m = \frac{1}{360};\ \frac{1}{6};\ \frac{1}{4};\ \frac{1}{3};\ \frac{1}{2};\ \frac{4}{5};\ \frac{4}{3};\ \frac{3}{2};\ 3;\ 4$	$n = 1$	[21]
7 dec.	$m = 1.$	$n = 2$	[1] [21]
7 dec.	$m = \frac{1}{4};\ 2;\ 4.$	$n = 2$	[21]
7 dec.	$m = 1.$	$n = 3;\ 4;\ 5$	[21]
7 dec.	$m = \frac{1}{4};\ 4.$	$n = 2$	[20]
7 dec.	$m = 1.$	$n = \frac{1}{2}$	[1] [21]
7 dec.	$m = \sqrt{2}.$	$n = \frac{1}{2}$	[20] [21]
7 dec.	$m = \frac{1}{\sqrt{2}}.$	$n = \frac{1}{2}$	[1]
7 dec.	$m = \frac{1}{\sqrt{2}};\ 2;\ 2\sqrt{2}.$	$n = \frac{1}{2}$	[21]
7 dec.	$m = 1.$	$n = \frac{1}{3};\ \frac{2}{3};\ -1$	[21]
7 dec.	$m = \frac{1}{\sqrt[3]{6}}.$	$n = \frac{1}{3}$	[21]
7 dec.	$m = 1.$	$n = \frac{4}{3}$	[22]
7 dec.	$m = \frac{1}{2};\ 1.$	$n = -1$	[1]
7 dec.	$m = \frac{1}{4};\ 2;\ 3;\ 6;\ 360.$	$n = -1$	[21]
7 dec.	$m = \frac{1}{2};\ 180;\ 360.$	$n = -1$	[20]

7 dec.	$m = \frac{1}{3};\ \frac{1}{2};\ \frac{3}{4};\ 4.$	$n = -1$	
7 dec.	$m = 1.$	$n = -\frac{1}{2};\ -2;\ -3$	
7 dec.	$m = \frac{1}{\sqrt{2}};\ \frac{1}{2};\ \sqrt{2};\ \sqrt{3};\ 2.$	$n = -\frac{1}{2}$	[21]
7 dec.	$m = 1;\ \sqrt[3]{\frac{3}{4}}.$	$n = -\frac{1}{3}$	
7 dec.	$m = 1.$	$n = -\frac{1}{2};\ -2$	[1]
7 dec.	$m = \frac{1}{\sqrt{2}};\ \sqrt{2}.$	$n = -\frac{1}{2}$	
7 dec.	$m = 1.$	$n = -\frac{1}{2}$	[20]
6 dec.	$m = 1.$	$n = -5$	
5 dec.	$m = \frac{1}{4};\ \frac{1}{3};\ \frac{1}{2};\ \frac{1}{\sqrt{2}};\ 1;\ \sqrt{2}.$	$n = 1$	[22]
5 dec.	$m = \frac{1}{4};\ 1;\ 4.$	$n = 2$	
5 dec.	$m = 1.$	$n = \frac{1}{2};\ \frac{1}{3};\ \frac{3}{2};\ 1;\ 3;\ 4;\ 5;\ -1;\ -2;\ -3;\ -4$	
5 dec.	$m = \frac{1}{\sqrt{2}};\ \sqrt{2}.$	$n = \frac{1}{2}$	
5 dec.	$m = \sqrt{2};\ \sqrt{3}.$	$n = \frac{1}{2}$	[22]
5 dec.	$m = \frac{1}{\sqrt[3]{4}};\ \frac{1}{\sqrt[3]{2}};\ \sqrt[3]{2}.$	$n = \frac{1}{3}$	
5 dec.	$m = \sqrt[3]{2};\ \sqrt[3]{3}.$	$n = -\frac{1}{3}$	
4 dec.	$m = 3.$	$n = 2$	[21]
4 dec.	$m = \frac{1}{360}.$	$n = 1$	[20]

lg mM

100 dec.	$m = 1$	[93]
50 dec.	$m = 1$	[97] [256]
27 dec.	$m = 2$	[133]
24 dec.	$m = \pi$	[59]
20 dec.	$m = \frac{\pi}{200}$	[77]
8 dec.	$m = 1$	[13]
7 dec.	$m = 1$	[21]
5 dec.	$m = 1$	[22]

$\lg \frac{1}{M}$

6 dec.	[21]
5 dec.	[22]

$n \ln m$

330—331 dec.	$n = 1$.	$m = 2; 3; 5; 7; 10; 17$	[210]
325 dec.	$n = 1$.	$m = 10$	[135]
290—329 dec.	$n = 1$.	$m = 11; 13; 17; 19; 23; 29; 31; 37; 101; 9901$	[212]
274 dec.	$n = 1$.	$m = 17$	[209]
273—274 dec.	$n = 1$.	$m = 2; 3; 5; 7; 10$	[226]
272 dec.	$n = 1$.	$m = 2; 3; 5; 7$	[97]
262 dec.	$n = 1$.	$m = 2; 3; 5; 7; 10$	[225]
262 dec.	$n = 1$.	$m = 500$	[97]
262 dec.	$n = 1$.	$m = 1000$	[161] [225]
213 dec.	$n = 1$.	$m = 71; 113$	[209]
205 dec.	$n = 1$.	$m = 2; 3; 5; 10$	[146] [223]
201 dec.	$n = 1$.	$m = 2; 3; 5; 10$	[219]
100 dec.	$n = 1$.	$m = 2; 3; 5; 10$	[218]
60 dec.	$n = \pm 1$.	$m = 2; 3; 5; 7; 10$	[116]
23 dec.	$n = 1\,(1)\,10$.	$m = 10$	[113]
20 dec.	$n = 1\,(1)\,20$.	$m = 10$	[18]
15 dec.	$n = 1\,(1)\,12$.	$m = 2$	[182]
12 dec.	$n = 1\,(1)\,20$.	$m = 2$	[172]

$n \ln m\pi$

213 dec.	$n = 1$.	$m = 1$	[209]
48 dec.	$n = 1$.	$m = 1$	[97] [206] [230]
35 dec.	$n = 1$.	$m = 1$	[51] [230]
32 dec.	$n = 1\,(1)\,9$.	$m = 1$	[97]
30 dec.	$n = 20; 22$.	$m = 1$	[230]
25 dec.	$n = \frac{1}{2}$.	$m = 2$	[196]
20 dec.	$n = 1$.	$m = \frac{1}{2}; \frac{1}{8}$	[9] [39]
15 dec.	$n = 1$.	$m = 1$	[100]
8 dec.	$n = 1$.	$m = 1$	[13]

$\ln M$

54 dec.	[97]

lg sin 1 & lg cos 1

dec. [25]

Miscellaneous constants

$\rho = 0{,}47693\,62762\,04469\,87338$

The constant ρ, which is widely used in probability theory, is defined by the expression

$$\frac{2}{\sqrt{\pi}} \int_0^{\rho} e^{-x^2}\, dx = \frac{1}{2}.$$

Values of the following constants and their logarithms are given to 23 decimals in /255/:

$$\rho;\ \frac{1}{\rho};\ \rho^2;\ \rho\sqrt{2};\ 2\rho\sqrt{\pi};\ \frac{2\rho}{\sqrt{\pi}};\ \frac{2\rho^2}{\sqrt{\pi}};\ \frac{1}{\rho\sqrt{\pi}};\ \frac{1}{2\rho\sqrt{\pi}};\ \rho\sqrt{\pi};\ \frac{1}{\rho\sqrt{2}};\ \rho\sqrt[6]{\pi};$$

$$\rho\sqrt[10]{\frac{1}{4}\pi};\ \rho\sqrt{\pi-2};\ \rho\sqrt{\frac{1}{36}(15\pi-8)};\ \rho\sqrt{\frac{1}{1600}(945\pi-128)};\ \rho\sqrt{\frac{4}{3}};$$

$$\rho\sqrt[4]{\frac{4}{3}};\ \rho\sqrt{\frac{113}{45}};\ \rho\sqrt[6]{\frac{8}{15}};\ e^{\rho^2};\ e^{\rho^2}\sqrt{\pi};\ e^{-\rho^2};\ \frac{2e^{-\rho^2}}{\pi}$$

Dubois-Reymond's constants C_m

Dubois-Reymond's constants are defined by the expression:

$$C_m = \int_0^{\infty} \left| \frac{d}{dt}\left(\frac{\sin t}{t}\right)^m \right| dt - 1.$$

Values of the following Dubois-Reymond constants have been calculated:

10 dec. $m = 2;\ 3;\ 4;\ 6$ [244]

Lesbesgue's constants $\rho_n = \frac{2}{\pi} \int_0^{\frac{1}{2}\pi} \frac{|\sin(2n+1)\theta|}{\sin\theta}\, d\theta$

14 dec. $n = \frac{9}{2}$ [243]

Values of the magnitudes C_n, by which Lesbesgue's constants may be expressed, are also given

$$\rho_n = \frac{2}{\pi} \int_0^{\frac{1}{2}\pi} \frac{|\sin(2n+1)\theta|}{\sin\theta}\, d\theta \sim \frac{4}{\pi^2}\left[\ln(2n+1) + C_0 + \sum_{r=1}^{\infty}(-1)^{r-1}\frac{C_r}{(2n+1)^{2r}}\right].$$

15 dec. $n = 0(1)7$ [243]

Constants of the lemniscate functions

Lemniscate constants are defined by the expressions

$$\frac{1}{2}\tilde{\omega} = \int_0^1 \frac{dx}{\sqrt{1-x^4}} = 1{,}31102\,87771\,46059\ldots$$

$$\frac{1}{2}\frac{\pi}{\tilde{\omega}} = \int_0^1 \frac{x^2\,dx}{\sqrt{1-x^4}} = 0{,}59907\,01173\,67796\ldots$$

$$\frac{1}{2}\tilde{\omega}$$

23 dec. [56]

$$\frac{1}{2}\frac{\pi}{\tilde{\omega}}$$

20 dec. [56]

$$\frac{\tilde{\omega}}{\pi}$$

21 dec. [86]

$$\sqrt{\frac{\tilde{\omega}}{\pi}}$$

26 dec. [56]

$$\lg \frac{1}{2}\tilde{\omega}$$

14 dec. [56]

$$\ln \tilde{\omega}$$

25 dec. [56]

$$\ln \frac{1}{2}\tilde{\omega}$$

24 dec. [56]

Stirling's numbers of the first and second kind

Stirling's numbers of the first kind S_n^ν

$n = 2(1)52.$	$\nu = n - 1$	[247]
$n = 3(1)65.$	$\nu = n - 2$	
$n = 4(1)51.$	$\nu = n - 3$	
$n = 5(1)50.$	$\nu = n - 4$	
$n = 6(1)51.$	$\nu = n - 5$	
$n = 1(1)12.$	$\nu = 1(1)6$	[249]
$n = 7(1)12.$	$\nu = 7(1)12$	

Stirling's numbers of the second kind G_n^ν

$n = 1(1)12.$	$\nu = 1(1)12$	[249]

Chapter 15

PRIMES, FACTORS AND PRODUCTS. QUOTIENTS AND FRACTIONS

Certain information on tables of prime numbers, factors, products and quotients is given in this chapter as a supplement to the present work.

Although multiplication and division tables are still being compiled, they have lost their former significance in mathematical calculations owing to the development of machine methods of calculation. Only a few of the most commonly used tables of this type are described here.

The chapter contains the following sections:

primes;
factors and divisors;
products;
division tables;
tables involving vulgar fractions.

Primes

In addition to these tables, primes may also be found from factor tables, information on which is given in this chapter.

Lists of primes

$1 < x < 400\,313$	[110]
$1 < x < 60\,000$	[46]
$1 < x < 12\,919$	[94]
$1 < x < 10\,000$	[1] [70] [95] [106]
$1 < x < 9\,997$	[93]
$1 < x < 9\,973$	[57]
$1 < x < 5\,987$	[19]
$1 < x < 5\,020$	[75]
$1 < x < 4\,000$	[17]

[1] See pages 532 and 550 for the references to Chapter 15.

$1 < x < 2\,129$	[50]
$1 < x < 1\,000$	[114]
$289 < x < 15\,250$	[76]
$1009 < x < 9\,973$	[18]
$3677 < x < 13\,999$	[125]
$1 < x < 217\,219$	[126]

Primitive roots of primes

From 5 to 11 981	[123]
From 1 to 5 000	[117]
From 1 to 4 000	[17]
From 1 to 3 000	[115]
" 103 to 1 009	[124]
" 3 000 to 5 000	[116]
" 5 000 to 10 000	[116] [117]

Powers of primes x^n

4—34 fig. $\left\{ \begin{matrix} 2 \leqslant x \leqslant 97. \quad n = 11\,(1)\,N \\ 16 \leqslant N \leqslant 120 \end{matrix} \right\}$ [100]

Powers of primes dividing $a^n - 1$

$a = 2; 3; 5; 6; 7; 10; 11; 12. \qquad n = 1\,(1)\,144$ [120]

Differences of successive primes

2—3 fig. $3 \leqslant x \leqslant 4652\,353$ [122]

Presentation of primes as sums of squares

Presentation of primes of the type $4n + 1$ as the sum of two squares

$5 \leqslant x \leqslant 11\,981$ [123]

Presentation of primes of the type $6n + 1$ as the sum of two squares

$7 \leqslant x \leqslant 12\,007$ [123]

Presentation of primes of the type $8n + 1$ as the sum of two squares

$17 \leqslant x \leqslant 5953$ [123]

Factors and divisors

Prime factors

$x = 4(1)10192$	[100]
$x = 1(1)10017$	[109]
$x = 1(1)10000$	[80] [97]
$x = 4(1)\ 5000$	[84]
$x = 1(1)\ 4850$	[70]
$x = 1(1)\ 2000$	[102]
$x = 1(1)\ 1000$	[28] [33] [107] [113]
$x = 1(2)999$	[13] [14]
$x = 1(1)969$	[52]
$x = 1(1)100$	[114]

Prime factors of numbers of the type $n! \pm 1$

$n = 1(1)22$ [127]

Prime factors of numbers of the type $n^{16} + 1$

$n = 2(1)13;\ 21;\ 27;\ 32$ [121]

Divisors

$1 \leqslant x \leqslant 60\,000$	[46]
$1 \leqslant x \leqslant 1000$	[9]
$11 \leqslant x \leqslant 1000$	[18]

Divisors of numbers $2^n + 1$

$n = 1(1)64$ [118]

Divisors of numbers $2^n - 1$

$n = 1(2)63$ [118]

Products

Products of type mn

$m = 1(1)1\,000\,000\,000.$	$n = 1(1)10$	[26]
$m = 1(1)1\,000\,000.$	$n = 1(1)1\,000\,000$	[74]
$m = 1(1)10\,000.$	n — any 9-digit number.	[64]
$m = 1(1)10\,000.$	$n = 1(1)100$	[21] [69] [88]

$m = 1\,(1)\,10\,000.$	$n = 1\,(1)\,10\,000$	[34]
$m = 1\,(1)\,10\,000.$	$n = 1\,(1)\,10$	[37]
$m = 1000\,(1)\,10\,000.$	$n = 1\,(1)\,10$	[59]
$m = 1\,(1)\,2000.$	$n = 1\,(1)\,2000$	[8]
$m = 1\,(1)\,1500.$	$n = 1\,(1)\,10$	[7]
$m = 1\,(1)\,1200.$	$n = 1\,(1)\,10$	[62]
$m = 1\,(1)\,1025.$	$n = 1\,(1)\,100$	[15]
$m = 1\,(1)\,1000.$	$n = 1\,(1)\,1000$	[36] [111]
$m = 1\,(1)\,1000.$	$n = 1\,(1)\,500$	[11]
$m = 1\,(1)\,1000.$	$n = 1\,(1)\,100$	[4] [5] [9] [40] [55] [92] [99] [103]
$m = 1\,(1)\,1000.$	$n = 1\,(1)\,10$	[20] [44] [49] [86]
$m = 3\,(1)\,1000.$	$n = 1\,(1)\,100$	[3]
$m = 10\,(1)\,1000.$	$n = 1\,(1)\,100$	[27] [51]
$m = 11\,(1)\,1000.$	$n = 1\,(1)\,100$	[66]
$m = 11\,(1)\,1000.$	$n = 1\,(1)\,30$	[45]
$m = 12\,(1)\,1000.$	$n = 1\,(1)\,10$	[67]
$m = 100\,(1)\,1000.$	$n = 1\,(1)\,10$	[53]
$m = 100\,(1)\,1000.$	$n = 11\,(1)\,100$	[60]
$m = 100\,(1)\,1000.$	$n = 100\,(1)\,990$	[87]
$m = 100\,(1)\,1000.$	$n = 100\,(1)\,1000$	[30]
$m = 1\,(1)100;\ 200;\ 500.$	$n = 2\,(1)\,100;\ 900;\ 1000$	[77]
$m = 1\,(1)\,100.$	$n = 1\,(1)\,10\,(10)\,100\,(100)\,1000\,(1000)\,10\,000\,(10\,000)\,100\,000$	[56]
$m = 1\,(1)\,100.$	$n = 1000\,(1)\,3000$	[79]
$m = 1\,(1)\,100.$	$n = 11\,(1)\,1000$	[16]
$m = 10\,(1)\,100.$	$n = 100\,(1)\,1000$	[58]
$m = 1\,(1)\,100.$	$n = 1\,(1)\,100$	[10] [22] [39] [43] [48] [68] [83] [93] [98]
$m = 10\,(10)\,90.$	$n = 10\,(0{,}2)\,99{,}8$	[119]
m is any multi-digit number.	$n = 11\,(1)\,100$	[91]
$m = 11\,(1)\,100.$	$n = 11\,(1)\,100$	[38] [72]
$m = 11\,(1)\,100.$	$n = 1\,(1)\,100$	[23] [65]
$m = 1\,(1)\,100.$	$n = 1\,(1)\,10$	[31] [105] [112]
$m = 1\,(1)\,50.$	$n = 11\,(1)\,50$	[114]
$m = 1\,(1)\,50.$	$n = 1\,(1)\,50$	[29]
$m = 11\,(0{,}5)\,51(1)\,100$	$n = 1\,(0{,}5)\,100$	[104]
$m = 11\,(1)\,25.$	$n = 1\,(1)\,25$	[12]
$m = 1\,(1)\,20.$	$n = 1\,(1)\,20$	[81]

	$m = 1\,(1)\,10.$	$n = 1\,(1)\,1000$	[32] [42]
	$m = 1\,(1)\,10.$	$n = 1\,(1)\,10$	[6] [54]
	m — is any number	$n = 1\,(1)\,100$	[89]
	m — is any number	n — a 10-fig. no.	[63]
	m — up to 7 sig. fig	n — is prime	[73]
	$m = 0{,}25\,(0{,}25)\,8{,}5.$	$n = 0{,}25\,(0{,}25)\,15{,}5$	[43]
	$m = 10\,(1)\,100.$	$n = 3\,(0{,}25)\,49{,}5$	[82]
3 dec.	$m = 0\,(10)\,1000.$	$n = 0\,(0{,}001)\,1$	[61]
3 fig.	$m = 11\,(1)\,30.$	$n = 11\,(1)\,30$	[2]
2 fig.	$m = 11\,(1)\,1000.$	$n = 11\,(1)\,100$	[78]
	$m + n < 20\,000.$		[111]

Products calculated by the formula

$$A \times B = Q\,(A + B) - Q\,(A - B)$$

10 fig. [96]

Products of type $m\,(10x + y)$

5 dec. $m = 11\,(1)\,1000.$ $10x = 0\,(10)\,90.$ $y = 0\,(1)\,10$ [66]

Products of type $2mn$

$m = 1\,(1)\,1000.$ $n = 1\,(1)\,10$ [24]

Products of type $3mn$

$m = 1\,(1)\,1000.$ $n = 1\,(1)\,10$ [24]

Products of type mn^2

$m = 1\,(1)\,100.$ $n = 1\,(1)\,100$ [108]

Products of prime numbers

$m = 1\,($ primes $)\,400.$ $n = 1\,($ primes $)\,400$ [25]

Division tables

Division tables $m : n$

m — is any multi-digit number	$n = 11\,(1)\,99$	[91]
$m = 1\,(1)\,100.$	$n = 1\,(1)\,500$	[11]
$m = 1\,(1)\,1000.$	$n = 1\,(1)\,100$	[5] [99]
$m = 10\,(1)\,1000.$	$n = 1\,(1)\,100$	[51]

	$m = 11\,(1)\,1000.$	$n = 1\,(1)\,30$	[45]
	$m = 1\,(1)\,1000.$	$n = 1\,(1)\,9$	[86]
	$m = 1\,(1)\,100;\ 200;\ 500.$	$n = 1\,(1)\,100;\ 900;\ 1000$	[77]
	$m = 1\,(1)\,100.$	$n = 1\,(1)\,100$	[39]
	$m = 1\,(1)\,9.$	$n = 1\,(1)\,1000$	[42]
	m is any number	n is a ten-figure no.	[63]
6—7 dec.	$m = 1\,(1)\,9.$	$n = 1\,(1)\,1000$	[32]
5 fig.	$m = 1\,(100)\,9900.$	$n = 101\,(0{,}5)\,200\,(1)\,400$	[35]
5 fig.	$m = 1\,(1)\,1000.$	$n = 1\,(1)\,100$	[9]
5 fig.	$m = 10\,(1)\,100.$	$n = 101\,(1)\,1000$	[90]
5 dec.	$m = 1\,(1)\,10.$	$n = 11\,(1)\,100$	[8]
4 dec.	$m = 2\,(2)\,30\,(5)\,100.$	$n = 1\,(1)\,100$	[93]
3 dec.	$m = 1\,(1)\,10.$	$n = 10\,(1)\,20$	[2]
2—3 fig.	$m = 1000\,(1)\,1\,000\,000.$	$n = 11\,(1)\,100$	[78]

Division tables $mn:m$

3 dec.	$m = 1\,(10)\,1000.$	$n = 0\,(0{,}001)\,1$	[61]

Division tables $m(10x+y):m$

5 dec.	$m = 11\,(1)\,1000.$	$10x = 0\,(10)\,90.$	$y = 0\,(1)\,90$	[66]

Division tables of square roots $\sqrt{m}:\sqrt{n}$

4 dec.	$m = 2\,(2)\,30\,(5)\,100.$ $m = 1\,(1)\,100.$	$n = 1\,(1)\,100$ $n = 2\,(2)\,30\,(5)\,100$	[93]

See also the tables of square roots of vulgar fractions on page 16.

Division tables, primes

$m = 1\,(1)\,100\,($ primes $)\,400.$ $n = 1\,(1)\,100\,($ primes $)\,400$ [25]

Tables involving vulgar fractions

Tables for converting vulgar fractions to decimals

8 dec.	$m = 1\,(0{,}1)\,32.$	$n = 1\,(1)\,32$	[93]
6 dec.	$m = 1.$	$n = 1\,(1)\,1000$	[41]
6 dec.	$1 \leqslant m \leqslant 16$	$2 \leqslant n \leqslant 17$	[80]

4 dec.	$1 \leqslant m \leqslant 15.$	$2 \leqslant n \leqslant 84$	[85]
3 dec.	$m = 1.$	$n = 1\,(1)\,100$	[2]

See also the tables of reciprocals in Chapter 1.

Irreducible fractions

6 dec.	$1 \leqslant m \;\&\; n \leqslant 20$	[71]

PART TWO

REFERENCES

References to Chapter I

POWERS, RATIONAL AND ALGEBRAIC FUNCTIONS

1. **Адамович С. В.** Формулы по арифметике, алгебре, геометрии и тригонометрии. Клабуков, С.-Петербург, 1904.

2. **Адлер А.** Пятизначные логарифмы с вспомогательными таблицами для графического расчета и таблицами часто встречающихся числовых величин. Берлин — Рига, 1923.

3. **Александров В. А.** Математические таблицы для быстрых подсчетов. «Новое время», М., 1916.

4. **Алексеев В. А.** Сборник таблиц для расчета вероятностей. Арт. акад. им. Дзержинского, М., 1944.

5. **Андреев П. П.** Математические таблицы. Госстатиздат, М., 1952.

6. **Андреев П. П.** Таблицы для процентных вычислений. Госстатиздат, М., 1950.

7. **Андрейчин Р. и Матеев А.** Петозначни логаритмични и други таблици. София, 1949.

8. **Антонов Л. М.** Таблицы квадратов, кубов, степеней и упрощенного умножения. Л. и И. Белолипские, Николаев, 1910.

9. **Артемьев Д.** Таблицы пятизначных логарифмов натуральных чисел и тригонометрических функций. Ладыжников, Берлин, 1924.

10. **Асатиани Л. Г.** Малые таблицы. Готовые проценты, умножение и деление. «Техника да шрома», Тбилиси, 1946.

11. **Асатиани Л. Г.** Карманные таблицы умножения и деления. ГИЗ, М.—Л., 1930.

12. **Безикович Я. С.** Таблицы тригонометрические и логарифмические. «Кубуч», Л., 1931.

13. **Беленький Н. С.** Карманная счетная таблица для технических расчетов и хозяйственных вычислений. ГТТИ, М.— Л., 1934.

14. **Беленький Н. С.** Таблицы обратных чисел (от 1 до 10 000). Союзоргучет, М., 1939.

15. **Беленький Н. С.** Счетные таблицы для хозяйственных вычислений. Госфиниздат, М., 1949.

16. **Березовский Б. Я.** Справочник по элементарной математике, механике и физике. ГИТТЛ, М.— Л., 1939.

17. **Бирзвалк К.** Таблицы умножения и таблицы натуральных чисел, тригонометрических функций с приложением справок и формул из низшей геодезии. «Лотос», С.-Петербург, 1910.

18. **Блюмберг Я.** Таблицы пятизначных логарифмов чисел и тригонометрических величин. Вальтер и Рап, Рига, 1921.

19. **Богомольний М. А.** Таблиці знаходити добутки та частки (множення та ділення трицифрових на двоцифрові та трицифрові менші від 500). Тех.-теор. вид-во, Харків — Київ, 1932.

20. **Борисов С. А.** Математический справочник. Госкиноиздат, М., 1938.

21. **Брадис В. М.** Четырехзначные математические таблицы. Учпедгиз, М., 1950.

22. **Брилинг С. Р.** Математические формулы и таблицы. «Макиз», М., 1928.

23. **Б р и л и н г С. Р.** Математический справочник. ГТТИ, М.— Л., 1933.

24. **Б р о н ш т е й н И. Н. и С е м е н д я е в К. А.** Справочник по математике. Гостехиздат М.—Л., 1945.

25. **В о р о н е ц А. М.** Таблицы логарифмов и справочник формул и вспомогательных таблиц по курсу элементарной математики. Д. Сытин, М., 1917.

26. **В ы г о д с к и й М. Я.** Справочник по элементарной математике. ГИТТЛ, М.— Л., 1951.

27. **Г л а з е н а п С. П.** Карманные таблицы логарифмов чисел и тригонометрических функций с четырьмя десятичными знаками. Изд-во АН СССР, Л., 1934.

28. **Г л а з е н а п С. П.** Пятизначные таблицы логарифмов. Изд-во АН СССР, Л., 1934.
МИ, БЛ, ГАИШ, ИТМиВТ

29. **Г о р б у н о в А. И.** Математические таблицы квадратов, кубов, квадратных и кубичных корней, длин окружностей... М. К. Ивановский, Петроград, 1916.

30. **Г р и н б е р г М.** Натуральная таблица для возвышения в степень (2-ю и 3-ю) и извлечение корня (квадратного и кубического). Фридланд, Могилев, 1890.

31. **Г у р в и ц Л. Е.** Таблицы обратных чисел. Центр. упр. нар. хоз. учета Госплана СССР, М., 1933.

32. **Г у т е р м а н А. В. и Б р а т а ш В. Н.** Математические таблицы для конструкторов машиностроителей. ГНТИ Маш. лит., Москва — Свердловск, 1944.

33. **Г у т е р м а н И. В. и И в а н о в а К. Г.** Математические таблицы, 1. Союзоргучет М., 1939.

34. **Д и б о л ь д Р.** Формулы для вычисления поверхностей и объемов и таблицы для вычисления прямоугольных треугольников. Николаевское отделение Имп. Рус. техн. об-ва, Николаев, 1898.

35. **Д и т р и х к е й т О.** Быстрое счисление. Семизначные логарифмы и антилогарифмы. Г. В. Гольстен, Петербург, 1904.

36. **Д о б р о х о т о в П. И.** Таблицы логарифмов и тригонометрических величин (для градусных мер и делений угломера). Госиздат УзССР, Ташкент, 1935.

37. **Д о л г у ш и н П. А.** Четырехзначные таблицы логарифмов чисел и тригонометрических функций с приложением элементарной теории табличных вычислений. «Сотрудник», Киев, 1911.

38. **Д у б б е л ь Г.** Справочник по математике для инженеров, студентов и преподавателей математики. ОНТИ, М.— Л., 1933.

39. **Ж у р а в л е в С. Д.** Умножение и возведение в квадрат всевозможных чисел, извлечение квадратного корня из чисел до 1 000 000 и деление на один и два знака. М.

40. **И в а н о в Н. Н. и Х р е н н и к о в Н. А.** Таблицы натуральных пятизначных тригонометрических величин и квадратов чисел. Авторы, Одесса, 1929.

41. **И в а н о в К. П.** Таблицы для вычисления многочленов. Гостехиздат, М.— Л., 1949.

42. **К а й г о р о д о в А. И.** Четырехзначные таблицы логарифмов, антилогарифмов, натуральных тригонометрических величин и их логарифмов, перевода градусов в радианы, квадратных корней, обратных величин и др. Белорус. гос. акад. с. х., Горки, 1926.

43. **К а й г а р а д а у А.** Чатырохзначныя табліцы. Бел. дзярж. выд. Вычпэдсэктар, Менск, 1932.

44. **К а з а н с к и й С. П.** Таблицы квадратов, кубов и других математических величин. «Кубуч», Томск, 1933.

45. **К а м е н ь щ и к о в Н.** Таблицы логарифмов с четырьмя десятичными знаками. ГИЗ, Москва - Петроград, 1922.

46. **Карманная книжка математики.** Собрание таблиц, правил и формул. Андерс, М.,1861.

47. **Квадраты чисел от 1 до 1 000.** 1853.

48. **К о б е ц Н.** Карманная справочная книга для инженеров и техников. «Светоч», Одесса, 1927.

49. **К о г а н Я. М.** Краткий математический справочник. «Кубуч», Л., 1936.

50. **К о р н а к о в В. П.** Извлечение квадратных и кубичных корней из чисел с приложением таблиц квадратов, кубов, квадратных и кубичных

корней от 1 до 1 000. Э. Арнгольд, Петербург, 1900.

51. **Кутник С. Е.** Таблицы вычислений для инженеров, строителей, студентов, техников и учащихся профтехшкол. Окрпрофуч, Харьков, 1926.

52. **Кутузов Н. И.** Таблицы для разбивки круговых кривых. Дориздат, М., 1950.

53. **Кущенко В. С.** Математические формулы и таблицы, 1. Л., 1948.

54. **Лорченко В. И. и Оглоблин Н. В.** Четырехзначные логарифмические таблицы. Корчак-Новицкий, Киев, 1910.

55. **Майер Б. А.** Математические формулы. Справочник по арифметике, алгебре, геометрии, тригонометрии, высшей математике. Киев, 1930.

56. **Максимов М. Г.** Сборник вспомогательных таблиц для землемеров-землеустроителей, лесоустроителей и техников по съемке и нивелировке городов. ГТТИ, Л.— М., 1932.

57. **Мар'янов Б. М.** Лічильні таблиці. Держ. наук.-техн. вид-во України, Київ — Харків, 1933.

58. **Машиностроение. Энциклопедический справочник,** I, кн. 1. ГНТИ маш. лит., М., 1947.

59. **Мечников В. В.** Полные трехзначные таблицы логарифмов. Арт. акад. РККА, Л., 1933.

60. **Митропольский А. К.** Краткие таблицы (с приложением очерка вычислительной техники). Изд-во Лесотехн. акад. им. С. М. Кирова, Л., 1950.

61. **Михайлов А.** Таблицы логарифмов с четырьмя десятичными знаками. Логарифмы чисел, тригонометрических величин, натуральные тригонометрические функции и другие таблицы, упрощающие вычисления. ГИЗ, Москва — Петроград, 1923.

62. **Москвічов.** Таблиці логаритмів, антилогаритмів, логаритмів тригонометричних функцій. ОНТВУ, Харків — Київ, 1932.

63. **Недзялковский А. А.** Собрание таблиц и формул для инженеров, архитекторов и механиков, 1. Колесов и Михин, Петербург, 1867.

64. **Нейшулер Л. Я.** Таблицы. Деление, логарифмы, точное умножение и квадраты. ГТТИ, М., 1933.

65. **Никулин М. А.** Таблицы для практических вычислений. Перевод русских мер в метрические, таблицы степеней и корней чисел, десятичных логарифмов, тригонометрических функций, а также другие математические таблицы, необходимые для техника и учащегося. ГТТИ, М.— Л., 1932.

66. **Нумеров Б. В.** Таблицы для трехзначного вычисления. Гос. вычисл. ин-т, Петроград, 1922.

67. **Оглоблин Д. Н.** Пятизначные таблицы натуральных значений тригонометрических функций для маркшейдеров. Металлургиздат, Свердловск — Москва, 1949.

68. **Орлов С. В.** Трехзначные таблицы логарифмов и тригонометрических функций. ГИЗ, М., 1920.

69. **Первый русский карманный счетчик.** Быков, М., 1902.

70. **Перельман Я. И.** Таблицы и правила для вычислений. Учпедгиз, М.— Л.. 1931.

71. **Подтягин М. Е.** Таблицы умножения четырехзначных чисел на двузначные. ГОНТИ, М.— Л., 1932.

72. **Попов И. Г.** Математические таблицы. Учпедгиз, М., 1952.

73. **Пржевальский Е.** Пятизначные таблицы логарифмов. ГИЗ, М.— Л., 1928.

74. **Придатко С. В.** Таблицы четырехзначных логарифмов чисел и антилогарифмов. Таблицы натуральных величин тригонометрических функций и их логарифмов. А. Бенке, Петроград, 1915.

75. **Пеевски В. и Венедиков М.** Пятзначни логаритмични таблици на числата и тригонометричните функции при новоградусно деление. «Наука и изкуство», София, 1949.

76. **Пятизначные таблицы логарифмов.** Воениздат, М., 1951.

77. **Пятизначные таблицы логарифмов чисел и тригонометрических величин.** Геодезиздат, М., 1949.

78. **Пятэцкі П. Д.** Справачнік па матэматыцы. Дзярж. выд. Беларуси, Менск, 1932.

79. **Райнов А. В.** Четырехзначные математические таблицы (связно-функциональные). ОНТИ, М.— Л., 1936.

80. **Родин А. Ф.** Пятизначные логарифмические таблицы чисел и тригонометрических функций с приложением гауссовых логарифмов и различных вспомогательных таблиц. Линдеберг, С.-Петербург, 1895.

81. **Сегал Б. И. и Семендяев К. А.** Пятизначные математические таблицы. Изд-во АН СССР, М.— Л., 1948.

82. **Справочник по математике для обогатителей.** Металлургиздат, М., 1951.

83. **Справочное руководство по машиностроению, I.** ГНТИ Украины, Харьков, 1937.

84. **Справочник по элементарной математике, механике и физике.** ГИТТЛ, М.— Л., 1943.

85. **Стойков Д.** Спомогателни таблици за изчисления по машинна техника, метална техника и электротехника. София, 1947.

86. **Таблицы Барлоу — квадратов, кубов, квадратных корней и обратных величин всех целых чисел до 12.500.** Изд-во иностр. лит., М., 1950.

87. **Таблицы для вычисления квадратных чисел с 1000 до 8000.** Уралмашзавод, Свердловск, 1936.

88. **Таблиці квадратів, кубів, логаритмів, тригонометричних і гіперболічних функцій, квадратових і кубових коренів та інших величин.** Техн.-теорет. вид-во, Харків, 1932.

89. **Таблицы квадратов трехзначных чисел, составных чисел до 10000, тригонометрических величин и пр.** А. А. Карцев, М., 1887.

90. **Таблицы логарифмов Вестфаля.** Гл. Морской штаб, С.-Петербург, 1834.

91. **Таблицы логарифмов комплексных чисел и перехода от декартовых координат к полярным.** Изд-во АН СССР, М., 1952.

92. **Таблицы степеней, корней, обратных величин и логарифмов.** Изд-во Сев.-Зап. обл. промбюро ВСНХ, Л., 1927.

93. **Таблицы математических и физических величин.** Научн. книгоизд., Петроград, 1923.

94. **Таланов А. В.** Таблицы вычислителя БЗР. Арт. акад. им. Дзержинского, М., 1941.

95. **Устрялов В. Ф.** Вспомогательные при решениях математических задач таблицы первоначальных множителей. Б. М. Вольф, С.-Петербург, 1901.

96. **Фейгельсон Е. Г.** Таблицы десятичных логарифмов и натуральных тригонометрических функций. Учпедгиз, М.— Л., 1931.

97. **Филоненко А. С.** Таблицы шестизначных натуральных величин тригонометрических функций для счетных машин. Госкартогеодезия, М.—Л. 1932.

98. **Франк М. Л.** Карманный справочник по математике. Т-во профессоров и преподавателей, С.—Петербург, 1912.

99. **Хохлов А. И.** Карманные математические таблицы. Пятизначные. Гостехиздат, М.— Л., 1947.

100. **Hütte.** Справочник для инженеров, техников и студентов, I. Госмашметиздат, М.— Л., 1934.

101. **Четырехзначные таблицы логарифмов.** Коллект. печ., Л., 1928.

102. **Чиханов Б.** Таблицы пятизначных логарифмов чисел и тригонометрических величин. А. И. Мамонтов, М., 1917.

103. **Чэркашин Л.** Чатырохзначныя матэматычныя табліцы. Дзярж. наук-тэхн. выд-во, Менск, 1934.

104. **Яковкин М. В.** Таблица квадратов четырехзначных чисел. Госстатиздат, М., 1950.

105. **Янке Е. и Эмде Ф.** Таблицы функций с формулами и кривыми. ГИТТЛ, М.—Л., 1949.

106. **Abbott P. W.** Mathematical tables and formulae. Green and Co., London, 1918.

107. **Allen E. S.** Six-place tables. McGraw — Hill book Co., New York—London, 1947.

108. **Attwood** E. L. Practical five-figure mathematical tables. Macmillan, London, 1950.

109. **Babington** J. Pyrotechnia. London, 1635.

110. **Bachheimer** R. Potenz-und Wurzeltafeln enthaltend die Quadrate, Kuben, Quadrat-und Kubikwurzeln der Zahlen 1—1000, sowie die Umfänge und Inhalte der Kreise mit den Durchmessern 1—1000. Leipzig—Wien, 1928.

111. **Barlow** P. New mathematical tables. G. and S. Robinson, London, 1814.

112. **Barlow** P. Tables of squares' cubes, square roots, cube roots and reciprocals of all integer numbers up to 12 500. Spon, London; Chemical Publishing Co., New York, 1947.

113. **Beyrodt** G. und **Reissner** A. Tabellenbuch zur Wirtschaftmathematik. Fachwissen Lehrbücher für die Berufsfach und Fachschule. «Volk und Wissen», Berlin — Leipzig, 1949.

114. **Blater** J. Tafel der Viertel-Quadrate aller ganzen Zahlen von 1 bis 200 000. A. Hölder, Wien, 1887

115. **Boebert** K. W. Tafeln der Quadratzahlen aller natürlichen Zahlen von 1 bis 25 200... Fleischer, Leipzig, 1812.

116. **Bohde** H. Mathematisch-technische Zahlentafeln. Springer, Berlin, 1933.

117. **Bojko** J. Neue Tafel der Viertelquadrate aller natürlichen Zahlen von 1 bis 20 000. Speidel, Zürich, 1909.

118. **Boll** M. Tables numériques universelles des laboratoires et bureaux d'étude. Dunod, Paris, 1947.

119. **Bremiker** C. Logarithmisch-trigonometrische Tafeln mit fünf Decimalstellen. Weidmann, Berlin, 1922.

120. **Bremiker** C. Tafeln vierstelliger Logarithmen. Weidmann, Berlin, 1907.

121. **Bremmer** H. Terrestrial radio waves. Theory of propagation. Elsevier Publ. Co., New York, 1949.

122. **Britisch Association for the Advancement of Science.** Mathematical tables, 9: Table of powers giving integral powers of integers. Cambridge Univ. Press, Cambridge, 1940.

123. **Buchner** J. P. Tabula radicum, Quadratorum, und Cuborum, in welcher von allen gegebenen Cubis und Quadratis du radix biss auf 12 000, also fort ohne eintziges rechnen sammt der darinnen steckender surdischen oder errational-zahl extrahiret zufinder; und wiederum von einen ieglichem Radice, biss auf 12 000 sein Quadratum und Cubus ausgerechnet zu sehen. Neurnberg, 1701.

124. **Card** M. A table for calculating the thicknesses of segments of spheres. Downing College, Cambridge, 1943.

125. **Carmichael** R. D. and **Smith** E. R. Mathematical tables and formulae. Ginn, Boston — New York, 1931.

126. **Castle** F. Five-figure logarithmic and other tables. Macmillan, London, 1910.

127. **Claessens** F. Tables des carrées, cubes, racines carrée et cubique. Gauthier-Villars, Paris, 1920.

128. **Clark** J. B. Mathematical and physical tables. Oliver and Boyd, Edinburgh, 1950.

129. **Claudel** J. Formules, tables, et renseignements pratiques; aide-mémoire des ingénieurs, des architecteurs. Paris, 1854.

130. **Claudel** J. Tables: 1. Des carres et des cubes des nombres entiers successifs... 2. Des longueurs des circonférences et des surfaces des cercles... 3. Des expressions trigonométriques naturelles des angles successifs, de minute en minute. Carilian-Goeury et V-or Dalmont, Paris, 1850.

131. **Comrie** L. J. Chambers's four-figure mathematical tables. W. and R. Chambers, Edinburgh — London, 1947.

132. **Dale** J. B. Five-figure tables of mathematical functions comprising tables of logarithms, powers of numbers, trigonometric, elliptic, and other transcendental functions. Arnold, London, 1949.

133. **Davenport** C. B. Statistical method with special reference to biological variation. New York, 1899.

134. **Dwight H. B.** Tables of integrals and other mathematical data. Macmillan, New York, 1947.

135. **Dwight H. B.** Mathematical tables. McGraw-Hill, New York — London, 1941.

136. **Eschke.** Mathematische Tabellen und Formeln. W. Mayer, Berlin, [193...].

137. **Fergusson J. C.** Fergusson's percentage unit of angular measurement, with logarithms. Longmans, London, 1912.

138. **Fisher R. A.** and **Yates F.** Statistical tables for biological, agricultural and medical research. Oliver and Boyd, London — Edinburgh, 1949.

139. **Formular matematic şi tehnic.** Pentru elevi, studenţi şi tehnicieni. Editura tehnica, Bucureşti, 1949.

140. **Garrett H. E.** Statistics in psychology and education. Longmans, New York — London, 1937.

141. **Gauss F. G.** Vierstellige logarithmische und trigonometrische Tafeln (Sexagesimal unterteiler Altgrad). «Volk und Wissen», Leipzig — Berlin, 1946.

142. **Gauss F. G.** Fünfstellige vollständige logarithmische und trigonometrische Tafeln. Strien, Halle, 1908.

143. **Gauss F. G.** Fünfstellige voll ständige trigonometrische und polygonometrische Tafeln für Machinenrechnen. Wittwer, Stuttgart, 1918.

144. **Gelin E.** Recueil de tables numériques et de formules. Author, Nuy, 1891—1892.

145. **Elznic V.** a **Valouch M.** Geoma. Pětimístné tabulky hodnot a logaritmů goniometrických funkcí v setinném dělení, tabulky geodetické a katastralni. Vydala jednota českych matematiků a fysiků. Praha, 1944.

146. **Glover J. W.** Tables of applied mathematics in finance, insurance and statistics. Wahr, Ann Arbor, Michigan, 1930.

147. **Guldinus P.** De Centro Gravitatis, I, Appendix. Vienna, 1635.

148. **Burington R. S.** Handbook of mathematical tables and formulas. Sandusky, Ohio, 1946.

149. **Hantschl J.** Logarithmisch-trigonometrisches Handbuch... Rohrmann und Schweigerd, Wien, 1833.

150. **Hawelka R.** Vierstellige Tafeln der Kreis- und Hyperbelfunktionen, sowie ihrer Umkehrfunktionen im Komplexen in Schritten von $\frac{\pi}{2}$ 0,02 und i 0,02. Vieweg, Braunschweig, 1931.

151. **Hayashi K.** Fünfstellige Funktionentafeln. Springer, Berlin, 1930.

152. **Hayashi K.** Sieben—und mehrstellige Tafeln der Kreis- und Hyperbelfunktionen und deren Produkte sowie der Gamma-funktion. Springer, Berlin, 1926.

153. **Hayashi K.** Tafeln der Besselschen-, Theta-, Kugel- und anderen Funktionen. Springer, Berlin, 1930.

154. **Hayashi K.** Tafeln für die Differenzenrechnung sowie für die Hyperbel,-Besselschen, elliptischen und anderen Funktionen. Springer, Berlin. 1933.

155. **Heckmann O.** Rechnenformeln und Fehlergrenzer. Städtisches Vermessungs und Liegenschaftsamt. Mannheim, 1937.

156. **Horton R. E.** Weir experiments, coefficients and formulae. U. S. geological survey, water-supply and irrigation paper, № 200. Washington, 1907.

157. **Hutton C.** Tables of the products and powers of numbers. Board of Longitude, London, 1781.

158. **Tavole logarithmiche a cinque cifre decimali.** Italy Istituto Idrografico della R. Marina. Genoa, 1916.

159. **Ives H. C.** Mathematical tables reprinted from searles and Ives' field engineering, with additions. Wiley, New York; Chapman and Hall, London, 1945.

160. **Ives H. C.** Seven place natural trigonometrical functions. Wiley, New York; Chapman and Hall, London, 1929.

161. **Jäderin E.** Tafeln für Interpolation und Ausgleichung. Stockholm, Norstedt, 1915.

162. **Jahn G. A.** Tafeln der Quadrat- und Kubikwurzeln aller Zahlen von 1 bis 25 500 der Quadratzahlen aller Zahlen von 1 bis 27 000 und der Kubik-

zahlen von 1 bis 24 000. Barth, Leipzig, 1839.

163. Jahnke E. und Emde F. Funktionentafeln mit Formeln und Kurven. Teubner, Leipzig — Berlin, 1933.

164. Jones G. W. Logarithmic tables. Macmillan, London; Author, Ithaca, New York, 1893.

165. Jordan W. Mathematische und geodätische Hilfstafeln. Hannover, 1895.

166. Kahlau R. Vierstellige Tafeln zum logarithmischen und numerischen Rechnen. Bonness und Hachfeld, Potsdam, 1944.

167. Kaye G. W. C. and Laby T.H. Four-figure mathematical tables. London, 1925.

168. Kelley T. L. Statistical method. Macmillan, New York, 1924.

169. Kelley T. L. The Kelley statistical tables. Macmillan, New York, 1938. ЛГУ, МИ

170. Kent W. and Kent R. T. Kent's mechanical engineers' handbook. Wiley, New York; Chapman and Hall, London, 1923.

171. Keturzenkles logaritmu, ir kitu dydziu lenteles. Matematines formules. Valstybine Pedagogines Literaturos Leidykla, Kaunas, 1946.

172. Köhler H. G. Logarithmisch-trigonometrisches Handbuch. Tauchnitz, Leipzig, 1876, 1898.

173. Kolb A. und Bauman F. Blitz. Rechnentafeln nebst den Tabellen: Kreis-Inhalte und Umfänge Quadrate, Kuben, Quadrat- und Kubik-Wurzeln. Fr. Lintz, Trier, 1903.

174. Kulik J. P. Tafeln der Quadrat- und Kubik-Zahlen aller natürlichen Zahlen bis Hundert Tausend nebst ihrer Anwendung auf die Zerlegung grosser Zahlen in ihre Faktoren. Fleischer, Leipzig, 1848.

175. Kulik J. P. Neue Multiplikationstafeln. Fleischer, Leipzig, 1851.

176. Küstner H. Fünfstellige Logarithmen. «Volk und Wissen». Berlin—Leipzig, 1952.

177. Lambert J. H. Zusätze zu den logarithmischen und trigonometrischen Tabellen. Haude und Spener, Berlin, 1770.

178. Lambert J. H., Felkel A. Supplementa tabularum logarithmicarum et trigonometricarum... Lisbon, 1798.

179. Laundy S. L. Table of quarter-squares of all integer numbers, up to 100 000. Layton, London, 1856.

180. Lohse O. Tafeln für numerisches Rechnen mit Maschinen. Engelmann, Leipzig, 1935.

181. Loky B. Negyjegyü függvenytablazatok. Budapest, 1950.

182. Lötzbeyer P. Vierstellige Tafeln zum Logarithmischen und Zahlenrechnen. Graphische Rechentafeln. Teubner, Leipzig—Berlin, 1928.

183. Ludolf L. J. Tetragonometria tabularia... Amsterdam, 1690.

184. Maseres F. The doctrine of permutations and combinations. London, 1795.

185. Merpaut J. M. Tables Arithmonomiques... Vannes, 1832.

186. Merriman M. Text-Book on the method of least squares. Willey, New York; Chapman and Hall, London, 1910.

187. Miller J. C. P. Tables for converting rectangular to polar coordinates. Scientific Computing Service, London, 1943.

188. Milne-Thomson L. M. Standard table of square roots. Bell, London, 1929.

189. Milne-Thomson L. M. and Comrie L. J. Standard four-figure mathematical tables. Including many new tables trigonometrical functions for radians, inverse trigonometrical and hyperbolic functions and an extended table of natural logarithms. Macmillan, London, 1948.

190. Miner J. R. Tables of $\sqrt{1-r^2}$ and $1-r^2$ for use in partial correlation and in trigonometry. Johns Hopkins Press, Baltimore, 1922.

191. Miscellaneous physical tables. Planck's radiation functions and electronic functions. New York, 1941.

192. Molterer J. Tabellen zur Zahlentheorie 52-stellige Briggische Logarithmen von 1—1100 und Potenzen der Zahl 2 bis 2^{400}. Welsermuhl, Wels, 1937.

193. Moore J. Arithmetic. London, 1650.

194. Mursi Z. Tables of Legendre associated functions. Cairo, 1941.

195. **Nautical Almanac Office. Planetary co-ordinates for the years 1800—1940.** L. J. Comrie, H. M. Stationery Office, London, 1933.

196. **Nautical Almanac Office. Seven-figure trigonometrical tables for every second of time.** L. J. Comrie, H. M. Stationery Office, London, 1939.

197. **P a t z W.** Tafel der Regelmässigen Kettenbrüche für die Quadratwurzeln aus den natürlichen Zahlen von 1—10.000. Becker und Erler, Leipzig, 1941.

198. **P e a r s o n K.** Tables of the incomplete Γ-Function. H. M. Stationery Office, London, 1934.

199. **P e a r s o n K.** Tables for statisticians and biometricians, I. Biometric Laboratory, University College, London, 1930.

200. **P e a r s o n K.** Tables for statisticians and biometricians, II. Biometric Laboratory, University College, London, 1931.

201. **P e t e r s C. F. W.** Astronomische Tafeln und Formeln. Mauke, Hamburg, 1871.

202. **P e t e r s J.** Zehnstellige Logarithmen der Zahlen von 1 bis 100.000 nebst einem Anhang mathematischer Tafeln, I. Reichsamt für Landesaufnahme, Berlin, 1922.

203. **P l a s s m a n n J.** Tafel der Viertel-Quadrate aller Zahlen von 1 bis 20009 zur Erleichterung des Multiplizierens vierstelliger Zahlen. Jänecke, Leipzig, 1933.

204. **P o t i n L.** Formules et tables numériques. Doin et Gauthier-Villars, Paris, 1925.

205. **P r y d e J.** Chamber's educational course. Mathematical tables. Chambers, London — Edinburgh, 1930.

206. **R a t a s s e p p K.** Matemaatilised tabelid. Pedagoogiline Kirjandus, Tallinn, 1948.

207. **R a y l e i g h, Lord,** (S t r u t J. W.). The theory of sound, I, Macmillan, London, 1894.

208. **R e u t e r J., S e r g e l i u s M.** Taulukot-Tabellen. Helsinki, 1945.

209. **L a r s e n H. D.** Rinehart mathematical tables, formulas and curves. Chapman and Hall, London, 1949.

210. **R i v a d r M.** Tables des sinus, tangentes, secantes et de leurs logarithmes, avec la construction de ces tables, et les problèmes de la trigonométrie rectiligne et sphérique. Desaint, Saillant, Paris, 1743.

211. **R o h r b a c h C.** Vierstellige logarithmisch-trigonometrische Tafeln. Ehlermann, Dresden, 1929.

212. **R o s e n b a c h I. B., W h i t m a n E. A., M o s k o v i t z D.** Mathematical tables. Ginn, Boston, 1946.

213. **R u s s e l l A.** A treatise on the theory of alternating currents, I Cambridge Univ. Press, Cambridge, 1914.

214. **S a l o m o n J. M.** Logarithmische Tafeln. Gerold, Vienna, 1827.

215. **S a s s e n f e l d H. M.** und **T s c h u n k o H. F. A.** Mathematische Tafeln für Mathematiker, Naturwissenschaftler, Ingenieure. Erster Teil. Elementare Funktionen. Fr. Lamade, Walldorf bei Heidelberg, 1949.

216. **S c h a d e C.** Taschenrechner «Fix». Die Rechnenmaschine in der Tasche. Engelhard-Reyher, Gotha, 1942.

217. **S c h u b e r t H.** Vierstellige Tafeln und Gegentafeln für logarithmisches und trigonometrisches Rechnen. W. Gruyter, Berlin, 1945.

218. **S c h ü l k e A.** Tafeln. Vierstellige Logarithmen Funktions- und Zahlenwerte. «Volk und Wissen», Leipzig, 1949.

219. **S c h u l t e - T i g g e s R. A.** Rechentafeln für höhere Lehranstalten insbesondere zu Mehlers Hauptsätzen der Elementar-Mathematik. W. Gruyter, Berlin — Leipzig, 1929.

220. **S c h u l z e J. C.** Neue und erweiterte Sammlung logarithmischer, trigonometrischer und anderer... Tafeln, 1 — 2. Mylius, Berlin, 1778.

221. **S e e l e y W. J.** Impedance computing tables. Engineers Publishing Co., Philadelphia, 1936.

222. **S e m i l l e r H.** und **S e m i l l e r A.** Vierstellige Logarithmen und Zahlentafeln. Springer, Berlin, 1938.

223. **S h a n k s W.** Contributions to mathematics. Comprising chiefly the rectification of the circle to 607 places

of decimals. Cambridge and Durham, London, 1853.

224. Sokal S. i Torbus W. Tablice matematyczno-fizyczne do użytku szkoł zawodowych. Państwowe wyd-wa szkolnictwa zawodowego. Warszawa, 1951.

225. Spencer R. C. and Reynolds G. E. A table of normalized parabolic coordinates and arc lengths. Air Force Cambridge Research Laboratories, publications E 4083. Cambridge, 1951.

226. Stalmasek J. Tabul'ky funkcií a konštant (z matematiky, fyziky a chemie). Štátne nakl., Bratislava, 1951.

227. Studnicka F. J. Kapesní logarithmické tabulky. Praha, 1870.

228. **Tabele şi formule matematice.** Ed. tehnica, Bucureşti, 1951.

229. **Tabellen der Quadrate, Kuben, Wurzeln, Kreisumfange und Inhalte, Kreisfunktionen und Logarithmen.** Leipzig.

230. **Table of reciprocals of the integers from 100.000 through 200.009.** Columbia Univ. Press, New York, 1943.

231. **Table of the first ten powers of the integers from 1 to 1.000.** Prepared by the Matematical Tables Project of the Work Projects Administration for the City, New York, 1939.

232. **Tables of fractional powers, I:** Values of a^x, II: Values of x^a. Columbia Univ. Press, New York, 1946.

233. Ledertzen P. Manuale Mathematicum, darinn begriffen die Tabulae Sinuum, Tangentium, Secantium... Straßburg, 1619.

234. **Tafelsammlung und Hilfsbuch für Maschinenbauer, Mechaniker, Werkzeugmacher und andere Metallarbeiter.** Leipzig, 1948.

235. Timpenfeld P. Tabellen der Quadrate von 1 bis 15.000, Kuben von 1 bis 3.000... Krügen, Dortmund, 1937.

236. Trautwine J. C. The Civil Engineer's Reference-Book. Chapman and Hall, London, 1937.

237. **United States Bureau of Reclamation. Hydraulic and excavation tables.** U. S. Department of Interior Bureau of Reclamation. 1940.

238. Vega G., Hülsse J. A. Sammlung mathematischer Tafeln. Weidmann, Leipzig, 1849.

239. Vlacq A. Tabellen der Sinuum, Tangentium und Secantium, wie auch der Logarithmorum vor die Sinus, Tangentes und die Zahlen von 1 bis 10 000. Franckfurt — Leipzig, 1738.

240. **Des vollstandigen Mathematischen,** 2. Leipzig, 1742.

241. Weigel H. W. $x^n + y^n = z^n$? Die elementare Lösung des Fermat-Problems. Author, Leipzig, 1933.

242. Weisbach J. Der Ingenieur Sammlung von Tafeln, Formeln und Regeln. Braunschweig, 1863.

243. Wolf R. Handbuch der Mathematik, Physik, Geodäsie und Astronomie, I. Zürich, 1870.

244. Lomnicki A. Tablice matematyczno-fizyczne czterocyfrowe. Dla użytku szkoł średnich. Ksiaznica-Atlas, Lwów—Warszawa, 1932.

245. Zacharias M. und Meth P. Vierstellige Logarithmentafeln. W. Gruyter, Berlin — Leipzig, 1927.

246. **Zahlentafeln für die Grundschule.** «Volk und Wissen», Berlin, 1950.

247. Zimmermann L. Vollständige Tafeln der Quadrate aller Zahlen bis 100.009. Berechnet und herausgegeben. Wichmann, Berlin, 1948.

248. Zimmerman L. Rechentafeln. Reiss, Liebenwerda, 1923.

249. **Abhandlungen der Königlich Böhmischen Gesellschaft der Wissenschaften** (Prague), 1860, (5), 11, 21—123.
Kulik J. P. Beiträge zur Auflösung höherer Gleichungen überhaupt und der kubischen Gleichungen insbesondere.

250. **Acta Societatis Scientiarum Fennicae** (Helsingfors), 1906, 32, № 6, 1—27.
Tallqvist H. J. Tafeln der Kugelfunctionen $P_n(x)$ und ihrer abgeleiteten Functionen.

251. **American Journal of Mathematics Pure and Applied,** 1878, I, 184—239.
Lucas E. Théorie des fonctions numériques simplement périodiques.

252. **American Mathematical Monthly,** 1925, 32, 379.
Lehmer D. H. Table of cube and fifthroots.

253. **Archiv der Mathematik und Physik** (Leipzig — Berlin), 1902, 2, 249—252.
H e r t z e r H. Periode des Dezimalbruches für $\frac{1}{P}$, wo P eine Primzahl.

254. **Archiv für Elektrotechnik** (Berlin), 1927, **17**, 79—101.
Potentialtheorie der Hängeisolatoren.

255. **Astronomische Nachrichten** (Kiel), 1936, **258**, 9—10.
M ö l l e r J. P. and R a s m u s s e n H. Q. Tafel der Funktion $x^{\frac{2}{3}}$ zur Verwendung bei parabolischer Bahnbestimmung nach der Methode von B. Strömgren.

256. **The Astrophysical Journal** (Chicago), 1950, 3, № 1, 142—156.
C h a n d r a s e k h a r S. and M ü n c h G. On the integral equation governing the distribution of the true and the apparent rotational velocities of stars.

257. **Journal für die reine und angewandte Mathematik** (Berlin), 1849, **38**, 216—240.
O e t t i n g e r. Untersuchungen über die analytischen Facultäten.

258. **Memoirs of Britisch Astronomical Association** (London), 1934, **30**, № 3, 107—148.
F o x e l l J. T. Lunar occultation maps.

259 **The Messenger of Mathematics** (Cambridge), 1877, **7**, 86—88.
D r a c h S. M. Cube roots of primes to 31 places.

260. **Proceedings of the American Academy of Arts and Sciences** (Boston), 1922, **57**, № 7, 173—191.
P i e r c e Q. W. A table and method of computation of electric wave propagation, transmission line phenomena, optical refraction and inverse hyperbolic functions of a complex variable.

261. **Proceedings of the Physical Society of London**, 1920, **32**, 252—264.
S m i t h T. On tracing rays through an optical system.

262. **Proceedings of the Royal Society of London**, 1873—1874, 22, 384—388.
S h a n k s W. On the number of figures in the period of the reciprocal of every prime number between 20 000 and 30 000.

263. **Proceedings of the Royal Society of London**, 1873—1874, 22, 200—210.
S h a n k s W. On the number of figures in the period of the reciprocal of every prime number below 20 000.

264. **Psychometrika** (Journal devoted to the development of psychology as a quantitative rational science Colorado Springs Psychometric Society), 1946, **11**, № 4, 223—237.
R e y n o l d s W. M. A. A prepunched master deck for the computation of square roots on IBM electrical accounting equipment.

265. **Scripta Mathematica** (New York), 1949, 15, № 3—4, 247—251.
U h l e r H. S. Table of exact values of high powers of 2.

266. **Zeitschrift für Mathematik und Physik** (Leipzig), 1909, 57, 337—373.
G r e i n e r R. Über des Fehlersystem der Kollektivmasslehre.

267. C o m r i e L. J. Chambers's six-figure mathematical tables, II. New-York, 1949.

268. V a l o u c h M.—V a l o u c h M. A. Sedmimístné logaritmy čisel od 1 do 120 000. Nakladatelstvi Československe Akademie Věd, Praha, 1953.

269. V a l o u c h M.—V a l o u c h M.A. Pětimístné logaritmy čisel. Nakladatelstvi Československe Akademie Věd, Praha, 1953.

270. R y b n e r J., S o r e n s e n K. S. Tabel til Brug ved Addition af Komplekse Tal. Gjellerups, København, 1948.

References to Chapter 2

TRIGONOMETRIC FUNCTIONS, VARIOUS VALUES CONNECTED WITH THE CIRCLE AND THE SPHERE

1. **А б о л ь н и к Б. С.** Карманные таблицы трехзначных логарифмов. «Красное знамя», Томск, 1937.

2. **А б о л ь н и к Б. С.** Трехзначные таблицы логарифмов. Транспечать, М., 1930.

3. **А д л е р А.** Пятизначные логарифмы с вспомогательными таблицами для графического расчета и таблицами часто встречающихся числовых величин. Берлин — Рига, 1923.

4. **Десятизначные таблицы логарифмов комплексных чисел и перехода от декартовых координат к полярным.** Изд-во АН СССР, М., 1952.

5. **А л е к с а н д р о в В. А.** Математические таблицы для быстрых подсчетов. «Новое время», М., 1916.

6. **А л е к с е е в В. А.** Сборник таблиц для расчета вероятностей. Арт. акад. им. Дзержинского, М., 1944.

7. **А н д р е й ч и н Р. и М а т е е в А.** Петозначни логаритмични и други таблици. София, 1949.

8. **А н и ч к о в Д.** Теоретическая и практическая тригонометрия. Типограф. компания, М., 1787.

9. **Б е з и к о в и ч Я. С.** Таблицы тригонометрические и логарифмические. «Кубуч». Л., 1931.

10. **Б е л е н ь к и й Н. С.** Счетные таблицы для хозяйственных вычислений. Госфиниздат. М., 1949.

11. **Б е р е з о в с к и й Б. Я.** Справочник по элементарной математике, механике и физике. ГИТТЛ. М.— Л., 1939.

12. **Б и р з в а л к К.** Таблицы умножения и таблицы натуральных чисел, тригонометрических функций с приложением справок и формул из низшей геодезии. «Лотос», С.-Петербург, 1910.

13. **Б л ю м б е р г Я.** Таблицы пятизначных логарифмов чисел и тригонометрических величин. Вальтер и Рап, Рига, 1921.

14. **Б о г о м о л ь н и й М. А.** Таблиці знаходити добутки та частки (множення та ділення трицифрових на двоцифрові та трицифрові менші від 500). Харків—Київ, Тех.-теор. видво. 1932.

15. **Б о р д ю с с о н а.** Угломер или таблицы хорд дуги каждого угла, считая от одной минуты до 180 градусов, выраженных в величинах радиуса, разделенного на 1000 равных частей, для употребления инженерами, занимающимися съемкой и черчением планов. Петербург, 1827.

16. **Б о р и с о в С. А.** Математический справочник. Госкиноиздат, М., 1938.

17. **Б р а д и с В. М.** Четырехзначные математические таблицы. Учпедгиз, М., 1950.

18. **Б р е м и к е р К.** Таблицы логарифмов чисел и тригонометрических функций с шестью десятичными знаками. Редбюро ГУГСК НКВД СССР, М., 1938.

19. **Брилинг С. Р.** Математический справочник. ГТТИ, М.—Л., 1933.

20. **Брилинг С. Р.** Математические формулы и таблицы. «Макиз», М., 1928.

21. **Бронштейн И. Н. и Семендяев К. А.** Справочник по математике. Гостехиздат, М.— Л., 1945.

22. **Брунс К.** Таблицы логарифмов чисел и тригонометрических функций с семью десятичными знаками. Редбюро ГУГК при СНК СССР, М., 1939.

23. **Ветчинкин В. П.** Новые формулы и таблицы эллиптических интегралов и функций. Изд-во Военновозд. акад. РККА, М., 1935.

24. **Виноградов В.** Таблицы тригонометрических величин, II. Житомир.

25. **Воронец А. М.** Таблицы логарифмов и справочник формул и вспомогательных таблиц по курсу элементарной математики. Сытин, М., 1917.

26. **Восьмизначные таблицы тригонометрических функций.** Геодезиздат, М., 1950.

27. **Выгодский М. Я.** Справочник по элементарной математике. ГИТТЛ, М.—Л., 1951.

28. **Гаусс Ф. Г.** Пятизначные таблицы натуральных значений тригонометрических функций. Геодезиздат, М., 1943.

29. **Гибшман Е. А.** Таблицы значений тригонометрических величин с 6 десятичными знаками через каждую минуту. А. И. Мамонтов, М., 1914.

30. **Глазенап С. П.** Карманные таблицы логарифмов чисел и тригонометрических функций с четырьмя десятичными знаками. Изд-во АН СССР, Л., 1934.

31. **Глазенап С. П.** Пятизначные таблицы логарифмов. Изд-во АН СССР, Л., 1934.

32. **Гопфен Б. В.** Логарифмы, антилогарифмы и тригонометрические величины в упрощенных таблицах. М., 1919.

33. **Горбунов А. И.** Математические таблицы квадратов, кубов, квадратных и кубичных корней, длин окружностей. М. К. Ивановский, Петроград, 1916.

34. **Гутерман А. В. и Браташ В. Н.** Математические таблицы для конструкторов машиностроителей. Г.Н.Т.И. Маш. лит., Москва — Свердловск, 1944.

35. **Гутерман И. В. и Иванова К. Г.** Математические таблицы, II. Союзоргучет, М., 1940.

36. **Двайт Г. Б.** Таблицы интегралов и другие математические формулы. Гос. изд-во иностр. лит., М., 1948.

37. **Дибольд Р.** Формулы для вычисления поверхностей и объемов и таблицы для вычисления прямоугольных треугольников. Николаевское отделение Имп. Рус. техн. об-ва, Николаев, 1898.

38. **Доброхотов П. И.** Таблицы логарифмов и тригонометрических величин (для градусных мер и делений угломера). Гос. изд-во Уз. ССР, Ташкент, 1935.

39. **Долгушин П. А.** Четырехзначные таблицы логарифмов чисел и тригонометрических функций с приложением элементарной теории табличных вычислений. «Сотрудник», Киев, 1911.

40. **Дуббель Г.** Справочник по математике для инженеров, студентов и преподавателей математики. ОНТИ, М.— Л., 1933.

41. **Жиронкин Е. И.** Таблицы деления и таблицы секансов. Автор, Л., 1929.

42. **Записки Государственного гидрологического института, VI.** Л., 1932.

43. **Иванов Н. Н. и Хренников Н. А.** Таблицы натуральных пятизначных тригонометрических величин и квадратов чисел. Авторы, Одесса, 1929.

44. **Известия Казахского филиала АН СССР** (Алма-Ата), Серия горного дела, вып.1946, 1(22).

45. **Казанский С. П.** Таблицы квадратов, кубов и других математических величин. «Кубуч», Томск, 1933.

46. **Кайгарадау А.** Чатырохзначныя табліцы. Бел. дзярж. выд-во Вычпэдсэктар, Менск, 1932.

47. **Кайгородов А. И.** Четырехзначные таблицы логарифмов, антилогарифмов, натуральных тригонометрических величин и их логарифмов, перевода градусов в радианы, квадратных корней, обратных величин и др. Белорус. Гос. акад. с. х., Горки, 1926.

48. **Каллет Ф.** Таблицы логарифмов чисел от 1 до 108 000, синусов, косинусов и тангенсов от секунды до секунды для первых пяти градусов и от десяти до десяти секунд для всех градусов четверти окружности. М., 1837.

49. **Каменьщиков Н.** Таблицы логарифмов с четырьмя десятичными знаками. ГИЗ, Москва — Петроград, 1922.

50. **Каменьщиков Н.** Трехзначные логарифмы. ГИЗ, М., 1921.

51. **Каменьщиков Н.** Четырехзначные логарифмы чисел и тригонометрических функций. ГИЗ, М., 1921.

52. **Кобец Н.** Карманная справочная книга для инженеров и техников. Одесса, «Светоч», 1927.

53. **Коган Я. М.** Краткий математический справочник. «Кубуч», Л., 1936.

54. **Кончин В. В.** Таблицы натуральных тригонометрических и редукционных величин. Геодезиздат, М., 1940.

55. **Кутник С. Е.** Таблицы вычислений для инженеров, строителей, студентов, техников и учащихся профтехшкол. Окрпрофуч, Харьков, 1926.

56. **Кутузов Н. И.** Таблицы для разбивки круговых кривых. Дориздат, М., 1950.

57. **Левшунов М. Т.** Таблицы произведений на tang'сы и cotg'сы для землеустроительных, топографических и других вычислений. Омск, 1930.

58. **Лорченко В. И.** и **Оглоблин Н. В.** Четырехзначные логарифмические таблицы. Корчак-Новицкий, Киев, 1910.

59. **Майер Б. А.** Математические формулы. Справочник по арифметике, алгебре, геометрии, тригонометрии, высшей математике. Киев, 1930.

60. **Максимов М. Г.** Сборник вспомогательных таблиц для землемеров-землеустроителей, лесоустроителей и техников по съемке и нивелировке городов. ГТТИ, Л.—М., 1932.

61. **Машиностроение. Энциклопедический справочник, I,** кн. 1. ГНТИ. Маш. лит., М., 1947.

62. **Мечников В. В.** Полные трехзначные таблицы логарифмов. Арт. акад. РККА, Л., 1933.

63. **Митропольский А. К.** Краткие таблицы (с приложением очерка вычислительной техники). Изд-во Лесотехн. акад. им. С. М. Кирова, Л., 1950.

64. **Михайлов А.** Таблицы логарифмов с четырьмя десятичными знаками. Логарифмы чисел, тригонометрических величин, натуральные тригонометрические функции и другие таблицы, упрощающие вычисления. ГИЗ, Москва — Петроград, 1923.

65. **Мореходные таблицы 1943 г.** Изд-во Гидрограф. упр. ВМС, 1949.

66. **Москвічов.** Таблиці логаритмів, антилогаритмів, логаритмів тригонометричних функцій. ОНТВУ, Харків — Київ, 1932.

67. **Недзялковский А. А.** Собрание таблиц и формул для инженеров, архитекторов и механиков, 1. Колесов и Михин, Петербург, 1867.

68. **Никонов М. П.** Таблица тригонометрических функций. М. П. Никонов, 1912.

69. **Никулин М. А.** Таблицы для практических вычислений. Перевод русских мер в метрические, таблицы степеней и корней чисел, десятичных логарифмов, тригонометрических функций, а также другие математические таблицы, необходимые для техника и учащегося. ГТТИ, М.— Л., 1932.

70. **Нумеров Б. В.** Таблицы натуральных значений тригонометрических функций с пятью десятичными знаками. ГТТИ, Л., 1933.

71. **Нумеров Б. В.** Таблицы для трехзначного вычисления. Гос. вычисл. ин-т, Петроград, 1922.

72. **Обломиевский Д.** Вспомогательные таблицы тригонометрических функций для малых дуг. Военно-топогр. отд. Глав. штаба, С-Петербург, 1875.

73. **Оглоблин Д. Н.** Пятизначные таблицы натуральных значений тригонометрических функций для маркшейдеров. Металлургиздат, Свердловск — Москва, 1949.

74. **Орлов П. М.** Таблицы для вычисления умножением приращений прямоугольных координат. Винер, Москва, 1914; Воронеж, 1930.

75. **Орлов С. В.** Трехзначные таблицы логарифмов и тригонометрических функций. ГИЗ, М., 1920.

76. **Павловский А.** Таблицы логарифмов по изданию Каллета, содержащие логарифмы чисел от 1 до 108000; также логарифмы синусов и тангенсов для каждой секунды первых пяти градусов и для каждых десяти секунд всех градусов четверти круга. А. Павловский, Харьков, 1820.

77. **Пеевски В. и Венедиков М.** Петозначни логаритмични таблици на числата и тригонометричните функции при ново градусно деление. «Наука и изкуство», София, 1949.

78. **Перельман Я. И.** Таблицы и правила для вычислений. Гос. учеб.-педагог. изд-во, М.— Л., 1931.

79. **Петерс И.** Семизначные таблицы тригонометрических функций (для десятичного деления квадранта). М., 1946.

80. **Петерс И.** Шестизначные таблицы тригонометрических функций, содержащие значения шести тригонометрических функций через каждые 10″ дуги квадранта, разделенного на 90°, и значения котангенсов и косекансов через каждую секунду дуги от 0° 0″ до 1°20′. ГТТИ, М.—Л., 1932.

81. **Попов И. Г.** Математические таблицы. Учпедгиз, М., 1952.

82. **Пржевальский Е.** Пятизначные таблицы логарифмов чисел и тригонометрических величин. Гос. учебн.-педагогич. изд., М., 1937.

83. **Придатко С. В.** Таблицы четырехзначных логарифмов чисел и антилогарифмов. Таблицы натуральных величин тригонометрических функций и их логарифмов. А. Бенке, Петроград, 1915.

84. **Пятэцкі П. Д.** Справачник па матэматыцы. Дзярж. выд-во Беларуси, Менск, 1932.

85. **Пятизначные таблицы логарифмов чисел и тригонометрических величин.** Геодезиздат, М., 1949.

86. **Пятизначные таблицы натуральных значений тригонометрических функций.** Упр. землеустройства Днепропетр. Облзо, Днепропетровск, 1940.

87. **Пятизначные таблицы тригонометрических функций.** Геодезиздат, М., 1953.

88. **Таблицы тригонометрических величин и их логарифмов для углов в делениях угломера.** Гос. изд-во оборонной промышл., М., 1954.

89. **Райнов А. В.** Четырехзначные математические таблицы (связно-функциональные). ОНТИ, М.—Л., 1936.

90. **Родин А. Ф.** Пятизначные логарифмические таблицы чисел и тригонометрических функций с приложением гауссовых логарифмов и различных вспомогательных таблиц. Линдеберг, С.-Петербург, 1895.

91. **Романов А. Д.** Таблицы обыкновенных четырехзначных логарифмов и антилогарифмов, а также таблицы тригонометрических величин. Автор, С.-Петербург, 1901.

92. **Сегал Б. И. и Семендяев К. А.** Пятизначные математические таблицы. Изд-во АН СССР. М.—Л., 1948.

93. **Справочник по математике для обогатителей.** Металлургиздат, М., 1951.

94. **Справочное руководство по машиностроению, 1.** ГНТИ Украины, Харьков, 1937.

95. **Справочник штурмана по математике.** Изд-во Гидрограф. упр. ВМС. 1948.

96. **Справочник по элементарной математике, механике и физике.** ГИТТЛ. М.— Л., 1943.

97. **Стойков Д.** Спомогателни таблици за изчисления по машинна техника, метална техника и электротехника. София, 1947.

98. **Щеглов Н. Т.** Таблицы Бригговых логарифмов чисел и триго-

нометрических линий о семи десятичных. Трей, С.-Петербург, 1856.

99. **Таблиці квадратів, кубів, логаритмів, тригонометричних і гіперболічних функцій, квадратових і кубових коренів та інших величин.** Техне-теоретичне вид., Харьків, 1932.

100. **Таблицы квадратов трехзначных чисел, составных чисел до 10 000, тригонометрических величин.** А. А.Карцев, М., 1887.

101. **Таблицы логарифмов.** Чуваш. гос. изд-во, Чебоксары, 1931.

102. **Таблицы логарифмов и синусов, тангенсов, секансов.** М., 1703.

103. **Таблицы логарифмов чисел от 1 до 10 000 и натуральных синусов, тангенсов и секансов с логарифмами.** С.-Петербург, 1746.

104. **Таблицы логарифмов на числа от единицы до 10000 и на синусы, тангенсы, косинусы и котангенсы всех градусов и минут четверти круга, с таблицею синусов, тангенсов и секансов на градусы четверти круга, полагая радиус 100000,00 равных частей.** Морской Шляхетский кадетский корпус, Петербург, 1761.

105. **Таблицы натуральных значений тригонометрических величин для углов, выраженных в тысячных дистанции.** АНИМИ УВМС РККА, Л., 1934.

106. **Таблицы обыкновенных логарифмов чисел и тригонометрических линий для употребления в морском кадетском корпусе.** Петербург, 1862.

107. **Таблицы синусов, тангенсов и секансов и логарифма синусов и тангенсов; тоже и чисел, еже есть от единого даже до 10 000.** М., 1716.

108. **Таблицы степеней, корней, обратных величин и логарифмов.** Изд.Сев.-зап. обл. Промбюро ВСНХ, Л., 1927.

109. **Таблицы тангенсов углов для вычисления превышений** (для десятичного деления квадранта). НИИ Военно-топ. сл. ВС СССР, М., 1948.

110. **Таблицы тригонометрических функций. Пособие при решении прямоугольных треугольников без помощи логарифмов.** Новгород, 1912.

111. **Таблицы тригонометрических и эвольвентной функций для расчета зубчатых колес.** Типолит. НАМИ, М., 1950.

112. **Таблицы математических и физических величин.** Научн. книгоизд-во, Петроград, 1923.

113. **Таблицы четырехзначных логарифмов и основных физических величин.** Глаголев М. М. и Пиотровский М. Ю., С.-Петербург, 1910.

114. **Тараденков И. Ф.** Таблицы натуральных величин sin-в и cos-в для вычисления приращений прямоугольных координат при помощи арифмометра. Центр. бюро секции специалистов Союза сельхозлесрабочих СССР, М.

115. **Труды Центрального научно-исследовательского института геодезии, аэросъемки и картографии** (Москва), вып.30, 1939.
Таблицы для астрономических вычислений.

116. **Упрощенные таблицы логарифмов и тригонометрических величин.** С.-Петербург, 1902.

117. **Федоров Г. Р.** Таблицы натуральных значений тригонометрических функций для артиллерийского деления окружности. Гос. воен. изд-во, М., 1934.

118. **Фейгельсон Е. Г.** Таблицы десятичных логарифмов и натуральных тригонометрических функций. Учпедгиз, М.—Л., 1931.

119. **Филатов А. П.** Таблицы пятизначных и четырехзначных логарифмов чисел и антилогарифмов от 1 до 1000 с прибавками от 1000 до 100 000, расположенные на одном листе. С приложением правил логарифмирования и других вспомогательных таблиц. «Общественная польза», М., 1909.

120. **Филоненко А. С.** Таблицы шестизначных натуральных величин тригонометрических функций для счетных машин. Госкартогеодезия, М. — Л., 1932.

121. **Франк М. Л.** Карманный справочник по математике. Т-во профессоров и преподавателей, С.-Петербург, 1912.

122. **Хохлов А. И.** Карманные математические таблицы. Пятизначные. Гостехиздат, М.—Л , 1947

123. **Хренов Л. С.** Пятизначные таблицы тригонометрических функ-

ций, содержащие натуральные значения шести тригонометрических функций через каждую минуту от 0° до 360° и значения котангенсов и косекансов через каждую секунду дуги от 0°00′ до 1°00′ и через каждые 10″ дуги от 1° до 10°. ГИТТЛ, М.–Л., 1954.

124. **Стойчев Д.** Шестозначни таблици за стойностите на гониометричните функции sin, cos, tg и cotg при старо и ново деление на кръга. «Народна просвета», София, 1953.

125. **Хренов Л. С.** Семизначные таблицы тригонометрических функций. ГИТТЛ, М.— Л., 1951.

126. **Хренов Л. С.** Пятизначные таблицы тригонометрических функций с аргументом, выраженном в часовой мере, содержащие значения шести тригонометрических функций через каждые 4^s от 0 до 24^h, значения котангенсов и косекансов через $0^s,1$ от $0^h 00^m$ до $0^h 08^m$ и через 1^s от $0^h 08^m$ до $0^h 40^m$. ГИТТЛ, М., 1954.

127. **Хренов Л. С.** Таблицы тригонометрических функций (четырехзначные sin,cos,tg, ctg, sec, cosec). Кабинет топографии и картографии Воронеж. Гос. пед. ин-та, Воронеж, 1940.

128. **Hütte.** Справочник для инженеров, техников и студентов. Госмашметиздат, М.— Л., 1934.

129. **Четырехзначные таблицы логарифмов.** Арт. акад. им. Дзержинского, Л., 1935.

130. **Чиханов Б.** Таблицы пятизначных логарифмов чисел и тригонометрических величин. А. И. Мамонтов, М., 1917

131. **Чэркашин Л.** Чатырохзначныя матэматычныя табліцы. Дзярж. навук.-тэхн. выд-во, Менск, 1934.

132. **Шестизначные таблицы тригонометрических функций.** Геодезиздат, М., 1952.

133. **Штейнбреннер Г.** Пятизначные таблицы тригонометрических функций. НИИ Военно-топ. сл. ВС СССР, М., 1946.

134. **Янке Е.** и **Эмде Ф.** Таблицы с формулами и кривыми. ГИТТЛ, М.—Л., 1949.

135. **Abbott P. W.** Mathematical tables and formulae. Longmans, Green and Co., London, 1918.

136. **Ageton A. A.** Manual of celestial navigation. Van Nostrand, New York, 1942.

137. **Albrecht C. T.** Formeln und Hilfstafeln für geographische Ortsbestimmungen. Leipzig, 1789; Engelmann, Leipzig, 1908.

138. **Allen E. S.** Six-place tables. Mc Graw-Hill, New York — London, 1947.

139. **Ambronn L.** und **Domke J.** Astronomisch-geodätische Hilfstafeln. Mittler, Berlin, 1909.

140. **Andoyer M. H.** Nouvelles tables trigonométriques fondamentales (Valeurs naturelles), I. Sines. Hermann, Paris, 1915.

141. **Andoyer M. H.** Nouvelles tables trigonométriques fondamentales (Valeurs naturelles), II. Tangents. Hermann, Paris, 1916.

142. **Andoyer M. H.** Nouvelles tables trigonométriques fondamentales (Valeurs naturelles), III. Secants. Hermann, Paris, 1918.

143. **Andrew J.** Astronomical and nautical tables. London, 1805.

144. **Aquino R.** Altitude and azimuth tables. Potter, London; Author, Rio de Janeiro, 1924.

145. **Aquino R.** Sea and air navigation tables. Author, Rio de Janeiro, 1927.

146. **Attwood E. L.** Practical five-figure mathematical tables. Macmillan, London, 1950.

147. **Bachheimer R.** Potenz- und Wurzeltafeln enthaltend die Quadrate, Kuben, Quadrat und Kubikwurzeln der Zahlen 1—1000, sowie die Umfänge und Inhalte der Kreise mit den Durchmessern 1—1000. Leipzig — Wien, 1928.

148. **Ball F.** Altitude tables. Lat. 31° to 60°. Dec. 0° to 24°. Potter, London, 1927.

149. **Balzer F.** und **Dettwiler H.** Fünfstellige natürliche Werte der Sinus- und Tangentenfunktionen neuer Teilung für Maschinenrechnen. Wittwer, Stuttgart, 1919.

150. **Banhagel E.** Logarithms and slide rule for practical use. Liff-Davis, New York, 1945.

151. **Bauschinger J.** Tafeln zur theoretischen Astronomie. Engelmann, Leipzig, 1901, 1934.

152. **Bauschinger J.** und **Peters J.** Logarithmisch-trigonometrische Tafeln mit acht Decimalstellen, 1 — 2. Engelmann, Leipzig, 1936.

153. **Becker G. F.** and **Van Orstrand C. E.** Hyperbolic functions. Smithsonian Institution, Washington, 1909.

154. **Beetle R. D.** Five-place logarithmic and trigonometric tables. McGraw-Hill, New York — London, 1933.

155. **Benson R. E.** Natural trigonometric functions containing the natural sine, cosine, tangent and contangent to seven decimal places for every ten seconds of arc from 0° to 90° semi-quadrantally arranged. Benson Book Co., Los-Angeles, 1941.

156. **Beyrodt G.** und **Reissner A.** Tabellenbuch zur Wirtschaftmathematik. Fachwissen Lehrbücher für die Berufsfach- und Fachschule. «Volk und Wissen», Berlin — Leipzig, 1949.

157. **Blackburne H. S.** Tables for azimuths, great-circle sailing, and reduction to the meridian... I. Lat. 60° N to 60° S. Dec. 90° N to 90° S. II. Lat. 60° to Poles. Dec. 90° N to 90° S. N. Z. Government, Wellington, 1923.

158. **Boileau J. T.** A new and complete set of traverse tables. Allen. London, 1931.

159. **Bogen-Tafeln für Teilung des Kreises in 400^g.** Berlin, 1938.

160. **Boll M.** Tables numériques universelles des laboratoires et bureaux d'étude. Dunod, Paris, 1947. 1). Remarques et compléments aux tables numériques universelles de M. Boll. Dunod, Paris, 1949.

161. **Bohde H.** Mathematisch-technische Zahlentafeln. Springer, Berlin, 1933.

162. **Bowditch N.** The new american practical navigator. Gov. Printing Office, Washington, 1874.

163. **Bowditch N.** Useful tables from the american practical navigator. U. S. Hydrographic Office, № 9, part II. Washington, 1911.

164. **Brandenburg H.** Hétjegyü trigonometrial táblázat. Honved terkepészeti intézet. Budapest, 1951.

165. **Brandenburg H.** Sechsstellige trigonometrische Tafel... Lorentz Leipzig, 1932.

166. **Brandenburg H.** Siebenstellige trigonometrische Tafel alter Kreisteilung für Berechnungen mit der Rechenmaschine. Lorentz, Leipzig, 1931.

167. **Brandenburg H.** Zwei trigonometrische Tafeln alter Kreisteilung zur Berechnung der Hülfs- oder Richtungsgrössen *a* und *b* bei Ausgleichung von Dreiecksnetzpunkten. Lorentz, Leipzig, 1932.

168. **Bremiker C.** Logarithmisch-trigonometrische Tafeln mit fünf Decimalstellen. Weidmann, Berlin, 1922.

169. **Bremiker C.** Tafeln vierstelliger Logarithmen. Weidmann, Berlin, 1907.

170. **Briggs H.** and **Gellibrand H.** Trigonometria Britannica. Gouda, 1633.

171. **British Association for the Advancement of Science. Mathematical Tables, I.** Circular and hyperbolic functions, exponential, sine and cosine integrals, factorial function and allied functions Hermitien probability functions. Prepared by the Committee for the Calculation of Mathematical Tables. Cambridge Univ. Press, 1951.

172. **British Association for the Advancement of Science. Mathematical Tables, II.** Emden Function. Cambridge Univ. Press, London, 1932.

173. **Brown's Nautical Almanac for 1935.** Brown, Son and Ferguson, Glasgow, 1935.

174. **Bruhns C. C.** Neues logarithmisch-trigonometrisches Handbuch auf sieben Decimalen. Leipzig, 1881.

175. **Buckingham E.** Manual of gear design. Section one. Mathematical tables. Machinery, New York, 1935.

176. **Burrau C.** Tafeln der Funktionen Cosinus und Sinus mit den natürlichen sowohl reellen als rein

imaginären Zahlen als Argument (Kreis und Hyperbelfunktionen). Reimer, Berlin, 1907.

177. **B u r t o n S. M.** A set of nautical tables. G. Philip, London, 1941.

178. **C a l l e t F.** Tables portatives de logarithmes, contenant les logarithmes des nombres, depuis 1 jusqu'a 108.000; les logarithmes des sinus et tangentes,...,F. Didot, Paris, 1795.

179. **C a m p b e l l J. W.** Numerical tables. The Douglas Print. Co., Edmonton, Alta, 1946.

180. **C a r e y F. S.** and **G r a c e S. F.** Four-place mathematical tables with forced decimals. Longmans, London, 1927.

181. **C a r m i c h a e l R. D.** and **S m i t h E. R.** Mathematical tables and formulas. Ginn, Boston — New York, 1931.

182. **C a s t l e F.** Five-figure logarithmic and other tables. Macmillan, London, 1910.

183. **C h a p p e l l E.** Five-figure mathematical tables. Chambers, London; Van Nostrand, New York, 1915.

184. **C h a u v e n e t W.** A manual of spherical and practical astronomy, I, II. Lippincott, Philadelphia, 1900.

185. **C h r e t i e n H.** Nouvelles tables des sinus naturels... Revue d'Optique, Paris, 1932.

186. **C l a r k D. K.** A manual of rules, tables and data for mechanical engineers. Blackie, London, 1891.

187. **C l a r k D. K.** The mechanical engineer's pocket-book of tables, formulae, rules and data. Crosby Lockwood, London, 1919.

188. **C l a r k J. B.** Mathematical and physical tables. Oliver and Boyd, Edinburgh, 1950.

189. **C l a u d e l J.** Formules, tables et renseignements pratiques; aide-mémoire des ingénieurs, des architecteurs. Paris, 1854.

190. **C l a u d e l J.** Tables: 1. Des carrés et des cubes des nombres entiers successifs de 1 à 10000. 2. Des longueurs des circonférences... 3. Des expressions trigonométriques. Carilian-Goeury et V-or Dalmont, Paris, 1850.

191. **C l o u t h F. M.** Tafeln zur Berechnung goniometrischer Coordinaten. Wittwer, Stuttgart, 1922.

192. **C o m r i e L. J.** Chambers's four-figure mathematical tables. W. and R. Chambers, Edinburgh — London, 1947.

193. **C o m r i e L. J.** Four-figure tables of the natural and logarithmic trigonometrical function with the argument in time. Author, London, 1931.

194. **C r e l l i n W.** Complete traverse table. Robertson, Melbourne, 1881.

195. **D a l e J. B.** Five-figure tables of mathematical functions, comprising tables of logarithms, power of numbers, trigonometric, elliptic and other transcendental functions. Arnold, London, 1949.

196. **D a n j o n A.** Tables des fonctions trigonométriques. Hachette, Paris, 1949.

197. **D a v i s H. T.** Tables of the higher mathematical functions, I. Principia Press, Bloomington, Indiana, 1933.

198. **D a v i s P. L. H.** Requisite tables... Potter, London, 1905.

199. **D a v i s R. A.** Tables of natural sines and radians. Marchant Calculating Machine Co., Oakland, California, 1941.

200. **D e c k e r t A.** Die Logarithmen der trigonometrischen Funktionen und der natürlichen Zahlen. Druck Kösel und Pustet.

201. **D r e i s o n s t o k J. Y.** Navigation tables for mariners and aviators. U. S. Hydrographic Office, Washington, 1935.

202. **D w i g h t H. B.** Mathematical tables. McGraw-Hill, New York — London, 1941.

203. **E n b e r g J.** and **L ä n g s t r ö m E.** Tables à six décimales des valeurs naturelles des fonctions trigonométriques pour la division centesimale. Haeggström, Stockholm, 1925.

204. **E n g e l E.** Zehnstellige Tafeln der Sinus, Cosinus und Tangenten für die dezimale Teilung Nonagesimalgrades. Generaldirektion des Grundsteuerkatasters, Vienna, 1920.

205. **E s c h k e.** Mathematische Tabellen und Formeln. W. Mayer, Berlin, [193...]

206. **F a r l e y R.** Natural versed sines for 0° to 125° and logarithmic versed sines from 0° to 135° of 0^h to 9^h used in computing lunar distances for the nautical almanac. London, 1856.

207. **F e r g u s s o n J. C.** Fergusson's percentage unit of angular measurement, with logarithms... Longmans, London, 1912.

208. **C i u r i l e a n u D. J.** Tabele trigonometrice. Editura tehnica, Bucureşti, 1950.

209. **Formular matematic şi tehnic.** Pentru eleve, studenţi şi tehnicieni. Editura tehnica, Bucureşti, 1949.

210. **G a r d i n e r W.** Tables of logarithms, for all numbers from 1 to 102.100 and for the sines and tangents to every ten seconds of the quadrant; as also for the sines of the first 72 minutes to every single second. London, 1742.

211. **G a u s s F. G.** Fünfstellige vollständige logarithmische und trigonometrische Tafeln. Halle, Strien, 1908.

212. **G a u s s F. G.** Fünfstellige logarithmisch-trigonometrische Tafeln für Dezimalteilung des Quadranten. Wittwer, Stuttgart, 1937.

213. **G a u s s F. G.** Fünfstellige vollständige trigonometrische und polygonometrische Tafeln für Machinenrechnen. Wittwer, Stuttgart, 1918.

214. **G a u s s F. G.** Vierstellige logarithmische und trigonometrische Tafeln (Sexagesimal unterteiler Altgrad). «Volk uns Wissen». Berlin — Leipzig, 1946.

215. **E l z n i c V. a V a l o u c h M.** Geoma. Pětimístné tabulky hodnot a logaritmů goniometrických funkcí v setinném dělení, tabulky geodetické a katastrálni. Vydala jednota českych matematikû a fysikû, Praha, 1944.

216. **G i f f o r d E.** Natural secants. Heywood, Manchester, 1929.

217. **G i f f o r d E.** Natural sines to every second of arc, and eight places of decimals. Heywood, Manchester, 1926.

218. **G i f f o r d E.** Natural tangents. Heywood, Manchester, 1927; Los Angeles, 1943.

219 **G i n g r i c h J. E.** Aerial and marine navigation tables. McGraw-Hill, New York — London, 1931.

220. **G o n g g r i j p B.** Logarithmische en goniometrische tafels en bijtafels. Noordhoff, Groningen, 1916.

221. **G o o d w i n H. B.** An equatorial azimuth table. Potter, London 1921

222. **G o s s a r t A.** Table des carrés de une à cent millions an moyen de laquelle on obtient des produits exacts... Paris, 1865.

223. **G o u s s i n s k y B.** Five-figure tables for checking traverse computations with the aid of a calculating machine. Survey of Palestine, Jaffa, 1943.

224. **G r a v e l i u s H.** Fünfstellige logarithmisch-trigonometrische Tafeln für die Decimalteilung des Quadranten. Reimer, Berlin, 1886.

225. **G r o n w a l d t A.** Sehnentafel... Quedlinburg, 1850.

226. **G u r d e n R. L.** Traverse tables... Griffin, London, 1906.

227. **H a l l W.** Tables and constants to four figures. Cambridge Univ. Press, Cambridge, 1905.

228. **B u r i n g t o n R. S.** Handbook of mathematical tables and formulas. Sandusky, Ohio, 1946.

229. **H a n n y n g t o n J. C.** Haversines natural and logarithmic used in computing lunar distances for the nautical almanac. H. M. Stationery Office, London, 1876.

230. **H a n t s c h l J.** Logarithmisch-trigonometrisches Handbuch... Rohrmann und Schweigerd, Wien, 1833.

231. **H a w e l k a R.** Vierstellige Tafeln der Kreis- und Hyperbelfunktionen, sowie ihrer Umkehrfunktionen im Komplexen in Schritten von $\frac{\pi}{2}$ 0,02 und i 0.02. Vieweg, Braunschweig, 1931.

232. **H a y a s h i K.** Fünfstellige Funktionentafeln... Springer, Berlin, 1930.

233. **H a y a s h i K.** Fünfstellige Tafeln der Kreis- und Hyperbelfunktionen, sowie der Funktionen e^x und e^{-x} mit den natürlichen Zahlen als Argument. Gruyter, Berlin, 1944.

234. **Hayashi K.** Sieben- und mehrstellige Tafeln der Kreis- und Hyperbelfunktionen und deren Produkte sowie der Gammafunktion. Springer, Berlin, 1926.

235. **Hayashi K.** Tafeln der Besselschen-, Theta-, Kugel—und anderen Funktionen. Springer, Berlin, 1930.

236. **_edrick E. R.** Logarithmic and trigonometric tables. Macmillan, New York, 1920.

237. **Hilfstafeln für Tachymetrie.** Stuttgart, Matzlersche Verlagsbuchhandlung, 1939.

238. **Hobert J. P. et Ideler L.** Nouvelles tables trigonométriques calculées pour la division décimale du quart de cercle. Berlin, 1799.

239. **Hof H.** Seven place full natural trigonometric tables. Sine. Michigan Univ. Research Association, Ann Arbor, 1946.

240. **Hof H.** Seven place full natural trigonometric tables. Cosine. Michigan Univ. Research Association, Ann Arbor, 1943.

241. **Hof H.** Seven place full natural trigonometric tables. Tangent. Michigan Univ. Research Association, Ann Arbor, 1943.

242. **Hof H.** Seven place full natural trigonometric tables.Cotangent. Michigan Univ. Research Association, Ann. Arbor, 1944.

243. **Hoüel J.** Tables de logarithmes à cinq décimales pour les nombres et les lignes trigonométriques, suivées des logarithmes d'addition et de saustraction ou logarithmes de Gauss et de diverses tables usuélles. Gauthier-Villars, Paris, 1864.

244. **Hoüel J.** Recueul de formules et de tables numériques. Gauthier-Villars, Paris, 1901.

245. **Hughes - Comrie L. J.** Tables for sea and air navigation. Henry Hughes and Son, London, 1938.

246. **Hudson L.** and **Mills E. S.** Natural trigonometric functions. Tables: sine, cosine, tangent, cotangent, secant and cosecant to eight decimal places with second differences to ten decimal places semi-quadrantally arranged. N. J., Monroe Calculating Machine Co., Orange, 1941.

247. **Huntington E. V.** Four-place tables of logarithms and trigonometric functions, with auxiliary tables. Mass., Harvard Co — op. Soc., Cambridge, 1931.

248. **Huszar G., Kreko B.** Táblázatok a gazdasagi számtanhoz. Tankönyvkiadó, Budapest, [195—].

249. **Hutton C.** Mathematical tables. London, 1811.

250. **Hütte.** Des Ingenieurs Taschenbuch, I. Ernst, Berlin, 1931.

251. **Tavole logaritmiche a cinque cifre decimali.** Jtaly istituto Idrografico della R. Marina, Genoa, 1916.

252. **Tavole logaritmiche a cinque cifre decimali, 6.** Jtaly istituto Idrografico della R. Marina. Genoa, 1941.

253. **Ives H. C.** Mathematical tables reprinted from Searles and Ives' Field Engineering, with additions. Wiley, New York; Chapman and Hall, London, 1945.

254. **Ives H. C.** Natural trigonometrical functions to seven decimal places for every ten seconds of arc. Wiley, New York; Chapman and Hall, London, 1945.

255. **Ives H. C.** Seven place natural trigonometrical functions. Wiley, New York; Chapman and Hall, London, 1929.

256. **Jadanza N.— Hammer E.** Tachymeter-Tafeln für zentesimale Winkelteilung. Wittwer, Stuttgart, 1909.

257. **Jahnke E.** und **Emde F.** Funktionentafeln mit Formeln und Kurven. Leipzig — Berlin, 1) 1909; 2) 1933; 3) 1938.

258. **Jansky F.** Fünfstellige Tabellen der trigonometrischen Funktionen zur Koordinatenberechnung der Polygonzugpunkte mit der Rechenmaschine. Trobor und Söhne, Königinhof an der Elbe, 1944.

259. **Jauget R. J.** Mathematical tables. For ascertaining the distance of an object. Sunderland, 1906.

260. **Jordan W.** Hilfstafeln für Tachymetrie. Metzler, Stuttgart, 1908.

261. **Jordan W.** Mathematische und geodätische Hilfstafeln. Hannover, 1895.

262. **Jordan W.** Opus Palatinum. Sinus und Cosinus Tafeln von 10″ zu 10″. Hahn, Hannover — Leipzig, 1929, 1936.

263. Jörgensen N. R. Undersögelser over frekvensfladér og korrelation. Busck, Copenhagen, 1916.

264. Jurisch C. L. H. M. Tables containing the natural sines and cosines to seven decimal figures of all angles between 0° and 90° to every ten seconds... Cape Town, 1884.

265. Kahlau R. Vierstellige Tafeln zum logarithmischen und numerischen Rechnen. Bonnes und Hachfeld, Potsdam, 1944.

266. Kaye G. W. C. and Laby T.H. Four-figure mathematical tables. London, 1925.

267. Kekkonen A., Järvi H. Matematiikan Perussäännöt ja Kaavat. Karjalan Kommunistinen Maatalouskorkeekakoulu, Petroskoi, 1936.

268. Kelvin, Lord. Tables for facilitating sumner's method at sea. Teubner and Francis for Author, London, 1876.

269. Kempe H. R. and Smith W. H. The Engineer's Year-book... for 1927. Grosby Lockwood, London, 1927.

270. Kepler J. Chilias logarithmorum... Marburg, 1624.

271. Keturzenkles logaritmu ir kitu dydziu lenteles. Matematines formules. Valstybine Pedagogines Literaturos Leidykla. Kaunas, 1946.

272. Kleyer A. Fünfstellige korrekte logarithmische Tafeln nebst einer trigonometrischen Tafeln und einer Anzahl von anderen Tabellen. J. Maier, Stuttgart, 1883.

273. Köhler H. G. Logarithmisch-trigonometrisches Handbuch. Tauchnitz, Leipzig, 1862, 1898.

274. Kolb A. und Bauman F. «Blitz». Rechentafeln nebst den Tabellen: Kreis-Inhalte und Umfänge Quadrate, Kuben, Quadrat- und Kubikwurzeln. Fr. Lints., Thier, 1903.

275. König A. Handbuch der Astrophysiks, 1, 502—559. Springer, Berlin, 1933.

276. Krovák J. Šestimístné tabulky přirozenych hodnot funkcí sinu a kosinu uhlu v setinném dělení. Praha, (19...)

277. Küstner H. Fünfstellige Logarithmen. «Volk und Wissen», Berlin—Leipzig, 1952.

278. Knott C. G. Four-figure mathematical tables. Chambers, London — Edinburgh, 1905.

279. Lambert J. H. Zusätze zu den logarithmischen und trigonometrischen Tabellen. Haude und Spener, Berlin, 1770.

280. Lambert J. H. — Felkel A. Supplementa tabularum logarithmicarum et trigonometricarum... Lisbon, 1798.

281. Lecky S. T. S. General utility tables. Philip, London, 1923.

282. Lecky S. T. S. *ABC* tables. Table D. London, 1892.

283. Legendre A. M. Exercices de calcul intégral, 3. Huzard-Courcier, Paris, 1816.

284. Legendre A. M. Traité des fonctions élliptiques, 2. Huzard-Courcier, Paris, 1826.

285. Lella A. Five-place table of natural trigonometric functions to hundredths of a degree. Wiley, New York; Chapman and Hill. London, 1934.

286. Ligowski W. Sammlung fünfstelliger logarithmischer, trigonometrischer, nautischer und astronomischer Tafeln (nebst Erklärungen und Formeln der Astronomie mit besonderer Rücksicht auf die Nautik). Toeche, Kiel, 1873.

287. Ligowski W. Tafeln der Hyperbelfunktionen und der Kreisfunktionen... Ernst, Berlin, 1890.

288. Littlehales G. W. The summer line of position furnished... U. S. Hydrographic Office, № 203, 1923, № 204, 1925, Washington.

289. Lohse O. Tafeln für numerisches Rechnen mit Maschinen. Leipzig, 1909; Engelmann, Leipzig, 1935.

290. Loky B. Negyjegyü függvénytáblázatok. Budapest, 1950.

291. Louis H. and Caunt G. W. Tacheometer tables. Arnold, London, 1919.

292. Louis H. and Caunt G. W. Traverse tables. Arnold, London, 1901.

293. Low D. A. and Low B. B. A pocket-book for mechanical engineers. Longmans, London, 1938.

294. Lötzbeyer P. Vierstellige Tafeln zum Logarithmischen und Zahlenrechnen. Graphische Rechentafeln. Teubner, Leipzig—Berlin, 1928.

295. Lynch J. K. Calculation of $\frac{\sin x}{x}$ tables. Commonwealth of Austra-

lia postmaster-general's department. Engineering branch research laboratory report № 3238. Melbourne, 1949.

296. **Lynn T.** Horary tables. London, 1827.

297. **Mackay A.** The theory and practice of finding the longitude. Longman, London, 1810.

298. **Massaloup J. V.** Logarithmisch-trigonometrische Hülfstafeln. Brockhaus, Leipzig, 1847.

299. **Meldola A.** Logarithmische Tafeln der Nummer-Logarithmen (1—10 000) der Sinus und Tangenten. Aue, Altona, 1840.

300. **Meyer U.** und **Deckert A.** Tafeln der Hyperbelfunktionen. Kösel und Pustet, Kempten, 1924.

301. **Miller F. H.** Calculus. Wiley, New York, 1946.

302. **Miller J. C. P.** Tables for converting rectangular to polar coordinates. Scientific Computing Service, London, 1943.

303. **Milne—Thomson L. M.** and **Comrie L. J.** Standard four-figure mathematical tables. Including many new tables trigonometrical functions for radians, inverse trigonometrical and hyperbolic functions and an extended table of natural logarithms. Macmillan, London, 1948.

304. **Myerscough W.** and **Hamilton W.** Rapid navigation tables. Pitman, London, 1939.

305. **Natural sines and cosines to eight decimal places.** Gov. print. off., Washington, 1942.

306. **Nassau J. J.** A textbook of practical astronomy. McGraw-Hill, New York — London, 1932.

307. **Nautucal Almanac (abridged). The nautical almanac abridged for the use of seamen for the year 1936.** H. M. Stationery Office, London, 1935.

308. **Nautical almanac office. Five-figure tables of natural trigonometrical functions.** War Office, London, 1947.

309. **Nautical almanac office. Seven-figure trigonometrical tables for every second of time.** H. M. Stationery Office, London, 1939.

310. **Neill W. T.** Natural sines and cosines for every ten seconds of arc and to seven places of decimals. New Zealand, Department of Lands and Survey, Wellington, 1927.

311. **Newcomb S.** Logarithmic and other mathematical tables. Holt, New York; Bell, London, 1895.

312. **Vlacq A.** Nieuwe konstige tafelen, sinium, tangentium et secantium. Amsterdam, 1658.

313. **Oberg E.** and **Jones F. D.** Machinery's handbook... Industrial Press, New York, 1936.

314. **Ogura S.** New altitude and azimuth tables... Hydrographic Department, Tokyo, 1924.

315. **Otho V.** Meteoroscopium... Neustadt, 1596.

316. **Oughtred W.** Trigonometria R. Stokes and A. Haughton, London, 1657.

317. **Pasquich J.** Tabulae logarithmico-trigonometricae Contractae cum novis accessionibus ad abbreviandos facilioresque reddendos calculos trigonometricos (Abgekürzte logarithmisch-trigonometrische Tafeln). Weidmann, Leipzig, 1817.

318. **Patz W.** Tafel der Regelmässigen Kettenbrüche für die Quadratwurzeln aus den natürlichen Zahlen von 1—10000. Becker und Erler, Leipzig, 1941.

319. **Peters C. F. W.** Astronomische Tafeln und Formeln. Mauke, Hamburg, 1871.

320. **Peters J.** Einundzwanzigstellige Werte der Funktionen Sinus und Cosinus. Reimer, Berlin, 1911.

321. **Peters J.** Siebenstellige Werte der trigonometrischen Funktionen von Tausendstel zu Tausendstel des Grades. Optische Anstalt C. P. Goerz. Berlin—Friedenau, 1918.

322. **Peters J.** Hilfstafeln zur Zehnstelligen Logarithmentafel. Preussische Landesaufnahme, Berlin, 1919.

323. **Peters J.** Zehnstellige Logarithmentafel, I: Zehnstellige Logarithmen der Zahlen von 1 bis 100 000 nebst einem Anhang mathematischer Tafeln. Reichsamt für Landesaufnahme, Berlin, 1922.

324. **Peters J.** Sechsstellige Tafel der trigonometrischen Funktionen, enthaltend die Werte der sechs trigonometrischen Funktionen von zehn zu zehn Bogensekunden... Dümmler, Berlin — Bonn, 1929.

325. **Peters J.** Sechsstellige Werte der Kreis- und Evolventen-Funktionen von Hundertstel zu Hundertstel des Grades. Köllmann-Werke A. G., Leipzig, 1937.

326. **Peters J.** Achtstellige Tafel der trigonometrischen Funktionen für jede Sexagesimalsekunde des Quadranten. Reichsamt für Landesaufnahme, Berlin, 1939.

327. **Peters J.** Sechsstellige Werte der trigonometrischen Funktionen von Tausendstel zu Tausendstel des Neugrades. Wichmann, Berlin, 1943.

328. **Pierce M. R.** Position tables for aerial and surface navigation, N 209. U. S. Hydrographic Office, Washington, 1931.

329. **Pinto J. C.** The «Simplex» navigation and· avigation tables. Azores, Fayal. Sold by Brown, Glasgow, 1933.

330. **Pitiscus B.** Thesaurus Mathematicus... Frankfurt, 1613.

331. **Potin L.** Formules et tables numériques. Doin et Gauthier-Villars, Paris, 1925.

332. **Prevost G.** Tables de fonctions sphériques et de leurs intégrales pour calculer les coefficients du développement en série de polynomes de Laplace d'une fonction de deux variables indépendantes. Soc. Sci. Phys. et Nat., Bordeaux; Gauthier-Villars, Paris, 1933.

333. **Prokes A.** Tabured. Tabellen zur Reduktion von schräg gemessenen Entfernungen für zentesimale Kreisteilung. Wichmann, Berlin, 1943.

334. **Pryde J.** Chambers's educational course. Mathematical tables. Chambers, London — Edinburgh, 1930.

335. **Raper H.** Nautical tables. Potter, London, 1920.

336. **Ratassepp K.** Matemaatilised tabelid. Pedagoogiline Kirjandus, Tallinn, 1948.

337. **Reuter J., Sergelius M.** Taulukot — Tabeller. Helsinki, 1945.

338. **Rheticus G.** Magnus Canon... in Opus Palatinum... V. Otho, Neustadt, 1596.

339. **Larsen H. D.** Rinehart mathematical tables, formulas and curves. Chapman and Hall, London, 1949.

340. **Rivard M.** Tables des sinus, tangentes, secantes et de leurs logarithmes, avec la construction de ces tables, et les problèmes de la trigonométrie rectiligne et sphérique. Desaint, Saillant, Paris, 1743.

341. **Rohrbach C.** Vierstellige logarithmisch-trigonometrische Tafeln. Ehlermann, Dresden, 1929.

342. **Rosenbach I. B., Whitman E. A., Moskovitz D.** Mathematical tables. Ginn, Boston, 1946.

343. **Roussilhe H.** and **Brandicourt.** Tables à 8 dècimals des valeurs naturelles des sinus, cosinus et tangentes... International Geodetic and Geophysical Union, Paris, 1933.

344. **Rubey H.** Engineering surveys-logarithmic and trigonometric tables. Macmillan, New York, 1941.

345. **Russell A.** A treatise on the theory of alternating currents, II. Cambridge Univ. Press, Cambridge, 1916.

346. **Sassenfeld H. M.** und **Tschunko H. F. A.** Mathematische Tafeln für Mathematiker, Naturwissenschaftler, Ingenieure. Erster Teil: Elementare Funktionen. Fr. Lamade, Walldorf bei Heidelberg, 1949.

347. **Schorr R.** Sammlung von Hilfstafeln der Hamburger Sternwarte in Bergedorf. Lucas Gräfe, Hamburg, 1916.

348. **Schorr R.** Sammlung von Hilfstafeln der Hamburger Sternwarte. 1) Publ. J., 1935; 2) Publ. K., 1935.

349. **Schubert H.** Fünfstellige Tafeln und Gegentafeln für logarithmisches und trigonometrisches Rechnen. Teubner, Leipzig, 1897.

350. **Schubert H.** Vierstellige Tafeln und Gegentafeln für logarithmisches und trigonometrisches Rechnen. W. Gruyter, Berlin, 1945.

351. **Schülke A.** Tafeln. Vierstellige Logarithmen Funktions und Zahlenwerte. «Volk und Wissen», Leipzig, 1949.

352. **Schulte-Tigges R. A.** Rechentafeln für höhere Lehranstalten insbesondere zu Mehlers Hauptsätzen der Elementar-Mathematik. W. Gruyter, Berlin — Leipzig, 1929.

353. **Schulze J. C.** Recueil de tables astronomiques, **1**, **2**, **3**. Acad. Roy. de Prusse, Berlin, 1776.

354. **Schuster A.** An introduction to the theory of optics. Arnold, London, 1909.

355. **Scott C. M. L.** Satellite station tables. Arnold, London, 1934.

356. **Seeley W. J.** Impedance computing tables. Engineers Publishing Co., Philadelphia, 1936.

357. **Semiller H.** und **Semiller A.** Vierstellige Logarithmen und Zahlentafeln. Springer, Berlin, 1938.

358. **Service Géographique de l'Armée. Tables des logarithmes à huit décimales...** Gauthier-Villars, Paris, 1891.

359. **Shortrede R.** Traverse tables... Layton, London, 1864.

360. **Smart W. M.** and **Shearme F. N.** Position line tables (sine method). Potter, London, 1924.

361. **Smoley C. K.** Segmental functions. C. K. Smoley and Sons, Scranton, 1937.

362. **Sokal S.** i **Torbus W.** Tablice matematyczno-fizyczne do uzytku szkoł zawodowych. Państwowe wyd-wa szkolnictwa zawodowego, Warszawa, 1951.

363. **Spence W.** An essay on the theory of the various orders of logarithmic transcendents. London — Edinburgh, 1809.

364. **Spenceley G. W.** and **Spenceley R. M.** Smithsonian elliptic functions tables. The Smithsonian Institution. Washington, 1947.

365. **Stalmasek J.** Tabul'ky funkcií a konštant z matematiky, fyziky a chemie. Státne nakl., Bratislava, 1951. БЛ

366. **Standard field tables and trigonometric formulas..** Gov. Print. Off., Washington, 1939.

367. **Steinbrenner G.** Fünfstellige trigonometrische Tafeln neuer Teilung (Dezimalteilung des Quadranten) zum Maschinenrechnen. Grimme, Natalis und Co, Braunschweig, 1914.

368. **Steinheil A.** und **Voit E.** Handbuch der angewandten Optik, I. Teubner, Leipzig, 1891.

369. **Strömgren B.** Optical sine-tables giving seven-figure values of $x-\sin x$ with arguments x and $\sin x$. Bianco Lunos Bogtrykkeri A/S, København, 1945.

370. **Studnicka F. J.** Kapesní logarithmické tabulky. Praha, 1870.

371. **Stumpff K.** Tafeln und Aufgaben zur harmonischen Analyse und Periodogrammrechnung. Springer, Berlin, 1939.

372. **Sust O.** Vierstellige Tafeln der natürlichen Werte der Sinus- und Tangensfunktionen für neue Teilung nebst einer Quadrattafel. Fachbuchverlag, Leipzig, 1952.

373. **Table şi formule matematice.** Ed. tehnica, Bucureşti, 1951.

374. **Tabellen der Quadrate, Kuben, Wurzeln, Kreisumfänge und Inhalte, Kreisfunktionen und Logarithmen.** Leipzig.

375. **Tabellen der sinuum, tangentium und secantium.** Amsterdam, 1689.

376. **Table of arcsin x.** Prepared by the Mathematical Tables Project conducted under the sponsorship of the National Bureau of Standards. Columbia Univ. Press, New York, 1945.

377. **Table of arctan x.** Prepared by the Mathematical Tables Project conducted under the sponsorship of the National Bureau of Standards. Columbia Univ. Press, New York, 1942.

378. **Tables of the Bessel function of fractional order.** Prepared by the Computation Laboratory of the National Applied Mathematics Laboratories, National Bureau of Standards. Columbia Univ. Press, New York, 1) I, 1948; 2) II, 1949.

379. **Tables of circular and hyperbolic sines and cosines for radian arguments.** Prepared by U. S. Department of Commerce, National Bureau of Standards. Washington, 1940, 1949.

380. **Table of circular and hyperbolic tangents and cotangents for radian arguments.** Columbia Univ. Press, New York, 1943.

381. **Robbins C. K.** and **Little N.** Tables for calculus. The Macmillan Co., New York, 1942.

382. **Tables of the functions $\frac{\sin\varphi}{\varphi}$ and of its first eleven derivatives.** The annals of the computation laboratory

of Harvard university, XXII. Cambridge, Massachusetts, Harward Univ. Press, New York, 1949.

383. **Tables relating to Mathieu functions. Charakteristic values, coefficients and Joining factors.** Cambridge Univ. Press, New York, 1951.

384. **Tafel der Sinus, Tangenten und Secanten, mit dem Opus Palatinum.** Hochhausen, Jena, 1846.

385. **Tables of sines, and cosines for radian arguments.** Prepared by the Mathematical Tables Project, National Bureau of Standards. Washington, 1940.

386. **Tables of sine, cosine and exponential integrals, II.** Prepared by the Mathematical Tables Project, National Bureau of Standards. Washington, 1940.

387. **Tables of sine, cosine, integrals for arguments from 10 to 100.** National Bureau of Standards. Washington, 1942.

388. **Table of sines and cosines to fifteen decimal places at hundeidths of a degree.** United States Government Printing Office, Washington, 1949.

389. **Ledertzen P.** Manuale Mathematicum, darinn begriffen die Tabulae Sinuum, Tangentium, Secantium. . Straßburg, 1619.

390. **Tafelsammlung und Hilfsbuch für Mashinenbauer, Mechaniker, Werkzeugmacher und andere Metallarbeiter.** Leipzig, 1948.

391. **Templeton W. — Hutton W. S.** The practical mechanic's workshop companion... Crosby Lockwood, London, 1900.

392. **Templeton W. — Maynard S.** The millwright and engineer's pocket companion... Simpkin, London, 1858.

393. **Todd C.** A Series of tables... Longman, London, 1853.

394. **Towson J. T.** Tables to facilitate the practice of great circle sailing. Hydrographic Office, London, 1912.

395. **Trautwine J. C.** The civil engineer's reference-book. Chapman and Hall, London, 1937.

396. **Turrell F. M.** Tables of surfaces and volumes of spheres and of prolate and oblate spheroids and spheroidal coefficients. Univ. of California Press, Berkeley — Los Angeles, 1946.

397. **Új Hétszamjegyu Logarithmus Kezikönyv.** Honved térképészeti intézet. Budapest, 1952.

398. **Piatkowski F. i Pilitowski T.** Tablice naturalnych wartości funkcji trygonometrycznych w podziale gradusowym od 0^g do 100^g. Państwowe Przedsiębiorstwo Wydawnictw kartograficznych, Warszawa, 1952.

399. **Valori naturali delle funzioni goniometriche.** Firenze istituto geografico militare, 1943.

400. **Vega G.** Thesaurus Logarithmorum Completus (Vollständige Sammlung grösserer logarithmisch-trigonometrischer Tafeln). Weidmann, Leipzig, 1794.

401. **Vega G. — Hülsse J. A.** Sammlung mathematischer Tafeln... Weidmann, Leipzig, 1849.

402. **Vega G.** 10 place Logarithms. Including Wolfram's tables of natural logarithms. Reprint of the rare of 1794. Stechert G. E., New York, 1946.

403. **Strauchii A.** Tabellen der Sinium, Tangentium, Logarithmorum... Amsterdam, 1700.

404. **Vlacq A.** Arithmetica logarithmica. Gouda, 1628.

405. **Vlacq A.** Tabulae sinuum, tangentium, et secantium et logarithmi sinuum, tangentium et numerorum ab unitate ad 10 000. Francofurti et Lipsiae, 1726, 1768.

406. **Vogel W. F.** Angular spacing tables. Vinco Corporation, Detroit, Michigan, 1943.

407. **Des vollständigen Mathematischen, 2.** Leipzig, 1742.

408. **Weems P. V. H.** Line of position book. Annapolis, 1927.

409. **Weisbach J.** Der Ingenieur Sammlung von Tafeln, Formeln und Regeln. Braunschweig, 1863.

410. **Wijdenes P.** Five-place tables. Noordhoff, Groningen, 1937.

411. **Wirtz C.** Tafeln und Formeln aus Astronomie und Geodäsie. Springer, Berlin, 1918.

412. **Wittke H.** Vademekum für Vermessungstechnik. Metzlersche Verlagsbuchhandlung, Stuttgart, 1948.

413. **Witkowski A. W.** Tablice matematyczno-fizyczne. Orgelbrand, Warszawa, 1904.

414. **Wolf R.** Handbuch der Mathematik, Physik, Geodäsie und Astronomie, 1. Zürich, 1870.

415. Wolfe J. H. and Phelps E. R. Natural trigonometric functions with differences and rapid and easy method of interpolation. Authors, Detroit, Michigan, 1942.

416. Woodward R. S. Smithsonian geographical tables. Smithsonian Institution, Washington, 1906.

417. Lomnicki A. Tablice matematyczno-fizyczne czterocyfrowe. Dla użytku szkoł średnich. Ksiaznica-Atlas, Lwów — Warszawa, 1932.

418. Zacharias M. und Meth P. Vierstellige Logarithmentafeln. W. Gruyter, Berlin — Leipzig, 1927.

419. **Zahlentafeln für die Grundschule**. «Volk und Wissen», Berlin, 1950.

420. Zimmermann H. Rechentafeln. Ausgabe A, ohne besondere Quadrattafel. Ernst, Berlin, 1929.

421. **Abhandlungen der Mathematisch-Physikalischen Classe der Königlich Bayerischen Akademie der Wissenschaften**, 1885, 15, 229—328.
Lommel E. C. J. Die Beugungserscheinungen einer kreisrunden Oeffnung und eines kreisrunden Schirmchens.

422. **Abhandlungen der Mathematisch-Physikalischen Classe der Königlich Bayerischen Akademie der Wissenschaften**, 1886, 15, 529—664.
Lommel E. C. J. Die Beugungserscheinungen geradlinig begrenzter Schirme.

423. **Annals of the Observatory of Lund** (printed in Sweden), 1936, 5.
Tillman E. Altitude tables for mariners and aviators.

424. **Archiv der Mathematik und Physik** (Leipzig), 1843, (1), 3, 27—34.
Bretschneider C. A. Berechnung der Grundzahl der natürlichen Logarithmen, sowie mehrerer anderer mit ihr zusammenhängender Zahlwerthe.

425. **The Astronomical Journal**, 1933, 42, 123—125; 196.
Herget P. A table of sines and cosines to eight decimal places.

426. * **British Association for the Advancemen Science. Reports**, 1916. 59—89.

427. **British Association for the Advancement of Science. Reports**, 1924, 275—295.

428. **British Association for the Advancement of Science. Reports**, 1928, 305—340.

429. **Electrotechnische Zeitschrift** (Berlin), 1891, 12, № 39, 509—520.
Tabelle für tg $\varphi \cdot \mathrm{tg}(x-\varphi)$.

430. **Handbook of Aeronautics** (London), 1938, 3, 78—84.
Pritchard J. L. Aircraft desing data and formulae.

431. **Ingenieur-Archiv** (Berlin), 1932, 3, 206—214.
Gradstein S. Erzwungene Torsionsschwingungen von Kurbelwellen.

432. **Journal des Mathématiques pures et appliquées** (Paris), 1894, (4), 10, 391—472.
Hamy M. Sur le développement approché de la fonction perturbatrice dans le cas des inéqualités d'ordre élevé.

433. **Le Journal de Phisique et le Radium**, 1946, 7, № 8, 231—243.
Colombani A. La théorie des filtres électriques et les polynomes de Tchebichef.

434. **Journal of the Washington Academy of Sciences**, 1922, 12, 423—436.
Van Orstrand C. E. and Shoultes M. A. Values of sine θ and cosine θ to 33 places of decimals for various values of θ expressed in sexagesimal seconds.

435. **Luftfahrt-Forschung** (München—Berlin), 1940, 17, 47—53.
Jaecuel K. Auflösung der Prandteschen Zirculationsgleichung durch trigonometrische Reihen.

436. **Mathematical and Physical Papers**, 1884, 2, 153.
Thomson W. On the forces concerned in the laying and lifting of deep-sea cables.

437. **Mathematical Tables and other Aids to Computation** (Washington), 1951, 5, № 33, 9—11.
Salzer H. E. Radix table for trigonometric functions and their inverses to high accuracy.

438. **Memoirs of the National Academy of Sciences**, 1925, 14, № 5, 3—79.
Van Orstrand C. E. Tables of the exponential function and

of the circular sine and cosine to radian argument.

439. **Monthly Notices of the Royal Astronomical Society** (London), 1902, 62, 633—703.
A l d i s W. S. Explanation of use of tables of $\frac{1}{2}(\theta + \cos\theta)$.

440. **The New Zealand Journal of Sciences and Technology.** General section B, 1944, 26, 156—159.
J o n e s W. M. Table of directions cosines to four figures at intervals of one degree in latitude and longitude over a zone of 60° of longitude and latitude from 0° to 40°.

441. **Philosophical Magazine** (London), 1935, (7), 20, № 134, 721—731.
A i r e y J. R. The circular and hyperbolic functions, argument $\frac{x}{\sqrt{2}}$.

442. **Philosophical Magazine** (London), 1935, (7), 20, № 134, 731—738.
A i r e y J. R. The circular sine and cosine functions, argument $\log_e x$.

443. **Philosophical Transactions of the Royal Society of London**, 1949, 242, 1—32.
F o x E. N. The diffraction of two dimensional sound pulses incident on an infinite uniform slit in a perfectly reflecting screen.

444. **Proceedings of the Institute of Radio Engineering** (New York), 1942, 30, 7, 335—349.
K i n g R. and B l a k e F. G. The self-impedance of a symmetrical antenna.

445. **Proceedings of the Institute of Radio Engineering** (New York), 1944,32, 114.
W h i n n e r y J. R. and J a m i e s o n H. W. Equivalent circuits for discontinuities in transmission lines.

446 **Proceedings of the Royal Society of London**, 1871, 19, 259—266.
K e l v i n, L o r d. On the determination of a ship's place from observations of altitude.

447. **Publication der Astronomischen Gesellschaft**, 1908, 22.
B ö r g e n C. Logarithmisch-trigonometrische Tafel, auf 11 (bezw. 10) Stellen.

448. **Reviews of Modern Physics** (Lancaster), 1938, 10, 264—269.
B a r t k y W. Numerical calculation of a generalized complete elliptic integral.

449. **Sitzungsberichte der Mathematisch-naturwissenschaftlichen Classe der Kaiserlichen Akademie der Wissenschaften** (Wien), 1848, 4, 165—181.
H e r r m a n n. Bestimmung der trigonometrischen Functionen aus den Winkeln und der Winkel aus den Funktionen, bis zu einer beliebigen Grenze der Genauigkeit.

450. **Terrestrial Magnetism and Atmospheric Electricity** (Baltimore), 1935, 40, 307—316.
R o m a n I. A Table of inverse trigonometric functions in radians.

451. **Terrestrial Magnetism and Atmospheric Electricity** (Baltimore), 1935, 40, 443—447.
Y o s t F. L., W h e e l e r J. A. and B r e i t G. Coulomb wave-functions.

452. **Tracts for Computers**, 1938, 23.
C o m r i e L. J. Tables of $\tan^{-1} x$ and $\log(1 + x^2)$ to assist in the calculation of the ordinates of a Pearson type IV curve.

453. **Veröffentlichungen des Astronomischen Rechen-Instituts zu Berlin—Dahlem**, 1929, 47.
P e t e r s J. Verwandlung von rechtwinkligen Platten-Koordinaten und sphärischen Koordinaten ineinander.

454. **Zeitschrift für Kristallographie** (Leipzig), 1933, (A), 85, № 5/6, 404—419.
S h e r m a n J. and B r o c k w a y L. Four-place table of $\frac{\sin x}{x}$.

455. **C o m r i e L. J.** Chambers's six-figure mathematical tables, I. New York, 1949.

456. **C o m r i e L. J.** Chambers's six-figure mathematical tables, II. New York, 1949.

457. **R a d v a n y i B.** Involut. Nehézipari Könyvkiadó, Budapest, 1954.

458. **V a l o u c h M. — V a l o u c h M. A.** Sedmimístné logaritmy čísel od 1 do 120 000. Nakladatelství Československé Akademie Věd, Praha. 1953.

459. **V a l o u c h M. — V a l o u c h M. A.** Pětimístné logaritmy čísel. Nakladatelství Československé Akademie Věd, Praha, 1953.

References to Chapter 3

EXPONENTIAL AND HYPERBOLIC FUNCTIONS

1. **Алексеев В. А.** Сборник таблиц для расчета вероятностей. Арт. акад. им. Дзержинского, М., 1944.

2. **Андреев П. П.** Математические таблицы. Госстатиздат, М., 1952.

3. **Брилинг С. Р.** Математический справочник. ГТТИ, М.— Л., 1933.

4. **Брилинг С. Р.** Математические формулы и таблицы. «Макиз», М., 1928.

5. **Бронштейн И. Н. и Семендяев К. А.** Справочник по математике. Гостехиздат, М.—Л., 1945.

6. **Ватсон Г. Н.** Теория бесселевых функций, 2. Изд-во иностр. лит., М., 1949.

7. **Вега Г.** Таблицы семизначных логарифмов. Геодезиздат, М., 1949.

8. **Ветчинкин В. Н.** Новые формулы и таблицы эллиптических интегралов и функций. Изд-во Военновозд. акад. РККА, М., 1935.

9. **Гнеденко Б. В.** Курс теории вероятностей. ГИТТЛ, М.— Л., 1950.

10. **Гутерман А. В. и Браташ В. Н.** Математические таблицы для конструкторов машиностроителей. ГНТИ Маш. лит., Москва—Свердловск, 1944.

11. **Гутерман И. В. и Иванова К. Г.** Математические таблицы. II. Союзоргучет, М., 1940.

12. **Двайт Г. Б.** Таблицы интегралов и другие математические формулы. Гос. изд-во иностр. лит., М., 1948.

13. **Дитрихкейт О.** Быстрое счисление. Семизначные логарифмы и антилогарифмы. Г. В. Гольстен, Петербург, 1904.

14. **Дуббель Г.** Справочник по математике для инженеров, студентов и преподавателей математики. ОНТИ, М.— Л., 1933.

15. **Кайгарадау А.** Чатырохзначныя таблицы. Бел. дзярж. выд. Вычпэдсэктар, Менск, 1932.

16. **Коган Я. М.** Краткий математический справочник. «Кубуч», Л., 1936.

17. **Машиностроение. Энциклопедический справочник,** I, кн. 1. ГНТИ Маш., лит., М., 1947.

18. **Мечников В. В.** Полные трехзначные таблицы логарифмов. Артил. акад. РККА, Л., 1933.

19. **Митропольский А. К.** Краткие таблицы (с приложением очерка вычислительной техники). Изд-во Лесотехн. акад. им. С. М. Кирова, Л., 1950.

20. **Сегал Б. И. и Семендяев К. А.** Пятизначные математические таблицы. Изд-во АН СССР, М.— Л., 1948.

21. **Справочник по математике для обогатителей.** Металлургиздат, М., 1951.

22. **Справочное руководство по машиностроению,** I. ГНТИ Украины, Харьков, 1937.

23. **Таблиці квадратів, кубів, логаритмів, тригонометричних і гіперболічних функцій, квадратових і кубових коренів та інших величин.** Технетеоретич. вид-во, Харків, 1932.

24. **Таблицы математических и зических величин.** Научн. книгоизд-во, Петроград, 1923.

25. **Тимофеев Н.** О наивыгоднейших результатах наблюдений. Униврситетская типогр., Киев, 1870.

26. **Хренов Л. С.** Таблицы натуральных значений тригонометрических функций в единицах времени. Тип. АН Казах. ССР, Алма-Ата, 1947.

27. **Hütte.** Справочник для инженеров, техников и студентов, 1. Госмашметиздат, М.—Л., 1934.

28. **Чэркашин Л.** Чатырохзначныя матэматычныя табліцы. Дзярж. наук.-тэхн. выд-ва, Менск, 1934.

29. **Янке Е., Эмде Ф.** Таблицы функций с формулами и кривыми. ГИТТЛ, М.—Л., 1949.

30. **Вестник инженеров и техников** (Москва), 1948, № 2—4, 114—117.
Карякин Н. Н. Метод узловых депланаций для расчета тонкостенных многопролетных стержней на кручение.

31. **Журнал технической физики** (Киев), 1932, 2, 97—117.
Уманский А. Таблицы гиперболико-тригонометрических функций.

32. **Труды Сибирского физико-технического института**, 1930.
Конусов В. Ф. Таблицы интегралов

$$\Phi_m(x) = \int_1^\infty \beta^m e^{-\beta x} d\beta.$$

33. **Abbott P. W.** Mathematical tables and formulae. Longmans, Green and Co., London, 1918.

34. **Allen E. S.** Six-place tables. McGraw-Hill Book Co., New York — London, 1947.

35. **Angot A.** Compléments de mathématiques à l'usage des ingénieurs de l'électrotéchnique et des télécommunications. Revue d'Optique, Paris, 1949.

36. **Attwood E. L.** Practical five-figure mathematical tables. Macmillan, London, 1950.

37. **Becker G. F.** and **Van Orstrand C. E.** Hyperbolic function. Smithsonian Institution, Washington, 1909.

38. **Blakesley T. H.** A table of hyperbolic cosines and sines. Physical Society, London, 1890.

39. **Bohde H.** Mathematisch-technische Zahlentafeln. Springer, Berlin, 1933

40. **Boll M.** Tables numériques universelles des laboratoires et bureaux d'étude. Dunod, Paris, 1947.

41. **British Association for the Advancement of Science. Mathematical Tables,** 6. Bessel functions. Functions of orders zero and unity, Cambridge, 1937.

42. **British Association for the Advancement of Science. Mathematical tables,** I: Circular and hyperbolic functions, exponential and sine and cosine integrals, factorial function and allied functions Hermitian, probability functions. Prepared by the Committee for the Calculation of Mathematical Tables. Cambridge Univ. Press. 1951.

43. **Bruns H.** Wahrscheinlichkeitsrechnung und Kollektivmasslehre. Teubner, Leipzig — Berlin, 1906.

44. **Burrau C.** Tafeln der Funktionen Cosinus und Sinus mit den natürlichen sowohl reellen als rein imaginären Zahlen als Argument (Kreis und Hyperbelfunktionen). Reimer, Berlin, 1907.

45. **Byerly W. E.** An elementary treatise on Fourier's series and spherical, cylindrical and ellipsoidal harmonics. Ginn, Boston, 1893.

46. **Campbell J. W.** Numerical tables. The Douglas Print Co., Edmonton, Alta, 1946.

47. **Carmichael R. D.** and **Smith E. R.** Mathematical tables and formulas. Ginn, Boston — New York, 1931.

48. **Clark J. B.** Mathematical and physical tables. Oliver and Boyd, Edinburgh, 1950.

49. **Comrie L. J.** Chambers's fourfigure mathematical tables. Edinburgh— London, 1947.

50. **Dale J. B.** Five-figure tables of mathematical functions, comprising tables of logarithms, powers of numbers, trigonometric, elliptic and other transcendental functions. Arnold, London, 1949.

51. **Dwight H. B.** Mathematical tables. McGraw-Hill, New York — London, 1941.

52. **F o w l e F. E.** Smithsonian physical tables. Smithsonian Institution, Washington, 1934.

53. **Formular matematic şi tehnic.** Pentru elevi, studenţi şi tehnicieni. Editura tehnica, Bucureşti, 1949.

54. **F o r t i A.** Nuove tavole delle funzioni iperboliche... Societa Laziale, Rome, 1892.

55. **G a u s s F. G.** Vierstellige logarithmische und trigonometrishe Tafeln (Sexagesimal unterteiler Altgrad). «Volk und Wissen», Berlin — Leipzig, 1946.

56. **V a l o u c h M.— V a l o u c h M. A.** Pětimístné logaritmy čisel. Nakladatelství Československé Akademie Věd, Praha, 1953.

57. **G r e e n h i l l A. G.** The applications of elliptic functions. Macmillan, London — New York, 1892.

58. **G u y o n E.** Les problèmes de navigation et la carte marine. Berger-Levrault, Paris — Nancy, 1896.

59. **H a l l W.** Tables and constants to four figures. Cambridge Univ. Press, Cambridge, 1905.

60. **B u r i n g t o n R. S.** Handbook of mathematical tables and formulas. Sandusky, Ohio, 1946.

61. **H a w e l k a R.** Vierstellige Tafeln der Kreis- und Hyperbelfunktionen, sowie ihrer Umkehrfunktionen im Komplexen in Schritten von $\frac{\pi}{2}$ 0,02 und i 0,02. Vieweg, Braunschweig, 1931.

62. **H a y a s h i K.** Fünfstellige Funktionentafeln. Springer, Berlin, 1930.

63. **H a y a s h i K.** Fünfstellige Tafeln der Kreis-und Hyperbelfunktionen sowie der Funktionen e^x und e^{-x} mit den natürlichen Zahlen als Argument. W. Gruyter and Co., Berlin, 1944.

64. **H a y a s h i K.** Sieben—und mehrstellige Tafeln der Kreis-und Hyperbelfunktionen und deren Produkte sowie der Gammafunktion. Springer, Berlin, 1926.

65. **H a y a s h i K.** Tafeln der Besselschen — Theta—, Kugel— und anderen Funktionen Springer, Berlin, 1930.

66. **H a y a s h i K.** Tafeln für die Differenzenrechnung sowie zur Hyperbel —,Bessel'schen, elipti en und anderen, Funktionen. Springer, Berlin, 1933.

67. **H o b s o n E. W.** A treatise on plane trigonometry. Cambridge Univ. Press, Cambridge, 1891.

68. **H o l t a p p e l H. W.** Tafels van e^x. Groningen, Noordhoff, 1938.

69. **H o ü e l J.** Recueil de ormules etf de tables numériques. Gauthier-Villars, Paris, 1868, 1901.

70. **H ü t t e.** Des Ingenieurs Taschenbuch' 1. Ernst, Berlin, 1920.

71. **J a h n k e E.** und **E m d e F.** Funktionentafeln mit Formeln und Kurven. Teubner, Leipzig — Berlin, 1933.

72. **K a y e G. W. C.** and **L a b y T. H.** Tables of physical and chemical constants and some mathematical functions. Longmans, London, 1943.

73. **K e m p e H. R.** and **S m i t h W. H.** The Engineer's Year-book... for 1927... Grosby Lockwood, London, 1927.

74. **K e n n e l l y A. E.** Tables of complex hyperbolic and circular functions. Harvard Univ. Press, Cambridge, Massachusetts, 1921.

75. **K ö h l e r H. G.** Logarithmisch-trigonometrisches Handbuch. Tauchnitz, Leipzig, 1898.

76. **K ü s t n e r H.** Fünfstellige Logarithmen. «Volk und Wissen», Berlin— Leipzig, 1950.

77. **L e g e n d r e A. M.** Exercices de calcul intégral, 3. Paris, 1816.

78. **L e g e n d r e A. M.** Traité des fonctions élliptiques, 2. Huzard-Courcier, Paris, 1826.

79. **L i g o w s k i W.** Tafeln der Hyperbelfunktionen und der Kreisfunktionen... Ernst, Berlin, 1890.

80. **L o k y B.** Négyjegyü függvénytáblázatok. Budapest, 1950.

81. **M c M a h o n J.** Hyperbolic functions. Wiley, New York; Chapman and Hall, London, 1906.

82. **M e l l o r J. W.** Higher mathematics for students of chemistry and physics. Longmans, London, 1931.

83. **M e y e r U. und D e c k e r t A.** Tafeln der Hyperbelfunktionen. Druck Kösel und Pustet, Kempten, 1924.

84. **M i l l e r W. L. and R o s e b r u g h T. R.** Numerical values of certain functions involving e^{-x}. University of Toronto Studies, Canada, 1904.

85. **M i l l e r F. H.** Calculus. Wiley, New York, 1946.

86. **M i l n e - T h o m s o n L. M. and C o m r i e L. J.** Standard four-figure mathematical tables. Macmillan, London, 1948.
ВГБПЛ, ГАПШ, МП, ПТМиВТ, БЛ

87. **M i l n e - T h o m s o n L. M.** Die elliptischen Funktionen von Jacobi. Springer, Berlin, 1931.

88. **M o r s e P. M.** Vibration and sound. McGraw-Hill, New York — London, 1936.

89. **N e w m a n F. W.** The higher trigonometry. Superrationals of second order. Macmillan and Bowes, Cambridge, 1892.

90. **P e a r s o n K.** Tables for statisticians and biometricians, I. Biometric Laboratory, University College, London, 1930.

91. **P e a r s o n K.** Tables of the incomplete Γ-function. Biometric Laboratory, University College, London, 1934.]

92. **P e i r c e B. O.** A Short table of integrals. Ginn, Boston—New York, 1910.

93. **P e r n o t F. E.** Abridged table of hyperbolic functions. Univ. Calif. Press, Berkeley, 1915.

94. **P e r n o t F. E.** Electrical phenomena in parallel conductors, I. Wiley, New York; Chapman and Hall, London, 1918.

95. **P e r n o t F. E. and W o o d s R. M.** Logarithms of hyperbolic functions to twelve significant figures. Univ. Calif. Publ. in Eng., 1918. 1, 13, 297—467.

96. **P e t e r s J.** Zehnstellige Logarithmentafel, I: Zehnstellige Logarithmen der Zahlen von 1 bis 100.000 nebst einem Anhang mathematischer Tafeln. Reichsamt für Landesaufnahme, Berlin, 1922.

97. **P o t i n L.** Formules et tables numériques. Doin et Gauthier-Villars, Paris, 1925.

98. **P r y d e J.** Chambers's educational course. Mathematical tables. Chambers, London — Edinburgh, 1930.

99. **R a p e r H.** Nautical tables. Potter, London, 1920.

100. **L a r s e n H. D.** Rinehart methematical tables, formules and curves. Chapman and Hall, London, 1949.

101. **R i v a r d M.** Tables des sinus, tangentes, secantes et de leurs logarithmes, avec la construction de ces tables, et les problèmes de la trigonomètrie rectiligne et sphérique. Desaint. Saillant, Paris, 1743.

102. **R o s e n b a c h I. B., W h i t m a n E. A., M o s k o v i t z D.** Mathematical tables. Ginn, Boston, 1946.

103. **R u s s e l l A.** A treatise on the theory of alternating currents, II. Cambridge Univ. Press, Cambridge, 1916.

104. **S a k a m o t o S.** Tables of Gudermannian angles and hyperbolic functions. Tetsudo-Tosho-Kyoku, Tokyo, 1934.

105. **S a s s e n f e l d H. M. und T s c h u n k o H. F. A.** Mathematische Tafeln für Mathematiker, Naturwissenchaftler, Ingenieure. Erster Teil: Elementare Funktionen. Fr. Lamade, Walldorf bei Heidelberg, 1949.

106. **S c h o r r R.** Sammlung von Hilfstafeln der Hamburger Sternwarte in Bergedorf. Lucas Gräfe, Hamburg, 1916.

107. **S c h ü l k e A.** Tafeln. Vierstellige Logarithmen Funktions- und Zahlenwerte. «Volk und Wissen», Leipzig, 1949.

108. **S c h u l t e - T i g g e s R. A.** Rechentafeln für höhere Lehranstalten insbesondere zu Mehlers Hauptsätzen der Elementar-Mathematik. W Gruyter, Berlin — Leipzig, 1929.

109. **S c h u l z e J. C.** Neue and erweiterte Sammlung logarithmischer, trigonometrischer und anderer ...Tafeln, 1 — 2. Mylius, Berlin, 1778.

110. **S i l b e r s t e i n L.** Synopsis of applicable mathematics, with tables. Bell, London, 1923.

111. **S o k a l S. i T o r b u s W.** Tablice matematyczno—fizyczne do użytku szkoł zawodowych. Państwowe wyd-wa szkolnictwa zawodowego, Warszawa, 1951.

112. Spenceley G. W. and Spenceley R. M. Smithsonian elliptic functions tables. The Smithsonian Institution, Washington, 1947.

113. Spencer R. C. and Reynolds G. E. A table of normalized parabolic coordinates and arc lengths. Air Force Cambridge Research Laboratories, publications E. 4083, Cambridge, 1951.

114. Stalmasek J. Tabul'ky funkcií a konštant z matematiky, fysiky a chemie. Štátne nakl., Bratislava, 1951.

115. Stumpff K. Tafeln und Aufgaben zur harmonischen Analyse und Periodogrammrechnung. Springer, Berlin, 1939.

116. **Table of circular and hyperbolic tangents and cotangents for radian arguments.** Columbia Univ. Press, 1943.

117. Robbins C. K. and Little N. Tables for calculus. The Macmillan Co., New York, 1942.

118. **Tables of circular and hyperbolic, sines and cosines for radian arguments.** Prepared by U. S. Department of Commerce. National Bureau of Standards, New York — Washington, 1940.

119. **Tables of the exponential function** e^x. Prepared by National Bureau of Standards, Computation laboratory, Washington; United States, Government Printing Office, New York, 1951.

120. **Tables of inverse hyperbolic functions.** The annals of the computation laboratory of Harvard University, XX. Harvard Univ. Press, Cambridge, 1949.

121. Vassal V. Nouvelles tables donnant avec cinq décimales les logarithmes vulgaires et naturels des nombres de 1 à 10 800 et des fonctions circulaires et hyperboliques, pour tous les degrés du quart de cercle de minute en minute. Gauthier-Villars, Paris, 1872.

122. Vega G. — Hülsse J. A. Sammlung mathematischer Tafeln... Weidmann, Leipzig, 1849. БЛ

123. Vega G. Logarithmisch-trigonometrische Tafeln... (Tabulae logarithmico-trigonometricae...), 1—2. Weidmann, Leipzig, 1797.

124. Vega G. Logarithmische, trigonometrische und andere... Tafeln und Formeln. Vienna, 1783.

125. Watson G. N. A treatise on the theory of Bessel functions. Cambridge Univ. Press, Cambridge, 1945.

126. Smith E. S. Binomial, normal and Poisson probabilities. Published by the author, Bel Air, Md., 1953.

127. **Archiv für Elektrotechnik** (Berlin), 1914, 2, 57—67.
Siegbahn M. Über die Darstellung der Vorgänge bei Kabeln durch Raumkurven.

128. **Archiv für Elektrotechnik** (Berlin), 1927, 17, 79—101.
Potentialtheorie der Hängeisolatoren.

129. **Archiv der Mathematik und Physik** (Leipzig), 1843, (1), 3, 27—34.
Bretschneider C. A. Berechnung der Grundzahl der natürlichen Logarithmen, sowie mehrerer anderer mit ihr zusammenhängender Zahlwerthe.

130. **Atti del Congresso Internazionale dei Matematici** (Bologna), 1930, 3, 357—371.
Milne-Thomson L. M. A seven-decimal table of inverse hyperbolic cotangents.

131. **Biometrika** (London), 1903, 2, 260—272.
Elderton W. P. Graduation and analysis of a sickness table.

132. **Biometrika** (London), 1942, 32, 243—262.
Beall G. The transformation of data from entomological field experiments so that the analysis of variance becomes applicable.

133. **British Association for the Advancement of Science. Reports**, 1926, 273—297.

134. **British Association for the Advancement of Science. Reports**, 1928, 305—340.

135. **Handbook of Aeronautics** (London), 1938, 3, 78—84.
Pritchard J. L. Aircraft desing data and formulae.

136. **Ingeniöurdenskabelige skriften, Danmarks naturvidenskabelige samfund**, 1933, 10.
Prohaska C. W. Simpler equations for catenarys.

137. **Jahrbuch der Radioaktivität und Elektronik**, 1906, 3, 120—133, 290.
G r u n e r P. Tables of exponential functions with negative exponents, $y = e^{-x}$.

138. **Journal of the Institute of Actuaries**, 1870, 15, 432—446.
M a k e h a m W. M. A table for determining the amounts etc. of continuous annuities certain.

139. **Journal of Physical Chemistry**, 1942, 46, 641—662.
S h e r m a n J. and E w e l l R.B. A six-place table of the Einstein functions.

140. **Journal of the Washington Academy of Sciences**, 1913, 3, 345—357.
V a n O r s t r a n d C. E. Table of the exponential functions e^x, e^{-x} for $x = 0{,}0$ *to* $x = 32.0$ to either 20 decimals or 20 significant figures.

141. **Journal für die reine und angewandte Mathematik**, 1831, 7, 72—96, 176—200.
G u d e r m a n n C. Theorie der potential- oder cyklisch-hyperbolischen Funktionen.

142. **Journal für die reine und angewandte Mathematik**, 1832, 8, 64—116, 194—212.
G u d e r m a n n C. Theorie der potential- oder cyklisch-hyperbolischen Funktionen.

143. **Journal für die reine und angewandte Mathematik**, 1832, 9, 81—96, 193—208, 297—304, 362—378.
G u d e r m a n n C. Theorie der potential- oder cyklisch-hyperbolischen Funktionen.

144. **Kongelige Danske videnskabernes selskab** (Kjobenhavn), 1884, (6), 2, № 6, 185—308.
G r a m J. P. Undersogelser angaaende maengden af primtal under en given graense.

145. **Memoirs of the National Academy of Sciences** (Washington), 1925, 14, № 5, 3—79.
V a n O r s t r a n d C. E. Tables of the exponential function and of the circular sine and cosine to radian argument.

146. **Memoirs Ryojun College of Engineering**, 1929, 2, № 2.
N o d a S. Table of $\mathrm{gd}^{-1}\varphi - \sin\varphi$ for φ at intervals of 10′ from 0° to 35°.

147. **Metron.** (**Rivista internationale di statistica**), 1921, 1, № 4, 3—32.
F i s h e r R. A. On the «Probable Error» of ı coefficient of correlation-deduced from a small sample.

148. **Philisophical Magazine**, 1910, (6), 19, 673—684.
P e r r y J. Telephone circuits.

149. **Philosophical Magazine**, 1921, (6), 41, № 241, 81—96.
D a r n l e y E. R. The transverse vibrations of Beams and the whirling of shafts supported at intermediate points.

150. **Philosophical Magazine**, 1923, (6), 45, 846—849.
W r i n c h H. E. H. and W r i n c h D. M. Table of the Bessel functions $I_n(x)$.

151. **Philosophical Magazine** (London), 1924, (6), 47, 62—65.
W r i n c h H. E. H. and W r i n c h D. M. Tables of Bessel functions.

152. **Philosophical Magazine**, 1935, (7), 20, № 134, 721—731.
A i r e y J. R. The circular sine and hyperbolic functions, argument $\frac{x}{\sqrt{2}}$.

153. **Proceedings of the Cambridge Philosophical Society**, 1950, 46, 164—181.
H o p k i n s H. G. Elastic deformations of infinite strips.

154. **Proceedings of the Physical Society of London**, 1911, 23, 344—345.
J o h n s t o n e A. Short table of circular and hyperbolic functions for complex values of the argument.

155. **Proceedings of the Royal Society of Edinburgh**, 1931—1932, 52, I, 236—250.
M i l n e - T h o m s o n L. M. The zeta functions of Jacobi. A seven-decimal table of $z(u/m)$ at interval 0.01 for u and for values of $m(=k^2)$ from 0.1 to 1.0.

156. **Quarterly Journal of Mathematics**, 1885, 20, 213—220.
C a y l e y A. On the orthomorphosis of the circle into the parabola.

157. **Scientific Papers of the Instituts of Physical and Chemical Research** (Tokyo), 1929, 3, 181—215.

Akahira T. Tables of $\frac{e^{-x}}{x}$ and $\int_x^\infty \frac{e^{-u}}{u}du$ from $x = 20$ to $x = 50$.

158. **Skandinavisk aktuarietidskrift,** 1938, 21, 47—71.

Steffensen J. F. A table of the function $G(x) = \frac{x}{1-e^{-x}}$ and its applications to problems in compound interest.

159. **Transactions of the Cambridge Philosophical Society,** 1883, 13, 243—272.

Glaisher J. W. L. Tables of the exponential function.

160. **Transactions of the Cambridge Philosophical Society,** 1883, 13, № 3, 145—241.

Newman F. W. Tables of the descending exponential function to 12 or 14 places of decimals.

161. **Transactions of the Cambridge Philosophical Society,** 1889, 14, № 3, 237—249.

Newman F. W. Tables of the exponential e^x to 12 places of decimals.

162. **Transactions of the Connecticut Academy of Arts and Science** (New Haven), 1937, 32, 381—434.

Uhler H. S. A new table of reciprocals of factorials and some derived numbers.

163. **Transactions Faculte of Actuaries in Scotland,** 1918, 8, 57—86.

164. **Transactions of the Royal Society of Canada,** 1903, (2), 9, 73—107.

Miller W. L. and Rosebrugh T. R. Numerical values of certain functions involving e^{-x}. (Tables of the integral $\int_x^\infty e^u u^n du$)

165. **Transactions of the Royal Society of Edinburgh,** 1887, 33, 95.

Clark J. Table calculated for «on the foundations of the kinetic theory of gases».

166. **Transactions of the Royal Society of Edinburgh,** 1889, 39, № 9, 257—321.

Burgess J. On the definite integral $\frac{2}{\sqrt{\pi}} \int_0^t e^{-t^2}dt$ with extended table of values, also $\frac{2}{\sqrt{\pi}}e^{-x^2}$.

167. **Zeitschrift für angewandte Mathematik und Mechanik** (Berlin), 1950, (5, 6), 30, 169—174.

Prandtl L. und Vandrey F. Fliessgesetze normalzäher Stoffe im Rohr. Ein Beitrag zur Rheologie.

168. **Comrie L. J.** Chambers's six-figure mathematical tables, I. New York, 1949.

169. **Comrie L. J.** Chambers's six-figure mathematical tables, II. New York, 1949.

170. **Таблицы e^x и e^{-x}.** Москва. Изд-во АН СССР, 1955.

References to Chapter 4

COMMON AND NATURAL LOGARITHMS

1. **Абольник Б. С.** Трехзначные таблицы логарифмов. Транспечать, М., 1930.

2. **Адлер А.** Пятизначные логарифмы с вспомогательными таблицами для графического расчета и таблицами часто встречающихся числовых величин. Берлин — Рига, 1923.

3. **Адрианов И. В.** Таблицы поправок за центрировку и редукцию на пунктах триангуляции. Углетехиздат, М., 1951.

4. **Алексеев В. А.** Сборник таблиц для расчета вероятностей. Арт. акад. им. Дзержинского, М., 1944.

5. **Андреев П. П.** Математические таблицы. Госстатиздат, М., 1952.

6. **Андрейчин Р. и Матеев А.** Петозначни логаритмични и други таблици. София, 1949.

7. **Аничков Д.** Теоретическая и практическая тригонометрия. Типогр. компания, 1787.

8. **Артемьев Д.** Таблицы пятизначных логарифмов натуральных чисел и тригонометрических функций. Ладыжников, Берлин, 1924.

9. **Афанасьев П. А.** Таблицы логарифмов простых чисел и тригонометрических линий. Департамент народного просвещения, С.-Петербург, 1848.

10. **Баушингер И. и Петерс И.** Таблицы логарифмов чисел и тригонометрических функций с восемью десятичными знаками. I и II. Геодезиздат, М. 1). I — 1942; 2) II—1944.

11. **Безикович Я. С.** Таблицы тригонометрические и логарифмические. «Кубуч», Ленинград, 1931.

12. **Безу.** Курс математики, 1—4. Университетская тип., М., 1806.

13. **Бень Я.** Таблицы обыкновенных логарифмов о пяти знаках. Вайсберг Б. Р., Винница, 1910.

14. **Березовский Б. Я.** Справочник по элементарной математике, механике и физике. Техн.-теор. лит., М.— Л., 1939.

15. **Бик А.** Четырехзначные логарифмы чисел, их антилогарифмы и логарифмы тригонометрических функций. Лисснер и Роман, М., 1895.

16. **Блюмберг Я.** Таблицы пятизначных логарифмов чисел и тригонометрических величин. Вальтер и Рап, Рига, 1921.

17. **Богомольний М. А.** Таблиці знаходити добутки та частки (множення та делення трицифрових на двоцифрові та трицифрові менші від 500). Техн.-теор. вид-во, Харків—Київ, 1932.

18. **Борисов С. А.** Математический справочник. Госкиноиздат, М., 1938.

19. **Брадис В. М.** Четырехзначные математические таблицы. Учпедгиз, М., 1950.

20. **Бремикер К.** Таблицы логарифмов чисел и тригонометрических функций с шестью десятичными знаками. Ред. бюро ГУГСК НКВД СССР, М., 1938.

21. **Брилинг С. Р.** Математические формулы и таблицы. «Макиз», М., 1928.

22. **Брилинг С. Р.** Математический справочник. ГТТИ, М.— Л., 1933.

23. **Бронштейн И. Н. и Семендяев К. А.** Справочник по математике. Гостехиздат, М.—Л., 1945.

24. **Брунс К.** Таблицы логарифмов чисел и тригонометрических функций с семью десятичными знаками. Ред. бюро ГУГК при СНК СССР, М., 1939.

25. **Буссе Ф. И.** Сокращенные таблицы обыкновенных логарифмов. Составлено по руководству Веги для употребления в учебных заведениях. Битепаж Ф. А.,С.-Петербург, 1885.

26. **Былинский А. И.** Пятизначные таблицы логарифмов целых чисел и углов в делениях артиллерийского угломера. Воениздат, М., 1939.

27. **Вега Г.** Таблицы семизначных логарифмов. Геодезиздат, М., 1949.

28. **Ветчинкин В. П.** Новые формулы и таблицы эллиптических интегралов и функций. Изд-во Военно-возд. акад. РККА, М., 1935.

29. **Воронец А. М.** Таблицы логарифмов и справочник формул и вспомогательных таблиц по курсу элементарной математики. Сытин Д., М., 1917.

30. **Выгодский М. Я.** Справочник по элементарной математике. ГИТТЛ, М.— Л., 1951.

31. **Гаусс Ф. Г.** Пятизначные полные логарифмические и тригонометрические таблицы. ГТТИ, М.— Л., 1933.

32. **Герц Н.** Семизначные логарифмы тригонометрических функций для каждой секунды времени. ГУЧЕК, НКВД СССР, М., 1938.

33. **Глазенап С. П.** Карманные таблицы логарифмов чисел и тригонометрических функций с четырьмя десятичными знаками. Изд-во АН СССР, Л., 1934.

34. **Глазенап С. П.** Пятизначные таблицы логарифмов с приложением других таблиц, упрощающих вычисления. Изд-во АН СССР, Л., 1932.

35. **Глазенап С. П.** Шестизначные таблицы логарифмов сумм и разностей по Гауссу. Изд-во АН СССР, М.—Л., 1935.

36. **Гнеденко Б. В.** Курс теории вероятностей. ГИТТЛ, М.—Л., 1950. *

37. **Гопфен Б. В.** Логарифмы, антилогарифмы и тригонометрические величины в упрощенных таблицах. Моск. гор. сов. нар. хоз., М., 1919.

38. **Гутерман А. В. и Браташ В. Н.** Математические таблицы для конструкторов машиностроителей. ГНТИ Маш. лит., Москва—Свердловск, 1944.

39. **Гутерман И. В. и Иванова К. Г.** Математические таблицы, I—II. М., 1939—1940.

40. **Двайт Г. Б.** Таблицы интегралов и другие математические формулы. Гос. изд-во иностр. лит., М., 1948

41. **Десятизначные таблицы логарифмов комплексных чисел и перехода от декартовых координат к полярным.** Изд-во АН СССР, М., 1952.

42. **Дитрихкейт О.** Быстрое счисление. Семизначные логарифмы и антилогарифмы. Гольстен Г. В., С.-Петербург, 1904.

43. **Доброхотов П. И.** Таблицы логарифмов и тригонометрических величин (для градусных мер и делений угломера). Гос. изд-во УзССР, Ташкент, 1935.

44. **Долгушин П. А.** Четырехзначные таблицы логарифмов чисел и тригонометрических функций. С приложением элементарной теории табличных вычислений. «Сотрудник», Киев, 1911.

45. **Дуббель Г.** Справочник по математике для инженеров, студентов и преподавателей математики. ОНТИ, М.— Л., 1933.

46. **Ермолаев С. И.** Таблицы для вычисления траекторий численным интегрированием. Военно-морск. акад. корабл. и вооружения им. А. Н. Крылова. Л., 1951.

47. **Закович М. С.** Сокращенные таблицы логарифмов и соответствующих им чисел. «Польза», М., 1906.

48. **Иордан В.** Логарифмо-тригонометрические таблицы для нового (десятичного) деления с шестью знаками. ГНТИ, М.—Л., 1931.

49. **Кайгарадау А.** Чатырохзначныя табліцы. Бел. Дзярж. Выд. Вычпэдсэктор, Менск, 1932.

50. **Кайгородов А. И.** Четырехзначные таблицы логарифмов, антилогарифмов, натуральных тригонометрических величин и их логарифмов, перевода градусов в радианы, квадратных корней, обратных величин и др. Белорус. гос. акад. с. х., Горки, 1926.

51. **Калугин Я. И.** Счетные таблицы (пособие для калькуляторов и составителей смет). Комбинат П.П.П., Ульяновск, 1925.

52. **Каллет Ф.** Таблицы логарифмов чисел от 1 до 108000, синусов, косинусов и тангенсов от секунды до секунды для первых пяти градусов и от десяти до десяти секунд для всех градусов четверти окружности. М., 1837.

53. **Каменьщиков Н.** Четырехзначные логарифмы чисел и тригонометрических функций. Госиздат, М., 1921.

54. **Каменьщиков Н.** Логарифмическая таблица. «Начатки знаний», Петроград, 1922.

55. **Каменьщиков Н.** Таблицы логарифмов с четырьмя десятичными знаками. ГИЗ, Москва — Петроград, 1922.

56. **Каменьщиков Н.** Трехзначные логарифмы. ГИЗ, М., 1921.

57. **Кобец Н.** Карманная справочная книга для инженеров и техников. «Светоч», Одесса, 1927.

58. **Коган Я. М.** Краткий математический справочник. «Кубуч», Л., 1936.

59. **Комендантов Н. В.** (ред.) Пятизначные таблицы логарифмов и тригонометрических величин для углов в дуге и во времени. ГТТИ, М.—Л., 1932.

60. **Краткий повторительный курс логарифмических чисел и тригонометрических величин.** С.-Петербург, 1883.

61. **Кутник С. Е.** Таблицы вычислений для инженеров, строителей, студентов, техников и учащихся профтехшкол. Окрпрофуч, Харьков, 1926.

62. **Кутузов Н. И.** Таблицы для разбивки круговых кривых. Дориздат, М., 1950.

63. **Кюстер Ф. В.** Таблицы логарифмов для химиков, фармацевтов, врачей и физиков. Госхимиздат, М.—Л., 1941.

64. **Левитус Д.** Четырехзначные таблицы логарифмов и антилогарифмов. Сытин, М., 1909.

65. **Летник А. Я.** Пятизначные таблицы логарифмов чисел от 1 до 10 000. Спиридонов В. С., М., 1911.

66. **Лорченко В. И. и Оглоблин Н. В.** Четырехзначные логарифмические таблицы. Корчак-Новицкий Н. Т., Киев, 1910.

67. **Лунский Н. С.** Вычисление по сокращенным таблицам многозначных логарифмов Gray's. Лисснер Г. и Собко Д., М., 1917.

68. **Майер Б. А.** Математические формулы. Справочник по арифметике, алгебре, геометрии, тригонометрии, высшей математике. Киев, 1930.

69. **Максимов М. Г.** Сборник вспомогательных таблиц для землемеров, землеустроителей и техников по съемке и нивелировке городов. ГТТИ, Л.—М., 1932.

70. **Маркович Б. А.** Начальные логарифмы. Стасюлевич, Петербург, 1912.

71. **Машиностроение. Энциклопедический справочник, 1,** кн. 1. ГНТИ. Маш. лит., М., 1947.

72. **Мечников В. В.** Полные трехзначные таблицы логарифмов. Арт. акад. РККА, Л., 1933.

73. **Минин С. В.** Настольная таблица логарифмов и антилогарифмов с 4 десятичными знаками без применения интерполяции. ГИТТЛ, 1952.

74. **Минин С. В.** Пятизначные таблицы логарифмов. ГИТТЛ, М.— Л., 1949.

75. **Митропольский А. К.** Краткие таблицы (с приложением очерка вычислительной техники). Изд-во Лесотехн. акад. им. С. М. Кирова, Л., 1950.

76. **Михайлов А.** Таблицы логарифмов с четырьмя десятичными знаками. Логарифмы чисел, тригонометрических величин, натуральные тригонометрические функции и другие таблицы, упрощающие вычисления. ГИЗ, Москва — Петроград, 1923.

77. **Михальський М. М.** Перше ознайомлення з логаритмами й праця за ними. Держ. вид-во Укр., Харків—Одесса, 1930.

78. **Михель В. М.** Четырехзначные таблицы логарифмов. Изд-во Главсевморпути, М.—Л., 1947.

79. **Мореходные таблицы, IV.** Изд-во Гидрограф. департамента Морского министерства. Петербург, 1905.

80. **Мореходные таблицы.** Изд-во Гидрограф. упр. ВМС, 1949.

81. **Морозов.** Таблица для получения логарифма по данному числу. Таблица для получения числа по данному логарифму. Шмидт П. И., Петербург, 1879.

82. **Москвічов.** Таблиці логаритмів, антилогаритмів, логаритмів тригонометричних функцій. Держтехвидав, Харків, 1931.

83. **Недзялковский А. А.** Собрание таблиц и формул для инженеров, архитекторов и механиков, 1. Колесов и Михин, Петербург, 1867.

84. **Нейшулер Л. Я.** Таблицы. Деление, логарифмы, точные умножения и квадраты. ГТТИ, М., 1933.

85. **Никулин М. А.** Таблицы для практических вычислений. Перевод русских мер в метрические, таблицы степеней и корней чисел, десятичных логарифмов, тригонометрических функций, а также другие математические таблицы, необходимые для техника и учащегося. ГТТИ М.— Л., 1932.

86. **Нумеров Б. В.** Таблицы десятичных логарифмов чисел от 1000 до 10 000. ГТТИ, Л., 1932.

87. **Обломиевский Д.** Вспомогательные таблицы тригонометрических функций для малых дуг. Военнотопогр. отд. Глав. штаба, С.-Петербург, 1875.

88. **Орбинский А. Р.** Таблицы четырехзначных логарифмов для школ и кабинета. «Mathesis», Одесса, 1919.

89. **Орлов С. В.** Трехзначные таблицы логарифмов и тригонометрических функций. ГИЗ, М., 1920.

90. **Отто Л. Р.** Таблицы логарифмов. Добровольный флот, М., 1924.

91. **Павловский А.** Таблицы логарифмов, по изданию Каллета, содержащие логарифмы чисел от 1 до 108 000; также логарифмы синусов и тангенсов для каждой секунды первых пяти градусов и для каждых десяти секунд всех градусов четверти круга. Павловский А., Харьков, 1820.

92. **Пеевски В. и Венедиков М.** Петозначни логаритмични таблици на числата и тригонометричните функции при ново градусно деление. «Наука и изкуство», София, 1949.

93. **Первый русский карманный счетчик.** Быков, М., 1902.

94. **Перельман Я. И.** Таблицы и правила для вычислений. Учпедгиз, М.— Л., 1931.

95. **Петерс И.** Шестизначные таблицы тригонометрических функций, содержащие значения шести тригонометрических функций через каждые 10″ дуги квадранта, разделенного на 90°, и значения котангенсов и косекансов через каждую секунду дуги от 0°0″ до 1°20′. Изд. геодез. и картогр. ГУГК при СНК СССР, М., 1941.

96. **Поляков П. Н.** Начальная алгебра и собрание алгебраических задач. Улитин В. И., М., 1881.

97. **Попов И. Г.** Математические таблицы. Учпедгиз, М., 1952.

98. **Придатко С. В.** Таблицы четырехзначных логарифмов чисел и антилогарифмов. Таблицы натуральных величин тригонометрических функций и их логарифмов. Бенке А., Петроград, 1915.

99. **Пржевальский Е.** Пятизначные таблицы логарифмов чисел и тригонометрических величин. ГИЗ, М.—Л., 1928.

100. **Пятизначные таблицы логарифмов и тригонометрических величин для углов в дуге и во времени.** ГТТИ, М.—Л. 1932.

101. **Пятизначные таблицы логарифмов.** Воениздат, М., 1953.

102. **Пятэцкі П. Д.** Справачнік па матэматыцы. Дзярж. выд-во Беларусі, Менск, 1932.

103. **Райнов А. В.** Четырехзначные математические таблицы (связно-функциональные). ОНТИ, М.—Л., 1936.

104. **Ратнер И. Н.** Таблицы обыкновенных логарифмов о двенадцати знаках. Варшава, 1900.

105. **Ратнер И. Н.** Таблицы обыкновенных логарифмов о десяти знаках. Бр. Шульдберг, Варшава, 1894.

106. **Родин А. Ф.** Пятизначные логарифмические таблицы чисел и тригонометрических функций с приложением гауссовых логарифмов и различных вспомогательных таблиц. Линдеберг, С.-Петербург, 1895.

107. **Рябков Г. З.** Таблицы пятизначных логарифмов чисел и тригонометрических величин. Автор, Одесса, 1904.

108. **Рябков Г. З.** Таблицы пятизначных логарифмов чисел от 1 до 10 000. Автор, Одесса, 1895.

109. **Слудский И.** Сокращенные четырехзначные таблицы логарифмов и некоторые другие математические, физические и астрономические таблицы. Лисснер Г. и Гешель Л., М., 1902.

110. **Сокращенные таблицы обыкновенных логарифмов.** (Составлено по руководству Веги). С.-Петербург, 1855.

111. **Справочник по математике для обогатителей.** Металлургиздат, М., 1951.

112. **Справочное руководство по машиностроению,** I. ГНТИ Украины, Харьков, 1937.

113. **Справочник по элементарной математике, механике, физике.** ГИТТЛ, М.— Л., 1943.

114. **Слюсарев Г. Г.** Четырехзначные таблицы логарифмов чисел, синусов и тангенсов. Детиздат, Л., 1938.

115. **Стойков Д.** Спомогателни таблици за изчисления по машинна техника, метална техника и электротехника. София, 1947.

116. **Субботин М. Ф.** Многозначные таблицы логарифмов. Изд-во АН СССР, М.— Л., 1940.

117. **Таблица логарифмов.** Московская ордена Ленина с.-х. акад. им. Тимирязева. Кафедра неорган. химии, М., 1950.

118. **Таблица логарифмов тригонометрических величин и логарифмов чисел с тремя знаками.** Пулково, 1900.

119. **Таблица логарифмов чисел от 1 до 10 000 и натуральных синусов, тангенсов и секансов с логарифмами.** Морская академия, С.-Петербург. 1746.

120. **Таблиці квадратів, кубів, логаритмів тригонометричних і гіперболічних функцій, квадратових і кубових коренів та інших величин.** Техне-теоретич. вид-во, Харків, 1932.

121. **Таблицы логарифмов.** Чувашское гос. изд-во, Чебоксары, 1931.

122. **Таблицы логарифмов Вестфаля.** Гл.морской штаб, С.-Петербург, 1834.

123. **Таблицы логарифмов и синусов, тангенсов, секансов.** М., 1703.

124. **Таблицы логарифмов на числа от 1 до 10 000 и на синусы, тангенсы, косинусы и котангенсы всех градусов и минут четверти круга с таблицею синусов, тангенсов и секансов на градусы четверти круга. полагая радиус 100000,00 равных частей.** Морской шляхетский кадетский корпус, Петербург, 1761.

125. **Таблицы логарифмов четырех десятичных знаков. Антилогарифмы.** Київ, 1891.

126. **Таблицы логарифмов чисел и тригонометрических величин, составленные Лаландом** (стереотипное изд. с таблиц Лаланда, Лейпциг, 1844). Бр. Салаевы, М., 1868.

127. **Таблицы логарифмов чисел и тригонометрических линий.** Изд. Главным гидрограф. упр. Морского министерства, 1901.

128. **Таблицы обыкновенных логарифмов чисел и тригонометрических линий** (для употребления в морском кадетском корпусе). Морская академия, Петербург, 1862.

129. **Таблицы пятизначных логарифмов чисел, тригонометрических функций, $\sin^2 {}^1/_2$ угла, сумм и разностей.** Гос. изд-во Отд. военной лит., Петроград, 1921.

130. **Таблицы синусов, тангенсов и секансов и логарифмы синусов и танген-**

сов, тоже и чисел еже есть от единого даже до 10 000. М., 1716.

131. **Таблицы степеней, корней, обратных величин и логарифмов.** Сев.-зап. обл. промбюро ВСНХ, Л., 1927.

132. **Таблицы математических и физических величин.** Научн. книгоизд-во, Петроград, 1923.

133. **Таблицы четырехзначных логарифмов и основных физических величин.** Глаголев М. М. и Покровский М. Ю., С.-Петербург, 1910.

134. **Таблицы четырехзначных логарифмов чисел от 1 до 10 000 и натуральных тригонометрических величин.** «Далекая окраина», Владивосток, 1919.

135. **Тарасов Н. П.** Логарифмы. Гос. техн. изд-во, М., 1927.

136. **Тимофеев Н.** О наивыгоднейших результатах наблюдений. Университетская типогр., Киев, 1870.

137. **Трехзначные логарифмы.** Дом занимат. науки, Л., 1940.

138. **Упрощенные таблицы логарифмов и тригонометрических величин.** С.-Петербург, 1902.

139. **Фейгельсон Е. Г.** Таблицы десятичных логарифмов и натуральных тригонометрических функций. Учпедгиз, М.— Л.,1931.

140. **Филатов А. П.** Таблицы пятизначных и четырехзначных логарифмов чисел и антилогарифмов от 1 до 1000 с прибавками от 1000 до 10 000, расположенные на одном листе. С приложением правил логарифмирования и других вспомогательных таблиц. «Общественная польза», М., 1909.

141. **Франк М. Л.** Карманный справочник по математике. Т-во профессоров и преподавателей, С.-Петербург, 1912.

142. **Хохлов А. И.** Карманные математические таблицы. Пятизначные. Гостехиздат, М.— Л., 1947.

143. **Hütte.** Справочник для инженеров, техников и студентов. Госмашметиздат, М.— Л., 1934.

144. **Цех И.** Семизначные таблицы логарифмов сумм и разностей. Геодезиздат, М., 1940.

145. **Чэркашин Л.** Чатырохзначныя матэматычныя таблицы. Дзярж. наук-тэхн. выд-во, Менск, 1934.

146. **Четырехзначные таблицы логарифмов.** Арт. Акад. РККА им. Дзержинского, Л., 1935.

147. **Чиханов Б.** Таблицы пятизначных логарифмов чисел и тригонометрических величин. Мамонтова А. М., М., 1917.

148. **Шейнессон С.** Таблицы логарифмов, вып. V. Госиздат Киргизской ССР.

149. **Швейцер Б.** Четырехзначные таблицы логарифмов. Готье Т. В. М., 1855.

150. **Щеглов Н. Т.** Таблицы бригговых логарифмов чисел и тригонометрических линий о семи десятичных. Трей Я., Петербург, 1856.

151. **Яновский А. С.** Таблицы разностей логарифмов или логарифмов отношений. «Мосполиграф», М.,1928.

152. **Записки императорской Академии наук (Петербург),** 1867, 12, кн. 2, № 4.
Буняковский В. Я. О суммовании численных таблиц по приближению.

153. **Труды Центрального научно-исследовательского института геодезии, аэросъемки и картографии (Москва),** 1939, 30.
Таблицы для астрономических вычислений.

154. **Abbott P. W.** Mathematical tables and formulae. Longmans, Green and Co., London, 1918.

155. **Ageton A. A.** Manual of celestial navigation. Van Nostrand, New York, 1942.

156. **Valouch M.—Valouch M. A.** Pětimístné logaritmy čísel. Nakladatelství československé Akademie Věd, Praha, 1953.

157. **Airy G. B.** Greenwich observations for 1837, Appendix III. London, 1838.

158. **Albrecht C. T.** Logarithmisch-trigonometrische Tafeln mit fünf Dezimalstellen. Stankiewicz, Berlin, 1884.

159. **Albrecht C. T.** Formeln und Hilfstafeln für geographische Ortsbestimmungen. Engelmann, Leipzig, 1908.

160. **Allen E. S.** Six-place tables. McGraw-Hill Book Co., New York — London, 1947.

161. **Ambronn L.** und **Domke J.** Astronomisch-geodätische Hilfstafeln. Mittler, Berlin, 1909.

162. **Andoyer M. H.** Nouvelles tables trigonométriques fondamentales (Logarithmes). Hermann, Paris, 1911.

163. **Andoyer M. H.** Tables logarithmiques à treize décimales. Hermann, Paris, 1922.

164. **Aquino R.** Altitude and azimuth tables. Author, Rio de Janeiro; Potter, London, 1924.

165. **Attwood E. L.** Practical five-figure mathematical tables. Macmillan, London, 1950.

166. **Banhagel E.** Logarithms and slide rule for practical use. Liff-Davis, New York, 1945.

167. **Barlow P.** New mathematical tables. G. and S. Robinson, London, 1814.

168. **Bauschinger J.** und **Peters J.** Logarithmisch-trigonometrische Tafeln mit acht Dezimalstellen, enthaltend die Logarithmen aller Zahlen von 1 bis 200 000 und die Logarithmen der trigonometrischen Funktionen für jede Sexagezimalsecunde des Quadranten neu berechnet und herausgegeben, 1—2. Engelmann, Leipzig, 1936.

169. **Becker E.** Logarithmisch-trigonometrisches Handbuch auf fünf Decimalen. Tauchnitz, Leipzig, 1928.

170. **Becker G. F.** and **Van Orstrand C. E.** Hyperbolic functions. Smithsonian Institution, Washington, 1909.

171. **Beetle R. D.** Five-place logarithmic and trigonometric tables. McGraw-Hill, New York — London, 1933.

172. **Benson R. E.** Natural trigonometric functions containing the natural sine, cosine, tangent and cotangent to seven decimal places for every ten seconds of arc from 0° to 90° semi-quadrantally arranged. Benson Book Co., Los Angeles, 1941.

173. **Beyrodt G.** und **Reissner A.** Tabellenbuch zur Wirtschaftmathematik. Fachwissen Lehrbücher für die Berufsfach und Fachschule. «Volk und Wissen», Berlin—Leipzig, 1949.

174. **Bohde.** Mathematisch-technische Zahlentafeln. Springer, Berlin, 1933.

175. **Boll M.** Tables numériques universelles des laboratoires et bureaux d'étude. Dunod, Paris, 1947.

176. **Borda C.** et **Delambre J. B. J.** Tables trigonométriques decimales ou table des logarithmes des sinus, secantes et tangentes, suivant la division du quart de cercle en 100 degrés. Paris, 1801.

177. **Brandenburg H.** Siebenstellige trigonometrische Tafel alter Kreisteilung für Berechnungen mit der Rechenmaschine. Lorentz, Leipzig, 1931.

178. **Briggs H.** Arithmetica logarithmica. London, 1624.

179. **Briggs H.** and **Gellibrand H.** Trigonometria Britannica. Gouda, 1633.

180. **Bremiker C.** Tafeln vierstelliger Logarithmen. Weidmann, Berlin, 1907.

181. **Bremiker C.** Logarithmisch-trigonometrische Tafeln mit sechs Decimalstellen. Berlin, 1909. (Англ. изд. 1924).

182. **Bremiker C.** Logarithmisch-trigonometrische Tafeln mit fünf Decimalstellen. Weidmann, Berlin, 1937.

183. **Brown's Nautical Almanac for 1935.** Brown, Son and Ferguson, Glasgow, 1935.

184. **Bruhns C. C.** Neues logarithmisch-trigonometrisches Handbuch auf sieben Decimalen. Leipzig, 1881.

185. **Burton S. M.** A set of nautical tables. Philip G., London, 1941.

186. **Byrne O.** Practical, short and direct method of calculating the logarithms of any given number and the number corresponding to any given logarithm. Appleton, London, 1849.

187. **Caillet** V. Tables des logarithmes et cologarithmes des nombres et des lignes trigonométriques à six décimales... Challamelaine, Paris, 1890.

188. **Callet** F. Tables portatives de logarithmes, contenant les logarithmes des nombres, depuis 1 jusqu'a 108 000; les logarithmes des sinus et tangentes... F. Didot, Paris, 1795.

189. **Carey** F. S. and **Grace** S. F. Four-place mathematical tables with forced decimals. Longmans, London, 1927.

190. **Carmichael** R. D. and **Smith** E. R. Mathematical tables and formulas. Ginn, Boston — New York, 1931.

191. **Chauvenet** W. A manual of spherical and practical astronomy, 1—2. Lippincott, Philadelphia, 1900.

192. **Chappell** E. Five-figure mathematical tables. Chambers, London; Van Nostrand, New York, 1915.

193. **Clark** J. B. Mathematical and physical tables. Oliver and Boyd, Edinburgh, 1950.

194. **Cohn** B. Tables of addition and substraction logarithms with six decimals. Scientific Computing Service, London, 1939.

195. **Comrie** L. J. Chambers's four-figure mathematical tables. Chambers W. and R., Edinburgh—London, 1947.

196. **Comrie** L. J. Four-figure tables of the natural and logarithmic trigonometrical function with the argument in time. Author, London, 1931.

197. **Crüger** P. Praxis trigonometriae logarithmicae cum logarithmorum tabulis. Amsterdam — Danzig, 1634.

198. **Dale** J. B. Five-figure tables of mathematical functions comprising tables of logarithms, powers of numbers, trigonometric, elliptic and other transcendental functions. Arnold, London, 1949.

199. **Danjon** A. Tables des fonctions trigonométriques. Hachette, Paris, 1949.

200. **Dase** Z. Tafeln der natürlichen Logarithmen der Zahlen. Sommer, Vienna, 1850.

201. **Davis** H. T. Tables of the higher mathematical functions, I. Indiana Principia Press, Bloomington, 1933.

202. **Davis** P. L. H. Requisite tables... Potter, London, 1905.

203. **Deckert** A. Die Logarithmen der trigonometrischen Funktionen und der natürlichen Zahlen. Druck Kösel und Pustet.

204. **Deprez** F. Tables pour le calcul à la machine des logarithmes à 13 décimales. Francke, Berne, 1939.

205. **Des Vollstandigen Mathematischen,** 2. Leipzig, 1742.

206. **Desvallees** H. R. Tables logarithmiques et trigonométriques à quatre décimales. Gauthier-Villars, Paris, 1920.

207. **Dietrichkeit** O. Siebenstellige Logarithmen und Antilogarithmen aller vierstelligen Zahlen und Mantissen von 1000—9999 bezw., 0000—,9999 mit Rand-Index und Interpolationseinrichtung für vier bis sieben-stelliges Schnellrechnen. Springer, Berlin, 1903.

208. **Dodson** J. The calculator: being correct and necessary tables for computation. London, 1747.

209. **Dodson** J. The antilogarithmic canon, being a table of numbers, consisting of eleven-places of figures, corresponding to all logarithms under 100 000. London, 1742.

210. **Domke** F. Nautische astronomische und logarithmische Tafeln nebst Erklärung und Gebrauchs-Anweisung für die Koniglich-Preussischen Navigations-Schulen. R. Decker, Berlin, 1892.

211. **Douglas** G. Mathematical tables, containing the logarithms of numbers; tables of sines. Edinburgh, 1809.

212. **Dreisonstok** J. Y. Navigation tables for mariners and aviators. U. S. Hydrographic Office, Washington, 1935.

213. **Duarte** F. J. Nouvelles tables de log n! à 33 décimales depuis $n=1$ jusqu'à $n=3000$. Kundig, Geneva; Index Generalis, Paris, 1927.

214. **Duarte** F. J. Nouvelles tables logarithmiques à 36 décimales. Gauthier-Villars, Paris, 1933.

215. **Duffield** W. W. Logarithms, their nature computation and uses, with logarithmic tables of numbers and circular functions to 10 decimal places. Washington, 1897.

216. Dupuis J. Tables de logarithmes à sept décimales. Hachette, Paris, 1885.

217. Dwight H. B. Tables of integrals and other mathematical data. Macmillan, New York, 1947.

218. Dwight H. B. Mathematical tables. McGraw-Hill, New York — London, 1941.

219. Eilmann M. Eine neue leichte Divisions- und Multiplikationsmethode, um Briggische Logarithmen in 13 Dezimalstellen oder ihre zugehörende Absolutzahlen eins durchs andere zu finden. Blothe, Osnabrück, 1804.

220. Eschke. Mathematische Tabellen und Formeln. Mayer W., Berlin [193...].

221. Farley R. Natural versed sines for 0° to 125° and logarithmic versed sines from 0° to 135° of 0^h to 9^h used in computing lunar distances for the nautical almanac. London, 1856.

222. Farley R. Tables of six-figure logarithms. Longman, London, 1840.

223. Ferrol F. Das Ferrol'sche neue Rechnungsverfahren. Huthmacher, Bonn, 1913.

224. Filipowski H. E. A table of antilogarithms. Bell, London, 1851.

225. Fisher R. and Yates F. Statistical tables for biological, agricultural and medical research. Oliver and Boyd, London — Edinburgh, 1949.

226. Formular matematic şi tehnic. Pentru elevi, studenti şi tehnicieni. Editura tehnica, Bucuresti, 1949.

227. Gardiner W. Tables of logarithms for all numbers from 1 to 102.100 and for the sines and tangents to every ten seconds of the quadrant: as also for the sines of the first 72 minutes to every single second. London, 1742.

228. Gauss K. F. Werke, 1876, 3, 244—246.
Table of addition and subtraction logarithms.

229. Gauss F. G. Fünfstellige vollständige trigonometrische und polygonometrische Tafeln für Maschinenrechnen. Wittwer, Stuttgart, 1925.

230. Gauss F. G. Vierstellige logarithmische- und trigonometrische Tafeln (Sexagesimal unterteiler Altgrad). «Volk und Wissen», Berlin — Leipzig, 1946.

231. Gauss F. G. Fünfstellige vollständige logarithmische und trigonometrische Tafeln. Strien, Halle, 1908.

232. Gauss F. G. Fünfstellige logarithmisch-trigonometrische Tafeln für Dezimalteilung des Quadranten. Wittwer, Stuttgart, 1937.

233. Elznic V. a Valouch M. Geoma. Pětimístné tabulky hodnot a logaritmů goniometrických funkcí v setinněm dělení. tabulky geodetické a katastrálni. Vydala jednota českych matematiků a fisicũ, Praha, 1944.

234. Gernerth A. Fünfstellige gemeine Logarithmen der Zahlen und der Winkelfunktionen von 10 zu 10 Sekunden nebst der Proportionaltheilen ihrer Differenzen. Beck, Wien, 1866.

235. Glaisher J. Factor table for the sixth million, containing the least factor of every number not divisible by 2.3 or 5 between 5 000 000 and 6 000 000. Taylor, London, 1883.

236. Glover J. W. Tables of applied mathematics in finance, insurance and statistics. Wahr, Ann Arbor, Michigan, 1930.

237. Gonggrijp B. Logarithmische en goniometrische tafels en bijtafels. Noordhoff, Groningen, 1916.

238. Granville W. A. Four-place tables of logarithms. Ginn, Boston, 1908.

239. Gravelius H. Fünfstellige logarithmisch-trigonometrische Tafeln für die Decimaltheilung des Quadranten. Reimer, Berlin, 1886.

240. Gray P. Tables for the formation of logarithms and antilogarithms to twenty-four or any less number of places. Layton, London, 1876.

241. Gray P. Tables and formulae for the computation of life contingencies. Longman, London, 1870.

242. Gray P. Tables for the formation of logarithms and antilogarithms to twelve places. Layton, London, 1865.

243. Gray P. A table for the easy formation of antilogarithms to 12

or any less number of places. Tyler and Reed, London, 1848.

244. Guillemin A. Tables de logarithmes à 3 quatrades et nombres correspondants avec 12—13 chiffres. Gauthier-Villars, Paris, 1912.

245. Gundelfinger S. Sechsstellige Gaussische und siebenstellige gemeine Logarithmen. Veit, Leipzig, 1902.

246. Gundelfinger S. und Nell A. M. Tafeln zur Berechnung neunstelliger Logarithmen. Bergsträsser, Darmstadt, 1891.

247. Hammer E. Sechsstellige Tafel der Werte $\log_{10} \frac{1+x}{1-x}$. Teubner, Leipzig, 1902.

248. Burington R. S. Handbook of mathematical tables and formulas. Sandusky, Ohio, 1946.

249. Hannyngton J. C. Haversines natural and logarithmic used in computing lunar distances for the nautical almanac. H. M. Stationery Office, London, 1876.

250. Hannyngton J. C. Table of logarithms and antilogarithms (four figures) 1 to 10 000. Layton, London, 1912.

251. Hantschl J. Logarithmisch-trigonometrisches Handbuch. Rohrmann und Schweigerd, Wien, 1833.

252. Hassler F. K. Tables logarithmiques et trigonométriques a sept décimales. Carvill, New York, 1834.

253. Hayashi K. Fünfstellige Funktionentafeln. Springer, Berlin, 1930.

254. Hayashi K. Sieben-und mehrstellige Tafeln der Kreis- und Hyperbelfunktionen und deren Produkte sowie der Gammafunktion. Springer, Berlin, 1926.

255. Hedrick E. R. Logarithmic and trigonometric tables Macmillan, New York, 1920.

256. Herz N. Siebenstellige Logarithmen der trigonometrischen Funktionen für jede Zeitsekunde. Teubner, Leipzig, 1885.

257. Hobert J. P. et Ideler L. Nouvelles tables trigonométriques calculées pour la division décimale du quart de cercle. Berlin, 1799.

258. Hof H. Seven place full natural trigonometric tables. Tangent. Michigan Univ. Research Association, Ann Arbor, 1943.

259. Holtappel H. W. Tafels van e^x. Noordhoff, Groningen, 1938.

260. Hoppe R. Tafeln zur dreißigstelligen logarithmischen Rechnung. Koch, Leipzig, 1876.

261. Hoüel J. Recueil de formules et de tables numériques. Gauthier-Villars, Paris, 1868; 1901.

262. Hoüel J. Tables de logarithmes à cinq décimales pour les nombres et les lignes trigonométriques. Gauthier-Villars, Paris, 1858.

263. Hudson L. and Mills E. S. Natural trigonometric functions. Tables: sine, cosine, tangent, cotangent, secant and cosecant to eight decimal places with second differences to ten decimal places semi quadrantally arranged. Monroe Calculating Machine Co, Orange, New York, 1941.

264. Hughes—Comrie L. J. Tables for sea and air navigation. Henry Hughes and Son, London, 1938.

265. Hunter J. Auxiliary tables of the Survey of India, II: Mathematical tables. Government of India, Dehra Dun, 1921.

266. Huszar G., Kreko B. Táblázatok a gazdasági szamtanhoz. Tankönyvkiadó, Budapest [195...].

267. Hütte. Des Ingenieurs Taschenbuch, I. Ernst, Berlin, 1931.

268. Hutton C. Mathematical tables. London, 1811.

269. Ingenieurs-schul, I. Nürenberg, 1637.

270. Tavole logaritmiche a cinque cifre decimali. Italy Istituto Idrografico della R. Marina. Genoa, 1916.

271. Tavole logaritmiche a cinque cifre decimali, 6. Italy Istituto Idrografico della R. Marina. Genoa, 1941.

272. Ives H. C. Mathematical tables reprinted from Searles and Ives field engineering, with additions. Wiley, New York; Chapman and Hall, London, 1945.

273. Ives H. C. Natural trigonometrical functions to seven decimal places for every ten seconds of arc. Wi-

ley, New York; Chapman and Hall, London, 1945.

274. **Ives H. C.** Seven place natural trigonometrical functions. Wiley, New York; Chapman and Hall, London, 1929.

275. **Jäderin E.** Tables abregées pour le calcul des logarithmes vulgaires à 12 décimales. Norstedt, Stockholm, 1914.

276. **Jahn G. A.** Tafeln der sechsstelligen Logarithmen für die Zahlen 1 bis 100 000 für die Sinus und Tangenten, I — II. Voigt und Fernau, Leipzig, 1844.

277. **Johnson F. W.** Non-interpolating logarithms, cologarithms and antilogarithms. The Simplified Series Publishing Co., San Francisco, 1930.

278. **Johnson F. W.** Easily interpolated trigonometric tables with non-interpolating logs, cologs and antilogs. The Simplified Series Publishing Co., San Francisco, 1933

279. **Jones G. W.** Logarithmic tables. Macmillan, London, 1893.

280. **Jordan W. — Eggert O.** Logarithmisch-trigonometrische Tafeln für neue (zentesimale) Teilung mit sechs Dezimalstellen. Jordan, Stuttgart, 1894.

281. **Jordan W.** Mathematische und geodätische Hilfstafeln. Hannover, 1895.

282. **Kahlau R.** Vierstellige Tafeln zum logarithmischen und numerischen Rechnen. Bonness und Hachfeld, Potsdam, 1944.

283. **Kaye G. W. C.** and **Laby T. H.** Four-figure mathematical tables. London, 1925.

284. **Kekkonen A., Järvi H.** Matematiikan Perussäännöt ja Kaavot. Karjalan Kommunistinen Maatalouskorkeakoulu, Petroskoi, 1936.

285. **Kepler J.** Chilias Logarithmorum... Marburg, 1624.

286. **Keturzenkles logaritmu ir kitu dydziu lenteles. Matematines formules.** Valstybine Pedagogines Literaturos Leidykla, Kaunas, 1946.

287. **Kleyer A.** Fünfstellige korrekte logarithmische Tafeln nebst einer trigonometrischen Tafeln und einer Anzahl von anderen Tabellen. Maier J., Stuttgart, 1883.

288. **Köhler H. G.** Logarithmisch-trigonometrisches Handbuch. Tauchnitz, Leipzig, 1876, 1898

289. **Knott C. G.** Four-figure mathematical tables. Chambers, London—Edinburgh, 1905.

290. **Küstner H.** Fünfstellige logarithmen. «Volk und Wissen» Berlin — Leipzig, 1952.

291. **La Lande J.** Logarithmisch-trigonometrische Tafeln. Tauchnitz, Leipzig, 1844.

292. **Lambert J. H.** Zusätze zu den logarithmischen und trigonometrischen Tabellen. Haude und Spener, Berlin, 1770.

293. **Lambert J. H. — Felkel A.** Supplementa tabularum logarithmicarum et trigonometricarum... Lisbon, 1798.

294. **Legendre A. M.** Exercices de calcul integral, 3. Huzard-Courcier, Paris, 1816.

295. **Legendre A. M.** Traité des fonctions élliptiques, 2. Huzard-Courcier, Paris, 1826.

296. **Leonelli G. Z.** Supplement logarithmique. Bordeaux, 1803.

297. **Lienard R.** Tables fondamentales a 50 décimales des sommes S_n, u_n, Σ_n. Centre de Documentation Universitaire, Paris, 1948.

298. **Ligowski W.** Sammlung fünfstelliger logarithmischer, trigonometrischer... Tafeln. Toeche, Kiel, 1873.

299. **Loky B.** Négyjegyü függvénytáblázatok. Budapest, 1950.

300. **Lötzbeyer P.** Vierstellige Tafeln zum logarithmischen und Zahlenrechnen. Graphische Rechentafeln. Teubner, Leipzig—Berlin, 1928.

301. **Lynn T.** Horary tables. London, 1827.

302. **Maseres F.** Scriptores logarithmici, or a collection of several curious tracts on the nature and construction of logarithms, 1—6. 1791—1807.

303. **Matthiessen E. A.** Tafeln zur bequemern Berechnung des Logarithmen der Summe oder Differenz... . Hammerich, Altona, 1824.

304. McLaren G. C. Improved four-figure logarithm table. Cambridge Univ. Press, 1915.

305. Meldola A. Logarithmische Tafeln der Nummer-Logarithmen (1—10 000), der Sinus und Tangenten. Aue, Altona, 1840.

306. Mendizabal (y) Tamborrel J. Tables des logarithmes à huit decimales des nombres de 1 à 125 000 et des fonctions goniométriques. Hermann, Paris, 1891.

307. **Metals Handbook**, 1948, 196—199. Smith C. S. The intercommunication of atomic and weight percentages.

308. Meyer U. und Deckert A. Tafeln der Hyperbelfunktionen. Druck Kösel und Pustet, Kempten, 1924.

309. Miller F. H. Calculus. Wiley, New York, 1946.

310. Milne-Thomson L. M. and Comrie L. J. Standard four-figure mathematical tables. Macmillan, London, 1948.

311. **Miscellaneous physical tables. Planck's radiation funkctions and electronic functions.** Prepared by the Mathematical Tables Project of the Work Projects Administration for the City, New York, 1941.

312. Molterer J. Tabellen zur Zahlentheorie 52-stellige Briggische Logarithmen von 1—1100 und Potenzen der Zahl 2 bis 2^{400}. Welsermuhl, Wels, 1937.

313. Montferrier A. S. L'arithmétique des logarithmes ou application usuelle des logarithmes aux calcules numériques et géométriques, avec une table des logarithmes des nombres de 1 à 10.000. Paris, 1853.

314. Müller O. — Rajna M. Tavole di logaritmi con cinque decimali. Hoepli, Milan, 1935.

315. Müller O. Hilfstafeln für praktische Messkunde nebst logarithmisch-trigonometrischen Tafeln. Zürich, 1897.

316. Namur A. and Mansion P. Tables de logarithmes à 12 décimales. Gauthier-Villars, Bruxelles, 1877.

317. Napier J. Mirifici Logarithmorum Canonis Descriptio. Edinburgh, 1614.

318. Napier J., Filipowski H. The wonderful canon of logarithms. Lizars, Edinburgh, 1857.

319. Napier J., Wright E. A description of the admirable table of logarithmes... . London, 1618.

320. **Nautical Almanac** (abridged). The nautical almanac abridged for the use of seamen for the year 1936. Stationery Office, London, 1936.

321. **Nautical Almanac Office** (NAO). Five-figure tables of natural trigonometrical functions. War Office, London, 1941.

322. **Extended tables of Fresnel integrals.** U. S. Navy, Naval Research Laboratory. Boston, 1948.

323. Neper J. Arithmétique logarithmique ou la construction et usage d'une table contenant des logarithmes de tous les nombres de puis l'unite jusques'a 100 000 Goude, 1628.

324. Newcomb S. Logarithmic and other mathematical tables. Holt, New York; Bell, London, 1895.

325. Newton J. Trigonometria Britannica: or the doctrine of triangles in two books... A table of logarithms to 100.000 there to annexed, with the artificial sines and tangents, to the hundredth part of every degree, and the first three degrees to a thousand parts. London, 1658.

326. Vlacq A. Nieuwe konstige tafelen, sinuum, tangentium et secantium... . Amsterdam, 1658.

327. Valouch M.—Valouch M. A. Sedmimístné logaritmy čísel od 1 do 120 000. Nakladatelství Československé Akademie Věd, Praha, 1953.

328. Oughtred W. Trigonometria. R. Stokes and A. Haughton, London, 1657.

329. Parkhurst H. M. Astronomical tables... . Author, New York, 1889.

330. Peirce B. O. A short table of integrals. Ginn, Boston, 1910.

331. Peters C. F. W. Astronomische Tafeln und Formeln. Mauke, Hamburg, 1871.

332. Peters J. Einundzwanzigstellige Werte der Funktionen Sinus und Cosinus. Reimer, Berlin, 1911.

333. Peters J. Logarithmic table to seven places of decimals of the trigonometrical functions for every second of the quadrant. Engelmann, Leipzig, 1911.

334. Peters J. Fünfstellige Logarithmentafel der trigonometrischen Funktionen für jede Zeitsekunde des Quadranten. Reimer, Berlin, 1912.

335. Peters J. Hilfstafeln zur Zehnstelligen Logarithmentafeln. Preussische Landesaufnahme, Berlin, 1919.

336. Peters J. Sechsstellige Logarithmen der trigonometrischen Funktionen von 0° bis 90° für jedes Tausendstel des Grades. Preussische Landesaufnahme, Berlin, 1921.

337. Peters J. Siebenstellige Logarithmen der trigonometrischen Funktionen von 0° bis 90° für jedes Tausendstel des Grades. Preussische Landesaufnahme, Berlin, 1921.

338. Peters J. Zehnstellige Logarithmentafel, I: Zehnstellige Logarithmen der Zahlen von 1 bis 100.000 nebst einem Anhang mathematischer Tafeln. Reichsamt für Landesaufnahme, Berlin, 1922.

339. Peters J. Zehnstellige Logarithmentafeln, II: Zehnstellige Logarithmen der trigonometrischen Funktionen von 0° bis 90° für jedes Tausendstel des Grades. Preussische Landesaufnahme, Berlin, 1919.

340. Peters J. Sechsstellige Tafel der trigonometrischen Funktionen, enthaltend die Werte der sechs trigonometrischen Funktionen von zehn zu zehn Bogensekunden.... Dümmler, Berlin — Bonn, 1929.

341. Peters J. Siebenstellige Logarithmentafel, I: Logarithmen der Zahlen, Antilogarithmen, Additions und Subtraktionslogarithmen... Reichsamt für Landesaufnahme, Berlin, 1940.

342. Peters J. Siebenstellige Logarithmentafel, II: Logarithmen der trigonometrischen Funktionen für jede zehnte Sekunde des Neugrades... Reichsamt für Landesaufnahme, Berlin, 1940.

343. Pineto S. Tables de logarithmes vulgaires à dix décimales construites d'après un nouveau mode. Acad. Imp. Sci., St. Petersbourg, 1871.

344. Pinto J. C. The «Simplex» navigation and avigation tables. Azores, Fayal, 1933.

345. Plummer H. C. Four-figure tables with mathematical formulae. Macmillan, London, 1941.

346. Potin L. Formules et tables numériques. Doin et Gauthier-Villars, Paris, 1925.

347. Pryde J. Chambers's educational course. Mathematical tables. Chambers, London — Edinburgh, 1930.

348. Prytz H. Tables d'antilogarithmes. Lehmann and Stage, Copenhagen, 1886.

349. Raper H. Nautical tables. Potter, London, 1920.

350. Ratassepp K. Matemaatilised tabelid. Pedagoogiline Kirjandus, Tallin, 1948.

351. Rheticus G. Magnus canon... in opus palatinum... . V. Otho, Neustadt, 1596.

352. Larsen H. D. Rinehart mathematical tables, formulas and curves. Chapman and Hall, London, 1949.

353. Rivard M. Tables des sinus, tangentes, secantes et de leurs logarithmes, avec la construction de ces tables, et les problèmes de la trigonométrie rectiligne et sphérique. Desaint, Saillant, Paris, 1743.

354. Roe N., Wingate E. Tabulae logarithmicae. London, 1633.

355. Rohr M. Die Logarithmen der Sinus und Tangenten für 0° bis 5° und der Cosinus und Cotangenten für 85° bis 90° von Tausendstel zu Tausendstel Grad. Weidmann, Berlin, 1900.

356. Rohrbach C. Vierstellige logarithmisch-trigonometrische Tafeln. Ehlermann L., Dresden, 1929.

357. Roman W. Tables numériques pour les calculs scientifiques. Rr. W. Junk — E. Ditcurs — La Haye, 1950.

358. Rosenbach J. B., Whitman E. A., Moskovitz D. Mathematical tables. Ginn, Boston, 1946.

359. Rubey H. Engineering surveys logarithmic and trigonometric tables. Macmillan, New York, 1941

360. **R ü h l m a n n M.** Logarithmisch-trigonometrische und andere... Tafeln. Arnold, Leipzig, 1837.

361. **S a l o m o n J. M.** Logarithmische Tafeln... Gerold, Vienna, 1827.

362. **S a n g E.** A new table of seven-place logarithms of all numbers from 20.000 to 200.000. Layton, London, 1915.

363. **S a s s e n f e l d H. M.** und **T s c h u n k o H. F. A.** Mathematische Tafeln für Mathematiker, Naturwissenschaftler, Ingenieure. Erster Teil: Elementare Funktionen. Fr. Lamade, Walldorf bei Heidelberg, 1949.

364. **S c h e u t z J.** and **E.** Speciments of tables, calculated, stereomoulded and printed by machinery. London, 1857.

365. **S c h r ö n L.** Siebenstellige gemeine Logarithmen der Zahlen von 1 bis 108.000 und der Sinus, Cosinus, Tangenten und Cotangenten. Vieweg, Braunschweig, 1875, 1878.

366. **S c h u b e r t H.** Vierstellige Tafeln und Gegentafeln für logarithmisches und trigonometrisches Rechnen. W. Gruyter, Berlin, 1945

367. **S c h ü l k e A.** Tafeln. Vierstellige Logarithmen Funktions- und Zahlenwerte. «Volk und Wissen», Leipzig, 1949.

368. **S c h u l t e - T i g g e s R. A.** Rechentafeln für höhere Lehranstalten insbesonders zu Mehlers Hauptsätzen der Elementar-Mathematik. W. Gruyter, Berlin — Leipzig, 1929.

369. **S c h u l z e J. C.** Neue und erweiterte Sammlung logarithmischer, trigonometrischer und anderer... Tafeln, 1 и 2. Mylius, Berlin, 1778.

370. **S c o t t E. E.** A short table of logarithms and antilogarithms to ten places... Layton, London, 1897.

371. **S c o t t E. E.** Tables of logarithms and antilogarithms to five places. Layton, London, 1912.

372. **S c o t t E. E.** Tables of logarithms and antilogarithms to five places... to which is added a table for finding logarithms and antilogarithms to ten places... Layton, London, 1870.

373. **S e m i l l e r H.** und **S e m i l l e r A.** Vierstellige Logarithmen- und Zahlentafeln. Springer, Berlin, 1938.

374. **Nouvelles tables de logarithmes à cinq décimales...** Service Géographique de l'armée. Gauthier-Villars, Paris. 1939.

375. **Tables des logarithmes à huit décimales...** Service Géographique de l'armée. Gauthier-Villars, Paris, 1891.

376. **S h a r p A.** Geometry improv'd... by A. S. Philomath (nom de plume). London, 1717.

377. **S h o r t r e d e R.** Logarithmic tables..., containing logarithmic sines and tangents... Black, Edinburgh, 1849.

378. **S h o r t r e d e R.** Logarithmic... tables. Black, Edinburgh, 1844.

379. **S h o r t r e d e R.** Logarithmic tables, containing logarithms to numbers... Black, Edinburgh, 1849.

380. **Siebenstellige Logarithmen und Antilogarithmen aller vierstelligen Zahlen und Mantissen von 1000—9999.** Springer, Berlin, 1903.

381. **S i l b e r s t e i n L.** Synopsis of applicable mathematics, with tables. Bell, London, 1923.

382. **S m a r t W. M.** and **S h e a r m e F. N.** Position line tables (line method). Potter, London, 1924.

383. **S m o l e y C. K.** Segmental functions. C. K. Smoley and Sons, Scranton, 1937.

384. **S o k a l S. i T o r b u s W.** Tablice matematyczno-fizyczne do użytku szkoł zawodowych. Państwowe wyd-wa szkolnictwa zawodowego, Warszawa, 1951.

385. **S p e n c e l e y G. W., S p e n c e l e y R. M.** and **E p p e r s o n E. R.** Smithsonian logarithmic tables to base *e* and base 10. Smithsonian Instn, Washington, 1952.

386. **S p e i d e l l J.** New logarithmes. London, 1619.

387. **S p e r o t t i E.** I logaritmi per i Ragionieri: Tavole dei logaritmi volgari a Otto decimali dei numeri interi da 1 a 100.000. Capelli, Rocca San Casciano, 1911.

388. **S t a l m a s e k J.** Tabul'ky funkcií a konstant z matematiky,

fyziky a chemie. Ŝtátne. nakl., Bratislava, 1951.

389. **Standard field tables and trigonometric formulas.** United States Government Printing Office, Washington, 1939.

390. **S t e g m a n n F.** Tafel der natürlichen Logarithmen auf fünf Decimalen und der Gaussischen Logarithmen. Koch, Marburg, 1856.

391. **S t e i n h a u s e r A.** Anhang zu allen deutschen Ausgaben von Logarithmen Tafeln... Beck, Vienna, 1857.

392. **S t e i n h a u s e r A.** Kurze Hilfstafel zur bequemen Berechnung fünfzehnstellige Logarithmen... Beck, Vienna, 1865.

393. **S t e i n h a u s e r A.** Hilfstafeln zur präcisen Berechnung zwanzigstelliger Logarithmen... Gerold, Vienna, 1880.

394. **S t u d n i c k a F. Y.** Kapesní logarithmické tabulky. Praha, 1870.

395. **Tabele şi formule matematice.** Ed. tehnica, Budapeşti, 1951.

396. **Tabellen der sinuum, tangentium und secantium.** Amsterdam, 1689.

397. **Tables de logarithmes pour les sinus et tangentes de toutes les minutes du quart de cercle, et pour tous es nombres naturels depuis 1 jusqu'à** 20.000. Desaint, Paris, 1768.

398. **R o b b i n s C. K.** and **L i t t l e N.** Tables for Calculus. The Macmillan Co., New York, 1942.

399. **La C a i l l e N. L., L a L a l a n d e M.** Tables de logarithmes pour les sinus et tangentes de toutes les minutes du quart de cercle, et pour tous les nombres naturels depuis 1 jusqu'à 21. 600. Barrois l'áiné. Paris, 1804.

400. **G r e v e A.** Fünfstellige logarithmische und trigonometrische Tafeln nebst einer grösseren Anzahl von Hilfstafeln. Velhagen und Klasing, Bielefeld — Leipzig, 1893.

401. **Tables des logarithmes à huit décimales, I — II.** I: Logarithmes des nombres entièrs de 1 à 120 000. II: Logarithmes des fonctions circulares. Paris, 1944.

402. **Table of natural logarithms, I.** Logarithms of the integers from 1 to 50 000. National Bureau of Standards, New York, 1941.

403. **Table of natural logarithms, II.** Logarithms of the integers from 50.000 to 100.000. National Bureau of Standards, New York, 1941.

404. **Table of natural logarithms, III.** Logarithms o. the decimal numbers from 0.0001 to 5.0000. National Bureau of Standards, New York, 1941.

405. **Table of natural logarithms, IV.** Logarithms of the decimal numbers from 5.0000 to 10.0000. National Bureau of Standards, New York, 1941.

406. **Table of the first ten powers of the integers from 1 to 1000.** Prepared by the Mathematical Tables Project of the Work Projects Administration for the City, New York, 1939.

407. **T h i e l e W.** Tafel der Wolframschen hyperbolischen 48-stelligen Logarithmen. Dünnhaupt, Dessau, 1908.

408. **T h o m a n F.** Tables de logarithmes à 27 décimales... Paris, 1867.

409. **U h l e r H. S.** Original tables to 137 decimal places of natural logarithms for factors of the form $1 \pm n \cdot 10^{-p}$ enhanced by auxiliary tables of logarithms of small integers. Author, New-Haven, 1942.

410. **Uj Hétszámjeqyu Logarithms Kezikönyv.** Honved Térképészeni Intézet, Budapest, 1952.

411. **Tables of *S* and *T* values to ten places.** United States Coast and Geodetic survey. Washington, 1934.

412. **U r s i n u s B.** Magnus canon triangulorum logarithmicus. Coloniae, 1624.

413. **U r s i n u s B.** Cursus mathematici practici... Coloniae, 1618.

414. **U r s i n G. F.** Logarithmi VI decimalium. Author, Copenhagen, 1827.

415. **V e g a G.** Thesaurus Logarithmorum Completus... (Vollständige Sammlung grösserer logarithmisch-trigonometrischer Tafeln...). Weidmann, Leipzig, 1794.

416. **V e g a G.** Logarithmisch-trigonometrische Tafeln... (Tabulae Logarithmico-trigonometricae...), **1—2.** Weidmann, Leipzig, 1797.

417. **V e g a G.** Logarithmische, trigonometrische und andere... Tafeln und Formeln. Vienna, 1783.

418. **V e g a G. — H ü l s s e J. A.** Sammlung mathematischer Tafeln.. Weidmann, Leipzig, 1849.

419. **V e g a G. — B r e m i k e r C.** Logarithmisch-trigonometrisches Handbuch. Weidmann, Berlin, 1856.

420. **V e g a G.** 10 place logarithms. Including Wolfram's tables of natural logarithms. Reprint of the rare of 1794. Stechert, New York, 1946.

421. **S t r a u c h i i A.** Tabellen der Sinuum, Tangentium, Logarithmorum... Amsterdam, 1700.

422. **Vierstellige Logarithmen der trigonometrischen Functionen in Zeit ausgedrückter Winkel.** Petersburg, 1882.

423. **V l a c q A.** Tabulae sinuum, tangentium, et secantium, et logarithmi sinuum, tangentium, et numerorum ab unitate ad 10 000. Francofurti et Lipsiae, 1726.

424. **V l a c q A.** Arithmetica Logarithmica... Gouda, 1628.

425. **W a c k e r b a r t h A. F. D.** Fern-ställiga Logarithm-Tabeller... Upsala, 1867.

426. **W e e m s P. V. H.** Line of position book. Annapolis, 1927.

427. **W e i d e n b a c h.** Tafel um den Logarithmen von $\frac{x+1}{x-1}$ zu finden. Copenhagen, 182 9.

428. **W e i s b a c h J.** Der Ingenieur Sammlung von Tafeln, Formeln und Regeln... Braunschweig, 1863.

429. **W e s t p h a l J. H.** Logarithmische Tafeln. Königsberg, 1821.

430. **W e s t r i c k F. A.** Fünfstellige Logarithmen. Münster, Westphalia, Aschendorff, 1892.

431. **W i j d e n e s P.** Five-place tables. Noordhoff, Groningen, 1937.

432. **W i t k o w s k i A. W.** Tablice matematyczno-fizyczne. Orgelbrand, Warszawa, 1904.

433. **W i t t k e H.** Vademekum für Vermessungstechnik. Metzlersche Verlagsbuchhandlung, Stuttgart, 1948.

434. **W i t t s t e i n T.** Siebenstellige Gaussische Logarithmen... Hahn, Hanover, 1866.

435. **W o l f R.** Handbuch der Mathematik, Physik, Geodäsie und Astronomie, I. Zürich, 1870.

436. **W o l f e J. H.** and **P h e l p s E. R.** Natural trigonometric functions with differences and rapid and easy method of interpolation. Authors, Detroit, Michigan, 1942.

437. **L o m n i c k i A.** Tablice matematyczno-fizyczne czterocyfrowe. Dla uzytku szkoł srednich. Ksiaznica-Atlas, Lwow — Warszawa, 1932.

438. **Z a c h a r i a s M.** und **M e t h P.** Vierstellige Logarithmentafeln. W. Gruyter, Berlin — Leipzig, 1927.

439. **Z e c h J.** Tafeln der Additions- und Substractions Logarithmen für sieben Stellen. Weidmann, Berlin, 1910.

440. **Academie Royale de Belgique. Bulletin classe des Sciences** (Bruxelles), 1875, (2), **40,** 74—139, 455—524.
H o u z e a u J.· C. Fragments sur le calcul numérique.

441. **Archiv für Electrotechnik,** 1927, **17,** 79—101.
Potentialtheorie der Hängeisolatoren.

442. **Archiv der Mathematik und Physik,** 1864, **41,** № 32, 305.
S p i t z e r S. Lösung einer Aufgabe der Variations-Rechnung.

443. **Astronomische Nachrichten** (Altona), 1826, **4,** 53.
V i e t z. Natürliche Logarithmen.

444. **Biometrics,** 1949, **5,** № 2, 127—143.
M a t h e r K. M. The analysis of extinction time data in bioassay.

445. **Bulletin Astronomique** (Paris), 1922, (2), **2,** 5—32.
A n d o y e r M. H. Tables fondamentales pour les logarithmes d'addition et de soustraction.

446. **The Collected Mathematical Papers,** 1897, **12,** 328—336.
C a y l e y A. On the orthomophosis of the circle into the parabola.

447. **Journal für die reine und angewandte Mathematik** (Berlin — Leipzig), 1832, **9,** 81—96, 193—208, 297—304, 362—378.
G u d e r m a n n C. Theorie der potential- oder cyklisch-hyperbolischen Funktionen.

448. **Journal of the Institute of Actuaries and Assurance Magazine,** 1884, **24,** 251.
M c K e n z i e D. J. On

an easy method of forming logarithms and antilogarithms correct to ten or eleven places, with the aid of Mr. Peter Gray's tables for the formation of logarithms and antilogarithms to twelve places.

449. **Journal of the Institute of Actuaries and Assurance Magazine**, 1887, 26, 55.
On logarithms correct to ten places of decimals.

450. **Journal of Physical Chemistry**, 1942, (6), 46, № 6, 641—642.
Sherman J. and Ewell R. B. A six-place table of the Einstein functions.

451. **Memoirs of British Astronomical Association** (London), 1934, **30**, № 3, 107—148.
Foxell J. T. Lunar occultation maps.

452. **Messenger of Mathematics.** 1873, 3,66—92.
Wace H. On the calculation of logarithms.

453. **Philosophical Magazine** (London), 1930, (7), **10**, 110—127.
Wright C. E. Note of the potential and attraction of rectangular bodies.

454. **Philosophical Magazine** (London), 1878, (5), 5, 241—267.
Ayrton W. E. and Perry J. Experiments on the heat-conductivity of Stone, based on Fourier's theorie de la Chaleur.

455. **Philosophical Transactions of the Royal Society**, 1714, 29, № 339, 52—55.
Long J. A new method for making logarithms and vice versa.

456. **Proceedings of the National Academy of Sciences**, 1943, 29, 319—325.
Uhler H. S. Natural logarithms of small prime numbers.

457. **Proceedings of the National Academy of Sciences**, 1938, 24, 23—30.
Uhler H. S. Log π and other basic constants.

458. **Proceedings of the Royal Society** (London), 1878, (A), 27, 88—94.
Adams J. C. Note on the value of Euler's constant, likewise the values of the Napierian logarithms of 2, 3, 5, 7 and 10 and of the modulus of common logarithms all carried to 260 places of decimals.

459. **Proceedings of the Royal Society**(London), 1899, (A), 64, №407, 203—223.
Aldis W. S. Tables for the solution equation $\frac{d^2y}{dx^2}+\frac{1}{x}\frac{dy}{dx}-(1+\frac{n^2}{x^2})y=0$.

460. **Publication der Astronomischen Gesellschaft**, 1908, 22.
Börgen C. Logarithmisch-trigonometrische Tafel auf 11 (bezw. 10) Stellen.

461. **Rees's Cyclopaedia** (Universal Dictionary of Arts, Sciences and Literature), 1819, 21: «Logarithms»; 28: «Hyperbolic Logarithms».

462. **Sitzungsberichte der Mathematisch-naturwissenschaftlichen Classe der Keiserlichen Akademie der Wissenschaften** (Wiener Sitzungsber.), 1848, 2, 175—193.
Hermann. Verbesserung der H. Callet'schen Tafel der gemeinen Logarithmen mit 20 Decimalen, nebst Vorschlägen für die weitere Förderung dieses Zweckes.

463. **Tracts for Computers**, 1924, **11**, IX.
Thompson A J. Logarithmetica britannica being a standard table of logarithms to twenty decimal places. Numbers 90 000 to 100 000.

464. **Tracts for Computers**, 1926, **13**, I.
Henderson J. Bibliotheca tabularum mathematicarum, logarithmic tables. A logarithms of numbers.

465. **Tracts for Computers**, 1927, **14**, VIII.
Thompson A. J. Logarithmetica britannica being a standard table of logarithms to twenty decimal places. Numbers 80 000 to 90 000.

466. **Tracts tor Computers**, 1928, **16**, IV.
Thompson A. J. Logarithmetica britannica being a standard table of logarithms to twenty decimal places. Numbers 40 000 to 50 000.

467. **Tracts for Computers**, 1931, **17**, V.
Thompson A. J. Logarithmetica britannica being a standard table of logarithms to twenty decimal places. Numbers 50 000 to 60 000.

468. **Tracts for Computers**, 1933, **18**, VI.
Thompson A. J. Logarithmetica britannica being a standard

table of logarithms to twenty decimal places. Numbers 60 000 to 70 000.

469. **Tracts for Computers,** 1934, **19,** I.
Thompson A. J. Logarithmetica britannica being a standard table of logarithms to twenty decimal places. Numbers 10 000 to 20 000.

470. **Tracts for Computers,** 1935, **20,** VII.
Thompson A. J. Logarithmetica britannica being a standard table of logarithms to twenty decimal places. Numbers 70 000 to 80 000.

471. **Tracts for Computers,** 1937, **21,** III.
Thompson A. J. Logarithmetica britannica being a standard table of logarithms to twenty decimal places. Numbers 30 000 to 40 000.

472. **Tracts for Computers,** 1938, **23.**
Comrie L. J. Tables of $\tan^{-1}x$ and $\log(1+x^2)$ to assist in the calculation of the ordinates of a Pearson type IV curve.

473. **Tracts for Computers,** 1952, **22,** II.
Thompson A. J. Logarithmetica britannica... Numbers 20 000 to 30 000.

474. **Transactions of the Cambridge Philosophical Society,** 1883, **13,** 243—272.
Glaisher J. W. L. Tables of the exponential functions.

475. **Veröffentlichungen des Kgl. Astronomischen Rechen-Instituts zu Berlin,** 1919, **43.**
Peters J. und Stein J. Zweiundfünfzigstellige Logarithmen.

476. **Zeitschrift für Mathematik und Physik** (Leipzig), 1903, **49,** 465—467.
Schnöckel J. Tafel der Antilogarithmen für die Basis 2.

477. **Berkeley L. M.** Addition-subtraction logarithms to five decimal places. City, White Book and Sypply Co., New York, 1930.

478. **Neuste Schriften der Naturforschenden Gesellschaft in Danzig,** 1862, 6, № 4.
Gronau J. F. W. Tafeln für die hyperbolischen Sectoren und für die Logarithmen ihrer Sinus und Cosinus.

479. **Comrie L. J.** Chambers's six-figure mathematical tables, **I.** New York, 1949.

480. **Comrie L. J.** Chambers's six-figure mathematical tables, **II,** New York, 1949.

References to Chapter 5

FACTORIALS, EULER INTEGRALS AND RELATED FUNCTIONS

1. **Абрамов А. А.** Таблицы $\ln\Gamma(z)$ в комплексной области. Изд-во АН СССР, М., 1953.

2. **Алексеев В. А.** Сборник таблиц для расчета вероятностей. Арт. Акад. им. Дзержинского, М., 1944.

3. **Андреев П. П.** Математические таблицы. Госстатиздат, М., 1952.

4. **Брилинг С. Р.** Математический справочник. ГТТИ, М.—Л., 1933.

5. **Бронштейн И. Н. и Семендяев К. А.** Справочник по математике. Гостехиздат, М.—Л., 1945.

6. **Вега Г.** Таблицы семизначных логарифмов. Геодезиздат, М., 1949.

7. **Глазенап С. П.** Пятизначные таблицы логарифмов. Изд-во АН СССР, Л., 1934.

8. **Двайт Г.Б.** Таблицы интегралов и другие математические формулы. Гос. изд-во иностр. лит., М., 1948.

9. **Каменьщиков Н.** Таблицы логарифмов с четырьмя десятичными знаками. ГИЗ, Москва—Петроград, 1922.

10. **Машиностроение. Энциклопедический справочник**, I, кн. 1. ГНТИ, Маш. лит. М., 1947.

11. **Митропольский А. К.** Краткие таблицы (с приложением очерка вычислительной техники). Изд-во Лесотехн. акад. им. Кирова, Л., 1950.

12. **Михайлов А.** Таблицы логарифмов с четырьмя десятичными знаками. Логарифмы чисел, тригонометрических величин, натуральные тригонометрические функции и другие таблицы, упрощающие вычисления. ГИЗ, М.— Л., 1923.

13. **Пржевальский Е.** Пятизначные таблицы логарифмов чисел и тригонометрических величин. ГИЗ, М.—Л., 1928.

14. **Пятизначные таблицы логарифмов.** Воениздат, М., 1953.

15. **Родин А. Ф.** Пятизначные логарифмические таблицы чисел и тригонометрических функций с приложением гауссовых логарифмов и различных вспомогательных таблиц. Линдеберг, С.-Петербург, 1895.

16. **Сегал Б. И. и Семендяев К. А.** Пятизначные математические таблицы. Изд-во АН СССР, М.— Л. 1948.

17. **Слуцкий Е. Е.** Таблицы для вычисления неполной Г-функции и функции вероятности χ^2. Изд-во АН СССР, М.— Л., 1950.

18. **Справочное руководство по машиностроению**, I. ГНТИ Украины, Харьков, 1937.

19. **Таблицы Барлоу-квадратов, кубов, квадратных корней и обратных величин всех целых чисел до 12 500.** Изд-во иностр. лит., М., 1950.

20. **Hütte.** Справочник для инженеров, техников и студентов, 1. Госмашметиздат, М.—Л., 1934.

21. **Шпильрейн Я. Н.** Таблицы специальных функций. Числовые значения, графики и формулы, 1. ГТТИ, М.—Л., 1933.

22. **Янке Е.** и **Эмде Ф.** Таблицы функций с формулами и кривыми. ГИТТЛ, М.— Л., 1949. *

23. **Allen E. S.** Six-place tables. McGraw-Hill Bock Co., New York — London, 1947.

24. **Alliaume M.** Tables jusqu'à $n = 1200$ des factorielles n! décomposées en facteurs premiers et en factorielles premières. Librairie Univ., Louvain, 1928.

25. **Andoyer M.** Nouvelles tables trigonométriques fondamentales (valeurs naturelles). 1 (sines). Hermann, Paris, 1915.

26. **Angot A.** Compléments de mathématiques à l'usage des ingénieurs de l'électrotêchnique et des télécommunications. Revue d'Optique, Paris, 1949.

27. **Attwood E. L.** Practical five-figure mathematical tables. Macmillan, London, 1950.

28. **Barlow P.** New mathematical tables.... G. and S. Robinson, London, 1814.

29. **Barlow P.** Tables of squares, cubes, square roots, cube roots, and reciprocals of all integer numbers up to 12 500. Chemical Publishing Co., New York; Spon, London, 1947.

30. **Beetle R. D.** Five-place logarithmic and trigonometric tables. McGraw Hill, New York — London, 1933.

31. **Bertrand J. L. F** Traité de calcul différentiel et de calcul intégral, 2: Calcul intégral. Gauthier-Villars, Paris, 1870.

32. **Beyrodt G.** und **Reissner A.** Tabellenbuch zur Wirtschaftmathematik Fachwissen Lehrbücher für die Berufsfach und Fachschule. «Volk und Wissen», Berlin — Leipzig, 1949.

33. **Boccardi J. C. M.** Tables logarithmiques des factorielles jusqu'à 10 000! Imprimerie Mistral, Cavaillon, 1932.

34. **Boll M.** Tables numériques universelles des laboratoires et bureaux d'étude. Dunod, Paris, 1947.

35. **Brandenburg H.** Hétjegyü trigonometrial táblázat. Honved térképészeti intézet. Budapest, 1951.

36. **Brandenburg H.** Sechsstellige trigonometrische Tafel. Lorentz, Leipzig, 1932.

37. **Brandenburg H.** Siebenstellige trigonometrische Tafel alter Kreisteilung für Berechnungen mit der Rechenmaschine. Lorentz, Leipzig, 1931.

38. **British Association for the Advancement of Science. Mathematical tables,** 1; Circular and hyperbolic functions, exponential and sine and cosine integrals, factorial function and allied functions, Hermitian probability functions., Cambridge Univ. Press, 1951.

39. **Carr G. S.** A synopsis of elementary results in pure mathematics, II. Hodgson, London, 1886.

40. **Comrie L. J.** Chambers's four-figure mathematical tables. W. and R. Chambers, Edinburgh—London, 1947.

41. **Dale J. B.** Five-figure tables of mathematical functions comprising tables of logarithms, powers of numbers, trigonometric, elliptic and other transcendental functions. Arnold, London, 1949.

42. **Davenport C. B.** Statistical method with special reference to biological variation. New York, 1899.

43. **Davis H. T.** Tables of the higher mathematical functions, I. Indiana Principia Press, Bloomington, 1933.

44. **Davis H. T.** Tables of the higher mathematical functions, II. Indiana Principia Press, Bloomington, 1935.

45. **Degen C. F.** Tabularum ad faciliorem et breviorem probabilitatis computationem utilium Enneas. Copenhagen, 1824.

46. **Duarte F. J.** Nouvelles tables de log n! á 33 décimales depuis $n = 1$ jusqu'á $n = 3000$. Kundig, Genewa, Index Generalis, Paris, 1927.

47. **Duarte F. J.** Nouvelles tables logarithmiques á 36 décimales. Gauthier-Villars, Paris, 1933.

48. **Dwight H.** Tables of integrals and other mathematical data. Macmillan, New York, 1947.

49. **Dwight H. B.** Mathematical tables. McGraw-Hill, New York — London, 1941.

50. Egersdörfer R. und Egersdörfer L. Formeln und Tabellen der zugeordneten Kugelfunktionen, 1. Art von $n = 1$ bis $n = 20$. Erster Teil: Formeln. Springer, Berlin, 1936.

51. Elderton W. P. Frequency curves and correlation Cambridge Univ. Press, Cambridge, 1938.

52. Encyclopaedia Metropolitana, 1845, 2, 393—490.
De Morgan A. Theory of probabilities.

53. Euler L. Introductio in ànalysin infinitorum, I. Lepsiae et Berolini, 1922.

54. Euler L. Opuscula analytica, 2. St. Peterburg, 1785.

55. Fisher R. and Yates F. Statistical tables for biological, agricultural and medical research. Oliver and Boyd, London — Edinburgh, 1949.

56. Formular matematic şi tehnic. Pentru elevi, studenţi şi tehnicieni. Editura tehnica, Bucuresti, 1949.

57. Fowle F. E. Smithsonian physical tables. Smithsonian Institution, Washington, 1934.

58. Fry T. C. Probability and its engineering uses. Van Nostrand, New York, 1928.

59. Gauss C. F. Werke (Göttingen), 1866, 3, 123—162.
Disquisitiones Generales Circa Seriem Infinitam.

60. Glover J. W. Tables of applied mathematics in finance, insurance and statistics. Wahr, Ann Arbor, Michigan, 1930.

61. Burington R. S. Handbook of mathematical tables and formulas. Sandusky, Ohio, 1946.

62. Hayashi K. Fünfstellige Funktionentafeln. Springer, Berlin, 1930.

63. Hayashi K. Sieben- und mehrstellige Tafeln der Kreis- und Hyperbelfunktionen und deren Produkte sowie der Gammafunktion. Springer, Berlin, 1926.

64. Hayashi K. Tafeln der Besselschen-, Theta-. Kugel- und anderen Funktionen. Springer, Berlin, 1930.

65. Hayashi K. Tafeln für die Differenzenrechnung... Springer, Berlin 1933.

66. Hobson E. W. A treatise on plane trigonometry. Cambridge Univ. Press, Cambridge, 1891.

67. Hoüel J. Recueil de formules et de tables numériques. Gauthier-Villars, Paris, 1868; 1901.

68. Hoüel J. Tables de logarithmes à cinq décimales pour les nombres et les lignes trigonométriques. Gauthier-Villars, Paris, 1858.

69. Jahn G. A. Tafeln der Quadrat- und Kubikwurzeln aller Zahlen von 1 bis 25.500, der Quadratzahlen aller Zahlen von 1 bis 27.000 und Kubikzahlen von 1 bis 24.000. Barth, Leipzig, 1839.

70. Jahnke E. und Emde F. Funktionentafeln mit Formeln und Kurven. Teubner, Leipzig — Berlin, 1) 1909, 2) 1938.

71. Jeffreys H. Theory of probability. Clarendon Press, Oxford, 1939.

72. Jones G. W. Logarithmic tables. Author, Ithaca, New Jork; Macmillan, London, 1893.

73. Jörgensen N. R. Undersögelser over frekvensflader og korrelation. Busck, Copenhagen, 1916.

74. Köhler H. G. Logarithmisch-trigonometrisches Handbuch. Tauchnitz, Leipzig, 1876.

75. Lambert J. H. — Felkel A. Supplementa tabularum logarithmicarum et trigonometricarum.. Lisbon, 1798.

76. Laska W. Sammlung von Formeln der reinen und angewandten Mathematik. Vieweg, Braunschweig, 1888—1894.

77. Legendre A. M. Exercices de calcul intégral, 1. Huzard-Courcier, Paris, 1811.

78. Legendre A. M. Exercices de calcul intégral, 2. Huzard-Courcier, Paris, 1814.

79. Legendre A. M. Traité des fonctions elliptiques, 2. Huzard-Courcier, Paris, 1826.

80. Loky B. Négyjegyü függvénytáblázatok. Budapest, 1950.

81. Lupton S. Numerical tables and constants in elementary science. Macmillan, London, 1906.

82. **M i l n e — T h o m s o n L. M.** and **C o m r i e L. J.** Standard four-figure mathematical tables. Including many new tables trigonometrical functions for radians, inverse trigonometrical and hyperbolic functions and an extended table of natural logarithms. Macmillan, London. 1948

83. **D e M o r g a n A.** The differential and integral calculus. London, 1842.

84. **P e a r s o n K.** Tables of the incomplete beta-function. Biometrica Office, University College, London, 1934.

85. **P e a r s o n K.** Tables of the incomplete Γ-function. Biometrika Office, University College, London, 1934.

86. **P e a r s o n K.** Tables for statisticians and biometricians, I. Biometric Laboratory. University College, London, 1930.

87. **P e a r s o n K.** Tables for statisticians and biometricians, II. Biometric Laboratory. University College, London, 1931.

88. **P e i r c e B. O.** A short table of integrals. Ginn, Boston — New York, 1910.

89. **P e t e r s J.** Zehnstellige Logarithmentafel 1: Zehnstellige Logarithmen der Zahlen von 1 bis 100.000 nebst einem Anhang mathematischer Tafeln. Reichsamt für Landesaufnahme, Berlin, 1922.

90. **P l u m m e r H. C.** Probability and frequency. Macmillan, London, 1940.

91. **P o t i n L.** Formules et tables numériques. Doin et Gauthier-Villars, Paris. 1925.

92. **L a r s e n H. D.** Rinehart mathematical tables, formulas and curves. Chapman and Hall, London, 1949.

93. **R o h r b a c h C.** Vierstellige logarithmisch trigonometrische Tafeln. Ehlerman, Dresden, 1929.

94. **R o s e n b a c h I. B., W h i t m a n E. A., M o s k o v i t z D.** Mathematical tables. Ginn, Boston, 1946.

95. **R u s s e l l A.** A treatise on the theory of alternating currents, I. Cambridge Univ. Press, Cambridge, 1914.

96. **S a s s e n f e l d H. M.** und **T s c h u n k o H. F. A.** Mathematische Tafeln für Mathematiker, Naturwissenschaftler, Ingenieure. Erster Teil: Elementare Funktionen. Fr. Lamade, Walldorf bei Heidelberg, 1949.

97. **S c h l ö m i l c h O. X.** Analytische Studien, 1 — 2. Engelmann, Leipzig, 1848.

98. **S c h o r r R.** Sammlung von Hilfstafeln der Hamburger Sternwarte in Bergedorf. Lucas Gräfe, Hamburg, 1916.

99. **S c h r ö n L.** Table d'interpolation pour le calcul des parties proportionnelles. Paris, 1869.

100. **S c h u m a n n W. O.** Elektrische Durchbruchfeldstärke von Gasen. Springer, Berlin, 1923.

101. **S i l b e r s t e i n L.** Synopsis of applicable mathematics, with tables. Bell, London, 1923.

102. **S t a l m a s e k Y.** Tabul'ky funkcií a konštant z matematiky, fyziky a chemie. Štátne nakl., Bratislava, 1951.

103. **S t a n l e y I. P.** and **W i l k e s M. V.** Table of the reciprocal of the gamma-function for complex argument. Computation Centre, Univ. of Toronto, 1950.

104. **Tabele şi formule matematice.** Ed. tehnica, Bucureşti, 1951.

105. **Tables of the error function and of its first twenty derivatives.** Harvard Univ. Press, Cambridge, Massachusetts, 1952.

106. **T h o m a n F.** Tables de logarithmes à 27 décimales. Paris, 1867.

107. **U h l e r H. S.** Exact values of the first 200 factorials. Author. New York, 1944.

108. **Valori naturali delle funzioni goniometriche.** Firenze istituto geografico militare, 1943.

109. **V e g a G.** Logarithmisch-trigonometrische Tafeln (Tabulae Logarithmico-trigonometricae), 1—2. Weidmann, Leipzig, 1797.

110. **V e g a G. — H ü l s s e J. A.** Sammlung mathematischer Tafeln ..Weidmann, Leipzig, 1849.

111. **W i l l i a m s o n B.** An elementary treatise on the integral calculus. Longmans, London, 1884.

112. **Acta Mathematica** (Uppsala), 1883, 2, 261—295.

Bourguet L. Sur les intégrales Euleriennes et quelques autres fonctions uniformes.

113. **Acta Mathematica** (Uppsala), 1931, 56, 273—354.
Ginzel I. Die Konforme Abbildung durch die Gammafunktion.

114. **Acta Mathematica** (Uppsala), 1933, 61, 263—322.
Bendersky L. Sur la fonction gamma généralisée.

115. **Acta Mathematica** (Uppsala), 1951, 85, 117—181.
Kreyszig E. Über den allgemeinen Integralsinus Si (z, α).

116. **American Journal of Mathematics pure and applied,** 1884, 6, 115—130.
Hill G. W. On certain possible abbreviations in the computation of the long-period inequalities of the moon's motion due to the direct action of the planets.

117. **Annales de l'Observatoire de Paris, Mémoires,** 1856, 2, 1—30.
Leverrier U. J. J. Formules pour le calcul des coefficients du développement de la fonction $(1 + \alpha^2 - 2\alpha \cos\theta)^-$.

118. **Archiv der Mathematik und Physik** (Leipzig), 1841, 1, 193—203.
Grunert J. A. Über die Bestimmung der Anzahl der verschiedenen Arten, auf welche sich ein n-eck durch Diagonalen in lauter m-ecke zerlegen, lässt, mit Bezug auf einige Abhandlungen der Herren Lame, Rodrigues, Binet, Catalan und Duhamel in dem Journal de Mathematiques..., III, IV.

119. **Archiv der Mathematik und Physik** (Leipzig), 1865, 43, 134—210.
Knar J. Die Harmonischen Reihen.

120. **Archiv der Mathematik und Physik** (Leipzig), 1887, 5, 279—344.
Siebel A. Exacte Trennung der reellen Wurzeln numerischer algebraischen und transcendenten Gleichungen.

121. **Atti della Accademia Nazionale dei Lincei. Rendiconti. Classe di scienze fisiche, matematiche e naturali** (Roma), 1947, (8), 3, № 3—4, 254—257.
Ghizzetti A. Tavola della funzione Euleriana Γ(z) per valori complessi dell'argomento.

122. **Atti dell'Instituto nazionale delle assizurazioni,** 1930, 2, 245—266.
Cassinis G. Sull'impiedo di alcune funzioni transcendenti nelle rappresentazioni empiriche.

123. **Biometrika** (London), 1902, 1, 155—163.
Elderton W. P. Table for testing the goodness of fit of theory of observation.

124. **Biometrika** (London), 1909, 7, 43—47.
Duffell J. H. Tables of the Γ-function.

125. **Biometrika** (London), 1914, 10, 208—214.
Lee A. Table of the Gaussian «Tail» function; when the «Tail» is larger than the body.

126. **Biometrika** (London), 1917, 11, № 4, 328—413.
Soper H. E., Young A. W., Cave B. M., Lee A. Pearson K. On the distribution of the correlation coefficient in small samples

127. **Biometrika** (London), 1932, 24, № 3—4, 293—350.
Pearson K., Stouffer S. A. and David F. N. Further applications in statistics of the $I_m(x)$ Bessel function.

128. **British Association for the Advancement of Science. Reports,** 1916, 123—126.

129. **British Association for the Advancement of Science. Reports,** 1929, 251—262.
Lodge A. Mathematical tables.

130. **Bulletin of the American Mathematical Society** (New York), 1943, 49, 58.
Salzer H. E. and Hillman A. P. Exact values of the first 120 factorials.

131. **United States Department of Commerce Bureau of Standards. Bulletin** (Washington), 1908, 5, 1—132.
Rosa E. B. and Cohen L. Formulae and tables for the calculation of mutual and self-inductance.

132. **United States Department of Commerce Bureau of Standards. Bulletin** (Washington), 1912, 8, 1—237.
Rosa E. B. and Grover F. W. Formulae and tables for the calculation of mutual and self-inductance.

133. **Elektrotechnik und Maschinenbau** (Wien), 1906, 24, 996.

E m d e F. Zur Berechnung der Elektromagnete.

134. **Konigsberger Archiv für Naturwissenschaft und Mathematik,** 1812, 1, 1—40, 239—240.
B e s s e l F. W. Values of li(x).

135. **Memoirs of the National Academy of Sciences** (Washington), 1925, **14**, № 5, 3—79.
V a n O r s t r a n d C. E. Tables of the exponential function and of the circular sine and cosine to radian argument.

136. **Mémoires de la Societé des Sciences Physiques et Naturelles de Bordeaux,** 1886, (3), **3**, № 1, 1—184.
B r u n e l G. Monographie de la fonction gamma.

137. **Mémorial de l'Office National Méteorologique de France** (Paris), 1925, **10.**
M o n t e s s u s (de B a l l o r e) R. et D u a r t e F. J. Table à 12 décimales de log n! pour toutes les valeurs de n de 1 à 1000.

138. **Memorie de Reale Istituto Veneto de Scienze, Lettere ed Arti** (Venezia), 1874, **18**, 125—162.
B e l l a v i t i s G. Logarithm-integral, exponential-integral, and other Eulerian integrals.

139. **Philosophical Magazine** (London), 1916, (6), **31**, 520—528.
A i r e y J. R. Bessel and Neumann functions of equal order and argument.

140. **Philosophical Magazine** (London), 1930, (7), 9, 184—196.
H i g u c h i S. On some closed algebraic curves and their application to dynamical problems.

141. **Philosophical Magazine** (London), 1942, (7), **33**, 115—130.
M i a n A. M. and C h a p m a n S. Approximate formulae for functions expressed as definite integrals.

142. **Philosophical Magazine** (London), 1942, (7), **33**, 754—761.
C o u l s o n C. A. and D u n c a n s o n W. E Some new values for the exponential integral.

143. **Philosophical Transactions of the Royal Society** (London), 1870, **160**, 367—388.
G l a i s h e r J. W. L. Tables of the numerical values of the sine-integral, cosine-integral, and exponential-integral.

144. **Physical Review**, 1931, (2), **37**, 507—531.
B a r t l e t t J. H. Orbital Valency.

145. **Physical Review**, 1931, (2), **38**, 255—276.
R o s e n N. Calculation of interaction between atoms with s-electrons.

146. **Physical Review**, 1931, (2), **38**, 2099—2114.
R o s e n N. The normal state of the hydrogen molecule.

147. **Physical Review**, 1942, (2), **61**, 528—530.
R o s e n b e r g P. Evaluation of functions related to Tait's mean free path.

148. **Proceedings of the Cambridge Philosophical Society**, 1940, **36**, 2, 173—182.
M i s r a R. D. On the stability of crystal lattices.

149. **Proceedings of the Cambridge Philosophical Society**, 1940, **36**, 466—478.
B o r n M. and M i s r a R. D Stability of crystal lattices.

150. **Proceedings of the London Mathematical Society**, 1871—1873, 4, 48—56.
G l a i s h e r J. W. L. On the constants that occur in certain summations by Bernoulli's series.

151. **Proceedings of the National Academy of Sciences** (Washington), 1942, **28**, 59—62.
U h l e r H. S. The coefficients of Stirling's series for log $\Gamma(x)$.

152. **Proceedings of the Physico-Mathematical Society of Japan**, 1938, (3), **20**, № 1, 1—70.
K o t a n i M., A m e m i y a A. and S i m o s e T. Tables of integrals useful for the calculations of molecular energies.

153. **Proceedings of the Royal Society of Edinburgh**, 1930, **50**, 225—242.
R o b e r t s O. F. T. Radiative diffusion in the atmosphere.

154. **Proceedings of the Royal Society of London**, 1909, (A), **82**, 43—70.
G o l d E. The isothermal layer of the atmosphere and atmospheric radiation.

155. **Quarterly Journal of Mathematics,** 1864, 6, 82—108.
J e f f e r y H. M. On the derivatives of the gamma functions.

156. **Quarterly Journal of Mechanics and applied Mathematics** (Oxford), 1948, 1.
H a r t r e e D. R. and J o h n s t o n S. On a function associated with the logarithmic derivative of the gamma-function.

157. **Technical Note,** 1949, **106**, 11.
L o t k i n M. Table of the first 200 factorials to 20 places.

158. **Tracts for Computers** (London), 1919, **1**.
P a i r m a n E. Tables of the digamma and trigamma functions.

159. **Tracts for Computers** (London), 1922, 8.
P e a r s o n E. S. Table of the logarithms of the complete Γ-function.

160. **Tracts for Computers,** 1923, 9.
B r o w n l e e J. Log $\Gamma(x)$ for $x=1$ to 50.9, by intervals of 0,01.

161. **Transactions of the Cambridge Philosophical Society,** 1883, **13**, 243—272.
G l a i s h e r J. W. L. Tables of the exponential function.

162. **Transactions of the Connecticut of Academy of Arts and Science** (New Haven) 1937, **32**, 381—434.
U h l e r H. S. A new table of reciprocals of factorials and some derived numbers.

163. **Transactions of the Royal Society of Canada,** 1903, (2), 9, 73—107.
M i l l e r W. L. and R o s e b r u g h F. R. Numerical values of certain functions involving e^{-x} (Tables of the integral $\int_x^{\infty} e^{-u}u^n du$).

164. **Transactions of the Royal Society of Edinburgh,** 1887, 33, 95.
C l a r k J. Table calculated for «On the foundations of the kinetic theory of gases».

165. **Transactions of the Royal Society of Edinburgh,** 1917, 52, 167—174.
R o b b i n s F. Factorials and allied products with their logarithms.

166. **C o m r i e L. J.** Chambers's six-figure mathematical tables, I. New York, 1949.

167. **C o m r i e L. J.** Chambers's six-figure mathematical tables, II. New York, 1949.

168. **V a l o u c h M.— V a l o u c h M. A.** Pětimístné logaritmy čísel. Nakladatelství. Československé Akademie Věd., Praha, 1953.

References to Chapter 6

SINE AND COSINE INTEGRALS, EXPONENTIAL AND LOGARITHMIC INTEGRALS AND RELATED FUNCTIONS

1. **Справочное руководство по машиностроению, I.** ГНТИ Украины, Харьков, 1937.
2. **Таблицы интегрального синуса и косинуса.** Изд-во АН СССР, М., 1954.
3. **Таблицы интегральной показательной функции.** Изд-во АН СССР, М, 1954.
4. **Татаринов В. В.** Трехзначные таблицы интегральных синусов и косинусов. Связьтехиздат, Л.— М., 1934.
5. **Шпильрейн Я. Н.** Таблицы специальных функций. Числовые значения, графики и формулы, 1. ГТТИ, М.— Л., 1933.
6. **Янке Е. и Эмде Ф.** Таблицы функций с формулами и кривыми. ГИТТЛ. М.— Л., 1949.
7. **Автоматика и телемеханика,** 1949, 10, № 5, 362—376.
   Солодовников В. В. О применении трапециодальных частотных характеристик к анализу качества систем автоматического регулирования.
8. **Tables of modified cosine-integral.** Aircraft Radiation Systems Laboratory. Stanford Research Institute, 1951.
9. **Angot A.** Complements de mathématiques à l'usage des ingéniéurs de l'électrotéchnique et des télécommunications. Revue d'Optique, Paris, 1949.
10. **Schelkunoff S. A.** Applied mathematics for engineers and scientists. Van Nostrand, New York, 1948.
11. **Boll M.** Tables numériques universelles des laboratoires et bureaux d'étude. Dunod, Paris, 1947.
12. **British Association for the Advancement of Science. Mathematical tables, I:** Circular and hyperbolic functions, exponential and sine and cosine integrals, factorial functions. Cambridge Univ. Press, 1951.
13. **Dale J. B.** Five-figure tables of mathematical functions comprising tables of logarithms, powers of numbers, trigonometric elliptic and other transcendental functions. Arnold, London, 1949.
14. **Fowle F. E.** Smithsonian physical tables. Smithsonian Institution, Washington, 1934.
15. **Hoüel J.** Recueil de formules et de tables numériques. Gauthier-Villars, Paris. 1868, 1901
16. **Jahnke E.** und **Emde F.** Functionentafeln mit Formeln und Kurven. Teubner, Leipzig — Berlin, 1938.
17. **Glaisher J.** Factor table for the sixth million containing the least factor of every number not divisible by 2,3 or 5 between 5 000 000 and 6 000 000. Taylor, London, 1883.
18. **Jörgensen N. R.** Undersögelser over frekvensflader og korrelation. Busck, Copenhagen, 1916.
19. **Laska W.** Sammlung von Formeln der reinen und angewandten Mathematik. Vieweg, Braunschweig, 1888—1894

20. **L e g e n d r e A. M.** Exercices de calcul intégral, 3. Huzard-Courcier, Paris, 1816.

21. **L e g e n d r e A. M.** Traité des fonctions élliptiques, 2. Huzard-Courcier, Paris, 1826.

22. **Carnegie Institute Publications** (Washington), 1914, 165.
L e h m e r D. H. List of prime numbers from 1 to 10 006 721.

23. **M i l l e r W. L. and R o s e b r u g h T. R.** Numerical values of certain functions involving e^{-x}. University of Toronto Studeis, Canada, 1904.

24. **S o l d n e r J.** Théorie et tables d'une nouvelle fonction transcendante. Munich, 1809.

25. **S t e n b e r g L.** Tabulae logarithmi-integralis, 1. Malmö, 1861.

26. **S t e n b e r g L.** Tabulae logarithmi-integralis, 2. Malmö, 1867.

27. **S t e n b e r g L.** Tabulae logarithmi-integralis, 3. Malmö, 1871.

28. **Table of sine and cosine integrals for arguments from 10 to 100.** National Bureau of Standards, Washington, 1942.

29. **Tables of the generalized exponential-integral functions.** Harvard Univ. Press, Cambridge, Massachusetts, 1949.

30. **Tables of generalized sine- and cosine-integral functions, 1—2.** Harvard Univ. Press, Cambridge, Massachusetts, 1949.

31. **Tables of sine, cosine and exponential integrals, I.** National Bureau of Standards, Washington, 1940.

32. **Tables of sine, cosine and exponential integrals, II.** National Bureau of Standards, Washington, 1940.

33. **T a n i K.** Tables of Si(x) and Ci(x) for the range $x = 0$ to $x = 50$. Naval Experimental and Research Establishment, Meguro, Tokyo, 1931.

34. **Abhandlungen der mathematisch-physikalischen Classe der Königlich Bayerischen Academie des Wissenschaften** (München), 1899, 19, 450—508.
L o m m e l E. Theorie Dämmerungsfarben.

35. **Acta Mathematica** (Uppsala), 1951, 85, 117—181.
K r e y s z i g E. Über den allgemeinen Integralsinus Si(z, α).

36. **Archiv der Mathematik und Physik** (Leipzig), 1843, (1), 3, 27—34.
B r e t s c h n e i d e r C. A Berechnung der Grundzahl der natürlichen Logarithmen, sowie mehrerer anderer mit ihr zusammenhängender Zahlwerthe.

37. **Arkiv för mathematik, astronomi och fysik** (Stockholm), 1933, (A), 24, № 5, 35—36.
H a l l e n E. General theory of electric oscillatory circuits.

38. **Australian Journal of Scientific Research** (Melbourne), 1948, (A), 1, № 1, 1—17.
G o d e r e y G. H. Diffraction of light from sources of finite dimensions.

39. **British Association for the Advancement of Science. Reports,** 1883, 118—125.

40. **British Association for the Advancement of Science. Reports,** 1927, 220—254.
Confluent hypergeometric function, Ei($\pm x$), Si(x), Ci(x); zeros of $J_{-\nu}(x)$, ber x, bei x, ker x, kei x, and their first derivatives.

41. **British Association for the Advancement of Science. Reports,** 1928, 305—340.
Sin θ, cos θ, sinh πx, cosh πx, Si(x), Ci(x), $\frac{\partial}{\partial \nu} J\nu(x)$ and integrals of the error integral.

42. **G a u s s C. F.** Werke, 1863, 2, 444—44.

43. **Ingeniorvidenskabelige skrifter** (Danmarsk naturvidenskabelige samfund), 1935, (A), 38.
P e d e r s e n P. O. Radiation from a vertical aerial over flat perfectly conducting earth.

44. **The Journal of the Acoustical Society of America,** 1949, 21, № 4, 348—359.
G r i f f i n g V. and F o x F. E. Theory of ultrasonic intensity gain due to concave reflectors.

45. **Journal für die reine und angewandte Mathematik** (Berlin — Leipzig), 1846, 33, 316—324.
Schlömilch O. X. Note sur quelques intégrales définies.

46. **Kongelige Danske videnskabernes selskab** (Kjobenhavn), 1884, (6). 2, № 6, 185 — 308.
Gram J. P. Undersogelser angaaende maengden af primtal under en given graense.

47. **Königsberger Archiv für Naturwissenschaft und Mathematik**, 1812, 1, 1—40, 239—240.
Bessel F. W. Values of li(x).

48. **Memorie del Reale Istituto Veneto de Scienze, Lettere ed Arti** (Venezia), 1874, 18, 125—162.
Bellavitis G. Logarithm-integral, exponential-integral and other Eulerian integrals.

49. **Philips Research Reports** (Eindhaven, Holland), 1949, 4, 179—188.
Bouwkamp C. Y On the effective length of a linear transmitting antenna.

50. **Philosophical Magazine** (London), 1942, (7), 33, 754—761.
Coulson C. A. and Duncanson W. E. Some new values for the exponential-integral.

51. **Philosophical Transactions of the Royal Society** (London), 1870, 160, 367—388.
Glaisher J. W. L. Tables of the numerical values of the sine-integral, cosine-integral, and exponential-integral.

52. **Philosophical Transactions of the Royal Society**, 1937, (A), 236, 381—422.
King L. V. On the radiation field of a perfectly conducting base insulated cylindrical antenna over a perfectly conducting plane earth and the calculation of radiation resistance and reactance.

53. **Proceedings of the Institute of Radio Engineering** (New York), 1942, 30, № 7, 335—349.
King R. and Blake F. G. The self-impedance of a symmetrical antenna.

54. **Proceedings of tde Physico-Mathematical Society of Japan**, 1938, (3), 20, № 1, 1—70.
Kotani M., Amemiya A. and Simose T. Tables of integrals useful for the calculations of molecular energies.

55. **Proceedings of the Physico-Mathematical Society of Japan**, 1940, (3), 22.
Kotani M., Amemiya A. and Simose T. Tables of integrals useful for the calculations of molecular energies, II.

56. **Proceedings of the Royal Society of London**, 1914, (A), 90, 318—323.
Rayleigh, Lord (Strutt J. W.) Some calculations in illustration of Fourier's theorem.

57. **Proceedings of the Royal Society of London**, 1917, (A), 93, 293—312.
Lamb H. On the deflection of the vertical by tidal loading of the earch's surface.

58. **Quarterly of Applied Mathematics** (Menasha), 1948, 5 № 4, 394—402.
Bouwkamp C. J. Concerning a new transcendent its tabulation and application in antenna theory.

59. **Scientific Papers of the Institute of Physical and Chemical Research** (Tokyo), 1929, 3, 181—215.
Akahira T. Tables of $\frac{e^{-x}}{x}$ and $\int_x^\infty \frac{e^{-u}}{u}\,du$ from $x = 20$ to $x = 50$.

60. **Transactions of the Royal Society of Canada**, 1903, (2), 9, 73—107.
Miller W. L. and Rosebrugh T. R. Numerical values of certain functions involving e^{-x} (Tables of the integral $\int_x^\infty e^{-u}u^n du$).

61. **Zeitschrift für Astrophysik** (Berlin), 1932, 4, 339—357.
Unsöld A. Zur Deutung der Intensitätsverteilung in den Fraunhoferschen Linien, II. Teil: Die Intensität der Linien flügel-approximative Lösung der Schwarzschildschen Integralgleichung für eine beliebig geschichtete Atmosphere.

62. **Zeitschrift für Mathematik und Physik**, 1861, 6, 127—139.
Bretschneider C. A. Über die Berechnung des Integrallogarithmen und einiger mit ihm zusammenhängenden anderen Funktionen.

References to Chapter 7

PROBABILITY INTEGRALS AND RELATED FUNCTIONS

1. **Алексеев В. А.** Сборник таблиц для расчета вероятностей. М., Арт. акад. им. Дзержинского, 1944.

2. **Андреев П. П.** Математические таблицы. Госстатиздат, М., 1952.

3. **Бронштейн И. Н. и Семендяев К. А.** Справочник по математике. Гостехиздат, М.—Л., 1945.

4. **Ватсон Г. Н.** Теория бесселевых функций, II. Изд-во иностр. лит., М., 1949.

5. **Гнеденко Б. В.** Курс теории вероятностей. ГИТТЛ, М.— Л., 1950.

6. **Двайт Г. Б.** Таблицы интегралов и другие математические формулы. Изд-во иностр. лит., М., 1948.

7. **Карпов К. А.** Таблицы функций $W(z)=e^{-z^2}\int_0^z e^{x^2}\,dx$ в комплексной области. Изд-во АН СССР, М., 1954.

8. **Машиностроение.** Энциклопедический справочник, 1, кн. 1. ГНТИ Маш. лит., М., 1947.

9. **Митропольский А. К.** Интегралы вероятностей. Лесотехн. акад. им. С. М. Кирова, Л., 1948.

10. **Сегал Б. И. и Семендяев К. А.** Пятизначные математические таблицы. Изд-во АН СССР, М.— Л., 1950.

11. **Слуцкий Е. Е.** Таблицы для вычисления неполной Г-функции и функции вероятности χ^2. Изд-во АН СССР, М.— Л., 1950.

12. **Справочное руководство по машиностроению,** I. ГНТИ Украины, Харьков, 1937.

13. **Таблицы интегралов Френеля.** Изд-во АН СССР, М., 1953.

14. **Тимофеев Н.** О наивыгоднейших результатах наблюдений. Университетская типограф., Киев, 1870.

15. **Фаддеева В. Н. и Терентьев Н. М.** Таблицы значений функции $W(z)=e^{-z^2}\left(1+\frac{2i}{\sqrt{\pi}}\int_0^z e^{t^2}dt\right)$ от комплексного аргумента. Гостехиздат, М., 1954.

16. **Хренов Л. С.** Таблицы натуральных значений тригонометрических функций в единицах времени. Изд-во АН Казах. ССР, Алма-Ата, 1947.

17. **Шпильрейн Я. Н.** Таблицы специальных функций. Числовые значения, графики и формулы, 1. ГТТИ, М.-Л., 1933.

18. **Янке Е. и Эмде Ф.** Таблицы функций с формулами и кривыми. ГИТТЛ, М.—Л., 1949.

19. **Airy G. B.** Mathematical tracts. Cambridge, 1858.

20. **Airy G. B.** On the algebraical and numerical theory of errors of observations... Macmillan, Cambridge — London, 1879.

21. **Aitken A. C.** Statistical mathematics. Oliver and Boyd, Edinburgh—London, 1945.

22. **Allen E. S.** Six-place tables. McGraw-Hill, New York — London, 1947.

23. **Angot A.** Compléments de mathématiques à l'usage des ingénieurs

de l'électrotéchnique et des télécommunications. Revue d'Optique, Paris. 1949.

24. Arenberg D. L. and Levin D. Table of Fresnel integrals and derived functions. Naval Research Laboratory Field Station, Boston, 1948.

25. Bauschinger J. Tafeln zur theoretischen Astronomie. Engelmann, Leipzig, 1934.

26. Bessel F. W. Fundamenta Astronomiae. Königsberg, 1818.

27. Bertrand J. L. F. Calcul des probabilités. Gauthier-Villars, Paris, 1888.

28. Beyrodt G. und Reissner A. Tabellenbuch zur Wirtschaftmathematik. Fachwissen Lehrbücher für die Berufsfach und Fachschule. «Volk und Wissen», Berlin — Leipzig, 1949.

29. Boll M. Tables numériques universelles des laboratoires et bureaux d'étude. Dunod, Paris, 1947.

30. Born M. Optik. Ein Lehrbuch der elektromagnetischen Lichttheorie. Springer, Berlin, 1933.

31. Bowley A. L. Elements of statistics P. S. King and Son, London; Scribner, New York, 1926.

32. British Association for the Advancement of Science. Mathematical tables, I: Circular and hyperbolic functions, exponential and sine and cosine integrals factorial function and allied functions, Hermitian probability functions. Cambridge Univ. Press, 1951.

33. British Association for the Advancement of Science. Mathematical tables, VII.
Sheppard W. F. The probability integral. Univ. Press, Cambridge, 1939.

34. Bruns H. Wahrscheinlichkeits rechnung und Kollektivmasslehre. Teubner, Leipzig — Berlin, 1906.

35. Brunt D. The combination of observations. Cambridge Univ. Press, Cambridge, 1931.

36. Burgess R. W. Introduction to the mathematics of statistics. Houghton Mifflin, Boston—New York, 1927.

37. Castelnuovo G. Calcolo delle probabilitita, 1—2. Bologna, 1933.

38. Charlier C. V. L. Vorlesungen über die Grundzüge der mathema tischen Statistik. Verlag «Scientia» Lund, 1920.

39. Chauvenet W. A manual of spherical and practical astronomy, 1—2. Lippincott, Philadelphia, 1900.

40. Czuber E. Theorie der Beobachtungsfehler. Teubner, Leipzig, 1891

41. Czuber E. Wahrscheinlichkeitsrechnung. Teubner, Leipzig — Berlin, 1941.

42. Dale J. B. Five-figure tables of mathematical functions comprising tables of logarithms, powers of numbers, trigonometric, elliptic and other transcendental functions. Arnold, London, 1949.

43. Darmois G. Statistique mathématique. Doin, Paris, 1928.

44. Davenport C. B. Statistical methods in biology, medicine and psychology. Wiley, New York; Chapman and Hall, London, 1936.

45. Davis H. T. and Nelson W. F. C. Elements of statistics with applications to economic data. Principia Press, Bloomington, Indiana, 1935.

46. Deltheil R. Erreurs et moindres carres (Traité du calcul des probabilités... par E. Borel..., tome I, Fasc, II). Gauthier - Villars, Paris, 1930.

47. Dwight H. B. Mathematical tables. McGraw-Hill, New York — London, 1941.

48. Eddington A. S. Stellar movements and the structure of the universe. Macmillan, London, 1914.

49. Edwards J. A treatise on the integral calculus, II. Macmillan, London, 1922.

50. Encyclopedia Britannica (Black, Edinburgh), 1839.
Galloway T. A treatise on probability.

51. Faa de Bruno F. Traité élementaire du calcul des erreurs, avec des tables stéréotypes. Gauthier-Villars, Paris, 1869.

52. **F e c h n e r G. T.** Elemente der Psychophysik. Breitkopf und Härtel, Leipzig, 1889.

53. **F i s h e r A.** The mathematical theory of probabilities. Macmillan, New York, 1922.

54. **F i s h e r R. A.** Statistical methods for research workers. Oliver and Boyd, Edinburgh — London, 1938.

55. **F i s h e r R. A.** and **Y a t e s F.** Statistical tables for biological, agricultural and medical research. Oliver and Boyd, London — Edinburgh, 1949.

56. **Fowle F. E.** Smithsonian physical tables. Smithsonian Institution, Washington, 1934.

57. **F r y T. C.** Probability and its engineering uses. Van Nostrand, New York, 1928.

58. **G a r r e t t H. E.** Statistics in psychology and education. Longmans, New York — London, 1937.

59. **G i b b s R. W. M.** The adjustment of errors in practical science. Oxford Univ. Press, 1929.

60. **G l o v e r J. W.** Tables of applied mathematics in finance, insurance and statistics. Wahr, Ann Arbor, Michigan, 1930.

61. **B u r i n g t o n R. S.** Handbook of mathematical tables and formulae. Sandusky, Ohio, 1946.

62. **H a y a s h i K.** Fünfstellige Funktionentafeln. Springer, Berlin, 1930.

63. **H o l z i n g e r K. J.** Statistical methods for students in education. Ginn, Boston, 1928.

64. **H o ü e l J.** Recueil de formules et de tables numériques. Gauthier-Villars, Paris, 1868, 1901.

65. **J a h n k e E.** und **E m d e F.** Funktionentafeln mit Formeln und Kurven. Teubner. Leipzig — Berlin, 1938.

66. **J e a n s J. H.** The dynamical theory of gases. Cambridge Univ. Press, 1925.

67. **J o h n s o n W. W.** The theory of errors and method of least squares. Wiley, New York; Chapman and Hall, London, 1892.

68. **J o n e s G. W.** Logarithmic tables. Author, New York; Macmillan, London, 1893

69. **J o n e s D. C.** A first course in statistics. Bell, London, 1924

70. **J o r d a n C.** Statistique mathématique. Gauthier-Villars, Paris, 1927.

71. **J ö r g e n s e n N. R.** Undersögelser over frekvensflader og korrelation. Busck, Copenhagen, 1916.

72. **K e l l e y T. L.** The Kelley statistical tables. Harvard Univ. Press, Cambridge, 1938.

73. **K e l l e y T. L.** Statistical method. Macmillan, New York, 1924.

74. **K e n n e y J. F.** Mathematics of statistics, 1—2. Chapman and Hall, London, 1940.

75. **K r a m p C.** Analyse des Refractions. Strasbourg — Leipzig, 1799.

76. **L e g e n d r e A. M.** Traité des fonctions élliptiques, 2. Huzard-Courcier, Paris, 1826.

77. **L e v y H.** and **R o t h L.** Elements of probability. Clarendon Press, Oxford, 1936.

78. **L o m m e l E. C. J.** Studien über die Bessel'schen Functionen. Teubner, Leipzig, 1868.

79. **L u p t o n S.** Notes on observations. Macmillan, London, 1898.

80. **M a r c u m J. J.** Tables of Hermite polynomials and the derivatives of the error function. Santa Monica, 1948.

81. **M a r k o f f A. A.** Table des valeurs de l'intégrale $\int_x^\infty e^{-t^2}dt$. St. Petersbourg, 1888.

82. **M a r k o f f A. A.** Wahrscheinlichkeitsrechnung. Teubner, Leipzig — Berlin, 1912.

83. **M a s c a r t E.** Traité d'optique, I. Gauthier-Villars, Paris, 1889.

84. **M a s c a r t E.** Traité d'optique, III. Gauthier-Villars, Paris, 1893.

85. **M e l l o r J. W.** Higher mathematics for students of chemistry and physics. Longmans, London, 1931.

86. **M e r r i m a n M.** Text-book on the method of least squares. Wiley, New York; Chapman and Hall, London, 1910.

87. **Milne-Thomson L. M.** and **Comrie L. J.** Standard four-figure mathematical tables. Including many new tables, trigonometrical functions for radians, inverse trigonometrical and hyperbolic functions and an extended table of natural logarithmis. Macmillan, London, 1948

88. **Mitchell A. C. G.** and **Zemansky M. W.** Resonance radiation and excited atoms. Univ. Press, Cambridge, 1934.

89. **De Morgan A.** An essay on probabilities. Longman, London, 1838.

90. **De Morgan A.** The differential and integral calculus. London, 1842.

91. **Extended tables of Fresnel integrals.** U. S. Navy, Naval Research Laboratory, Boston, 1948.

92. **Oppolzer T.** Lehrbuch zur Bahnbestimmung der Kometen und Planeten, II. Engelmann, Leipzig, 1880.

93. **Pearson K.** Tables for statisticians and biometricians, 1. Biometric Laboratory, University College, London, 1930.

94. **Pearson K.** Tables for statisticians and biometricians, 2. Biometric Laboratory, University College, London, 1931.

95. **Peirce B. O.** A short table of integrals. Ginn, Boston—New York, 1910.

96. **Peters C. C.** and **Van Voorhis W. R.** Statistical procedures and their mathematical bases. Mc Graw-Hill, New York — London, 1940.

97. **Plummer H. C.** Probability and frequency. Macmillan, London, 1940.

98. **Potin L.** Formules et tables numériques. Doin et Gauthier-Villars, Paris, 1925.

99. **Preston T.** The theory of light. Macmillan, London, 1890.

100. **Rider P. R.** An introduction to modern statistical methods. Wiley, New York; Chapman and Hall, London, 1939.

101. **Rietz H. L.** Handbook of mathematical statistics. Houghton Mifflin, Boston—New York, 1924.

102. **Larsen, H. D.** Rinehart mathematical tables, formulas and curves. Chapman and Hall, London, 1949.

103. **Rivard M.** Tables des sinus, tangentes, secantes et de leurs logarithmes, avec la construction de ces tables, et les problèmes de la trigonométrie rectiligne et sphérique. Desaint J., Saillant Ch., Paris, 1743.

104. **Rosenbach I. B.**, **Whitman E. A.**, **Moskovitz D.** Mathematical tables. Ginn, Boston, 1946.

105. **Rosser J. B.** Theory and application of $\int_0^z e^{-x^2}dx$ and $\int_0^z e^{-p^2y^2}dy\int_0^y e^{-x^2}dx$, I: Methods of Computation. Brooklyn, New York, 1948.

106. **Schorr R.** Sammlung von Hilfstafeln der Hamburger Sternwarte in Bergedorf. Lucas Gräfe, Hamburg, 1916.

107. **Schumann W. O.** Elektrische Durchbruchfeldstärke von Gasen. Springer, Berlin, 1923.

108. **Silberstein L.** Synopsis of applicable mathematics, with tables. Bell, London, 1923.

109. **Smart W. M.** Stellar dynamics. Univ. Press, Cambridge, 1938.

110. **Stalmasek J.** Tabul'ky funkcií a konštant z matematiky, fyziky a chemie. Štátne nakl., Bratislava, 1951.

111. **Stumpff K.** Tafeln und Aufgaben zur harmonischen Analyse und Periodogammrechnung. Springer, Berlin, 1939.

112. **Tables of the error function and of its first twenty derivatives.** Harvard Univ. Press, Cambridge, Massachusetts, 1952.

113. **Tables of the probability functions, I:** $H'(x) = \frac{2}{\sqrt{\pi}} e^{-x^2}$ and $H(x) = \frac{2}{\sqrt{\pi}}\int_0^x e^{-t^2} dt$. National Bureau of Standards, Washington, 1941.

114. **Tables of the probability functions, II:** $\frac{1}{\sqrt{2\pi}} e^{-\frac{1}{2}x^2}$ and $\frac{1}{\sqrt{2\pi}}\int_x^\infty e^{-\frac{1}{2}t^2} dt$.

National Bureau of Standards, Washington, 1942.

115. **Thiele T. N.** Theory of observations. Layton, London, 1903.

116. **Uspensky J. V.** Introduction to mathematical probability. McGraw-Hill, New York — London, 1937.

117. **Van Deuren P.** Leçons sur le calcul des probabilités, I: La théorie des probabilités. Gauthier-Villars, Paris — Namur, 1934.

118. **Wijngaarden Adrian van and Scheen W. L.** Table of Fresnel integrals. Akademie van wetenschappen, Amsterdam, 1949.

119. **Watson G. N.** A treatise on the theory of Bessel functions. Univ. Press, Cambridge, 1945.

120. **Witkowski A. W.** Tablice matematyczno-fizyczne. Orgelbrand, Warszawa, 1904.

121. **Worthing A. G.** Treatment of experimental data. John Wiley, New York; Chapman and Hall, London, 1944.

122. **Yule G. U. and Kendall M. G.** An introduction to the theory of statistics. Griffin, London, 1937

123. **Abhandlungen der Mathematisch-Physikalischen Classe der Königlich Bayerischen Akademie der Wissenschaften** (München), 1886, **15**, 529—664.
Lommel E. C. J. Die Beugungserscheinungen geradlinig begrenzter Schirme.

124. **Acta Universitatis Lundensis** (Lund), 1906, (5), **1**.
Charlier C. V. L. Researches into the theory of probability.

125. **Acta Universitatis Lundensis** (Lund), 1912, (2), 8.
Charlier C. V. L. Studies in stellar statistics, I.

126. **Annalen der Physik** (Leipzig), 1907, (4), **23**, 875—904.
Ignatowsky W. V. Diffraction und Reflexion abgeleitet aus den Maxwellschen Gleichungen.

127. **Annales der Physik** (Leipzig), 1935, (5), **24**, 49—83.
Baumann H. Zur Verwendung von Operatoren in der Kontinuums-dynamik.

128. **Annalen der Physik und Chemie**, 1882, (3), **15**, 49—80.
Struve H. Fresnel's Interferenzerscheinungen; theoretisch und experimentell.

129. **Annalen der Physik und Chemie**, 1882, (3), **17**, 720—725.
Lindstedt A. Zur Theorie der Fresnel'schen Integrale.

130. **Annales de l'Observatoire de Paris. Memoires**, 1885, (B), **18**.
Radau R. Tables de l'intégrale $\psi(z) = e^{z^2}\int_z^\infty e^{-t^2}dt$.

131. **The Astrophysical Journal** (Chicago), 1938, **88**, 508—515.
Hjerting F. Tables facilitating the calculation of line absorption coefficients.

132. **The Astrophysical Journal** (Chicago), 1950, **111**, 142—156.
Chandrasekhar S. and Münch G. On the integral equation governing the distribution of the true and the apparent rotational velocities of stars.

133. **Berliner Astronomisches Jahrbuch for 1834.** 1832, 249—312.
Encke J. F. Über die Methode der kleinisten Quadrate.

134. **Biometrika** (London), 1902, **1**, 155—163.
Elderton W. P. Table for testing the goodness of fit of theory of observation.

135. **Biometrika** (London), 1903, **2**, 174—190.
Sheppard W. F. New tables of the probability integral.

136. **Biometrika** (London), 1907, **5**, 404—406.
Sheppard W. F. Table of derivate of the normal curve.

137. **Biometrika** (London), 1908, **6**, 59—68.
Pearson K. and Lee A. On the generalized probable error in multiple normal correlation.

138. **Biometrika** (London), 1910, **7**, 437—451.
Everitt P. F. Tables of the tetrachoric functions for fourfold correlation tables.

139. Biometrika (London), 1912, **8**, 385—395.
Everitt P. F. Supplementary tables for finding the correlation coefficient from tetrachoric groupings.

140. Biometrika (London), 1914, **10**, 208—214.
Lee A. Table of the Gaussian «Tail» function; when the «Tail» is larger than the body.

141. Biometrika (London), 1917, **11**, 284—291.
Lee A. Further supplementary tables for determining high correlations from tetrachoric groupings.

142. Biometrika (London), 1921, **13**, 426—428.
Pearson K. and Soper H. E. Tables of ordinates of the normal curve for each permille of frequency.

143. Biometrika (London), 1925, **17**, 343—354.
Lee A. Table of the first twenty tetrachoric functions to seven decimal places.

144. Biometrika (London), 1925, **17**, 364—387.
Tippett L. H. C. On the extreme individuals and the range of samples taken from a normal population.

45. Biometrika (London), 1926, **18**, 395—400.
Mills J. P. Table of the ratio: area to bounding ordinate, for any portion of the normal curve.

146. Biometrika (London), 1927, **19**, 354—404.
Lee A. Supplementary tables for determining correlation for tetrachoric groupings.

147. Biometrika (London), 1930, 22, 1—35.
Moul M., Elderton E. M., Fieller E. C., Pretorius J. and Church A. E. R. On the remaining tables for determining the volumes of a bi-variate normal surface.

148. Biometrika (London), 1931, 22, 368—376.
Kondo T. and Elderton E. M. Tables of normal curve functions to each permille of frequency.

149. Biometrika (London), 1942, **32**, 301—310.
Pearson E. S. and Hartley H. O. The probability integral of the range in samples of n-observations from a normal population.

150. Biometrika (London), 1950, **37**, 168—172.
Hartley H. O. and Pearson E. S. Table of the probability integral of the t-distribution.

151. British Association for the Advancement of Science. Reports, 1926, 273—297.
Fresnel integrals, $\sinh x$, $\cosh x$, and the confluent hypergeometric function.

152. British Association for the Advancement of Science. Reports, 1928, 305—340.
$\sin\theta$, $\cos\theta$, $\sinh \pi x$, $\cosh \pi x$, $\mathrm{Si}(x)$, $\mathrm{Ci}(x)$, $\frac{\partial}{\partial \nu} J_\nu(x)$ and integrals of the error integral.

153. Encyclopedia Britannica, 1911, **22**, 376—403.
Edgeworth F. Y. Probability.

154. Encyclopaedia Metropolitana, 1845, 2, 393—490.
De Morgan A. Theory of probabilities.

155. Handbuch der Physik, 1906, **6**, 1032—1119.
Pockels F. Beugung des Lichtes.

156. Journal of Educational Psychology, (Baltimore), 1938, 29, 491—500.
Krause R. H. and Conrad H. S. Students' tables of the unit normal curve, for abscissae expressed in terms of the probable error or PE : I. Areas corresponding to abscissae; II. Abscissae corresponding to areas.

157. Journal of Experimental Education, 1937, 5, 278—285.
Conrad H. S. and Krause R. H. An extension of the Kelley-Wood and Konds-Elderton tables of abscissae of the unit normal curve for areas $(\frac{1}{2}\alpha)$ between 0.4500 and 0.49999 99999.

158. Journal of the Franklin Institute, 1944, 237, 495—497.
Terrill H. M. and Sweeny L. An extension of Damson's table of the integral of e^{x^2}.

159. **Journal of the Franklin Institute,** 1944, 238, 220—222.
Terrill H. M. and Sweeny L. Table of the integral of e^{x^2}.

160. **Journal of the Institute of Actuaries and Assurance Magazine** (London), 1873, **17**, 50.
Woolhouse W. S. B. On the phylosophy of statistics.

161. **Journal of Physical Chemistry,** 1931, 35, 2785—2884.
Miller W. L. and Gordon A. R. Numerical evaluation of infinite series and integrals which arise in certain problems of linear heat flow, electrochemical diffusion.

162. **Journal of Psychology,** 1938, 5, 397—424.
Krause R. H. and Conrad H. S. New and extended tables of the unit normal curve: I. Areas $\left(\frac{1}{2}\alpha\right)$ corresponding to abscissae from.00 to 10.0 PE; II. Abscissae $\left(\frac{x}{PE}\right)$ corresponding to areas $\left(\frac{1}{2}\alpha\right)$ from·000 to ·49999 99999».

163. **Leipziger Berichte** (Berichte über die Verhandlungen der Kgl. Sächsischen Gesellschaft der Wissenschaften), 1928, 80, 217—223.
Sakamoto S. Tafel der Funktionen

$$y = \frac{e^{-x^2}}{x} \int_0^{\infty} e^{a^2} da \text{ und } z = e^{-x^2} \int_0^{\infty} e^{a^2} da$$

im intervall $x=0$ bis $x=10$.

164. **Mathematics and Physics Papers,** 1890, 3, 484—515.
Kelvin, Lord. Ether, electricity, and ponderable matter

165. **Mémoires de l'Academie des Sciences de l'Institut de France** (Paris), 1826, 5, 339—475.
Fresnel A. J. Mémoire sur la diffraction de la lumière.

166. **Mémoires Couronnés et autres mémoires publiés par l'Academie Royale des Sciences, des Lettres et des beaux Arts de Belgique** (Bruxelles), 1863, **31**, 1—52.
Gilbert P. Recherches analytiques sur la diffraction de la lumière.

167. **Memoirs and Proceedings of the Manchester Literary and Philosophical Society,** 1936, 80, 85—102.
Hartree D. R. Some properties and applications of the repeated integrals of the error function.

168. **Metron** (Rivista internationale di statistica), 1925, (3), 5, 105—108; 113—120.
Gosset W. S. («Student»). New tables for testing the significance of observations.

169. **Philips Research Reports** (Eindhoven, Netherlands), 1952, 7, 45—53.
Diemer G. and Dyjkgraaf H. Langmuir's ξ,η tables for the exponential region of the $I_a - V_a$ characteristic.

170. **Philosophical Magazine** (London), 1871, (4), 42, 421—436.
Glaisher J. W. L. On a class of definite integrals.

171. **Philosophical Magazine** (London), 1918, (6), 35, 30—49.
Silberstein L. Light distribution round the focus of a lens, at various apertures.

172. **Philosophische Studien** (Wundt), (Leipzig), 1894, 9, 145—150.
Kämpfe B. Tafel des Integrals $\varphi(\gamma) = \frac{2}{\sqrt{\pi}}\int_0^{\gamma} e^{-t^2}dt$.

173. **Physical Review,** 1930, (2), 36, 219—238.
Zemansky M. W. Absorption and collision broadening of the mercury resonance line.

174. **Physical Review** (Lancaster—New York), 1936, (2), 49, 78—97.
Nottingham W. B. Thermionic emission from tungsten and thoriated tungsten filaments.

175. **Physical Review** (Lancaster—New York), 1942, (2), 61, 528—530.
Rosenberg P Evaluation of functions related to Tait's mean free path.

176. **Proceedings of the Cambridge Philosophical Society,** 1897, 9, 259—269.
Mayall R. H. D. On the diffraction pattern near the focus of a telescope.

177. **Proceedings of the London Mathematical Society,** 1898, (1), 29, 519—522.
D a w s o n H. G. On the numerical value of $\int_0^h e^{x^2}\,dx$.

178. **Proceedings of the London Mathematical Society,** 1905, (2), 2, 371—400.
L a m b H. On deep-water waves.

179. **Proceedings of the Physico-Mathematical Society of Japan,** 1935, 17, 101—118.
O k a y a T. Sur le phénomène de Liesegang.

180. **Proceedings of the Royal Society of London,** 1916, (A), 92, 57—81.
T e r a z a w a K. On deep-sea water waves caused by a local disturbance on or beneath the surface.

181. **Proceedings of the Royal Society of London,** 1950, (A), **204**, № 1078, 396—405.
R h o d e s P. Fermi-Dirac functions of integral order.

182. **Psychometrika** (Journal devoted to the development of psychology as a quantitative rational Science Colorado Springs Psychometric Society), 1937, (1), 2, 55—66.
K r a u s e R. H. and C o n r a d H. S. A seven-decimal table of the area (α) under the unit normal curve for abscissae expressed in terms of *PE*.

183. **Quarterly Journal of Mechanics and applied Mathematics** (Oxford), 1948, 1, 319—326.
G o o d w i n E. T. and S t a t o n J. Table of $\int_0 e^{-u^2}du/(u+x)$.

184. **Science Reports of Tohoku Imperial University,** 1917, (1), 6, № 4, 169—181.
T e r a z a w a K. On the oscillations of the deep-sea surface caused by a local disturbance.

185. **Transactions of the Royal Society of Edinburgh,** 1887, **33**, 95.
C l a r k J. Table calculated for «on the foundations of the kinetic theory of gases».

186. **Transactions of the Royal Society of Edinburgh,** 1897—1898, **39**, № 9, 257—321.
B u r g e s s J. On the definite integral $\frac{2}{\sqrt{\pi}}\int_0^t e^{-t^2}dt$ with extended table of values, also $\frac{2}{\sqrt{\pi}}e^{-x^2}$.

187. **Verhandelingen K. Akademie van wetenschappen** (Amsterdam), 1931, 14, № 4.
B o u m a n J., J o n g W. F. Grafische bepaling van buigingsfiguren.

188. **Zeitschrift für angewandte Mathematik und Mechanik** (Berlin), 1943, 23, № 1.
S t ä b l e i n F. und S c h l ä f e r R. Numerische Berechnung von $y(x) = e^{-x^2}\int_0^x e^{t^2}dt$.

189. **Zeitschrift für angewandte Mathematik und Mechanik** (Berlin), 1951, 31, № 11—12 393—394.
E m e r s l e b e n O. Numerische Werte des Fehlerintegrals für $\sqrt{n\pi}$.

190. **Zeitschrift für angewandte Mathematik und Mechanik** (Berlin), 1949, 29, 289—300.
S c h w e i k e r t G. Zur Theorie des Gasdrucks gegen eine bewegte Wand.

191. **Zeitschrift für Physik** (Berlin), 1927, 45, 114—134.
K a p z o v N. und G w o s d o w e r S. Über die verschiedenen Schwingungsarten die von einer Elektronenröhre in der Schaltung von Barkhausen und Kurz erreugt werden.

192. **C o m r i e L. J.** Chambers's six-figure mathematical tables, II. Van Nostrand, New York, 1949.

References to Chapter 8

ELLIPTIC INTEGRALS AND ELLIPTIC FUNCTIONS

1. **Бронштейн И. Н.** и **Семендяев К. А.** Справочник по математике. Гостехиздат, М.— Л.,1945.

2. **Ветчинкин В. П.** Новые формулы и таблицы эллиптических интегралов и функций. Изд-во Военно-возд. акад. РККА, М., 1935.

3. **Двайт Г. Б.** Таблицы интегралов и другие математические формулы. Гос. изд-во иностр. лит., М., 1948.

4. **Записки Государственного гидрологического института** (Ленинград), 1932, VI.

5. **Машиностроение. Энциклопедический справочник**, I, кн. 1. ГНТИ Маш. лит., М. 1947.

6. **Милн-Томсон Л.** Эллиптические функции Якоби. Держ. наук техн. вид-во Укр., Харьков, 1933.

7. **Самойлова-Яхонтова Н. С.** Таблицы эллиптических интегралов ОНТИ, М.— Л. 1935.

8. **Сегал Р. П.** и **Семендяев К. А** Пятизначные математические таблицы. Изд-во АН СССР, М.— Л. 1948.

9. **Сикорский Ю. С.** Элементы теории эллиптических функций с приложениями к механике. ОНТИ, М.— Л., 1936.

10. **Справочное руководство по машиностроению**, I. ГНТИ Украины, Харьков, 1937.

11. **Hütte.** Справочник для инженеров, техников и студентов. Госмашметиздат, М.—Л., 1934.

12. **Шпильрейн. Я. Н.** Таблицы специальных функций. Числовые значения, графики и формулы, II. ГТТИ, М.—Л., 1934.

13. **Янке Е.** и **Эмде Ф.** Таблицы функций с формулами и кривыми. ГИТТЛ, М.—Л., 1949.

14. **Adams E. P.** and **Hippisley R. L.** Smithsonian mathematical formulae and tables of elliptic functions. Smithsonian Institution, Washington, 1947.

15. **Allen E. S.** Six-place tables. McGraw-Hill, New York — London, 1947.

16. **Bertrand J. L. F.** Traité de calcul différentiel et de calcul intégral, II. Gauthier-Villars, Paris 1870.

17. **Boll M.** Tables numériques universelles des laboratoires et bureaux d'étude. Dunod, Paris, 1947.

18. **Dale J. B.** Five figure tables of mathematical functions comprising tables of logarithms, powers of numbers trigonometric, elliptic and other transcendental functions. Arnold, London, 1949.

19. **Dwight H. B.** Mathematical tables. McGraw-Hill, New York—London, 1941.

20. **Fowle F. E.** Smithsonian physical tables. Smithsonian Institution, Washington, 1934.

21. **Greenhill A. G.** The applications of elliptic functions. Macmillan, London—New York, 1892.

22. **Hancock H.** Elliptic integrals. Wiley, New York; Chapman and Hall, London, 1917.

23. **Burington R. S.** Handbook of mathematical tables and formulae. Sandusky, Ohio, 1946.

24. **Hayashi K.** Fünfstellige Funktionentafeln. Springer, Berlin, 1930.

25. **Hayashi K.** Tafeln der Besselschen-,Theta-, Kugel- und anderen Funktionen. Springer, Berlin, 1930.

26. **Hayashi K.** Tafeln für die Differenzenrechnung... Springer, Berlin, 1933.

27. **Hoüel J.** Recueil de formules et de tables numériques. Gauthier-Villars, Paris, 1868, 1901.

28. **Hütte.** Des Ingenieurs Taschenbuch, I. Ernst, Berlin, 1931.

29. **Jahnke E.** und **Emde F.** Funktionentafeln mit Formeln und Kurven. Teubner, Leipzig — Berlin, 1938.

30. **Laska W.** Sammlung von Formeln der reinen und angewandte Mathematik. Vieweg, Braunschweig, 1888—1894.

31. **Legendre A. M.** Exercices de calcul intégral, 1. Huzard-Courcier, Paris, 1811.

32. **Legendre A. M.** Exercices de calcul intégral, 3. Huzard-Courcier, Paris, 1816.

33. **Legendre A. M.** Traité des fonctions élliptiques, 2. Huzard-Courcier, Paris, 1826.

34. **Levy L.** Precis élémentaire de la théorie des fonctions élliptiques avec tables numériques et applications. Gauthier-Villars, Paris, 1898.

35. **Low A. R.** Normal elliptic functions. University of Toronto Press, Toronto, 1950.

36. **Maxwell J. C.** A treatise on electricity and magnetizm, II. Clarendon Press, Oxford, 1893.

37. **Meissel E. D. F.** Sammlung mathematischer Tafeln. Iserlohn. im Selbstverlag, 1860.

38. **Milne-Thomson L. M.** Die elliptischen Funktionen von Jacobi. Springer, Berlin, 1931.

39. **Milne-Thomson L. M.** Jacobian elliptic function tables. Dover Publications Including, New York, 1950.

40. **Oberhettinger F.** und **Magnus W.** Anwendung der elliptischen Functionen in Physik und Technik (Die Grundlehren der Mathematischen Wissenschaften in Einzeldarstellungen), 55. Berlin, 1948.

41. **Peirce B. O.** A short table of integrals. Ginn, Boston—New York, 1910.

42. **Pidduck F. B.** A treatise on electricity. Cambridge Univ. Press, 1925.

43. **Potin L.** Formules et tables numériques. Doin et Gauthier-Villars, Paris, 1925.

44. **Rosenbach J. B., Whitman E. A., Moskovitz D.** Mathematical tables. Ginn, Boston 1946.

45. **Runkle J. D.** New tables for determining the values of the coefficients, in the perturbative function of planetary motion, which depend upon the ratio of the mean distances. Smithsonian Contributions, 9, Appendix, 1856.

46. **Russell A.** A treatise on the theory of alternating currents, I. Cambridge Univ. Press, 1914.

47. **Schlömilch O. X.** Vorlesungen über einzelne Theile der höheren Analysis. Vieweg, Braunschweig, 1879.

48. **Schmidt J. G.** System elliptischer Bogen. Reimer, Berlin, 1842.

49. **Silberstein L.** Synopsis of applicable mathematics, with tables. Bell, London, 1923.

50. **Spenceley G. W.** and **Spenceley R. M.** Smithsonian elliptic functions tables. Smithsonian Institution, Washington, 1947.

51. **Strayhorne D. A.** A study of an elliptic function. Chicago, 1946.

52. **Verhulst P. F.** Traité élémentaire des fonctions élliptiques. Brussels, 1841.

53. **Witkowski A. W.** Tablice matematyczno-fizyczne. Orgelbrand, Warszawa, 1904.

54. **Archiv für Electrotechnik** (Berlin), 1915, 3, 187—202.
Spielrein J. Die Induktivität eisenfreier Kreisringspulen.

55. **Archiv der Mathematik** (Basel—Stuttgart), 1949—1950, 2, № 2, 117—125.

R i e g e l s F. Formeln und Tabellen für ein in der räumlichen Potentialtheorie auftretendes elliptisches Integral.

56. **Astronomical Papers prepared for the Use the American Ephemeris and Nautical Almanac** (Washington), 1882, 1, 315—361.
H i l l G. W. On Gauss's method of computing secular perturbations, with an application to the action of Venus on Mercury.

57. **Astronomical Papers prepared for the Use of the American Ephemeris and Nautical Almanac** (Washington), 1891, 3, 1—200.
N e w c o m b S. Development of the perturbative function and its derivatives, in sines and cosines of multiples of the eccentric anomalies, and in powers of the eccentricities and inclinations.

58. **Bihang till Kgl. Svenska vetenskaps akademiens hadlingar** (Stockholm), 1900, 25, № 7.
B o h l i n K. Tables des fonctions elliptiques.

59. **British Association for the Advancement of Science. Reports,** 1912, 39—68.

60. **British Association for the Advancement of Science. Reports,** 1913, 115—130.

61. **British Association for the Advancement of Science. Reports,** 1919, 43—81.

62. **Electrical Engineering** (New York), 1935, 54, 709—711.
D w i g h t H. B. Elliptic integrals of large moduli.

63. **Electrotechnik und Maschinenbau** (Wien), 1912, 30, 221—226; 246—251; 267—271.
E m d e F. Die Berechnung eisenfreier Drosselspulen.

64. **Journal für die reine und angewandte Mathematik** (Berlin — Leipzig), 1848, 36, 1—74.
P l a n a J. Nouvelles formules pour reduire d'integrale $v=\int\left(\frac{T}{x^{1/2}}\right)dx$.

65. **Journal für die reine und angewandte Mathematik** (Berlin — Leipzig), 1843, 26 93—114.
J a c o b i C. G. J. Zur Theorie der elliptischen Functionen.

66. **Journal of the College of Science** (Imperial University of Tokyo), 1909, 27, № 6.
N a g a o k a H. The inductance coefficients of solenoids.

67. **Journal of the London Mathematical Society,** 1930, 5, 148—149.
M i l n e - T h o m s o n L. M. Ten decimal table of the nome q.

68. **Journal of Mathematics and Physics,** 1941, 20, 127—206; 336.
N e u m a n C. Tables of complete elliptic integrals.

69. **Journal of Mathematics and Physics,** 1948, 26, 234—245.
C a m b i E. Complete elliptic integrals of complex Legendrian modulus.

70. **Journal of Mathematics and Physics,** 1950, 29, № 2, 69—75.
K a p l a n E. L. Multiple elliptic integrals.

71. **Monthly Notices of the Royal Astronomical Society** (London), 1877, **37,** 366—386.
G l a i s h e r J. W. L. On an elliptic function solution of Kepler's problem.

72. **Monthly Notices of the Royal Astronomical Society** (London), 1902, **62,** 494—503.
I n n e s R. T. A. Jacobi's note (q) in astronomical formulae with numerical tables.

73. **Monthly Notices of the Royal Astronomical Society** (London), 1909, **69,** 648—650.
R o b b i n s F. Notes on certain coefficients appearing in the algebraical development of the perturbative functions.

74. **Monthly Notices of the Royal Astronomical Society** (London), 1939, **99,** 259—265.
F l e t c h e r A. Tables of the two chief Laplace coefficients.

75. **Philosophical Magazine** (London), 1935, (7), 19. 177—188.
A i r e y J R Toroidal functions and the complete elliptic integrals.

76. **Philosophical Magazine** (London), 1940, (7), 30, 516—519.

Fletcher A. A table of complete elliptic integrals.

77. **Philosophical Transactions of the Royal Society** (London), 1884, (A), 175, 411—460.
Rayleigh Lord and Sidgwick H. On the electro-chemical equivalent of silver.

78. **Philosophical Transactions of the Royal Society** (London). 1929. (A), 228, 275—329.
Rosenhead L. The Karman street of vortices in a channel of finite breadth.

79. **The Physical Review**, 1945, (2), 67, № 11 и 12, 351—357.
Osborn J. A. Demagnetizing factors of the general ellipsoid.

80. **Proceedings of the London Mathematical Society**, 1887, (1), 17, 262—287.
Greenhill A. G. Solution of the cubic and quartic equations by means of Weierstrass's elliptic functions.

81. **Proceedings of the London Mathematical Society** (London), 1932, (2), 33, 160 — 164.
Milne-Thomson L. M. Ten-figure table of the complete elliptic integrals K, K', E, E' and a table of $\frac{1}{\theta_3^2(0/\tau)}; \frac{1}{\theta_3^2(0/\tau')}$.

82. **Proceedings of Royal Artillery Institute**, 1889, 17, 181—216.
Greenhill A. G. The trajectory of a projectile for the cubic law of resistance.

83. **Proceedings of the Royal Society of Edinburgh**, 1931—1932, 52, №1, 236—250.
Milne-Thomson L. M. The Zeta functions of Jacobi, a seven-decimal table of z(u/*m*) at interval 0.01 for u and forvalues of m(= =k²) from 0.1 to 1.0.

84. **The Quarterly Journal of Mechanics and Applied Mathematics** (Oxford), 1951, 4, № 2, 223—235.
Fletcher A. Tables of two integrals and of Spielrein's inductance function.

85. **Scientific Papers of the U. S. Bureau of Standards** (Washington), 1924, 19, 541—576.
Curtis H. L. and Sparks C. M. Formulae, tables and curves for computing the mutual inductance of two coaxial circles.

86. **Scientific Papers of the U. S. Bureau of Standards** (Washington), 1924, 20, 1—18.
Grover F. W. Tables for the calculation of the mutual inductance of circuits with circular symmetry about a common axis.

87. **Scientific Papers of the Institute of Physical and Chemical Research** (Tokyo), 1922, 1, 1—67.
Nagaoka H. and Sakurai S. Tables of theta functions, elliptic integrals *K* and *E*, and associated coefficients in the numerical calculation of elliptic functions.

88. **Scientific Papers of the Institute of Physical and Chemical Research** (Tokyo), 1927, 2, 69—180.
Nagaoka H. and Sakurai S. Tables for facilitating the calculation of self-inductance of circular coil and of mutual inductance of coaxial circular currents.

89. **United States Department of Commerce, Bureau of Standards. Bulletin** (Washington), 1908, 5, 1—132.
Rosa E. B. and Cohen L. Formulae and tables for the calcula tion of mutual and self-inductance.

90. **United States Department of Commerce, Bureau of Standards. Bulletin** (Wachington) 1912, 8, 1—237.
Rose E. B. and Grover F. W. Formulae and tables for the calculation of mutual and self-inductance.

91. **United States Department of Commerce, Bureau of Standards. Bulletin** (Washington), 1912, 8, 269—393.
Rose E. B., Dorsey N. E. and Miller J. M. A determination of the international ampere in absolute measure.

92. **Zeitschrift für angewandte Mathematik und Mechanik** (Berlin), 1925, 5, 319—331.
Bergmann S. Über die Berechnung des magnetischen Feldes in einem Einphasen-Transformator.

93. **Proceedings of the Physico-Mathematical Society of Japan**, 1946, 25, № 1, 26—36.

References to Chapter 9

LEGENDRE FUNCTIONS AND POLYNOMIALS

1. **Бронштейн И. Н.** и **Семендяев А. К.** Справочник по математике. Гостехиздат, М.— Л., 1945.

2. **Марков А. А.** Исчисление конечных разностей. «Mathesis», Одесса, 1911.

3. **Машиностроение. Энциклопедический справочник**, I, кн. 1. ГНТИ Маш. лит., М. 1947.

4. **Справочное руководство по машиностроению.** I. ГНТИ Украины, Харьков, 1937.

5. **Шпильрейн Я. Н.** Таблицы специальных функций. Числовые значения, графики и формулы, II. ГТТИ, М.— Л., 1934.

6. **Янке Е.** и **Эмде Ф.** Таблицы функций с формулами и кривыми. ГИТТЛ, М.— Л., 1949.

7. **Angot A.** Compléments de mathématiques à l'usage des ingénieurs de l'électrotéchnique et des télécommunications. Revue d'Optique, Paris, 1949.

8. **Bauschinger J.** Tafeln zur theoretischen Astronomie. Engelmann, Leipzig, 1934.

9. **Bertrand J. L. F.** Traité de calcul différentiel et de calcul intégral, II. Gauthier-Villars. Paris, 1870.

10. **Boll M.** Tables numériques universelles des laboratories et bureaux d'étude. Dunod, Paris, 1947.

11. **British Association for the Advancement of Science. Mathematical Tables.** Legendre polynomials. Part-volume A, Cambridge, 1946.

12. **Byerly W. E.** An elementary treatise on Fourier's series and spherical, cylindrical and ellipsoidal harmonics. Ginn, Boston, 1893.

13. **Carr G. S.** A synopsis of elementary results in pure mathematics. Hodgson, London, 1886.

14. **Dale J. B.** Five-figure tables of mathematical functions comprising tables of logarithms, powers of numbers, trigonometric, elliptic — other transcendental functions. Arnold, London, 1949.

15. **Doodson A. T.** and **Warburg H. D.** The admiralty tide tables, III. H. M. Stationery Office, London, 1936.

16. **Dwight H. B.** Mathematical tables. McGraw-Hill New-York — London, 1941.

17. **Egersdörfer R** und **Egersdörfer L.** Formeln und Tabellen der zugeordneten Kugelfunktionen, 1. Art von $n=1$ bis $n=20$. Erster Teil : Formeln. Springer, Berlin, 1936.

18. **Fowle F. E.** Smithsonian physical tables. Smithsonian Institution, Washington, 1934.

19. **Hayashi K.** Sieben-und mehrstellige Talfen der Kreis-und Hyperbelfunktionen. Springer, Berlin, 1926.

20. **Hayashi K.** Fünfstellige Funk'tionentafeln. Springer, Berlin, 1930.

21. **Hayashi K.** Tafeln der Bessel'schen-, Theta-, Kugel- und anderen Funktionen. Springer, Berlin, 1930.

22. Heine E. Handbuch der Kugelfunktionen, II, ч. I. Reimer, Berlin, 1881.

23. Hobson E. W. The theory of spherical and ellipsoidal harmonics. Cambridge Univ. Press, 1931.

24. Jahnke E. und Emde F. Funktionentafeln mit Formeln und Kurven. Teubner, Leipzig — Berlin, 1938.

25. Markoff A. A. Differenzenrechnung. Teubner, Leipzig, 1896.

26. Mises R. Verzeichnis berechneter Funktionentafeln. Erster Teil. Bessel'sche-, Kugel- und elliptische Funktionen. VDI-Verlag, Berlin, 1928.

27. Moors B. P. Valeur approximative d'une intégrale définie. Gauthier-Villars, Paris, 1905.

28. Mursi Z. Tables of Legendre associated functions. Abbasia, Cairo 1941.

29. Prevost G. Tables de fonctions sphériques et de leurs intégrales pour calculer les coefficients du développement en serie de polynomes de Laplace d'une fonction de deux variables indépendantes. Soc. Sci. Phys. et Nat., Bordeaux; Gauthier-Villars, Paris, 1933.

30. Rosenbach I. B., Whitman E. A., Moskovitz D. Mathematical tables. Ginn, Boston, 1946.

31. Runge C. und König H. Vorlesungen über numerisches Rechnen. Springer, Berlin, 1924

32. Scarborough J. B. Numerical mathematical analysis. Johns Hopkins Press, Baltimore; Oxford Univ. Press, Milford, London, 1930.

33. Schmidt A. Tafeln der normierten Kugelfunktionen... Engelhard-Reyher, Gotha, 1935.

34. Silberstein L. Synopsis of applicable mathematics, with tables. Bell, London, 1923.

35. Schorr R. Sammlung von Hilfstafeln der Hamburger Sternwarte in Bergedorf. Lucas Gräfe, Hamburg, 1916.

36. **Tables of associated Legendre functions.** National Bureau of Standards. Columbia Univ. Press, New York, 1945.

37. Tallqvist H. J. Grunderna of teorin for sferiska funktioner... Helsingfors, 1905.

38. Todhunter I. An elementary treatise on Laplace's functions, Lame's functions and Bessel's functions. Macmillan, London, 1875.

39. Whittaker E. T. and Robinson G. The calculu of observations. Blackie, Londons — Glasgow, 1926.

40. **Abhandlungen der Kgl. Gesellschaft der Wissenschaften zu Göttingen,** 1922, 11, № 1.
Prey A. Darstellung der Höhen- und Tiefenverhaltnisse der Erde durch eine Entwicklung nach Kugelfunctionen bis zur 16. Ordnung.

41. **Abhandlungen Preussisches Meteorologisches Institut** (Berlin), 1925, 8, № 2.
Schmidt A. Allgemeine Formeln zur Vereinfachung häufig wiederholter Potentialberechnungen durch Benutzung fester Stationsgruppen.

42. **Acta Mathematica** (Uppsala), 1930, 54, 185—204.
Nyström E. J. Über die practische Auflösung von Integralgleichungen mit Anwendungen auf Randwertaufgaben.

43. **Acta Mathematica** (Uppsala), 1931, 56, 273—354.
Ginzel I. Die Konforme Abbildung durch die Gamma-funktion.

44. **Acta Societatis Scientiarum Fennicae**, 1906, 32, № 6, 1—27.
Tallqvist H. Tafeln der Kugelfunktionen $P_n(x)$ und ihrer abgeleiteten Funktionen.

45. **Acta Societatis Scientiarum Fennicae**, 1906, 33, № 4.
Tallqvist H. Tafeln der Kugelfunktionen P_n (cos θ).

46. **Acta Societatis Scientiarum Fennicae**, 1908, 33, № 9.

Tallqvist H. Tafeln der abgeleiteten und zugeordneten Kugelfunktionen, 1.

47. **Acta Societatis Scientiarum Fennicae**, 1937, 2, № 4, 43.
Tallqvist H. Sechsstellige Tafeln der 16 ersten Kugelfunktionen $P_n(x)$.

48. **Acta Societatis Scientiarum Fennicae**, 1938, (A), 2, № 11.
Tallqvist H. Sechsstellige Tafeln der 32 ersten Kugelfunktionen $P_n(\cos\theta)$.

49. **Annales de Géophysique** (Paris), 1948, 4, № 2, 96—108.
Lecolazet R. et Pluvinage P. États de régimes permanents électrodynamiques dans l'atmosphère.

50. **Biometrika** (London), 1938, **30**, 193—194.
Bayly B. F. Gauss'quadratic formula with twelve ordinates.

51. **British Association for the Advancement of Science. Reports** (London), 1879, 46—57.

52. **Bulletin of the American Mathematical Society**, 1942, 48, № 10, 739—743.
Lowan A. N., Davids N. and Levenson A. Table of the zeros of the Legendre polynomials of order 1—16 and the weight coefficients for Gauss' mechanical quadrature formula.

53. **Coll. Scientific Papers**, 1900, 2, № 2,
Adams J. C. Terrestrial magnetism.

54. **Societas Scientiarum Fennica, Commentationes Physico-Mathematicae** (Helsinki), 1932, 6, № 3.
Tallqvist H. Tafeln der 24 ersten Kugelfunktionen $P_n(\cos\theta)$.

55. **Societas Scientiarum Fennica, Commentationes Physico-Mathematicae** (Helsinki), 1932, 6, № 10.
Tallqvist H. Tafel der Kugelfunktionen $P_{25}(\cos\theta)$ bis $P_{32}(\cos\theta)$.

56. **Societas Scientiarum Fennica, Commentationes Physico-Mathematicae** (Helsinki), 1932, 6, № 11.
Tallqvist H. Potentialtafeln für die Kreislinie, Kreisfläche und Kreisscheibe.

57. **Comptes Rendus de l'Academie des Sciences**, 1880, 90, 913—915.
Radau R. Remarques sur la formule de quadrature de Gauss.

58. **Iowa State College Journal of Science**, 1938 12, № 2, 263—274.
Smith E. R. and Higdon A. Zeros of the Legendre polynomials.

59. **Journal of Mathematics and Physics**, 1951, 29, 292—299.
Carrus P. A. and Treuenfels C. G. Tables of roots of incomplete integrals of associated Legendre functions of fractional orders.

60. **Journal de Mathématiques**, 1880, (3), 6, 283—336.
Radau R. Étude sur les formules d'approximation qui servent à calculer la valeurs numérique d'une intégrale définie.

61. **Mémoires de la Societé des Science Physiques et Naturelles de Bordeaux**, 1886, (3), 3, № 1, 1—184.
Brunel G. Monographie de la fonction gamma.

62. **The Messenger of Mathematics**, 1887, 17, 21—23.
Cayley A. Note on the Legendrian coefficients of the second kind.

63. **The Messenger of Mathematics**, 1922, 52, 71—78.
Bateman H. Some problems in potential theory.

64. **Philosophical Magazine and Journal of Science**, 1891, (5), 32, 512—523.
Holland C. E., Jones P. R. and Lamb C. G. Table of zonal spherical harmonics. Short explanation and some illustrations of its use, by professor John Perry.

65. **Philosophical Transactions of the Royal Society** (London), 1903 (A), 200, 181—223.
Schuster A. On some definite integrals and a new method of reducing a function of spherical coordinates to a series of spherical harmonics.

66. **Philosophical Transactions of the Royal Society** (London), 1904, (A), 203, 87—110.
Lodge A. Values of Legendre's functions from P_0 to P_{20} at intervals of 5 degrees.

67. **Physical Review**, 1931, (2), 37, 507—531.
Bartlett J. H. Orbital valency.

68. **Physical Review,** 1931, (2), 38, 255—276.
R o s e n N. Calculation of interaction between atoms with s-electrons.

69. **Physical Review,** 1931, (2), 38, 2099—2114.
R o s e n N. The normal state of the hydrogen molecule.

70. **Proceedings of the National Academy of Sciences,** 1919, 5, 79—82.
W e b s t e r A. G. and F i s h e r W. Tables of the zonal spherical harmonic of the second kind $Q_1(z)$ and $Q_1'(z)$.

71. **Proceedings of the Physico-Mathematical Society of Japan,** 1938, (3) 20, № 1, 1—70.
K o t a n i M., A m e m i y a A. and S i m o s e T. Tables of integrals useful for the calculations of molecular energies.

72. **Proceedings of the Royal Society** (London), 1899, 64, 192—202.
F a r r C. C. On some expressions for radial and axial components of magnetic force.

73. **Quarterly Journal of Mathematics,** 1891, 25, 200—202.
P e r o t t J. The Gaussian interpolation theory formulae for $n =$ = 7,8 and 9.

74. **Terrestrial Magnetism and Atmospheric Electricity,** 1896, I, 73—76.
S c h m i d t A. Logarithmen der Kugelfunctionen der ersten fünf Ordnungen von fünf zu fünf Grad.

75. **Tracts for Computers** (London), 1923, 10.
I r w i n J. O. On quadrature and cubature...

76. **Zeitschrift für angewandte Mathematik und Mechanik** (Berlin), 1937, 17, 48—55.
F o u q u e t W. Dipolare Koordinaten und Kugelfunktionen.

77. **Zeitschrift für angewandte Mathematik und Mechanik** (Berlin), 1940, 20, 277—279.
V a n d r e y F. Tafel der acht ersten Kugelfunktionen zweiter Art.

78. **Zeitschrift für Mathematik und Physik,** 1899, 44, 327—338.
S c h m i d t A. Formeln zur Transformation der Kugelfunktionen bei linearer Änderung des Koordinatensystems.

79. **Кармазина Л. Н.** Таблицы полиномов Якоби. Изд-во АН СССР, М., 1954.

80. **Quarterly of Applied Mathematics,** 1953, 11, 311—318.
G r a y M. C. Legendre functions of fractional order.

81. **Tables des fonctions de Legendre associees.** Revue d'Optique, Paris, 1952.

References to Chapter 10

CYLINDER FUNCTIONS

1. Бронштейн И. Н. и Семендяев А. К. Справочник по математике. Гостехиздат, М.— Л., 1945.

2. Ватсон Г. Н. Теория бесселевых функций, II. Изд-во иностр. лит., М., 1949.

3. Глазенап С. П. Математические и астрономические таблицы в двух частях. Изд-во АН СССР, Л., 1932.

4. Грей Э. и Мэтьюз Г. Б. Функции Бесселя и их приложения к физике и механике. Изд-во иностр. лит., М., 1953.

5. Двайт Г. Б. Таблицы интегралов и другие математические формулы. Изд-во иностр. лит., М., 1948.

6. Люстерник Л. А., Акушский И. Я., Диткин В. А. Таблицы Бесселевых функций. Гостехиздат, М.— Л., 1949.

7. Машиностроение. Энциклопедический справочник, I, кн. 1. ГНТИМаш. лит., М., 1947.

8. Сегал Б. И. и Семендяев К. А. Пятизначные математические таблицы. Изд-во АН СССР, М.—Л., 1948.

9. Справочное руководство по машиностроению, I. ГНТИ Украины, Харьков, 1937.

10. Таблицы значений функций Бесселя от мнимого аргумента. Изд-во АН СССР, М.—Л., 1950.

11. Фаддеева В. Н., Гавурин М. К. Таблицы функций Бесселя $J_n(x)$ целых номеров от 0 до 120. ГИТТЛ, М.—Л., 1950.

12. Фок В. А. Таблицы функций Эйри Инф. отд. НИИ-108, М., 1946.

13. Шпильрейн Я. Н. Таблицы специальных функций. Числовые значения, графики и формулы, ч. I. ГТТИ, М.—Л., 1933.

14. Янке Е. и Эмде Ф. Таблицы функций с формулами и кривыми. ГИТТЛ, М.—Л., 1949.

15. Всеукраїнська Академія Наук. Природничо-технічний відділ. Журнал математичного циклу (Київ), 1932.
Динник А. Таблицы функций Бесселя дробного порядка.

16. Труды физико-математического института им. Стеклова, 2, №1. Изд-во АН СССР, Л., 1931.
Bursian V. and Fock V. Table of the functions $\int_x^\infty K_0(x)\,dx$; $\int_0^x I_0(x)\,dx$; $e^x \int_x^\infty K_0(x)\,dx$; $e^{-x}\int_0^x I_0(x)\,dx$.

17. Ученые записки Ленинградского университета. Серия физических наук, I, № 1, 1935, стр. 4—9.
Бурсиан В. Р. Таблицы значений функции $I_{1/3}$.

18. Airy G. B. Mathematical tracts. Cambridge, 1858.

19. Anding E. Sechsstellige Tafeln der Bessel'schen Funktionen imaginären Argumentes. Engelmann, Leipzig, 1911.

20. Angot A. Complements de mathématiques à l'usage des ingénieurs de

l'électrotéchnique et des télécommunications. Édition de la Revue d'Optique, Paris, 1949.

21. **B o l l M.** Tables numériques universelles des laboratoires et bureaux d'étude. Dunod, Paris, 1947.

22. **B r e m m e r H.** Terrestrial radio waves. Theory of propagation. Elsevier Publ. Co., New York, 1949.

23. **British Association for the Advancement of Science. Mathematical tables, VI.** Bessel functions, I: Functions of orders zero and unity. Cambridge, 1937.

24. **British Association for the Advancement of Science. Mathematical Tables, VII.**
S h e p p a r d W. F. The probability integral. Cambridge, 1939.

25. **B y e r l y W. E.** An elementary treatise on Fourier's series and spherical, cylindrical and ellipsoidal harmonics. Ginn, Boston, 1893.

26. **B y e r l y W. E.** Harmonic functions. Wiley, New York; Chapman and Hall, London, 1906.

27. **C a m b i E.** Eleven- and fifteenplace tables of Bessel functionen of the first kind to all significant orders. Dover, New York, 1948.

28. **D a l e J. B.** Five- figure tables of mathematical functions, comprising tables of logarithms, powers of numbers, trigonometric, elliptic and other transcendental functions. Arnold, London, 1949.

29. **D w i g h t H. B.** Mathematical tables. McGraw-Hill, New York — London, 1941.

30. **D w i g h t H. B.** Tables of integrals and other mathematical data. Macmillan, New York, 1947.

31. **F i s c h e r W.** Manuscript table of Bessel functions. 1926.

32. **F o w l e F. E.** Smithsonian physical tables. Smithsonian Institution, Washington, 1934.

33. **G r a y A.** Absolute measurements in electricity and magnetism, II. Macmillan, London, 1921.

34. **G r a y A.** and **M a t h e w s G. B.** A treatise on Bessel functions. Macmillan, London, 1931.

35. **G u m p r e c h t R. O.** and **S l i e p c e v i c h C. M.** Tables of Riccati—Bessel functions for large arguments and orders. Edwards, Ann Arbor, Michigan, 1951.

36. **H a n s e n P. A.** Ermittelung der absoluten Störungen in Ellipsen von beliebiger Excentricität und Neigung (Schriften der Sternwarte Seeberg. I). Gotha, 1843.

37. **H a y a s h i K.** Fünfstellige Funktionentafeln. Springer, Berlin, 1930.

38. **H a y a s h i K.** Tafeln der Besselschen-, Theta-, Kugel- und anderer Funktionen. Springer, Berlin, 1930.

39. **H a y a s h i K.** Tafeln für die Differenzenrechnung. Springer. Berlin, 1933.

40. **H u m p h r e y s W. J.** Physics of the air. Lippincott for Franklin Institute, Philadelphia, 1920.

41. **J a h n k e E.** und **E m d e F.** Funktionentafeln mit Formeln und Kurven. Teubner, Leipzig und Berlin, 1) 1909; 2) 1933; 3) 1938.

42. **L o m m e l E. C. J.** Studien über die Bessel'schen Functionen. Teubner, Leipzig, 1868.

43. **M a s c a r t H.** et **J o u b e r t J.** Leçons sur l'électricité et le magnetisme, I. Masson et Gauthier Villars, Paris, 1896.

44. **M c L a c h l a n N. W.** Bessel functions for engineers. Clarendon Press, Oxford, 1934.

45. **M e i s s e l E. D. F.** Programm 1889—1890 der Oberrealschule. Kiel, 1890.

46. **M i l l e r J. C. P.** The Airy integral giving tables of solution of the differential equation $y'' = xy$. Cambridge Univ. Press. 1946

47. **M o r g a n S. P.** Tables of Bessel functions of imaginary order and imaginary argument. Pasadena, Institute of Technology California, 1947.

48. **M o r s e P. M.** Vibration and sound. McGraw-Hill, New York — London, 1936.

49. **P e a r s o n K.** Tables for statisticians and biometricians, 2. Biometric Laboratory, University College, London, 1931.

50. P e r n t e r J. M. Meteorologische Optik. Braumüller, Vienna — Leipzig, 1906.

51. P r a n d t l L. Kipp-Erscheinungen. Koch, Nuremberg, 1901.

52. R a y l e i g h, Lord (S t r u t t J. W.). The theory of sound, I. Macmillan, London, 1894.

53. R a y l e i g h, Lord (S t r u t t J. W). The theory of sound, II. Macmillan, London, 1896.

54. R e i n s t e i n E. Untersuchungen über die Transversalschwingungen. Leipzig, 1911.

55. R o s e n b a c h J. B., W h i t m a n E. A., M o s k o v i t z D. Mathematical tables. Ginn, Boston, 1946.

56. R u s s e l l A. A treatise on the theory of alternating currents, I. Cambridge Univ. Press, 1914.

57. R u s s e l l A. A treatise on the theory of alternating currents, II. Cambridge Univ. Press, 1916.

58. S c h o r r R. Sammlung von Hilfstafeln der Hamburger Sternwarte in Bergedorf. Lucas Gräfe, Hamburg, 1916.

59. S c h w e r d F. M. Die Beugungserscheinungen. Schwan und Goetz. Mannheim, 1835.

60. S i l b e r s t e i n L. Synopsis of applicable mathematics, with tables. Bell, London, 1923

61. S m a r t W. M. Stellar dynamics. Cambridge Univ. Press, 1938.

62. **Table of the Bessel functions $J_0(z)$ and $J_1(z)$ for complex arguments.** Columbia Univ. Press, New York, 1950.

63. **Table of the Bessel functions $Y_0(z)$ and $Y_1(z)$ for complex arguments.** Prepared by the Computation Laboratory, National Bureau of Standards, Columbia Univ. Press, New Yo k, 1950.

64. **Tables of the Bessel functions $Y_0(x)$ $Y_1(x)$, $K_0(x)$, $K_1(x)$: $0 \leqslant x \leqslant 1$.** Gov. Print. Off, Washington, 1948.

65. **Tables of the Bessel functions of fractional order.** Prepared by the Computation Laboratory of the National Applied Mathematics Laboratories, National Bureau of Standards. 1) **I**, 1948, 2) **II**, 1949. Columbia Univ. Press, New York, 1948.

66. **Tables of the Bessel functions of the first kind of orders zero and one.** By the Staff of the Computation Laboratory, **III**. Harvard Univ. Press, Cambridge, Massachusetts, 1947.

67. **Tables of the Bessel functions of the first kind of orders two and three.** By the Staff of the Computation Laboratory, **IV**. Harvard Univ. P ess, Cambridge, Massachusetts, 1947.

68. **Tables of the Bessel functions of the first kind of orders four, five, and six.** By the Staff of the Computation Laboratory, **V**. Harvard Univ. Press, Cambridge, Massachusetts, 1947.

69. **Tables of the Bessel functions of the first kind of orders seven, eight and nine.** By the Staff of the Computation Laboratory, **VI**. Harvard Univ. Press, Cambridge, Massachusetts 1947.

70. **Tables of the Bessel functions of the first kind of orders ten, eleven and twelve.** By the Staff of the Computation Laboratory, **VII**. Harvard Univ. Press, Cambridge, Massachusetts. 1947.

71. **Tables of the Bessel functions of the first kind of orders thirteen, fourteen and fifteen.** By the Staff of the Computation Laboratory, **VIII**. Harvard Univ. Press, Cambridge, Massachusetts, 1947.

72. **Tables of the Bessel functions of the first kind of orders sixteen through twenty-seven.** The Annals of the Computation Laboratory of Harvard University, **IX**. Harvard Univ. Press, Cambridge, Massachusetts, 1948.

73. **Tables of the Bessel functions of the first kind of orders twenty-eight through thirty-nine.** The Annals of the Computation Laboratory of Harvard University, **X**. Harvard Univ. Press, Cambridge, Massachusetts, 1948.

74. **Tables of the Bessel functions of the first kind of orders forty through fifty-one.** The Annals of the Computation Laboratory of Harvard University, **XI**. Harvard Univ. Press, Cambridge, Massachusetts, 1948.

75. **Tables of the Bessel functions of the first kind of orders fifty-two through sixty-three.** The Annals of the Computation Laboratory of Harvard University, XII. Harvard Univ. Press, Cambridge, Massachusetts, 1949.

76. **Tables of the Bessel functions of the first kind of orders sixty-four through seventy-eight.** The Annals of the Computation Laboratory of Harvard University, XIII. Harvard Univ. Press, Cambridge, Massachusetts, 1949.

77. **Tables of the Bessel functions of the first kind of orders s^venty-nine through one hundred thirty-five,** The Annals of the Computation Laboratory of Harvard University, XIV. Harvard Univ. Press, Cambridge, Massachusetts, 1951.

78. **Tables of the modified Hankel functions of order one-third and of their deiivatives.** The Annals of the Computation Laboratory of Harvard University, II. Harvard Univ. Press, Cambridge, Massachusetts, 1945.

79. **Tables of spherical Bessel functions I.** Prepared by the Mathematical Tables Project, National Bureau of Standards, Columbia Univ. Press, New York, 1947.

80. **Tables of spherical Bessel functions, II.** Prepared by the Mathematical Tables Project, National Bureau of Standards, Columbia Univ. Press, New York, 1947.

81. **Thomson J. J.** Notes on recent researches in electricity and magnetism. Clarendon Press, Oxford, 1893.

82. **Thornton D. L.** Mechanics applied to vibrations and balancing. Chapman and Hall, London, 1939.

83. **Tölke F.** Bessel'sche und Hankel'sche Zylinderfunktionen nullter bis dritter Ordnung vom Argument $r\sqrt{i}$. Wittwer, Stuttgart, 1936.

84. **Shibogaki W.** Table of the modified Bessel functions. Kyushu University Physics Department, 1946.

85. **Watson C. G.** Catenarian functions (Selected Engineering Papers, № 179). Institution of Civil Engineers, London, 1935.

86. **Watson G. N.** A treatise on the theory of Bessel functions. Cambridge Univ. Press, 1945.

87. **Witkowski A. W.** Tablice matematyczno-fizyczne. Orgelbrand, Warszawa, 1904.

88. **Abhandlungen Akademie Berlin,** 1824. 1826, 1—52.
Bessel F. W. Untersuchung des Theils der planetarischen Störungen, welcher aus der Bewegung der Sonne ensteht.

89. **Abhandlungen der Königlichen Akademie der Wissenschaften zu Berlin, 1888.** Mathematische Abhandlungen. 1889, 1—23.
Meissel E. D. F. Tafel der Bessel'schen Funktionen I_k^0 und I_k^1 von $k = 0$ bis $k = 15{,}5$ berechnet.

90. **American Mathematical Monthly,** 1930, **37**, 491—492.
Schelkunoff S. A. A note on a paper concerning a Bessel function.

91. **Annalen der Physik** (Leipzig), 1906, (4), **19**, № 1, 80—115.
Kalähne A. Elektrische Schwingungen in ringförmigen Metallröhren.

92. **Annalen der Physik** (Leipzig), 1925, (4), **78**, № 1, 1—34.
Gans R. Mikroskopische Probleme.

93. **Annalen der Physik** (Leipzig), 1929, (5), **2**, № 5, 649—675.
Franke G. Erzwungene Schwingungen einer eingespannten kreisförmigen Platte.

94. **Annalen der Physik und Chemie** (Leipzig), 1882, (3), **17**, 1008—1016.
Struve H. Beitrag zur Theorie der Diffraction an Fernröhren.

95. **Annales Scientifiques de l'école Normale Supérieure,** 1866, 3, 55—95.
Bourget J. Mémoire sur le mouvement vibratoire des membranes circulaires.

96. **British Association for the Advancement of Science.** Table of Bessel function, II. Committee of Mathematical Tables. Cambridge, 1952.

97. **Short table of Bessel functions** $I_{n+1/2}$; $K_{n+1/2}$. Royal Society Mathema-

tical Tables Committee, Cambridge, 1952.

98. **Archiv für Elektrotechnik** (Berlin), 1929, 23, 162—180.
Ollendorff F. Choking coils with saturated iron cores.

99. **Archiv der Mathematik und Physik**, 1911, (3), 18, 337—338.
Dinnik A. Tafeln der Besselschen Funktionen $J_{\pm\frac{1}{3}}$ und $J_{\pm\frac{2}{3}}$.

100. **Archiv der Mathematik und Physik**, 1913, (3), 20, 238—240.
Dinnik A. Tafeln der Bessel'schen Funktionen $J_{\pm\frac{1}{2}}$, $J_{\pm\frac{3}{2}}$, $J_{\pm\frac{5}{2}}$.

101. **Archiv der Mathematik und Physik**, 1913, (3), 20, 289—294.
Airey J. R. The vibrations of cylinders and cylindrical shells.

102. **Archiv der Mathematik und Physik**, 1913, (3), 21, 324—326.
Dinnik A. Tafeln der Bessel'schen Funktionen $J_{\pm\frac{1}{4}}$ und $J_{\pm\frac{3}{4}}$.

103. **Archiv der Mathematik und Physik**, 1914, (3), 22, 30—43.
Airey J. R. Tables of the Neumanns functions or Bessel functions of the second kind.

104. **Archiv der Mathematik und Physik** (Berlin), 1914, (3), 22, 226—227, 376.
Dinnik A. Tafeln der Bessel'schen Funktionen $J_{\pm\frac{1}{3}}(ix)$ und $J_{\pm\frac{2}{3}}(ix)$.

105. **Archiv der Mathematik und Physik**, 1916, (3), 24, 239—250.
Emde F. Zur Berechnung der reellen Nullstellen der Bessel'schen Zylinderfunktionen.

106. **Arkiv för Matematik, Astronomi och Fysik** (Stockholm), 1918, 13, № 14.
Westgren A. Die Veränderungsgeschwindigkeit der lokalen Teilchenkonzentration in kolloiden Systemen.

107. **Arkiv för matematik, astronomi och fysik** (Stockholm), 1923, 17, № 12, 50.
Hogner E. A contribution to the theory of ship waves.

108. **Arkiv för matematik, astronomi och fysik** (Stockholm), 1933, (A), 24, № 5, 1—36.
Hallen E. Allgemeine Theorie des elektrischen Schwingungskreises.

109. **Arkiv för matematik, astronomi och fysik** (Stockholm), 1934, (A), 24, № 9, 1—50
Köhler H. Chlorine distribution and groups of drops in fog. Calculation of the colour of the corona in white light. Critical remarks on the coagulation theory of fog drops.

110. **Astronomische Nachrichten** (Kiel), 1891, 128, 153—156.
Meissel E. D. F. Abgekürzte Tafel der Bessel'schen Functionen $J_{\kappa}^{(n)}$.

111. **Astronomische Nachrichten** (Kiel), 1891, 128, 435—438.
Meissel E. D. F. Beiträge zur Theorie der Bessel'schen Funktionen.

112. **Astronomische Nachrichten** (Kiel), 1892, 129, 281—284.
Meissel E. D. F. Neue Entwicklungen über die Bessel'schen Functionen.

113. **The Astrophysical Journal** (Chicago), 1920, 51, 73—102.
Nagaoka H. Diffraction of a telescope objective in the case of a circular source of light.

114. **Australian Journal of Scientific Research** (Melbourne), 1948, (A), 1, № 1 1—17.
Goderey G. H. Diffraction of light from sources of finite dimensions.

115. **The Bell System Technical Journal** (New York), 1947, 26.
Kinzer J. P. and Wilson I. G. End plate and side wall currents in circular cylinder cevity resonator. Appendix, integration of $\int_0^x \frac{J_n(x)}{J'_n(x)}\,dx$, tables calculated by Miss F. C Larkey

116. **Biometrika** (London), 1929, 21, 194—201.
Elderton E. M. Table of the product moment I_m function

117. **Biometrika** (London), 1932, 24, № 3—4, 293—350.
Pearson K., Stouffer S. A. and David F. N. Further appli-

cations in statistics of the $I_m(x)$ Bessel function.

118. **Biometrika** (London), 1949, **36**, 142—148.
Gumbel E. J. Probability tables for the range.

119. **British Association for the Advancement of Science. Reports** (London), 1889, 28—32.

120. **British Association for the Advancement of Science. Reports** (London), 1893, 227—279.

121. **British Association for the Advancement of Science. Reports** (London), 1896, 98—149.

122. **British Association for the Advancement of Science. Reports** (London), 1907, 94—97.

123. **British Association for the Advancement of Science. Reports** (London), 1909, 33—36.

124. **British Association for the Advancement of Science. Reports** (London), 1911, 73—78.

125. **British Association for the Advancement of Science. Reports** (London), 1912, 56—68.
Webster A. G. Bessel functions, II.

126. **British Association for the Advancement of Science. Reports** (London), 1913, 115—130.

127. **British Association for the Advancement of Science. Reports** (London), 1914, 87—102.
Doodson A. T. Bessel functions of half-integral order.

128. **British Association for the Advancement of Science. Reports** (London), 1915, 27—38.

129. **British Association for the Advancement of Science. Reports** (London), 1916, 59—89.

130. **British Association for the Advancement of Science. Reports** (London), 1916, 92—96.

131. **British Association for the Advancement of Science. Reports** (London), 1916, 97—107.
Doodson A. T. Bessel functions of half-integral order.

132. **British Association for the Advancement of Science. Reports** (London), 1916, 108—122.
Savidge H. G. Tables of the ber and bei and ker and kei functions with further formulae for their computation.

133. **British Association for the Advancement of Science. Reports** (London), 1922, 263—272.
Riccati-Bessel functions and zeros of $J_n(x)$.

134. **British Association for the Advancement of Science. Reports** (London), 1923, 287—298.

135. **British Association for the Advancement of Science. Reports** (London), 1924, 275—295.

136. **British Association for the Advancement of Science. Reports** (London), 1925, 221—253.

137. **British Association for the Advancement of Science. Reports** (London), 1927, 220—254.
Confluent hypergeometric function, Ei $(\pm x)$, Si (x), Ci (x); zeros of $J_{-\nu}(x)$, ber x, bei x, ker x, kei x, and their first derivatives.

138. **British Association for the Advancement of Science. Reports** (London), 1928, 305—340.
Sin θ, cos θ, sinh πx, cosh πx, si(x), ci (x), $\frac{\partial}{\partial\nu} J_\nu(x)$ and integrals of the error integral.

139. **Bulletin of the American Mathematical Society,** 1897, (2), **3**, 153—155.
Willson R. W. and Peirce B. O. Table of the first forty roots of the Bessel equation $J_0(x) = 0$ with the corresponding values of $J_1(x)$.

140. **Bulletin of the American Mathematical Society,** 1927, **33**, 760—772.
Davis H. T. and Kirkham W. J. A new table of the zeros of the Bessel functions $J_0(x)$ and $J_1(x)$.

141. **Canadian Journal of Research,** 1931, (A), **5**, 149—155.
Ruedy R. Propagation of longitudinal waves in cylindrical rods.

142. **Canadian Journal of Research** (Ottawa), 1935, (A), **13**, № 1, 10—15.
Ruedy R. Anomalous dispersion of sound in solid cylindrical rods.

143. **Canadian Journal of Research** (Ottawa), 1938, (A), **16**, 138—148.
Ruedy R. Vibrations of power lines in a steady wind.

144. **Canadian Journal of Research** (Ottawa), 1941, (B), **19**, 261—274.
Heatley A. H. Reaction rates of the oxidation of liquid acetaldehyde.

145. **Encyclopedia Britannica** (London), 1888, **24**, 421—459.
Rayleigh, Lord (Strutt J. W). Wave theory of light.

146. **Gerlands Beiträge zur Geophysik** (Leipzig), 1929, **24**, № 4, 386—436.
Wagner A. Frequency distribution of wind elements.

147. **Industrial and Engineering Chemistry** (New York), 1942, **34**, 874—877.
Olson F. C. W. and Schultz O. T. Temperatures in solids during heating or cooling.

148. **Iowa State College Journal of Science**, 1938, **12**, № 2, 263—274.
Smith E. R. and Higdon A. Zeros of the Legendre polynomials.

149. **Iowa State College Journal of Science**, 1938. **12**, 471—474.
Bates J. M. Zeros of a class of polynomials associated with Bateman's *k*-function.

150. **Jahrbuch der drahtlosen Telegraphie und Telephonie**, 1911, **4**, 501—515.
Pedersen P. O. Wirbelstromverluste in und effektiver Widerstand von geraden, runden Metallzylindern.

151. **Japanese Journal of Physics**, 1937, **12**, 9—25.
Okaya F. Liesegang Rings.

152. **The Journal of the Acoustical Society of America**, 1949, **21**. 617—625.
Pachner J. Pressure distribution in the acoustical field excited by a vibrating plate.

153. **Journal of the Aeronautical Sciences** (New York), 1951, **18**, 478—483.
Like Y. L. and Dengles M. A. Tables of the Theodorsen circulation function for generalized motion.

154. **Journal of the American Institute of Electrical Engineers**, 1890, **18**, 4—37.
Kelvin, Lord. Ether, electricity and ponderable matter.

155. **Journal of the American Institute Electrical Engineers**, 1923, **42**, 827—831
Dwight H. B. A precise method of calculation of skin in isolated tubes.

156. **Journal of Applied Physics** (Lancaster), 1940, **11**, 208—219.
Muskat M., Morgan F. and Meres M. W. Studies in lubrication. III. The lubrication of plane sliders of finite Width.

157. **Journal of Applied Physics** (Lancaster), 1942, **13**, 496—502.
Bertram S. Calculation of axially symmetric fields.

158. **Journal of College of Science, Imperial University Japan**, 1891, **4**, 301—322.
Nagaoka H. Diffraction phenomena produced by an aperture on a curved surface.

159. **Journal of Franklin Institute**, 1919, **188**, 433—488, 607—674.
Humphreys W. J. Optics of the air.

160. **Journal of the Franklin Institute**, 1944, **237**, 301—303.
Smith D. B., Rodgers L. M. and Traub E. H. Zeros of Bessel functions.

161. **Journal of Mathematics and Physics**, 1937, **16**, 76.
Clapp M. M. Four place table of the Bessel function.

162. **Journal of Mathematics and Physics**, 1943, **22**, 2—12.
Lowan A. N. and Abramowitz M. Tables x of the integrals $\int_0^x J_0(t)dt$ and $\int_0^x Y_0(t)\,dt$.

163. **Journal of Mathematics and Physics**, 1943, **22**, 51—57.
Lowan A. N., Blanch G. and Abramovitz M. Tables of $Ji_0(x) = \int_x^\infty \frac{J_0(t)\,dt}{t}$ and related functions.

164. **Journal of Mathematics and Physics**, 1943, **22**, № 4, 208—209.
Lowan A. N. and Hillman A. P. A short table of the first five zeros of the transcendental equation $J_0(x)Y_0(kx) - J_0(kx)Y_0(x) = 0$.

165. **Journal of Mathematics and Physics**, 1945, **24**, № 1, 30—50.
Corrington M. S. and Michle W. Tables of Bessel functions $J_n(x)$ for large arguments.

166. **Journal of Mathematics and Physics**, 1945, **24**, № 3—4, 144—147.
Corrington M. S. Tables of Bessel functions $J_n(1000)$.

167. **Journal of Mathematics and Physics,** 1946, **24**, 252—259.
Table of the Struve functions $L_\nu(x)$ and $H_\nu(x)$.

168. **Journal of Mathematics and Physics,** 1947, **26**, № 1, 22—28.
Infeld L., Smith V. G. and Chien W. Z. On some series of Bessel functions.

169. **Journal of Mathematics and Physics,** 1948, **27**, 210—219.
Jaeger J. C. Repeated integrals of Bessel functions and the theory of transients in filter circuits.

170. **Journal of Mathematics and Physics,** 1950, **29**, 49—51.
Abramowitz S. Tables integrals of Struve functions.

171. **Journal of Mathematics and Physics,** 1950, **29**, 56—59.
Horton C. W. A short table of Struve functions and of some integrals involving Bessel and Struve functions.

172. **Journal of Mathematics and Physics,** 1951, **30**, № 2, 102—105.
Bogert B. P. Some roots of an equation involving Bessel functions.

173. **Journal of the Physical Society of Japan** (Tokyo), 1949, **4**, № 2, 73—77.
Kotani M. and Takahashi H. Numerical tables of functions useful for the calculation of resonant frequencies of cavity magnetron.

174. **L'Enseignement Mathématique,** 1923. **23**, 193—202.
Brasey E. Table des fonctions de Bessel.

175. **Luftfahrtforschung** (München-Berlin), 1944, **20**, 341—372.
Schwarz L. Untersuchung einiger mit den Zylinderfunktionen nullter Ordnung verwandter Funktionen.

176. **Mathematical Tables and other Aids to Computation** (Washington), 1944, **1**, № 5, 133—135.
Note on the computation of the Bessel function $I_n(x)$.

177. **Mathematical Tables and other Aids to Computation** (Washington), 1945, **1**, № 9, 353—354.
Zeros of certain Bessel functions of fractional order.

178. **Mathematical Tables and other Aids to Computation** (Washington), 1945, **1**, № 9, 337—351.
Machines for solving algebraic equations.

179. **Mathematical Tables and other Aids to Computation** (Washington), 1949, **3**, № 25, 351—352.
Complex zeros of $Y_0(z)$, $Y_1(z)$ and $Y'_1(z)$.

180. **Mathematics and Physics Papers,** 1890, **3**, 484—515.
Kelvin, Lord. Ether, electricity and ponderable matter.

181. **Mathematische und Naturwissenschaftliche Berichte aus Ungarn,** 1894, **11**, 362—373.
Steiner L. Intensitäts-verhältnisse der Beugungserscheinung durch eine kreisförmige Öffnung.

182. **Memoirs and Proceedings of the Manchester Literary and Philosophical Society,** 1904, **19**, № 48.
Isherwood J. G. Tables of the Bessel functions for pure imaginary values of the argument.

183. **Memoirs and Proceedings of the Manchester Literary and Philosophical Society,** 1939, **83**, 183—188.
Hartree D. R. and Johnston S. Note on the function

$$X(x) = \int_0^\infty e^{-(x-\omega^2)^2}\, d\omega.$$

184. **The Messenger of Mathematics,** 1897, **26**, 98—101.
Smith B. A. Table of Bessel's functions Y_0 and Y_1.

185. **The Messenger of Mathematics,** 1897, **27**, 20—98, 152—182.
Glaisher J. W. L. On the definite integrals connected with the Bernoullian function.

186. **Monthly Notices of the Royal Astronomical Society** (London), 1908, **68**, 588—605.
Eddington A. S. On the mathematical theory of two-star-drifts, and on the systematic motions of zodiacal stars.

187. **Monthly Notices of the Royal Astronomical Society,** 1919, **79**, 575—593
Conrady A. E. Star-discs

188. **Monthly Notices of the Royal Astronomical Society,** 1921, **81**, 547—565.
Buxton A. Star-discs.

189. **Monthly Notices of the Royal Astronomical Society,** 1933, **94**, 14—35.
Chandrasekhar S. The solar chromosphere

190. **Munich Abhandlungen** (Abhandlungen der Mathematisch-Physikali

schen Klasse der Königlich-Bayerischen Akademie der Wissenschaften), 1885, 15, 229—328.
Lommel E. C. J. Die Beugungserscheinungen einer kreisrunden Oeffnung und eines kreisrunden Schirmchens.

191. **Munich Abhandlungen** (Abhandlungen der Mathematisch-Physikalischen Klasse der Königlich-Bayerischen Akademie der Wissenschaften), 1886, 15, 529—664.
Lommel E. C. J. Die Beugungserscheinungen geradlinig begrenzter Schirme.

192. **Munich Abhandlungen** (Abhandlungen der Mathematisch-Physikalischen Klasse der Königlich Bayerischen Akademie der Wissenschaften), 1899, 19, 383—448.
Seeliger H. Die scheinbare Vergrösserung des Erdschattens bei Mondfinsternissen.

193. **Philips Research Reports** (Holland), 1949, 4, 179—188.
Bouwkamp C. J. On the effective length of a linear transmitting antenna.

194. **Philosophical Magazine** (London), 1841, (3), 18, 1—10.
Airy G. B. On the diffraction of an annular aperture.

195. **Philosophical Magazine** (London), 1879, (5), 8, 403—411.
Rayleigh, Lord (Strutt J. W.). Investigations in optics, with special reference to the spectroscope.

196. **Philosophical Magazine** (London), 1891, (5), 31, 87—99.
Rayleigh, Lord (Strutt J. W.). On pinhole photography.

197. **Philosophical Magazine** (London), 1891, (5), 32, 512—523.
Holland C. E., Jones P. R. and Lamb C. G. Table of zonal spherical harmonics. Short explanation and some illustrations of its use, by professor John Perry.

198. **Philosophical Magazine** (London), 1892, (5), 34, 177—180.
Rayleigh, Lord (Strutt J. W.). On the instability of cylindrical fluid surfaces.

199. **Philosophical Magazine** (London), 1898, (5), 45, 122—123.
Smith B. A. Tables following paper by J. H. Michell: «Tables of $kJ_0(x) - Y_0(x)$ and $kJ_1(x) - Y_1(x)$».

200. **Philosophical Magazine** (London) 1909, (6), 17, 524—552.
Russell A. The effective resistance and inductance of a concentric main and methods of computing the ber and bei allied functions.

201. **Philosophical Magazine** (London), 1909, (6), 18, 6—17.
Nicholson J. W. On the relation of Airy's integral to the Bessel functions.

202. **Philosophical Magazine** (London), 1910, (6), 19, 49—58.
Savidge H. G. Tables of the ber and bei and ker and kei functions with further formulae for their computation.

203. **Philosophical Magazine** (London), 1910, (6), 20, 591—610.
Russell A. The convection of heat from a body cooled by a stream of fluid.

204. **Philosophical Magazine** (London), 1911, (6), 21, 736—742.
Airey J. R. The oscillations of chains and their relation to Bessel and Neumann functions.

205. **Philosophical Magazine** (London), 1911, (6), 22, 658—663.
Airey J. R. Tables of the Neumann functions $G_n(x)$ and $Y_n(x)$.

206. **Philosophical Magazine** (London), 1913, (6), 25, 85—106.
Ward P. F. The transverse vibrations of a rod of varying cross section.

207. **Philosophical Magazine** (London), 1915, (6), 30, 459—481.
Paris E. T. On the polarization of light scattered by spherical metal particles.

208. **Philosophical Magazine** (London), 1916, (6), 31, 520—528.
Airey J. R. Bessel and Neumann functions of equal order and argument.

209. **Philosophical Magazine** (London), 1916, (6), 32, № 187, 7—14.
Airey J. R. The roots of Bessel and Neumann functions of high order.

210. **Philosophical Magazine** (London), 1918, (6), 35, 30—49.
Silberstein L. Light distribution round the focus of a lens, at various apertures.

211. **Philosophical Magazine** (London), 1918, (6), **35**, 364—370.
W a t s o n G. N. Bessel functions of equal order and argument

212. **Philosophical Magazine** (London), 1918, (6), **36**, 129—141.
W e b b H. A. and A i r e y J. R. The practical importance of the confluent hypergeometric function.

213. **Philosophical Magazine** (London), 1918, (6), **36**, 234—242.
A i r e y J. R. The addition theorem of the Bessel function of zero and unit orders.

214. **Philosophical Magazine** (London), 1921, (6), **41**, 200—205.
A i r e y J. R. Bessel function of small fractional order and their application to problems of elastic stability.

215. **Philosophical Magazine** (London), 1923, (6), **45**, 846—849.
W r i n c h H. E. H. and W r i n c h D. M. Tables of the Bessel functions $I_n(x)$.

216. **Philosophical Magazine** (London), 1923, (6), **45**, 1026—1049.
B i s a c r e F. F. D. The calculation of the skin effect in electrical conductors.

217. **Philosophical Magazine** (London), 1924, (6), **47**, 62—65.
W r i n c h H. E. H. and W r i n c h D. M. Tables of Bessel functions.

218. **Philosophical Magazine** (London), 1924, (6), **47**, 736.
O w e n S. P. Table of values of the integral $\int_0^x K_0(t)\,dt$.

219. **Philosophical Magazine** (London), 1925, (6), **50**, 1261—1264.
C a r r i n g t o n H. The frequencies of vibration of flat circular plates fixed at the circumference.

220. **Philosophical Magazine** (London), 1934, (7), **18**, 610—624.
M c L a c h l a n N. W. and M e y e r s A. L. The polar form of the ker and kei functions with applications to eddy current heating.

221. **Philosophical Magazine** (London), 1935, (7), **19**, 236—243.
A i r e y J. R. The Bessel derivatives $\frac{\partial J\nu(x)}{\partial x}$ and $\frac{\partial^2 J\nu(x)}{\partial x^2}$.

222. **Philosophical Magazine** (London), 1935, (7), **19**, 230—235.
A i r e y J. R. Bessel functions of nearly equal order and argument.

223. **Philosophical Magazine** (London), 1935, (7), **20**, № 132, 343—347.
B i c k l e y W. G. and N a y l e r J. A short table of the functions $Ki_n(x)$, from $n = 1$ to $n = 16$.

224. **Philosophical Magazine** (London), 1937, (7), **24**, 1041—1055.
C o l w e l l R. C. and H a r d y H. C. The frequencies and nodal systems of circular plates.

225. **Philosophical Magazine** (London), 1944, (7), **35**, 582—588.
B e l l R. P. Eigen-values and eigen-functions for the operator $\frac{d^2}{dx^2} - |x|$.

226. **Philosophical Magazine** (London), 1944, (7), 35, 693—705.
B l o c h A. On the temperature coefficient of air-cored self-inductances.

227. **Philosophical Magazine** (London), 1944, (7), **35**, 812—818.
C a r s t e n H. R. F. and M c K e r r o w N. W. The tabulation of some Bessel functions $K_\nu(x)$ and $K_\nu^{\bullet}(x)$ of fractional order.

228. **Philosophical Transactions of the Royal Society of London**, 1914, (A), **214**, 373—422.
K i n g L. V. On the convection of heat from small cylinders in a stream of fluid determination of the correction constants of small platinum wire with applications to hot-wire anemometry.

229. **Philosophical Transactions of the Royal Society of London**, 1921, (A), **222**, 57—100.
B u t t e r w o r t h S. Eddy-current losses in cylindrical conductors, with special applications to the alternating current resistances of short coils.

230. **Philosophical Transactions of the Royal Society of London**, 1926 (A), **226**, 135—155.
T o p p i n g J. and L u d l a m A. E. Tables of log $K_0(x)$ over

the range $x = 2$ to $x = 12$ at intervals of 0.001.

231. **Philosophical Transactions of the Royal Society of London,** 1951, (A), 243, 221—249.
Barnett M. P. and Coulson C. A. The evaluation of integrals occuring in the theory of molecular structure. Parts I and II.

232. **Physical Review,** 1936, (2), 49, 174—189.
Yost F. L., Wheeler J. A. and Breit G. Coulomb wave functions in repulsive fields.

233. **Proceedings of the American Academy of Arts and Sciences** (Boston), 1906, 41, 505—540.
Peirce B. O. On the manner of growth of a current in the coil of a nearly-closed electromagnet as influenced by the width of the air gap.

234. **Proceedings of the American Academy of Arts and Sciences** (Boston), 1907, 43, 99—182.
Peirce B. O. Rise of the current in the coil of an electromagnet.

235. **Proceedings of the American Institute of Electrical Engineers** (New York), 1915, 34, 1749—1814.
Kennelly A. E., Laws F. A. and Pierce P. H. Experimental researches on skin effect in conductors.

236. **Proceedings of the Cambridge Philosophical Society,** 1949, 45, 554—557.
Hartree D. R. The tabulation of Bessel functions for large argument.

237. **Proceedings Koninklijke Nederlandse Akademie van wetenschappen** (Amsterdam), 1950, 53, № 6, 869—881.
A. van Wijngaarden. A table of partitions into two squares with an application to rational triangles.

238. **Proceedings of the London Mathematical Society,** 1878, (1). 10, 4—13.
Rayleigh, Lord (Strutt J. W.). On the instability of jets.

239. **Proceedings of the London Mathematical Society,** 1884, (1), 15, 139—149.
Lamb H. On the induction of electric currents in cylindrical and spherical conductors.

240. **Proceedings of the London Mathematical Society,** 1912,(2), 11, 104-126.
Nicholson J. W. The pressure of radiation on a cylindrical obstacle.

241. **Proceedings of the Physico-Mathematical Society of Japan,** 1939, (3), 21, 287—298.
Okaya T. Complementary functions of modified Bessel functions.

242. **Proceedings of the Physical Society of London,** 1909, 21, 581—614.
Russell A. The effective resistance and inductance of a concentric main and methods of computing the ber and bei and allied functions.

243. **Proceedings of the Physical Society of London,** 1910, 22, 105—114.
Savidge H. G. Tables of the ber and bei and ker and kei functions with further formulae for their computation.

244. **Proceedings of the Physical Society of London,** 1910, 22, 432—452.
Russell A. The convection of heat from a body cooled by a stream of fluid.

245. **Proceedings of the Physical Society of London,** 1911, 23, № 3, 219—224.
Airey J. R. The roots of the Neumann and Bessel functions.

246. **Proceedings of the Physical Society of London,** 1911, 23, № 3, 225—232.
Airey J. R. The vibrations of circular plates and their relation to Bessel functions.

247. **Proceedings of the Physical Society of London,** 1913, 25, № 4, 294—297.
Butterworth S. On the evaluation of certain combinations of the ber, bei and allied functions.

248. **Proceedings of the Physical Society of London,** 1933, 45, 462—468.
Nancarrow H. A. Tables to facilitate the calculation of the temperature-distribution in a cylinder.

249. **Proceedings of the Physical Society of London,** 1943, 55, 286—291.
Selwyn E. W. H. The limit of visual resolution.

250. **Proceedings of the Physical Society of London,** 1948, 60, 71—82.
Collie C. H., Hasted J. B. and Ritson D. M. The cavity resonator

method of measuring the dielectric constants of polar liquids in the centimetre band.

251. **Proceedings of the Royal Society of Edinburgh,** 1942, (A), 61, 229—230.
Jaeger J. C. and Jaeger M. C. A short table of $\int_0^\infty \frac{e^{-xu^2}}{J_0^2(u)\,Y_0^2(u)}\frac{du}{u}$.

252. **Proceedings of the Royal Society of London,** 1899, (A), 64, № 407, 203—223.
Aldis W. S. Tables for the solution of the equation $\frac{d^2y}{dx^2}+\frac{1}{x}\frac{dy}{dx}-\left(1+\frac{n^2}{x^2}\right)y=0$.

253. **Proceedings of the Royal Society of London,** 1900, (A), 66, 32—43.
Aldis W. S. On the numerical computation of the functions $G_0(x)$, $G_1(x)$ and $J_n(x\sqrt{i})$.

254. **Proceedings of the Royal Society of London,** 1910, (A), 84, 25—46.
Rayleigh, Lord. The incidence of light upon a transparent sphere of dimensions comparable with the wavelength.

255. **Proceedings of the Royal Society of London,** 1917, (A), 93, № 651, 293—312.
Lamb H. On the deflection of the vertical by tidal loading of the earth's surface.

256. **Proceedings of the Royal Society of London,** 1918, (A), 94, 190—206.
Watson G. N. The zeros of Bessel functions.

257. **Proceedings of the Royal Society of London,** 1918, (A), 94, 307—314.
Airey J. R. The Lommel-Weber function and its application to the problem of electric waves on a thin anchor ring.

258. **Proceedings of the Royal Society of London,** 1922, (A), 101, 493—508.
Wrinch D. M. On the lateral vibrations of bars of conical type.

259. **Proceedings of the Royal Society of London,** 1925, (A), 108, 77—92.
Havelock T. H. Studies in wave resistance: the effect parallel middle body.

260. **Proceedings of the Royal Society of London,** 1929, (A), 123, 440—465.
Goldstein S. On the vortex theory of screw propellers.

261. **Proceedings of the Royal Society of London,** 1930, (A), 130, 29—37.
Watson G. N. The use of series of Bessel functions in problems connected with cylindrical wind-tunnels.

262. **Proceedings of the Royal Society of London,** 1951, (A), 204, № 1079, 533—548.
Wolf E. Light distribution near focus in an error-fre diffraction image.

263. **Quarterly Journal of Mathematics,** 1936, 7, 124—133.
Knight R. C. The potential of a sphere inside an infinite circular cylinder.

264. **Quarterly Journal of Mechanics and Applied Mathematics** (Oxford), 1948, 1, № 2, 220—224.
Goodwin E. T. and Staton J. Table of $J_0(j_{0,n}r)$.

265. **Quarterly Journal of Mechanics and Applied Mathematics** (Oxford), 1949, 2, 212—227.
Rothman M. Table of $\int_0^x I_0(x)\,dx$ for 0(0.1) 20(1) 25.

266. **Quarterly Journal of Mechanics and Applied Mathematics** (Oxford), 1949, 2, № 4, 452—457.
Olver F. W. J. Transformation of certain series occurring in aerodynamic interference calculations.

267. **Quarterly Journal of Mechanics and Applied Mathematics** (Oxford), 1951, 4, № 1, 106—115.
Gordon A. N. The field induced by an oscillating magnetic dipole outside a semi-infinite conductor.

268. **Atti della R. Accademia Nazionale dei Lincei, Rendiconti.** Classe di scienze Fisiche, Matematiche e Naturale (Roma), 1938, (6), 27, 525—528.
Watson G. N. Tabellazione di una particolare funzione definita da un integrale improprio.

269. **Rendiconti del Seminario matematico. Universita di Roma,** 1939, (4), 3, 271—279.
Badellino M. Sul colcolo delle funzioni di Bessel.

270. **Sankhya** (Indian Journal of Statistics), 1947, 8, 235—248.
Bose P. K. On recursion formulae, tables and Bessel function populations associated with the distribution of classical D^2-statistic.

271. **Science Reports of the Tohoku Imperial University,** 1939, (1), 28, 20—84.
Takagi M. On a statistical domain theory of ferro-magnetic crystals, I. Magnetization and magnetostriction.

272. **Terrestrial Magnetism and Atmospheric Electricity** (Baltimore), 1935, 40, 443—447.
Yost E. L., Wheeler J. A. and Breit G. Coulomb wave-functions.

273. **Mascart** E. Traité d'Optique, 1. Gauthier-Villars, Paris, 1889.

274. **Mascart** E. Traité d'Optique, 3. Gauthier-Villars, Paris, 1893.

275. **Transactions of the American Institute of Electrical Engineers** (New York), 1915, 34, 1953—2018.
Kennelly A. E., Laws F. A. and Pierce P. H. Experimental researches on skin effect in conductors.

276. **Transactions of the American Institute of Electrical Engineers** (New York), 1923, 42, 850—859.
Dwight H. B. Proximity effect in wires and thin tubes.

277. **Transactions of the American Institute of Electrical Engineers** (New York), 1929, 48, 812—820.
Dwight H. B. Bessel functions for $A - C$ problems.

278. **Transactions of the American Institute of Electrical Engineers** (New York), 1939, 58, 787—790.
Dwight H. B. Values of the Bessel functions ber x and bei x and their derivatives.

79. **Transactions of the American Institute of Electrical Engineers** (New York), 1941, 60, 135—136.
Dwight H. B. A five-figure table of the Bessel function $I_n(x)$.

280. **Transactions of the American Society of Mechanical Engineers** (New York), 1935, 57, 96—98.
Newman A. B. and Church A. H. The temperature history during the constant-rate heating of a solid circular cylinder.

281. **Transactions of the Cambridge Philosophical Society,** 1835, 5, 283—291.
Airy G. B. On the diffraction of an object-glass with circular aperture.

282. **Transactions of the Cambridge Philosophical Society,** 1838, 6, 379—402.
Airy G. B. On the intensity of light in the neighborhood of a caustic.

283. **Transactions of the Cambridge Philosophical Society,** 1849, 8, 595—599.
Airy G. B. On the intensity of light in the neighborhood of a caustic.

284. **Transactions of the Cambridge Philosophical Society,** 1851, 9, 166—187.
Stokes G. G. On the numerical calculation of a class of definite integrals and infinite series.

285. **Transactions of the Cambridge Philosophical Society,** 1891, 15, № 2, 139—266.
Chree C. On some compound vibrating systems.

286. **Transactions of the Royal Society of Canada,** 1943, (3), 37, 13—29.
Heatley A. H. A short table of the Toronto function.

287. **United States Department of Commerce Bureau of Standards. Bulletin** (Washington), 1912, 8 1—237.
Rosa E. B. and Grover F. W. Formulae and tables for the calculation of mutual and self-inductance.

288. **Wiener Sitzungsberichte** (Sitzungsberichte der Mathematisch-naturwissenschaftlichen Classe der Kaiserlichen Akademie der Wissenschaften) (Wien), 1936, (II A), 145, 797—826.
Karas K. Eigenschwingungen inhomogener Saiten.

289. **Zeitschrift für angewandte Mathematik und Mechanik** (Berlin), 1925, 5, 80—83.
Ikeda Y. Über die Nullstellen der Besselschen Funktionen.

290. **Zeitschrift für angewandte Mathematik und Mechanik** (Berlin), 1936, 16, № 4, 248—252.
Karas K. Tabellen für Bessel'sche Funktionen erster und zweiter Art mit den Parametern $\nu = \pm \frac{2}{3}$; $\pm \frac{1}{4}$; $\pm \frac{3}{4}$.

291. **Zeitschrift für angewandte Mathematik und Mechanik** (Berlin), 1936, 16, 321—335.
Biezeno C. B. und Koch J. J. Die Berechnung des statisch-unbestimmt gestützten geschlossenen Kreisringes.

292. **Zeitschrift für angewandte Mathematik und Mechanik** (Berlin), 1939, 19, 36—54.
Müller R. Über die zahlenmässige Beherrschung und Anwendung einiger der Bessel'scher verwandten Funktionen nebst Bemerkungen zum Gebiet der Bessel-Funktionen.

293. **Zeitschrift für angewandte Mathematik und Mechanik** (Berlin), 1939, 19, № 6, 372—379.
Brixy E. Logarithmus der Bessel'schen Funktionen reellen positiven Argumentes.

294. **Zeitschrift für angewandte Mathematik und Mechanik** (Berlin), 1952, 32, № 1, 27—30.
Hagen G. B. Über iterierte Integration von Bessel-Funktionen.

295. **Zeitschrift für Mathematik und Physik** (Leipzig), 1857, 2, №1, 137—165.
Schlömilch O. X. Über die Bessel'schen Funktion.

296. **Zeitschrift für Mathematik und Physik** (Leipzig), 1870, 15, 141—169.
Lommel E. C. J. Über die Anwendung der Bessel'schen Funktionen in der Theorie der Beugung.

297. **Zeitschrift für Mathematik und Physik** (Leipzig), 1907, 54, 55—86.
Kalähne A. Über die Wurzeln einiger Zylinderfunktionen und gewisser aus ihnen gebildeter Gleichungen.

298. **Zeitschrift für Mathematik und Physik**, 1914, 63, 1—66.
Weinreich W. Über den Temperaturverlauf in stromdurchflossenen Drähten besonders in Fall von Wechselstrom.

299. **Zeitschrift für Physik** (Berlin), 1926, 38, 308.

300. **Zeitschrift für technische Physik** (Leipzig), 1920, 1, 213—221.
Hort W. Die Geschwindigkeitsverteilung im Innern rotierender zäher Flüssigkeiten.

301. **Gumprecht R. O., Sliepcevich C. M.** Tables of light scattering functions for spherical particles. Engineering Research Institute University of Michigan, 1951.

302. **Сообщения Всероссийского Центрального Исполнительного Комитета**, 1922, 121—126.
Динник А. Таблицы функций Бесселя комплексного аргумента.

303. **Commentationes** (Pontifica Akademia Scientiarum), 1945, 9, № 3, 87—113.
Bordoni P. G. Sulle funzioni di Stokes.

304. **Journal of the Iron and Steel Institute**, 1952, **172**, 387—402.
Manterfield D., Cresswell J. D. and Herne H. The quickimmersion thermo-couple for liquid steel.

305. **Tables of Bessel-Clifford functions of orders zero and one.** Applied Mathematics Series, № 28. U.S. Government Printing Office, Washington, 1953.

306. **Proceedings of the Cambridge Philosophical Society,** 1953, 49, 446—448.
Chandrasekhar S. and Elbert D. The roots of $J_{-\left(l+\frac{1}{2}\right)}(\lambda\eta)\, J_{l+\frac{1}{2}}(\lambda) - J_{l+\frac{1}{2}}(\lambda\eta) \times J_{-\left(l+\frac{1}{2}\right)}(\lambda) = 0$.

307. **Proceedings of the Cambridge Philosophical Society,** 1954, **50**, 266—268.
Chandrasekhar S. and Elbert D. The roots of $Y_n(\lambda\eta) \times J_n(\lambda) - J_n(\lambda\eta)\, Y_n(\lambda) = 0$.

308. **Quarterly Journal of Mechanik and Applied Mathematik,** 1954, 7, № 1, 1—7.
Rothman M. The problem of an infinite plate under an inclined loading with tables of the integrals of Ai $(\pm x)$ and Bi $(\pm x)$.

309. **Quarterly of applied Mathematik,** 1952, 10, 255—262.
Thiruvenkatachar V. R. and Ramakrishna B. S. A case of combined radial and axial heat blow in composite cylinders.

310. **Tables of spherical wave functions.** Tokyo Numerical Computation Bureau, Report № 3. Kyohdoh, Setagaya, Tokyo, 1950.

311. **Knudsen H. L.** Bidrag til teorien for antennsystemer med hel Eller Delvis rotations — symmetri. I Komission hos Teknisk Forlag, København, 1953.

312. **Butler T.** and **Pohlhausen K.** Tables of definite integrals involving Bessel functions of the first kind. WADC Technical Report 54—420, Wright Air Development Center. 1954.

References to Chapter 11

CERTAIN SPECIAL FUNCTIONS AND INTEGRALS

1. **Алексеев В. А.** Сборник таблиц для расчета вероятностей. Арт. акад. им. Дзержинского, М., 1944.

2. **Ватсон Г. Н.** Теория бесселевых функций. Изд-во иностр. лит., М., 1949.

3. **Ветчинкин В. П.** Новые формулы и таблицы эллиптических интегралов и функций. Изд-во Воен.-возд. акад. РККА, М., 1935.

4. **Кармазина Л. Н.** Таблицы полиномов Якоби. Изд-во АН СССР, М., 1954.

5. **Митропольский А. К.** Интеграл вероятностей. Изд-во Лесотехн. акад. им. Кирова, Л. 1948.

6. **Шпильрейн Я. Н.** Таблицы специальных функций. Числовые значения, графики и формулы, ч. II. ГТТИ, М.— Л., 1934.

7. **Янке Е. и Эмде Ф.** Таблицы функций с формулами и кривыми. ГИТТЛ, М.— Л., 1949.

8. **Журнал экспериментальной и теоретической физики**, 1948, 18, 5, 457—461.
Угаров В. А. К теории страт.

9. **Труды Сибирского физико-технического института**, 1930.
Конусов В. Ф. Таблицы интегралов $\Phi_m(x) = \int_1^\infty \beta^m e^{-\beta m}\, d\beta$.

10. **Ballistic Research Laboratories.** Tables of ballistic functions $\xi(\theta)$, $c(\xi)$ $s(\xi)$, $X(\beta,b)$, $Y(\beta,b)$, $T(\beta,b)$. Aberdeen Proving Ground, 1949.

11. **Boll M.** Tables numériques universelles des laboratoires et bureaux d'étude. Dunod, Paris, 1947.

12. **British Association for the Advancement of Science. Mathematical Tables, 9:** Table of powers giving integral powers of integers. Cambridge Univ Press, 1950.

13. **Davis H. T.** Tables of the higher mathematical functions, II. Principia Press, Bloomington, Indiana, 1935.

14. **Delury D. B.** Values and integrals of the orthogonal polynomials up to $n = 26$. Published for Ontario Research Foundation by University of Toronto Press, Toronto, 1950.

15. **Fisher R. A. and Yates F.** Statistical tables for biological, agricultural and medical research. Oliver and Boyd, London — Edinburgh, 1949.

16. **Hartree D. R.** Note on the function $X(x) = \int_0^\infty e^{-(x-w^2)^2} dw$. Manchester, 1939.

17. **U. S. National Advisory Committee for Aeronautics, Technical Note,** 1948, № 1716.
Huckel V. Tables of hypergeometric functions for use in compressible-flow theory.

18. **Jahnke E. und Emde F.** Funktionentafeln mit Formeln und Kurven. Teubner, Leipzig — Berlin, 1938.

19. **Jörgensen N. R.** Undersögelser over frekvensflader og korrelation. Busck, Copenhagen, 1916.

20. **Milne-Thomson L. M.** The calculus of finite differences. Macmillan, London, 1933.

21. **Munk M. and Rawling.** Tables of Chaplygin functions. Naval

Ordnance Laboratory, Memorandum 10816. 1950.

22. **Newman F. W.** The higher trigonometry. Superrationals of second order. Macmillan and Bowes, Cambridge, 1892.

23. **Nörlund N. E.** Vorlesungen über Differenzenrechnung. Springer, Berlin, 1924.

24. **Peters C. C.** and **Van Voorhis W. R.** Statistical procedures and their mathematical bases. Mc-Graw-Hill, New-York—London, 1940.

25. **Prevost G.** Tables de fonctions sphériques et de leurs intégrales pour calculer les coefficients de développement en série de polynomes de Laplace d'une fonction de deux variables indépendantes. Soc. Sci. Phys. et Nat., Bordeaux; Gauthier-Villars, Paris, 1933.

26. **Renfer H.** Die Definitionen der Bernoullischen Funktionen... Wyss, Bern, 1900.

27. **Sasuly M.** Trend analysis of statistics. Brookings Institution, Washington, 1934.

28. **Spence W.** An essay on the theory of the various orders of logarithmic transcendents. London — Edinburgh, 1809.

29. **Steffensen J. F.** Interpolation. Bailliere, Tindall and Cox, Baltimore, 1927.

30. **Stratton J. A., Morse P. M., Chu L. J.** and **Hutner R. A.** Elliptic cylinder and spheroidal wave functions. Wiley, New York; Chapman and Hall, London, 1941, 1944.

31. **Tables of the confluent hypergeometric function $F(\frac{n}{2}, \frac{1}{2}, x)$ and related functions** (National Bureau of Standards, Applied Mathematics Series, № 3). United States Government Printing Office, Washington, 1949.

32. **Tables relating to Mathieu functions. Characteristic values, coefficients and Joining factors.** Cambridge Univ. press, New York, 1951.

33. **Watson G. N.** A treatise on the theory of Bessel functions. Cambridge Univ. Press, 1945.

34. **Wiener N.** Extrapolation, interpolation and smoothing of stationary time series with engineering applications. John Wiley and sons, New York, 1949.

35. **Acta Mathematica**, 1903, 27, 289—304.
Gram J. P. Note sur les zeros de la fonction $\zeta(s)$ de Riemann.

36. **Acta Mathematica**, 1926, 48, 393—400.
Walther A. Anschauliches zur Riemann'schen Zetafunktion.

37. **Acta Polytechnica** (Stockholm), 1948, 18.
Westberg R. On the harmonic and biharmonic problems of a region bounded by a circle and two parallel planes.

38. **American Mathematical Monthly**, 1936, 43, 354—358.
Smith E. R. Zeros of the Hermitian polynomials.

39. **Annalen der Physik**, 1913, (4), 43, 49—95.
Debye P. Interferenz von Röntgenstrahlen und Wärmebewegung.

40. **Annals of Mathematics**, 1907—1908, (2) 9, 167—172.
Gilman F. Evaluation of the probability integral.

41. **The Annals of Mathematical Statistics** (Baltimore), 1947, 18, № 3, 384—412.
Gumbel E. J. The distribution of the range.

42. **Arkiv för matematik, astronomi och fysik** (Stockholm), (A), 1943, 29, № 29.
Reiz A. On the numerical solution of certain types of integral equations.

43. **The Astrophysical Journal** (Chicago), 1940, 92, 27—49.
Morse P. M. The opacity of gas mixtures in stellar interiors.

44. **The Astrophysical Journal** (Chicago), 1949, 110, 355—374.
Krogdahl M. K. On broabening of hydrogen lines in stellar spectra — I.

45. **The Astrophysical Journal** (Chicago), 1950, 111, № 1, 142—156.
Chandrasekhar S. and Münch G. On the integral equation governing the distribution of the true and the apparent rotational velocites of stars.

46. **Biometrika** (London), 1923, 14, 281—310.
Pearson K. and Elderton E. M On the variate difference method.

47. **Biometrika** (London), 1931. 22, 362—367.
Pearson K. Tables of the mean value and squared standard deviation of the square of a multiple correlation coefficient.

48. **British Association for the Advancement of Science. Reports,** 1926 273—297.
Fresnel integrals, sinh x, cosh x, and the confluent hypergeometric function.

49. **British Association for the Advancement of Science. Reports,** 1927, 220—254.
Confluent hypergeometric function, $Ei(\pm x)$, $Si(x)$, $Ci(x$; zeros of $J_{-\nu}(x)$, ber x, bei x, ker x, kei x and their first derivatives.

50. **Bulletin of the American Mathematical Society,** 1948, 54, 765—769.
Greenwood R. E. and Miller J. J. Zeros of the Hermite polynomials and weights for Gauss'mechanical quadrature formula.

51. **Bulletin of the American Mathematical Society,** 1949, 55, № 10, 1004—1012.
Salzer H. E. and Zucker R. Table of the zeros and weight factors of the first fifteen Laguerre polynomials.

52. **Bulletin des Sciences Mathématiques** (Paris), 1938, (2), 62, 171—182.
Ser J. Formules et tables pour le calcul numérique de certaines séries divergentes.

53. **Canadian Journal of Research** (Ottawa), 1936, (A),14, 181—193.
Bronson H. L. and Wilson A. J. C. The heat capacities of silver, nickel, zinc, cadmium and lead from —80° to 120° C. БАН, БЛ

54. **Canadian Journal of Research** (Ottawa), 1950, (A), 28, № 2.
Table of the function $e^{-\alpha z/\gamma} M(\alpha,\gamma,z)$.

55. **Commentationes** (Pontificia Academia Scientiarum), 1945, 9, № 3, 87—113.
Bordoni P. G. Sulle funzioni di Stokes.

56. **Comptes Rendus,** 1887, 104, 1156—1159.
Jensen J. L. W. V. Sur la fonction $\zeta(s)$ de Riemann.

57. **Ingenieur-Archiv** (Berlin), 1937, 8, 99—103.
Olsson R. G. Tabellen der konfluenten hypergeometrischen Funktion erster und zweiter Art.

58. **Ingenieur-Archiv** (Berlin), 1937, 8, 373—380.
Olsson R. G. Über einige Lösungen des Problems der rotierenden Scheibe.

59. **Ingenieur-Archiv** (Berlin), 1938, 9, 126—137.
Gammel R. Das Durchschlagen von Kreisringen.

60. **Journal of Applied Physics** (Lancaster), 1950, 21, № 11, 1108—1114.
Bartberger C. L. The magnetic field of a plane circular loop.

61. **Journal of Mathematics and Physics,** 1926, 6, 1—32.
Beattie J. A. Six-place tables of the Debue energy, and specific heat functions.

62. **Journal of Mathematics and Physics,** 1933, 12, 291—297.
Russel J. B. A table of Hermite functions.

63. **Journal of Mathematics and Physics,** 1937, 16, 1—42.
Sutton C. S. An investigation of the average distribution of twin prime numbers.

64. **Journal of Mathematics and Physics,** 1938, 17, 123—199.
Lanczos C. Trigonometric interpolation of empirical and analytical functions.

65. **Journal of Mathematics and Physics,** 1942, 21, 264—283.
Lowan A. N. and Horenstein W. On the function $H(m, a, x) = \exp(-ix) F(m + 1 - ia; 2m + 2, ix)$.

66. **Journal of Mathematics and Physics,** 1943, 22, 136—147.
Lowan A. N. and Laderman J. Tables of Fourier coefficients.

67. **Journal of Mathematics and Physics,** 1949, 28, 183—191.
Mac Donald A. D. Properties of the confluent hypergeometric function.

68. **Journal of Mathematics and Physics,** 1950, 29, 146—178.
Clark R. A. On the theory of thin elastic toroidal shells.

69. **Journal of Mathematics and Physics**, 1951, **30**, № 1, 40—43.
Kruse U. E. and Ramsey N. F. The integral $\int_0^\infty y^3 \exp\left(-y^2 + \frac{ix}{y}\right) dy$.

70. **Journal of Mathematics and Physics**, 1952, **30**, № 4, 214—225.
Grossley F. R. E. A hyperelliptic function as a non-linear oscillation.

71. **Journal für die reine und angewandte Mathematik**, 1832, **8**, 298—300.
Clausen T. Über die Funktion $\sin \varphi + \frac{1}{2^2} \sin 2\varphi + \frac{1}{3^2} \sin 3\varphi + \ldots$

72. **Journal of Research of the National Bureau of Standards** (Washington), 1951, **47**, 288—290.
Abramowitz M. Tables of the functions $\int_0^\varphi \sin^{1/3} x dx$ and $\left(\frac{4}{3}\right) \sin^{-4/3} \varphi \times \times \int_0^\varphi \sin^{1/3} x dx$.

73. **Kongelige Danske videnskabernes selskab** (København), 1925, (8), **10**, № 3, 311—325.
Gram J. P. Tafeln für die Riemannsche Zetafunktion.

74. **Le Journal de Physique et le Radium**, 1946, (8), **7**, № 8, 231—243.
Colombani A. La théorie des filtres électriques et les polinomes de Tchebichef.

75. **Mathematical Gazette** (London), 1912, **6**, 332—336.
Sheppard W. F. The power-sum formula and the Bernoullian function.

76. **Memoirs of the Imperial Marine Observatory** (Kobe, Japan), 1936, **6**, № 2, 137—157.
Hidaka F. Tables for computing the Mathieu functions of odd order and their derivatives.

77. **Messenger of Mathematics** (Cambridge), 1871, **1**, 106—111.
Glaisher J. W. L. On the integrals $\int_0^\infty \sin (x^n) dx$ and $\int_0^\infty \cos (x^n) dx$.

78. **Messenger of Mathematics** (Cambridge), 1897, **27**, 20—98.
Glaisher J. W. L. On the definite integrals connected with the Bernoullian function.

79. **Mitteilungen der Naturforschenden Gesellschaft** (Bern), 1898, 1—56.
Otti H. Eigenschaften Bessel'schen Funktionen II-ter Art.

80. **Monthly Notices of the Royal Astronomical Society** (London), 1907, **67**, 444—447.
Robbins F. Computation of secular perturbations: Appendix.

81. **Monthly Notices of the Royal Astronomical Society** (London), 1908, **68**, 605—608.
Merfield C. J. Tables of the two hypergeometrical functions $F\left(\frac{1}{6}, \frac{5}{6}, 2, \sin^2 \frac{1}{2}\right)$ and $F\left(-\frac{1}{6}, \frac{7}{6}, 2, \sin^2 \frac{1}{2}\right)$ between the limits iota equals 90 and 180 degrees.

82. **Oversikt af Finska vetenskaps-societeten forhandlingar** (Helsingfors), 1912, (A), **54**, № 3.
Backlund R. J. Einige numerische Rechnungen, die Nullprodukte der Riemann'schen ζ-Funktion betreffend.

83. **Oversigt over det Kongelige Danske Videnskabernes Selskabs Forhandlingar** (København), 1902, 1—15.
Gram J. P. Note sur les zeros de la fonction ζ (s) de Riemann.

84. **Oversigt over det Kongelige Danske videnskabernes selskabs forhandlingar** (København), 1895, 303—308.
Gram J. P. Note sur le calcul de la fonction ζ(s) de Riemann.

85. **Philosophical Magazine** (London), 1918, (6), **36**, 129—141.
Webb H. A. and Airey J. R. The practical importance of the confluent hypergeometric function.

86. **Philosophical Magazine** (London), 1938, (7), **25**, 273—282.
Airey J. R. The radiation integrals $\int_0^\infty \frac{dx}{x^\alpha (e^x - 1)}$.

87. **Philosophical Magazine** (London), 1939, (7), **27**, 229—239.
Squire H. B. The lift and drag of a rectangular wing spanning a free jet of circular section.

88. **Philosophical Magazine** (London), 1943, (7), **34**, 600—607.

Powell E. O. An integral related to the radiation integrals.

89. **Philosophical Magazine** (London), 1947, (7), 38, № 280, 331—339.
Salzer H. Table of coeffecients for repeated integration with differences.

90. **Philosophical Magazine** (London), 1949, (7), **40**, № 302, 351—368.
Mitchell K. Tables of the function $\int_0^z \frac{-\log|1-y|}{y} dy$ with an account of some properties of this and related functions.

91. **Philosophical Transactions of the Royal Society** (London), 1939, (A), 237, 67—104.
McDougall J. and Stoner E. C. The computation of Fermi-Dirac functions.

92. **Philosophical Transactions of the Royal Society** (London), 1949, (A), 242, 1—32.
Fox E. N. The diffraction of two dimensional sound pulses incident on an infinite uniform slit in a perfectly reflecting screen.

93. **Proceedings of the Cambridge Philosophical Society**, 1938, 34, 412—423.
Wang J. S. The kinetics of absorption with long-range interaction between absorbed particles.

94. **Proceedings of the Cambridge Philosophical Society**, 1942, 38, 34—39.
Orr W. J. C. Expansions for a particular class of exponential-logarithmic integrals.

95. **Proceedings of the Cambridge Philosophical Society**, 1953, 49, 612—622.
Slater L. J. On the evaluation of the confluent hypergeometric function.

96. **Proceedings of the London Mathematical Society**, 1871—1873, 4, 48—56.
Glaisher J. W. L. On the constants that occur in certain summations by Bernoulli's series.

97. **Proceedings of the National Academy of Sciences** (Washington), 1940, 26, 491—496.
Bateman H. The polynomial of Mittag-Leffler.

98. **Proceedings of the Royal Society of Edinburgh**, 1925, 46, 20—29.
Ince E. L. Researches into the characteristic numbers of the Mathieu equation.

99. **Proceedings of the Royal Society of Edinburgh**, 1926, 46, 316—322.
Ince E. L. Researches into the characteristic numbers of the Mathieu equation.

100. **Proceedings of the Royal Society of Edinburgh**, 1927, 47, 294—301.
Ince E. L. Researches into the characteristic numbers of the Mathieu equation.

101. **Proceedings of the Royal Society of Edinburgh**, 1932, 52, № 1, 355—433.
Ince E. L. Tables of the elliptic-cylinder functions.

102. **Proceedings of the Royal Society of Edinburgh**, 1940, 60, 47—63.
Ince E. L. The periodic Lame functions.

103. **Proceedings of the Royal Society of Edinburgh**, 1940, 60, 83—99.
Ince E. L. The periodic Lame functions.

104. **Proceedings of the Royal Society of London**, 1920, (A), 98, 414—427.
Milne E. A. and Fowler R. H. Siren harmonics and a pure tone siren.

105. **Proceedings of the Royal Society of London**, 1935, (A), **151**, 234—255.
Titchmarsh E. C The zeros of the Riemann zeta-function.

106. **Proceedings of the Royal Society of London**, 1936, (A), **157**, 261—263.
Titchmarsh E. C. The zeros of the Riemann zeta-function.

107. **Proceedings of the Royal Society of London**, 1937, (A), **163**, 298—317.
Rosenhead L. and Miller J. C. P. Electrosmosis between plane parallel walls produced by high-frequency alternating currents.

108. **Proceedings of the Royal Society of London**, 1950, (A), **201**, № 1067, 586—599.
Frank F. C. Radially symmetric phase growth controlled by diffusion.

109. **Proceedings of the Royal Society of London**, 1950, (A), **204**, № 1078, 396—405.

Rhodes P. Fermi-Dirac functions of integral order.

110. **Publications of the Dominion Astrophysical Observatory** (Ottawa), 1931, 4, № 15, 209—260.
Zanstra H. An application of the quantum theory to the luminosity of diffuse nebulae.

111. **Quarterly of Applied Mathematics** (Menasha), 1943, 1, 215—236.
Lubkin S., Stoker J. J. Stability of columns and strings under periodically varying forces.

112. **Quarterly of Applied Mathematics** (Menasha), 1946, 4, № 1, 45—99.

Schoenberg I. J. Contributions of the problem of approximation of equidistant data by analytic functions Part A: On the problem of smoothing or graduation. A first class of analytic approximation formulae.

113. **Quarterly Journal of Mechanics and applied Mathematics** (Oxford), 1950, 3, № 2, 236—240.
Conolly B. W. A short table of the confluent hypergeometric function $M(\alpha,\gamma,x)$.

114. **Quarterly Journal of Mathematics** (Oxford), 1896, 28, 1—174.
Glaisher J. W. L. Products and series involving prime numbers only.

115. **Quarterly Journal of Mathematics** (Oxford), 1898, 29, 1—168.
Glaisher J. W. L. The Bernoullian function.

116. **Quarterly Journal of Mathematics** (Oxford), 1899, 30, 166—204.
Glaisher J. W. L. On the sums of the series $1^n + 2^n + \ldots x^n$ and $1^n - 2^n + \ldots \pm x^n$.

117. **Quarterly Journal of Mathematics** (Oxford), 1915, 46, 33—51.
Joffe S. A. Sums of like powers of natural numbers.

118. **Quarterly Journal of Mathematics** (Oxford), 1937, 8, 75—80.
Pidduck F. B. Electrical notes x. Multiple orbits in the magnetron.

119. **Tables of Whittaker functions.** Report № 8 of Numerical Computation Bureau. Published by the Tsuneta Yano Memorial Society, Tokyo, 1954.

120. **Quarterly Journal of Mechanics and Applied Mathematics** (Oxford), 1950, 3, 385—410.
Price A. T. Electromagnetic induction in a semi-infinite conductor with a plane boundary.

121. **Quarterly Journal of Mechanics and Applied Mathematics** (Oxford), 1950, 3, 107—112.
Scorer R. S. Numerical evaluation of integrals of the form $I = \int_{x_1}^{x_2} f(x) e^{\varphi(x)} dx$ and the tabulation of the function

$$Gi(z) = \frac{1}{\pi} \int_0^x \sin\left(uz + \frac{1}{3}u^3\right) du.$$

122. **Quarterly Journal of Mechanics and Applied Mathematics** (Oxford), 1952, 5, № 1, 109—115.
Godwin H. J. A method for the evaluation of

$$\int_0^\infty x^m \left[\sqrt{\left(\frac{2}{\pi}\right)} \int_x^\infty \exp\left(-\frac{1}{2}t^2\right) dt \right]^n dx.$$

123. **Quarterly Journal of Mechanics and Applied Mathematics** (Oxford), 1952, 5, № 1, 116—123.
Powell E. O. A table of the generalized Riemann zeta function in a particular case.

124. **Quarterly Journal of Mechanics and Applied Mathematics** (Oxford), 1952, 5, № 2, 196—205.
Chester W. The reflection of a transient pulse by a parabolic cylinder and a paraboloid of revolution.

125. **Atti della Accademia Nazionale dei Lincei. Rendiconti. Classe di scienze ficiche, matematiche e naturalı** (Roma), 1938, (6), 27, 2.
Watson G. N., Levi-Civita T. Tabellazione di una particolare funzione definita da un integrale improprio.

126. **Atti della Accademia Nazionale dei Lincei. Rendiconti. Ciasse di scienze fisiche, matemateche e naturali** (Roma), 1948, (8), 5, 339—344.
Stankiewicz L. Sul calcolo della piastra poggiata su suolo elastico.

127. **Reviews of Modern Physics,** 1933, 5, 169—202.

Blake F. C. On the factors affecting the reflection intensities by the several method of x-Ray analysis of crystal structures.

128. Sankhya (The Indian Journal of Statistics. Calcutta), 1951, **11**, № 2, 153—166.
Nath P. Confluent hypergeometric function.

129. Transactions of the American Mathematical Society, 1925, 27, 49—60.
Hutchinson J. I. On the roots of the Riemann zeta function.

130. Transactions of the Cambridge Philosophical Society, 1927, **23**, 303—336.
Goldstein S. Mathieu functions.

131. Transactions of the Royal Society of Canada (Ottawa), 1943, (3), **37**, 13—29.
Heatley A. H. A short table of the Toronto function.

132. Zeitschrift für Astrophysik (Berlin), 1931, 2, № 1, 1—29.
Zanstra H. Untersuchungen über planetarische Nebel.

133. Zeitschrift für Astrophysik (Berlin), 1932, 4, 118—152.
Strömgren B. The opacity of stellar matter and the hydrogen content of the stars.

134. Tables of Chebyshev Polynomials $S_n(x)$ and $C_n(x)$. Applied Mathematics Series, № 9. Government Printing Office, Washington, 1952.

135. Annali di matematica pura ed applicata, 1949, (4), 28, 263—289.
Tricomi F. G. Sul comportamento asintotico dei polinomi di Laguerre.

136. Успехи математических наук, 1952, (нов. сер.), 7, № 6, 3—96.
Дородницын А. А. Асимптотические законы распределения собственных значений для некоторых особых видов дифференциальных уравнений второго порядка.

References to Chapter 12

SOLUTIONS OF CERTAIN EQUATIONS

1. Битковский И. И. Таблицы для решения уравнения Гаусса $\sin(z-q) = m \sin^4 z$. Астр. обс. Новосибирского ун-та.

2. Осипов В. С. Таблицы для решения кубических уравнений. Изд-во ВИА РККА, Москва, 1937.

3. Щумягский Б. М. Таблицы для решения кубических уравнений методом основ. Гостехиздат, М.-Л., 1950.

4. Янке Е. и Емде Ф. Таблицы функций с формулами и кривыми. ГИТТЛ, М.— Л., 1949.

5. Adams E. P. and Hippisley R. L. Smithsonian mathematical formulae and tables of elliptic functions. Smithsonian Institution, Washington, 1947.

6. Boll M. Tables numériques universelles des laboratoires et bureaux d'étude. Dunod, Paris, 1947.

7. British Association for the Advancement of Science. Mathematical Tables, 2. London, 1932.
   Emden functions.

8. Brockmeyer E. Sandsynligbedsregningens anvendelse i telefonien. København 1949.

9. Chandrasekhar S. An introduction to the study of stellar structure. Chicago Univ. Press, 1939.

10. Eddington A. S. The internal constitution of the stars. Cambridge Univ. Press, 1930.

11. Emden R. Gaskugeln. Teubner, Leipzig—Berlin, 1907.

12. Hayashi K. Fünfstellige Funktionentafeln. Springer, Berlin, 1930.

13. Jahnke E. und Emde F. Funktionentafeln mit Formeln und Kurven. Teubner, Leipzig — Berlin. 1) 1909; 2) 1933.

14. Katz D. Pocket tables for cubics. A systematic method for algebraic treatment of cubic equations. South Milwaukee, Wisconsin, 1933.

15. McLachlan N. W. Complex variable and operational calculus with technical applications. Cambridge Univ. Press, 1939.

16. Pernter J. M. Meteorologische Optik. Braumüller, Vienna — Leipzig, 1906.

17. Rayleigh, Lord. The theory of sound, I. Macmillan, London, 1894.

18. Rayleigh, Lord. The theory of sound, II. Macmillan, London, 1896.

19. Silberstein L. Synopsis of applicable Mathematics, with tables. Bell, London, 1923.

20. Verdet E. Leçons d'optique physique, I. Masson, Paris, 1869.

21. Abhandlungen der Mathematisch-Physikalischen Klasse der Königlich-Bayerischen Akademie der Wissenschaften (München), 1885, 15, 229—328.
   Lommel E. C. J. Die Beugungserscheinungen einer kreisrunden Oeffnung und eines kreisrunden Schirmchens.

22. Abhandlungen der Mathematisch-Physikalischen Classe der Königlich-Bayerischen Akademie der Wissenschaften (München), 1886, 15, 529—664.

L o m m e l E. C. J. Die Beugungserscheinungen geradlinig begrenzten Schirme.

23. **Acta Polytechnica**, 1948, 18.
W e s t b e r g R. On the harmonic and biharmonic problems of a region bounded by a circle and two parallel planes.

24. **Annalen der Physik**, 1921, (4), 65, 41—50.
C o l l o J. B. Ein sender elektrischer Wellen im Innern einer metallischen Hülle.

25. **Archiv der Mathematik**, 1949—1950, 2, 186—191.
H o d a p p W. Über die Hermite'schen Funktionen zweiter Art von reellem und rein imaginärem Argument.

26. **Archiv der Mathematik und Physik**, 1887, 5, 279—344.
S i e b e l A. Exacte Trennung der reellen Wurzeln numerischer algebraischen und transcendenten Gleichungen.

27. **The Astrophysical Journal** (Chicago), 1939, 89, № 1, 116—118.
C h a n d r a s e k h a r S. Lane-Emden function $\theta_{3.25}$.

28. **Astronomische Nachrichten** (Altona), 1872, 78, 177—198.
G a s p a r i s A. Table pour resoudre l'équation $m \sin^4 z = \sin(z-q)$; m, q positifs.

29. **The Cyclopaedia of Universal Dictionary of Arts, Sciences and Literature** (London), 1860, 4, 951—967.
De M o r g a n A. Involution and evolution.

30. **Encyclopaedia Britannica** (London), 1888, 24, 421—459.
R a y l e i g h, L o r d. Wave theory of light.

31. **Handbuch der Astrophysik** (Berlin), 1930, 3, 1, 187.
M i l n e E. A. Thermodynamics of the stars.

32. **Journal of Mathematics and Physics**, 1950, 29, 146—178.
C l a r k R. A. On the theory of thin elastic toroidal shells.

33. **Journal of Mathematics and Physics**, 1945, 24, № 1, 51—64.
W e l l s C. P. and S p e n c e R. D. The parabolic cylinder functions.

34. **Journal of Mathematics and Physics**, 1945, 24, 109—125.
T e a P. L. A graphical method for the numerical solution of Fredholm's integral equation of the second kind.

35. **Journal of Mathematics and Physics**, 1947, 26, № 2, 79—92.
B o u w k a m p C. J. On spheroidal wave functions of order zero.

36. **Kongliga Svenska vetenskaps akademies handlingar** (Stockholm), 1913, 51, № 5.

37. **Mathematical Tables and other Aids to Computation** (Washington), 1944, I, 202—203.
W r e n c h J. W. A root of the equation 10 log $x=x$.

38. **Mathematician**, 1850, 3, 289—291.

39. **The Messenger of Mathematics**, 1879, 9, 93—95.
L e w i s T. C. Some cases of vortex motion.

40. **Monthly Notices of the Royal Astronomical Society** (London), 1931, 91, 55—63.
F a i r c l o u g h N. Numerical integration of Emden's polytropic equation of index three.

41. **Monthly Notices of the Royal Astronomical Society** (London), 1932, 93, № 1, 40—49.
F a i r c l o u g h N. Numerical integration of Emden's polytropic equation of index three.

42. **Oeuvres Complétes**, 1882, (1), 1, 3—318.
C a u c h y A. L. Théorie de la propagation des ondes à la surface d'un fluide pesant d'une profondeur indéfinie.

43. **Oregon State Monographs** (math. and stat. ser.), 1935, I.
M i l n e W. E. Tables of derivatives for damped vibrations.

44. **Philosophical Magazine** (London), 1878, (5), 6, 241—263.
B l a k e J. F. On the measurement of the curves formed by cephalopods and other mollusks.

45. **Philosophical Magazine** (London), 1908, (6), 16, 1—23; 687—711.
K e l v i n, L o r d and G r e e n G. The problem of a spherical gaseous nebula.

46. **Philosophical Magazine** (London), 1935, (7), 20, 528—531.
Silberstein L. The roots of cos $z=z$.

47. **Philosophical Magazine** (London), 1943, (7), 34, 575.
Hillman A. P. and Salzer H. E. Roots of sin $z=z$.

48. **Philosophical Magazine** (London), 1948, (7), 39, № 299, 1004—1005.
Robbins C. J. and Smith R. C. T. A table of roots of sin $z=-z$.

49. **Physical Review**, 1923, (2), 22, 347—356.
Langmuir I. and Blodgett K. B. Currents limited by space charge between coaxial cylinders.

50. **Physical Review**, 1930, (2), 36, 630—647.
Baker E. B. The application of the Fermi-Thomas statistical model to the calculation of potential distribution in positive ions.

51. **Physical Review**, 1931, (2), 38, 1898—1902.
Bush V. and Caldwell S. H. Thomas-Fermi equation solution by the differential analyzer.

52. **Proceedings of the American Society of Civil Engineers**, 1920, 46, 375—425.
Smith B. A. Arched dams.

53. **Proceedings of the Cambridge Philosophical Society**, 1938, 34, 365—374.
Wilson A. H. The binding energies of the hydrogen isotopes.

54. **Proceedings of the Edinburgh Mathematical Society**, 1891, 9, 80—83.
Miller T. H. On the numerical values of the roots of the equation cos $x=x$.

55. **Proceedings Koninkijke Nederlandsche Academie vom wetenschappen** (Amsterdam), 1946, 49, № 7, 822.
Schouten J. A. The product of n homogeneous linear forms.

56. **Proceedings of the London Mathematical Society**, 1872, (1), 4, 93—103.
Rayleigh, Lord. On the vibrations of gas contained within a rigid spherical envelope.

57. **Proceedings of the London Mathematical Society**, 1882, (1), 13, 182—212.
Lamb H. On the vibration of an elastic sphere.

58. **Proceedings of the London Mathematical Society**, 1929, (2), 29, 293—308.
Watson G. N. Theorems stated by Ramanujan (v): approximations connected with e^x.

59. **Proceedings of the Physical Society of London**, 1932, 44, 17—24.
McKay A. T. Diffusion for the infinite plane sheet.

60. **Proceedings of the Royal Society of Edinburgh**, 1908, 28, 259—302.
Kelvin, Lord and Green G. The problem of a spherical gaseous nebula.

61. **Proceedings of the Royal Society of London**, 1949, (A), 200, № 1060, 20—23.
Chandrasechar S. On Heisenberg's elementary theory of turbulence.

62. **Proceedings of the Royal Society of London**, 1949, (A), 200, № 1060, 100—113.
Davies C. N. The sedimentation and diffusion of small particles.

63. **Quarterly Journal of Mathematics**, 1933, 4, 140—146.
Watson G. N. Du Bois-Reymond's constants.

64. **Quarterly Journal of Mechanics and applied Mathematics**, 1951, 4, № 1, 42—63.
Lock R. C. The velocity distribution in the laminar boundary layer between parallel streams.

65. **Rendiconti dell'Istituto Lombardo di Scienze e Letter**, 1939, 72, 333—361
Casale F. Su di una equazione collegata a quella di Keplers, I, II.

66. **Simon Stevin**, 1950, 27, 177—192.
Thijsen W. P. Over de geitereerde machtsverheffing.

67. **Technical Report Aeronautical Research Committee 1934—1935**, 1934, I, № 1632, 320—375.
Howarth L. On the calculation of steady flow in the boundary layer near the surface of a cylinder in a stream.

68. **Transactions of the American Electrochemical Society** (New York), 1935. 66, 345—358.
Newmann A. B. and Green L. The temperature history and rate of heat loss of an electrically heated slab.

69. **Transactions of the American Society of Civil Engineers** (New York), 1921, 83, 2027—2093.
Smith B. A. Arched dams.

70. **Transactions of the American Society of Mechanical Engineers** (Journal of Applied Mechanics section), 1937, 59.

71. **Transactions of the Cambridge Philosophical Society**, 1864, 10, 331—358.

72. **University of California, Publications in Engineering**, 1944, 5, № 3, 59.
Perry R. L. and Berggren W. P. Transient heat conduction in hollow cylinders after sudden changes of innersurface temperature.

73. **University of Oregon Publications** (Eugene, Oregon), 1929, I, № 1.
Milne W. E. Tables of damped vibrations.

74. **University of Oregon Publications** (Eugene, Oregon), 1923, 2, № 2.
Milne W. E. Damped vibrations.

75. **Zeitschrift für angewandte Mathematik und Mechanik**, 1943, 23, № 2, 101—117.
Buchholz H. Die konfluente hypergeometrische Funktion mit besonderer Berücksichtigung ihrer Bedeutung für die Integration der Wellengleichung in den Koordinaten eines Rotationsparaboloides.

76. **Proceedings of the Royal Society of London**, 1950, (A), 201, 159—174; 175—186.
Taylor G. The formation of a blast wave by a very intense explosion. I: Theoretical discussion. II: The atomic explosion of 1945.

77. **Панов Д. Ю.** Справочник по численному решению дифференциальных уравнений в частных производных. ОГИЗ-Гостехиздат, М., 1943.

78. **Диткин В. А. и Кузнецов П. И.** Справочник по операционному исчислению. ГИТТЛ, М.-Л., 1951.

79. **Камке Э.** Справочник по обыкновенным дифференциальным уравнениям. Изд-во иностр. лит., М., 1951.

80. **Mathematical Tables and other Aids to Computation** (Washington), 1953, 7, № 43, 165—168.
Iverson, K. E. The zeros of the partial sums of *ez*.

References to Chapter 13

SUMS AND VALUES CONNECTED WITH FINITE DIFFERENCES

1. **Андреев П. П.** Математические таблицы. Госстатиздат, М., 1952.

2. **Брадис В. М.** Четырехзначные математические таблицы, Учпедгиз, М., 1950.

3. **Гнеденко Б. В.** Курс теории вероятностей, ГИТТЛ, М.—Л., 1950.

4. **Гутерман А. В. и Браташ В. Н.** Математические таблицы для конструкторов-машиностроителей ГНТИ Маш. лит., Москва—Свердловск, 1944. **МГУ, ПБ, БЛ**

5. **Иванов К. П.** Таблицы для вычисления многочленов. Гостехиздат, М.—Л., 1949.

6. **Каллет Ф.** Таблицы логарифмов чисел от 1 до 108000, синусов, косинусов и тангенсов от секунды до секунды для первых пяти градусов, и от десяти до десяти секунд для всех градусов четверти окружности. М., 1837.

7. **Карпов К. А.** Таблицы коэффициентов интерполяционной формулы Лагранжа. Приложение к таблицам функции $W(z)=e^{-z}\int_0^z e^{x^2}\,dx$ в комплексной области. Изд-во АН СССР, М., 1954.

8. **Марков А. А.** Исчисление конечных разностей. «Mathesis», Одесса, 1911.

9. **Машиностроение.** Энциклопедический справочник, т. I, кн. 1. ГНТИ Маш. лит., М., 1947.

10. **Таблицы Барлоу-квадратов, кубов, квадратных корней и обратных величин для целых чисел до 12 500.** Изд-во иностр. лит., М., 1950.

11. **Фаддеева В. Н. и Гаврин М. К.** Таблицы функций Бесселя $J_n(x)$ целых номеров от 0 до 120. ГИТТЛ, М.— Л., 1950.

12. **Hütte.** Справочник для инженеров, техников и студентов, Госмашметиздат, М.— Л., 1934.

13. **Прикладная математика и механика,** 1948, 12, вып 2, 233—240.
Китовер К. А. Таблицы сумм некоторых бесконечных тригонометрических рядов.

14. **Adams E. R.** and **Hippisley R. L.** Smithsonian mathematical formulae and tables of elliptic functions. Smithsonian Institution, Washington, 1947.

15. **Albrecht C. T.** Formeln und Hilfstafeln für geographische Ortsbestimmungen. Engelmann, Leipzig, 1908.

16. **Barbette E.** Les sommes de $p^{\text{ièmes}}$ Puissances distinctes égales à une $p^{\text{ième}}$ Puissan e, suivie d'une table des 5000 premiers nombres triangulaires. Gauthier-Villars, Liège—Paris, 1910.

17. **Barlow P.** New mathematical tables. G. and S. Robinson, London, 1814.

18. **Barlow P.** Tables of squares, cubes, square roots, cube roots. Spon, London; Chemical Publishing Co, New York, 1947.

19. **Bauschinger J.** Tafeln zur theoretischen Astronomie Engelmann, Leipzig, 1934.

20. **Bertrand J. L. F.** Traité de calcul différentiel et de calcul intégral, II: Calcul Intégral. Gauthier-Villars, Paris, 1870.

21. **B o c c a r d i J. C. M.** Tables logarithmiques des factorielles jusqu'à 10 000! Imprimerie Mistral, Cavaillon, 1932.

22. **B o l l M.** Tables numériques universelles des laboratoires et bureaux d'étude. Dunod, Paris, 1947.

23. **British Association for the Advancement of Science. Mathematical Tables, I.**
Circular and hyperbolic functions, exponential and sine and cosine integrals, factorial function and allied functions, Hermitian probability functions. Cambridge Univ. Press, 1951.

24. **British Association for the Advancement of Science. Mathematical Tables, II.**
Emden function. London, 1932.

25. **British Association for the Advancement of Science. Mathematical tables, IV.**
Bessel functions. Part 1: Functions of orders zero and unity. Cambridge Univ. press, 1937.

26. **British Association for the Advancement of Science. Mathematical tables, IX.**
Table of powers giving integral powers of integers. Cambridge, 1940.

27. **B r u n s H.** Grundlinien des wissenschaftlichen Rechnens. Teubner, Leipzig, 1903.

28. **C a l l e t F.** Tables portatives de logarithmes, contenant les logarithmes des nombres, depuis 1 jusqu'à 108 000 les logarithmes des sinus et tangentes... F. Didot. Paris. 1795.

29. **C a r r G. S.** A synopsis of elementary results in pure mathematics. 1; 2. Hodgson, London, 1880; 1886.

30. **C h a p p e l l E.** A table of coefficient to facilitate interpolation by means of the formulae of Gauss. Bessel and Everett. Author, London, 1929.

31. **D a l e J. B.** Five-figure table of mathematical functions comprising tables of logarithms. powers of numbers trigonometric, elliptic and other transcendental functions. Arnold, London, 1949.

32. **D a v i s H. T.** Tables of the higher mathematical functions, I. Indiana, Principia Press, Bloomington, 1933.

33. **D a v i s H. T.** Tables of the higher mathematical functions, II. Indiana Principia Press, Bloomington, 1935.

34. **D e M o r g a n A.** The differential and integral calculus. London, 1842.

35. **D e s v a l l e e s H. R.** Tables logarithmiques et trigonométriques à quartre décimales. Gauthier-Villars, Paris, 1920.

36. **D w i g h t H. B.** Mathematical tables. McGraw-Hill, New York — London, 1941.

37. **E d w a r d s J.** A treatise on the integral calculus, II. Macmillan, London, 1922.

38. **E u l e r L.** Introductio in analysin infinitorum, I. Lipsiae — Berolini, 1922.

39. **E u l e r L.** Institutiones calculi differentialis, 2. Petersbourg, [1789].

40. **E u l e r L.** Opuscula analytica, II. Petersburg, 1785.

41. **F r y T. C.** Probability and its engineering uses. Van Nostrand, New York, 1928.

42. **G a u s s C. F.** Werke (Göttingen), 1866, 3, 163—196.
Methodus nova integralium valores per approximationem inveniendi.

43. **G l o v e r J. W.** Tables of applied mathematics in finance, insurance and statistics. Wahr, Ann Arbor, Michigan, 1930.

44. **C o u l d e n C. H.** Methods of statistical analysis. Wiley, New York; Chapman and Hall, London, 1939.

45. **B u r i n g t o n R. S.** Handbook of mathematical tables and formulae. Sandusky, Ohio, 1946.

46. **H a n t s c h l J.** Logarithmisch-trigonometrisches Handbuch. Rohrmann und Schweigerd, Wien, 1833.

47. **H a n s e n P. A.** Tables de la lune, construites d'après le principe Newtonian de la graviation universelle. Admiralty, London, 1857.

48. Hayashi K. Fünfstellige Funktionentafeln. Springer, Berlin, 1930.

49. Hindenburg C. F. Sammlung combinatorisch-analytischer Abhandlungen, II. Leipzig, 1800.

50. Hütte. Des Ingenieurs Taschenbuch, I. Ernst, Berlin, 1931.

51. Tavole logaritmische a cinque cifre decimali. Italy. Istituto Idrografico della R. Marina, Genoa, 1916.

52. Jäderin E. Tafeln für Interpolation und Ausgleichung. Norstedt, Stockholm, 1915.

53. Joncourt E. De la nature et des principaux usages de la plus simple espèce de nombres trigonaux. The Hague, 1762.

54. Jones G. W. Logarithmic tables. Macmillan, London; Author, Ithaca, New York, 1893.

55. Kelley T. L. The Kelley statistical tables. Macmillan, New York, 1938.

56. Köhler H. G. Logarithmisch-trigonometrisches Handbuch. Tauchnitz, Leipzig, 1898.

57. Lacroix S. F. Traité des différences et des séries... Duprat, Paris, 1800.

58. Lambert J. H.—Felkel A. Supplementa tabularum logarithmicarum et trigonometricarum. Lisbon, 1798.

59. Laska W. Sammlung von Formeln der reinen und angewandten Mathematik. Vieweg, Braunschweig, 1888—1894.

60. Legendre A. M. Essai sur la théorie des nombres. Huzard-Courcier, Paris, 1808.

61. Legendre A. M. Traité des fonctions elliptiques, 2. Huzard-Courcier, Paris, 1826.

62. Legendre A. M. Exercices de calcul intégral, 2. Huzard-Courcier, Paris, 1814.

63. Lienard R. Tables fondamentales à 50 décimales des sommes S_n, U_n, Σ_n. Centre de Documentation Universitaire, Paris, 1948.

64. Lindow M. Numerische Infinitesimalrechnung. Dümmler, Berlin — Bonn, 1928.

65. Little A. S. A table of interpolation multipliers, for obtaining through the means of calculating machines, intermediate rates bond values at yield intervals of one ten-thousandth per cent. Mass. Financial Publishing Company, Boston; Routledge, London, 1927.

66. Loomis E. An introduction to practical astronomy, with a collection of astronomical tables. Harper, New York, 1873.

67. Markoff A. A. Differenzenrechnung. Teubner, Leipzig, 1896.

68. Milne-Thomson L. M. The calculus of finite differences. Macmillan, London, 1933.

69. Milne-Thomson L. M. and Comrie L. J. Standard four-figure mathematical tables. Including many new tables, trigonometrical functions for radians, inverse trigonometrical and hyperbolic functions and an extended table of natural logarithms. Macmillan, London, 1948.

70. Moors B. P. Valeur approximative d'une intégrale définie. Gauthier-Villars, Paris, 1905.

71. Mursi Z. Tables of Legendre associated functions. Cairo, 1941.

72. Nautical Almanac. Interpolation tables. Reprinted from the Nautical Almanac for 1931. H. M. Stationery Office, London, 1931.

73. Nautical Almanac. Interpolation tables. Reprinted from the Nautical Almanac for 1935. H. M. Stationery Office, London, 1935.

74. Nautical Almanac. Interpolation and allied tables. Reprinted from the Nautical Almanac for 1937. H. M. Stationery Office, London, 1937.

75. Nautical Almanac Office. Planetary coordinates for the years 1800—1940. H. M. Stationery Office, London, 1933.

76. Nautical Almanac Office. Planetary coordinates for the years 1940—1960. H. M. Stationery Office, London. 1939.

77. Newman F. W. The higher trigonometry. Superrationals of second order. Macmillan and Bowes, Cambridge, 1892.

78. Pearson K. Tables for statisticians and biometricians, II. Biometric Laboratory, University College, London, 1931.

79. Peters C. F. W. Astronomische Tafeln und Formeln. Mauke, Hamburg, 1871.

80. **P e t e r s J.** Zehnstellige Logarithmentafeln, I: Zehnstellige Logarithmen der Zahlen von 1 bis 100 000 nebst einem Anhang mathematischer Tafeln. Reichsamt für Landesaufnahme, Berlin, 1922.

81. **P o t i n L.** Formules et tables numériques. Doin et Gauthier-Villars, Paris, 1925.

82. **P r y d e J.** Chamber's educational course. Mathematical tables. Chambers, London—Edinburgh, 1930.

83. **R i c e H. L.** The theory and practice of interpolation. Nichols, Lynn, Mass., 1899.

84. **R u n g e C.** und **K ö n i g H.** Vorlesungen über numerisches Rechnen. Springer, Berlin, 1924.

85. **S a l z e r H. E.** Table of coefficients for obtaining the first derivative without differences. Gov. print. off., Washington, 1948.

86. **S a s u l y M.** Trend analysis of statistics. Brookings Institution, Washington, 1934.

87. **S c h o r r R.** Sammlung von Hilfstafeln der Hamburger Sternwarte in Bergedorf. Lucas Gräfe, Hamburg, 1916.

88. **S i l b e r s t e i n L.** Synopsis of applicable mathematics, with tables. Bell, London, 1923.

89. **S p e n c e W.** An essay on the theory of the various orders of logarithmic transcendents. London — Edinburgh, 1809.

90. **S t a l m a s e k J.** Tabul'ky funkcií a konstant (z matematiky, fyziky a chemie). Štátne nakl., Bratislava, 1951.

91. **S t e f f e n s e n J. F.** Interpolation. Bailliere, Tindall and Cox, Baltimore, 1927.

92. **Table of arcsin x.** National Bureau of Standards. Columbia Univ. Press, New York, 1945.

93. **Table of arctan x.** Prepared by the Mathematical Tables Project, Work Project Administration by the Federal Works Agency. Columbia Univ. Press, New York, 1942.

94. **Table of the Bessel functions $J_0(z)$ and $J_1(z)$ for complex arguments.** Columbia Univ. Press, New York, 1950.

95. **Tables of the binomial probability distribution.** National Bureau of Standards. Applied mathematics, series 6. Washington, 1949.

96. **Table of circular and hyperbolic tangents and cotangents for radian arguments.** Columbia Univ. Press, New York, 1943.

97. **Table of sine and cosine integrals for arguments from 10 to 100.** National Bureau of Standards, Washington, 1942.

98. **Table of sines and cosines to fifteen decimal places at hunderdths of a degree.** Gov. print. off., Washington, 1949.

99. **Table of 5-point Lagrangean interpolation coefficients.** Prepared by the Mathematical Tables Project of the Work Projects Administration for the City of New York, 1942.

100. **Tables of the Bessel function $Y_0(z)$ and $Y_1(z)$ for complex arguments.** Columbia Univ. press, New York, 1950.

101. **Tables of the Bessel functions of fractional order.** Columbia Univ. press, New York. 1) I: 1948; 2) II: 1949.

102. **Tables of Lagrangean interpolation coefficients.** Columbia Univ. press, New York, 1944.

103. **Tables of sine, cosine and exponential integrals: I, II.** Prepared by the Mathematical Tables Project. National Bureau of Standards, New York, 1940.

104. **Tables relating to Mathieu functions. Characteristic values, coefficients and joining factors.** Cambridge Univ. press, New York, 1951

105. **V e g a G.** Logarithmisch-trigonometrische Tafeln... (Tabulae Logarithmico-Trigonometricae), 1 — 2. Weidmann, Leipzig, 1797.

106. **V e g a G. — H ü l s s e J. A.** Sammlung mathematischer Tafeln. Weidmann, Leipzig, 1849.

107. **W e i g e l H.** W. $x^n+y^n=z^n$? Die elementare Lösung des Fermat-Problems. Author, Leipzig, 1933.

108. **W h i t t a k e r E. T.** and **R o b i n s o n G.** The calculus of observations. Blakie, London — Glasgow, 1926.

109. **Willers F. A.** Numerische Integration. Sammlung Göschen, № 864. De Gruyter, Leipzig — Berlin, 1923.

110. **Wirtz G.** Tafeln und Formeln aus Astronomie und Geodäsie. Springer, Berlin, 1918.

111. **Acta Mathematica,** 1887, **10,** 299—302.

Stieltjes, T. J. Table des valeurs des sommes $s_k = \sum_{1}^{\infty} n^{-k}$.

112. **Annals of Mathematical Statistics** (Baltimore), 1931, **2,** 355—359.

Bailey J. L. A table to facilitate the fitting of certain logistic curves.

113. **Annals of Mathematical Statistics** (Baltimore), 1932, **3,** 257—357.

Jordan. C. Approximation and graduation according to the principle of least squares by orthogonal polynomials.

114. **Annals of Mathematical Statistics** (Baltimore), 1948, **19,** 279—281.

Smirnov N. V. Table for estimating the goodness of fit of empirical distributions.

115. **Archiv der Mathematik und Physik,** 1856, **26,** 1—42.

Oettinger L. Beiträge zur Summiring der Reihen.

116. **Archiv der Mathematik und Physik,** 1914, (3), **22,** 223.

Airey J. R. The asymptotic expansions of Bessel and other functions.

117. **Astronomiska jakttagelser och undersökningar a Stockholms observatorum** (Uppsala), 1877, (3), **1.**

Gylden H. Recueil de tables contenant les développements numériques a employer dans le calcul des perturbations des cométes.

118. **Atti del Congresso Internazionale dei Mathematici,** Bologna, 1932, **6.** 157—177.

Jordan C. Sur une formula d'interpolation.

119. **British Association for the Advancement of Science. Reports,** 1879, 46—57.

120. **British Association for the Advancement of Science. Reports,** 1880, 321—378.

Merrifield C. W. Report on the present state of knowledge of the application of quadratures and interpolation to actual data.

121. **Bulletin of the National Research Council** (Washington), 1933, **92.**

Bennet A. A., Milne W. E. and Bateman H. Numerical integration of differential equations.

122. **Communications on Pure and Applied Mathematics,** 1950, **3,** 177—199.

Friedman B. Theory of underwater explosion bubbles.

123. **Comptes Rendus,** 1864, 59, 618—620.

Catalan E. Analyse mathématique. Mémoire sur la transformation des séries et sur quelques intégrales définies.

124. **Comptes Rendus,** 1867, **64,** 1138—1141.

Bresse. Sur la recherche des fonction auxiliaires, dans l'application de la méthode de Kummer à la sommation des séries.

125. **Compte Rendu de l'Association Française pour l'Avancement des Sciences** (Paris), 1877, **6,** 172—175.

Glaisher J. W. L. Euler's table of $\Sigma_n = \frac{1}{2^n} + \frac{1}{3^n} + \frac{1}{5^n} + \frac{1}{7^n} + \frac{1}{11^n} + \ldots$, corrected.

126. **Encyclopedia Britannica** (London), 1910, **14,** 706—710.

Sheppard W. F. Interpolation.

127. **Handbuch der Physik** (Berlin), 1928, **3,** 13—15. Mathematische Hilfsmittel in der Physik.

Mader K Ausgleichsrechnung, graphisches Rechnen, numerisches Rechnen.

128. **Journal des mathematiques pures et appliquées** (Paris), 1874, (2), **19,** 19—34.

129. **Journal für die reine und angewandte Mathematik** (Berlin), 1830, **5,** 380—382.

Clausen T. Über die Summe der Reihen $1 + \frac{1}{2^3} + \frac{1}{3^3} + \frac{1}{4^3} + \ldots$ und $1 + \frac{1}{7^2} + \frac{1}{13^2} + \frac{1}{19^2} \ldots$

130. **Journal für die reine und angewandte Mathematik,** 1830, **6,** 1—39.

Gudermann C. Theorie der potenzial-oder cyklisch-hyperbolischen Funktionen.

131. **Journal of Mathematics and Physics,** 1930, **9,** 166—180.

Rutledge G. and Crout P. D. Tables and methods of extending tab.

les for interpolation without differences.

132. **Journal of Mathematics and Physics**, 1939, 18, 153—173.
Epstein L. F. A function related to the series for e^x.

133. **Journal of Mathematics and Physics**, 1940, 19, 34—92.
Crout P. D. An application of polinomial approximation to the solution of integral equations arising in physical problems.

134. **Journal of Mathematics and Physics**, 1941, 20, 310—335.
Hildebrand F. B. and Crout P. D. A least square procedure for solving integral equations by polinomial approximation.

135. **Journal of Mathematics and Physics**, 1943, 22, 204—207.
Blanch G. and Rhodes I. Seven-point Lagrangean integration formulae.

136. **Journal of the Institute of Actuaries** (London), 1874, 18, 305—317.
Sprague T. B. On Lubbock's formula for approximating to the value of a life annuity.

137. **Journal of Research of the National Bureau of Standards**, 1950, 44, 89—102.
Forsythe G. E. Solution of the telegrapher's equation with boundary conditions on only one characteristic.

138. **Kongelige Danske videnskabernes selskab** (Kjobenhavn), 1884, (6), 2, № 6, 185—308.
Gram J. P. Undersogelser angaaende maengden of primtal under en given graense.

139. **Mathematical Gazette** (London), 1939, 23, 352—359.
Bickley W. G. Formulae for numerical integration.

140. **Mémoires de la Société des Science Physiques et Naturelles de Bordeaux**, 1886, (3), 3, № 1, 1—184.
Brunel G. Monographie de la fonction gamma.

141. **The Messenger of Mathematics** (Cambridge), 1876, 6, 71—76.
Glaisher J. W. L. On a numerical continued product.

142. **The Messenger of Mathematics** (Cambridge), 1877, 6, 189—192.
Glaisher J. W. L. Further note on certain numerical continued products.

143. **The Messenger of Mathematics** (Cambridge), 1878, 7, 192.
Glaisher J. W. L. Numerical value of a series.

144. **The Messenger of Mathematics** (Cambridge), 1898, 28, 1—17.
Glaisher J. W. L. Table of the values of $\frac{1}{2}\cdot\frac{2}{3}\cdot\frac{4}{5}\cdots\frac{(x-1)}{x}$, the denominators being the series of prime numbers.

145. **The Messenger of Mathematics** (Cambridge), 1903, 33, 20—30.
Glaisher J. W. L. On the series $1-\frac{1}{3^4}+\frac{1}{5^4}-\frac{1}{7^4}+\frac{1}{9^4}-\ldots$

146. **The Messenger of Mathematics** (Cambridge), 1912—1913, 42, 35—49.
Glaisher J. W. L. Numerical values of the series $1-\frac{1}{3^n}+\frac{1}{5^n}-\frac{1}{7^n}+\frac{1}{9^n}-$ etc.

147. **The Messenger of Mathematics** (Cambridge), 1912—1913, 42, 50—58.
Glaisher J. W. L. Numerical values of the series $1-\frac{1}{3^n}+\frac{1}{5^n}-\frac{1}{7^n}+\frac{1}{9^n}-$ etc. for $n=2, 4, 6$.

148. **The Messenger of Mathematics** (Cambridge), 1912—1913, 42, 94—106.
Glaisher J. W. L. Numerical values of the series $1-\frac{1}{3}-\frac{1}{5}+\frac{1}{7}+\frac{1}{9}-\frac{1}{11}-\frac{1}{13}+\ldots$ and $1+\frac{1}{3^2}-\frac{1}{5^2}-\frac{1}{7^2}+\frac{1}{9^2}+\frac{1}{11^2}-\frac{1}{13^2}-\ldots$

149. **The Messenger of Mathematics** (Cambridge), 1917, 47, 97—107.
Glaisher J. W. L. Table of binomial theorem coefficients.

150. **Monthly Notices of the Royal Astronomical Society**, 1928, 88, 506—523.
Comrie L. J. On the construction of tables by interpolation.

151. **Opera Omnia**, 1913, (1), 10, 349.
Euler L. Institutiones calculi differentialis.

152. **Proceedings of the American Academy of Arts and Sciences** (Boston), 1929, 63, 421—437.

Huntington E. V. Tables of Lagrangean coefficients for interpolating without differences.

153. **Proceedings of the Institute of Radio Engineering**, 1944, 32, 114.
Whinnery J. R. and Jamieson H. W. Equivalent circuits for discontinuities in transmission lines.

154. **Proceedings of the London Mathematical Society**, 1871—1873, 4, 48—56.
Glaisher J. W. L. On the constants that occur in certain summations by Bernoulli's series.

155. **Proceedings of the London Mathematical Society**, 1877, 8, 140—145.
Glaisher J. W. L. Numerical values of the first twelve of π, of their reciprocals, and of certain other related quantities.

156. **Proceedings of the London Mathematical Society**, 1877, 8, 200—204.
Glaisher J. W. L. On the numerical value of a certain series.

157. **Proceedings of the London Mathematical Society**, 1935, (2), 39, 272—281.
Knight R. C. The potential of a circular cylinder between two infinite planes.

158. **Proceedings of the Royal Society of London**, 1881, 33, 4—10.
Merrifield C. W. The sums of the series of the reciprocals of the prime numbers and of their powers.

159. **Quarterly Journal of Mathematics**, 1891, 25, 347—362.
Glaisher J. W. L. On the sums of the inverse powers of the prime numbers.

160. **Quarterly Journal of Mathematics**, 1891, 25, 369—375.
Glaisher J. W. L. On the series $\frac{1}{2}+\frac{1}{3}+\frac{1}{5}+\frac{1}{7}+\frac{1}{11}+$ etc.

161. **Quarterly Journal of Mathematics**, 1891, 25, 375—383.
Glaisher J. W. L. On the series $\frac{1}{3}-\frac{1}{5}+\frac{1}{7}+\frac{1}{11}-\frac{1}{13}-$ etc.

162. **Quarterly Journal of Mathematics**, 1893, 26, 33—47.
Glaisher J. W. L. On the series $\frac{1}{3^2}-\frac{1}{5^2}+\frac{1}{7^2}+\frac{1}{11^2}-\frac{1}{13^2}-\dots$

163. **Quarterly Journal of Mathematics**, 1893, 26, 48—65.
Glaisher J. W. L. On the series $\frac{1}{2}+\frac{1}{5}-\frac{1}{7}+\frac{1}{11}-\frac{1}{13}+$ etc.

164. **Quarterly Journal of Mathematics**, 1899, 30, 18—46.
Sheppard W. F. On the relations between Bernoulli's and Euler's numbers.

165. **Quarterly Journal of Mathematics**' 1900, 31, 338.

166. **Quarterly Journal of Mathematics**, 1914, 45, 141—158.
Glaisher J. W. L. Tables of $1 \pm 2^{-n} + 3^{-n} \pm 4^{-n} +$ etc. and $1 + 3^{-n} + 5^{-n} + 7^{-n} +$ etc. to 32 places of decimals.

167. **Quarterly Journal of Mathematics**, 1915, 46, 52—65.
Johnson W. W. On Cotesian numbers: their history, computation and values to $n = 20$.

168. **Tracts for Computers** (London), 1920, 2.
Pearson K. On the construction of tables and on interpolation. Part I. Uni-variate tables.

169. **Tracts for Computers** (London), 1943, 5.
Thompson A. J. Table of the coefficients of Everett's central-difference interpolation formula.

170. **Transactions of the American Society of Mechanical Engineer** (Journal of Applied Mechanics Section), New York, 1935, 57, A12—A16, A109—A110.

171. **Панов Д. Ю.** Справочник по численному решению дифференциальных уравнений в частных производных. ОГИЗ-Гостехиздат, М., 1943. *

172. **Valouch M. — Valouch M. A.** Sedmimístné logaritmy čísel od 1 do 120 000. Nakladatelství Československé Akademia Věd. Praha, 1953.

References to Chapter 14

MATHEMATICAL CONSTANTS

1. **Андреев П. П.** Математические таблицы. Госстатиздат, М., 1952.

2. **Брилинг С. Р.** Математический справочник для студентов, инженеров и техников. ГТТИ, М.— Л., 1933.

3. **Брунс К.** Таблицы логарифмов чисел и тригонометрических функций с семью десятичными знаками. Редбюро ГУГК при СНК СССР, М., 1939.

4. **Воронец А. М.** Таблицы логарифмов и справочник формул и вспомогательных таблиц по курсу элементарной математики. Сытин, М., 1917.

5. **Глазенап С. П.** Пятизначные таблицы логарифмов с приложением других таблиц, упрощающих вычисления. Изд-во АН СССР, Л., 1934.

6. **Глазенап С. П.** Карманные таблицы логарифмов чисел и тригонометрических функций с четырьмя десятичными знаками. Изд-во АН СССР, Л., 1934.

7. **Гутерман А. В. и Браташ В. Н.** Математические таблицы для конструкторов машиностроителей. Москва — Свердловск, ГНТИ Маш. лит., 1944.

8. **Гутерман И. В. и Иванова К. Г.** Математические таблицы, II. Союзоргучет, М., 1940.

9. **Каллет Ф.** Таблицы логарифмов чисел от 1 до 108000, синусов, косинусов и тангенсов от секунды до секунды для первых пяти градусов, и от десяти до десяти секунд для всех градусов четверти окружности. М., 1837.

10. **Павловский А.** Таблицы логарифмов по изданию Каллета, содержащие логарифмы чисел от 1 до 108 000; также логарифмы синусов и тангенсов для каждой секунды первых пяти градусов и для каждых десяти секунд всех градусов четверти круга. Павловский А., Харьков., 1820.

11. **Пржевальский Е.** Пятизначные таблицы логарифмов чисел и тригонометрических величин. ГИЗ, М.— Л., 1928.

12. **Пятизначные таблицы логарифмов.** Сост. по заказу Арт. упр. КА (обраб. М. Я. Кичигиным). Воениздат, М., 1942.

13. **Пятизначные таблицы логарифмов.** Воениздат, М., 1946; 1953.

14. **Ратнер И. Н.** Таблицы обыкновенных логарифмов о двенадцати знаках. Варшава, 1900.

15. **Родин А. Ф.** Пятизначные логарифмические таблицы чисел и тригонометрических функций с приложением гауссовых логарифмов и различных вспомогательных таблиц. Линдеберг, С.-Петербург, 1895.

16. **Сегал Б. И. и Семендяев К. А.** Пятизначные математические таблицы. Изд-во АН СССР, М.— Л., 1948.

17. **Записки императорской Академии Наук по физико-математическому отделению, 1905, (8), 16, № 10.** Серебренников С. З. Таблицы первых девяноста чисел Бернулли.

18. **Субботин М. Ф.** Многозначные таблицы логарифмов. Изд-во АН СССР, М.— Л., 1940.

19. Таблицы Барлоу-квадратов, кубов квадратных корней и обратных величин всех целых чисел до 12 500. Изд-во иностр. лит., М., 1950.

20. Хренов Л. С. Пятизначные таблицы тригонометрических функций, содержащие натуральные значения шести тригонометрических функций через каждую минуту от 0° до 360° и значения котангенсов и косекансов через каждую секунду дуги от 0° до 1° и через каждые 10″ дуги от 1° до 10°. 1954.

21. Хренов Л. С. Семизначные таблицы тригонометрических функции. ГИТТЛ, М.— Л., 1951.

22. Hütte. Справочник для инженеров, техников и студентов, 1. Госмашметиздат, М.— Л., 1934.

23. Янке Е. и Эмде Ф. Таблицы функций с формулами и кривыми. ГИТТЛ, М.— Л., 1949.

24. Adams E. P. and Hippisley R. L. Smithsonian mathematical formulae and tables of elliptic functions. Smithsonian Institution, Washington, 1947.

25. Andoyer M. H. Nouvelles tables trigonométriques fondamentales (logarithmes). Hermann, Paris, 1911.

26. Andoyer M. H. Nouvelles tables trigonométriques fondamentales (valeurs naturelles), 1 (Sines). Hermann, Paris, 1915.

27. Andoyer M. H. Cours de mécanique céleste, I. Gauthier-Villars, Paris, 1923.

28. Attwood E. L. Practical five-figure mathematical tables. Macmillan, London, 1950.

29. Barlow P. Tables of squares, cubes, square roots, cube roots, and reciprocals of all integers numbers up to 12.500. Spon, London, 1947.

30. Bauschinger J. und Peters J. Logarithmisch-trigonometrische Tafeln mit acht Dezimalstellen, enthaltend die Logarithmen aller Zahlen von 1 bis 200 000 und die Logarithmen der trigonometrischen Funktionen für jede Sexagesimalsecunde des Quadranten, neue berechnet und herausgegeben, 1—2. Engelmann, Leipzig, 1936.

31. Becker G. F. and van Orstrand C. E. Hyperbolic functions. Smithsonian Institution, Washington, 1924.

32. Bernoulli J. Ars Conjectandi. Impensis Thurnisiorum, Basileae, 1713.

33. Beyrodt G. und Reissner A. Tabellenbuch zur Wirtschaft-mathematik. Fachwissen Lehrbücher für die Berufsfachleute und Fachschule. «Volk und Wissen», Berlin—Leipzig, 1949.

34. Boll M. Tables numériques universelles des laboratoires et bureaux d'étude. Dunod, Paris, 1947.

35. Boole G. A treatise on the calculus of finite differences. Macmillan, London, 1880.

36. Brandenburg H. Siebenstellige trigonometrische Tafel alter Kreisteilung für Berechnungen mit der Rechenmaschine. Lorentz, Leipzig, 1931.

37. Brandenburg H. Hétjegyü trigonometrial táblázat. Honved térképészeti intézet, Budapest, 1951.

38. Bruhns C. C. Neues logarithmisch-trigonometrisches Handbuch auf sieben Decimalen. Leipzig, 1881.

39. Callet F. Tables portatives de logarithmes, contenant les logarithmes des nombres depuis 1 jusqu'à 108 000, les logarithmes des sinus et tangentes... F. Didot, Paris, 1795.

40. Chrystal G. Algebra, II. Black, Edinburgh, 1906.

41. Claudel J Tables: 1. Des carres et des cubes des nombres entiers successifs de 1 à 10 000. 2. Des longueurs des circonférences et des surfaces des cercles. 3. Des expressions trigonométriques naturelles des angles successifs de minute en minute. Carilian - Goeury et V-or Dalmont, Paris, 1850.

42. Dale J. B. Five figure tables of mathematical functions comprising tables of logarithms, powers of numbers, trigonometric, elliptic and other transcendental functions. Arnold, London, 1949.

43. Davis H. T. Tables of the higher mathematical functions, II.

Principia Press, Bloomington, Indiana, 1935.

44. **Davis H. T.** and **Nelson W. F.C.** Elements of statistics with applications to economic data. Principia Press, Bloomington, 1935.

45. **De Morgan A.** A budget of paradoxes. Principia Press, Bloomington, Indiana; London, 1872.

46. **Duarte F. J.** Nouvelles tables de log $n!$ à 33 décimales depuis $n=1$ jusqu'à $n = 3000$. Kundig, Geneva; Index Generalis, Paris, 1927.

47. **Duarte F. J.** Nouvelles tables logarithmiques à 36 décimales. Gauthier-Villars, Paris, 1933.

48. **Dwight H.** Tables of integrals and other mathematical data. Macmillan, New York, 1947.

49. **Dwight H. B.** Mathematical tables. McGraw-Hill, New York — London, 1941.

50. **Egersdörfer R.** und **Egersdörfer L.** Formeln und Tabellen der zugeordneten Kugelfunktionen 1. Art von $n=1$ bis $n=20$, I, № 6. Springer, Berlin, 1936.

51. **Euler L.** Opuscula analytica, II. Petropoli, 1785.

52. **Fisher R.** and **Yates F.** Statistical tables for biological, agricultural and medical research. Oliver and Boyd, London — Edinburgh, 1949.

53. **Formular matematic şi tehnic.** Pentru eleve, studenţi şi tehnicieni. Editura tehnica, 1949.

54. **Fowle F. E.** Smithsonian physical tables. Smithsonian Institution, Washington, 1934.

55. **Gauss C. F.** Werke (Göttingen), 1866, **3**, 123—162.
Disquisitiones generales circa series infinitam.

56. **Gauss C. F.** Werke (Göttingen), 1866, **3**, 413—432.
De curva lemniscata II.

57. **Gauss F. G.** Vierstellige logarithmische und trigonometrische Tafeln (Sexagesimal interteiler Altgrad). «Volk und Wissen», Berlin—Leipzig, 1946.

58. **Glover J. W.** Tables of applied mathematics in finance, insurance and statistics. Wahr, Ann Arbor, Michigan, 1930

59. **Gray P.** Tables for the formation of logarithms and antilogarithms to twenty-four or any less number of places. Layton, London, 1876.

60. **Burington R. S.** Handbook of mathematical tables and formulae. Sandusky, Ohio, 1946.

61. **Hantschl J.** Logarithmisch-trigonometrisches Handbuch... Rohrmann und Schweigerd, Wien, 1833.

62. **Hayashi K.** Sieben und mehrstellige Tafeln der Kreis- und Hyperbelfunktionen und deren Produkte, sowie der Gammafunktion.
Springer, Berlin, 1926.

63. **Hayashi K.** Tafeln der Besselschen Theta-, Kugel- und anderer Funktionen. Springer, Berlin, 1930.

64. **Hayashi K.** Fünfstellige Funktionentafeln. Springer, Berlin, 1930.

65. **Hindenburg C. F.** Sammlung combinatorisch-analytischer Abhandlungen, II. Leipzig, 1800.

66. **Hobson E. W.** A treatise on plane trigonometry. Cambridge Univ. Press, 1891.

67. **Holtappel H. W.** Tafels van e^x. Noordhoff, Groningen, 1938.

68. **Hoppe R.** Tafeln zur dreißigstelligen logarithmischen Rechnung. Koch, Leipzig, 1876.

69. **Houël J.** Tables de logarithmes à cinq décimales pour les nombres et les lignes trigonométriques. Gauthier-Villars, Paris, 1858.

70. **Houël J.** Recueil de formules et de tables numériques. Gauthier-Villars, Paris, 1868, 1901.

71. **Hütte.** Des Ingenieurs Taschenbuch, I. Ernst, Berlin, 1931.

72. **Hutton C.** Mathematical tables. London, 1811.

73. **Tavole logarithmische a cinque cifre decimali.** Italy istituto idrografico della R. Marina, Genoa, 1941.

74. **Ives H. C.** Natural trigonometrical functions to seven decimal places for every ten seconds of arc. Wiley, New York; Chapman and Hall, London, 1945.

75. **Jahnke** E. und **Emde** F. Funktionentafeln mit Formeln und Kurven. Teubner, Leipzig — Berlin, 1923, 1938.

76. **Jones** G. W. Logarithmic tables. Macmillan, London, Author, New York, 1893.

77. **Jordan** W. — **Eggert** O. Logarithmisch-trigonometrische Tafeln für neue (zentesimale) Teilung mit sechs Dezimalstellen. Jordan, Stuttgart, 1894.

78. **Klügel** G. S. Wörterbuch. **Grunert** J. A. Bernoullische Zahlen. Leipzig, 1833.

79. **Kulik** J. P. Tafeln der Quadrat- und Kubik-Zahlen aller natürlichen Zahlen bis Hundert Tausend nebst ihrer Anwendung auf die Zerlegung grosser Zahlen in ihre Faktoren. Fleischer, Leipzig, 1848.

80. **Lacroix** S. F. Traité des différences et des séries. Duprat, Paris, 1800.

81. **Legendre** A. M. Exercices de calcul intégral, 1. Huzard-Courcier, Paris, 1811.

82. **Legendre** A. M. Exercices de calcul intégral, 3. Huzard-Courcier, Paris, 1816.

83. **Leonelli** G. Z. Supplement logarithmique. Bordeaux, 1803.

84. **Loran** J. Logaritmos de 13 decimales. Tables y Nuevos Procedimientos de Calculos. Imprenta López, Buenos Aires, 1934.

85. **Low** D. A. and **Low** B. B. A pocket-book for mechanical engineers. Longmans, London, 1938.

86. **Markov** A. A. Differenzenrechnung. Teubner, Leipzig, 1896.

87. **Milne-Thomson** L. M. and **Comrie** L. J. Standard four-figure mathematical tables. Including many new tables trigonometrical functions for radians, inverse trigonometrical and hyperbolic functions and an extended table of natural logarithms. Macmillan, London, 1948.

88. **Molterer** J. Tabellen zur Zahlentheorie 52-stellige Briggische Logarithmen von 1—1100 und Potenzen der Zahl 2 bis 2^{400}. Welsermuhl, Wels, 1937.

89. **Newman** F. W. The higher trigonometry. Superrationals of second order. Macmillan and Bowes, Cambridge, 1892.

90. **Nielsen** N. Handbuch der Theorie der Gammafunktion. Teubner, Leipzig, 1906.

91. **Nielsen** N. Traité élementaire des nombres de Bernoulli. Gauthier-Villars, Paris, 1923.

92. **Nörlund** N. E. Vorlesungen über Differenzenrechnung. Springer, Berlin, 1924.

93. **Parkhurst** H. M. Astronomical tables... Author, New York, 1889.

94. **Pearson** K. Tables for statisticians and biometricians, II. Biometric Laboratory, University College. London, 1931.

95. **Pernter** J. M. Meteorologische Optik. Braumüller, Vienna — Leipzig, 1906.

96. **Peters** J. Einundzwanzigstellige Werte der Funktionen Sinus und Cosinus. Reimer, Berlin, 1911.

97. **Peters** J. Zehnstellige Logarithmentafel, I: Zehnstellige Logarithmen der Zahlen von 1 bis 100,000 nebst einem Anhang mathematischer Tafeln. Reichsamt für Landesaufnahme, Berlin, 1922.

98. **Peters** J. Sechsstellige Werte der Kreis- und Evolventen Funktionen von Hundertstel zu Hundertstel des Grades... Köllmann-Werke, Leipzig, 1937.

99. **Peters** J. Sechstellige Werte der trigonometrischen Funktionen von Tausendstel zu Tausendstel des Neugrades. Wichmann, Berlin, 1943.

100. **Potin** L. Formules et tables numériques. Doin et Gauthier-Villars, Paris, 1925.

101. **Rietz** H. L. Handbook of mathematical statistics. Houghton Mifflin, Boston — New York, 1924

102. **Rohrbach** C. Vierstellige logarithmisch-trigonometrische Tafeln. Ehlermann, Dresden, 1929.

103. **Rosenbach** I. B., **Whitman** E. A., **Moskovitz** D. Mathematical tables. Ginn, Boston, 1946.

104. **Saalschütz** L. Vorlesungen über die Bernoullischen Zahlen. Springer, Berlin, 1893.

105. **Sassenfeld** H. M. und **Tschunko** H. F. A. Mathematische

Tafeln für Mathematiker, Naturwissenschaftler, Ingenieure. Erster Teil: Elementare Funktionen. Fr. Lamade, Walldorf bei Heidelberg, 1949.

106. S a s u l y M. Trend analysis of statistics. Brooking Institution, Washington, 1934.

107. S c a r b o r o u g h J. B. Numerical mathematical analysis. John Hopkins Press, Baltimore; Milford, Oxford Univ. Press, London, 1930.

108. S c h e r k H. F. Mathematische Abhandlungen. Reimer, Berlin, 1825.

109. S c h r ö n L. Table d'interpolation pour le calcul des parties proportionnelles. Paris, 1869.

110. S c h r ö n L. Siebenstellige gemeine Logarithmen der Zahlen von 1 bis 108 000 und der Sinus, Cosinus, Tangenten und Cotangenten. Vieweg, Braunschweig, 1875.

111. S h a n k s W. Contributions to mathematics, comprising chiefly the rectification of the circle to 607 places of decimals. Cambridge and Durham, London, 1853.

112. S o l d n e r J. Théorie et tables d'une nouvelle fonction transcendante. Munich, 1809.

113. S p e n c e l e y G. W., S p e n c e l e y R. M. and E p p e r s o n E. R. Smithsonian logarithmic tables to base *e* and base 10. Smithsonian inst., Washington, 1952.

114. S t a l m a s e k J. Tabul'ky funkcií a konstant z matematiky, fyziky, a chemie Štátne nakl., Bratislava, 1951.

115. S t e i n h a u s e r A. Hilfstafeln zur präcisen Berechnung zwanzigstelliger Logarithmen. Gerold, Vienna, 1880.

116. S t u c k e y J. J. Table of compound interest at $^1/_8$ per cent and of antilogarithms to base 1.00125. Allen and Unwin, London, 1914.

117. S t u d n i c k a F. J. Lectures on monoperiodic functions. Prague, 1892.

118. S t u m p f f K. Tafeln und Aufgaben zur harmonischen Analyse und Periodogrammrechnung. Springer, Berlin, 1939.

119. **Tabele si formule matematice.** Editura tehnica, Bucureşti, 1951.

120. **Tabellen der Quadrate, Kuben, Wurzeln, Kreisumfänge und Inhalte, Kreisfunktionen und Logarithmen.** Leipzig.

121. **Tables des logarithmes à huit décimales, I—II. I:** Logarithmes des nombres entiers de 1 à 120 000. **II:** Logarithmes des fonctions circulares. Paris, 1944.

122. **Table of arc sin** x. Columbia Univ. Press, New York, 1945.

123. **Tables of Bessel funktions of fraktional order.** Columbia Univ. Press, New York, 1) **I.** 1948. 2) **II.** 1949.

124. **Table of circular and hyperbolic tangents and cotangents for radian arguments.** Columbia Univ. press, New York, 1943.

125. **Tables of fractional powers.** Part **I:** Values of a^x. Part II: Values of x^a. Columbia Univ. Press, New York, 1946.

126. **Table of natural logarithms, I:** Logarithms of the integers from 1 to 50 000. National Bureau of Standards, New York, 1941.

127. **Table of natural logarithms, II:** Logarithms of the integers from 50 000 to 100000. National Bureau of Standards, New York, 1941.

128. **Table of natural logarithms, III:** Logarithms of the decimal numbers from 0.000 1 to 5.0000. National Bureau of Standards, New York, 1941.

129. **Table of natural logarithms, IV:** Logarithms of the decimal numbers from 5.0000 to 10.0000. National Bureau of Standards, New York, 1941.

130. **Table of sine and cosine integrals for arguments from 10 to 100.** National Bureau of Standards, New York, 1942.

131. **Tables relating to Mathieu functions. Characteristic values, coefficients and Joining factors.** Cambridge Univ. Press, New York, 1951.

132. **Tables of sine, cosine and exponential integrals, I—II.** National Bureau of Standards, New York, 1940.

133. T h o m a n F. Tables de logarithmes à 27 décimales. Paris, 1867.

134. T e m p l e t o n W.—M a y n a r d S. The millwright and engineer's pocket companion... Simpkin, London, 1858.

135. **Uhler H. S.** Original tables to 137 decimal places of natural logarithms for factors of the form $1 \pm n \cdot 10^{-p}$ enchanced by auxiliary tables of logarithms of small integers. Author, New Haven, 1942.

136. **Valori naturali delle funzioni goniometriche.** Firenze istituto geografico militare, 1943.

137. **Vega G.** Logarithmisch-trigonometrische Tafeln... (Tabulae logarithmico-trigonometricae), I, II. Weidmann, Leipzig, 1797.

138. **Vega G.—Hülsse J. A.** Sammlung mathematischer Tafeln. Weidmann. Leipzig, 1849.

139. **Veröffentlichungen des Astronomischen Rechen-Instituts zu Berlin**, 1919, 43.
Peters J. und Stein J. Zweiundfünfzigstellige Logarithmen.

140. **Vogel W. F.** Angular spacing tables. Vinco Corporation, Detroit, Michigan, 1943.

141. **Witkowski A. W.** Tablice matematyczno-fizyczne. Orgelbrand, Warszawa, 1904.

142. **Zimmermann H.** Rechentafeln. Ausgabe A, ohne besondere Quadrattafel Ernst, Berlin, 1929.

143. **Abhandlungen der Mathematisch-Physicalischen Klasse der Königlich Bayerischen Akademie der Wissenschaften** (München), 1885, 15, 229—328.
Lommel E. C. J. Die Beugungerscheinungen einer kreisrunden Oeffnung und eines kreisrunden Schirmchens.

144. **Abhandlungen der Mathematisch-Physikalischen Klasse der Königlich Bayerischen Akademie der Wissenschaften** (München), 1886, 15, 529—664.
Lommel E. C. J. Die Beugungserscheinungen geradlinig begrenzter Schirme.

145. **Abstracts of the Papers Communicated to the Royal Society of London**, 1850—1854, 6, 273—275.
Gassiot J. P. On the extension of the value of the ratio of the circumference of a circle to its diameter.

146. **Abstracts of the Papers Communicated to the Royal Society of London**, 1850—1854, 6, 396—398.
Sabine C. E. On the extension of the value of the base of Napier's logarithms of the Napierian logarithms of 2, 3, 5 and 10; and of the modulus of Briggs's or the common system of logarithms; all to 205 places of decimals.

147. **Allgemeine Literatur-Zeitung** (Halle—Leipzig), 1817, 63, 503.
Rothe H. A. Bekanntmachung für Mathematiker.

148. **American Mathematical Monthly**, 1921, 28, 114—116.
Uhler H. S. Eight values, to 54 D, of $e^{\pi/n}$ including i^i.

149. **American Mathematical Monthly**, 1921, 28, 116—121.
Archibald R. C. Historical notes on the relation $e^{-\frac{\pi}{2}} = i^i$

150. **American Journal of Mathematics** (Baltimore), 1926, 48, 139—143.
Lehmer D. H. On the value of the Napierian base.

151. **The Analyst**, 1882, 9, 180.

152. **Annals of Mathematics** (Princeton), 1935, (2), 36, 637—649.
Lehmer D. H. Lacunary recurrence formulae for the numbers of Bernoulli and Euler.

153. **Archiv der Mathematik und Physik**, 1841, (1), 1, 9—11.
Paucker G. Weitere Berechnung verschiedener auf das Kreisverhältniss π begründeter Zahlen.

154. **Archiv der Mathematik und Physik**, 1843, (1), 3, 27—34.
Bretschneider C. A. Bechnung der Grundzahl der natürlichen Logarithmen, sowie mehrerer anderer mit ihr zusammenhängender Zahlwerte.

155. **Archiv der Mathematik und Physik**, 1853, 21, 121—174.
Lehmann W. Beitrag zur Berechnung der Zahl π, welche das Verhältniss des Kreis-Durchmessers zum Umfang ausdrückt.

156. **Archiv der Mathematik und Physik**, 1856, 26, 1—42.
Oettinger L. Beiträge zur Summiring der Reihen.

157. **Astronomische Nachrichten** (Altona), 1826, 4, 53.
Vietz. Natürliche Logarithmen.

158. **Astronomische Nachrichten** (Altona), 1847, 25, 207—210.
Clausen T. Über die Zahl π, die das Verhältniss des Durchmessers zum Umfange des Kreises ausdrückt.

159. **British Association for the Advancement of Science. Reports** (London), 1871, Trans. Sect., 16—18.
Glaisher J. W. L. On the calculation of e (the base of the Napierian logarithms) from a continued fraction.

160. **British Association for the Advancement of Science. Reports** (London), 1877. Math. and Physics Section, 8—14.
Adams J. C. On the calculation of Bernoulli's numbers up to B_{62} by means of Staudt's theorem.

161. **British Association for the Advancement of Science. Reports** (London), 1877, 14—15.

162. **Cambridge England University Astronomical Observatory**, 1890, 22.
Adams J. C. On the calculation of the Bernoullian numbers from B_{32} to B_{62}.

163. **Comptes Rendus de l'Académie des Sciences** (Paris), 1860, 50, 905—906.
Thoman F. Logarithmes des 40 premiers nombres de Bernoulli.

164. **The Cyclopaedia or Universal Dictionary of Arts, Sciences and Literature**, 1860, 5, 999.

165. **Duke Mathematical Journal** (Durham), 1936, 2, 460—464.
Lehmer D. H. An extension of the table of Bernoulli numbers.

166. **Encyclopedie des Sciences Mathématiques, pures et appliquées** (Paris), 1906, I, № 4, 47—160.
Andoyer M. H. Calcul des differences et interpolation.

167. **Encyclopedia Britannica** (London), 1911, 16, 868—877.
Glaisher J. W. L. Logarithms.

168. **The Encyclopaedia of Pure Mathematics**, 1847.
Hall T. G. Calculus of finite differences.

169. **Geodaetisk institut meddelelse** (København), 1940, 14.
Pedersen P. O. Über die numerische Berechnung der Kettenbrüche nebst einer Berechnung der Grundzahl der natürlichen Logarithmen.

170. **Jahrbuch über die Fortschritte der Mathematik** (Berlin), 1894, 23, 440—441.
Studnicka F. J. Über die Berechnung der transcendenten Zahl.

171. **Jahrbuch über die Fortschritte der Mathematik** (Berlin), 1897, 25, (for 1893—1894), 736.
Studnicka F. J. Vorträge über monsperiodische Funktionen.

172. **Journal für die reine und angewandte Mathematik**, 1830, 6, 1—39.
Gudermann C. Theorie der potenzial-oder cyklish-hyperbolischen Funktionen.

173. **Journal für die reine und angewandte Mathematik**, 1830, 6, 162—194.
Gudermann C. Theorie der potenzial-oder cyklisch-hyperbolischen Funktionen.

174. **Journal für die reine und angewandte Mathematik**, 1830, 6, 311—396.
Gudermann C. Theorie der potenzial-oder cyklisch-hyperbolischen Funktionen.

175. **Journal für die reine und angewandte Mathematik**, 1832, 9, 81—96, 193—208, 297—304, 362—378.
Gudermann C. Theorie der potenzial- oder cyklisch-hyperbolischen Funktionen.

176. **Journal für die reine und angewandte Mathematik**, 1840, 20, 11—12.
Rothe H. A.— Ohm M. Etwas über die Bernoullischen Zahlen.

177. **Journal für die reine und angewandte Mathematik**, 1862, 60, 375—376.
Oettinger L. Ueber die richtige Werthbestimmung der Constant der Integrallogarithmes.

178. **Journal für die reine und angewandte Mathematik**, 1875, 79, 67—98.
Stern M. A. Zur Theorie der Eulerschen Zahlen.

179. **Journal für die reine und angewandte Mathematik**, 1878, 85, 269—272.
Adams J. C. Table of the values of the first sixty-two numbers of Bernoulli.

180. **Journal of the Indian Mathematical Society** (Madras), 1911, 3, 219—234.
Ramanujan S. Some properties of Bernoulli's numbers.

181. **Kongelige Danske videnskabernes selskab** (Kjobenhavn), 1925, (8), **10**, № 3, 311—325.
Gram J. P. Tafeln für die Riemann'sche Zetafunktion.

182. **Königsberger Archiv für Naturwissenschaft und Mathematik** (Königsberg), 1812, **1**, 1—40, 239—240.
Bessel F. W. Values of li (x).

183. **L'Intermédiaire des Mathématiciens**, 1920, **27**, 65.
Bastien L. Relations entre e et π.

184. **Mathematical Gazette** (London), 1912, **6**, 332—336.
Sheppard W. F. The power-sum formula and the Bernoullian function.

185. **Mathematical Magazine** (Washington), 1884, **1**, 207—208.
Boorman J. M. Square-root notes.

186. **Mathematical Tables and other Aids to Computation** (Washington), 1943, **1**, № 1, 31.
Lehmer D. H. Queries-Replies.

187. **Mathematical Tables and other Aids to Computation** (Washington), 1950, **4**, № 29, 11—15.
Reitwiesner G. W. An ENIAC determination of π and e to more than 2000 decimal places.

188. **Mathematical Tables and other Aids to Computation** (Washington), 1954, **8** № 45, 49—50.
Wrench J. W. A new approximation to the reciprocal of π.

189. **Memoirs of the National Academy of Sciences** (Washington), 1921, **14**, № 5, 3—79.
Van Orstrand C. E. Tables of the exponential function and of the circular sine and cosine to radian argument.

190. **The Messenger of Mathematics**, 1877, **7**, 51—54.
Gray P. Verification and extension of the value of the cube root of 2.

191. **The Messenger of Mathematics**, 1877, **7**, 50—51.
Martin A. Computation of the cube root of 2.

192. **The Messenger of Mathematics**, 1877, **6**, 105—106.
Gray P. Values of the trigonometrical quadratic surds.

193. **The Messenger of Mathematics**, 1897, **26**, 20—98, 152—182.
Glaisher J. W. L. On the definite integrals connected with the Bernoullian function.

194. **The Messenger of Mathematics**, 1898, **28**, 36—79.
Glaisher J. W. L. Classes of recurring formulae involving Bernoullian numbers.

195. **The Messenger of Mathematics**, 1912, **42**, 94—106.
Glaisher J. W. L. Numerical values of the series $1-1/3-1/5+1/7+1/9-1/11-1/13+$ etc. and $1+1/3^2-1/5^2-1/7^2+1/9^2+1/11^2-1/13^2-$etc.

196. **Novi Commentarii Academiae Scientiarum Imperialis Petropolitanae**, 1770, **14**, 129—167.
Euler L. De Partitione numerorum.

197. **Opera Omnia**, 1913, (1), **10**, 349.
Euler L. Institutiones calculi differentialis.

198. **Penny Cyclöpaedia**, 1840, **16**, 366

199. **Philosophical Magazine** (London), 1923, (6), **45**, 846—849.
Wrinch H. E. H. and Wrinch D. M. Tables of the Bessel functions $I_n(x)$.

200. **Philosophical Magazine** (London), 1935, (7), **20**, № 134, 721—731.
Airey J. R. The circular and hyperbolic functions, argument $\frac{x}{\sqrt{2}}$.

201. **Philosophical Magazine** (London), 1935, (7), **20**, № 134, 731—738.
Airey J. R. The circular sine and cosine functions, argument $\log_e x$.

202. **Philosophical Transactions of the Royal Society** (London), 1870, **160**, 367—388.
Glaisher J. W. L. Tables of the numerical values of the sine-integral, cosine-integral, and exponential-integral.

203. **Proceedings of the Cambridge Philosophical Society**, 1940, 36, 283—287.
Miller J. C. P. On a criterion for oscillating solutions of a linear differential equation of the second order.

204. **Proceedings of the London Mathematical Society** (London), 1873, 4, 308—312.
Glaisher J. W. L. On the calculation of the theoretical unit-angle to a great number of decimal places.

205. **Proceedings of the London Mathematical Society**, 1877, 8, 140—145.
Glaisher J. W. L. Numerical values of the first twelve of π, of their reciprocals, and of certain other related quantities.

206. **Proceedings of the London Mathematical Society**, 1883, 14, 134—139.
Glaisher J. W. L. Calculation of the hyperbolic logarithm of π.

207. **Proceedings of the London Mathematical Society**, 1899, 31, 216—235.
Glaisher J. W. L. On a set of coefficients analogous to the Eulerian numbers.

208. **Proceedings of the London Mathematical Society**, 1899, (1), 31, 423—488.
Sheppard W. F. A method for extending the accuracy of certain mathematical tables.

209. **Proceedings of the National Academy of Sciences** (Washington), 1938, 24, 23—30.
Uhler H. S. Log π and other basic constants.

210. **Proceedings of the National Academy of Sciences** (Washington), 1940, 26, 205—212.
Uhler H. S. Recalculation and extension of the modulus and of the logarithms of 2, 3, 5, 7 and 17.

211. **Proceedings of the National Academy of Sciences** (Washington), 1942, 28, 59—62.
Uhler H. S. The coefficients of Stirling's series for $\log \Gamma(x)$.

212. **Proceedings of the National Academy of Sciences** (Washington), 1943, 29, 319—325.
Uhler H. S. Natural logarithms of small prime numbers.

213. **Proceedings of the National Academy of Sciences** (Washington), 1951, 37, № 7, 63—67.
Uhler H. S. Many-figure approximations to $\sqrt{2}$, and distribution of digits in $\sqrt{2}$ and $\frac{1}{\sqrt{2}}$.

214. **Proceedings of the Royal Society of London**, 1853, 6, 273—275.
Rutherford W. On the extension on the value of the ratio of the circumference of a circle to its diameter.

215. **Proceedings of the Royal Society of London**, 1854, 6, 397—398.
Shanks W. On the extension to the value of the base of Napier's logarithms; of the Napierian logarithms of 2, 3, 5, 10; and of the modulus of Briggs, or the common system of logarithms; all to 205 places of decimals.

216. **Proceedings of the Royal Society of London**, 1867, 15, 429—432.
Shanks W. On the numerical value of Euler's constant.

217. **Proceedings of the Royal Society of London**, 1868, 16, 154.
Shanks W. Second supplementary paper on the calculation of the numerical value of Euler's constant.

218. **Proceedings of the Royal Society of London**, 1871, 19, 514—524.
Glaisher J. W. L. On the calculation of Euler's constant.

219. **Proceedings of the Royal Society of London**, 1872, 20, 27—29.
Shanks W. On the numerical values of e, $\log_e 2$, $\log_e 3$, $\log_e 5$ and $\log_e 10$, also on the numerical value of M, ..., all to 205 decimals.

220. **Proceedings of the Royal Society of London**, 1872, 20, 29—34.
Shanks W. On the numerical value of Euler's constant.

221. **Proceedings of the Royal Society of London**, 1873, 21, 318—319.
Shanks W. On the extension of the numerical value of π.

222. **Proceedings of the Royal Society of London**, 1873, 22, 45—46.
Shanks W. On certain discrepancies in the published value of π.

223. **Proceedings of the Royal Society of London**, 1873—1874, 22, 27—29.
Shanks W. On the numerical values of e, $\log_e 2$, $\log_e 3$, $\log_e 5$ and $\log_e 10$; also on the numerical value of M, the modulus of the common system of logarithms, all to 205 decimals.

224. **Proceedings of the Royal Society of London**, 1873—1874, 22, 29.
Shanks W. On the numerical value of Euler's constants and on the summation of the harmonic series employed in obtaining such value.

225. **Proceedings of the Royal Society of London**, 1878, (A), 27, 88—94.
Adams J. C. Note on the value of Euler's constant likewise the values of the Napierian logarithms of 2, 3, 5, 7 and 10, and of the modulus of common logarithms, all carried to 260 places of decimals.

226. **Proceedings of the Royal Society of London**, 1887, 42, 22—25.
Adams J. C. Supplementary note on the values of the Napierian logarithms of 2, 3, 5, 7 and 10 and of the modulus of common logarithms.

227. **Proceedings of the Royal Society of London**, 1899, (A), 64. № 407, 203—223.
Aldis W. S. Tables for the solution of the equation

$$\frac{d^2y}{dx^2} + \frac{1}{x} \cdot \frac{dy}{dx} - \left(1 + \frac{h^2}{x^2}\right) y = 0.$$

228. **Quarterly Journal of Mathematics**, 1875, 13, 255—259.
Glaisher J. W. L. Numerical values of certain continued fractions.

229. **Quarterly Journal of Mathematics**, 1878, 15, 125—127.
Glaisher J. W. L. An approximate numerical theorem involving e and π.

230. **Quarterly Journal of Mathematics**, 1891, 25, 362—368; 384.
Glaisher J. W. L. Calculation of the hyperbolic logarithm of π to thirty decimal places.

231. **Quarterly Journal of Mathematics**, 1893, 26, 48—65.
Glaisher J. W. L. On the series $\frac{1}{2} + \frac{1}{5} - \frac{1}{7} + \frac{1}{11} - \frac{1}{13} + \text{etc}$

232. **Quarterly Journal of Mathematics**, 1896, 28, 1—174.
Glaisher J. W. L. Products and series involving prime numbers only.

233. **Quarterly Journal of Mathematics**, 1898, 29, 1—168.
Glaisher J. W. L. The Bernoullian function.

234. **Quarterly Journal of Mathematics**, 1899, 30, 18—46.
Sheppard W. F. On the relations between Bernoulli's and Euler's numbers.

235. **Quarterly Journal of Mathematics**, 1913, 44, 105—112.
Glaisher J. W. L. On the last two figures in certain coefficients-analogous to the Eulerian numbers.

236. **Quarterly Journal of Mathematics**, 1914, 45, 1—51.
Glaisher J. W. L. On the Eulerian numbers (formulae residues, end-figures), with the values of the first twenty-seven.

237. **Quarterly Journal of Mathematics**, 1914, 45, 187—222.
Glaisher J. W. L. On the coefficient in the expansion of $\frac{\cos x}{\cos 2x}$ and $\frac{\sin x}{\cos 2x}$.

238. **Quarterly Journal of Mathematics**, 1914, 45, 350—372.
Ramanujan S. Modular equations and approximations to π.

239. **Quarterly Journal of Mathematics**, 1915, 46, 33—51.
Joffe S. A. Sums of like powers of natural numbers.

240. **Quarterly Journal of Mathematics**, 1916, 47, 103—126.
Joffe S. A. Calculation of the first thirty-two Eulerian numbers from central differences of zero.

241. **Quarterly Journal of Mathematics**, 1916, 47, 311—333.
Schott G. A. and Watson G. W. Asymptotic formulae occuring in electron theory.

242. **Quarterly Journal of Mathematics**, 1920, 48, 193—271.
Joffe S. A. Calculation of eighteen more, fifty in all, Eulerian numbers from central differences of zero.

243. **Quarterly of Journal Mathematics**, 1930, **1**, 310—318.
Watson G. N. The constants of Landau and Lebesque.

244. **Quarterly Journal of Mathematics**, 1933, **4**, 140—146.
Watson G. N. Du Bois Reymond's constants.

245. **Scripta Mathematica**, 1949, **15**, 94—96.
Uhler H. S. The arabian Nights, factorial and the weighted mean factorial.

246. **Sitzungsberichte der Mathematisch-naturwissenschaftlichen Klasse der Keiserlichen Akademie der Wissenschaften** (Wien), 1848. **I**, 465—481.
Herrmann. Bestimmung der trigonometrischen Functionen aus den Winkeln und der Winkel aus den Funktionen, bis zu einer belibigen Grenze der Genauigkeit.

247. **Skoplje, Yugoslavia, Universitet, Filosofski Fakultet, Prirodno-matematichki oddel, Godishen Zbornik**, 1948, **I**, 49—95.
Mitrinovich D. S. O Stirling-ovim brojevima.

248. **Sphinx** (Paris), 1932, **6**, Juin, 94.
Vatriquant S. Courrier du Sphinx.

249. **The Tôhoku Mathematical Journal** (Sendai, Japan), 1933, **37**, 254—278, 481.
Jordan C. On Stirling's numbers.

250. **Tracts for computers**, 1924, **11**, IX.
Thompson A. J. Logarithmetica Britannica being a standard table of logarithms to twenty deci mal places numbers 90 000 to 100 000.

251. **Transactions of the American Mathematical Society**, 1925, **27**, 49—60.
Hutchinson J. I. On the roots of the Riemann zeta function.

252. **Transactions of the Cambridge Philosophical Society**, 1873, **12**, 384—391.
Glaisher J. W. L. Tables of the first 250 Bernoulli numbers to nine figures and their logarithms to ten figures.

253. **Transactions of the Cambridge Philosophical Society**, 1883, **13**, 243—272.
Glaisher J. W. L. Tables of the exponential function.

254. **Transactions of the Connecticut Academy of Science** (New Haven), 1937, **32**, 381—434.
Uhler H. S. A new table of reciprocals of factorials and some derived numbers.

255. **Transactions of the Royal Society of Edinburgh**, 1898, **39**, № 9, 257—321.
Burgess J. On the definite integral $\frac{2}{\sqrt{\pi}}\int_0^t e^{-t^2}\,dt$ with extended table of values, also $\frac{2}{\sqrt{\pi}}\,e^{-x^2}$.

256. **Veröffentlichungen des Kgl. Astronomischen Rechen-Instituts zu Berlin**, 1919, **43**.
Peters J. und Stein J. Zweiundfünfzigstellige Logarithmen.

257. **Zeitschrift für Mathematik und Physik**, 1861, **6**, 127—139.
Bretschneider C. A. Über die Berechnung des Integrallogarithmen und einiger mit ihm zusammenhangenden anderen Funktionen.

258. **Comrie L. J.** Chambers s six-figure mathematical tables, I. New York, 1949.

259. **Записки Императорской Академии наук по физико-математическому отделению** (Петербург), 1906, (8), **19**, № 4.
Серебренников С. З. Новый способ вычисления чисел Бернулли.

260. **Scripta Mathematica**, 1952, **18**, № 2 173—176.
Uhler H. S. Many-figure approximations for $\sqrt[3]{2}$, $\sqrt[3]{3}$, $\sqrt[3]{4}$ and $\sqrt[3]{9}$ with χ^2 data.

References to Chapter 15

PRIMES, FACTORS AND PRODUCTS. QUOTIENTS AND FRACTIONS

1. **Адамович С. В.** Формулы по арифметике, алгебре, геометрии и тригонометрии. Клобуков, С.-Петербург, 1904.
2. **Александров В. А.** Математические таблицы для быстрых подсчетов. «Новое время», М., 1916.
3. **Асатиани Л. Г.** Большие счетные таблицы для механического и быстрого умножения и деления. Планхозгиз, М., 1931.
4. **Асатиани Л. Г.** Карманные таблицы умножения и деления. ГИЗ, М.—Л., 1930.
5. **Асатиани Л. Г.** Малые таблицы, готовые проценты, умножение и деление. „Техника да шрома", Тбилиси, 1946.
6. **Афанасьев М. А.** Пифагорова таблица умножения. Кнебель, Москва, [1911].
7. **Барзаковский Н.** Вспомогательные таблицы при умножении простых чисел с одним, двумя и более знаками. Евстифеев Н., С.-Петербург, 1894.
8. **Беленький Н. С.** Карманная счетная таблица для технических расчетов и хозяйственных вычислений. ГТТИ, М.— Л., 1934.
9. **Беленький Н. С.** Счетные таблицы для хозяйственных вычислений. Госфиниздат, М., 1949.
10. **Бирзвалк К.** Таблицы умножения и таблицы натуральных чисел, тригонометрических функций с приложением справок и формул из низшей геодезии. «Лотос», С.-Петербург, 1910.
11. **Богомольний М. А.** Таблиці знаходити добутки та частки (множення та ділення трицифрових на двоцифрові та трицифрові менші від 500). Техн.-теор. вид-во, Харків-Київ, 1932.
12. **Большая таблица умножения.** С.-Петербург, 1854.
13. **Брилинг С. Р.** Математические формулы и таблицы. «Макиз», М., 1928.
14. **Брилинг С. Р.** Математический справочник. ГТТИ, М.— Л., 1933.
15. **Бурачевский А. К.** Вспомогательные таблицы умножений и делений. С.-Петербург, 1912.
16. **Васильев А. П.** Пространная таблица умножения. Тифлис, 1872.
17. **Виноградов И. М.** Основы теории чисел. ГИТТЛ, М.—Л., 1949.
18. **Воронец А. М.** Таблицы логарифмов и справочник формул и вспомогательных таблиц по курсу элементарной математики. Сытин, М., 1917.
19. **Выгодский М. Я.** Справочник по элементарной математике. ГИТТЛ, М.— Л., 1951.
20. **Гарбовский К.** Таблица умножения неограниченных чисел. Портовая таможня, Л., 1925.
21. **Гольденберг И. В.** Карманный арифмометр. Автор, М., 1925.
22. **Гольденберг И. В.** Общедоступные счетные таблицы «Самосчет». Готовые результаты по таблицам умножения, деления и вычисления процентов. Автор, М., 1927.

23. **Горбунов А. И.** Карманный множитель. Федотов П. А., Петроград, 1915.

24. **Гринберг М.** Натуральная таблица для возвышения в степень (2-ю и 3-ю) и извлечения корня (квадратного и кубического). Фридланд, Могилев, 1890.

25. **Грузон И. Ф.** Описание таблиц всеобще полезных для умножения и деления изобретенных. Шнор, Петербург, 1808.

26. **Гусар А. Г.** Автоматическая таблица умножения (или деления) «Микст». Феокритов, Саратов, 1914.

27. **Гутерман И. В.** Таблицы умножения трехзначных чисел на двузначные. Союзоргучет, М., 1939.

28. **Гутерман А. В. и Браташ В. Н.** Математические таблицы для конструкторов машиностроителей. ГНТИ, Маш. лит., Москва — Свердловск, 1944.

29. **Деценко К. В.** Краткие сведения из арифметики. Нестеров А. К., 1911.

30. **Дзуцев С. М.** Таблицы умножения и возведения в квадрат трехзначных чисел и извлечения квадратного корня из чисел, не превосходящих 999 000. Союзоргучет, М., 1939.

31. **Дитрихкейт О.** Быстрое счисление. Семизначные логарифмы и антилогарифмы. Гольстен Г. В., С.-Петербург, 1904.

32. **Думанов М. Т.** Сметачни таблици за умножение и деление с трицифрени числа. «Наука и изкуство», София, 1952.

33. **Дынин З. и Мороз Я.** Таблица множителей чисел первой тысячи. Паровая типо-лит. Безобразова, Борисоглебск, 1915.

34. **Езерский Ф. В.** Готовое умножение и деление в таблицах на всякую желаемую величину. Автор, С.-Петербург, 1875.

35. **Жиронкин Е. П.** Таблицы деления и таблицы секансов. Автор, Л., 1929.

36. **Журавлев С. Д.** Умножение и возведение в квадрат всевозможных чисел, извлечение квадратного корня из чисел до 1 000 000 и деление на один и два знака. М.

37. **Ивончик И. М.** Таблица умножения. Шайнберг, Севастополь, 1912.

38. **Казачков Н. Г.** Таблица умножения (двузначных чисел на двузначные). Союзоргучет. М., 1938.

39. **Карманов Н. Н.** Книжка для скорого и безошибочного перемножения и деления всяких чисел (деление на один и два знака). Автор, Гомель, 1895.

40. **Карманов Н. Н.** Счетные настольные таблицы готовых произведений всех чисел первой сотни на все числа первой тысячи. Автор, Гомель, 1895.

41. **Карманная книжка математики.** Собрание таблиц, правил и формул. Андерс, М., 1861.

42. **Кирюшин Е. Д.** Умножение и деление при помощи особых таблиц. Шовский А. А., М., 1910.

43. **Клебанюк М. П.** Таблица для моментального умножения целых чисел и десятичных дробей и нахождения процентов и результатов процентных складок. Швец Х. М., Житомир, 1916.

44. **Книзе В.** Таблицы, облегчающие производство арифметического умножения простых (отвлеченных) чисел любой величины. Вольф, Б. М. С.-Петербург, 1894.

45. **Козырев С. Д.** Таблица для быстрого и безошибочного умножения и деления. Козырев, С. Д., Нижний Новгород, 1912.

46. **Крайчик М.** Таблицы первоначальных чисел до 60000 и наименьших делителей составных. Соськин Ш. А., Минск, 1904.

47. **Купарадзе Г. З.** Таблицы произведений. «Техника да шрома», Тбилиси, 1952.

48. **Левасем А. и Цукерштейн А. П.** Счетчик автоматический (умножение, деление и вычисление процентов). Мосполиграф, М., 1925.

49. **Лунский Н. С.** Таблицы, сводящие умножение любых чисел к сложению, а деление — к вычитанию, с подробной инструкцией. Типо-лит. Центросоюза, М., 1928.

50. **Майер Б. А.** Математические формулы. Справочник по арифметике алгебре, геометрии, тригоно-

метрии, высшей математике. Киев, 1930.

51. **Мар'янов Б. М.** Лічильні таблиці. Держ. наук.-техн. вид-во України, Київ-Харків, 1933.

52. **Мень И. М.** Таблицы всех первоначальных множителей, входящих в состав чисел от 1 до 1000, кроме кратных 2-м и 5-ти. Корчак-Новицкий, Киев, 1915.

53. **Митропольский А. К.** Краткие таблицы (с приложением очерка вычислительной техники). Изд-во Лесотехн. акад. им. Кирова, Л., 1950.

54. **Михайловский И.** Портативная таблица умножения. Ташкент, 1924.

55. **Мурашов К. В.** Практическая таблица вычислений. Швец, Бар, 1903.

56. **Наливайко А. К.** Таблицы умножения и деления. Автор, Гришино, 1927.

57. **Недзялковский А. А.** Собрание таблиц и формул для инженеров, архитекторов и механиков, 1. Колесов и Михин, С.-Петербург, 1867.

58. **Нейшулер Л. Я.** Таблицы. Деление, логарифмы, точное умножение и квадраты. ГТТИ, М., 1933.

59. **Нейшулер Л. Я.** Серия таблиц умножения, № 1. Нижполиграф, Нижний Новгород, 1929.

60 **Нейшулер Л. Я.** Таблицы умножения многозначных чисел. Донск. политехн. ин-т, Новочеркасск, 1930.

61. **Нумеров Б. В.** Таблицы для трехзначного вычисления. Гос. вычисл. инст., Петроград, 1922.

62. **Общедоступная таблица умножения от 1-го до 1200 на однозначные числа.** Воронеж, 1905.

63. **Орлов Е. И.** Таблицы для механического производства умножения и деления всевозможных чисел на всевозможные. Петровский, С.-Петербург, 1877.

64. **Орлов П. М.** Таблицы произведений (умножения). Венер, М., 1910.

65. **О'Рурк А. Н.** Сомножитель. Томск. отд. Сиб. обл. гос. изд-ва, Томск, 1923.

66. **О'Рурк А. Н.** Таблицы умножения. Госстатиздат, М., 1953. *

67. **Первый русский карманный счетчик.** Быков, М., 1902.

68. **Писаревский И.** Быстрый счетчик — умножение, деление, вычисление процентов. «Наука и школа», Л., 1928.

69. **Подтягин М. Е.** Таблицы умножения четырехзначных чисел на двузначные. ОГИЗ, М.— Л., 1932.

70. **Попов И. Г.** Математические таблицы. Учпедгиз, М., 1952. *

71. **Сегал Б. И.** и **Семендяев К. А.** Пятизначные математические таблицы. Изд-во АН СССР, М.—Л., 1948.

72. **Снипишский Л. И.** Таблицы для облегчения умножения двузначных чисел. Берман Л. и К°, С.-Петербург, 1874.

73. **Способ упрощенного умножения и таблица произведений на простые множители всех, произвольной величины, чисел от нуля до бесконечности.** Голике Р., С.-Петербург, 1893.

74. **Субботин П.** Таблица для умножения и деления многозначных чисел. Союзоргучет, М., 1937.

75. **Схема таблицы простых чисел.** Тип. Губернского правления, Казань, 1891.

76. **Щеглов Н. Т.** Таблицы Бригговых логарифмов чисел и тригонометрических линий о семи десятичных. Трей Я., С.-Петербург, 1856.

77. **Таблицы готового счета для умножения и деления разных чисел.** Камерштейн, Ростов-на-Дону, 1911.

78. **Таблицы деления на двузначные числа.** Сост. в Стат. отделе Моск. гор. управы, М., 1901.

79. **Таблицы для умножения, проверенные на арифметической машине Томсоном Петриком.** Берлин — Либава, 1875.

80. **Таблицы квадратов трехзначных чисел, составных чисел до 10 000, тригонометрических величин и пр.** Карцев А. А., М., 1887.

81. **Таблицы умножения.** Швабе А., Фридрихштат, 1893.

82. **Таблицы умножения с дробными множителями.** Яковлев С. П., М., 1898.

83. **Ускор или образец сотенного умножения, в котором находятся произведения до 10 000.** Киев, 1852.

84. **Устрялов В. Ф.** Вспомогательные при решениях математических задач таблицы первоначальных множителей. Вольф Б. М., С.-Петербург, 1901.

85. **Франк М. Л.** Карманный справочник по математике. Т-во профессоров и преподавателей, С.-Петербург, 1912.

86. **Числов Е. Е.** Практический самосчетовод. Готовое умножение и деление. Поплавский А. П., М., 1903.

87. **Чудов А. А.** Таблицы умножения трехзначных чисел на трехзначные. Контора расчет. изд. ЦУНХУ Госплана СССР. М., 1940.

88. **Чудов А. А.** Таблицы умножения четырехзначных чисел на двузначные числа. Союзоргучет, М., 1940.

89. **Чудов А. А.** Таблицы деления и умножения многозначных чисел на двузначные. ОНТИ, М.— Л., 1935.

90. **Эйдельнант М. И.** Счетные таблицы для деления любого числа с точностью до 5 знач. цифр. НИХИ, Ташкент, 1930.

91. **Яковкин М. В.** Карманные таблицы умножения и деления многозначных чисел на двузначные. Гостехиздат, М., 1952.

92. **Яковкин М. В.** Таблицы умножения двузначных чисел на трехзначные и многозначные. Госстатиздат, М., 1950.

93. **Boll M.** Tables numériques universelles des laboratoires et bureaux d'étude. Dunod, Paris, 1947.

94. **Comrie L. J.** Chambers's six-figure mathematical tables, II. Van Nostrand, New York, 1949.

95. **Formular matematic şi tehnic.** Pentru eleve, studenti şi tehnicieni. Editura tehnica, Bucureşti, 1949.

96. **Gouritin P.** Les calcules cinématiques opérations à 10 chiffres exacts. Gauthier—Villars, Paris, 1948.

97. **Grosscurth et Godilo-Godlewsky.** Reduction des nombres 1—10 000 en nombres premiers. Moscov, 1880.

98. **Jordan W.** Mathematische und geodätische Hilfstafeln. Hannover, 1895.

99. **Kolb A. und Bauman F.** Blitz. Rechnentafeln nebst den Tabellen: Kreis-Inhalte und Umfänge, Quadrate, Kuben, Quadrat— und Kubikwurzeln. Fr. Lintz, Trier, 1903.

100. **Peters J.** Zehnstellige Logarithmen der Zahlen von 1 bis 100 000 nebst einem Anhang mathematischer Tafeln, I. Reichsamt für Landesaufnahme, Berlin. 1922.

101. **Rayleigh, Lord** (Strutt J. W.). The theory of sound, II. Macmillan, London, 1896.

102. **Sassenfeld H. M. und Tschunko H. F. A.** Mathematische Tafeln für Mathematiker, Naturwissenschaftler, Ingenieure. Erster Teil: Elementare Funktionen. Fr. Lamade, Walldorf bei Heidelberg, 1949.

103. **Schade C.** Taschenrechner «Fix». Die Rechnenmaschine in der Tasche. Engelhard-Reyher. Gotha, 1942.

104. **Der Schnellrechner in der Tasche.** Ein Hilfsbuch für Lohnbüros, Gutskanzleien, Forstverwaltungen für Schule und Haus, I—II. Hofmann, Zella, 1946.

105. **Siebenstellige Logarithmen und Antilogarithmen aller vierstelligen Zahlen und Mantissen von 1000—9999.** Springer, Berlin, 1903.

106. **Tabele şi formule matematice.** Ed. tehnica, Bucureşti, 1951.

107. **Tafelsammlung und Hilfsbuch für Maschinenbauer, Mechaniker, Werkzeugmacher und andere Metallarbeiter.** Leipzig, 1948.

108. **Tariffa Kircheriana id est inventum aucthoris novum.** Romae, 1679. БАН

109. **Törzstényezö táblázat.** „Népszava", Budapest, 1951.

110. **Vega G.—Hülsse J. A.** Sammlung mathematischer Tafeln. Weidman, Leipzig, 1849.

111. **Voisin A.** Tables des multiplications, ou logarithmes des nombres entiers depuis 1 jusqu'à 20.000. Paris, 1817.

112. **Weisbach J.** Der Ingenieur Sammlung von Tafeln, Formeln und Regeln. Braunschweig, 1863.

113. **Wolf R.** Handbuch der Mathematik, Physik, Geodäsie und Astronomie, I. Zürich, 1870.

114. **Zählentafeln für die Grundschule.** «Volk und Wissen», Berlin, 1950. БЛ

115. **Acta Mathematica** (Stockholm), 1893, **17**, 315—319.
Wertheim G. Tabelle der kleinsten primitiven Wurzeln *g* aller ungeraden Primzahlen *p* unter 3000.

116. **Acta Mathematica** (Stockholm), 1897, **20**, 153—157.
Wertheim G. Tabelle der kleinsten primitiven Wurzeln *g* aller Primzahlen *p* zwischen 3000 und 5000.

117. **Acta Mathematica** (Stockholm), 1912, **35**, 193—231.
Posse C. Expose succint des résultats principaux du mémoire posthume de Korkine, avec une table des racines primitives et des caractères, qui s'y rapportent, calculle par bui pour les nombres premiers inférieurs à 4000 et prolongés jusqu'à 5000.

118. **American Journal of Mathematics pure and applied** (Baltimore), 1878, **1**, 184—239.
Lucas E. Theorie des fonctions numériques simplement périodiques.

119. **Archiv der Mathematik und Physik,** 1858, **31**, 335—448.
Arndt F. Tabellarische Berechnung der reducierten binären kubischen Formen und Klassification derselben für alle successiven negativen Determinanten($-D$) von $D = 3$ bis $D = 2000$.

120. **The Duodecimal Bulletin,** 1948, **4**, 20—26.
Anon. List of primes 12855 (=25409) and powers of primes 5954(=10000) with periods not greater than 100 (=144).

121. **Euclides,** 1950, **10**, 74.
Gloden A. Resolution de la congruence $x^4 + 1 \equiv 0 \pmod{p^3}$ avec une table.

122. **The Journal of the London Mathematical Society,** 1934, **9**, 276—278.
Western A. E. Note on the magnitude of the difference between successive primes.

123. **Journal für die reine und angewandte Mathematik,** 1846, **30**, 166—182.
Jacobi C. G. J. Über die Kreistheilung und ihre Anwendung auf die Zahlentheorie.

124. **Journal für die reine und angewandte Mathematik,** 1853, **45**, 55—81.
Kulik J. P. Über die Tafel primitiver Wurzeln.

125. **The Messenger of Mathematics,** 1925—1926, **55**, 17—26.
Beeger N. G. W. H. On the congruence $2^{p-1} \equiv 1 \pmod{p^2}$ and Fermat's last theorem.

126. **Ree's Newcyclopaedia** (Cyclopaedia: or a new universal dictionary of arts and sciences), 1816, **28**.
Prime Numbers.

127. **Scripta Mathematica,** 1948, **14**, № 1, 24—26.
Kraitchik M. On the divisibility of factorials.

128. **Transactions of the Royal Society of Edinburgh,** 1879, **28**, 287—298.
Sang E. On the tabulation of all fractions having their values between two prescribed limits.

AUTHOR INDEX [1]

[1] Chapters are indicated in Roman numerals and the serial numbers of works by the author in Arabic.

TRANSLATIONS OF TITLES GIVEN IN CYRILLIC SCRIPT

For the convenience of users of this **Guide** who are unable to read titles printed in Cyrillic script, translations of these titles from the reference section are given below. Some of these titles are in themselves translations into Russian and this is usually apparent from the publishing house (GITTL - State Publishing House for Foreign Technical Literature and Inostranizdat - State Publishing House for Foreign Literature). The references are given with the same serial numbers as in the original (pages 411 to 536).

Chapter 1

1. Adamovich, S.V. Formulas for arithmetic, algebra, geometry and trigonometry. Klabukov, St.P., 1904.
2. Adler, A. Five-figure logarithms with auxiliary tables for graphic calculation and tables of frequently encountered numerical values. Berlin Riga, 1923.
3. Aleksandrov, V.A. Mathematical tables for rapid calculations. "Novoye vremya", M., 1916.
4. Alekseyev, V.A. A collection of tables for probability calculations. Dzerzhinskii Artillery Academy, M., 1944.
5. Andreev, P.P. Mathematical tables. Gosstatizdat, M., 1952.
6. Andreev, P.P. Tables for percentage calculations. Gosstatizdat, 1950.
7. Andreichin, R. & Mateev, A. Five-figure logarithmic and other tables. (Text in Bulgarian). Sofia, 1949.
8. Antonov, L.M. Tables of squares, cubes, powers and simplified multiplication. L. & I. Belolinskie, Nikolaev, 1910.

9. Artem'ev, D. Tables of five-figure logarithms of natural numbers and trigonometric functions. Ladyzhnikov, Berlin, 1924.

10. Asatiani, L.G. Small tables. Prepared percentages, multiplication and division. "Tekhnika da shroma", Tbilisi, 1946.

11. Asatiani, L.G. Pocket multiplication and division tables. GIZ, M. - L., 1930.

12. Bezikovich, Ya. S. Trigonometric and logarithmic tables. "Kubuch", L., 1931.

13. Belen'kii, N.S. A pocket calculator for technical and commercial calculations. GTTI, M. - L., 1934.

14. Belen'kii, N.S. Tables of reciprocals (from 1 to 10,000). Soyuzorguchet, M., 1939.

15. Belen'kii, N.S. Tables for commercial calculations. Gosfinizdat, M., 1949.

16. Berezovskii, B.Ya. Handbook of elementary mathematics, mechanics and physics. GITTL, M. - L., 1939.

17. Birzvalk, K. Multiplication tables and tables of natural numbers and trigonometric functions, with a supplement of corrections and formulas from elementary geodesy. "Lotos", St.P., 1910.

18. Blyumberg, Ya. Five-figure logarithms of numbers and trigonometric values. Val'ter and Rap, Riga, 1921.

19. Bogomol'nii, M.A. Tables for the multiplication of three-digit numbers by two-digit numbers and three-digit numbers less than 500. (Text in Ukrainian). Tekhteoretizdat, Kharkov - Kiev, 1932.

20. Borisov, S.A. Mathematical handbook. Goskino-izdat, M., 1938.

21. Bradis, V.M. Four-place mathematical tables. Uchpedgiz, M., 1950.

22. Briling, S.R. Mathematical formulas and tables. "Makiz", M., 1928.

23. Briling, S.R. Mathematical handbook. GTTI, M. - L., 1933.

24. Bronshtein, I.N. & Semendyaev, K.A. Handbook of mathematics. Gostekhizdat, M. - L., 1945.

25. Voronets, A.M. Logarithm tables and a reference to the formulas and auxiliary tables of the elementary mathematics course. D.Sytin, M., 1917.

26. Vygodskii, M.Ya. A handbook of elementary mathematics. GITTL, M. - L., 1951.

27. Glazenap, S.P. Pocket four-decimal tables of the logarithms of numbers and trigonometric functions. AN SSSR, L., 1934.

28. Glazenap, A.P. Five-figure logarithm tables. AN SSSR, L., 1934.

29. Gorbunov, A.I. Mathematical tables of squares, cubes, square and cube roots and circumferences M.K.Ivanovskii, Petrograd, 1916.

30. Grinberg, M. A natural table for raising to a power (2nd and 3rd) and deriving a root (square and cube). Fridland, Mogilev, 1890.

31. Gurvits, L.E. Tables of reciprocals. Tsentr. upr. nar. khoz. ucheta Gosplana SSSR, M., 1933.

32. Guterman, A.V. & Bratash, V.N. Mathematical tables for mechanical engineering design staff. GNTI Mash. lit., M. - Sverdlovsk, 1944.

33. Guterman, I.V. & Ivanova, K.G. Mathematical tables, pt. 1. Soyuzorguchet, M., 1939.

34. Dibol'd, R. Formulas for calculating surfaces and volumes and tables for calculating right angle triangles. Nikolaevsk Department, the Imperial Russian Technical Society, Nikolaevsk, 1898.

35. Ditrikheit, O. Rapid reckoning. Seven-figure logarithms and antilogarithms. G.V.Gol'sten, P., 1904.

36. Dobrokhotov, P.I. Tables of logarithms and trigonometric values (in degrees and protractor divisions). Gosizdat Uz.SSR, Tashkent, 1935.

37. Dolgushin, P.A. Four-figure tables of logarithms of numbers and trigonometric functions with a supplement on the elementary theory

37. of table computations. "Sotrudnik", Kiev, 1911.

38. Dubbel', G. A handbook of mathematics for engineers, students and mathematics teachers. ONTI, M. - L., 1933.

39. Zhuravlev, S.D. Multiplication and squaring of numbers, derivation of the square roots of numbers up to 1,000,000 and division by one-, two- and three-digit numbers. M.

40. Ivanov, N.N. & Khrennikov, N.A. Tables of natural five-figure trigonometric values and squares of numbers. Published by the authors. Odessa, 1929.

41. Ivanov, K.P. Tables for calculating polynomials. Gostekhizdat, M. - L., 1949.

42. Kaigorodov, A.I. Four-figure tables of logarithms, antilogarithms, natural trigonometric values and their logarithms, conversion of degrees to radians, square roots, reciprocals etc. Belorus. gos. akad. s. kh., Gorki, 1926.

43. Kaigaradau, A. Four-figure tables. (Text in Byelorussian). Minsk, 1932.

44. Kazanskii, S.P. Tables of squares, cubes and other mathematical values. "Kubuch", Tomsk, 1933.

45. Kamen'shchikov, N. Four-decimal tables of logarithms. GIZ., M. - Petrograd, 1922.

46. A mathematics pocket book. A collection of tables, rules and formulas. Anders, M., 1861.

47. Squares of numbers from 1 to 1,000. 1853.

48. Kobets, N. A pocket book for engineers and technicians. "Svetoch", Odessa, 1927.

49. Kogan, Ya.M. A short mathematical handbook. "Kubuch", L., 1936.

50. Kornakov, V.P. The derivation of square and cube roots with a supplement of tables of squares, cubes and square and cube roots from 1 to 1,000. E.Arngol'd, Petersburg, 1900.

51. Kutnik, S.E. Calculation tables for engineers, building technicians, students, technicians and pupils at trade schools. Okrprofuch., Kharkov, 1926.

52. Kutuzov, N.I. Tables for the decomposition of circular curves. Dorizdat, M., 1950.

53. Kushchenko, V.S. Mathematical formulas and tables, I. L., 1948.

54. Lorchenko, V.I. & Ogloblin, N.V. Four-figure logarithms. Korchak-Novitskii, Kiev, 1910.

55. Maier, B.A. Mathematical formulas. A handbook on arithmetic, algebra, trigonometry and higher mathematics. Kiev, 1930.

56. Maksimov, M.G. A collection of auxiliary tables for rural surveyors, forestry workers and technicians engaged on surveying and cartography in towns. GTTI, L. - M., 1932.

57. Maryanov, B.M. Computation tables. Gostekhizdat, Kiev-Kharkov, 1933. (Text in Ukr.).

58. Machine engineering. An encyclopaedic reference work. Vol. I, pt. I. GNTI mash. lit., M., 1947.

59. Mechnikov, V.V. Complete three-figure logarithm tables. Art. akad RKKA, L., 1933.

60. Mitropol'skii, A.K. Short tables (with a supplement on computing technique). Izd. Lesotekhn. akad. im. S.M.Kirova, L., 1950.

61. Mikhailov, A. Four-decimal tables of logarithms. Logarithms of numbers, trigonometric values, natural trigonometric functions and other tables which facilitate calculations GIZ, M.-P.,1923.

62. Moskvichov. Tables of logarithms, antilogarithms and logarithms of trigonometric functions. (Text in Ukrainian). GNTVU, Kharkov - Kiev, 1932.

63. Nedzyalkovskii, A.A. A collection of formulas for engineers, architects and mechanics. I. Kolesov & Mikhin, Petersburg, 1867.

64. Neishuler, L.Ya. Tables. Division, logarithms exact multiplication and squares. GTTI, M., 1933.

65. Nikulin, M.A. Tables for practical calculatio

65. Conversion of Russian measures to metric, tables of powers and roots, decimal logarithms, trigonometric functions and other mathematical tables necessary for the technician and student. GTTI, M. - L., 1932.

66. Numerov, B.V. Tables for three-place calculations. Gos. vychisl. inst., Petrograd, 1922.

67. Ogloblin, D.N. Five-figure tables of natural values of trigonometric functions for surveyors. Metallurgizdat, Sverdlovsk - Moscow, 1949.

68. Orlov, S.V. Three-figure tables of logarithms and trigonometric functions. GIZ, M., 1920.

69. The first Russian pocket calculator. Bykov, M., 1902.

70. Perel'man, Ya.I. Tables and rules for calculations. Uchpedgiz, M. - L., 1931.

71. Podtyagin, M.E. Tables for the multiplication of four-digit by two-digit numbers. GONTI, M. - L., 1932.

72. Popov, I.G. Mathematical tables. Uchpedgiz, M., 1952.

73. Przheval'skii, E. Five-figure logarithm tables. GIZ, M. - L., 1928.

74. Pridatko, S.V. Tables of four-figure logarithms of numbers and antilogarithms. Tables of natural values of trigonometric functions and their logarithms. A.Benke, Petrograd, 1915.

75. Peevski, V & Venedikov, M. Five-figure logarithm tables of numbers and trigonometric functions. (Text in Bulgarian). Sofia, 1949.

76. Five-figure tables of logarithms. Voenizdat, M., 1951.

77. Five-figure tables of logarithms of numbers and trigonometric functions. Geodezizdat, M., 1949.

78. Pyatetski, P.D. A handbook of mathematics. (Text in Byelorussian). Minsk, 1932.

79. Rainov, A.V. Four-figure functionally related mathematical tables. ONTI, M. - L., 1936.

80. Rodin, A.F. Five-figure tables of logarithms of numbers and trigonometric functions with a supplement of Gaussian logarithms and various auxiliary tables. Lindeberg, St.Pb., 1895.

81. Segal, B.I. & Semendyaev, K.A. Five-figure mathematical tables. AN SSSR, M. - L., 1948.

82. A handbook of mathematics for ore dressers. Metallurgizdat, M., 1951.

83. A guide to mechanical engineering, I. GNTI Ukraini, Kharkov, 1937.

84. A handbook of elementary mathematics, mechanics and physics. GITTL, M. - L., 1943.

85. Stoikov, D. Auxiliary tables for calculations in mechanical engineering, metallurgy and electric technology. (Text in Bulgarian). Sofia, 1947.

86. Barlow's tables of squares, cubes, square roots and reciprocals of all integers to 12,500. Inostranizdat, M., 1950.

87. Tables for calculating squares of numbers from 1,000 to 8,000. Uralmashzavod, Sverdlovsk,1936

88. Tables of squares, cubes, logarithms, trigonometric and hyperbolic functions, square and cube roots and other values. (Text in Ukrainian). Tekhteoretizdat, Kharkov, 1932.

89. Tables of squares of three-digit numbers, composite numbers to 10,000, trigonometric functions, etc. A.A.Kartsev, M., 1887.

90. Tables of Westphalian (Gaussian) logarithms. Gl. Morskoi shtab, St.Pb., 1834.

91. Tables of logarithms of complex numbers and of conversions from Cartesian to polar co-ordinates. AN SSSR, M., 1952.

92. Tables of powers, roots, reciprocals and logarithms. Izd. Sev.-Zap.obl.prombyuro VSNKh, L., 1927.

93. Tables of mathematical and physical values. Nauchn. knigoizd., Petrograd, 1923.

94. Talanov, A.V. Tables for calculators in sound-ranging batteries. Dzerzhinskii Artillery Academy, M., 1941.

95. Ustryalov, V.P. Tables of prime factors used in the solution of mathematical problems. B.M.Wolf, St.Pb., 1901.

96. Feigel'son, E.G. Tables of decimal logarithms and natural trigonometric functions. Uchpedgiz, M. - L., 1931.

97. Filonenko, A.S. Tables of six-figure natural values of trigonometric functions for calculating machines. Goskartogedeziya, M. - L., 1932.

98. Frank, M.L. A mathematical pocket book. T-vo professorov i prepodovatelei, St.Pb.,1912.

99. Khokhlov, K.I. Pocket mathematical tables. Five-figure. Gostekhizdat, M. - L., 1947.

100. Hütte. A handbook for engineers, technicians and students, I. Gosmashmetizdat, M. - L.,1934.

101. Four-figure tables of logarithms. Collective print shop, L., 1928.

102. Chikanov, B. Five-figure tables of logarithms of numbers and trigonometric values. A.I.Mamontov, M., 1917.

103. Cherkashin, L. Four-figure mathematical tables. (Text in Byelorussian). Minsk, 1934.

104. Yakovkin, M.V. Tables of squares of four-digit numbers. Gosstatizdat, M., 1950.

105. Jahnke & Emde. Tables of functions with formulas and curves. GITTL, M. - L., 1949.

Chapter 2

1. Abol'nik, B.S. Pocket tables of three-figure logarithms. "Krasnoe znamya", Tomsk, 1937.

2. Abol'nik, B.S. Three-figure tables of logarithms. Transpechat', M., 1930.

3. Same as Ch. 1/2.

4. Ten-figure logarithm tables of complex numbers and conversions from cartesian to polar co-ordinates. AN SSSR, M., 1952.

5. Same as Ch. 1/3.

6. Same as Ch. 1/4.

7. Same as Ch. 1/7.

8. Anichkov, D. Theoretical and practical trigonometry. Tipograf. kompaniya, M., 1787.

9. Same as Ch. 1/12.

10. Belen'kii, N.S. Tables for commercial Calculations. Gosfinizdat, M., 1949.

11. Same as Ch. 1/16.

12. Same as Ch. 1/17.

13. Same as Ch. 1/18.

14. Same as Ch. 1/19.

15. Bordyussona. The protractor or tables of the chords of arc for each angle from one minute to 180 degrees expressed in values of the radius divided into 1,000 equal parts, for the use of engineers engaged on surveying and the construction of plans. Petersburg, 1827.

16. Borisov, S.A. Mathematical handbook, M., 1938.

17. Bradis, V.M. Four-place mathematical tables. Uchpedgiz, M., 1950.

18. Bremiker, K. Tables of the logarithms of numbers and trigonometric functions to six decimal places. Redbyuro GUGSK NKVD SSSR, M., 1938.

19. Same as Ch. 1/23.

20. Same as Ch. 1/22.

21. Same as Ch. 1/24.

22. Bruns, K. Tables of logarithms of numbers and trigonometric functions to seven decimal places. Redbyuro GUGK pri SNK SSSR, M., 1939.

23. Vetchinkin, V.P. New formulas and tables of elliptic integrals and functions. Military Air Academy RKKA, M., 1935.

24. Vinogradov, V. Tables of trigonometric values, II. Zhitomir.

25. Same as Ch. 1/25.

26. Eight-figure tables of trigonometric functions. Geodezizdat, M., 1950.

27. Same as Ch 1/26.

28. Gauss, F.G. Five-figure tables of natural values of trigonometric functions. Geodezizdat, M., 1943.

29. Gibshman, E.A. Six-decimal tables of the values of trigonometric functions for each minute.' A.I.Mamontov, M., 1914.

30. Same as Ch. 1/27.

31. Same as Ch. 1/28.

32. Gopfen, B.V. Logarithms, antilogarithms and trigonometric values in simplified tables. M., 1919.

33. Same as Ch. 1/29.

34. Same as Ch. 1/32.

35. Guterman, I.V. & Ivanova, K.G. Mathematical tables, II. Soyuzorguchet, M., 1940.

36. Dwight, G.V. Tables of integrals and other mathematical formulas. Gosizdat. inostr. lit., M., 1948.

37. Same as Ch. 1/34.

38. Same as Ch. 1/36.

39. Same as Ch. 1/37.

40. Same as Ch. 1/38.

41. Zhironkin, E.I. Division tables and secant tables. The author. L., 1929.

42. Notes of the State Hydrological Institute, VI, L., 1932.

43. Same as Ch. 1/40.

44. Bulletin of the Kazakh branch of the USSR Academy of Sciences (Alma-Ata), Mining series, fasc. 1,(22), 1946.

45. Same as Ch. 1/44.

46. Same as Ch. 1/43.

47. Same as Ch. 1/42.

48. Callet, F. Tables of the logarithms of numbers from 1 to 108,000, sines, cosines and tangents by seconds for the first five degrees and by tens of seconds for all degrees in a quarter circle. M., 1837.

49. Same as Ch. 1/45.

50. Kamenshchikov, N. Three-decimal logarithms. GIZ, M., 1921.

51. Kamenshchikov, N. Four-figure logarithms of numbers and trigonometric functions. GIZ, M., 1921.

52. Same as Ch. 1/48.

53. Same as Ch. 1/49.

54. Konchin, V.V. Tables of natural trigonometric values. Geodezizdat, M., 1940.

55. Same as Ch. 1/51.

56. Same as Ch. 1/52.

57. Levshunov, M.T. Tables of the products of tangents and cotangents for improving, topographical and other calculations. Omsk, 1930.

58. Same as Ch. 1/54.

59. Same as Ch. 1/55.

60. Same as Ch. 1/66.

61. Same as Ch. 1/58.

62. Same as Ch. 1/59.

63. Same as Ch. 1/60.

64. Same as Ch. 1/61.

65. Navigation tables for 1943. Izdat. Gidrograf. upr. VMS, 1949.

66. Same as Ch. 1/62.

67. Same as Ch. 1/63.

68. Nikonov, M.P. Tables of trigonometric functions. M.P.Nikonov, 1912.

69. Same as Ch. 1/65.

70. Numerov, B.V. Five-decimal tables of natural values of trigonometric functions. GTTI, L., 1933.

71. Same as Ch. 1/66.

72. Oblomievskii, D. Auxiliary tables of trigonometric functions for small arcs. Voenno-topogr. otd. Glav. shtaba, St.Pb., 1875.

73. Same as Ch.1/67.

74. Orlov, P.M. Tables for calculating increments in rectangular coordinates by multiplication. Winer, M., 1914; Voronezh, 1930.

75. Same as Ch. 1/68.

76. Pavlovskii, A. Tables of logarithms in Callet's edition, comprising logarithms of numbers from 1 to 108,000; also logarithms of sines and tangents for each second of the first five degrees and for each ten seconds for all degrees up to a right angle. A.Pavlovskii, Kharkov, 1820.

77. Same as Ch. 1/75.

78. Perel'man, Ya.I. Tables and rules for calculations. Uchpedgiz, M.-L., 1931.

79. Peters, I. Seven-figure tables of trigonometric functions (for tenths of a quadrant). M., 1946.

80. Peters, I. Six-figure tables of trigonometric functions, containing values of six trigonometric functions for each 10" of arc of the quadrant divided into 90° and values of cotangents and cosecants for each second of arc from $0^\circ 0''$ to $1^\circ 20'$. GTTI, M. - L., 1932.

81. Popov, I.G. Mathematical tables. Uchpedgiz, M., 1952.

82. Przheval'skii, E. Five-figure logarithm tables of numbers and trigonometric values. Uchpedgiz, M., 1937.

83. Pridatko, S.V. Tables of four-figure logarithms of numbers and antilogarithms. Tables of natural values of trigonometric functions and their logarithms. A.Benke, Petrograd, 1915.

84. Pyatetski, P.D. A handbook of mathematics. (Text in Byelorussian). Minsk, 1932.

85. Five-figure tables of logarithms of numbers and trigonometric values. Geodezizdat, M., 1949.

86. Five-figure tables of natural values of trigonometric functions. Upr. zemleustroistva Dnepropetr. Oblzo, 1940.

87. Five-figure tables of trigonometric functions. Geodezizdat, M., 1953.

88. Tables of trigonometric values and their logarithms for angles and protractor divisions. Gosizdat oboronnoi prom., M., 1954.

89. Same as Ch. 1/79.

90. Same as Ch. 1/80.

91. Romanov, A.D. Tables of common four-figure logarithms and antilogarithms and also tables of trigonometric values. The author, St.Pb., 1901.

92. Same as Ch. 1/81.

93. Same as Ch. 1/82.

94. Same as Ch. 1/83.

95. Pilot's handbook of mathematics. Izd. Gidrograf. upr. VMS, 1948.

96. Same as Ch. 1/84.

97. Same as Ch. 1/85.

98. Shcheglov, N.T. Tables of Briggs' logarithms of numbers and trigonometric functions to seven decimals. Trei, St.Pb., 1856.

99. Same as Ch. 1/88.

100. Same as Ch. 1/89.

101. Tables of logarithms. Chuvash gosizd., Cheboksary, 1931.

102. Tables of logarithms and sines, cosines and tangents. M., 1703.

103. Tables of logarithms of numbers from 1 to 10,000 and natural sines, tangents and secants with logarithms. St.Pb., 1746.

104. Tables of logarithms for numbers from unity to 10,000 and for the sines, tangents, cosines and cotangents of all degrees and minutes of a right angle, with a table of sines, tangents and secants for a quarter circle assuming a radius of 100,000.00 units. Morskoi Shlyakhetskii kadetskii korpus, Petersburg, 1761.

105. Tables of natural values of trigonometric values for angles expressed in thousandths of the interval. ANIMI, UVMS, RKKA, L., 1934.

106. Tables of common logarithms of numbers and trigonometric lines for use in the Corps of Marine Cadets. Petersburg, 1862.

107. Tables of sines, tangents and secants and of the logarithm of sines and tangents; also of the numbers from unity to 10,000. M., 1716.

108. Tables of powers, roots, reciprocals and logarithms. Izd. Sev.-Zap. obl. prombyuro VSNKh, L., 1927.

109. Tables of tangents of angles for calculating excesses (in tenths of a quadrant). NII Voenno top. sl. VS SSSR, M., 1948.

110. Tables of trigonometric functions. An aid for the solution of right-angle triangles without having recourse to logarithms. Novgorod, 1912.

111. Tables of trigonometric and involute functions for gear calculations. Tipolit. NAMI, M., 1950.

112. Tables of mathematical and physical values. Nauchn. knigoizd. Petrgrad, 1923.

113. Tables of four-figure logarithms and fundamental physical values. Glagolev, St.Pb., 1910.

114. Taradenkov, I. F. Tables of natural values

114. of sines and cosines for calculating increments in rectangular coordinates using a calculating machine. Central Bureau of the specialists' section of the U.S.S.R. Unio of Agricultural Workers. M.

115. Works of the Central Research Institute for Geodesy, Aerosurveying and Cartography. (Moscow), fasc. 30, 1939. (Tables for astronomical calculations)

116. Simplified tables of logarithms and trigonometric values. St.Pb., 1902.

117. Fedorov, G.R. Tables of natural values of trigonometric functions for division of a circle in artillery. Voenizdat, M., 1934.

118. Same as Ch.1/96.

119. Filatov, A.P. Tables of five-figure and four-figure logarithms of numbers and antilogarithms from 1 to 1,000 with additions from 1,000 to 100,000 arranged on the one page. With a supplement on the rules for using logarithms and other auxiliary tables. "Obshchestvennaya pol'za", M., 1909.

120. Same as Ch. 1/97.

121. Same as Ch. 1/98.

122. Same as Ch. 1/99.

123. Khrenov, L.S. Five-figure tables of trigonometric functions, containing values of six trigonometric functions for each minute from 0 - 360° and values of cotangents and cosecants for each second of arc from 0°00' to 1°00' and for each 10 seconds from 1 to 10°. GITTL, M. - L., 1954.

124. Stoichev, D. Six-figure tables of sine, cosine, tangent and cotangent for the old and new divisions of the arc. (Text in Bulgarian). Sofia, 1953.

125. Khrenov, L.S. Five-figure tables of trigonometric functions. GITTL, M. - L., 1951.

126. Khrenov, L.S. Five-figure tables of trigonometric functions with argument in time containing values of the six trigonometric functions for each 4 sec. from 0 to 24 hrs, values of cotangents for each 0.1 sec. from 0 hrs, 00 min. to 0 hrs. 08 min. and for each sec. from 0 hr, 08 min. to 0 hr, 40 min. GITTL, M., 1954.

127. Khrenov, L.S. Tables of trigonometric functions (four-figure sine, cosine, tangent, cotangent, secant, cosecant). Gos. ped. inst., Voronezh, 1940.

128. Same as Ch. 1/100.

129. Four-figure tables of logarithms. Dzerzhinskii Artillery Academy, L., 1935.

130. Same as Ch. 1/102.

131. Same as Ch. 1/103.

132. Six-figure tables of trigonometric functions. Geodezizdat, M., 1952.

133. Shteinbrenner, G. Five-figure tables of trigonometric functions. NII Voenno-top. sl. VS SSSR, M., 1946.

134. Same as Ch. 1/105.

Chapter 3

1. Same as Ch. 1/4.

2. Same as Ch. 1/5.

3. Same as Ch. 1/23.

4. Same as Ch. 1/22.

5. Same as Ch. 1/24.

6. Watson, G.N. The theory of Bessel functions, Vol. 2. Inostranizdat, M., 1949.

7. Vega, G. Seven-figure tables of logarithms. Geodezizdat, M., 1949.

8. Same as Ch. 2/23.

9. Gnedenko, B.V. A course of probability theory. GITTL, M. - L., 1950.

10. Same as Ch. 1/32.

11. Same as Ch. 2/35.

12. Same as Ch. 2/36.

13. Same as Ch. 1/35.

14. Same as Ch. 1/38.

15. Same as Ch. 1/43.

16. Same as Ch. 1/49.

17. Same as Ch. 1/58.

18. Same as Ch. 1/59.

19. Same as Ch. 1/60.

20. Same as Ch. 1/81.

21. Same as Ch. 1/82.

22. Same as Ch. 1/83.

23. Same as Ch. 1/88.

24. Same as Ch. 1/93.

25. Timofeev, N. The most advantageous results of observations. Universit. tipogr., Kiev, 1870.

26. Khrenov, L.S. Tables of natural values of trigonometric functions in units of time. Tip. AN Kazakh.SSR, Alma-Ata, 1947.

27. Same as Ch. 1/100.

28. Same as Ch. 1/103.

29. Same as Ch. 1/105.

30. Engineers' and Technicians' Gazette (Moscow), 1948, No 2 - 4, 114-117.
Karykin, N.N. The nodal deplanation method in calculating torsion in thin-section multispan bars.

31. The Journal of Technical Physics (Kiev), 1932, 2, 97-117.
Umanskii, A. Tables of hyperbolic trigonometric functions.

32. Trans. of the Siberian Physical and Technical Institute. 1932.
Konusov, V.F. Tables of the integrals
$\Phi_m(x) = \int_1^\infty \beta^m e^{-\beta x} d\beta$.

170. Tables of e^x and e^{-x}. M., AN SSSR, 1955.

Chapter 4

1. Same as Ch. 2/2.

2. Same as Ch. 1/2.

3. Adrianov, I.V. Tables of corrections for centring and reduction at triangulation points. Ugletekhizdat, M., 1951.

4. Same as Ch. 1/4.

5. Same as Ch. 1/5.

6. Same as Ch. 1/7.

7. Same as Ch. 2/8.

8. Same as Ch. 1/9.

9. Afanas'ev, P.A. Tables of logarithms of prime numbers and trigonometric functions. Department of Public Education, St.Pb., 1848.

10. Baushinger, I & Peters, I. Tables of logarithms of numbers and trigonometric functions with eight decimals. I & II. Geodezizdat, M., I - 1942, II - 1944.

11. Same as Ch. 1/12.

12. Bezu. A course of mathematics, 1 - 4. Univ. tip., M., 1806.

13. Ben', Ya. Tables of common logarithms to five figures. Vaisberg, Vinnitsa, 1910.

14. Same as Ch. 1/16.

15. Bik, A. Four-figure logarithms of numbers, their antilogarithms and logarithms of trigonometric functions. Lissner and Roman, M., 1895.

16. Same as Ch. 1/18.

17. Same as Ch. 1/19.

18. Borisov, S.A. Mathematical handbook, Goskinoizdat, M., 1938.

19. Same as Ch. 1/21.

20. Same as Ch. 2/18.

21. Same as Ch. 1/22.

22. Same as Ch. 1/23.

23. Same as Ch. 1/24.

24. Same as Ch. 2/22.

25. Busse, F.I. Shortened tables of common logarithms. Compiled after Vega for use in teaching establishments. Bitepazh, F.A. St.Pb., 1885.

26. Bylinskii, A.N. Five-figure tables of logarithms of integers and angles in the divisions of an artillery protractor. Voenizdat, M., 1939.

27. Same as Ch. 3/7.

28. Same as Ch. 3/8.

29. Same as Ch. 1/25.

30. Same as Ch. 1/26.

31. Gauss, F.G. Complete five-figure logarithmic and trigonometric tables. GTTI, M. - L., 1933.

32. Gerts, N. Seven-figure logarithms of trigonometric functions for each second of time. GUChEK, NKVD SSSR, M., 1938.

33. Same as Ch. 1/27.

34. Glazenap, S.P. Five-figure logarithms with a supplement of other tables which facilitate calculations. AN SSSR, L., 1932.

35. Glazenap, S.P. Six-figure Gaussian tables of logarithms of sums and differences. AN SSSR, M. - L., 1935.

36. Same as Ch. 3/9.

37. Same as Ch. 2/32.

38. Same as Ch. 1/32.

39. Same as Ch. 1/33.

40. Same as Ch. 2/36.

41. Ten-figure tables of logarithms of complex numbers and conversions from cartesian to polar coordinates. AN SSSR, M., 1952.

42. Same as 1/35.

43. Same as Ch. 1/36.

44. Same as Ch. 1/37.

45. Same as Ch. 1/38.

46. Ermolaev, S.I. Tables for calculating trajectories by numerical integration. Krylov Naval Academy, L., 1951.

47. Zakovich, M.S. Shortened tables of logarithms and corresponding numbers. "Pol'za", M., 1906.

48. Six-place logarithm and trigonometric tables for the new division (tenths). GNTI, M. - L., 1931.

49. Same as Ch. 1/43.

50. Same as Ch. 1/42.

51. Kalugin, Ya. I. Calculation tables for computers and those who compile estimates. Combinat PPP, Ul'yanovsk, 1925.

52. Same as Ch. 2/48.

53. Same as Ch. 2/51.

54. Kamenshchikov, N. A table of logarithms. "Nachatki znanii", Petrograd, 1922.

55. Same as Ch. 1/45.

56. Kamenshchikov, N. Three-figure logarithms. GIZ, M., 1921.

57. Same as Ch. 1/48.

58. Same as Ch. 1/49.

59. Komendantov, N.V.(edit.). Five-figure tables of logarithms and trigonometric values for angles in an arc and in time. GTTI, M. - L., 1932.

60. A short revision course of logarithmic n[illegible]s and trigonometric values. St.Pb., 1883.

61. Same as Ch. 1/51.

62. Same as Ch. 1/52.

63. Kyuster, F.V. Tables of logarithms for chemists, pharmaceutists, surgeons and physicists. Goskhimizdat, M. - L., 1941.

64. Levitus, D. Four-figure tables of logarithms and antilogarithms. Sytin, M., 1909.

65. Letnik, A.Ya. Five-figure tables of logarithms of numbers from 1 to 10,000. Spiridonov, V.S., M., 1911.

66. Same as Ch. 1/54.

67. Lunskii, N.S. Computation using Gray's shortened tables of multi-place logarithms. Lisner & Sobko, M., 1917.

68. Same as Ch. 1/55.

69. Same as Ch. 1/56.

70. Markovich, B.A. Basic logarithms. Stasyulevich, Petersburg, 1912.

71. Same as Ch. 1/58.

72. Same as Ch. 1/59.

73. Minin, S.V. A desk table of logarithms and antilogarithms to four decimals with the use of interpolation. GITTL, 1952.

74. Minin, S.V. Five-figure logarithm tables. GITTL, M. - L., 1949.

75. Same as Ch. 1/60.

76. Same as Ch. 1/61.

77. Mikhal'skii, M.M. A first approach to logarithms and work with them (Text in Ukrainian). Gosizdat Ukraini, Kharkov - Odessa, 1930.

78. Mikhel', V.M. Four-figure tables of logarithms. Izd. glavsevmorputi, M. - L., 1947.

79. Navigational tables, IV. Izd. gidrograf. Dep. Morskogo m-tva, Petersburg, 1905.

80. Navigational tables. Izd. Gidrograf. upr. VMS, 1949.

81. Morozov. Tables for obtaining the logarithm of a given number. Tables for obtaining a number from a given logarithm. Schmidt, Petersburg, 1879.

82. Moskvichov. Tables of logarithms, antilogarithms, and logarithms of trigonometric functions. (Text in Ukr.). Kharkov, 1931.

83. Same as Ch. 1/63.

84. Same as Ch. 1/64.

85. Same as Ch. 1/65.

86. Numerov, B.V. Tables of the decimal logarithms of numbers from 1,000 to 10,000. GTTI, L., 1932.

87. Same as Ch. 2/72.

88. Orbinskii, A.P. Tables of four-figure logarithms for schools and offices. "Mathesis", Odessa, 1919.

89. Same as Ch. 1/68.

90. Otto, L.R. Tables of logarithms. Volunteer Fleet. M., 1924.

91. Same as Ch. 2/76.

92. Same as Ch. 2/77.

93. Same as Ch. 1/69.

94. Same as Ch. 1/70.

95. Same as Ch. 2/80, edition GUGK pri SNK SSSR, M., 1941.

96. Polyakov, P.N. A beginner's algebra and a collection of algebraic problems. Ulitin, M., 1881.

97. Popov, I.G. Mathematical tables. Uchpedgiz, M., 1932.

98. Same as Ch. 1/74.

99. Same as Ch. 2/82, edition GIZ, M. - L., 1928.

100. Five-figure tables of logarithms and trigonometric values for angles in an arc and in time. GTTI, M. - L., 1932.

101. Five-figure tables of logarithms. Voenizdat, M., 1953.

102. Same as Ch. 1/78.

103. Same as Ch. 1/79.

104. Ratner, I.N. Tables of common logarithms to 12 places. Warsaw, 1900.

105. Ratner, I.N. Tables of common logarithms to 10 places. Br. Shul'dberg, Warsaw, 1894.

106. Same as Ch. 1/80.

107. Ryabkov, G.Z. Tables of five-figure logarithms of numbers and trigonometric values. The author, Odessa, 1904.

108. Ryabkov, G.Z. Tables of five-figure logarithms of numbers from 1 to 10,000. The author, Odessa, 1895.

109. Sludskii, I. Shortened four-figure tables of logarithms and certain other mathematical, physical and astronomical tables. Lissner & Geshel', L. - M., 1902.

110. Shortened tables of common logarithms (compiled after Vega). St.Pb., 1855.

111. Same as Ch. 1/82.

112. Same as Ch. 1/83.

113. Same as Ch. 1/84.

114. Slusaryev, G.G. Four-figure tables of logarithms of numbers, sines and tangents. Detizdat, L., 1938.

115. Same as Ch. 1/85.

116. Subbotin, M.F. Multi-place tables of logarithms. AN SSSR, M. - L., 1940.

117. A table of logarithms. Timiryazev Agricultural Academy, M., 1950.

118. A table of logarithms of trigonometric values and logarithms of numbers to three figures. Pulkovo, 1900.

119. Same as Ch. 2/103.

120. Same as Ch. 1/88.

121. Tables of logarithms. Chuvash Gosizdat., Cheboksary, 1931.

122. Same as Ch. 1/90.

123. Tables of logarithms and sines, cosines and tangents. M., 1703.

124. Same as Ch. 2/104.

125. Tables of logarithms of four decimal places. Antilogarithms. Kiev, 1891.

126. La Lande's tables of logarithms of numbers and trigonometric values. (Stereotype edit. from La Lande's Tables, Leipzig, 1844). Br. Salaevy, M., 1868.

127. Tables of logarithms of numbers and trigonometric lines. Morskoe min., 1901.

128. Same as Ch. 2/106.

129. Tables of five-figure logarithms of numbers, trigonometric functions, $\sin^2$ of half the angle, sums and differences. Gosizdat. Otd. voennoi lit., Petrograd, 1921.

130. Same as Ch. 2/107.

131. Same as Ch. 2/108.

132. Same as Ch. 1/93.

133. Same as Ch. 2/113.

134. Tables of four-figure logarithms of numbers from 1 to 10,000 and natural trigonometric values. "Dalekaya okraina", Vladivostok, 1919.

135. Tarasov, N.P. Logarithms. Gostekhizdat, M., 1927.

136. Same as Ch. 3/25.

137. Three-figure logarithms. Dom zanimat. nauki, L., 1940.

138. Simplified tables of logarithms and trigonometric values. St.Pb., 1902.

139. Same as Ch. 1/96.

140. Same as Ch. 2/119.

141. Same as Ch. 2/121.

142. Khokhlov, A.I. Same as Ch. 2/122.

143. Same as Ch. 1/100.

144. Zech, I. Seven-figure tables of logarithms of sums and differences. Geodezizdat, M., 1940.

145. Same as Ch. 2/131.

146. Four-figure tables of logarithms. Dzerzhinskii Artillery Academy, RKKA, L., 1935.

147. Same as Ch. 2/130.

148. Sheinesson, S. Tables of logarithms, fasc. V. Gosizdat Kirg.SSR.

149. Shveitzer, B. Four-figure tables of logarithms. Got'e, T.V., M., 1855.

150. Same as Ch. 2/98.

151. Yanovskii, A.S. Tables of the differences of logarithms or of logarithms of ratios. "Mospoligraf", M., 1928.

152. <u>Notes of the Imperial Academy of Sciences (Petersburg)</u>, 1867, 12, Bk 2, No 4.
Bunyakovskii, B.Ya. The summing of numerical tables by approximation.

153. Same as Ch. 2/153.

<u>Chapter 5</u>

1. Abramov, A.A. Tables of $\ln\Gamma(z)$ in the field of complex arguments. AN SSSR, M., 1953.

2. Same as Ch. 1/4.

3. Andreev, P.P. Mathematical tables. Gosstatizdat, M., 1952.

4. Briling, S.R. Mathematical handbook. GTTI, M. - L., 1933.

5. Same as Ch. 1/24.

6. Vega, G. Tables of seven-figure logarithms. Geodezizdat., M., 1949.

7. Same as Ch. 1/28.

8. Same as Ch. 2/36.

9. Same as Ch. 2/49.

10. Same as Ch. 2/61.

11. Same as Ch. 2/63.

12. Same as Ch. 2/64.

13. Przheval'skii, E. Five-figure tables of

13. logarithms of numbers and trigonometric values. GIZ, M. - L., 1928.

14. Five-figure tables of logarithms. Voenizdat, M., 1953.

15. Same as Ch. 2/90.

16. Same as Ch. 1/81.

17. Slutskii, E.E. Tables for calculating the incomplete gamma function and the probability function χ^2. AN SSSR, M. - L., 1950.

18. Same as Ch. 1/83.

19. Same as Ch. 1/86.

20. Same as Ch. 1/100.

21. Shpil'rein, Ya.N. Tables of special functions. Numerical values, graphs and formulas, I. GTTI, M. - L., 1953.

22. Same as Ch. 1/105.

Chapter 6

1. Same as Ch. 1/83.

2. Tables of sine and cosine integrals. AN SSSR, M., 1954.

3. Tables of exponential integral functions. AN SSSR, M., 1954.

4. Tatarinov V.V. Three-figure tables of sine and cosine integrals. Svyaz'tekhizdat, L. - M., 1934

5. Same as Ch. 5/21.

6. Same as Ch. 1/105.

7. Automation and Telemechanics, 1949, 10, No 5, 362-376.
   Solodovnikov, V.V. The use of trapezoid frequency characteristics in the analysis of automatic control systems.

Chapter 7

1. Same as Ch. 1/4

2. Same as Ch. 5/3

3. Same as Ch. 1/24.

4. Watson, G. The theory of Bessel functions, II. Inostranizdat, M., 1949.

5. Same as Ch. 3/9.

6. Same as Ch. 3/12.

7. Karpov, K.A. Tables of the functions $W(z)=e^{-z^2}\int_0^z e^{x^2} d$ in the field of complex arguments. AN SSSR, M., 1954.

8. Same as Ch. 2/61.

9. Mitropol'skii, K.A. Probability integrals. Kirov Forestry Academy, L., 1948.

10. Same as Ch. 1/81.

11. Same as Ch. 5/17.

12. Same as Ch. 1/83.

13. Tables of Fresnel integrals. AN SSSR, M., 195

14. Same as Ch. 3/25.

15. Faddeeva, V.N. & Terent'ev, N.M. Tables of values of the function $W(z)=e^{-z^2}\left(1+\frac{2i}{\sqrt{\pi}}\int^z e^{t^2}dt\right)$ for complex arguments. Gostekhizdat, M., 1954.

16. Same as Ch. 3/26.

17. Same as Ch. 5/21.

18. Same as Ch. 1/105.

Chapter 8

1. Same as Ch. 2/21.

2. Vetchinkin, V.P. New formulas and tables of elliptic integrals and functions. Military Air Academy RKKA, M., 1935.

3. Same as Ch. 3/12.

4. Notes of the State Hydrological Institute (Leningrad), 1932, VI.

5. Same as Ch. 2/61.

6. Milne-Thomson, L. Jacobian elliptic functions Gostekhizdat Ukraini, Kharkov, 1933.

7. Samoilova-Yakhontova, M.S. Tables of elliptic integrals. ONTI, M. - L., 1935.

8. Same as Ch. 1/81.

9. Sikorskii, Yu.S. Elements of the theory of elliptic functions as applied to mechanics. ONTI, M. - L., 1936.

10. Same as Ch. 1/83.

11. Same as Ch. 1/100.

12. Same as Ch. 5/21.

13. Same as Ch. 1/105.

Chapter 9

1. Same as Ch. 2/21.

2. Markov, A.A. The calculation of finite differences. "Mathesis", Odessa, 1911.

3. Same as Ch. 2/61.

4. Same as Ch. 1/83.

5. Same as Ch. 5/21.

6. Same as Ch. 1/105.

7. Karmazina, L.N. Tables of Jacobian polynomials. AN SSSR, M., 1954.

Chapter 10

1. Same as Ch. 1/24.

2. Same as Ch. 7/4.

3. Glazenap, S.P. Mathematical and astronomical tables in two parts. AN SSSR, L., 1932

4. Grey, E & Mathews, G.B. Tables of integrals and other mathematical formulae. Inostranizdat, M., 1953

5. Same as Ch. 3/12.

6. Lyusternik, L.A., Akushskii, I.Ya., Ditkin, V.A. Tables of Bessel functions. Gostekhizdat, M. - L., 1949.

7. Same as Ch. 2/61.

8. Same as Ch. 1/81.

9. Same as Ch. 1/83.

10. Tables of values of Bessel functions of imaginary argument. AN SSSR, M. - L., 1950.

11. Faddeeva, V.N., Gavurin, M.K. Tables of the Bessel functions $J_n(x)$ of whole numbers from 0 to 120. GITTL, M. - L., 1950.

12. Fok, V.A. Tables of Airy functions. Inf. otd. NII-108, M., 1946.

13. Same as Ch. 5/21.

14. Same as Ch. 1/105.

15. All-Ukrainian Academy of Sciences. Natural and Technical Sciences Section. Journal of Mathematics, Kiev, 1932.

Dinnik, A. Tables of Bessel functions of fractional order.

16. Trans. of the Steklov Institute of Physics and Mathematics, 2, No 1, AN SSSR, L., 1931.
Bursian, V & Fok, V. Tables of the functions $\int_x^\infty K_0(x)\,dx$; $\int_0^x I_0(x)\,dx$; $e^x \int_x^\infty K_0(x)\,dx$; $e^{-x} \int_0^x I_0(x)\,dx$.

17. Proceedings of Leningrad University, Physical Sciences Section, 1, No 1, 1935, pp 4 - 9.
Bursian, V.R. Tables of values of the function $I_{1/3}$.

302. Communications of the All-Russian Central Executive Committee, 1922, 121-126.
Dinnik, A. Tables of Bessel functions of complex argument.

Chapter 11

1. Same as Ch. 1/4.

2. Same as Ch. 7/4.

3. Same as Ch. 8/2.

4. Karmazina, L.N. Tables of Jacobian polynomials. AN SSSR, M., 1954.

5. Mitropol'skii, A.K. The probability integral. Kirov Forestry Institute, L., 1948.

6. Same as Ch. 5/21.

7. Same as Ch. 1/105.

8. Journal of experimental and theoretical physics, 1948, 18, 5, 457-461.
Ugarov, V.A. On the theory of strata.

9. Works of the Siberian Physical and Technical Institute, 1930.
Konusov, V.F. Tables of the integrals $\Phi_m(x) = \int_1^\infty \beta^m e^{-\beta m} d\beta$.

136. Progress in Mathematics, 1952, (new series), 7, No 6, 3-96.
Dorodnitsyn, A.A. Asymptotic laws of the distribution of characteristic values for certain special cases of differential equations of the second order.

Chapter 12

1. Vitkovskii, I.I. Tables for solution of the Gaussian equation $\sin(z-q) = m \sin^4 z$. Novosibirsk University Astronomical Observatory.

2. Osipov, V.S. Tables for solution of cubic equations. Izd. VIA, RKKA, Moscow, 1937.

3. Shchumyagskii, B.M. Tables for the solution of cubic equations by the root method. Gostekhizdat, M. - L., 1950.

4. Same as Ch. 1/105.

77. Panov, D.Yu. A handbook on the numerical solution of differential equations in partial derivatives. OGIZ - Gostekhizdat, M., 1943.

78. Ditkin, V.A. & Kuznetsov, P.I. A handbook on operational calculation. GITTL, M. - L., 1951.

79. Kamke, E. A handbook of ordinary differential equations. Inostranizdat, M., 1951.

Chapter 13

1. Andreev, P.P. Mathematical tables. Gosstatizdat, M., 1952.

2. Bradis, V.M. Four-figure mathematical tables. Uchpedgiz, M., 1950.

3. Same as Ch. 3/9.

4. Same as Ch. 1/32.

5. Same as Ch. 1/41.

6. Same as Ch. 2/48.

7. Karpov, K.A. Tables of the coefficients of the Lagrange interpolation formula. Supplement to tables of the function $W(z)=e^{-z}\int_0^z e^{x^2}\,dx$ in the field of complex argument. AN SSSR, M., 1954.

8. Markov, A.A. The calculation of finite differences. "Mathesis" Odessa, 1911.

9. Same as Ch. 2/61.

10. Same as Ch. 1/86.

11. Same as Ch. 10/11.

12. Same as Ch. 1/100.

13. Applied Mathematics and Mechanics, 1948, 12, fasc. 2, 233-240.
Kitover, K.A. Tables of sums of certain infinite trigonometric series.

Chapter 14

1. Andreev, P.P. Mathematical tables. Gosstatizdat, M., 1952.

2. Briling, S.R. A mathematical handbook for students, engineers and technicians. GTTI, M. - L., 1933.

3. Bruns, K. Tables of logarithms of numbers and trigonometric functions to seven decimal places. Redbyuro GUGK pri SNK SSSR, M., 1939.

4. Same as Ch. 1/25.

5. Same as Ch. 4/34.

6. Same as Ch. 1/27.

7. Same as Ch. 1/32.

8. Same as Ch. 2/35.

9. Same as Ch. 2/48.

10. Same as Ch. 2/76.

11. Same as Ch. 5/13.

12. Five-figure tables of logarithms. Voenizdat, M., 1942.

13. Five-figure tables of logarithms. Voenizdat, M., 1946; 1953.

14. Same as Ch. 4/104.

15. Same as Ch. 1/80.

16. Same as Ch. 1/81.

17. Notes of the Imperial Academy of Sciences for the physics and mathematics section, 1905, (8), 16, No 10.
Serebrennikov, S.Z. Tables of the first 90 Bernoulli numbers.

18. Subbotin, M.F. Multiplace tables of logarithms. AN SSSR, M. - L., 1940.

19. Same as Ch. 1/86.

20. Same as Ch. 2/123.

21. Khrenov, L.S. Seven-figure tables of trigonometric functions. GITTL, M.-L., 1951.

22. Same as Ch. 1/100.

23. Same as Ch. 1/105.

259. Notes of the Imperial Academy of Sciences for the physics and mathematics section, (Petersburg), 1906, (8), 19, No 4.
Serebrennikov, S.Z. A new method for calculating Bernoulli numbers.

Chapter 15

1. Same as Ch. 1/1.

2. Same as Ch. 1/3.

3. Asatiani, L.G. Large calculating tables for rapid mechanical multiplication and division. Plankhozgiz, M., 1931.

4. Same as Ch. 1/11.

5. Same as Ch. 1/10.

6. Afanas'ev, M.A. Pythagorian multiplication table. Knebel', M., /1911/.

7. Barzakovskii, N. Auxiliary tables for the multiplication of one-, two- and three-digit prime numbers. Evstifeev, St.Pb., 1894.

8. Same as Ch. 1/13.

9. Same as Ch. 1/15.

10. Same as Ch. 1/17.

11. Same as Ch. 1/19.

12. Large multiplication table. St.Pb., 1854.

13. Same as Ch. 1/22.

14. Briling, S.R. Mathematical handbook, GTTI, M. - L., 1933.

15. Burachevskii, A.K. Auxiliary multiplication and division tables. St.Pb., 1912.

16. Vasil'ev, A.P. An extended multiplication table. Tiflis, 1872.

17. Vinogradov, I.M. Fundamentals of the theory of numbers. GITTL, M. - L., 1949.

18. Same as Ch. 1/25.

19. Same as Ch. 1/26.

20. Garbovskii, K. A table for the multiplication of unlimited numbers. L., 1925.

21. Gol'denberg, I.V. A pocket calculator. The author, M., 1925.

22. Gol'denberg, I.V. Popular calculating tables. Prepared results for tables of multiplication, division and the calculation of percentages. The author, M., 1927.

23. Gorbunov, A.I. The pocket multiplier. Fedotov, P.A. Petrograd, 1915.

24. Same as Ch. 1/30.

25. Gruzon, I.F. A description of generally useful tables for multiplication and division. Shnor, Petersburg, 1808.

26. Gusar, A.G. An automatic multiplication (or division) table. "Mikst", Feokritov, Saratov, 1914.

27. Guterman, I.V. Tables for the multiplication of three-digit by two-digit numbers. Soyuzorguchet, M., 1939.

28. Same as Ch. 1/32.

29. Detsenko, K.V. Brief notes on arithmetic. Nesterov, A.K., 1911.

30. Dzutsev, S.M. Tables for the multiplication and squaring of three-digit numbers and for

30. deriving the square roots of numbers not exceeding 999,000. Soyuzorguchet, M., 1939.

31. Same as 1/35.

32. Dumanov, M.T. Multiplication and division tables for three-digit numbers (Text in Bulgarian). Sofia, 1952.

33. Dynin, Z. & Moroz, Ya. Tables of the factors of the first thousand numbers. Bezobrazov, Borisoglebsk, 1915.

34. Ezerskii, F.V. Prepared multiplication and division in tables for any desired quantity. The author, St.Pb., 1875.

35. Same as Ch. 2/41.

36. Same as Ch. 1/39.

37. Ivonchik, I.M. A multiplication table. Shainberg, Sevastopol, 1912.

38. Kazachkov, N.G. A multiplication table (two digits by two digits). Soyuzorguchet, M., 1938.

39. Karmanov, N.N. A little book for rapid and accurate multiplication and division of all numbers (division by one and two digits). The author, Gomel', 1895.

40. Karmanov, N.N. Desk tables of prepared products of all numbers in the first hundred times all numbers in the first thousand. The author, Gomel', 1895.

41. A mathematics pocket book. Anders, M., 1861.

42. Kiryushin, E.D. Multiplication and division using special tables. Shovskii, A.A. M., 1910.

43. Klebanyuk, M.P. A table for the instant multiplication of integers and decimals and for finding percentages and the results of percentage additions. Shvets, Zhitomir, 1916.

44. Knize, V. Tables which simplify the multiplication of prime numbers of any value. Wolf, B.M., St.Pb., 1894.

45. Kozyrev, S.D. A rapid and accurate multiplication and division table. Kozyrev, Nizhnii Novgorod, 1912.

46. Kraichik, M. Tables of prime factors in the numbers to 60,000 and lowest common denominators of the components. Sos'kin, Minsk, 1904.

47. Kuparadze, G.Z. Tables of products. "Tekhnika da shroma", Tbilisi, 1952.

48. Levasem, A. & Tsukershtein, A.I. The automatic calculator (multiplication, division and percentage calculations). Mospoligraf, M., 1925.

49. Lunskii, N.S. Tables which reduce the multiplication of any number to addition and its division to subtraction, with detailed instructions. Tipo-lit.Tsentrosoyuza, M., 1928.

50. Same as Ch. 2/59.

51. Maryanov, B.M. Calculating tables.(Text in Ukrainian). Kiev-Kharkov, 1933.

52. Men', I.M. Tables of all prime factors in the numbers from 1 to 1,000 except 2nd and 5th multiples. Korchak-Novitskii, Kiev, 1915.

53. Same as Ch. 1/60.

54. Mikhailovskii, I. A handy multiplication table. Tashkent, 1924.

55. Murashov, K.V. A practical calculation table. Shvets, Bar, 1903.

56. Nalivaiko, A.K. Multiplication and division tables. The author, Grishino, 1927.

57. Same as Ch. 1/63.

58. Same as Ch. 1/64.

59. Neishuler, L.Ya. A series of multiplication tables, No 1. Nizhpoligraf, Nizhnii Novgorod, 1929.

60. Neishuler, L.Ya. Multiplication tables for multi-digit numbers. Donsk. politekh. inst. Novocherkassk, 1930.

61. Same as Ch. 1/66.

62. Popular multiplication table from 1 to 1,200 multiplied by single digit numbers. Voronezh, 1905.

63. Orlov, E.I. Tables for the mechanical multiplication and division of any number by any number. Petrovskii, St.Pb., 1877.

64. Tables of products (multiplication). Vener, M., 1910.

65. O'Rurk, A.N. A multiplier. Tomsk, 1923.

66. O'Rurk, A.N. Multiplication tables. Gosstatizdat, M., 1953.

67. Same as Ch. 1/69.

68. Pisarevskii, I. A rapid calculator - multiplication, division and percentage calculations. "Nauka i shkola", L., 1928.

69. Same as Ch. 1/71.

70. Same as Ch. 1/72.

71. Same as Ch. 1/81.

72. Snipishkii, L.I. A table for easy multiplication of two-digit numbers. Berman & co., St.Pb., 1874.

73. A simplified method of multiplication and a table of products of prime factors for all numbers of arbitrary value from nought to infinity. Golike, R. St.Pb., 1893.

74. Subbotin, P. A table for the multiplication of multi-digit numbers. Soyuzorguchet, M., 1937.

75. A draft table of prime numbers. Kazan', 1891.

76. Same as Ch. 4/150.

77. Tables of prepared calculation for the multiplication and division of various numbers. Kamershtein, Rostov-on-Don, 1911.

78. Table for division by two-digit numbers. Compiled in the statistical department of the Moscow City Administration. M., 1901.

79. Multiplication tables verified on an arithmetical machine by Thomson and Petrick. Berlin-Libava(Libau), 1875.

80. Tables of the squares of three-digit numbers up to 10,000 and of trigonometric values, etc. Kartsev, M., 1887.

81. Multiplication tables. Schwabe, Friedrichstadt, 1893.

82. Multiplication tables with decimal factors. Yakovlev, M., 1898.

83. The accelerator, or an example of multiplication by hundreds which gives products up to 10,000. Kiev, 1852.

84. Same as Ch. 1/95.

85. Same as Ch. 1/98.

86. Chislov, E.E. A practical calculator. Prepared multiplication and division. Poplavskii, A.P., M., 1903.

87. Chudov, A.A. Tables for the multiplication of four-digit by three-digit numbers. Soyuzorguchet, M., 1940.

88. Chudov, A.A. Tables for the multiplication of four-digit by two-digit numbers. Soyuzorguchet, M., 1940.

89. Chudov, A.A. Tables for the multiplication and division of multi-digit by two-digit numbers. ONTI, M. - L., 1935.

90. Eidel'nant, M.I. Tables for the division of any number with an accuracy of five figures. NIKhI, Tashkent, 1930.

91. Yakovkin, M.V. Pocket tables of multiplication and division of multi-digit by two-digit numbers. Gostekhizdat, M., 1952.

92. Yakovkin, M.V. Tables for the multiplication of two-digit by three- and multi-digit numbers. Gosstatizdat, M., 1950.